Oceanography and Marine Biology

An Introduction to Marine Science

Oceanography and Marine Biology

An Introduction to Marine Science

David W. Townsend

University of Maine

Sinauer Associates, Inc. • Publishers
Sunderland, Massachusetts U.S.A.

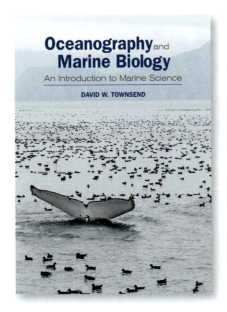

About the Cover

A humpback whale (*Megaptera novaeangliae*) dives amid flocking seabirds, Aleutian Islands, Alaska. © Flip Nicklin/Minden Pictures, Inc.

Address editorial correspondence and orders to:
Sinauer Associates
23 Plumtree Road
Sunderland, MA 01375 U.S.A.
Fax: 413-549-1118
Email: publish@sinauer.com
Internet: www.sinauer.com

Library of Congress Cataloging-in-Publication Data

Townsend, David W.
 Oceanography and marine biology : an introduction to marine science / David W. Townsend.
    p. cm.
 Includes index.
 ISBN 978-0-87893-602-1
 1. Marine biology. 2. Oceanography. I. Title.
 QH91.T69 2012
 577.7--dc23

                                                          2012003553

Printed in USA
5 4 3 2

To Roberta

Brief Contents

Contents

Chapter 4 Water: Its Chemical and Physical Properties 112

Chapter 5 Atmospheric Circulation and Ocean Currents 150

Chapter 6 Waves and Tides 192

Chapter 11
The Fishes   360

Chapter 12   Marine
Environments   390

Chapter 13
Marine Reptiles, Birds,
and Mammals   420

Preface

In recent years, college level textbooks in the marine sciences have fallen into one or the other of two overarching themes: oceanography or marine biology, for both of which there are several excellent books already available. Oceanography books focus much of their content on the geological history of the Earth and its oceans, and the physical and chemical processes that affect the nature of seawater, ocean currents, and the like. Unfortunately, once these topics have been addressed, their treatment of marine life is cut far too short. On the other hand, marine biology textbooks take the opposite tack, surveying the various groups of marine flora and fauna against a backdrop of basic biological and ecological principles, hardly discussing any oceanography at all. I have found this to be a problem; whether marine science or ocean science, it should include and give equal weight to all of these topics. In teaching my introductory ocean science class at the University of Maine for the past 15 years, I have tried to balance these two themes—aspects of marine life with the basics of physical, chemical, and geological processes that govern how life fits into the oceans in the first place. But over the years I became frustrated by the lack of any one book that I could assign to my class. So I decided to write a new textbook.

My intent in this book is to preserve the basic disciplinary elements and their order of presentation found in most oceanography texts, but to include how it all relates to life in the sea, adding specific chapters on life in the sea to do that. This book takes a more balanced approach to the broad topic of marine science, bringing marine biology and the traditional oceanographic disciplines closer together. Doing so more effectively introduces fundamentals of the physical sciences by showing students how such concepts form the oceanographic foundation upon which life in the sea depends.

Throughout the book, my goal is to hold the student's interest, to make the reading lively, and to make the various subjects relevant to a student's everyday life by using common examples wherever possible, often in side boxes that provide more in-depth explanation. These boxes are, in written form, some of the topics I have found that achieve moments when I truly have the class's interest and attention—when students' pens drop and eyes lift—moments every professor strives for in the classroom.

I also bring life to some of the more interesting historical figures, especially in the early chapters, in order to develop a sense in the reader of how scientific ideas have progressed. Another goal in these early chapters is to explain in an interesting way what science is and how it is done. Throughout the book, I use either specific examples or thought experiments to explain physical phenomena, and weave into the discussion how living organisms fit in. At the same time, I show how fascinating these various forms of marine life are in their own right by devoting five chapters to discussions of specific groups of marine organisms.

In the closing chapters, I build on the basic principles of oceanography and marine biology to show how commercial fisheries operate—to show why it is so frustratingly difficult to manage fisheries, and why we must begin to take aquaculture seriously. And, of course, I conclude with discussions of some of the issues facing the oceans—and the students who read this book.

I hope that this new approach to teaching some of the basics of marine science to undergraduates is successful, because that is all I am after: a more informed and scientifically literate person who appreciates how the Earth system operates and how we and society fit in.

Acknowledgments

This book would never have been written had the University of Maine not given me the privilege of developing a new class, "Introduction to Ocean Sciences," which I began teaching back in 1997 when we first started to offer an undergraduate major in the marine sciences. I am grateful to my various bosses back then who entrusted me with such a responsibility, in particular, the late Professor Bruce Sidell. He believed me when I said that I thought I could pull it off. I have been teaching that class ever since, and I thank all of my students for sitting through all those lectures and still recommending the class to their friends.

There are always long lists of people that deserve thanks for helping with such a large project as writing a textbook, and the deeper one reads into such an acknowledgment section, the more tedious and boring it becomes, I know. But please bear with me as I extend my sincere thanks to the many colleagues who have helped me in so many ways.

As is the case with so many of us who are in the latter half of our careers, we often think of those who have influenced us along the way—people we want to mention if and when we ever get the chance to do so, such as in the event that we write a textbook. And so I am. These people have shaped the way I have come to think about the oceans, and who have, probably without knowing it at the time, influenced the directions that my career has taken: Charlie Yentsch, Joe Graham, Tom Berman, Patrick Holligan, Bob Wall, Leon Cammen, and the late Hugh DeWitt—great men all; I truly am fortunate to have worked with them and to have known them.

Because this book crosses so many disciplines, far more than I could ever pretend to have expertise in, I must acknowledge, and sincerely thank, my colleagues for their patience in explaining things to me, in pointing me to references I had missed, and just being there when they must have been saying to themselves: "Uh oh, here comes Townsend with another dumb question…": I thank my University of Maine colleagues, Neal Pettigrew, Emmanuel Boss, Larry Mayer, Andy Pershing, Bob Steneck, Huijie Xue, Andy Thomas, Malcolm Shick, Kevin Eckelbarger, Jeff Runge, Maura Thomas, William Ellis, Fei Chai, Lee Karp-Boss, Susan Brawley, Yong Chen, Irv Kornfield, Gayle Zydlewski, Jim McCleave, Lew Incze, and Paul Rawson. In addition, I want to thank a number of colleagues who may not be aware of the help they have given me: Dennis McGillicuddy, Jeff Turner, Don Anderson, Bruce Keafer, Jeff Brown, Ted Durbin, Dan Lynch, Gary Shepherd, Bernie McAlice, David Mountain, Les Watling, Peter Wiebe, and Mike Sieracki.

I also want to thank the numerous colleagues who graciously allowed me to publish their original photographs and data, many of whom I have never met; and I thank the many reviewers of my book proposal and earlier draft chapters for their candid insights and valuable criticisms.

A word here about my publisher: It was only after several years of feeling my way through the seemingly endless variety of topics for inclusion in my lectures that I began to realize that my notes had grown to the point where they resembled a possible outline for a new textbook. But selling yet another marine science, oceanography, or marine biology textbook to a publisher is no easy task, I soon learned. Then I discovered Sinauer Associates, and Andy Sinauer, who was the first person to give this idea a second look. He, too, believed I could pull it off, and I guess only time will tell if he was right. Throughout this process, from inception to completion, the staff at Sinauer have been no less than amazing. I can honestly say that I have never worked with such a group of dedicated professionals. In particular, I want to acknowledge and thank my copy editor, Carrie Compton; she spotted so many awkward phrases, bad sentences and sections with unclear logic, and made so many helpful suggestions. I have never worked with anyone any better at her job—except my production editor, Azelie Aquadro; day after day she never failed to impress me with her skill, her talent, her dedication to this book, and her patience with me on so many details. The rest of the staff at Sinauer are no slouches, either: Chris Small, with his artist's eye for how to make this book hang together visually; Jen Basil-Whitaker, the one responsible for the book's layout, exceeded my expectations—what a joy it has been to see each chapter's page proofs for the first time, to see this work as students will; and photo editor David McIntyre, who knows what photographs I need better than I do, and who can find anything we asked of him. Quite simply, everyone with whom I have had even fleeting contact with at Sinauer Associates has left me with the same impression: it was my lucky day when I decided to send my book proposal to Andy.

Finally, I thank my family: My brother Terry, who promises to read every word; we'll see. My daughters Karen and Kristy, and my sons-in-law, Chris and Derek, for their support and encouragement as I began to think about writing a college textbook (the thought still scares me a bit). But above all others, I thank my wife of 36 years, my best friend, Roberta. She had to put up with me for all these months, and through it all she never stopped encouraging me to do the best job I could. It is to her that I dedicate this work.

David W. Townsend
University of Maine
March 8, 2012

Media and Supplements

to accompany **Oceanography and Marine Biology:
An Introduction to Marine Science**

eBook

(ISBN 978-0-87893-883-4)
www.coursesmart.com
Oceanography and Marine Biology is available as an eBook via CourseSmart, at
a substantial discount off the price of the printed textbook. The CourseSmart
eBook reproduces the look of the printed book exactly, and includes convenient tools for searching the text, highlighting, and note-taking. The eBook is
viewable in any Web browser, and via free apps for iPhone/iPad, Android,
and Kindle Fire.

Instructor's Resource Library

(ISBN 978-0-87893-887-2)
Available to qualified adopters, the *Oceanography and Marine Biology* Instructor's
Resource Library includes all of the textbook's figures and tables in a variety
of formats, making it easy for instructors to incorporate figures into lectures
and other course materials. All of the figures have been optimized for use in
the classroom and are provided as both low-resolution and high-resolution
JPEGs, as well as ready-to-use PowerPoint slides. The Instructor's Resource
Library also includes suggested answers to the end-of-chapter Discussion
Questions, written by the author.

To the Student

If you are taking the time to continue reading this continuation of the Preface you can count yourself among the very few who ever bother to do so. Preface materials in college textbooks are almost never assigned reading for a class and, after all, this book is intended to be a college textbook; so why are you still taking up your valuable time by continuing to read into this, yet another sentence? The most likely explanation is you are taking a class in ocean science and you happen to be one of those students who reads everything in "the book." On the other hand, it is possible that you are taking an ocean science class for reasons that have nothing to do with your college major or your career plans; the class for which you bought this book may just satisfy a requirement. And, of course, you may be considering majoring in one of the environmental or earth sciences such as oceanography or marine biology. Regardless, if you are still reading this Preface you must have some level of interest in the oceans and its creatures, and you probably saw this book or the class that assigned it as an opportunity to explore that interest further. Well, in this case you are not alone—you are in the majority, because most people are indeed keenly interested in, and even fascinated by, the oceans. But have you ever wondered just why that is, what it is about the oceans that seems to capture our interest? Please keep reading.

Ask a group of people to describe what the oceans mean to them and you are likely to get either a blank stare or some variation of, "Well, we need the oceans for the planet to survive." Interestingly, it is perhaps the blank stare that is more telling because in reality we all have some sense of the oceans' importance, it's just that we're not quite sure why that is. For one thing, everyone is at least aware that there are oceans, even though millions of people have never seen one. Certainly, most of us have a feel for how big the oceans are and that, for instance, it takes a pretty big boat to safely cross one. And many of us have at one time or another enjoyed being at a warm, sunny beach somewhere. Maybe that's it: that the oceans and their shorelines are breathtakingly beautiful places to visit every so often, and for many, that alone is reason enough to be curious about and interested in learning more about the oceans. Throughout history our choices of where to live would seem to support such a claim. Indeed, about half of the population of the United States lives within 50 miles of a coast and that fraction is expected to increase in future years. Still, there seems to be something else that draws us in. Maybe it's the myriad of interesting sea creatures that inhabit the oceans, creatures we've all marveled at either at a public aquarium or in shows we watch on cable TV. In fact, one could reasonably argue that the popularity of such television shows is a testament to the interest generated by the oceans and their creatures. Marine life is a powerful reminder of just how different the oceans are from our own terrestrial environment, and we humans are naturally curious about those creatures—why they look the way they do; why they inhabit different regions and great depths of the oceans the way they do; and what it is that allows them to survive in what, to us, is a hostile environment.

But what about the notion that "we need the oceans for our planet to survive"? Is there some merit to that sentiment? There is, and it weighs on peoples' minds because the oceans are more than just objects for our passive enjoyment. There are many extremely important aspects of the oceans' biology, physics, chemistry, and geology which do in fact affect the planet as a whole. This brings up the question: Are we and our activities harming the oceans in any way? What about pollution? Climate change? Overfishing? These are important issues that we hear about almost every day, but how many of us really understand enough about the oceans to discuss in an intelligent way how best to deal with them?

Whether it is the oceans' beaches or their magnificent life forms, there is so much to know and there are so many wonders to explore before any one of us can truly appreciate all there is that makes the oceans so important to our planet—and scientists are continually adding to that already-vast body of knowledge. At times it can all seem overwhelming, even to those of us who make a living studying the oceans.

Oceanographers and marine scientists (two terms you will often see, but which are synonymous) often speak of the World Ocean as one system. It is divided into great oceans and smaller seas that are connected with one another by a system of ocean currents in a global circulation pattern. This interconnectedness of the world's oceans and seas was made apparent after the *Apollo 8* astronauts photographed Earth while they were on the Moon and in transit in December of 1968. The poet James Dickey popularized the term "Blue Planet" after seeing the *Apollo* mission's "Earthrise" photographs, and for good reason: the oceans clearly stand out, with the planet as a whole shining a brilliant blue. Well, the name has stuck. Our oceans were seen, maybe for the first time, as what they truly are: big and magnificent.

The oceans are indeed big. They encompass almost three quarters (70.78%) of the Earth's surface area, which means, of course, that most of the world's surface area is water, not land. And the oceans are deep. Their average depth is nearly two and half miles (3796 meters), but the deepest depths are nearly seven miles. (The deepest spot is the Marianas Trench, which is just over 11,000 meters.) Compared with the average altitude of land on Earth (about 840 meters) the oceans are some 4.5 times as deep as the land is high; if all that water were spread evenly over a smooth Earth, it would still be 1.6 miles deep.

The oceans are both huge, and at the same time, almost insignificant. That is, even as deep and big as the World Ocean is, the oceans are still only "paper thin" (just over 2 miles) when compared with the diameter of the Earth, which is about 8000 miles. So even though they cover almost three quarters of the surface of the earth, the oceans are only 0.02% of Earth's mass and 0.13% of its volume. And interestingly, the majority of the oceans' surface area is not evenly distributed: there is more ocean area in the Southern Hemisphere, while the Earth's land masses are more concentrated in the Northern Hemisphere—a feature that is important to ocean currents, weather, and climate. Finally, most of the world's water supply is in the oceans, which have more than 97% of all the water on Earth. Rivers, lakes, groundwater, and the atmosphere have less than 1%, and the remainder, just more than 2%, is locked up in ice in Greenland and Antarctica.

Ok, so what? After all, this is just a long list of trivia, and you're probably still wondering: Why do we care? Why are the oceans important? Why study them? Well, that's what this book is all about. But to start, here are some points to consider: The oceans are responsible for a significant fraction of the world's food production—although they produce far less than most people realize. The oceans are important in controlling our weather and climate—in ways that are far more important than most realize. The oceans constitute the most efficient means of global transportation of bulk goods. And finally, the oceans are just plain *interesting*, which we have just argued is why most of us want to learn more about them.

The purpose of this book is really very simple: I wanted to provide an introductory exposure to and appreciation for the marine sciences, to provide a roadmap, of sorts, that will help to bring you through layers of understanding of and appreciation for the science of the oceans—what makes the oceans "work." This is a tall order, but perhaps by using some of our experiences as examples, I have illustrated basic principles a little better; I hope that my additional focus on marine biology, more so than other currently available ocean or marine science textbooks, works, and that I have maintained a balance with the basic oceanographic disciplines. And so I truly hope that you will get as much from this book as I have throughout my career as a research oceanographer and as someone who enjoys teaching the subject.

Oceanography and Marine Biology

An Introduction to Marine Science

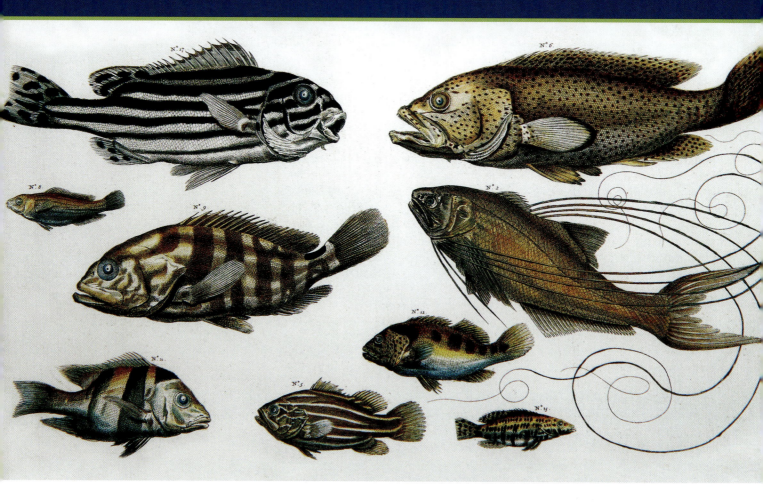

Contents

And so we begin our exploration of the science of the oceans. Much of what we know today we owe to the earliest ocean explorers who ventured across large expanses of open oceans, not necessarily to discover the oceans' secrets, but to seek out and explore distant lands, and in the process we began to accumulate knowledge about how our great oceans work. And we have no doubt been curious about marine life forms for a long time, both as a source of food and because they are just interesting to study. Certainly that was an opinion held by Albertus Seba (1665–1736), a Dutch naturalist who collected zoological specimens and had artists illustrate them, such as these from his *Cabinet of Natural Curiosities* published in 1758.

Early Foundations of the Ocean Sciences

As we begin our exploration of the marine, or ocean, sciences, we are going to adopt a historical perspective, tracing how our ancestors' appreciation and understanding of the world's oceans developed and matured over time. This way we can extend our understanding of fundamental scientific principles, an understanding that tends to govern how we "see" the oceans and the world. By following a few selected scientific developments of the near and distant past, we can build up, layer by layer, our own scientific understanding. Of course the challenge with this approach is: Where do we begin?

Selecting our historical starting point is a bit arbitrary, but it should probably be based on when marine science became *scientific*. In ancient times, much of our ancestors' understanding of the natural world was steeped in mythology and ideas of the supernatural. In about the sixth century B.C., the Greeks began to abandon that perspective and focus more on how natural causes shaped the physical world around them. In fact, it has been argued that science itself was "invented," or "discovered," about 600 B.C. by the Greek philosopher **Thales of Miletos**, who might have been the first person to formally abandon the supernatural approach. Quite simply, Thales believed that there were laws controlling nature, and that those laws could be discovered. A few hundred years later, **Aristotle** (384–322 B.C.), whom we might consider to be one of the first marine scientists, offered in his treatise *Meteorologica* a number of explanations for how the oceans were formed and why they are salty. He even attempted to explain why there are ocean tides. The Roman natural philosopher **Pliny the Elder** (ca. A.D. 23–79) provided numerous improvements on earlier marine observations, especially on tides and the ocean's saltiness, in his classic work, *Natural History*. The advancement of science was underway, but it was to be interrupted in the West for a thousand years. Only a few hundred years after Pliny's contribution, Europe was plunged into the Dark Ages, which lasted from the decline of the Roman Empire (ca. A.D. 450) to the European Renaissance (between the fourteenth and seventeenth centuries A.D.). Except for a treatise on tides by the **Venerable Bede** (A.D. 673–735) published in A.D. 730, very few scientific advancements were made in Europe (at least few were recorded for posterity).

But we know that long before the time of Thales or Aristotle, ancient mariners had successfully undertaken remarkable open ocean voyages, which, it would seem, must have required some level of understanding of "ocean science." Though their understanding was not truly scientific by today's standards or even by those of the great Greek philosophers, their feats were nonetheless remarkable.

The Early Ocean Voyagers

It is probably safe to say that interest in the oceans is as old as humankind; after all, we humans inhabit every major island and land mass on the globe except Antarctica. If modern civilization had a starting place, then we must somehow have ventured across oceans to expand to where we are today. And, of course, there must have been a motivation, some force that drove our ancestors to wander about and colonize the world. Long ago that motivation must have been born of simple necessity; our ancestors most certainly relied on the oceans for food, for early forms of trade and commerce, and, as history has shown, as an efficient means to wage wars across large distances. And so we have been living by the oceans throughout recorded history, extracting their resources and developing efficient ways to move about on them in ever more sophisticated ships. In fact, we have evidence of truly remarkable voyages that predate human history and led to the development of the ocean sciences.

FIGURE 1.1 (A) Ancient map of the world according to Herodotus, a Greek historian who lived in the fifth century B.C. (ca. 484–425 B.C.). (B) A modern map showing the same region and its relation to the rest of the globe.

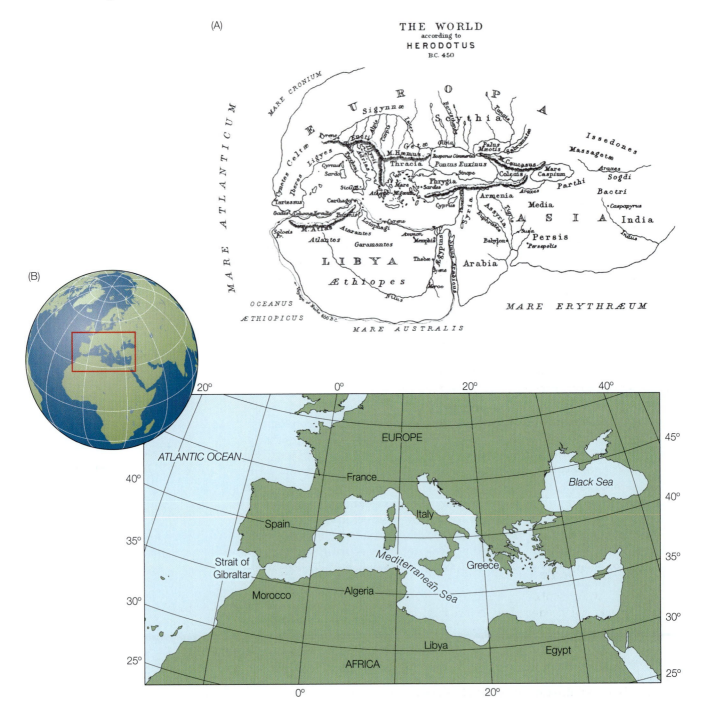

By 900–700 B.C., the ancient Greeks are known to have ventured as far west as the Strait of Gibraltar, the narrow passage that connects the Mediterranean Sea with the Atlantic Ocean, a distance of some 750 miles from their home. But that's as far as they dared to venture, because before them lay the open Atlantic Ocean, which, we now know, dwarfs the Mediterranean Sea. And as any sailor today without navigational aids should be, the ancient Greeks were fearful of venturing very far beyond sight of land. For this reason, early maps of the ancient world were remarkably accurate as far as the coastline of the Mediterranean Sea goes, but Gibraltar is where their maps of the world ended (**FIGURE 1.1**).

Because the Strait of Gibraltar has very strong ocean currents (for reasons we will see later), the ancient Greeks and Romans thought that on the other side was not an ocean, but a great river, a river they thought encircled the world. In fact, the word *ocean*, from the Greek *okeanos*, originally meant "great river."

Predating the ancient Greeks, but not nearly as well recorded in history, were the ancient Polynesian voyagers, who were known to be highly skilled navigators. As far back as 5000 years ago, they were somehow able to navigate the island-dotted South Pacific Ocean by using, it is thought, the positions of the Sun and stars, cloud formations, the patterns of surface waves on the ocean, wind directions, the flight paths of migratory birds, and even the smell of the water!

Even before that, the Polynesians had managed to settle throughout the vast expanse of the mid-South Pacific known as Oceania. The map in **FIGURE 1.2** shows the expansion of these ancient wanderers through the 10,000 or so islands in the western Pacific Ocean; they populated New Guinea 30,000 years ago, the Philippines 20,000 years ago, and Hawaii—the island system farthest away from any major land area—about 1500 years ago. These over-water migrations of the Polynesians (and the Greeks much later) were feats of ship building and physical courage, but they were fundamentally dangerous because there was no way to map the territory explored. In order for exploration of the world ocean to proceed beyond the haphazard and dangerous way of simply wandering around the oceans and hoping to bump into a land mass before their ship's rations are exhausted—in order for exploration to become more scientific—mariners needed a way to make maps that showed locations of oceans and land masses, and they needed a way to determine the precise position of their ships. They needed to develop basic navigational principles to know where they were and where they were going. As simple as this must sound, it was actually a big problem that only sorted itself out in the mid to late 1700s.

The Principles of Navigation

The problem of scale

In the early development of principles of navigation, two major problems needed to be solved. First, early mariners (as well as explorers on land) had to have some idea of *scale*. That is, they had to determine the relative size, or scale, of the Earth, its oceans, and its land masses. So step one was to determine how big the Earth is. About 200 B.C., enter the Greek astrono-

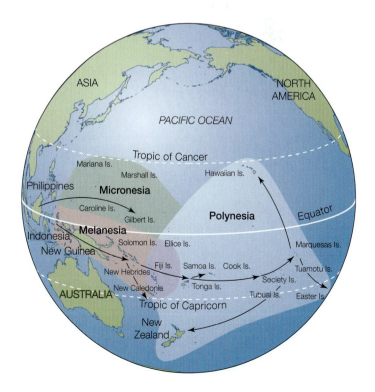

FIGURE 1.2 Oceania and the presumed order of colonization of regions of the island-dotted South Pacific Ocean.

mer, geographer, mathematician, and third librarian of the Great Library at Alexandria, **Eratosthenes of Cyrene** (he was actually from the region that is now Libya, in northern Africa). Eratosthenes was the first to calculate with any degree of accuracy the circumference of the Earth. The way he did this was quite clever and simple (**FIGURE 1.3**).

The ancient Greeks and Egyptians had known since the time of **Pythagoras** (about 600 B.C.) that the Earth is a sphere (this was, of course, long before all the trouble in the Middle Ages and the belief in a flat Earth that must have given Chistopher Columbus headaches). With that knowledge, Eratosthenes put together two observations: First, every year on the summer solstice (June 21) at high noon in the ancient city of Syene (known today as Aswan, Egypt), the Sun shone directly down a deep dry well. That is, the Sun was directly overhead. Of course, today we know that this is because June 21 is the longest day of year in the Northern Hemisphere—the first day of summer—when the Sun is at its highest point in the sky, which is directly over latitude 23.5° north (more about this below). And because Aswan, Egypt, is at about latitude 24° north, the Sun was very close to being directly over Aswan at 12:00 noon.

Second, Eratosthenes determined that in Alexandria, Egypt, which is some distance north of Aswan, the Sun was not directly overhead at the same time on June 21 that it was in Aswan. Instead it was some 7° shy of being directly overhead, which means that at the solstice, a long upright pole in Alexandria casts a shadow at an angle of about 7°. Eratosthenes not only knew that the Earth is a sphere, he also knew, or assumed, that the Sun is very, very far away—so far that the rays of sunlight are essentially parallel to one another as they arrive on the surface of the Earth. So his problem could be reduced to simple geometry: because there are 360 degrees in a circle, 7 degrees is about 1/50 of a circle (it is actually 1/50.43 of 360 degrees); therefore, the distance from Aswan to Alexandria is about 1/50.43 of the circumference of the Earth. So far, so good. Now for the most interesting part of all this: the ancient Egyptians knew that the distance from Aswan to Alexandria was just about 785 kilometers, or 491 miles (actually, the units they used were *stadia*). They calculated this distance based on uncannily accurate estimates of the walking speed of camels! Thus, Eratosthenes calculated the Earth's circumference as 491 × 50.43 = 25,252 miles—which is very close to the actual value. Eratosthenes was, indeed, an astute individual; he also determined the tilt of Earth's axis to within a degree, calculated that there are 365¼ days in a year, and proposed leap years, the addition of one day to the calendar every four years. But it would be two centuries before Julius Caesar made that modification to the calendar.

Eratosthenes' calculation of the Earth's circumference was a big breakthrough, and cartography (map making) took off after this, with maps on globes showing up in the Great Library at Alexandria. But alas, another Greek, **Ptolemy**, repeated the geometry exercise, and somehow messed up, concluding that the Earth was some 30% smaller in circumference than it actually is. And when the Library at Alexandria was burned under Roman Rule in A.D. 415, Eratosthenes's work was lost and we were left with Ptolemy's

FIGURE 1.3 Simple geometric method used by Eratosthenes to estimate the circumference of the Earth.

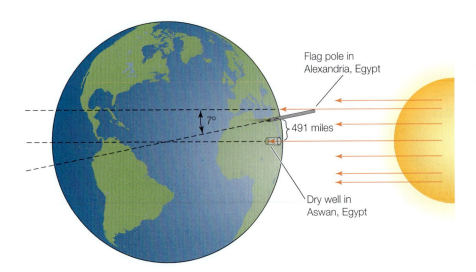

Flag pole in Alexandria, Egypt

7°

491 miles

Dry well in Aswan, Egypt

mistake throughout the Dark Ages (ca. A.D. 400–1400). Fortunately, Eratosthenes's work had been recorded by others, and was later rediscovered.

The problem of locating one's position

Once we had a good idea how big the Earth was, we had to devise a way to determine where we were on that Earth, and if we were traveling, where we were in relation to our destination. So the second problem that ancient mariners had to solve was how to establish some kind of coordinate system to allow maps to be drawn to scale. The answer to this problem is the system we know today as the grid of **latitude** and **longitude**, which looks just like an ordinary graph with an abscissa and an ordinate (or *x*- and *y*-axes). With two measurements (*x* and *y*) you can define, or fix, your position anywhere on the Earth (**FIGURE 1.4**).

This is straightforward when we are dealing with a flat plane, or a small area, such as a map of your local town or your college campus, and it even works quite well at larger scales, such a map of your state. But at still larger scales, because Earth is a sphere, there were problems to be solved before the grid could be established.

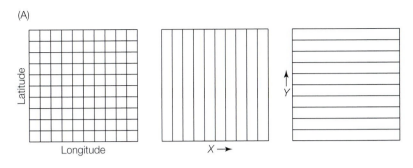

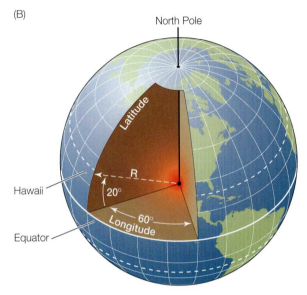

FIGURE 1.4 (A) The system we use for latitude and longitude functions just like an ordinary graph using *x* and *y* coordinates. Lines of longitude (*x*), running north–south, determine position in the east–west direction, and lines of latitude (*y*), running east–west, determine position in the north–south direction. (B) Lines of latitude are measured as angles from a point in the center of the Earth; lines of longitude are measured as angles from the axis of rotation.

Dealing with Spherical Geometry of the Earth

Latitude

On the spherical Earth, lines of latitude are based on angles, as shown in **FIGURE 1.5**. The angles are measured along the north–south axis from an imaginary point in the center of the Earth, and they transcribe lines that ring the Earth, much like the lines of the *y*-axis on the flat piece of paper just discussed. These angles give us the "lines of latitude." We can draw any number of such lines of latitude (an infinite number, really). And because latitude is a fraction of the Earth's circumference (a fraction of a 360° circle), 1 degree (1°) of latitude anywhere on Earth is an exact distance: it is exactly 1/360 the circumference of the Earth (**BOX 1A**). This gives us several latitudes that are important to keep in mind: the Equator = 0°; the North Pole = 90°N, and the South Pole = 90°S. There are several other major lines of latitude on the Earth, the lines we see drawn prominently on all globes, that are related to the annual changes in the relative positions of the tilted Earth orbit and the Sun's angle as its rays of light hit Earth, which give Earth its seasons.

THE SEASONS   The reason we have seasons is that the axis of the Earth's rotation (that is, the imaginary line that passes through the Earth from pole to pole, around which the Earth spins one rotation each day) is tilted at an

(A)

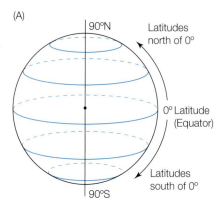

(B)

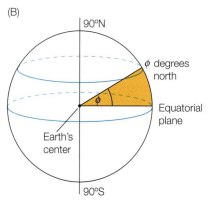

FIGURE 1.5 (A) Lines of latitude are measured from a point in the exact center of the Earth. Any angle drawn from this point to a point on the surface of the Earth will "transcribe" a line of latitude that encircles the globe. These lines of latitude give one's position in the north–south direction. (B) Zero degrees is the Equator, and we can measure from 0 to 90 degrees north or south of the Equator. 90°N is the North Pole, and 90°S is the South Pole.

angle of 23.5° relative to the plane on which the Earth orbits the Sun each year (**FIGURE 1.6**). The following is a detailed explanation of what happens over the course of one year in the Northern Hemisphere.

What happens over the course of a year is this: first, when the Sun is directly over the Equator (the top position in Figure 1.6), the Sun's rays of light hit all parts of the Earth equally, from pole to pole, as the Earth rotates once each day. This is the **autumn equinox**. As the Earth orbits the Sun for three months (which equals one-fourth of a year) to a new position one quarter the way around its orbit of the Sun, the Sun will be directly over latitude 23.5°S, which is the Tropic of Capricorn. The day this occurs is known as the **winter solstice**, when the Sun does not rise above the horizon at any time of the day north of latitude 66.5°N (the Arctic Circle). It is the shortest day of the year and the start of winter in the Northern Hemisphere. Also, notice in Figure 1.6 that at this time, there is always daylight below the Antarctic Circle—the Sun doesn't set all day. Allowing the Earth to continue orbiting the Sun for another three months, we are back to a position where the Sun is once again directly over the Equator and all of the Earth receives equal sunlight—12 hours of daylight and 12 hours of darkness—from pole to pole. This is the **vernal** (or **spring**) **equinox**, the first day of spring in the Northern Hemisphere. The fall and spring equinoxes occur when the Sun is halfway between its positions for the winter and summer solstices—that is, directly over the Equator. Three months after that, the Sun is directly over latitude 23.5°N, which is the Tropic of Cancer. This is the **summer solstice**, or the first day of summer in the Northern Hemisphere. On this day, the Sun never sets at latitudes above the Arctic Circle (66.5°N) and never rises at latitudes below 66.5°S. After another three months, a total of one year, we are back to the starting position in Figure 1.6.

FIGURE 1.6 Latitudinal changes in the angle of the Sun as a function of season.

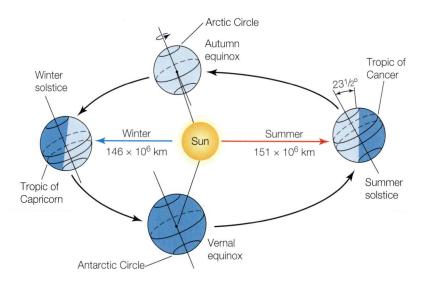

BOX 1A   Degrees of Latitude in Terms of Miles

Because latitude is a fraction of the Earth's circumference (some fraction of a 360° circle), then:

- 1 degree of latitude anywhere on Earth is an exact distance, and
- 1° = 1/360 the circumference of the Earth; and there are 360° in a circle, 60 minutes in a degree, and 60 seconds in a minute.

By definition: "1 minute of arc of a great circle of the Earth equals 1 nautical mile."

- 1 nautical mile = 1 minute of latitude
- 1 minute of latitude = 1852 meters (6076.1 feet)

Therefore, the circumference of the Earth equals:

- 360° × 60 min/degree = 21,600 minutes = 21,600 nautical miles (nmi)
- 21,600 nmi × 1852 m = 40,003.2 kilometers (km)
- 21,600 nmi × 1.11577 statute miles/n mi = 24,856 statute miles (mi)

Remember this general rule of thumb at sea: "A mile is a minute" (a nautical mile, that is).

It is important to note that the Southern Hemisphere experiences exactly the opposite throughout the year, but the terminology and the seasons are reversed. For example, the autumn equinox in the Northern Hemisphere is the vernal equinox in the Southern Hemisphere, and when it is the winter solstice in the Northern Hemisphere, the Southern Hemisphere is experiencing their summer solstice (the longest day of their year and the beginning of summer).

These seasonal phenomena are the basis for these prominent lines of latitude on all globes: the Arctic Circle (66.5°N), the Antarctic Circle (66.5°S), the Tropic of Cancer (23.5°N), the Tropic of Capricorn (23.5°S) and, of course, the Equator (0°). But more importantly, it is because of these differences in the Sun's angle relative to the Earth's surface over the course of year that we have differences in the heat input from the Sun's rays. When the Sun is at a low angle, such as in the Northern Hemisphere winter, the Earth's surface does not acquire much heat, and it cools down. The opposite occurs in summer. We'll be discussing more of this phenomenon in the chapter on atmospheric circulation.

DETERMINING ONE'S LATITUDE   Because latitude is an angle measured from a fixed point in the center of the Earth, it is fairly easy to determine one's latitude, or at least to get a good approximation. For example, based on the above discussion, one way is simply to estimate the Sun's angle and correct for the time of year. Even with the naked eye we have a pretty good idea where we are at any particular time of year, just by checking the Sun's angle at high noon, and making a mental correction for the time of year. Most of us notice that the Sun climbs higher in the sky as winter transitions into spring and summer (in the Northern Hemisphere, that is). Likewise, if we were to drive north or south any significant distance, we would notice a similar change in the Sun's angle relative to our starting point.

The technique of determining latitude based on the angle from the horizon to the Sun at its highest point of the day, and then correcting for season, was known at least as early as about A.D. 400, when mariners used early versions

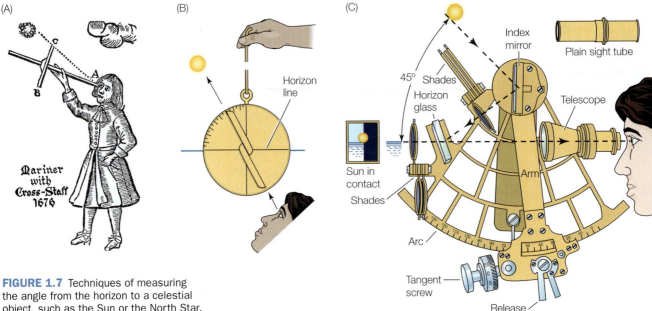

(A)

Mariner
with
Cross-Staff
1676

(B)

Horizon
line

(C)

Index
mirror

Plain sight tube

45° Shades

Horizon
glass

Telescope

Sun in
contact

Shades

Arm

Arc

Tangent
screw

Release

FIGURE 1.7 Techniques of measuring the angle from the horizon to a celestial object, such as the Sun or the North Star, Polaris; (A) cross-staff and (B) astrolabe, both in use as long ago as A.D. 400; (C) a modern sextant.

of a cross-staff and astrolabe to measure angles. Later, in the 1600s and 1700s the much more accurate sextant was developed (**FIGURE 1.7**).

An easier way than using the Sun to estimate latitude is to measure the angle to the North Star, Polaris. The angle from any given point in the Northern Hemisphere to Polaris is constant—it does not change with the seasons—because Polaris is almost exactly on a line in space that is an extension of the Earth's axis of rotation (**FIGURE 1.8**). This method only works in the Northern Hemisphere; unfortunately, there is no analog to the North Star in the Southern Hemisphere.

Longitude

Like latitude, longitude is also an angle measured from the center of the Earth, except that it is measured from an imaginary *line* that runs from pole to pole—along the Earth's axis of rotation through the center of the Earth—not from a *point* in the center (**FIGURE 1.9**). Because the Earth is rotating about that line, our east–west position, unlike our north–south position on the Earth's surface, is constantly changing over a 24-hour period as we spin through space. That is, if you do not move across the surface of the Earth, but are merely "on for the ride" as Earth rotates, you won't change position with respect to the angle measured from a central point, so your latitude will be constant; but, relative to the fixed stars out in space, your apparent longitude will be constantly changing.

So, what is needed is some reference line on the surface of the Earth, analogous to the Equator, that can serve as a zero-degree starting point for east–west calculations. That artificial reference line is the **Prime Meridian**, a line running from pole to pole on the surface of the Earth through one specific point. The Prime Meridian could have been placed anywhere, but the British had some influence at the time all this was being discussed and debated, and they declared that the Prime Meridian, longitude 0°, would run through Greenwich, England, where, not coincidentally, it already was and had been for a long time. All longitudes would be measured from 0° to 180° west of Greenwich, and 0° to 180° east of Greenwich, thus circling the globe.

(A)

(B)

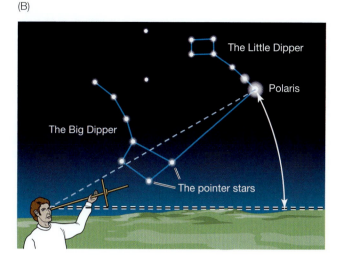

(C)

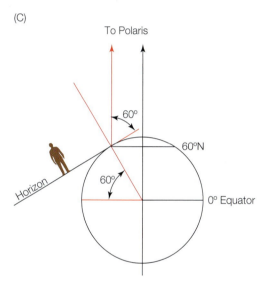

FIGURE 1.8 (A) A mariner using a sextant to measure the angle to the Sun to take a noon fix. (B) Polaris, the North Star, lies almost directly on Earth's axis of rotation, and its angle from the horizon can be measured using a cross staff. (C) The geometry showing that the angle from the horizon to Polaris is equal (very nearly so) to one's latitude.

(A)

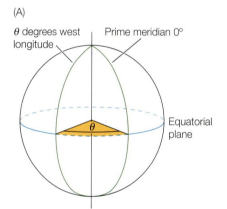

(B)

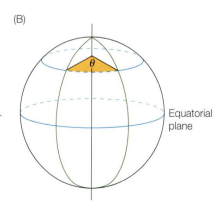

So now we had a coordinate system of latitude and longitude. And latitude, as we just discussed, was fairly easy to determine, even if you were at sea somewhere out of sight of land, for example. But longitude was another matter altogether. In fact, it was not until the mid-1700s that there was a reliable and relatively straightforward way to fix your longitude.

Unlike latitude, we could not use the positions of heavenly bodies to determine longitude—not easily, at least.[1] Imagine how disconcerting it must have been to be sailing east or west across a big ocean without any way of knowing precisely how far you had sailed. Sure, you could determine how far away from the Equator or the North Pole you were at any particular time, but how far east or west had you ventured from your starting point?

---

[1] One rather complicated method involved the tabulation of the Moon's position relative to a fixed star or the Sun, and changes in that position with time and with changes in longitude.

FIGURE 1.9 Longitude is also based on an angle (A), but it is a function of time. It is measured as a angle anywhere up or down along the Earth's axis of rotation. (B) The angle between any two meridians remains the same regardless of latitude north or south.

BOX 1B   Longitude and Time

Knowing that 1 hour is equivalent to 15° longitude (in one hour the Earth rotates 15°, or 1/24 of a complete rotation), then at your ship's present position, when the Sun is at its highest point in the sky (i.e., when it is exactly Local Apparent Noon, LAN), you can determine your longitude by comparing your LAN with the time it is in Greenwich (0° longitude). You thus need to keep an accurate clock on board (a chronometer) that is set to Greenwich Time, called Greenwich Mean Time, or just GMT.

For example, if your ship's chronometer says it is 2:00 p.m. GMT at the exact moment you determine that it is LAN at your ship's current position, then you and your ship are 2 hours "behind" Greenwich (i.e., you are to the west of Greenwich, in the sense that the Earth rotates west-to-east). Therefore your position is 2 × 15°, or longitude 30° west (west of Greenwich); you are somewhere out in the Atlantic Ocean.

Note that "behind" Greenwich is where *post meridiem*, or p.m., comes from. "Before" Greenwich, or *ante meridiem*, is the origin of a.m.

Because the Earth rotates to the east once each 24 hours, every hour it moves 360 degrees / 24 hours = 15° of longitude. This is why we see on virtually all globes these prominent lines of longitude, **meridians**, drawn every 15° east and west of Greenwich. Longitude is thus a function of *time—the difference between the local time at one's position and the time it is in Greenwich!* This difference in time (in hours) will dictate your angular distance from Greenwich. For example, at **Local Apparent Noon (LAN)** (12:00 o'clock local time), when the Sun is at its highest point in the sky at your present position, you need to know what time it is in Greenwich. To get your longitude, you simply multiply the time difference in hours by 15° (**BOX 1B**). And this was precisely the problem throughout much of the history of ocean exploration until the mid to late 1700s: in order to know the exact time at a prime meridian, to compare with your "local" time, you needed to bring a clock to sea with you—an accurate clock, set to the time of the prime meridian. It had to be accurate, because if it was fast or slow by only one minute, you would be off course by plus or minus 1/60 of 15°, which at the Equator is as much as 17 miles. And, of course, early clocks were good only on land, since they depended primarily on pendulums, and pendulum clocks don't work aboard a moving ship at sea!

In the days before there was any reliable way to determine longitude, voyagers would simply sail toward the latitude of their destination, usually unaware of their exact longitude, or of their east–west progress. They could compute latitude easily from the Sun and Polaris, based on the time of year (number of days after the winter solstice, etc.). So they would sail east or west for a given amount of time, or until they reached a known landmark, and then go north or south, in a course that amounted to a *square wave* pattern (**FIGURE 1.10**), which, of course, is not very efficient.

An accurate method to determine longitude was needed, and in 1714 the British Parliament established the Longitude Board,

FIGURE 1.10 Example of sailing in square waves from Plymouth, England to Jamestown, Virginia.

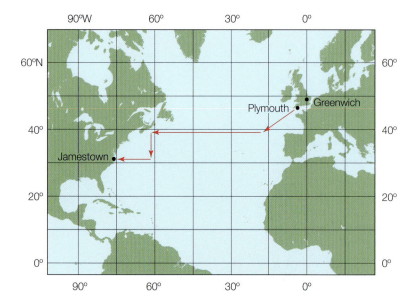

(A)

(B)

(C)

FIGURE 1.11 (A) John Harrison. Harrison's chronometers, like this version H5 (B), were used in the 1700s and early 1800s, when marine chronometers (C) took over until the mid to late 1900s. This marine chronometer, in a gimbaled case, was used aboard the HMS *Beagle*.

which offered a monetary prize to anyone who could find a practical way to determine longitude at sea. Several years later, the British Royal Admiralty offered a reward of £20,000 (a fortune in those times) to anyone who could build a clock that could be taken on long voyages at sea. **John Harrison** (1693–1776), a British clockmaker, took up the task in 1730, and over a span of some 40 years he developed five different versions, each one an improvement over the previous; his last, version H5, was shown to be accurate to about 1/3 of a second per day. While Harrison did manage to get financial support for his efforts from the Admiralty, he never received the £20,000 reward, nor did anyone else. At age 80, just three years before his death, Harrison received an award of £8,750 from the British Parliament in recognition of his accomplishment. By the late 1700s and early 1800s, various versions of Harrison's **chronometer** were routinely in use (**FIGURE 1.11**).

Today, even the cheapest department-store quartz watches keep very accurate time, on the order of that possible with Harrison's H4 and H5 models. However, we no longer rely on such "noon fixes" of the Sun to determine longitude; instead, we use GPS, a "Global Positioning System" based on a fleet of Earth-orbiting satellites that can triangulate one's geographic location to within a few meters!

Coming Out of the Dark Ages: The Renaissance Europeans

A more scientific and better organized approach to ocean exploration began about the time that Europe was coming out of the Dark Ages, during the Age of Discovery (late fifteenth century to the early seventeenth century). It was perhaps **Henry the Navigator** (1394–1460), a Prince of the Kingdom of Portugal during the early part of the Portuguese Empire, who was responsible for stimulating Europeans' interest in exploration of the world. He organized sailing explorations down the West African coast for the purpose of making navigational maps for military operations and for commercial purposes—especially, to discover a sea route to the Indies and the Orient and capture a share of the rich spice trade (*the Indies* referred to South and Southeast Asia, later known as the East Indies). Between 1444 and 1446 Henry sent some 30 ships on missions of discovery down the west coast of Africa, although he never went along himself. Prior to these voyages and dating back to the ancient Greeks, mariners had been fearful of sailing too far south of Gibraltar; legends spoke of sea monsters and a Sun so hot that ships would burn.

Henry is credited by some with the establishment of a Center for the Study of Navigation and Marine Science, and thus the development of organized and scientific approaches to ocean exploration. In addition, while the earliest use of a **magnetic compass** (**BOX 1C**) was apparently by Zheng He between 1405 and 1433, it is Prince Henry who is credited with is routine use.

Between the years 1405 and 1433, during the Ming Dynasty, the Chinese explorer **Zheng He** (1371–1433) from the Chinese province of Yunnan commanded a series of voyages throughout Southeast Asia and lands bordering the Indian Ocean as far west as East Africa. With a fleet of more than 300 ships, some longer than 130 meters, Zheng He spread word of Chinese cultural and scientific advances to distant lands. His fleet had the benefit of the magnetic compass and capabilities that included distilling fresh water from sea water and growing vegetables at sea, which allowed them to launch voyages that lasted months at sea. But following Zheng He's death in 1433, the Chinese abandoned their seagoing explorations for reasons that are not entirely clear today, except that the political climate had changed and the Chinese were becoming more isolated.

Of course no discussion of ocean explorers of the 1400s is complete without mention of **Christopher Columbus**, who, while also searching for a sea route to the Indies, discovered ("by accident") the New World. He actually landed in the Bahamas; exactly which island in the Bahamas that was remains unresolved. While he was not the first European to sail to the New World, his achievement is nonetheless significant in that he set into motion the exploration and colonization of the New World by Europeans.

As most of us will recall learning as school children, it was the **Vikings** who first came to North America, as many as two dozen times, around 950 A.D.

BOX 1C The Magnetic Compass

While the exact date when the magnetic compass was first used in navigating a ship is unknown, it is thought that the ancient Greeks, and certainly the Chinese, were aware of the north-seeking properties of a needle rich in loadstone (magnetite).

The needle orients itself in the Earth's magnetic field such that one end points to the Earth's magnetic North Pole, which may be 10–20 degrees off from true north, depending on location. Nonetheless, it was one of the first scientific breakthroughs in the science of voyaging, allowing a ship to orient its direction without relying on either celestial bodies or land-based reference points.

(Left) Magnetic compass (ca. 1920) with a delicately-balanced magnetic needle that always points to the magnetic North Pole, which deviates from the true North Pole, depending on your exact location. This instrument is marked showing a magnetic deviation of some 19° west of true north. (Right) A gimbaled marine compass, with a marked plate rather than a balanced needle, and immersed in oil to dampen the ship's motion.

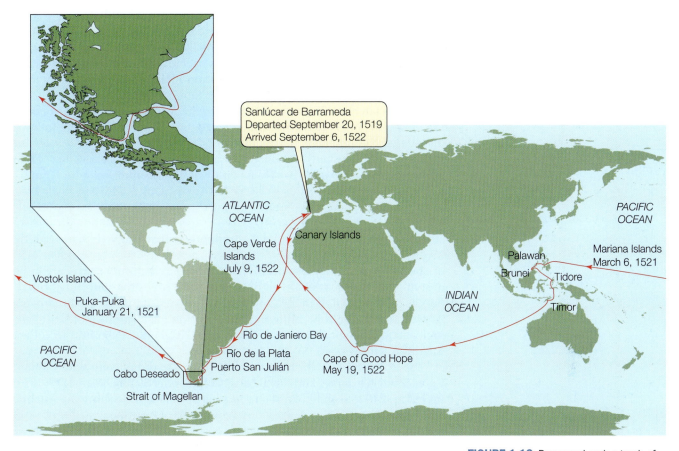

FIGURE 1.12 Presumed cruise track of Magellan's circumnavigation of the globe and a map showing detail of the Strait of Magellan and his route through it.

Nonetheless, Columbus was among the first to articulate the idea that if the Earth were a sphere, then the Indies could be reached by sailing west from Europe, across the Atlantic Ocean and around the world to Asia. And it was only after Columbus that charts of the world's oceans were drawn showing the New World. The age of ocean exploration was underway.

One of the greatest achievements in the history of ocean explorations has to be that of **Ferdinand Magellan** (1480–1521) of Portugal, who is credited with being the first to circumnavigate the globe. Well, his men did, anyway.

Like Columbus, Magellan believed that he could sail west to reach the Orient, and in 1519 he set out with 270 men and 5 ships. He discovered a shortcut that avoided the dangers of icebergs and notoriously heavy weather in what we now know as the Drake Passage, the narrow stretch of ocean separating Cape Horn at the southern tip of South America from Antarctica (Magellan was unaware that Antarctica existed, however; that discovery came much later). His shortcut is named, aptly, the Strait of Magellan (**FIGURE 1.12**).

Three years after they set out, 18 men in 1 ship returned to Portugal—without Magellan, who had been killed by natives in the Philippines.

Post-Renaissance Ocean Exploration

Some 125 years after Magellan, British naval officer **Captain James Cook** (1728–1779) led three major voyages of discovery. In 1768, Cook and the HMS *Endeavor* left Plymouth, England on a round-the-world voyage to document the transit of Venus across the Sun, in addition to exploring the South Pacific (**FIGURE 1.13**). Along the way he found and charted New Zealand and Australia's Great Barrier Reef. Cook was among the first to realize the importance of diet on long voyages. He used citrus extract to avoid **scurvy**, a debilitating and eventually fatal disease that results from Vitamin C deficiency that had theretofore wreaked havoc with crews on long voyages. He returned home three years later, in 1771, only to set out again on another three-year voyage the next year, in 1772. During this voyage, he is thought by some to have discovered the Antarctic continent, in 1773 (but see page 19). Cook's third voyage left England in 1776 to look for a **Northwest Passage** (see Box 1D) to get around North America. While exploring the Pacific Ocean he "found" Hawaii, charted the west coast of North America, and returned to Hawaii, where he and his crew apparently violated some taboos and the Hawaiians killed him.

Cook benefited from the new chronometer that Harrison had recently invented; he could determine his longitude with some precision, which, of course, is extremely important if accurate navigational charts are to be drawn. He also used lead weights attached to ropes for sounding bottom depths, and he coated the lead weights with wax so that they would catch material on the bottom, allowing him to bring back samples of both geological and biological material. Thus, Cook was one of the first to collect scientific samples.

The search for the Northwest Passage (**BOX 1D**) did not end with Captain Cook. Scottish explorer and Royal Navy Commander **Sir John Ross** (1777–1856) set out in 1818 to search for a such a northern sea lane connecting the Atlantic and Pacific Oceans. He left London in command of two ships, the *Isabella* and the *Alexander*, in what was to be the first of a series of Arctic expeditions aimed at resolving the question of whether a Northwest Passage existed. Along the way he made scientific observations regarding ocean currents, tides, and bottom depths, and like Cook, he collected specimens. While he did not discover a passage, he nonetheless sounded to a depth of 1919 meters near Greenland, recovering a bottom sample from that depth. We'll discuss this more later, but

FIGURE 1.13 (A) Painting of Captain James Cook by Nathaniel Dance, 1775. (B) Painting by Thomas Luny, 1790, of Cook's ship, HMS *Endeavour*, leaving Whitby Harbour in 1768.

(A)

(B)

sounding more than a mile deep was no mean feat. Ross's nephew, **Sir James Clark Ross** (1800–1862), also a British naval officer and Arctic explorer, famous for his discovery of the Ross Sea and Victoria Land in Antarctica, did even better than his uncle: in the South Atlantic, Ross sounded to a depth of 4893 meters—about 3 miles deep!

Known less for its contributions to marine science than for those to the development of the theory of evolution was the five-year voyage of **Charles Darwin** (1809–1882) aboard HMS *Beagle* from 1831 to 1836 (**FIGURE 1.14**). The voyage of the *Beagle* was primarily a mapping expedition, on which Charles Darwin served as the ship's naturalist. As we will learn later, this cruise led to a number of Darwin's significant contributions to our understanding of marine geology and biology.

A little-known but highly significant voyage of discovery was that of the United States Exploring Expedition (1838–1842), nicknamed the "U.S. Ex. Ex." The expedition was led by U.S. Navy Lt. **Charles Wilkes** (1798–1877) who was in command of the *Vincennes* (the flagship) and five other ships on a four-year, round-the-world voyage (**FIGURE 1.15**). This expedition was significant in that its crew included a number of natural scientists—botanists, taxidermists, and mineralogists. But, almost paradoxically, the expedition has been all but forgotten in history. While much of what is known is based on the detailed, five-volume "*Narrative*" published by Wilkes in 1845, journal accounts written by "Passed-Midshipman" (equivalent to today's Annapolis graduate) **William Reynolds** place the expedition in a different context. Apparently, Reynolds, like most of the crew, did not get along very well with his superior officer, Lt. Wilkes, according to one modern account.[2] But as for the legacy of this nonetheless monumental voyage, it was almost relegated to historical obscurity. Part of the reason is because the six year voyage spanned three presidential administrations: the eighth President of the United States, Martin Van Buren (1837–1841), a Jacksonian Democrat, under whom the expedition was launched, the ninth President William Henry Harrison (1841), a Whig who died of complications

---

[2] Nathaniel Philbrick, 2003. *Sea of Glory: America's Voyage of Discovery; The U.S. Exploring Expedition* (New York: Penguin).

FIGURE 1.14 (A) Watercolor portrait of the young Charles Darwin by George Richmond in the late 1830s. (B) HMS *Beagle* in the Strait of Magellan.

(A)

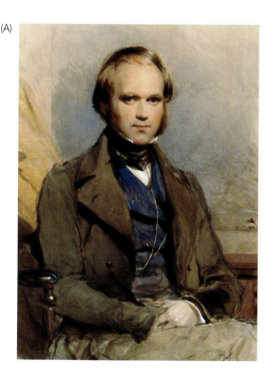

(B)

(A)

(B)

FIGURE 1.15 (A) The USS *Vincennes* in Disappointment Bay, 1844. (B) Photograph of the *Vincennes*'s controversial commander, U.S. Navy Lt. Charles Wilkes.

stemming from a cold in only his thirty-second day in office, and the tenth President John Tyler (1841–1845), a longtime Democratic-Republican, who was elected as Tyler's Vice President on the Whig ticket. Upon taking the oath of office in succeeding Harrison, Tyler became a bit of a rebel and took positions that were counter to his party's platform; most of his cabinet resigned and, his party, the Whigs, expelled him. In carving out his own place in history there was little political incentive for him to even acknowledge—much less show enthusiasm and support for—the successes of an initiative launched by a previous administration.

Although the accomplishments of the expedition were eclipsed by the political changes occurring in the U.S. during the years that the expedition was at sea, today we understand that, indeed, the U.S. Ex. Ex. accomplished a great deal. The expedition "discovered" Antarctica and mapped the Pacific Northwest, including San Francisco Bay and the treacherous mouth of the Columbia River. In four years, the expedition that began with 6 ships and 346 men sailed more than 87,000 miles, exploring and mapping more than 800 miles of coastline in the Pacific Northwest and more than 1500 miles of the Antarctic continent. While the primary mission was surveying and making navigational charts of distant coastlines—some of the 180 maps created were still being used during World War II—the crew also collected thousands of marine and terrestrial biological and geological samples and made significant contributions to anthropology. Perhaps the best outcome of the expedition was that it made it possible for a scientist to earn a living in the United States, which hadn't really been possible before, science being considered a "gentleman's hobby." The U.S. Ex. Ex., however, helped to change all that. Indeed, in part because of the tremendous volume of samples that needed to be catalogued and stored, the Ex. Ex. was the driving force that led to the creation of the Smithsonian Institute.

One result of all these world voyages of discovery was the realization that the oceans had powerful currents that, if not understood, could impede a ship's progress and set it well off course, but those open ocean currents were a bit enigmatic. After all, how could anything but a river have currents? (We'll learn how in a later chapter.) Significant insights into ocean currents

BOX 1D   The Search for the Northwest Passage

The search for a Northwest Passage, a commercial sea route to the north around North America that connects the Atlantic and Pacific Oceans, has been sought by explorers for centuries. In 1906, the Norwegian explorer **Roald Amundsen** (1872–1928) became the first to successfully navigate a Northwest Passage when he sailed a 47-ton steel-constructed seal-hunting vessel, the *Gjøa*, from Greenland to Alaska. But to complete the trip, Amundsen and his crew of six men ended up spending three winters locked in the ice in what today is northern Canada. Eventually they were able to navigate their way to the Bering Sea, to the north of Alaska, and to the Pacific Ocean. But the route they used could never become commercially viable because of the extremely shallow depths—just over a meter deep in spots. Larger, commercial vessels require much deeper depths. Nonetheless, Amundsen's feat was monumental, and was subsequently duplicated several times by others in specially fortified ships. Normally such a trip is blocked by sea ice and until just recently the waters of the Arctic Ocean have been frozen year-round to the extent that reliable passage of commercial ships between the Atlantic and Pacific Oceans was not possible, and so the only viable route remained the long way around South America.

In recent years, however, the Arctic ice has been melting more and more during late summer, as a result of global warming (which we will discuss later) such that a Northwest Passage is becoming a reality. By August 21, 2007, for example, the Northwest Passage opened up enough that large ships could pass without the need of an icebreaker. This image shows the sea ice extent as of September 9, 2011.

were first offered by **Matthew Fontaine Maury** (1806–1873), a U.S. Navy officer, who was among the first to begin to make sense of the world's ocean currents and to recognize that there was a worldwide pattern to them. His book, *The Physical Geography of the Sea*, first published in 1855, was the first "oceanography" book, although the term had not yet been coined. In it he discussed aspects of coastlines, ocean basins, winds, tides, ocean currents,

(A)

(B)

FIGURE 1.16 (A) Matthew Fontaine Maury (1806–1873). (B) Detail of his Plate IV, showing currents in the North Atlantic Ocean, from his *Physical Geography of the Sea*, originally published in 1855 and considered to be the first "oceanography" book.

and marine life. His summary maps of the major ocean currents (**FIGURE 1.16**) were remarkably accurate, as we'll see in Chapter 5. His analyses were quite elementary: he simply pored over a huge number of observations made by thousands of mariners, in the form of ships' logs and captains' journals "obtained from old sea chests," as he explained. For this work Maury is often referred to as the Father of Physical Oceanography. But it was not until a while after Maury's landmark book that the fields of oceanography and marine biology began in earnest.

The Birth of Oceanography

The Challenger *expedition*

At exactly 11:30 A.M. on Saturday morning, December 21, 1872, the refitted British steam corvette HMS *Challenger* under the command of Captain George S. Nares and his crew of 23 naval officers and 240 tars and bluejackets (deckhands), along with a 6-man scientific party, pushed off from the jetty at Portsmouth, England, and set sail for a three-and-a-half-year round-the-word expedition "to investigate the physical and biological condition of the great ocean basins."[3] With this relatively modest and, at the time, unheralded beginning, the modern science of oceanography was born.

The events leading up to the *Challenger*'s momentous voyage began in the mid-1800s amid a veritable marine biology craze that was sweeping Victorian England. But that bourgeoning interest in marine biology, by amateur naturalists for the most part, was primarily focused on collections and identifications of marine animal and plant specimens from near-shore and coastal waters. An exception was the oceanic collections made by **Thomas Henry Huxley** (1825–1895)

---

[3] Charles Wyville Thomson, 1878. *The Voyage of the "Challenger,"* vol. I. (New York: Harper & Brothers). p. 81.

on the voyage of the HMS *Rattlesnake* from 1846 to 1850. By the early 1870s, Britain was in the position, both politically and economically, to take the leadership role in a push to understand the oceans beyond the coastline by way of a concerted scientific study of the world's oceans. At least that was the gist of arguments made to the British Admiralty and the Royal Society of London by **Charles Wyville Thomson** (1830–1882), of Scotland's Edinburgh University (**FIGURE 1.17**), and **William Benjamin Carpenter** (1813–1885), an officer in the Royal Society. Wyville Thomson's and Carpenter's enthusiastic arguments for such a major voyage were in response to the recent successes of several local and much smaller-scale ocean surveys conducted between 1868 and 1871 by HMS *Lightning*, *Porcupine*, and *Shearwater,* and they were urging a much bolder follow-on expedition. It was they who first coined the term "oceanography" in its present context.

A particular motivation for the expedition was to resolve a simple question, which at the time was quite controversial: What life, if any, exists in the deep sea? The question was prompted by the preeminent scholar **Edward Forbes** (1815–1854), also an Edinburgh University professor, who had found during his biological surveys of the Mediterranean Sea in 1842 aboard HMS *Beacon* that different species of marine animals seemed to be aggregated according to depth. He defined a number of such depth zones along with their associated fauna down to a depth some 300 fathoms (ca. 550 meters) below which he believed the ocean to be devoid of life. It is important to stress that in those times this "discovery" was big news. After all, Forbes' **azoic theory**, as it became known, was entirely reasonable—it made sense that the great pressure, absence of light and cold temperatures in the deep sea must surely act to preclude the existence of life forms. It was an attractive idea that took hold—even though years earlier, John Ross and later his nephew James Clark Ross had found life forms from depths greater than 1000 fathoms (1828 meters). And certainly life in the deep sea was known at the time the first transatlantic telegraph cable was being laid between 1857 and 1866; during maintenance procedures, sections of the cable would come to the surface encrusted with various life forms. In the end, even though Wyville Thomson was aware of these contradictory findings and did not subscribe to the

FIGURE 1.17 (A) Charles Wyville Thomson, chief scientist of the *Challenger* expedition. (B) HMS *Challenger*.

(A)

(B)

deep sea azoic theory, he was nonetheless persuaded by William Carpenter that the idea deserved further investigation. But rationalizing a three-and-a-half-year round-the-world expedition based on the azoic theory was weak at best, and in fact there were other compelling justifications, including one that was much more controversial: to test the ideas of Charles Darwin.[4]

Darwin had published his *Origin of Species* just 13 years prior to the planning for the *Challenger* expedition, and in it he had speculated that marine organisms found on land only as fossils would be found alive in the oceans. Furthermore, he thought that gaps in the fossil record, the *missing links*, might be found in the deep sea, as well. But these ideas appear to have been downplayed at the time, and the search for deep sea life forms—to test the azoic theory—remained at the forefront of arguments in favor of the voyage. Indeed, Darwin and his ideas were quite controversial, to say the least. But his ideas were powerful nonetheless, and may have carried weight with the council of the Royal Society that was overseeing the expedition proposal. That council was headed by none other than T. H. Huxley, Darwin's staunchest supporter, who had himself been involved in the study of marine invertebrates during his voyage aboard the HMS *Rattlesnake*. And, of course, it was Huxley who 12 years earlier had waged a verbal battle with Bishop Wilberforce during a famous exchange at a meeting of the Royal Society, in which the bishop asked Huxley whether the apes from which he descended were on his grandfather's or his grandmother's side of the family. But Huxley's interests went beyond proving Darwin correct; he was an early proponent of the idea that there was, in fact, life in the great depths of the ocean.

Only a few years earlier, in 1868, Huxley had studied a sample of bottom sediment collected in 1857 from the Atlantic Ocean and found that it contained life forms that appeared to be protozoan cells within a slimelike substance that he described as containing veinlike threads. Thinking he had discovered a new life form, he named it *Bathybius haeckelii*, after the German naturalist **Ernst Haeckel** (1834–1919), who had earlier speculated that there must exist a primordial slimelike protoplasm from which all life had begun. Huxley believed that his *Bathybius* could be that primordial organism, and that samples collected from the bottom of the oceans by the *Challenger* would prove him right. As it turned out, of the more than 350 bottom samples collected by *Challenger*, no *Bathybius* was found. It was later determined, in 1875 by the chemist on board the *Challenger*, John Buchanan, that the slime was a precipitate that formed as a result of a chemical reaction between sea water and the alcohol used to preserve biological specimens. Huxley was notified almost immediately, and graciously acknowledged his mistake.

Although it did not confirm Huxley's pet idea, the *Challenger* expedition was hugely successful. It expanded our understanding of the ocean's chemistry, its bottom topography, and, especially, its biology. Under the direction of chief scientist Wyville Thomson, the expedition collected and described more than 4500 species that were new to science (**FIGURE 1.18**). The results and descriptions were published in 50 volumes, written by 76 authors, that took some 23 years to finish. Wyville Thomson's log was published in two volumes in 1878 and became a "best seller." The discoveries of the *Challenger* expedition stimulated the proliferation in marine laboratories (field stations) around the world.

A note on technology and sampling problems of the times

The technology available to scientists in the late nineteenth century made scientific work at sea difficult by today's standards, of course. The work-

[4]Richard Corfield, 2003. *The Silent Landscape: The Scientific Voyage of HMS Challenger* (Washington: Joseph Henry Press).

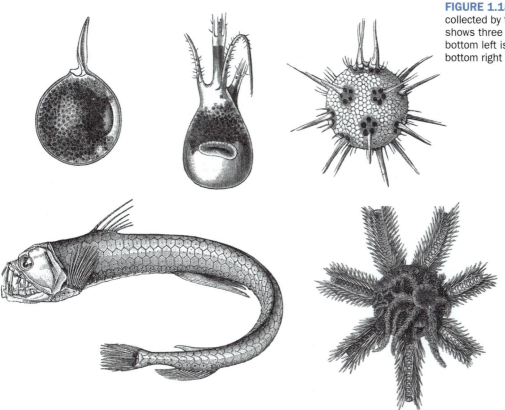

FIGURE 1.18 Woodcuts of organisms collected by the *Challenger*. The top row shows three forms of *Challengeria*, the bottom left is *Chauliodus sloanii*, and the bottom right is *Ophiocoma vivipara*.

ing spaces aboard the *Challenger* were limited and cramped, as **FIGURE 1.19** reveals for the chemistry laboratory and **FIGURE 1.20** for the working deck. While not much has changed in the last 140 years with respect to roominess at sea (see Figure 1.28), the analytical capabilities available aboard today's modern oceanographic research ships are a far cry from those illustrated in the woodcuts in Wyville Thomson's log. But the woodcuts do illustrate an important point to keep in mind as we review the history of the ocean sciences: the accomplishments of the nineteenth-century marine scientists were truly remarkable, given the low level of technology available to them as they sought answers to the most basic questions.

We have already discussed the first major question facing the early ocean explorers: Where am I and my ship? This problem was solved by the development of accurate chronometers, which could keep time on a moving ship at sea, and allow the determination of longitude. Three simple additional questions that constituted major technological barriers to the development of oceanography and marine biology might have been: How deep is the ocean under the ship? How does one get uncontaminated samples of animals and water? What is the water temperature at a given ocean depth?

The question of bottom depth may seem to be elementary, but it was not. For example, Matthew Fon-

FIGURE 1.19 The chemical laboratory aboard the *Challenger*.

FIGURE 1.20 The working deck of the *Challenger*. This is a small space for more than 200 deck hands to operate!

FIGURE 1.21 (A) The sounding machine (left), bottom dredge (center), and trawl (left) used aboard *Challenger*. (B) "Stopcock" bottles used to collect water samples from a specific depth.

taine Maury wrote in his 1855 book, *Physical Geography of the Sea*, that the matter was not straightforward at all:

[Others had attempted] to fathom the deep sea, some with silk threads, some with spun-yarn (coarse hemp threads twisted together), and some with the common lead and line of navigation. All of these attempts were made upon the supposition that when the lead reached the bottom, either a shock would be felt, or the line, becoming slack, would cease to run out.

The series of systematic experiments recently made upon the subject shows that there is no reliance to be placed on such a supposition, [for the shock caused by striking bottom can not be communicated through very great depths]. p. 200.

Maury goes on to describe attempts to measure bottom depths using the sounding-line method, including one which resulted in a report of the bottom reached at "forty-six thousand feet"; and in another report "Lieutenant J. P. Parker ... let go his plummet and saw a line fifty thousand feet long [nearly 10 miles!] run out after it as though the bottom had not been reached."

The solution that was finally reached, and used aboard the *Challenger*, involved the development of a relatively simple mechanical device, a "sounding machine" (**FIGURE 1.21A**), that could detect the slight change in tension on the sounding wire once the lead weights reached bottom and were mechanically released from the line. Using only this simple technology to sound the ocean depths, they were able to produce the remarkably accurate bathymetric (from the Greek *bathys*, deep) map of the Atlantic Ocean shown in **FIGURE 1.22**. Prior to this, almost nothing was known of the bathymetry of the oceans except that provided by Muary in his 1855 book. This map showed quite clearly the existence of the Mid-Atlantic Ridge, which we will be learning about in a later chapter. That submarine feature was not really understood until the 1960s, despite its discovery so long ago.

(A)

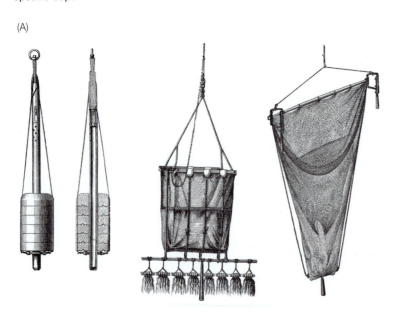

(B)

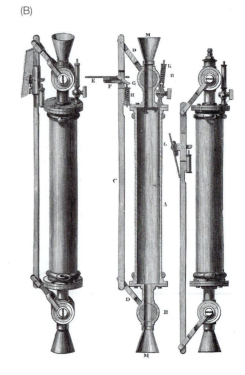

Collecting uncontaminated water from a particular depth for chemical analysis was accomplished using bottles that would close at the desired depth, thus trapping a volume of water for its return trip to the surface. Several clever devices were developed over the years for this purpose; examples of the samplers developed for use aboard the *Challenger* are shown in Figure 1.21B.

Of course, the *Challenger* scientists also wanted to know what the temperature was at the depth from which water samples were collected, as well as at other depths. This presented a bigger challenge to the instrument makers of the time. At first, scientists used simple, readily available thermometers that would record the maximum or minimum temperature encountered as they were lowered below the surface. These worked fine as long as the temperature increased or decreased continuously relative to the surface temperature, which, it turns out, is not always the case. Partway through the *Challenger* expedition, early versions of *reversing thermometers* became available and were sent to the ship. A reversing thermometer was ingeniously designed such that inverting it at the desired depth would cause a break in the mercury column, trapping a volume of mercury corresponding to the temperature at the target depth such that it would remain unchanged as it returns to the surface.

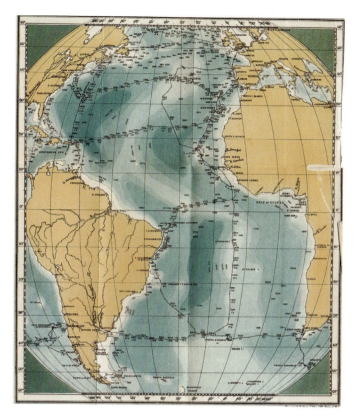

FIGURE 1.22 Cruise track with depth soundings and the resulting bathymetric contour map of the Atlantic Ocean produced by the *Challenger* expedition.

Oceanography and Marine Biology: Late Nineteenth to Early Twentieth Century

Fridtjof Nansen

No discussion of the history of oceanography and marine biology can be complete without an account of the Norwegian Arctic explorer (and first "professor of oceanography,") Fridtjof Nansen (1861–1930). Nansen holds claims to fame by virtue of any number of his achievements—he was, by all accounts, quite a Renaissance man. Not necessarily relevant, but interesting nonetheless: he was apparently a phenomenal athlete, and was instrumental in popularizing "modern" alpine skiing. In his youth he developed a reputation for being highly skilled at skiing down slopes at breakneck speeds on crude cross-country skis, which were not intended for that purpose, of course. As a scientist, Nansen began his career in neurobiology, and made his mark in that field. In his later years he was awarded the Nobel Peace Prize for his work relocating refugees after World War I. He is best known, however, as an Arctic explorer, who was the first to cross the Greenland ice sheet (by dogsled), but failed in his attempt to reach the North Pole. Ironically, that failed attempt to reach the Pole may have been his greatest achievement, especially with respect to its impact on the ocean sciences.

Nansen was driven to be the first person to reach the North Pole and in his attempt to do so, he applied an idea about the circulation of ice in the Arctic that would, in theory, assist him. His idea was based on observations of driftwood of Siberian tree species that arrived, somehow, on the icy shores in Norway. Nansen suspected that such was possible only if there were an ocean under

FIGURE 1.23 Fridtjof Nansen

the Arctic ice with currents that brought the driftwood—and ice—from Siberia to Norway. Whether or not there was an ocean under the Arctic ice was not known at the time. But if he were correct, he reasoned that by drifting with the ice sheets he could get a free ride to the North Pole.

He commissioned a specially designed ship, the *Fram*, a 38 meter schooner built with a 1.2-meter-thick oak hull (**FIGURE 1.23**), with the intention that it could withstand being frozen in the ice. Its shape and strength would allow it to be lifted upward by the pressure of the ice surrounding it, and not be crushed as ordinary ships would be. With a crew of 13 men and provisions for five years, he set out from Norway in the summer of 1893 with the intention of locking *Fram* in the ice off Siberia for a planned three years. But she was stuck in the ice for nearly four years (**FIGURE 1.24**)!

After sailing the open coastal waters of the Laptev Sea and around the New Siberian Islands off the coast of Russia (**FIGURE 1.25**), the *Fram* turned to the North. Eventually she became frozen in the ice at about latitude 78°N, or about 1200 kilometers (750 miles) from the Pole. The ship drifted with the ice at an average rate of 2 km per day, and came within 400 kilometers (ca. 250 miles) of the North Pole.

Once Nansen realized that his planned trajectory would not bring the ship over the Pole, he and **Frederick Johansen** left the *Fram* and set out from latitude 86°14′N by dogsled, hoping to get there before winter set in. They didn't make it; they ended up having to overwinter on the ice, spending five months in a snow cave, in total darkness and bitter cold (they slept together in a single sleeping bag to conserve their body heat), and surviving almost exclusively on walrus meat. The following spring, realizing that another assault on the Pole would be fruitless, they gave up and headed south. Some 14 months after leaving the *Fram*, Nansen and Johansen were rescued by a British seal-hunting expedition, on Franz Josef Land, north of Norway. Meanwhile, the *Fram*'s crew stayed with the ship, which finally broke free and sailed home. The *Fram* today is in a museum in Oslo, Norway.

Nansen never did reach the North Pole—that distinction would be claimed by Robert Peary on April 6, 1909, although this has been disputed. But his expedition provided strong evidence for his hypothesis that there was an ocean and not a continent beneath the Pole.

FIGURE 1.24 (A) The *Fram* in Bergen, Norway, en route to Siberia in the summer of 1893 and (B) frozen in the Arctic ice.

(A)

(B)

FIGURE 1.25 Track of the *Fram* expedition to the North Pole. The red line represents the track of the ship on open water, the blue line represents the track of the ship with the Arctic ice, and the black line represents Nansen and Johansen's trek across the ice to Franz Josef Land, where they were rescued.

Growth of marine research

The importance of the marine, or ocean, sciences was certainly spreading in the late 1800s leading up to Nansen's *Fram* expedition, not just in Europe and Victorian England, but around the world. In the United States, recognition of the importance of understanding how and why our commercial fisheries fluctuate led to the creation by act of Congress in 1871 of the U.S. Commission of Fish and Fisheries, located in Woods Hole, Massachusetts. The "Fish Commission" as it was better known, was the first conservation agency to be established by the federal government. In 1970 it became the National Oceanic and Atmospheric Administration (NOAA). The original mission of the Fish Commission was simple, if not in actual practice: "protection, study, management, and restoration of fish." Spencer Fullerton Baird (1823–1887), the Assistant Secretary of the Smithsonian, was appointed by President Ulysses S. Grant in 1871 to be the first Commissioner of Fisheries. One of the Commission's first research ships, the USS *Albatross* (**FIGURE 1.26**), believed to be the first ship specifically designed and built for fisheries research, was launched in 1882.

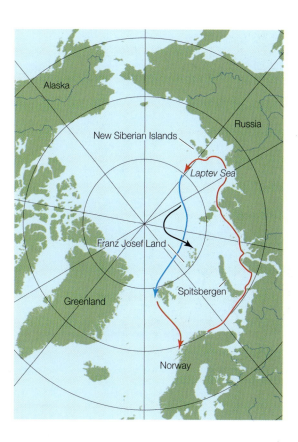

(A)

(B)

(C)

FIGURE 1.26 (A) Spencer Fullerton Baird, the first Commissioner of the U.S. Fish Commission. (B) The fisheries laboratory in Woods Hole, Massachusetts about 1885. (C) The USS *Albatross*.

(A)

Abb. 57. „Meteor" nach dem Umbau.

(B)

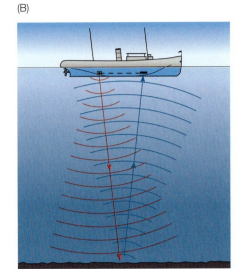

(C)

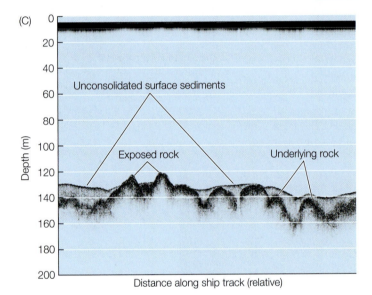

Depth (m)

Unconsolidated surface sediments

Exposed rock

Underlying rock

Distance along ship track (relative)

FIGURE 1.27 (A) The German research vessel *Meteor*. (B) A diagram illustrating the principle of echo sounding. (C) A modern computer monitor with echo sounder image (from a "precision depth recorder," or PDR, as echo sounders are now called) as the ship passes over bottom of differing depths and composition (soft sediment and rock).

Interest in the marine sciences led to advances in technology, and vice versa. For instance, the German *Meteor* expedition, from 1925 to 1927, pioneered the use of an echo sounder (**FIGURE 1.27**). With its complement of 123 officers, scientists, and crew, the *Meteor* criss-crossed the South Atlantic Ocean collecting some 67,400 echo soundings in addition to measurements of currents, temperature, salinity, and oxygen at more than 300 "stations." The echo soundings provided the first detailed snapshot of the ocean bathymetry, including details of the mid-ocean ridge and rift system first described by the *Challenger* expedition, which we will discuss in more detail in Chapter 3.

The principle of echo sounders has been known for some time, certainly since the early 1800s. The idea is to send out a sound pulse that travels to the bottom, reflects off the bottom and comes back up to a hydrophone on the ship. Knowing the speed of sound in water, and measuring the interval of time between the pulse and its return signal, one can easily compute the depth of the bottom. But the electronics required to make the technique feasible would not become available until the early twentieth century.

(A)

(B)

FIGURE 1.28 (A) Interior view of a laboratory aboard a modern oceanographic vessel, the research vessel *Endeavor*. One of the newest additions to the U.S. oceanographic fleet is the research vessel *Henry B. Bigelow* (B), commissioned in 2011; she will carry out oceanographic and fisheries research for NOAA, the National Oceanic and Atmospheric Administration.

Along with advances in technology came more modern research ships (**FIGURE 1.28**) and modern institutions devoted to the study of the oceans and their life forms. In 1888 the Marine Biological Laboratory was founded in Woods Hole, Masachusetts, next door to the Fish Commission's Fisheries Laboratory; and later, in 1939, the Woods Hole Oceanographic Institution was founded on the same small parcel of land (**FIGURE 1.29**). The Scripps Institution of Oceanography was founded in San Diego in 1892. At about the same time, there was a growing recognition that funds are required to support marine research. But not until we learned that the oceans are important to nations at war with one another did spending on the ocean sciences skyrocket. For example, in 1946, following World War II, the U.S. Congress created the Office of Naval Research with the stated mission of "planning, fostering, and encouraging scientific research in recognition of its paramount importance as related to the maintenance of future naval power and the preservation of national security." Today, it is still the largest funding agency for ocean sciences. Then in 1956, the National Science Foundation was created, followed two years later by NASA, the National Aeronautics and Space Administration, each of which is a major source of funding for marine scientific research. Federal government spending for ocean science research has led to a revolution in the way we view our planet—even our solar system.

FIGURE 1.29 A view from across Eel Pond harbor at Woods Hole, Massachusetts, of a portion of the Woods Hole Oceanographic Institution and the Marine Biological Laboratory.

Chapter Summary

- The ocean sciences began more than 3000 years ago, with the early Greek and Polynesian ocean explorers.

- The scientific approach to ocean exploration did not come about until reliable methods of navigation were developed, including map making, which requires a knowledge of the relative sizes of the Earth, its oceans, and its land masses.

- The first accurate calculation of the size of the Earth was made by some 1800 years ago by the Greek mathematician **Eratosthenes**. The magnetic compass had been available to explorers since the time of the ancient Greeks, so even before accurate maps were available, a ship would at least know where it was heading.

- To navigate at sea, one must be able to identify the **latitude** and **longitude** of one's own position. Determining latitude at sea was always relatively easy based on angles between the horizon and the Sun or the North Star. In the late 1700s, determining one's position east or west (longitude) became possible with the development of the **chronometer**, which can keep time aboard a moving ship, enabling one to know the difference in time of day between a predefined, fixed location and one's own position.

- Many important voyages were launched between those of **Columbus** and **Magellan** in the Age of Discovery in the late 1400s and those of **Cook** and **Darwin** in the 1700s and 1800s.

- In 1855, **Matthew Fontaine Maury** published the first modern study of the oceans, which detailed important features of ocean currents, such as the Gulf Stream. However, the HMS *Challenger* expedition from 1872–1876 led to the modern scientific fields of both oceanography and marine biology. Sampling technologies were developed at about the same time, including sounding machines, sampling bottles, and reversing thermometers, among others.

- Twentieth-century marine science was led by **Fridjhof Nansen**. In the early 1900s, he proved that there is an ocean under the ice of the Arctic. In the 1920s, another ship, the German *Meteor*, was the first to use an echo sounder to map bottom depths.

- The first marine research institutions began to be built at the end of the nineteenth century and the beginning of the twentieth century.

- Today, with modern ships and technologies, marine research is pursued at hundreds of colleges, universities, and private institutions, as well as by numerous government agencies.

Discussion Questions

1. What were some of the difficulties that hampered ocean exploration during the time of the ancient Greeks and Polynesians?

2. Why was it important to have determined the circumference of the Earth?

3. Why was it a technological difficulty to keep accurate time at sea? Why did most clocks in the eighteenth century work only on land?

4. Why is so little known about the voyages of the Chinese? The Vikings? What are some of the reasons for the relative obscurity and lack of appreciation for the significance of the U.S. Exploring Expedition?

5. What are some of the highlights of the post-Renaissance ocean voyages? What motivated those voyages?

6. What is the significance of the search for a Northwest Passage?

7. What are some of the basic sampling problems that had to be solved before there could be scientific studies of deep ocean waters, bottom sediments, and bottom animals? Why was it so difficult to determine bottom depth?

8. What were some of the scientific questions and political issues surrounding the promotion of the *Challenger* expedition?

Further Reading

Corfield, Richard, 2003. *The Silent Landscape: The Scientific Voyage of HMS Challenger*. Washington: Joseph Henry Press.

Deacon, Margaret, 1997. *Scientists and the Sea. 1650–1900: A Study of Marine Science*, 2nd ed. London: Academic Press.

Desmond, Adrian, 1994. *Huxley: From Devil's Disciple to Evolution's High Priest*. Reading, MA: Addison Wesley.

Huntford, Roland, 1997. *Nansen: The Explorer as Hero*. London: Duckworth. This biography of Frijdhof Nansen offers a fascinating account of the *Fram* expedition.

Philbrick, Nathaniel, 2003. *Sea of Glory: America's Voyage of Discovery; The U.S. Exploring Expedition*. New York: Penguin.

Rehbock, P. F., 1975. Huxley, Haeckel, and the Oceanographers: The case of *Bathybius Haeckelii*. *Isis* 66: 504–533.

Sobel, Dava, 1995. *Longitude: The True Story of a Lone Genius Who Solved the Greatest Scientific Problem of His Time*. New York: Penguin Press.

Thomson, Charles W., 1878. *The Voyage of the "Challenger,"* vol. I. New York: Harper & Brothers.

Contents

The Orion Nebula is 1500 light years from Earth, and is visible to the naked eye as a fuzzy-looking star. But it isn't a star. As this dramatic close-up image from the Hubble Space Telescope shows, it is actually a vast expanse of interstellar dust and gas where new stars are continually being born, with planets orbiting those stars, and, perhaps, oceans on those planets. Those same processes created our solar system, our Earth, and our oceans.

Origins and Connections: Science, the Universe, Earth, and Life

How were the oceans formed? For that matter, how was the Earth itself formed? And what about *life*? How did it first come about? These are profound questions that lie at the heart of any book about oceanography and marine biology and we will attempt to answer them in this chapter. But in order for us to appreciate the various scientific ideas and discoveries that underlie our still-growing understanding of these topics, it is helpful first to review the nature of science. And so we open this chapter with a discussion of what science itself is all about and how scientists "do science." A better grasp of these fundamental points will be useful as we attempt to draw a number of *connections*—connections that link such far-ranging phenomena as the early formation of the Universe and the various processes that produced the geological features of today's Earth, including the oceans and the life forms they support. Making these links, seeing how our scientific thinking has evolved and produced insights into seemingly unrelated phenomena, is a big challenge, because, quite frankly, some of the ideas and scientific findings we are about to discuss are not only fascinating, they are amazing, and they continue to offer surprises even to the scientists who study them.

What Is Science?

In our opening chapter we reviewed some of the highlights that stand out in the history of the marine sciences. But an important point we skipped over is: What exactly *is* science? That is, what is it about *marine science* that makes it *science*? These simple questions are actually more difficult to answer than it may seem at first. As we alluded to at the start of this chapter, we are about to venture into some pretty wild stuff, and without a fundamental appreciation of what science is all about, much of what we are about to discuss might seem more fanciful than scientific. So, before we go deeper into our target subset, the marine sciences, we are first going to discuss briefly the nature of science itself. To begin, let's explore what science is and is not.

Science and technology

At the outset, we need to stress an important point: although most people equate the two, technology and science are not one and the same.[1] Technology involves the design and production of something useful, such as a new tool. It predates science; it is a practice that has been with humans throughout our history. During our early history, we developed technologies such as agriculture and metal-working, which were fundamental advancements that did not depend on science. In fact, many technological advances arose even earlier in our history without the aid of science; examples include the use of fire in cooking food and the development of the wheel. Not to diminish the

---

[1] Much of the following discussion is based on Lewis Wolpert's excellent 1992 book, *The Unnatural Nature of Science* (Harvard University Press).

intellectual superiority of early humans, but simple technologies are used even by nonhumans: chimpanzees, for example, use tools; they use twigs to dig ants and other insects out of logs.

While technology seeks solutions to practical problems, science is an organized quest to discover nature's laws. Its objective is the pursuit of knowledge with which to better understand the world and how it works. And, as we pointed out in the last chapter, science was invented (or discovered) in the not-too-distant past, relative to the four million years or so during which hominids evolved, by the philosopher Thales of Miletos about 600 B.C. Thales is thought to have been the first person to try to explain the world in physical terms as opposed to the more abstract or supernatural terms that were far more common in those times. This, it has been argued, marked the birth of science. Thales's new natural philosophy, his early science, was really quite simple and straightforward: there are laws that control nature, and those laws can be discovered. This is what science is all about.

Interesting though it must have been back in Thales's time to ponder such a new philosophy of nature, there was a curious aspect of this *science*: it was really quite useless in a strictly pragmatic sense. Science was useless in the sense that the product of scientific inquiry is usually just a new idea, not a better tool or a better weapon. Even today we could argue that ideas are useless in a pragmatic sense, because they do not necessarily make life easier or more efficient in and of themselves. The reward of science in Thales's time, and even today, is intellectual gratification. Ideas do not normally lead to tangible rewards or riches (unless you win a Nobel Prize); you cannot get a patent for an idea or a new level of understanding, only for an invention or a new process. But the results of basic scientific research often lead to new technologies which, of course, sometimes yield big payoffs. The intriguing aspect to all this is that it is impossible to predict which ideas will turn out to be important or have life-changing ramifications for human beings.

Science was slow to develop as compared with technology, and even today it continues to be viewed with apprehension by some of us. One reason for this is that science is not always intuitive; it can even be counterintuitive at times, and it is often filled with surprises. Because science routinely deals with big numbers, such as extraordinarily large and extraordinarily small measures of time and space, values that exceed those which we deal with in our everyday lives, it can seem to violate our common sense. Here's an example of common-sense-defying scales, both large and small: if we go to the beach and dump a glass of fresh water into the ocean, and then wait long enough so that all of the molecules of fresh water have completely dispersed throughout the world ocean system, and then dip that same glass into the ocean—any ocean, anywhere in the world—there is an excellent chance that the glass will contain at least one of the original freshwater molecules. This is because there are more molecules of water in a single glass of water than there are glasses of water in all the world's oceans!

Another example of where our common sense fails us can be illustrated by a simple thought experiment: imagine a barrel, an ordinary 55-gallon oil drum, for example (**FIGURE 2.1**). You tie a string around it tightly, and notice that to do so, you need a string about 75 inches long. That is, the circumference of the barrel is 75 inches. Now, you want to insulate the barrel, so you decide to wrap some 6-inch-thick strips of Styrofoam insulation all around the barrel, and secure these strips with another piece of string. You find that this second string needs to be about 38 inches longer than the first. OK, now try the same thought experiment with the entire Earth; that is, tie a string around the Earth at the Equator. It will be about 25,000 miles long, the circumference of the Earth. Now add a 6-inch layer of insulation around the Equator and

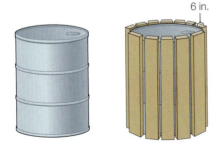

6 in.

FIGURE 2.1 An ordinary 55-gallon oil drum, around which you tie a string. Then you add a six-inch layer of insulation to the barrel, and again tie a string around the now-insulated barrel. How much longer does the string have to be?

tie a string around that. How much longer does this string have to be? The answer is: about 38 inches. No matter how large the circumference of the original circle is—25 feet, 25,000 miles, or 25 billion miles—the addition of *6 inches* to the radius will always increase the circumference by about *38 inches.* Do the math yourself, using different values for the radius, *r*:

$$\text{Circumference of the original circle} = 2\pi r$$
$$\text{Circumference of the padded circle} = 2\pi(r + 6")$$
$$[2\pi(r + 6")] - 2\pi r = 37.68"$$

These examples illustrate a fundamental point: that to practice science we cannot necessarily count on our common sense. After all, common sense is a reflection of our experiences (and our education); without any experience, or context, within which to view these two examples, we could draw incorrect conclusions about them. More importantly, common sense, even if it is misdirected, is very difficult for any of us to dismiss. This curious phenomenon was evident at the time of Aristotle (ca. 350 B.C.), the great philosopher whose early theory of gravity posited that heavier bodies fall faster than lighter ones. According to his theory, two objects of the same material, one larger and therefore heavier than the other, would fall at different speeds. Because of the great respect people had for Aristotle, his idea was accepted without question—*for 1800 years.* It wasn't challenged until the time of Galileo in the late 1500s. Quite simply, Aristotle's ideas had become an integral part of people's common sense. What has fascinated historians of science is that neither Aristotle nor, apparently, anyone else bothered to test the idea, even with the simplest thought experiments. One such thought experiment is shown in **FIGURE 2.2**. Imagine two rocks of the same material but of different sizes; if heavier rocks fall faster than light ones, the larger one might fall at a speed of, say, 8 m/sec, and the smaller one might fall at half that speed.[2] If we were to tie the two rocks together, we might predict, according to Aristotle's logic, that the larger rock would increase the speed of the smaller rock, and that the smaller rock would impede the speed of the larger one. The resulting speed of the two tied together, then, would be 6 m/sec. However, another logical possibility is that because the two rocks together weigh more than either one alone, together they will fall faster than either one alone. Either interpretation predicts preposterous results. But it took some 1800 years before this was pointed out!

After Galileo's new theory of gravity became widely known, our common sense became forever altered. There was no going back. Science would advance by building on this fundamental theory, and indeed, in the 1700s Isaac Newton quantified Galileo's theory of gravity. This is how science, and the understanding of science, advance: in layers of understanding that are built up over time. As Lewis Wolpert writes in his 1992 book, *The Unnatural Nature of Science*, "In all branches of science there is a great deal that must first be learned and understood at a deep level, so that the right questions are posed, before the generation of new thoughts can even be contemplated." Stated another way: we have to quash the natural tendency to rely on our common sense, if it is not supported by sufficient experience. And that "sufficient experience" is, to a large degree, the basic concepts we learn as students; our "experience" is our education.

So where does this leave us? We have briefly discussed how science differs from technology and how science is not

8 m/sec   4 m/sec   6 m/sec?

FIGURE 2.2 Thought experiment to test Aristotle's theory of gravity. Two rocks of the same material, but one larger than the other, are allowed to fall; Aristotle stated that the larger, and hence heavier, rock would fall faster. But what if the two are tied together? At what speed would they fall then?

---

[2] Today we know that objects don't just fall with a certain speed, all objects experience a gravitational acceleration, which on Earth is 9.8 meters per second per second, or m/sec².

always intuitive, but you may be wondering: How, then, do we do science? Well, as scientists, we want to understand the natural world and how it works, and so we often ask questions and devise orderly ways to answer them. For example, one of the questions posed by the *Challenger* expedition was, how deep is the ocean? In this case, they had to develop a new technology, a sounding machine, to answer the question. This is an example of how science can benefit from new technology. But there is more to doing science.

Ways of doing science

Most of us have been taught about the **scientific method**, whereby scientific inquiry is done in a stepwise process: initial *observation*, which leads to the formulation of a *hypothesis*, which is tested by an *experiment*. Based on the outcome and reproducibility of that experiment, we might then elevate our hypothesis to a *theory*, with which we can make *predictions*. But the scientific method does not always proceed through all these steps in order; there are times when strict adherence to the method may not be warranted. This point will, hopefully, become clearer if we present it in historical context.

Not many people realize that the concept of the scientific method as a process of "hypothesis testing" is based on ideas offered by the twentieth century philosopher of science **Sir Karl Popper** (1902–1994; **FIGURE 2.3**). In his book *The Logic of Scientific Discovery* (originally written in German in 1934 and translated into English in 1959), Popper was trying to discredit the rampant pseudoscience that prevailed in his youth in Austria, especially Marxism, astrology, and the psychology of Sigmund Freud. Popper leaned heavily on Einstein's Theory of Relativity as a model for what he considered to be good science because it made predictions about the real world that could be tested with experiments. His central point was that scientists cannot prove a theory no matter how many times they repeat their observations or experiments. They can only disprove, or falsify, a theory by finding a contradiction, thus sending the theory back to the drawing board or out the window altogether. The significance in all this is that Popper's philosophy prescribed a way to structure the conduct of scientific research in what today has become known as The Scientific Method. Popper argued that once we have a theory, we must subject it to the most vigorous scrutiny, trying hard to find a flaw in it—that is, to falsify it. As many philosophers of science have argued, a good scientist is a skeptic.

While Popper described a way to structure the conduct of scientific inquiry, he really didn't define science itself as well as others have. Perhaps Edward O. Wilson of Harvard University defined it best: "Science is the systematic enterprise of gathering knowledge about the world and organizing and condensing that knowledge into testable laws and predictions."[3] Wilson's definition is eloquent in that it can be reduced to two simple questions that will determine whether an idea, or theory, is scientific: Is it possible to devise an experimental test? Does it make the world more predictable? If the answer to either question is no, it isn't science.[4]

Much of what we are taught about the scientific method does not take into consideration this subtle bent that Wilson offered not that many years ago. And so we are taught that science is done by constructing testable, falsifiable hypotheses. As straightforward as that argument may seem, there is a fundamental problem with the way the scientific method is often presented: when most of us learned about the scientific method, we weren't taught much about what constitutes a "scientific hypothesis," whether it is a good one, or

FIGURE 2.3 Sir Karl Popper

[3] Edward O. Wilson, 1998. *Consilience: The Unity of Knowledge* (New York: Knopf).

[4] Robert L. Park, 2000. *Voodoo Science: The Road from Foolishness to Fraud* (New York: Oxford University Press).

even how one arrives at a hypothesis that is worthy of testing. It is all rather vague, isn't it? For example, imagine a situation where we have observed that a certain species of marine snail appears to feed exclusively on a single species of algae. Based on this observation, we might pose a hypothesis something like, "that particular species of snail eats only that species of algae." To test our hypothesis we would set up an experiment in which the snail is presented with alternative food sources in addition to what we suspect is its preferred alga. If our experiment shows that the snail shows no preference, then we would reject our hypothesis. That is, we look for a contradiction to our hypothesis. If we cannot find a contradiction, we accept our hypothesis, for the time being, anyway. But one day, we, or someone else, may find a contradiction which would require revising our hypothesis or starting over with an entirely new one.

In the previous example, the scientific method seems logical and sound. But what about the *Challenger* expedition? They could have tested the hypothesis that the oceans have no bottom; after all, as we mentioned in Chapter 1, there had been numerous well-documented but failed attempts to fathom the depths of the sea. But this hypothesis is clearly absurd, and it illustrates an inherent weakness of strict adherence to the scientific method. Where would falsifying that hypothesis get us? Instead, the question the *Challenger* expedition needed to answer was: How do depths vary throughout the oceans?

Hypothesis testing is, of course, extremely important in scientific research. It is the only way to search for cures to disease, to test the efficacy of new drugs, and so forth; we could go on to list many examples. And it becomes especially invaluable in cases where statistical significance is required to determine the final result of an experiment, which happens a lot. It is indeed a very useful and indispensible way to do science. It just has to be applied properly!

But, there is more—much more—to scientific research than the scientific method would seem to prescribe, such as exploration and discovery, for which few, if any, rules exist. This general line of thinking has led some philosophers of science to suggest that there is more than one scientific method.[5] Often, for example, answers to specific questions are not amenable to hypothesis testing but are more dependent on monitoring or observation. Observation, of course, is the first step in the scientific method, but isn't *it* scientific? There are still questions about our planet that cannot be properly answered by direct experimentation or hypothesis testing, such as: How fast is a particular ocean current? What new species of deep sea fishes remain to be discovered? What changes to marine ecosystems can we expect to see as a result of pollution or from unintentional introductions of non-indigenous species? Questions such as these are clearly scientific in nature, but, to answer them, we need to put a greater emphasis on the process of observation.

On "seeing" in science

Regardless of the specific method used, whether hypothesis testing, pure exploration, or some mix of the two, science requires us to have imagination, an ability to "see." And for this vision to be most effective, we need to have mastered certain concepts by way of our education and experience and to dismiss, for the most part, our common sense. We need to be prepared to discover.

This line of reasoning, on "seeing" in science, was developed by another philosopher of science, **Thomas Kuhn** (1922–1996), who wrote about "perceptual *gestalt*," from the German word meaning "the whole."[6] For example, we

---

[5] Gerald Holton, 1995. *Einstein, History, and Other Passions* (Woodbury, NY: American Institute of Physics Press).

[6] Thomas S. Kuhn, 1970. *The Structure of Scientific Revolutions*, 2nd ed. (Chicago: Chicago University Press).

might recognize a person from across a room, but if pressed to explain what it is we recognize, most of us would have some trouble. We just know that person's "gestalt." Other examples of perceptual gestalt might include: the cartographer's instant interpretation of the contour lines on a topographic map; the musician's ability to read musical notation; and the elite athlete's ability to *see* the playing field, to *read* the defense.

Kuhn explained that science does not advance by small increments over long periods of time, but rather by great leaps in understanding, leaps he called "**paradigm shifts**." He then showed how these paradigm shifts affect the way we "see" in science; in other words, once we learn to view the world in a different light, it is never again what it once was. We have already shown how in Galileo's time (late 1500s), the concept of gravity underwent a paradigm shift; after that time, nobody took Aristotle's theory of gravity seriously again.

This ability of a researcher to "see," which is a learned ability, supported by a healthy dose of initial talent, has been important to scientific discovery throughout history. As Kuhn puts it:

> What a man sees depends both upon what he looks at and also upon what his visual-conceptual experience has taught him to see. In the absence of such training, there can only be ... a bloomin' buzzin' confusion. (p. 113)[6]

Many initial scientific findings, discoveries, and experimental results have significant ramifications that extend far beyond those originally sought by the researcher. Yet that researcher often "sees" the significance of his results in an instant. For example, the discovery of penicillin by Alexander Fleming in the 1920s was purportedly an accident. But he noticed right away that the *Penicillium* mold growing all over his experiment had antibiotic properties, and the way was clear to discover other miracle drugs.

As we saw in our review of the history of marine science, the field advanced significantly as a result of simple exploration and discovery, every bit as much as it did by strict hypothesis testing. You might be telling yourself that this may have been the case a hundred years ago or during the Age of Exploration, but surely not now. Yet in fact, even today, many of our most interesting and important scientific discoveries in oceanography are being made in an exploratory or observational mode, not just in the experimental mode.

So, with this brief overview of a philosophy of science as background, we should be prepared to move on to some thought-provoking developments in the broad field of marine science that have shaped our current understanding of the initial formation of the Universe, the Earth and its oceans, and life itself. As you will see, some of these ideas and discoveries certainly challenge our common sense.

Origins: Where Did the Earth and Its Oceans Come From?

The "Big Bang"

Before about 13.7 billion years ago, there was nothing, and there was no such thing as time. And then something happened: time and matter began in an instant from a single point and exploded into an expanding cloud of matter—all the matter in the Universe today. Known as the **Big Bang**, this expansion of the Universe continues today.

The idea of the Big Bang dates back to 1927, when **Georges Lemaître** (1894–1966; **FIGURE 2.4A**) a Belgian priest and scientist, proposed that the Universe began with the explosion of what he called a primeval atom. He noted that astronomers had observed a "red shift" (**BOX 2A**) in distant regions of space, and he related that to a model of the Universe based on Einstein's Theory of Relativity. Some years later, **Edwin Hubble** (1889–1953; **FIGURE 2.4B**) found supporting evidence for such an explosion, and for the idea that distant galaxies in all directions are flying away from us at speeds proportional to their distance.

Earlier versions of the Big Bang theory posited that the expansion of the Universe would eventually slow down as it succumbed to the inward pull of gravity and that it would then begin to contract on itself. This line of reasoning predicted a sort of reverse Big Bang, or a Big Contraction, back to a single point at which another massive explosion would occur (and by this logic, maybe there *was* something prior to the Big Bang). But recent observations are showing not only that the distant galaxies are continuing to move away from one another, but that the speed of that expansion is *accelerating*!

FIGURE 2.4 (A) Monsignor Georges Le-maître, priest and scientist, ca. 1933. (B) American astronomer Edwin Hubble.

The first observations of an accelerating expansion of the Universe were reported by scientists in 1998, but they were at a loss to account for the mysterious source of energy driving it, which left them to hypothesize some sort of **dark energy** behind it all, yet to be discovered. Those initial observations were corroborated by subsequent studies by other scientists (in the field of *cosmology*, the study of the cosmos). Interestingly, if the accelerating expansion continues (and why shouldn't it?), all galaxies beyond our own (the Milky Way) will continue to "red shift" to the point that eventually the distant Universe will turn dark. We won't be able to see the stars of distant galaxies, not because they will accelerate to the point where they outrun the speed of light, but because their increasing wavelengths will make starlight invisible to us. The light waves reaching us will be red-shifted to the longer wavelengths that are outside the visible spectrum. This is wild stuff, and it isn't even science fiction; in fact it's better—because it's true.

Back to the Big Bang: about a million years after the massive explosion, the young Universe had cooled down enough that subatomic particles of matter could begin to coalesce to form elementary particles, **atoms**, of the lighter elements such as hydrogen and helium. Clouds of these particles began to form, and under their own mutual gravitational attraction, they condensed to form great masses. This gravitational process continued until the masses eventually became glowing **stars**. The immense heat and pressure inside the stars began to fuse hydrogen atoms together to form helium atoms, a process known as **nuclear fusion**, which releases huge amounts of energy in the form of even more heat and light. This, of course, is what a hydrogen bomb does; i.e., H + H → He + energy, lots of it!

These nuclear fusion reactions also make the heavier elements, such as iron, through continued fusion of the lighter elements. But eventually stars use up their hydrogen fuel (and after that their helium fuel), at which point, if they are large enough, they may explode,[7] creating a **supernova**, which blasts their mass of heavier elements into huge glowing clouds of interstellar

---

[7] If the star is of insufficient mass, it may first transition to a red giant or a white dwarf.

BOX 2A   Red Shift

Astronomers often refer to the "red shift" of light that is emitted by a star or galaxy and received here on Earth, or received by the Hubble Space Telescope as it orbits Earth. The degree to which light from that object is *red shifted* indicates the speed at which it is moving away from us.

The red shift principle is based on the **Doppler Effect**, an effect familiar to most of us. We have all heard the siren of an emergency vehicle (e.g., an ambulance, police car, or fire truck) or even a train whistle approaching and then moving away from us. As it approaches, the sound is higher in pitch, indicating a higher-frequency (shorter-wavelength) sound wave. As the vehicle passes by us we immediately hear a distinct change in tone to a lower-frequency (longer-wavelength) sound. The same principle can be applied to light waves—in particular, the light arriving on Earth from distant stars and galaxies.

For example, if a star or galaxy is emitting mostly yellow light and it is moving toward an observer, the wavelength of that yellow light will be compressed somewhat to a shorter wavelength; and the degree to which it is compressed is directly proportional to the object's speed. The light will be Doppler-shifted toward the blue end of the spectrum, or in other words, it will be "blue-shifted" as seen by the observer.

If the star or galaxy is moving away from the observer, then the wavelengths of light being emitted will be stretched to longer wavelengths, toward the red end of the spectrum, and hence, "red-shifted."

All stars and galaxies outside our own Milky Way are moving away from us—and away from everything else—and they have been since the Big Bang. As in any explosion of matter, stars and galaxies that are farther away from us are moving away from us faster

than those closer to us. For example, in this simple diagram of the Big Bang explosion, the dashed arrows represent some possible trajectories for objects propelled by the explosion, and the solid arrows represent the trajectories of three objects: the Milky Way (our galaxy), and galaxies 1 and 2. The distance between our Milky Way and galaxy 1, *a*, and the distance between our Milky Way and galaxy 2, *b*, are both increasing. However, from our perspective in the Milky Way, because our Milky Way and galaxy 1 are closer together, *a* increases more slowly than *b* increases. Therefore, because galaxy 2 is farther away from us than galaxy 1, it is moving away from us faster than galaxy 1 is. Galaxy 2 will therefore be red-shifted more than galaxy 1, and so we will observe it as having a redder color than galaxy 1.

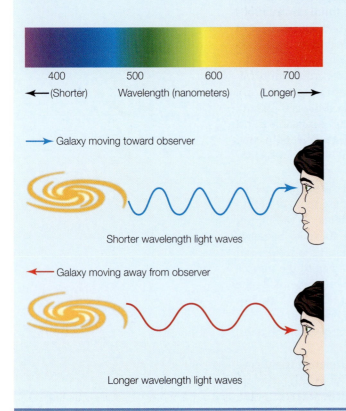

400   500   600   700
←(Shorter)   Wavelength (nanometers)   (Longer)→

→ Galaxy moving toward observer

Shorter wavelength light waves

← Galaxy moving away from observer

Longer wavelength light waves

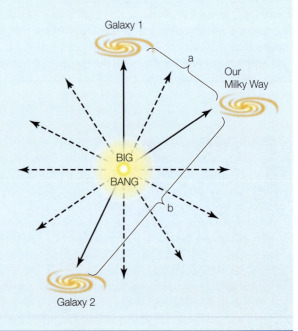

Galaxy 1

a

Our Milky Way

BIG BANG

b

Galaxy 2

(A)

(B)

FIGURE 2.5 (A) The Orion Nebula is visible to the naked eye, appearing as a blurry star. This photo was taken without magnification. (B) A higher magnification, high-resolution photo of the Orion Nebula taken by the Hubble Space Telescope.

dust and gases, called **nebulae** (singular, *nebula*). We can see several of these nebulae today in the night sky using even a cheap telescope or binoculars. For example, **FIGURE 2.5** shows the **Orion Nebula**, which, although some 1500 light years away from Earth, is visible to the naked eye as a "blurry star" in the constellation Orion, just below the three bright stars that form Orion's Belt. But of course, it isn't a single star.

Supernovae and the resulting nebulae do not mark the end of the process, however. Matter in nebulae also co-attract and condense and can eventually form new stars or combinations of stars, **planets** that orbit around stars, and **moons** that orbit around the planets. When huge numbers of stars co-attract one another, they can form **galaxies** (**FIGURE 2.6**), such as our own Milky Way galaxy (**FIGURE 2.7**).

FIGURE 2.6 Like our own galaxy, the galaxy shown here, Messier 101 (nicknamed the Pinwheel Galaxy), is a spiral galaxy. It is in the constellation Ursa Major (The Great Bear), also known as The Big Dipper. The Pinwheel Galaxy is 25 million light-years from Earth. This is one of the highest-resolution photographs of a spiral galaxy and was taken by the Hubble Space Telescope. It is composed of 51 individual Hubble exposures, in addition to elements from images from ground-based photos.

FIGURE 2.7 Photograph of our own Milky Way galaxy over a high desert in Chile with cacti in the foreground. Because we are part of the Milky Way and are inside it, all we can see is this side view, showing a dense strip of stars and interstellar dust and gases across the night sky, rather than the spiral shape we would see if we were above or below it, as in Figure 2.6.

The number of stars in a galaxy is truly remarkable. On a clear night we can see a few thousand stars with the unaided eye, but there are many more stars out there that we cannot see without using a telescope—several hundred million times more, just in our own galaxy, the Milky Way. The actual number of stars in the Milky Way is not known with certainty; current estimates range from 200 billion to 400 billion. And there are a lot of other galaxies in the Universe besides our own, as we learned shortly after the **Hubble Space Telescope** became operational.

Named for Edwin Hubble, the Hubble Space Telescope was carried into space and placed into Earth orbit by the Space Shuttle in April of 1990. During a 10-day period in December 1995, Hubble was pointed to a dark region of space near the outer stars of the handle of the Big Dipper. This area was chosen to allow the telescope to peer out into deep space in an area where there are few stars to get in the way. A series of 342 images were taken of a very small field of view within that region, a field roughly equivalent to 1/30 the diameter of the Moon; stated another way, it was like peering through a hole the size of a period on a sheet of paper held at arm's length. The spectacular images that resulted were released to the public at a meeting of the American Astronomical Society in 1996. They showed about 3000 faint but distinct galaxies (**FIGURE 2.8**)—not individual stars, but *3000 galaxies* of stars, with each galaxy containing some 200–400 billion stars!

A follow-up image was taken in the southern sky, in an area to the southwest of the constellation Orion, but this time it was of a slightly larger region; known as the Hubble Ultra Deep Field, it was equivalent to about 1/10 the diameter of the Moon, or three millionths of the sky. Data were collected over a 16-month period in 2003 and 2004 as Hubble orbited Earth, giving the image in **FIGURE 2.9**. Astronomers have estimated that this image contains some 10,000 galaxies—again, each with 200–400 billion stars! (What an amazing example these images are of exploration and discovery!)

Based on these new results from Hubble, cosmologists estimate that the *visible* Universe (that part of the Universe that can be seen with a telescope) has about

FIGURE 2.8 The Hubble Deep Field image taken by the Hubble space telescope in 1995. It shows nearly 3000 galaxies in an area equal to that of a period on a sheet of paper held at arm's length, which is equivalent to only two-millionths of the sky.

100 billion galaxies; but there are more out there. Estimates that include parts of the Universe not visible to us range as high as 500 billion galaxies. So, how many stars are there in the Universe? Well, assuming that the lower estimate of 200 billion stars (2×10^{11}) in the Milky Way is representative, and that there are (conservatively) 200 billion galaxies in the Universe, this would give us 4×10^{22} stars in the Universe, or 40,000 billion billion. This is a big number, of course, which makes more sense if we can relate it to something here on Earth. There are many estimates one could make, based on any number of assumptions, of course; but one estimate that has been used is the number of grains of sand on all the world's beaches, which turns out to be something on the order of 7×10^{18} or 7 billion billion.[8] This means that there are some 1000 to 10,000 times more stars in the Universe than there are grains of sand on all the world's beaches.

Early formation of planet Earth

After the Big Bang at the beginning of the Universe, as some stars ran out of their nuclear fusion fuel and exploded as supernovae, they created huge nebulae, the clouds of interstellar gases and dust particles that we have just discussed. Eventually, in our corner of the Milky Way, some of the interstellar dust and gases, especially hydrogen, came together under mutual gravitational attraction to form our Sun, but it didn't begin to shine right away; it would take a while before the young star would reach a critical point of temperature and pressure and ignite. Orbiting around the newly formed and yet-to-ignite Sun were disk-like rings of particles, similar to the rings of Saturn. Slowly those particles in the rings coalesced as a result of turbulent collisions, creating larger and larger particles, which eventually became **planetesimals**, spheroidal clumps of matter approximately a kilometer across (**FIGURE 2.10**). At this point, the gravity associated with the planetesimals was sufficient that they attracted one another to create even larger planetesimals, or **protoplanets**, which would eventually become our planet Earth and the other planets of our solar system.

As planetesimals continued to collide and the mass of the proto-Earth grew, it created a lot of heat, in addition to the heat of compression and heat

---

[8] Reported on the CNN Tech website in "Star survey reaches 70 sextillion," July 23, 2003.

FIGURE 2.9 Hubble Ultra Deep Field image, revealing 10,000 galaxies in a dark region of space equivalent to an area 1/10 the diameter of the Moon.

FIGURE 2.10 Diagram illustrating the protoplanetary disk around the Sun and the formation of planetesimals and, eventually, the planets.

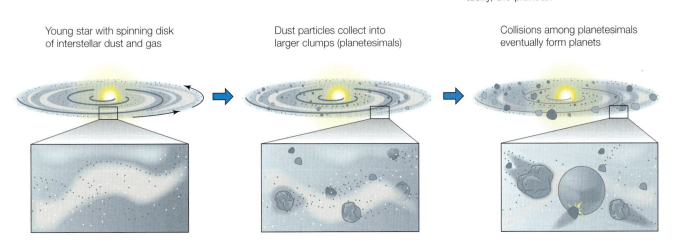

Young star with spinning disk of interstellar dust and gas

Dust particles collect into larger clumps (planetesimals)

Collisions among planetesimals eventually form planets

from radioactive decay of some of its elements. As early as 10 million years after the proto-Earth began to form, the inner protoplanet melted, and the molten material began sorting out into its heavier and lighter elements, with the heaviest elements, especially iron, settling to the center of mass. This gave the young Earth its metallic core, which is responsible for the Earth's magnetic field, and it produced an internal layered structure.

At some point during this process, the Sun reached its critical point and ignited, an event that released so much energy that it blew away most of the gaseous part of the disk surrounding it. The larger pieces of matter that remained, including the planetesimals, continued to coalesce into a system of planets (including the Earth) in orbit around the Sun. At about this point, an especially large planetesimal, perhaps as big as Mars, is thought to have struck Earth. It struck with such force that a large chunk of Earth, including some of its mantle, was blown into space and collected into another planetesimal that is now orbiting Earth, the Moon. The impact was so powerful that it is also thought to have altered the Earth's axis of rotation, producing the 23.5° tilt that is responsible for our seasons, and it may have sped up the Earth's rotation. This rather fantastic idea is supported by at least three powerful lines of evidence that came together following the Apollo moon missions in the 1960s and 1970s. Radiometric dating of moon rocks brought back to Earth by the Apollo astronauts showed them to be younger than the other celestial bodies in our solar system, including Earth, by some 30–55 million years. In other words, the Moon was formed later than Earth and the other planets that formed in the initial protoplanetary disk around the Sun. In addition, a reanalysis of the moon rocks in the late 1990s found them to be remarkably similar in composition to Earth's mantle and crust. These results, coupled with the estimated overall density of the Moon, which is relatively low, indicating that it does not have a metallic core, means that it did not form the same way the Earth and the other planets did. It seems that the Moon is made of part of the Earth's mantle and outer crustal layers that were blown out of the young Earth about four and a half billion years ago. This idea has come to be known as the "giant impact hypothesis."

It is not known whether the giant impact melted the entire Earth a second time, but there is consensus among geologists that a large portion of the Earth did melt and that it subsequently reorganized in layers of differing density. In this stratification, the lighter elements, such as silicon (Si), magnesium (Mg), aluminum (Al), and oxygen-bonded compounds, floated closer to the Earth's surface, where they cooled and eventually formed the Earth's crust, while the heavier elements coalesced in the center, or core, of the Earth.

It is important to realize that these processes were not confined to our minuscule portion of the Universe; that is, ours is not the only solar system in the Universe in which planets orbit around stars. In fact, we can identify new stars forming inside nebulae, and protoplanetary disks around those stars. For example, **FIGURE 2.11** shows a protoplanetary disk in the Orion Nebula, as imaged by the Hubble Space Telescope. And we have even found planets outside our own solar system. NASA and Caltech's Jet Propulsion Laboratory have an ongoing search called Planet Quest. As of late 2011, more than 650 **exoplanets**, planets outside our solar system, had been identified, the vast majority having been detected using indirect methods, rather than actual imaging. The intriguing question is: How many planets might there be in our galaxy with conditions suitable to support life? That number multiplied by the number of

FIGURE 2.11 Hubble Space Telescope picture of a protoplanetary disk in the Orion Nebula, viewed edge-on as it spins around a newborn star in the center of the disk. Earth probably formed out of just such a disk more than 4.5 billion years ago.

galaxies in the Universe (ca. 200 billion) would give a large number of planets indeed on which life could have evolved.

Origins of Earth's water

Finding other planets in the Universe is one thing, but finding planets with oceans of liquid water, like our own Earth, is another matter altogether. It is the presence of liquid water, after all, that is thought to have allowed life to evolve here on Earth, and it is the major determinant of whether life can exist on other planets. So a fundamental question is: Where did all the water on Earth come from?

Relatively recent studies have shown that as early as about 4.4 billion years ago, or roughly 150 million years after the Earth's formation, Earth already had its oceans; but where the water filling those oceans came from is a bit of a mystery and the subject of continued discussion and debate among scientists. There are generally two schools of thought on the matter, which have given rise to (1) the *extraterrestrial* theory and (2) the *outgassing* theory. The extraterrestrial theory is based on the fact that both meteorites and comets contain water (both chemically bound, and in solid [ice] form), and that both were colliding with Earth during and after its formation. One class of meteorites, the carbonaceous chondrites, which bombarded Earth early in its formation, contains between 3% and 22% water. And comets are mostly dust and water (ice); some contain up to 80% water. Could enough of these comets have struck Earth to account for a significant fraction of water currently present on Earth? Apparently, no. When scientists analyzed the isotopic ratios of hydrogen in the tails of three comets—Halley's, Hyakutake, and Hale–Bopp—they were shown to be markedly different from the hydrogen isotopic ratios in ocean water, suggesting that comets were not the source of the oceans' water.[9] In addition, scientists now believe that comets may not contain as much water as once thought. Nonetheless, while not accounting for the bulk of the oceans' water, comets could have provided some (less than 10%) of the water on Earth today. Meteorites and large asteroids are also thought to have contributed to Earth's water, but all the extraterrestrial sources combined do not account for the quantity of water on Earth. This leaves the outgassing theory.

The outgassing theory is based on the idea that water was part of Earth's original structure, and that as Earth formed, volcanic activity would have released water vapor to the surface, where it would have condensed. This volcanic process is still going on today. While both terrestrial and extraterrestrial sources were involved in filling the oceans with water, most of Earth's water was already here.

Whatever the exact sources of water to the young Earth, it is likely that the giant impact that formed the Moon must have blown away any atmosphere that the early Earth had. Now that there was no atmosphere to hold in the heat, the planet cooled relatively quickly (over a period of millions of years) to form a solid outer crust. Eventually, a new atmosphere formed with water vapor and clouds of tiny droplets of liquid water that fell as rain. And when Earth cooled enough that the rain no longer boiled away as soon as it struck the surface, the rain filled the oceans.

There are still holes in all of these theories, and details about where all of Earth's water came from and when and how the oceans were filled are far from resolved. Definitive answers await further research.

---

[9] The hydrogen isotopes analyzed were deuterium, or "heavy hydrogen," which has a nucleus containing one neutron and one proton, and protium, or "normal" hydrogen, with a nucleus containing only one proton.

Water elsewhere in the Universe

The exploration of space by the various NASA-launched space probes over the past couple of decades has greatly increased our understanding of our own Earth's formation, and has provided a few surprises. One such surprise came from data sent back by the Galileo spacecraft, which was launched in 1989 on a six-year journey to Jupiter. It arrived at Jupiter on December 7, 1995, and spent some time studying that planet before beginning a two-year extended mission to study one of Jupiter's moons, Europa. The results of that study, released in 1998, fueled speculation that Europa might have had, or may still have, an ocean which today is covered by a layer of ice, as hinted at by the presence of pressure ridges (**FIGURE 2.12**).

Given the probable sources of water on Earth, it stands to reason that evidence of water outside our own planet, such as that on Europa, should not come as a surprise. In fact, recent data suggest that even our Moon has an icy makeup. In 2009, NASA launched the Lunar Crater Observation and Sensing Satellite (LCROSS; **FIGURE 2.13**) to determine the presence of ice deposits trapped beneath the soil in the permanently shadowed craters of the Moon's polar regions. The plan was to collect and relay data from an impact and

FIGURE 2.12 Jupiter's moon, Europa, which is about the same size as our moon, is thought to be covered by a frozen ocean. (A–C) Photographs taken by the Galileo spacecraft during a mission to Jupiter and its moons in the late 1990s. (A) This image, taken as the Galileo spacecraft approached Europa on September 7, 1996, shows the moon's approximate natural color. (B) Closer view of Europa's surface showing pressure ridges in the surface ice and "freckles," or pock marks, that are thought to be the result of plumes of warmer water from beneath. (C) Photograph taken on February 20, 1997 showing the icy surface of the moon. This photo shows an area about 34 × 42 km, and shows plates of ice. These plates are apparently moving about and forming pressure ridges, similar to ice flows here on Earth. (D) The probable internal structure of Europa and its ice-covered ocean.

(A)

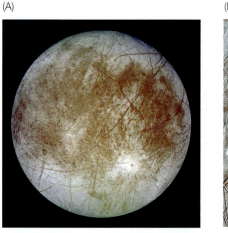

(B)

(C)

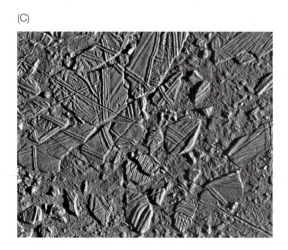

(D)

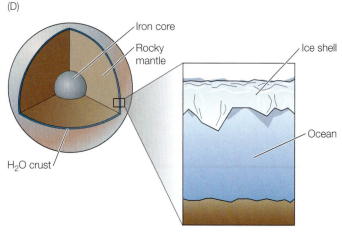

FIGURE 2.13 NASA's Lunar Crater Observation and Sensing Satellite (LCROSS) launched in 2009 to test whether there is water ice on the moon. There is.

debris plume that was to result when the launch vehicle's spent upper stage crashed into the surface of the Moon at a speed of some 10,000 kilometers per hour (6200 miles per hour). The mission was a success. It confirmed the presence of water ice on the Moon, although the water concentration was less than that found in soils of the driest deserts here on Earth.

In May of 2008, the Phoenix Lander (**FIGURE 2.14A**) touched down on the Martian surface to begin its search for traces of water. The conditions on Mars are not conducive to the presence of liquid water, at least not for very long. The temperatures are so low that any water would be in the solid form, ice, and the average atmospheric pressure is so low that if ice were exposed to the atmosphere, it would sublimate from a solid to a gas, skipping the intermediate liquid state. Nonetheless, there is evidence that in the past there was indeed abundant liquid water flowing on the surface of Mars, and that at one time there were even oceans similar to our own. The question that remains to be answered is: Where did all that water go? The most likely answer is that it is locked up in the permafrost at the poles. The Phoenix Lander confirmed this with an interestingly simple experiment. On June 19, 2008, NASA announced that clumps of bright material in depressions dug in the North Pole of Mars by the Phoenix Lander's robotic arm were water ice. After having been exposed, it sublimated (vaporized and began to disappear) over a four-day period (**FIGURE 2.14B**).

With these fascinating scientific ideas and findings about the early formation of the Universe and the solar system as a foundation, we move on in the next section to a more detailed discussion of the geophysics and chemistry of Earth itself.

FIGURE 2.14 (A) The Phoenix Lander settling to the surface of Mars in 2008. (B) The disappearance of water ice following four days' exposure in depressions dug by the Lander's robotic arm. Look closely at the white ice in the upper right portion of the impression mark; there is less ice there after four days. The box shows the area of the inset photos.

(A)

(B)

Day 20

2/3 in.

Day 24

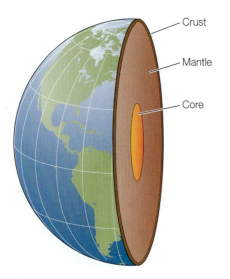

FIGURE 2.15 A cross section of the Earth showing the major internal layers that formed more than four billion years ago when a once-molten Earth slowly separated, stratifying its mass into layers of different densities as a result of their different chemical compositions.

Earth's Internal Structure

Exactly when and to what degree Earth melted early in its formative stages are still subjects of debate. Was the Earth partially or completely melted at the time of its initial formation? Did all or part of Earth melt following the "giant impact" that resulted in the formation of the Moon 35–55 million years later? Geologists are not certain. Nonetheless, the internal structure of the Earth today reveals a layering that required it to have had a molten state at one time or another long ago. We alluded earlier to the simplest model of Earth's internal layering, that of a metallic core, a mantle, and an outer crust (**FIGURE 2.15**).

Geologists have further differentiated these layers based on their physical and chemical characteristics. The chemical makeup of the Earth's interior reveals the nature of the planet's initial formation billions of years ago, and is the basis for the physical structure of the Earth's layers today. But to understand how these layers behave, which is fundamental to understanding, for example, continental drift and earthquakes, we have to consider the relative rigidity and fluidity of these layers, how those properties change with depth below the surface, and how they respond to stress.

Chemical characteristics of Earth's interior

The Earth's interior can be described as a series of layers differentiated by their chemical characteristics (**FIGURE 2.16**), as revealed by seismic analyses. The outermost layer is the **crust**, a major part of the rigid layer that completely covers the surface of the Earth. Relatively speaking, the crust is an extremely thin, lightweight, and brittle layer of rock, sort of like the crust of a spherical loaf of bread that has been baked a little too long. There are two types of crust, defined by the types of rock that predominate in them. **Continental crust**, which makes

FIGURE 2.16 A side-by-side comparison of the thicknesses of Earth's internal layers, based on their chemical composition (left) and their physical properties (whether they are rigid or fluidlike [right]). The depths do not line up. Note that the chemically distinct crust is very thin, but beneath it is a layer of chemically distinct upper mantle which has similar physical properties.

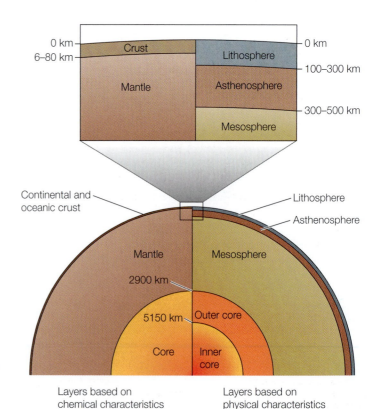

up the major land forms on Earth, is mostly **granite**, a relatively lightweight rock made of compounds of oxygen, silicon, and aluminum, generally referred to simply as *silicates*. **Oceanic crust** makes up the ocean floor and is primarily **basalt**, which is heavier (denser) than granite. It, too, is made up of compounds of oxygen and silicon (silicates), but it contains more metals, especially magnesium and iron, giving it a darker gray or black color (**FIGURE 2.17**). The crust varies in thickness, from about 6 to 80 km with the thickest crustal layers found beneath mountains; compared with Earth's diameter (about 13,000 kilometers, or 8000 miles), it is thus paper-thin. Beneath the crust is the much thicker **mantle**, about 2900 kilometers (ca. 1800 miles) thick. It is similar in chemical composition to oceanic crust in that it is made up of compounds of oxygen and silicon, but it has more of the metals magnesium and iron. The core, which is made up of an **inner** and an **outer core**, is mostly iron.

FIGURE 2.17 Granite (left), the main rock form that makes up continental crust, and basalt (right) the rock that makes up the oceanic crust.

No actual samples have ever been collected from beneath the crust, by the way. In fact, the deepest hole ever drilled was in Siberia, and it went down only about 12 km (7.5 miles) into the crust. The temperature at that depth was 245°C (more than 450°F), which is hotter than a pizza oven. Instead, we learned about the Earth's internal structure from seismic waves (see page 54, Passage of seismic waves through the Earth).

Considering changes in the physical properties at various depths inside the Earth, we have a different "model" of its layers, one that considers changes in temperature and pressure with depth inside the Earth, but which also relates to how the various layers respond to physical stress: do they crack or flow? That is, these layers may be rigid and brittle, cracking under stress, or they may behave more like a plastic or liquid, with fluid-like properties, thus flowing under stress. Those characteristics as a function of depth inside Earth give a different layering system, as illustrated in the right side of Figure 2.16.

Physical characteristics of Earth's interior

Still referring to Figure 2.16, we see that the **lithosphere** (from the Greek *lithos*, stone) is the cool, rigid outer layer, which includes the crust and the outermost part of the mantle. That is, the lithosphere includes that part of upper mantle which is chemically distinct from the crust, but which is nonetheless rigid and brittle and thus possesses crust-like physical characteristics. Therefore, the lithosphere is a bit thicker than the crust, on the order of 100 km (62 miles) thick.[10]

Just beneath the lithosphere is the **asthenosphere** (from the Greek *asthenes*, which means weak, or soft). This is the thin, hot, and slowly flowing upper layer of mantle. While not strictly a liquid, it is not exactly a solid either; it can deform like a plastic under stress, like white-hot piece of metal that can be shaped by a blacksmith's hammer. The asthenosphere is relatively thin; it only extends to a depth of about 300 km (187 miles). The lithosphere literally floats on top of the asthenosphere, and can move about if subjected to a sufficiently powerful force.

---

[10] The two terms, *crust* and *lithosphere*, have lost their distinction and become almost synonymous; some authors simply use the term *crust* to mean the lithosphere, but the lithosphere includes the crust and the chemically distinct upper part of the mantle. The distinction becomes more important when we discuss lithospheric plates.

Below the asthenosphere is the appropriately-named **mesosphere** (from the Greek *mesos*, middle). This is the less fluid, rigid part of the middle and lower mantle. It extends to a depth of about 2900 km (1800 miles). The great pressures at these depths have compressed it and made it more solid-like than the asthenosphere.

In the center of the Earth, we have the **inner** and **outer core**. The outer core is an extremely dense viscous liquid (that it is a liquid was verified in 1964, as we will discuss below), a molten mass that is much hotter than the mesosphere layer just above it—as hot as 5500°C (9900°F). Its density is four times that of the crust. The inner core is even hotter than the outer core (ca. 6000°C, similar to the surface of the Sun), but because of the great pressures which compress it into a solid phase, it is rigid, not liquid.

An Introduction to Geophysics

Heat sources and heat flow

In order to appreciate the forces at work inside the Earth, today and long ago, we need to go over a few basic principles of geophysics. For one thing, you must be wondering: Why are the deep layers of Earth so hot? And, even though it melted long ago, hasn't enough time passed that it should have cooled off by now? To answer those questions, we have to appreciate the nature of the sources of Earth's internal heat. First of all, during Earth's initial formation it was being bombarded by meteors and much larger planetesimals; these impacts generated a lot of heat. In addition, as all that mass accreted, as Earth's mass grew, the pressure from compaction as gravity pulled everything inward generated a lot of heat. (You can experience heat generated by compression for yourself: squeeze a copper penny in a bench vise over and over, and you can feel the penny getting hotter.) A third source of heat is that from **radioactive decay** of the heavier elements. The reason that Earth's interior is still so hot today is (mostly) that radioactive decay has never stopped; it continues today in the mantle.[11] This means that there is still a source of heat maintaining Earth's inner temperatures, temperatures that are hotter than the surface of the Sun. Because the Earth is a sphere, some of that heat flows outward, where it dissipates, but much of it is focused inward toward the center of the Earth where it has nowhere to go, and so the heating is concentrated (**FIGURE 2.18**).

There are two kinds of heat flow we need to consider: conduction and convection. **Conduction** is heat passing through a substance. Conduction is not the same as radiant heat transfer (which you can feel when you are standing in direct sunlight, or when you hold your hand near your face; the heat you feel from your hand, for example, is the infrared radiation emanating from your warm hand striking your skin and being converted into heat energy). A good example of heat conduction is what happens when you put a stainless steel spoon into a cup of hot tea; the tea's heat is first conducted into the submerged end of the spoon, and is then conducted along its entire length, causing even the far end of the spoon to get warmer, even though it is sticking up in the air.

Convection is the flow of heated material itself (water, air, or molten rock). For example, this is what happens when warm air rises. Heat is not being passed between nonmoving layers of air; what is moving is the air itself.

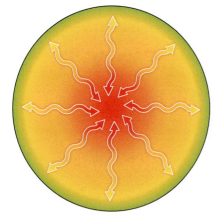

FIGURE 2.18 Radioactive decay continues to generate heat, even today, deep inside the Earth's mantle. That heat radiates away from the source in all directions, as represented by vectors (arrows): some heat radiates away from the center of the sphere and toward the surface, while some heat radiates inward. Because the Earth is a sphere, uniform heating in all directions produces more heat radiating inward, as illustrated by the arrowheads closer together in the center than at the outer edge. This focusing of heat toward the center of the Earth over long periods of time maintains the extremely high interior temperatures.

---

[11] First described by physicists in 1896, radioactive decay is the process whereby the nucleus of an unstable atom spontaneously emits particles of itself, and at the same time loses energy in the form of radiation. The result: an unstable atom of one element transforms into a stable atom of a different element while radiation is emitted, which heats the Earth's interior.

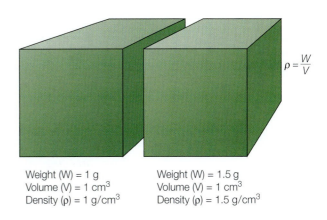

FIGURE 2.19 Example of two cubes, each the same size (volume, V) and shape, but of different chemical compositions, which results in different densities (ρ). Mass is measured as weight (W) in grams (g).

$$\rho = \frac{W}{V}$$

Weight (W) = 1 g	Weight (W) = 1.5 g
Volume (V) = 1 cm³	Volume (V) = 1 cm³
Density (ρ) = 1 g/cm³	Density (ρ) = 1.5 g/cm³

Concepts of mass, weight, density, buoyancy

As we mentioned above, the lithosphere floats on the plastic asthenosphere; because the lithosphere is less dense (its mass in a given volume is less) than the asthenosphere, it is buoyant. Let's review these principles of mass, density, and buoyancy in the context of geology here. (We'll be coming back to them again in a later chapter when we discuss how water behaves.)

To begin, anything that has **mass** has **weight** in the Earth's gravity; that weight is (usually) measured in grams. This distinction between mass and weight becomes clearer with a simple thought experiment (but don't actually try it). In a gravity field, such as here on the surface of the Earth, an ordinary carpenter's hammer is heavy compared with a sponge, which is light. But in space, aboard the International Space Station, for example, where there is zero gravity, they would both be "weightless." But even in a weightless environment, they each still possess mass (and if that mass is moving, it also possesses momentum). So, imagine being hit with each one; you would quickly learn that although it is weightless, the hammer still has mass.

Density is the mass per unit volume (**FIGURE 2.19**); its symbol is the Greek letter *rho* (ρ), and is usually expressed as grams per cubic centimeter (g/cm³). For example, water has a density of 1.0 g/cm³, while granite rock has a density of 2.7 g/cm³.

Buoyancy is the degree to which one object, or a substance, is less dense than another, and therefore whether it floats in the other, and if so, how high. This phenomenon is sometimes presented as **Archimedes' Principle**, which states that an object immersed in a fluid is *buoyed up* by a force equal to the weight of the fluid that the object displaces. An example often used to illustrate this is that of two identical ships, one loaded with cargo and one that is not. In each case, the water displaced is equal to the weight of the ship, so the ship that is loaded with cargo displaces more water and floats deeper; it is said to "draw more water."

In geology we refer to these phenomena in terms of the principle of **isostasy**, which is the state of gravitational equilibrium (flotation) between the rigid lithosphere and the liquid-like asthenosphere. That is, the lithosphere "floats" in the asthenosphere at an elevation that is dependent on its thickness and density—its total weight. For example, parts of the lithosphere that have mountain chains will float on the asthenosphere with a significant fraction of the mountains' mass sinking into the asthenosphere, such that it is in **isostatic equilibrium** (**FIGURE 2.20**), just like the ship that sinks into the water to its equilibrium point. Antarctica, for example, has been ice-free in the past and at those times it did not sink as deep into the asthenosphere as it does today, with its

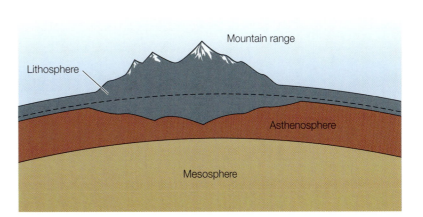

FIGURE 2.20 Vertically exaggerated schematic representation of isostatic equilibrium, whereby a mountain range, which is part of the lithosphere, floats deeper in the asthenosphere than the surrounding featureless and lighter lithosphere. Thus, the Earth's crust, or lithosphere, is thickest beneath mountain ranges.

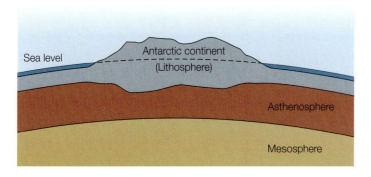

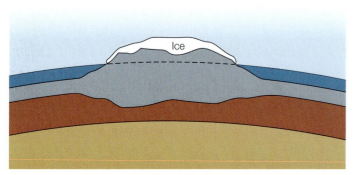

FIGURE 2.21 Isostatic sea level in Antarctica when it was ice free millions of years ago (and floated higher on the asthenosphere) compared to the sea level today, with ice cover depressing the continent deeper into the asthenosphere.

heavy load of ice (**FIGURE 2.21**). Relative sea level around Antarctica thus changed; sea level is higher today than it was long ago when it was ice-free, because the ice is depressing the continent into the asthenosphere. A world map showing the thickness of the lithosphere is given in **FIGURE 2.22**. The thickest crust (which is only part of the lithosphere, remember) is greater than 70 km and is located beneath the Himalayan Mountains; on the other hand, the thinnest crust (and thinnest lithosphere) is the ocean floor (the oceanic crust) which is on the order of 10 km.

The asthenosphere responds to stress (e.g., the weight of the Himalaya Mountains, or the coming and going of ice in Antarctica) by deforming, similar to the way water "deforms" around a ship, but not nearly so easily, because the asthenosphere is more "plastic" than "liquid." The lithosphere, unlike the asthenosphere, cannot deform this way; it is too rigid and brittle. The response of the lithosphere to stress is to fracture along the plane of weakness, creating a fault, which is one cause of earthquakes.

Seismic waves: Two types

The energy released by an earthquake is transmitted through the Earth as **seismic waves**, which are huge, low-frequency energy pulses (from the Greek *seismos*, earthquake). While powerful earthquakes can be very destructive and result in tragic human suffering, they also provide geologists with information that has proven useful in understanding the nature of the Earth's deep interior. In fact, is was just such a tragic earthquake, in Alaska in 1964, that provided the best evidence to date that the Earth's outer core is a liquid. We'll discuss that event in a minute; first, some more on the nature of seismic waves.

There are two kinds of seismic waves. Pressure waves, or **P waves** (also called compression waves), are waves of compression, and are akin to sound waves; shear waves, or **S waves**, are transverse, or side-to-side waves, and are akin to surface-water waves on the ocean. Earthquakes generate both types of waves, which travel through the Earth (in some cases, all the way through to the other side of the Earth). An illustration of P and S waves generated by a large

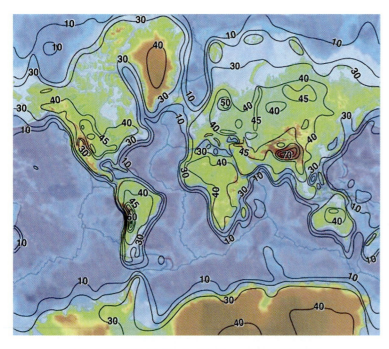

Thin   Thick

Crustal thickness

FIGURE 2.22 Computer-modeled map of crustal thickness in kilometers.

spring and a rope is given in **FIGURE 2.23**. These two different seismic waves have quite different characteristics, which are: (1) P waves travel faster than S waves; (2) P waves travel faster in a solid than a liquid; and (3) S waves cannot pass through a liquid. A couple of thought experiments help to illustrate these differences.

First, if we were to take a solid length of steel railroad track, a half-mile long, and hit the butt end with a hammer, we would send a compression wave, a P wave, down its length at a very high speed. The steel rail conducts the compression wave very well. If we were to pluck the rail like a guitar string, making a transverse wave, or S wave, we would notice that the rail also conducts the transverse wave, the S wave, but more slowly. If the rail were made of wood, rather than steel, it would still transmit each type of wave, but not as fast as steel does. The speed of a seismic wave is inversely proportional to the compressibility of the medium,[12] and since steel is less compressible than wood, it transmits waves faster.

Next, we line up hundreds of wood blocks, all perfect cubes the same size, so that each one is touching the one next to it, and repeat our experiment. First, we hit the butt end of the line with a hammer. We again notice that a compression wave, a P wave, travels along the length of the blocks quite well, sending the last block in the line flying. But when we pluck the row of blocks, nothing happens. Oh sure, the block we plucked moves sideways, but the others lined up next to it don't move, of course, because they aren't connected to one another. The blocks cannot transmit an S wave, only a P wave. In terms of their ability to transmit seismic waves, the wood blocks are analogous to a liquid; S waves cannot pass through a liquid, and P waves pass through liquid far less rapidly than they do through metal.

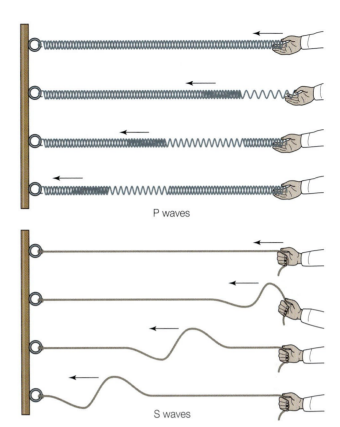

FIGURE 2.23 Examples of P waves and S waves using a large spring and a rope. The spring will propagate a compression wave along its length, as the person holding one end alternately pushes and pulls it. When tugged side to side, the rope will propagate an S wave, or transverse wave, along its length.

Refraction of seismic waves

When waves encounter a medium with a different level of compressibility, the wave speed increases or decreases accordingly, which causes the wave to bend. This "bending" is called wave **refraction**. Both types of seismic waves, P waves and S waves, traveling through the Earth change speed (and therefore, refract) depending on the compressibility of each new medium they encounter.

To understand refraction, first we need to understand how to represent waves, wavefronts, and the direction of wave travel. To start, let's use water as an example. If we were to toss a rock into a calm pond, we would, of course, see ripples (surface waves) which move away in a pattern of concentric circles from the spot where the rock hit the water (**FIGURE 2.24**). Some distance away from the center, as the concentric circles become quite large, you would

FIGURE 2.24 Example of wave propagation, based on surface water waves created by tossing a rock into a pond. As the waves move away in concentric circles, the diameters of the circular wave fronts become larger and larger, such that the waves, if we focus only on a small section, appear to become almost straight lines.

---

[12]Some authors explain that the speed of a seismic wave is directly proportional to the density of the medium—that is, the speed is faster in denser materials. Because the density of the Earth increases with depth (and the compressibility decreases), we can relate the speed of seismic waves to density, but the actual dependence is on compressibility, not density.

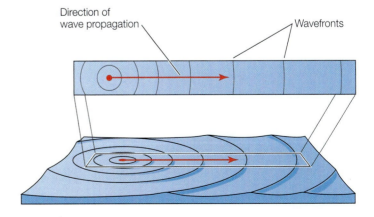

Direction of wave propagation

Wavefronts

FIGURE 2.25 Schematic diagram of light waves passing from air through glass, both (A) at an angle and (B) perpendicular to the glass surface. The speed of light is slower in glass than in air, so the waves in air tend to pile up upon being slowed in the glass, almost like cars at a turnpike toll booth; the wavelengths shorten inside the glass, like the spacing between cars. And when light strikes the glass at an angle, the slowing effect causes the light rays to refract, bending into the glass. When the light waves leave the glass, they refract in the opposite direction, as they speed up again in air.

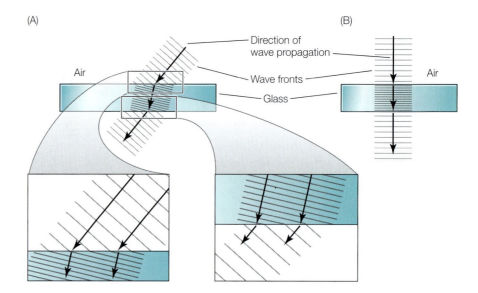

notice that those lines eventually appear to be straight. Those lines are the wavefronts, and they are essentially parallel to one another. Because of this phenomenon, when we draw wavefronts, we can draw them simply as straight, parallel lines. If you imagine the wavefront moving farther and father from the point where the rock entered the pond, you see that the direction of the wave propagation (or the direction the wave is traveling) is always 90° perpendicular to the wavefront, which is drawn as an arrow. While the example in Figure 2.24 is for surface water waves (which we'll discuss in Chapter 6), the same is true for sound waves, light waves, and seismic waves.

FIGURE 2.25A illustrates refraction in light waves passing from air into glass and then into air again. The speed of light is slower in glass than in air. Notice that the wavefronts are drawn as small, consecutive, parallel lines, while the direction of the wave is indicated by an arrow. As the wavefronts of light arrive at the glass, the part of the wavefront that first encounters the glass is slowed down, as illustrated, and the light is refracted. Note that this is only true for light that arrives at an angle; when a wavefront is parallel to the medium it encounters, all of the wavefront will meet the new medium at the same time, so no refraction will occur. The speed will be slowed uniformly across the wavefronts, but the light won't refract (**FIGURE 2.25B**). As the waves enter air again, they speed up, which bends the wave the opposite way.

Passage of seismic waves through the Earth

It is because of the predictability with which both S and P waves refract when they encounter a new medium that geologists have been able to determine the internal layering of the Earth. To illustrate, we'll use a couple thought experiments:

First, imagine that were are going to set off a dynamite charge on top of a rectangular block made of a material of uniform consistency throughout. The resulting seismic waves, both S and P waves, will travel through it in all directions, with spherical wave fronts as shown in **FIGURE 2.26**.

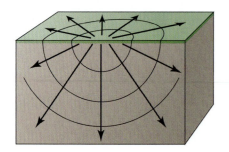

FIGURE 2.26 Schematic representation of seismic waves passing through a solid, such as the Earth. Curved lines represent the wave fronts, and arrows indicate the directions of wave propagation. Imagine wave fronts like layers of an onion—they radiate in three dimensions from the point of generation into the medium.

Now imagine two Earths, which, unlike our rectangular block, are spheres, of course (**FIGURE 2.27**). One of the spheres has uniform consistency throughout, similar to our block, but the other model Earth changes consistency with depth from the surface to the center, being least compressible in the center. If we were to set off a large explosion at the North Pole on these two imaginary Earths, we would observe the following: in the first Earth, which is of uniform consistency throughout, there would be no change in the speeds of seismic waves, so the waves would travel through the Earth in straight lines. But in the second Earth, the waves would be refracted toward the region where speed is slower, which means that those waves will refract back toward the surface, as illustrated.

The Great Alaska Earthquake in 1964 provided an opportunity to apply these principles, in that seismograph stations set up around the globe were in position to monitor the strength and arrival times of the earthquake-generated S and P waves as they propagated through the Earth. Analyses of those results revealed an interesting pattern (**FIGURE 2.28**). Both S and P waves from the earthquake were recorded over more than half the globe, but beyond that, on the opposite side of the globe, there was a curious feature: there was a band around the Earth, a shadow zone, where no seismic waves were recorded at all. And beyond that band, which was directly on the opposite side of the world from Alaska, only P waves were recorded—no S waves. Scientists recognized from this pattern of seismic wave transmission that the internal structure of the Earth must look something like that shown in **FIGURE 2.29**. The Earth's core must have an outer layer that is liquid, but an inner core that is solid. A liquid core would explain why no S waves were found opposite the earthquake's epicenter, because liquid does not conduct S waves. When P waves encounter the liquid outer core they are transmitted, but they are slowed down, thus bending (refracting) slightly downward. But because inside the liquid core the speed also increases with depth, the waves again refract outward, as illustrated. Interestingly, observations of an apparent acceleration of P waves that passed through the very center of the Earth had been recorded in 1935, so it had been known since then that the central core must be solid. (Remember, P waves travel faster in a solid than in a liquid).

The recordings of the S and P waves from the Great Alaska Earthquake provided direct physical evidence that Earth has both a solid inner core and a liquid outer core. The identification of Earth's various layers and the development of a fundamental appreciation of how they behave in a geophysical sense would

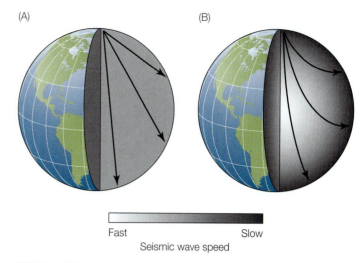

FIGURE 2.27 Cutaway diagrams of the Earth showing the directions of seismic waves (both S and P waves) generated by an earthquake at the North Pole as they pass through the Earth's interior under two theoretical conditions: assuming that (A) there is no change in seismic wave speed with depth in the Earth; or (B) that the velocity of seismic waves increases with increasing depth in the Earth. In the first instance, the waves travel in straight lines; there is no refraction. In the latter case, the waves refract toward the medium, in this case depth, of slower speed, which will bend the waves back toward the surface. The real Earth is more like this second case.

FIGURE 2.28 The pattern of seismic waves received by the worldwide network of seismic wave recording stations immediately following the 1964 Alaska earthquake. The earthquake epicenter is located by the arrow at the top of the globe. While both S and P waves were recorded over more than half the Earth's surface, there was a band around the Earth on the opposite side from Alaska where no S or P waves were recorded. Directly on the other side of the world from Alaska, no S waves were recorded, but P waves were.

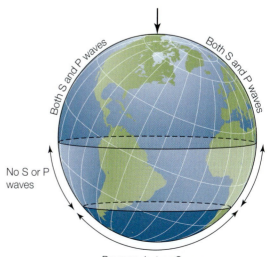

P waves, but no S waves

FIGURE 2.29 Interpretation of the seismic data recorded following the 1964 Alaska earthquake in terms of the physical properties of the Earth. The diagram shows the pattern of S and P waves predicted if the central core is a solid and the outer core is a liquid. Both S and P waves refract, but only P waves can propagate through the liquid core. The refraction of S and P waves in the mantle produce the shadow zone where no S or P waves were recorded.

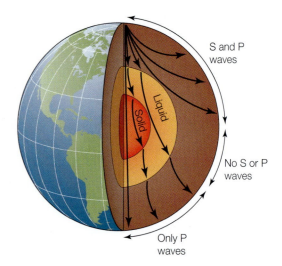

prove to be of paramount importance as scientists pieced together the various bits of evidence that would, eventually, lead to a general acceptance of the concept of continental drift.

So let's step back for a moment and review where we are. We have a Universe that was created by the Big Bang; stars in the Universe that fuse lighter elements into heavier elements; solar systems with planets, including Earth, that are made up of those elements after the stars explode and those elements are free to coalesce; and oceans that formed as the young Earth cooled. Sometime after all this, life appeared on Earth, perhaps originating in the ancient oceans. But when, and how?

The Origin of Life

Details of how life originated on Earth remain unknown, but scientists have determined that whatever happened did so between 4.2 and 3.7 billion years ago. And like the ideas about where all the water on Earth came from, ideas about the origin of life on Earth also fall into two general schools: one holds that life arrived here from space, and the other, that life originated here. In some ways the extraterrestrial theory is moot, because life from another celestial body would still have had to originate somewhere, somehow, sometime after the Big Bang. And, of course, where that might have happened, and under what conditions, would remain a total mystery. Nonetheless, we will briefly discuss the evidence for an extraterrestrial source of life here on Earth later in this section; but first we will focus our review on how life might have arisen here.

It was once thought that complex living organisms can be generated out of decaying organic matter, in a process of **spontaneous generation**, or **abiogenesis**. In the seventeenth century, various experiments demonstrated that when care was taken to prevent contaminating organisms, such as flies, from laying eggs on decaying matter, nothing happened. As more and more experiments produced similar results, a consensus developed toward **biogenesis**, the idea that all living things come from pre-existing living things. By the nineteenth century, abiogenesis from spontaneous generation had become a thing of the past, which certainly represented a scientific advance. But, without spontaneous generation (at least once in Earth's history), we were left with no explanation for how life first arose on Earth.

While Charles Darwin was silent on the issue in his *Origin of Species*, he did begin to ponder the question some years later. In a letter to a colleague, he suggested that life might have begun in a "warm little pond, with all sorts of ammonia and phosphoric salts, lights, heat, electricity, etc., present, so that a protein compound was chemically formed ready to undergo still more complex changes" and that today "such matter would be instantly devoured and absorbed, which would not have been the case before living creatures were formed."

Darwin's second point was reemphasized in 1924 by the Russian biochemist **Alexander Oparin** (1894–1980; **FIGURE 2.30**) in his book, *The Origin of Life*, that spontaneous generation had to have occurred at least once but that it was now impossible because life itself has wiped out the conditions in which it arose billions of years ago. Oparin argued that in an atmosphere devoid of molecular oxygen, a primeval soup of organic molecules could be formed by the action of sunlight, and that those molecules could further combine to form droplets that would "grow" by coming into contact with other droplets, and "reproduce" by dividing into smaller droplets.

At about the same time that Oparin published his book in Russian (it was not translated into English until 1938), British biologist **J. B. S. Haldane** (1892–1964) suggested in a 1929 paper that Earth's primitive atmosphere was exposed to a lot of lightning and ultraviolet radiation—much more so than today. He speculated that those processes interacting with the high levels of carbon dioxide and ammonia in the atmosphere could have produced some of the complex organic (carbon-based) molecules that make up living things. These two similar ideas have been collectively referred to as the *Oparin-Haldane Hypothesis*.

Then, in 1953, a graduate student at the University of Chicago, **Stanley Miller** (1930–2007), published a paper describing an experiment that he and his academic advisor, Nobel Prize-winning chemist **Harold Urey** (1893–1981), had performed. In this experiment, they mixed together water (H_2O), methane (CH_4), ammonia (NH_3), and hydrogen (H_2), substances presumed to be present in the early Earth's atmosphere (**FIGURE 2.31**). After passing an electric current through it for about a week they saw a number of complex molecules formed,

FIGURE 2.30 Alexander Oparin (right) and Andrei Kursanov in the laboratory in 1938.

FIGURE 2.31 (A) Stanley Miller in his laboratory. (B) Schematic design of the experimental apparatus used for the Miller-Urey experiment, in which Miller and his advisor simulated conditions of the early Earth's atmosphere and successfully synthesized a number of complex molecules, including amino acids, the building blocks of proteins.

(A)

(B)

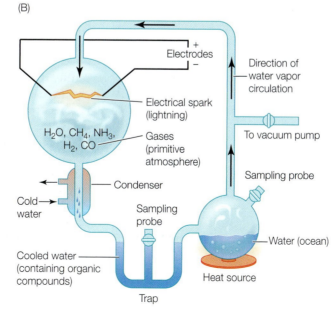

especially **amino acids**, the building blocks of proteins. However, they did not create any living thing.

The simplest imaginable living thing is a self-replicating molecule encased in some sort of cell membrane. It has been argued that simple cell membranes could have been made from phospholipids, which are known to form layers spontaneously when added to water; these *lipid bilayers* are basic components of modern cell membranes. But the origin of the first self-replicating molecule remains an open question. Modern cells achieve replication through the cooperative actions of proteins and nucleic acids, and it is thought that RNA molecules may have arisen from random polymerization of nucleotides. At some point, however, DNA took over the task of replication in modern cells.

Although the "proof of concept" experiment, in which a self-replicating protocell is synthesized in the laboratory, has not yet been achieved, there seems to be a growing consensus among scientists that these chance reactions at one time long ago did, in fact, result in a cell membrane encasing a self-replicating molecule. But this still leaves the question: Where did it all happen? In a pond, as suggested by Darwin? Or in the sea? Earlier ideas held that because nearly all living things depend upon saline water in their cells, life probably evolved in the sea—somehow. One idea was that when tide pools evaporated, leaving briny soups, amino acids could have come together against some surface, such as tiny bubbles of air, and with sunlight providing the energy, a self-replicating biochemical system might have formed. That idea lost support when it was learned that the Sun's UV radiation four billion years ago would have been too destructive for the complex molecules to have formed. Furthermore, the early oceans were probably perennially frozen over because the Sun was much weaker than it is today.

FIGURE 2.32 A "black smoker" in the vicinity of a deep sea hydrothermal vent, where seawater seeps into cracks near an undersea volcano and becomes heated to >350°C, dissolving minerals that, upon encountering the cold water immediately precipitate out, producing the black smoke. The first life forms on Earth are thought to have originated in such an ocean environment some four billion years ago. This image was shot at Sully Vent in the Main Endeavour Vent Field in the northeast Pacific Ocean. A bed of tube worms cover the base of the black smoker, and an acoustic hydrophone and sampling probes can be seen around the vent.

Today scientists are focused on the deep sea as the most likely environment for the initial evolution of life in Earth. In the late 1970s, deep submersibles investigating active undersea volcanoes came across areas where hot seawater, hotter than 350°C, was spewing out of cracks. The water was prevented from boiling because of the immense pressure at those depths, and it was rich in dissolved minerals that quickly precipitated out of solution upon encountering the cold surrounding waters, creating what have become known as "black smokers" (**FIGURE 2.32**). Entire biological communities were discovered that were being sustained by chemical energy from redox reactions involving hydrogen and carbon dioxide. These environments were common on the ancient sea floor at the time of the oceans' first formation, and it is now thought that early life forms first appeared here. Eventually these early life forms led to modern forms that established themselves in shallower depths of the oceans. But as Darwin pointed out long ago, whatever did happen four billion years ago could not happen again today, at least not at the surface of the oceans. The reason: life itself has changed the atmosphere and the oceans into a system not consistent with a new origin of life. We have too much oxygen in the atmosphere now (we'll come back to this issue shortly).

Of course, the other idea is that life may have been brought to Earth long ago from outer space, as farfetched as that may seem. The extraterrestrial hypothesis holds that life probably evolved first on Mars, and that a chunk of the Martian surface was blown into space by a large meteorite, and that eventually that chunk hit Earth as a meteorite, bringing with it some early protocell life form. An attractive aspect of this hypothesis is that life would not have to have evolved independently on separate planets, but only once, somewhere in the solar system, and then spread from there. Support for this idea came several years ago when a Martian meteorite found in Antarctica was analyzed and found to have evidence of primitive life (**FIGURE 2.33**). More recently, in 2009, NASA scientists announced that they had found, for the first time, the amino acid glycine in material collected by the Stardust space probe from the tail of a comet. These findings and the various conclusions that were offered, however, remain very controversial.

Clearly, there are bits and pieces of evidence that the building blocks of life are present, or were present at some time in the past, throughout the solar system, and if conditions were right (correct atmosphere and temperatures, etc.) then it should not surprise us that life may have evolved somewhere other than, or in addition to, here on Earth. Whether such early extraterrestrial life forms ever made the trip to Earth on a meteorite, for example, would seem to be a stretch, but who knows what new finds await us as we continue to explore our own solar system and extreme environments here on our own planet Earth, such as deep sea hydrothermal vents.

This broad way of thinking led Thomas Gold (1920–2004; **FIGURE 2.34**), a Cornell University astronomy professor, to argue that life may reside deep below the surface of Earth at the depths of oil wells (as deep as 8 km). His ideas were based on his controversial theory that oil and natural gas are not the product of decaying biomass from millions of years earlier in Earth's history, but that they have been around since the beginning. Astronomers know that hydrocarbons are widespread in the solar system and in distant nebulae, and as we just discussed, they have been found in meteorites. As part of Gold's overall idea, he posited that microorganisms deep below the Earth's surface might derive nutrition from petroleum, and that those organisms might explain the presence of more complex organic molecules commonly found in petroleum. Indeed, certain anaerobic (not requiring oxygen), thermophilic (heat-loving) microorganisms have been cultured from oil-well samples. Unfortunately, Gold's idea runs counter to the long-held and prevailing theory that oil and gas fields are the products of ancient biomass. But the reports of evidence of past life on Mars have stimulated research into "life in extreme environments," the results of which are truly intriguing. For example, themophilic microorganisms are known from hot springs in Yellowstone National Park (**FIGURE 2.35**) and deep sea hydrothermal vents. One bacterial ecosystem has been described which uses radioactivity as a source of energy!

We can expect this field to take off as more interest is stimulated. In the summer of 2010, a team of researchers reported that they had successfully constructed the first self-replicating synthetic bacterial cell.[13]

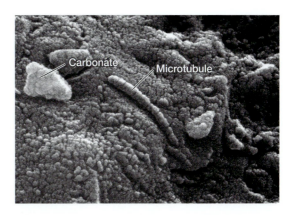

FIGURE 2.33 Electron micrograph of carbonate and a microtubule ca. 0.5 micrometers (µm) in diameter, from a Martian meteorite. Believed to be a fossil of a Martian organism from more than 3.6 billion years ago, it, and the meteorite that contained it, was blown from the surface of Mars about 16 million years ago and was recovered in Antarctica, where it fell about 13,000 years ago.

FIGURE 2.34 Professor Thomas Gold (1920–2004) of Cornell University, just before the Apollo moon mission in 1968. Known for his controversial ideas, especially those proposing the abiogenic origins of petroleum hydrocarbons, and microbial life at the depths of oil wells (several kilometers), he predicted in 1955 that future moon explorations would discover a fine coating of dust on the lunar surface, which was contrary to ideas at that time. Gold was right.

---

[13] Gibson, D. G. et al., 2010. Creation of a bacterial cell controlled by a chemically synthesized genome. *Science* 329: 52–56.

FIGURE 2.35 Example of a hot spring in Yellowstone National Park. Thermophiles, such as those coloring the area of this hot spring, thrive at temperatures as high as 80°C (176°F) in hot springs and in deep sea hydrothermal vents.

Chapter Summary

- We discussed how science and technology are different yet interdependent, how scientists do research, and the value and shortcomings of the **scientific method** as it is sometimes strictly interpreted. Science, not always intuitive, is often full of surprises, and doesn't always agree with our common sense. We introduced these topics in order to prepare to learn about the ideas and findings related to the formation of the Universe and the planet Earth, its oceans, and life forms.

- The formation of the Universe began with **The Big Bang** some 13.7 billion years ago.

- More recently, about 4.6 billion years ago, Earth formed. Our Moon was probably expelled from the early Earth upon impact with a **planetesimal** roughly the size of Mars.

- Earth formed by accretion of interstellar gas and dust particles. It melted under the heat of compaction and bombardment by meteors and asteroids, and only slowly cooled, sorting into internal layers as it did so. Studies of geophysical properties of the Earth and how **seismic waves** pass through the Earth have shown that it has a solid **inner core**, a molten **outer core**, a solid **mantle**, and the **continental** and **oceanic crusts**.

- By about 4.4 billion years ago, the Earth was sufficiently cool for oceans to exist, created by condensation of water on Earth's surface from outgassing water vapor. We know from recent NASA missions that water is common in our solar system.

- The interior of the Earth is still very hot today—6000°C in the inner core. The high temperature is maintained by the **radioactive decay** of heavier elements.

- How life first originated in Earth remains unknown, but evidence suggests it began between 4.2 and 3.7 billion years ago. Because the sun was weaker so long ago, Earth was probably frozen over at the time life began, and the first life forms probably evolved in deep-sea hydrothermal vents or similarly extreme terrestrial environments. Today's atmosphere has so much oxygen from photosynthesis that whatever did happen that formed life long ago could not happen now.

- Conceptual models and simulation experiments have shown how complex organic molecules may have been formed early in Earth's history. While no living thing has been produced in the laboratory, researchers have synthesized DNA that produced another self-replicating cell when inserted into a bacteria cell. (However, the team did not synthesize the new cell membrane.)

Discussion Questions

1. What are some examples of technology? Of science? How do they differ from one another? How can it be that we have always made technological advances throughout human history, but science only came into existence about 600 B.C.? Do you accept this point of view, or do you think that we have overly compartmentalized science to be purely the generation of new ideas or the discovery of natural laws?

2. Why do you think Aristotle's theory of gravity held sway for some 1800 years, until the time of Galileo? Can you explain the current theory of gravity?

3. What can you think of that are examples of exploration and discovery? Can you give some examples of Kuhn's "perceptual gestalt"?

4. Explain how we know that an object in space that is "red shifted" is moving away from us?

5. Why do scientists now think that the Universe is continuing to expand at a speed that is accelerating?

6. What are nebulae, and how do stars form in them?

7. How do scientists search for extrasolar planets in our galaxy? Can they, or do they, also search for planets in other galaxies?

8. Where did the water on Earth come from? What are (were) some of the competing ideas to explain where water came from?

9. How did geologists know that the Earth is layered inside? How is it possible that the Earth is still hot in its deep interior? How deep does one have to go before the temperatures become hotter than a pizza oven?

10. What is the asthenosphere? How thick is it? How did we determine that there is an asthenosphere? How does the thickness of the asthenosphere vary by location on Earth?

11. Why do seismic waves refract as they pass though the Earth's deep interior?

12. Where do most scientists think life first originated on Earth? What were the conditions likely to have been that promoted the origin of that early life? What is a definition of life? Has life ever been produced in the laboratory?

Further Reading

Gold, T., 1997. An unexplored habitat for life in the Universe? *American Scientist* 85: 408–411.

Holton, Gerald, 1995. *Einstein, History, and Other Passions.* Woodbury, NY: American Institute of Physics Press.

Kuhn, Thomas S., 1970. *The Structure of Scientific Revolutions*, 2nd ed. Chicago: Chicago University Press.

Miller, S. L., Schopf, J. W., and Lazcano, A., 1997. Oparin's "Origin of Life": Sixty years later. *Journal of Molecular Evolution* 44: 351–353.

Park, Robert L., 2000. *Voodoo Science: The Road from Foolishness to Fraud.* New York: Oxford University Press.

Riess, A. G., and Turner, M. S., 2004. From slowdown to speedup (the expanding universe). *Scientific American* 290: 62–67.

Sagan, Carl, 1996. *The Demon-Haunted World: Science as a Candle in the Dark.* New York: Ballentine Press.

Wilson, Edward O., 1998. *Consilience: The Unity of Knowledge.* New York: Knopf.

Wolpert, Lewis, 1992. *The Unnatural Nature of Science.* Harvard University Press.

Zimmerman, Michael, 1995. *Science, Nonscience, and Nonsense: Approaching Environmental Literacy.* Baltimore, MD: Johns Hopkins University Press.

Contents

Hawaii's active volcano, Kilauea (shown here), has been erupting continuously for nearly three decades, having started about the time that Mauna Loa last erupted in 1984. Then, in 1996, a new undersea volcano, Loihi, began to erupt. Hawaii's volcanoes provide direct, present day evidence of the forces operating deep inside the Earth that, over hundreds of millions of years, gave us the current shapes and geological features of our oceans and continents.

The Ocean Floor: Its Formation and Evolution

In the last chapter, when we discussed the enormity of the Universe, we took a slight risk: that we might, by extension, speculate on how insignificant we and our Earth are in comparison. Well, in comparison, we are small—minuscule, in fact—but if we confine our frame of reference to Earth, and not the entire Universe, we can still be very impressed by the power and scale of some of its wonders. For example, the present shapes of the oceans and land masses and their positions relative to one another were determined by extremely powerful forces at work deep inside the Earth, forces that are so powerful that they were literally unbelievable until as recently as the mid-1960s. It was then that we recognized what may be our latest scientific revolution—a general acceptance of **continental drift**. As we will see in this chapter, the forces that have pushed around entire continents as if they were just floating sheets of ice on a lake were not only important in the early formation of Earth billions of years ago, they continue to be important today.

In this chapter we will build on our overview of Earth's internal structure and discuss how a number of seemingly wild ideas about continental drift, some of

---

BOX 3A Scientific Revolutions

Thomas Kuhn explained that science does not advance by small increments over long periods of time, but by great leaps in understanding. These *scientific revolutions* bring about a wholesale change in basic scientific assumptions. Suddenly the world is seen in a different light.

Following the Dark Ages in Europe (ca. 400–1400 A.D.) when virtually all scientific discovery slowed or stopped altogether, there came a series of scientific revolutions. While philosophers of science are not in complete agreement on which scientific advances were truly revolutionary, we can nonetheless list a few major events here.

- 1543: Nicolaus Copernicus posited that the Earth was not the center of the universe, an idea that was met with stiff opposition and was not fully accepted until the early 1800s.
- Early 1600s: Johannes Kepler showed that the planets move in elliptical orbits.
- Mid-1600s: Galileo Galilei proposed the nature of forces acting on celestial bodies.
- Early 1700s: Isaac Newton described Galileo's forces quantitatively.
- 1859: Charles Darwin published *On the Origin of Species*, thus giving us the Theory of Evolution.
- Early 1900s: Several scientists developed the ideas of Relativity and Quantum Mechanics.
- 1950s: DNA was discovered and molecular biology was born.
- 1960s: Continental drift became generally accepted.

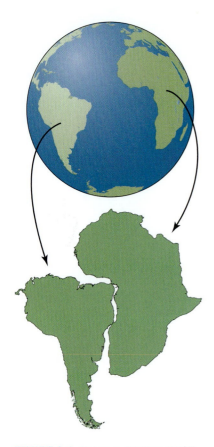

them offered hundreds of years ago, became accepted into mainstream science (**BOX 3A**). We will examine the progression of ideas and discoveries made over the past century as they slowly but surely led to a sound theoretical framework that describes how the ocean floor is continuously being created and destroyed, and how it and the continents are floating about on the Earth's surface. This revolutionary geological perspective will provide the context for our concluding discussion of the carpet-like covering that sits on the seafloor—the marine sediments. Sediments are in many ways the "timekeepers" of past geological and biological events in recent Earth history, and they help shape the structure and dynamics of many of today's marine ecosystems. But to make sense of this top layer of the ocean floor, we first need to understand the nature of continental drift.

FIGURE 3.1 Abraham Ortelius (1527–1598) was the first to speculate on the fit between South America and Africa, which became apparent when maps of the world showing South America appeared following Christopher Columbus's famous voyage of 1492.

Continental Drift

An emerging suspicion

Not long after Christopher Columbus's voyage of 1492, when maps began to show the New World, the rumors started. Once people could see the shape of South America they noticed the curious way it seemed to "fit" against Africa. (As we'll see later, the fit of the undersea continental shelves is even better than the fit of the coastlines.) The philosopher **Francis Bacon** (1561–1626) famously speculated on this in writing in 1620, and Benjamin Franklin is said to have pondered the nature of this fit in the 1700s. Earlier than either of them, **Abraham Ortelius** (1527–1598), a Flemish cartographer, not only recognized the fit, but also proposed the movements of continents as an explanation for it in his *Thesaurus Geographicus*, published in 1587 (**FIGURE 3.1**).

Over the next four centuries, a number of other, more-scientific observations began to accumulate, beyond the simple geometry of South America and Africa. Taken together, those observations suggested a very interesting geological history of our planet—that perhaps once, and maybe still, the continents were adrift, moving around on the surface of the Earth like boats on a pond. The story of how this scientific revolution—the acceptance of the theory of continental drift—developed and eventually became accepted is a captivating one, filled with ideas and observations offered by some of history's most interesting thinkers, many of whom were ahead of their time.

Early evidence

One of the first to come close to proposing continental drift was **Eduard Suess** (1831–1914), a British-born geologist who spent much of his career studying the geological history of the Alps (**FIGURE 3.2**). In 1857, while a professor of geology at the University of Vienna, Austria, Suess developed the idea that Europe and North Africa had once been connected to one another, but that they had since become separated by an ancient ocean. He named it the **Tethys Ocean** (after the Greek sea goddess Tethys), of which today's Mediterranean Sea is a remnant. His evidence was based on the distributions of marine fossils high up in the Alps, which led him to conclude that those mountains had been, at one time, the floor of the Tethys Ocean. His proposed explanation was that long ago, the slabs of crust that are now the Alps had been flooded by the sea; but then, somehow, the entire mountain range rebounded upward—pos-

FIGURE 3.2 Eduard Suess

sibly, he argued, as a result of **isostasy**, the tendency of continental crust to float on a fluid-like layer beneath it (see Chapter 2).

Vertical motions of mountain ranges was one thing, but horizontal movements of entire continents was quite another, and Suess stopped short of any such speculation. He did not actually propose continental drift. But he must have considered the idea; he reported in the first volume of his classic text, *The Face of the Earth* (published in four volumes between 1885 and 1909), that fossils of the *Glossopteris* fern had been found in South America, Africa, and India (they are also found in Antarctica and Australia, as reported much later; **FIGURE 3.3**).[1] His explanation was that widely separated Southern Hemisphere land masses had once been united in a single supercontinent, which he named **Gondwana-land**.[2]

It is clear that Suess had been thinking hard about this unusual correspondence of fossil distributions. This sliver of evidence, was just one piece of the great scientific puzzle that would occupy scientists for the better part of the next century. In fact, at the very outset of his four-volume masterpiece— in the first sentence on the first page of the first volume—Suess was already pondering what we would later learn was the breakup of an original master continent:

> If we imagine an observer to approach our planet from outer space, and, pushing aside the belts of red-brown clouds which obscure our atmosphere, to gaze for a whole day on the surface of the earth as it rotates beneath him, the feature beyond all others most likely to arrest his attention would be the wedge-like outlines of the continents as they narrow away to the South.

> This is indeed the most striking character presented by our map of the world, and has been so regarded ever since the chief features of our planet have become known to us. It recurs in the most diverse latitudes: Cape Horn, the Cape of Good Hope, Cape Comorin in the East Indies, Cape Farewell in Greenland are some of the best known examples. (p. 1)[1]

The "wedge-like" shapes of continents on their southern ends which Suess wrote of might be imagined as analogous to pieces of a cut-up cherry pie. But he never came to the conclusion that the continents are adrift. Instead, he proposed that the Earth is like an apple that shrinks as it ages, leaving its skin covered with wrinkles and folds, with the folds being mountain ranges. Suess

FIGURE 3.3 (A) Fossil of the *Glossopteris*, a genus of seed-bearing ferns that became extinct by the end of the Permian period, some 250 million years ago. It is named for its tongue-shaped leaves. This specimen was found in Australia. (B) Map of the distributions of *Glossopteris* fossils. Yellow areas represent reports given in Suess's *Faces of the Earth*; red areas represent later discoveries.

---

[1] Eduard Suess, 1909. *The Face of the Earth*, trans. Hertha B. C. Sollas (Oxford: Clarendon Press, 1924).

[2] Today it is simply called Gondwanaland (no hyphen) and sometimes Gondwana—you will see both terms used. Gondwana is the region in India known for its extensive fossil record of the *Glossopteris* flora.

FIGURE 3.4 One of the first depictions of the opening of the Atlantic Ocean due to the separation of the Eastern and Western Hemispheres published in 1858 in Snider-Pellegrini's *La Création et ses mystères dévoilés* (*Creation and Its Mysteries Unveiled*).

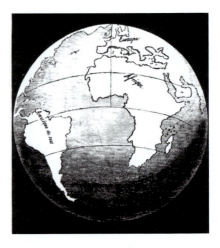

proposed that the oceans eventually flooded the spaces between those ridges. This idea became known as the Contraction Hypothesis. Suess was wrong about the mechanism of continental movements and mountain formation, but his fossil observations helped to stimulate thinking about the problem.

Some 17 years earlier, in 1858, **Antonio Snider-Pellegrini** had published a book in France, entitled *La Création et ses mystères dévoilés* (*Creation and Its Mysteries Unveiled*), which speculated that Africa and Europe were once connected to the Americas (**FIGURE 3.4**). His is perhaps the first published account of the opening of the Atlantic Ocean, by way of two great continents somehow drifting apart. Snider-Pellegrini based his argument on evidence that went well beyond the simple geometric fit: he showed that identical plant fossils could be found on both sides of the Atlantic. His proposed mechanism for that separation was the Great Flood described in the Bible.

By the end of the nineteenth century it was beginning to appear to some scientists that Earth's huge continents had indeed once been connected. But very few among them dared to speculate on such an outlandish idea, and it would be a generation later, in the early twentieth century, before the idea of drifting continents would be proposed by reputable scientists.

A modern history of an old idea

While the original seed of the idea of continental drift was planted more than four centuries ago by Abraham Ortelius, it would not begin to bear fruit for another three centuries, following Snider-Pellegrini's 1858 book and three other pivotal, and nearly simultaneous, presentations and publications.

The first was a talk given by amateur American geologist **Frank Bursley Taylor** (1860–1938) in 1908 and later published in 1910, in which he proposed actual movements of the continents.[3] Taylor argued that the shallow region that runs the length of the central Atlantic Ocean (which we now known as the Mid-Atlantic Ridge; see page 72) indicated the point where Africa and South America had once been joined, and from which they separated. Taylor's proposed mechanism was that some 100 million years ago, the Moon was captured by Earth's gravity and came so close that its gravitational pull on the Earth's surface dragged the continents toward the Equator (he offered no actual evidence, however). He also proposed that the continents, plowing through the ocean floor toward the Equator, had pushed up crustal material on their Equator-facing edges, thus building up both the Himalayan Mountains

---

[3] Taylor, F. B., 1910. Bearing on the tertiary mountain belt on the origin of the Earth's plan. *Bulletin of the Geological Society of America* 21: 179–226.

and the Alps (which was incorrect, as we now know). While many found Taylor's ideas intriguing, his rationale involving the tidal pull of the Moon was unconvincing, and in time his ideas were forgotten.

The second was an article published in 1911 by **Howard B. Baker** in the *Detroit Free Press* (a newspaper, of all places), in which Baker proposed that an original supercontinent had broken apart to give us today's continents. Unfortunately for Baker, his ideas were also doomed to obscurity, at least partly because he took such a bizarre approach in explaining the movements of the continents: the "Atlantic Rift" running down the center of his published diagram resulted, he explained, when the orbits of Earth and Venus coincided such that their gravitational attractions and the resulting tidal distortions on Earth stripped a large chunk of crustal matter from the Pacific Ocean, which subsequently became the Moon![4] (As discussed in Chapter 2, the Moon's initial formation is believed to have been the result of a massive collision, which is itself a strange story, but Baker's scenario is a bit fantastic.)

FIGURE 3.5 Photo of Alfred Wegener taken in 1930, during a German expedition to Greenland, shortly before his death.

The third, and most significant, was a book published in 1915 by German scientist **Alfred Wegener** (1880–1930) that seems to have started the scientific debate in earnest (**FIGURE 3.5**). Unfortunately, despite the abundance of detailed evidence Wegener presented in his book, most geologists at the time rejected the idea outright.[5]

Anyone who reads Wegener today can see that he was well ahead of his time. Not a geologist, but trained in astronomy, Wegener took Eduard Suess's early ideas and observations a step further, and with great audacity he proposed a bold new idea in a talk in 1912. He suggested the existence of an original supercontinent that he named **Pangaea**, which broke apart sometime between 250 and 350 million years ago into unequal pieces—today's continents—which slowly moved to their present positions. Moreover, he argued that the continents are still moving! He coined a term for this phenomenon: continental drift. He later detailed the evidence for his idea in *The Origin of Continents and Oceans*, originally published in German in 1915. It was revised several times in the years that followed, and translated into French and English in 1924.[6]

The story of Alfred Wegener and his theory is a fascinating one. Wegener was not only a brilliant scientist, he was also an interesting character—as is almost always the case with such historically significant figures. In 1904, at the age of 24, he received his PhD in Astronomy from the University of Berlin and became a Lecturer of Astronomy and Meteorology at the University of Marburg, Germany, where he earned a reputation for being an outstanding teacher. He focused his

---

[4] The notion that the Moon was formed this way, as a chunk of crust torn from the Pacific Ocean, was not all that bizarre a century ago. Osmond Fisher, in his *Physics of the Earth's Crust* (1881) speculated that the breakup of South America from Africa was in response to the gap left behind after that event (thus he proposed an early version of drifting continents). And in 1907, William H. Pickering wrote: "When the moon was separated from us, three-quarters of this crust was carried away and it is suggested that the remainder was torn in two to form the eastern and western continents. These then floated on the surface like two large ice floes." Pickering, W. H., 1907. The place of origin of the moon—the volcanic problem. *Popular Astronomy* 15: 274–287.

[5] Wegener, Taylor, and Baker were contemporaries. But in all likelihood they each proposed their ideas independently, unaware of each other's viewpoints. (In later editions of his book, Wegener pointed out that he was unaware of Taylor's 1910 paper when he wrote his first edition, published in 1915.)

[6] Alfred Wegener, 1927. *The Origin of Continents and Oceans*, 4th ed., trans. John Biram (repr., New York: Dover Publications, 1966).

(A)

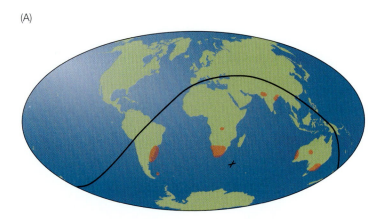

(B)

FIGURE 3.6 (A) Map based on Wegener's original map of glacial scour patterns and probable position of the South Pole (cross) and the Equator (line) at the time those scour marks were made. The orange areas are scour locations today, which all orient toward a probable starting point, the center of which is the South Pole. Glaciers could only have crept a finite distance away from the South Pole, not nearly as far as the scrape marks on continents are in their current locations, suggesting that it was the continents that moved away from the pole. This means that the geographic areas highlighted here were once on the Equator, but have since moved to their present locations. (B) Glacial scour left on a rock surface in Maine. The marks are oriented in the north–south direction and reflect the direction of glacial movement as small rocks and gravel in the glacial ice ground against the bedrock.

research on meteorology, not astronomy, and established himself in that field. Part of his research involved flying hot-air balloons to study the upper atmosphere, and in 1906 he set a world endurance record by staying up for more than 52 hours. At the age of 30, he published *The Thermodynamics of the Atmosphere* (1911) which became a popular textbook in Germany, helping to secure his position in the upper echelons of European science.

Like Fridjof Nansen, Wegener was an Arctic explorer. In 1906 he participated in the Danish expedition to Greenland, where he spent two years on the ice. He co-led another expedition to Greenland in 1912, the first to cross Greenland at its widest point (Nansen had taken a shortcut across southern Greenland). On his third expedition to Greenland, from 1929 to 1930, he documented the thickness of the ice cover: 1800 meters (more than a mile!)

Wegener's 1915 book, *The Origin of Continents and Oceans*, was extremely insightful in that it brought together a number of observations reported by others in support of the conclusion that the continents were indeed adrift. Wegener explained how scrape marks left on rock surfaces by ancient glaciers passing over them revealed a pattern of movement that could be traced back to a probable starting point common to several continents, a point away from which those continents had been drifting (**FIGURE 3.6**). He deduced common features among widely distributed mountain ranges, and suggested, for example, that mountain ranges on both sides of the Atlantic Ocean had once been a single range that split apart (**FIGURE 3.7**). With respect to the fit between Africa and South America and the continuous mountain ranges that appear to extend uninterrupted across both continents, he wrote: "It is just as if we were to refit the torn pieces of a newspaper by matching their edges and then check whether the lines of print ran smoothly across. If they do, there is nothing left but to conclude that the pieces were in fact joined in this way" (p. 77).[6] Wegener suggested, based on geological evidence, that the Andean Mountain Range had originally extended around South America and connected to South Africa. He also suggested—in contrast to Suess's Contraction Hypothesis—that the western North American Coastal and Cascade Ranges and the South American Andes mountains were the result of the continents crinkling as they were pushed to the west. Wegener was well aware of the principle of isostasy, that continental crust, at least, was supported by some kind of "plastic" material.[7] And he seized upon

[7] The term *isostasy* was coined in 1882 by Clarence. E. Dutton. Dutton, C. E., 1882. Some of the greater of physical geography. *Bulletin of the Philosophical Society of Washington* 11: 51–64.

Ernest Shackleton's 1908 report of coal in Antarctica. Coal is ancient plant biomass, but there isn't much plant life where Antarctica is now, and so Wegener reasoned that the Antarctic continent had not always been at the South Pole. In the fourth and last edition (1927) of his *Origin of Continents and Oceans*, Wegener reported on two sets of longitude determinations that others had made in Washington, DC, and Paris, France—the first made between 1913 and 1914 and the second in 1927. The comparison showed that in the 13-year period, the two cities had drifted about 4.3 meters away from each other, at a rate of just less than 32 centimeters per year (both were overestimates, we now know).[8] Comparative longitude determinations across the ocean were repeated often by various investigators during the 1920s and 1930s using radio tranmission times, and though the errors in these investigations were also large and the estimated rates of movement varied widely, the direction of that movement was always the same: the two continents were moving away from one another.

Wegener's book was controversial when it first appeared in 1915, and it was met with stiff resistance by the scientific establishment. He made extensive changes to subsequent editions in an attempt to assuage those criticisms, but the controversy continued. An example of the criticisms, and their tenor, is this quote taken from the opening of an address entitled "Wegener's Hypothesis of continental drift," given by Philip Lake at a meeting of the Royal Geographical Society (Great Britain) in 1923. Lake's contempt for Wegener and his ideas is a tad obvious in this opening passage:

> Wegener's views are now so widely known that a very brief intro-duction will be sufficient. He imagines that the continental masses are patches of lighter rock floating and moving in a layer of denser rock, of unknown thickness; and this denser rock forms the floor of the oceans. Following, with a slight alteration, the terminology of Suess he calls the lighter material the Sial and the denser layer the Sima. Suess's words are Sal and Sima, and there is no advantage in the change. Suess thinks, however, that the Sal is continuous, cover-ing the globe completely, and this is a fundamental difference.

> Wegener does not suppose the Sima to be actually liquid, but he believes it to be plastic enough to yield slowly under the strains to which it is subjected, much as a stick of sealing-wax supported at its ends will gradually bend without ever losing its apparent rigidity.

> In this paper, I shall not discuss the possibility of Wegener's conception. He does not profess to explain completely why the continents should move, but he does claim to have proved conclusively that such movement has taken place. It is the evidence on which he relies, and more particularly the geo-logical evidence, that I propose to examine.[9]

Unfortunately, Alfred Wegener died in 1930 without ever seeing his theory accepted; that acceptance would take another 40 years or so. After Wegener's death, a South African geologist, **Alexander du Toit** (1878–1948) took up the cause of continental drift in an influential book published in 1937, *Our Wandering Continents*. But this work also failed to convince the establishment. And so the theory of continental drift seemed to linger for decades after Wegener's book first appeared, waiting for more convincing evidence to accumulate. As recently

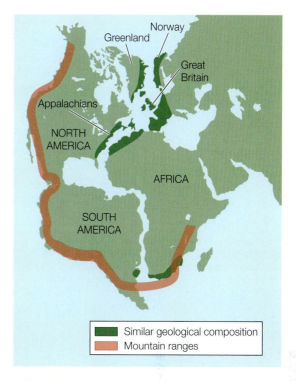

FIGURE 3.7 Wegener's interpretation of how mountain ranges on both sides of today's Atlantic Ocean were once connected but were pulled apart as the Atlantic Ocean opened up, and how western North American and South American mountain ranges were formed as the continents were pushed to the west, crinkling them upward into mountains. He also suggested that the Andean Range originally was continuous around South America to Africa.

---

[8] It was later shown that these estimates were too high by a factor of 10; much more recent spreading rate estimates for the Atlantic basin are on the order of 2–4 centimeters per year.
[9] Lake, P., 1923. Wegener's hypothesis of continental drift. *Geographical Journal* 61: 179–187.

as 1966, the description on the back cover of a paperback English translation of Wegener's fourth edition read, "Even today this important question remains undecided, and geologists are divided into strongly opposed groups about the Wegener hypothesis."[6]

To place that statement in historical context, remember: *men were walking on the Moon just three years later, in 1969*. This giant gap between our technological capabilities in the 1960s and our understanding of basic earth science, is, in hindsight, a bit shocking (and scholars have since tried to unravel the various issues that perpetuated the resistance to accepting continental drift.)[10,11]

So, what was wrong with Wegener's idea? Why did scientists resist for so long his idea that continents could be moving about on the surface of the Earth? Well, the reasons are many—some factual, and some cultural. We now know that his was indeed a beautiful idea in that it pulled together so many disparate lines of evidence, from geology to biology. But most scholars agree that his presentation of the concept suffered from a fundamental flaw: Wegener opened himself to attack by offering two competing explanations for the force that drives what he called "the flight from the poles," the apparent drift of continents away from the poles, where he thought they had once been, and toward the Equator. One of Wegener's proposed mechanisms for that "flight" was the centrifugal force of the rotating Earth, which he said would draw everything on its surface toward the Equator. This was not only a weak argument, as he readily admitted, but it was also wrong. He also suggested that tidal forces due to the gravitational attractions of the moon and sun might be at work. Again, this was not a convincing argument: at the time, tidal forces were thought to be far too weak to have any significant effect.[12] Wegener conceded that he would leave the issue of a driving force for others to discover; nonetheless, he emphasized that *the continents do move*.

Today, we wonder why the theory was held captive for so long for want of that plausible mechanism, that driving force. After all, earth scientists had long before accepted the idea of global Ice Ages, even though they had no explanation for the mechanism that caused them. Why was continental drift held to a higher standard?

Some authors have written that the biggest obstacle to acceptance of Wegener's hypothesis was the view of the mantle at that time—that it was solid rock. But, even a cursory reading of Wegener's book reveals that he and his predecessors already knew about isostasy and the "floating" continents. In fact, in his fourth edition of *The Origin of Continents and Oceans*, Wegener discussed suggestions made by other scientists that there might be convection currents in the viscous upper mantle that carry slabs of the Earth's crust along for a ride.

Wegener's ideas may also have fallen victim to bad timing: *The Origin of Continents and Oceans* was published in 1915—the same year that RMS *Lusitania*, the British luxury liner, was sunk by a German U-boat, killing more than 1100 people, including more than 150 Americans. It was this act that brought the United States into World War I against Germany. While the scientific community generally rises above the political din, this was a terrible war, and Germany (and Germans) were not very popular. Wegener fought in that war (and was shot twice). How much did nationalism enter into this debate about continental drift? One can only speculate.

[10] John A. Stewart, 1990. *Drifting Continents & Colliding Paradigms: Perspectives on the Geoscience Revolution* (Bloomington: Indiana University Press).

[11] Naomi Oreskes, 1999. *The Rejection of Continental Drift: Theory and Method in American Earth Science* (New York: Oxford University Press).

[12] Although still controversial, tidal forces have very recently been reconsidered as a potential factor in continental drift. Scoppola, B. et al., 2006. The westward drift of the lithosphere: A rotational drag? *Geological Society of America Bulletin* 118: 199–209.

(A) 200 million years ago

(B) 65 million years ago

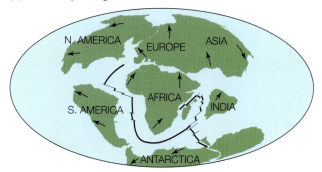

(C) Present

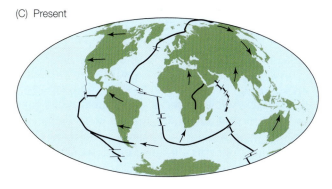

FIGURE 3.8 (A) The breakup of Pangaea some 225 million years ago, resulting in the formation of the supercontinents Laurasia and Gondwanaland. (B,C) The "drift" to today's positions. Spreading ridges are indicated by solid lines, transform faults are indicated by hatch marks, and arrows give the directions of plate movements. This modern scenario differs from Wegener's interpretation in that he did not account for the two supercontinents being separated by the Tethys, and he did not have India breaking away from Antarctica and moving to its present position.

For several decades after Wegener's death, most geologists seemed to ignore the idea altogether. For a long time even Wegener's own countrymen didn't embrace his theory. He even left Germany, where he had been a lecturer at the University of Hamburg, in 1924 to take the position of professor of meteorology and geophysics at the Graz University in Austria. Today, he is venerated. Not only did Germany issue a commemorative stamp bearing his portrait, but they founded the prestigious Alfred-Wegener Institute for Polar and Marine Research in 1980.

What Wegener had theorized nearly a century ago has become common knowledge today—that there was once an ancient supercontinent called Pangaea, which we now know began to break apart about 225 million years ago, initially forming two huge subcontinents, **Laurasia** and **Gondwanaland**. These two continents became separated by a new ocean, Eduard Suess's Tethys Ocean (today's Mediterranean Sea). The two subcontinents then each broke apart, resulting in the present-day configuration of the world's land masses and oceans. This process continues today (**FIGURE 3.8**).

So, how did Wegener's theory come to be accepted? What changed after his death? And just what were the lines of evidence that eventually led the international scientific community to embrace the notion of drifting continents?

Evidence begins to accumulate

In the early to mid-1900s, geologists were generating reams of data on the passage of seismic waves (from earthquakes) through the Earth, which supported the much-earlier-proposed idea of isostasy. That is, the data indicated that the upper mantle is not made of rigid, brittle stone, but has plastic properties, as Wegener proposed. Like white-hot iron, it can be deformed; we now call this plastic layer the asthenosphere.

In 1935, not long after Wegener's death, a Japanese scientist, **Kiyoo Wadati** (1902–1995), pointed out the locations and numbers of earthquakes and volcanoes around Japan and argued that they were likely the direct result of continental drift—that they might constitute direct evidence that great slabs of

(A)

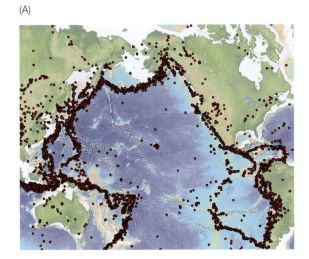

(B)

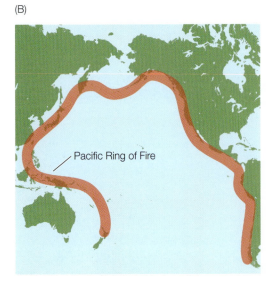

Pacific Ring of Fire

FIGURE 3.9 (A) Locations of all earthquakes of magnitude 5.0 and greater for the period 1973 to 2009. (B) A simplification of the pattern of those earthquakes, which illustrates the Pacific Ring of Fire.

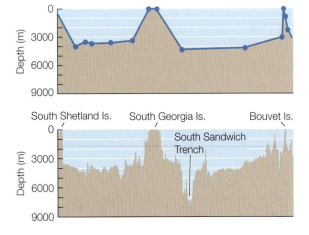

Earth's crust are moving about somehow. In 1940, seismologist **Hugo Benioff** (1899–1968) plotted the locations of earthquakes around the edge of the entire Pacific Ocean, revealing a pattern that became known as the **Pacific Ring of Fire** (**FIGURE 3.9**). The map in Figure 3.9 shows that earthquakes are going on in this region all the time, even today. Worldwide, there are more than a million earthquakes each year, of which more than 1000 are greater than magnitude 5 (**BOX 3B**).

Slowly, other supportive data began to accumulate. Radiometric dating of marine sediments after World War II showed (to the surprise of many) that the maximum age of sediments that blanket the bottom of the oceans is only about 200 million years, which is very young compared with the age of the Earth. This was a real eye-opener, because some rocks found on the continents are much, much older—on the order of 4 billion years old. Why are the sediments on the bottom of the oceans so young?

The ever-growing body of data collected after World War II using the newly developed echo sounders provided important new details of the ocean floor, especially a feature that had gone largely unnoticed: the **mid-ocean ridge** systems. In particular, the shape of the Mid-Atlantic Ridge was found to conform nicely to the positions of the continents—it runs right down the middle of the North Atlantic and South Atlantic Oceans. And running along that ridge is a deep valley, now known as the **Rift Valley**, which, as we will see shortly, came to be one of the key features in support of the idea of continental drift.

The existence of the Mid-Atlantic Ridge had been known since the time of the *Challenger* expedition, based on a relatively few depth soundings using just a wire-and-lead weight (each of which could take the better part of a day for the ship's crew to complete, depending on how deep it was). Charles Wyville Thomson's resulting bathymetric map showed the feature quite clearly (see Figure 1.22); and though he described this "axial ridge," he did not offer any explanation for its existence.

By the early twentieth century, the feature was well known; in his 1937 book, Alexander Du Toit listed a number of explanations offered by others for its existence, including Eduard Suess's Contraction Hypothesis, but he could not offer a convincing explanation of his own. Even in a classic oceanography text from 1942, continental drift was not mentioned, and the geological significance of "the Atlantic ridge," as they called it, was all but ignored.[13] The authors presented two depth profiles of a section of the South Atlantic Ocean (**FIGURE 3.10**), the first based

---

[13] Harald U. Sverdrup, Martin W. Johnson, and Richard H. Flemming, 1942. *The Oceans: Their Physics, Chemistry and General Biology* (New York: Prentice-Hall).

FIGURE 3.10 The German *Meteor* expedition (1925–1927) provided some of the first high-resolution echo soundings of the Atlantic Ocean. Bottom depths along an east–west section across the South Atlantic Ocean as determined by 13 wire soundings (top) were compared with more than 1300 echo soundings (bottom) by the *Meteor*. The South Sandwich Trench was clearly shown by the echo soundings. It was formed, we now know, by the collision of two slabs of oceanic crust, one sliding west beneath the other, creating the deep trench as well as the island arc that includes South Georgia Island.

The seismograph (or seismometer) was invented in the 1880s by British geologist John Milne (1850–1913) and his colleagues while working in Japan at Imperial College. The instrument has a very simple design: a stylus is suspended above a base that is firmly attached to the ground, such that the base moves with seismic waves (FIGURE). A paper recorder on the base slides beneath the freely suspended stylus, recording movements of the base—and the Earth that the base is sitting on. The system of assigning a magnitude to those seismic wave signals received by seismographs was developed much later. In 1935, Charles Francis Richter (1900–1985), an American seismologist, in collaboration with Beno Gutenberg (1889–1960) at Caltech, came up with what is commonly known as the Richter magnitude scale, which quantifies the size (or energy) of earthquakes. The scale is based on the difference in arrival times of S and P waves generated during an earthquake, which gives the distance between the seismic recorder and the disturbance; the amplitude of the S waves and the distance are plotted on a scale (as shown in the Figure, right) to give a magnitude for the disturbance. Some examples of Richter magnitudes and TNT explosions of comparable magnitude are listed here, along with example phenomena.

Richter magnitude	TNT for seismic energy yield	Examples
1.0	30 pounds	Large blast at a construction site
2.0	0.5 ton	Large quarry or mine blast
3.0	30 tons	Sufficient to blow up an area the size of a city block
4.0	1000 tons	Small nuclear weapon
5.5	80,000 tons	Virginia and Washington, DC, earthquake, 2011
6.5	5 million tons	Northridge, CA, earthquake, 1994
7.0	32 million tons	Largest nuclear bomb
8.0	1 billion tons	San Francisco, CA, earthquake, 1906
9.0	32 billion tons	Chilean earthquake, 1960
10.0	1 trillion tons	No known earthquake or explosion examples
12.0	60 trillion tons	No known earthquake or explosion examples

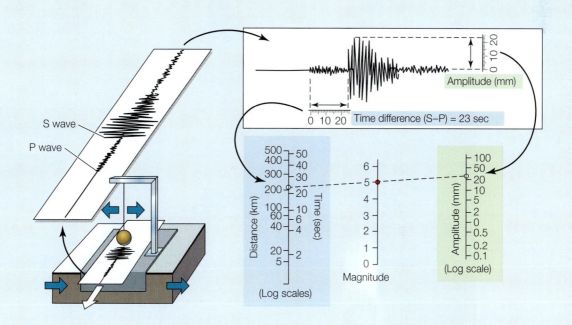

S wave

P wave

Amplitude (mm)

0 10 20 Time difference (S–P) = 23 sec

Distance (km)

Time (sec)

Magnitude

Amplitude (mm)

(Log scale)

(Log scales)

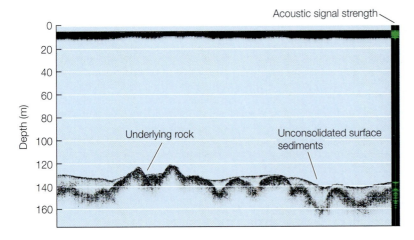

Acoustic signal strength

FIGURE 3.11 A modern echo sounding graph of the Northwest Atlantic continental shelf off New England, showing the depths of both solid rock and surface sediments. In places, a layer of bottom sediment 20 meters thick covers underlying rock; in other places, the rocks have been swept clean by tidal currents.

FIGURE 3.12 World map of sediment thickness on the bottom of the oceans. Data from actual bottom samples and sonic profiles were used in constructing the map. Thicknesses vary greatly, from as much as 2000 m directly adjacent to several continents, to less than 100 m in the centers of the ocean basins.

on soundings using a wire-and-lead weight, and the second on the echo soundings by the German *Meteor* expedition between 1925 and 1927 (see Figure 1.27A). Although the depth of the South Sandwich Trench, discovered by the *Meteor*, is striking, the authors merely commented that "the regions in which these deep sea trenches occur are sites of volcanic and seismic activity" (p. 29).[13]

Not only did echo sounders provide a detailed picture of trenches, mid-ocean ridge systems, and their central rift valleys, but because sound waves penetrate the ocean bottom, echo sounders also provide a measure of the thickness of the unconsolidated sediment on the seafloor (**FIGURE 3.11**). Those data, along with core samples taken of the sediments, revealed that sediment thickness on the ocean floor is quite variable. The thickest layers (some greater than 15 km) are adjacent to the continents, far removed from the center of the ocean basins, and the thinnest layers (on the order of 10 meters or less) are nearest the mid-ocean ridges.

These two data sets—one showing the *young age* of marine sediments relative to the age of the Earth and the other showing the *increasing thickness* of marine sediment on the ocean floor with increasing distance away from the mid-ocean ridges—suggested a radical conclusion: that new ocean floor is created continually in the vicinity of the ridge systems.

Modern data collected using cores and echo sounders were used to generate the map in **FIGURE 3.12**. Those data show that virtually everywhere in the oceans,

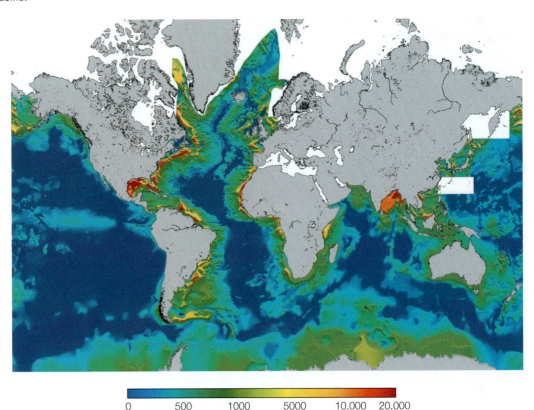

0   500   1000   5000   10,000   20,000
Sediment thickness (m)

sediment thicknesses are greatest nearest the continents, where the ocean floor is oldest—that is, sediments are thicker and older away from the mid-ocean ridges, where we now know that new ocean floor is created. Notice that there is little sediment in the eastern Pacific Ocean off North and South America (see page 83).

By the late 1940s and 1950s scientists were accumulating more and more data from echo sounders, and from seismic wave studies of the ocean floor. Those data allowed them to map the detailed features of the Mid-Atlantic Ridge and the Rift Valley running down its center (**BOX 3C**). Eventually, a picture of the worldwide mid-ocean ridge system, which we now know is more than 37,000 miles in length, came into focus (**FIGURE 3.13**).

Also in the late 1940s, scientists discovered that the layer of crustal material beneath the marine sediments—the oceanic crust—is much thinner than the continental crust (see Figure 2.22) and is composed of basalt. (Continental crust is made up of granite; see Chapter 2.) Because basalt is of volcanic origin,

(A)

(B)

FIGURE 3.13 (A) Relief map showing ocean depths and terrestrial elevations, produced using both sounding data and satellite altimetry data (see Box 3C). (B) The mid-ocean ridge systems are traced in yellow.

BOX 3C Satellite Altimetry

High-resolution maps of the ocean's depths were greatly improved after the *Meteor* expedition between 1925 and 1927, when echo sounders were first used. But even with echo sounders, bathymetric maps were still limited by the ships' coverage, leaving large gaps in the record. Resolution of the bathymetry of the seafloor reached a new level in the 1980s after the advent of satellite altimetry, a technique whereby details of the ocean floor can be resolved by Earth-orbiting satellites that map the flatness of the sea surface.

The surface of the oceans is not perfectly flat; in fact, it has many departures from flatness that result because (1) the Earth is rotating, making the diameter 43 km greater at the equator than the poles, and because (2) there are minute differences in the Earth's gravity at specific locations that produce bulges, or hills where sea level is elevated—up to tens of meters. Most of these bulges are caused by undersea mountains, which, because of their great mass, create locally increased gravity. (Remember: anything that has mass has gravity,

and mountains have a lot of mass.) For example, a person standing beside Mount Everest is not actually standing straight up, but is slightly tilted away from the mountain to compensate for its gravitational pull. Similarly, the additional component of gravity produced by undersea mountains causes water to be pulled laterally toward a location directly above them, and hence, the water piles up. Sea level is a bit higher there.

Despite these variations in height, at any point on the ocean—at the top of a local variation (a "hill") or at the base of one—the pull of gravity is the same. The surface of the oceans is a near-perfect "equipotential surface" in terms of Earth's gravity, having adjusted to variations in gravity created by features on the ocean floor. Therefore, if we were to place a basketball on the sea surface (assume the water is solid and we can do this), it wouldn't roll anywhere (**FIGURE A**). There is no "downhill" in the sense that gravity would not make the ball roll because the gravity everywhere on the surface of the ocean is the same. Because the sea surface has adjusted to the local

changes in gravity, the bulges are not easily detected, but their elevation above the *average* sea surface can be measured by satellites, which orbit in a perfect arc relative to the Earth's center of gravity (**FIGURE B**).

In 1985, the GEOSAT satellite was launched by the U.S. Navy to map departures in sea surface height from the overall average ocean surface height using a high-resolution radar altimeter that had a vertical resolution of 3 cm (see Appendix A). In 1995 the government declassified those data, making it possible to map fine-scale features on the ocean floor, giving us breathtaking relief maps of the ocean floor as shown here for the Atlantic (**FIGURE C**, top). Compare that image with what we would have using only ships and echo sounders (Figure C, bottom). Ship tracks where echo sounder data were available are plotted; notice the empty spaces.

Other radar altimetry satellites were launched after GEOSAT: TOPEX/Poseidon (1992–1995), GEOSAT-Follow-On (1998), Jason (2001), and most recently, a European Space Agency satellite.

FIGURE A

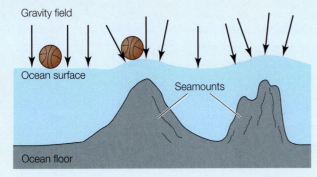

FIGURE B

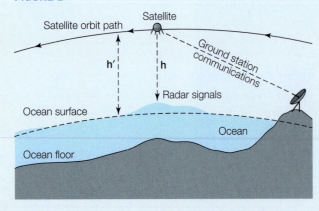

FIGURE C

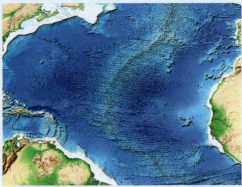

this finding suggested that the sources of continental and oceanic crust are very different, which supported the growing suspicion that new ocean floor might be created at the mid-ocean ridges by way of volcanic action.

By the 1960s, various lines of scientific evidence were coming together from a wide range of subdisciplines of geology, all of which supported the idea that new ocean floor is indeed being created in the vicinity of the mid-ocean ridges. At the same time that we were accumulating more and more data on the oceans' depths and the thicknesses of the overlying sediments and underlying crust, seismograph stations were being set up all around the word, first as part of the International Geophysical Year of 1957, and later as part of the network of stations to monitor the Nuclear Test Ban Treaty signed in 1963. Analyses of all these data streams showed that there is a layer in the upper mantle where seismic waves are slowed. This discovery provided confirmation that there is indeed an asthenosphere, a plastic-like layer beneath the rock-solid lithosphere. This unequivocal evidence of the existence of the asthenosphere was very compelling, and it was becoming apparent to a lot of scientists that Wegener's theory of continental drift just might have some merit; slowly but surely, people were coming around.

But the question remained: What drives this process? That is, if we assume that the continents are in fact adrift, what is it that makes them move? We know that Wegener's explanations (centrifugal and tidal forces) had already been dismissed. So what is it that actually pushes and pulls these great slabs of the Earth's crust?

In 1960, **Harry Hess** (1906–1969), a professor of geology at Princeton University, presented a conceptual model of how the seafloor might be in the process of continually being created.[14] He proposed that new oceanic crust is formed at the mid-ocean ridges,[15] and that it spreads outward laterally away from the rift valleys, thus creating new ocean floor. Hess explained that the mantle, though "solid," nonetheless flows, the way antique glass thickens at the bottom and thins at the top as it slides slowly downward under the pull of gravity, and that there are likely deep convection currents flowing within it (**FIGURE 3.14**). At the mid-ocean ridges, mantle material that rises all the way to the surface solidifies into basaltic rock and becomes new ocean floor. Hess wrote: "The continents do not plow though oceanic crust impelled by unknown forces; rather they ride passively on mantle material as it comes to the surface at the crest of the ridge and then moves laterally away from it" (p. 609).[14] He also wrote that the moving slabs of oceanic crust don't complete the convection cycle by following the downward-moving limb of mantle material; they stay at the surface, "thickening…a continent by deformation…Initially a mountain system and much larger root are formed" (p. 616).[14]

The idea of convection currents in the mantle was not new in the 1960s; it had been discussed much earlier by Wegener in the 1927 edition of his book. But even more significant is the similarity between Hess's model (as well as the versions published by Bruce Heezen in 1960 and Robert Deitz in 1961[15])

(A)

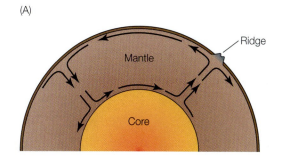

(B)

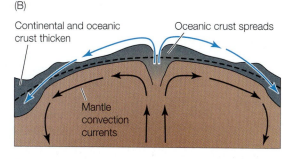

FIGURE 3.14 (A) Schematic diagram of convection cells in the mantle. (B) A diagram illustrating key aspects of Hess's theory of seafloor spreading, as interpreted from his 1962 paper, whereby new ocean floor is created at the mid-ocean ridge systems and spreads laterally away.

---

[14] Hess proposed his ideas in a talk in 1960, which was based on a grant report written that same year. He later published his ideas in a chapter, "History of ocean basins." Hess, H. H., 1962. "History of ocean basins," in *Petrologic Studies: A Volume to Honor A. F. Buddington*, ed. A. E. J. Engel, H. L. James, and B. F. Leonard (Geological Society of America). pp. 599–620.

[15] In 1960 Bruce C. Heezen also proposed that new ocean floor is being created and is spreading laterally away from the mid-ocean ridge systems in the Arctic, Indian, and Atlantic Oceans. (Heezen, B. C., 1960. The rift in the ocean floor. *Scientific American* 203: 98–110.) In 1961, Robert Dietz published a paper proposing seafloor spreading, in which he coined the term. (Deitz, R. S., 1961. Continent and ocean basin evolution by spreading of the sea floor. *Nature* 190: 854–857.) Dietz later acknowledged that Hess deserved credit for coming up with the theory first.

(A)

(B)

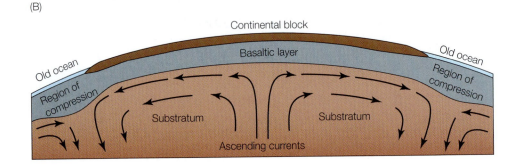

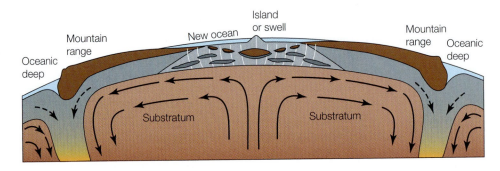

FIGURE 3.15 (A) Arthur Holmes (1890–1965) and (B) reproductions of his two figures illustrating his conceptual model of convection cells in the mantle (which he first proposed in 1928) and the resulting formation and spreading of new ocean floor, including the subduction of that seafloor (although he did not use the term).

with the proposed mechanism published in 1944 by British geologist **Arthur Holmes** (1890–1965; **FIGURE 3.15**).[16] Only Hess acknowledged Holmes's 1944 book, and even so, it was not a formal citation.

The new theory, which became known as the **Theory of Seafloor Spreading**, was firmly installed in the mainstream of scientific thinking by 1962. But it was not without its own shortcomings. In his explanation for how to counterbalance this continuing production of new ocean floor, Hess chose not to take the position that Heezen had in his paper published two years earlier—that the production of new ocean floor necessitates an "expanding Earth." Instead, Hess argued that the continents were being compacted and were thickening over time, we now know that Hess was only partially correct. Not only does oceanic crustal material thicken as he explained it, but, as we now know, some is also subducted. He asserted that the new ocean floor and the continents adjacent to it would simply move as one away from the ridge system, riding on top of the convection cell of flowing mantle. When the leading edge of a continent and the new seafloor moving with it reach the downward-moving limb of the convection cell, the crustal material doesn't follow the direction of the convection cell. Rather, the crustal material piles up—growing thicker and building mountain ranges. Balancing what comes up with something going down would seem to make sense, but Hess argued that continental crust is not dense enough and does not sink. However, he did imply some kind of a balance when he wrote that "the whole ocean is swept clean (replaced by new mantle material) every 300 to 400 million years."[17]

Today we know that as new ocean floor is *created*, it is also being *destroyed* in subduction zones, where one piece of oceanic crust—a **plate**—slides beneath another plate and downward, deep into the mantle. Arthur Holmes had proposed this back in 1944; notice in Figure 3.15 how Holmes's "basaltic layer" flows back down from whence it came, along with the descending

---

[16] Arthur Holmes, 1944. *Principles of Physical Geology* (New York: Thomas Nelson and Sons).
[17] Hess, H. H., 1962. "History of ocean basins," in *Petrologic Studies: A Volume to Honor A. F. Buddington*, ed. A. E. J. Engel, H. L. James, and B. F. Leonard (Geological Society of America) p. 617.

arm of the mantle convection cell. Hess alluded to such a phenomenon in his paper, but he did not make the point explicitly; that came later, in a paper he published in 1963. Most of the papers published at about this time seem to have ignored the possibility of subduction as well, but that process is important—because it turns out that a lot happens as slabs of crust meet one another (as we'll see below).

Hess's Theory of Seafloor Spreading, which it eventually came to be called, was indeed an important step forward in advancing the theory of continental drift—even if his model would seem to have been derived from Holmes's earlier model. Hess's paper answered a lot of questions and tied up a lot of loose ends, such as: If new seafloor is being created at the spreading centers in the mid-ocean ridges from volcanic material rising up out of the asthenosphere, the crust there should be hot. Indeed, observations showed that it is. Furthermore, if this newly formed crust cools as it slowly spreads and moves away, it should shrink in volume, thus becoming more dense, and over time sink progressively deeper into the asthenosphere with increasing distance away from the spreading ridges. This in turn should mean that the oceans are deeper father away from the spreading centers; as it turns out, the oceans are indeed deeper farther from the ridges. Using the same logic, sediment at the edges of the ocean basins should be *thicker* than in the centers of the oceans, reflecting the longer period of time for the accumulation of sediment (just as the dust on your furniture at home gets thicker the longer you wait between dustings). Therefore, marine sediments at the edges of the ocean basins should also be *older*; and, indeed, they are both thicker and older.

The theory of seafloor spreading still had a couple of problems: one was how to balance the rates of new ocean floor creation and old ocean floor destruction. The other was that because the rate of seafloor spreading is so slow (about 2–4 cm per year in the Atlantic, roughly the same as the rate your fingernails grow) it was virtually impossible to observe or measure the spreading directly.

One explanation for what happens to new ocean floor was given in 1962 in a paper by **Robert R. Coats** (1910–1995) of the U.S. Geological Survey on the origin of the Aleutian Islands off Alaska. He argued that a *megathrust*—the thrusting of one slab of ocean floor beneath another slab of oceanic crust at a steep angle (30°)—creates a deep sea **trench** system where the plates meet. Then, as material associated with the sinking plate is heated, volcanoes form. Finally, as the lava spewed from these volcanoes cools, islands form along the arc-shaped length of colliding plates of ocean floor (**FIGURE 3.16**).[18] Thus, one of the two colliding slabs of oceanic crust is destroyed, in a process we now call **subduction**. Curiously, Coats's paper seems to have been all but ignored for years after its publication, and the idea of **subduction zones**, in which ocean floor is destroyed, was only later added to the general Theory of Plate Tectonics.

We should point out here that there were a lot of ideas and discoveries in this field being published in the 1950s and 1960s, and it was not uncommon for some authors to overlook or otherwise miss important contributions reported by others. It is unfortunate, because much of the motivation that drives scientific research, besides curiosity, is the eventual recognition by one's peers. Being first to publish an idea or discovery, and being cited by future workers,

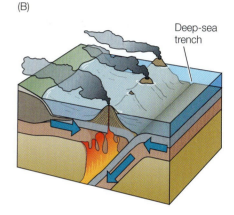

(A)

(B)

FIGURE 3.16 (A) The Aleutian Islands. (B) Model showing Robert Coats's theory for the mechanism of how island arcs are formed. According to Coats's theory, when a drifting slab of new ocean floor, created at one of the mid-ocean ridge systems, encounters another slab of oceanic crust, the denser one will slide beneath the lighter one, deforming the ocean floor as it does so and creating a deep sea trench. Sediments and crustal rock in the sinking slab melt and form volcanoes that extend upward off the ocean bottom, creating a line of islands.

---

[18] Coats first presented his ideas in 1961 in a talk which was published the next year. Robert R. Coats, 1962. "Magma type and crustal structure in the Aleutian arc," in *The Crust of the Pacific Basin*, ed. G. MacDonald and H. Kuno. *American Geophysical Union Monograph* 6: 92–109.

is how that recognition, that reward, is bestowed. It is especially unfortunate that the pioneering work of Holmes and Coats is often left out of this history of continental drift.

Not only was the field of marine geology itself advancing rapidly in the 1950s and 1960s, so too was the technology available to scientists. The end result was the coming together of numerous lines of new evidence in apparent support of the idea of continental drift. Wegener's theory, his wild idea of drifting continents, was beginning to seem plausible.

Paleomagnetism: The final proof?

Beginning in the 1950s and continuing through the 1960s, while Hess, Dietz, and Heezen were still formulating their ideas about seafloor spreading, scientists were using **magnetometers**—instruments that detect metallic objects and reveal aspects of their magnetic properties—to study the ocean floor. Developed for use in detecting enemy submarines in World War II, these magnetometers revealed curious patterns of magnetism in the ocean floor (**FIGURES 3.17**). Because the basalt that makes up the ocean floor is not just iron-rich, but also contains magnetite, a highly magnetic mineral, it records the polarity (the north-south magnetic pole orientation) and intensity of Earth's magnetic field upon its formation, as it solidifies after being extruded as volcanic lava. (This phenomenon of a "magnetic seafloor" had been known since the 1800s; it tends to distort compass readings in and around Iceland, for example, and generally plays havoc with marine navigation in that region.)

Earlier applications of **paleomagnetic** studies on land were revealing patterns of "remanent" magnetism—magnetism left over in rocks after their initial formation—in continental rocks that showed what appeared to be a "wandering of the poles." That is, when geologists reconstructed the directions of just where the North Pole should have been when the rocks first formed, the pole appeared to have been in motion. We now know that a significant part of this remanant magnetism was due to movement of the continents where those rocks formed, not to movement of the poles.[19]

---

[19] The word *remanence* (adj. *remanent*) specifically refers to magnetism remaining after the magnetizing force is removed, but you will sometimes see the term *remnant magnetism* used instead.

FIGURE 3.17 (A) The U.S. Coast and Geodetic Survey ship *Pioneer* passing under the Golden Gate Bridge. In August 1955, the *Pioneer* deployed the first towed marine magnetometer, which was developed by the Scripps Institution of Oceanography. (B) A later-model magnetometer being deployed from the stern of the *Pioneer* in 1965. (C) Modern versions of these instruments are still towed behind ships just off the bottom, where they measure magnetic properties of the ocean floor and send the data up the tow wire to the ship's computers for processing.

(A)

(B)

(C)

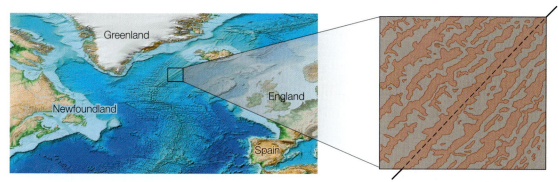

FIGURE 3.18 Striped pattern of magnetic reversals in the seafloor basalt on either side of the Mid-Atlantic Ridge southwest of Iceland. Pink regions represent "normal" magnetic orientation of the seafloor basalt, where it solidified at a time when the Earth's magnetic North Pole was north, and brown areas represent times when the orientation of the magnetic poles was reversed.

Remanent magnetism is created when basaltic magma flows from the volcanic mid-ocean ridges to the surface of the seafloor where it cools and solidifies in the cold bottom waters of the ocean. As it solidifies, tiny particles of iron-bearing minerals act like compass needles and align themselves with the Earth's magnetic field (they point north). For example, if we were to melt a piece of iron and then allow it to solidify here in the Earth's magnetic field, it would acquire the properties of a weak magnet. So, as the molten lava of basalt cools and solidifies, it "freezes" the orientation of the Earth's magnetic field into the rock where it remains unaltered in either strength or orientation (**BOX 3D**). When the magnetic properties of the seafloor were mapped by scientists towing magnetometers just above the bottom, a peculiar pattern emerged. Magnetic variations were revealed as stripes in the seafloor—alternations in the polarity of the magnetism—that ran parallel to the mid-ocean ridges, as shown in **FIGURE 3.18** for the Mid-Atlantic Ridge southwest of Iceland. Why there was a striped pattern of alternating polarity was a bit of a mystery.

In 1963 two marine geologists, Frederick Vine and Drummond Matthews, published a paper proposing an interpretation of those alternating magnetic stripes on the ocean floor.[20] They argued that the stripes were a reflection of past changes in the Earth's magnetic field whereby the polarity of the magnetic field was reversed. If this were true, then the striped pattern was powerful evidence that, indeed, new ocean floor was being created at the mid-ocean ridge and that it was spreading away from it.

That the Earth's magnetic field changes its polarity, whereby the magnetic North Pole becomes the magnetic South Pole and vice versa, was discovered in the early 1900s by geologists and archeologists who found that volcanic rocks, as well as bricks and earthenware, when heated and then cooled acquired a weak but stable magnetization that is oriented with the Earth's magnetic field. They found that rock samples from recent lava flows possessed a similar natural magnetization that was oriented very close to that of the Earth's current magnetic field, which, they deduced, was acquired when the lava flows originally cooled. However, they noticed that rocks that were much older, on the order of a half million years old, were magnetized with an orientation that was about 180 degrees from that of the Earth's present magnetic field—in other words, just the opposite of what was expected. The explanation for this phenomenon was that there must have been periodic reversals of the Earth's magnetic field—which means that a compass needle would reverse directions; what was once magnetic North would become magnetic South. This is all related to changes in the Earth's giant magnet: its liquid, iron-rich outer core. Every once in a great while, that giant magnet undergoes an internal restructuring that produces a flip-flop in polarity.

---

[20] Vine, F. J., and Matthews, D. H., 1963. Magnetic anomalies over oceanic ridges. *Nature* 199: 947–949.

BOX 3D Paleomagnetism and Reversal of the Earth's Magnetic Field

The magnetic poles are not the same as the geographic poles, which mark the axis of Earth's rotation; instead they are located some distance away from the true North and true South Poles. That difference in degrees is called *declination*. An ordinary compass needle will point to the magnetic North Pole, which is usually several degrees away from the geographic, or true, North Pole (**FIGURE A**). Examples of compass declination at several points on the Earth during a period of normal and reversed polarity are shown in Figure C. Because the Earth's magnetic field is three dimensional, as shown in **FIGURES B,C**, a properly balanced and oriented magnet needle will also indicate the *inclination* of the magnetic field, which causes the

needle to point upward or downward, depending on where the compass is on the surface of the Earth.

Magnetic rocks in the oceanic crust have recorded a near-permanent record of these features of the Earth's magnetic field at the time and location of the rocks' initial formation. Upon solidifying from molten lava, they acquire and lock in the declination, inclination, and intensity of the Earth's magnetic field. That magnetic record in the oceanic crust is revealed when magnetometers are towed just off the bottom. Not only will the record of declination, inclination, and intensity of the Earth's magnetic field in the ocean floor be revealed, but also whether the Earth was in a normal or reversed magnetic phase at the time the volcanic

basalt solidified. Thus, magnetic rocks in the Earth's crust, such as volcanic basalt—both on land and on the ocean floor—will exhibit a declination and inclination depending on where the rocks are sampled, and whether the Earth was in a normal or reversed magnetic phase at the time the volcanic basalt solidified.

FIGURE C

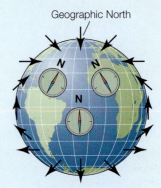

Geographic North

Normal magnetic field

Geographic North

Reversed magnetic field

FIGURE A

Magnetic North   True North

D

Declination

Horizontal

I

Inclination

FIGURE B

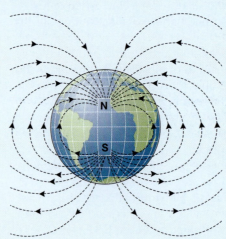

By 1964, scientists had determined that these magnetic reversals happen every few hundred thousand years, and they documented with some precision the dates of seven reversals that had occurred over the last 4 million years.[21] Today we have records of other reversals extending back in time some hundreds of millions of years.

FIGURE 3.19 illustrates how spreading magma (lava) extruded from the mid-ocean ridges would behave as it flowed outward, locking into itself a record of the Earth's magnetic field at the time it solidified. The pattern that results is much

[21] Cox, A., Doell, R. R., and Dalrymple, G. B., 1964. Reversals of the Earth's magnetic field. *Science* 144: 1537–1543.

(A)

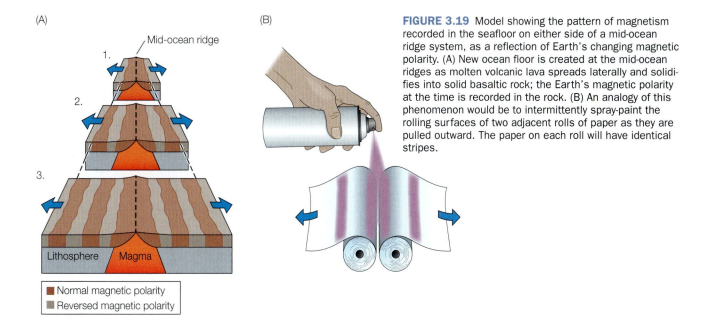

(B)

FIGURE 3.19 Model showing the pattern of magnetism recorded in the seafloor on either side of a mid-ocean ridge system, as a reflection of Earth's changing magnetic polarity. (A) New ocean floor is created at the mid-ocean ridges as molten volcanic lava spreads laterally and solidifies into solid basaltic rock; the Earth's magnetic polarity at the time is recorded in the rock. (B) An analogy of this phenomenon would be to intermittently spray-paint the rolling surfaces of two adjacent rolls of paper as they are pulled outward. The paper on each roll will have identical stripes.

the same as what you would see if a couple of side-by-side rolls of paper towels were being pulled out in opposite directions as you periodically spray-painted the space where the rolls met. A pattern of stripes would emerge, comparable to the stripes of magnetism seen on opposite sides of mid-ocean ridges.

The pattern of paleomagnetism that became locked into the basalt of the ocean floor provided scientists not only with clear evidence of the spreading of new ocean floor away from the ridge systems, but also with a way to estimate the *age* of oceanic crust (because the dates of past reversals had been determined). By the mid-1970s, much of the Pacific and Atlantic Ocean basins had been mapped by towed magnetometers, revealing a global pattern of stripes of magnetic reversals that ran parallel to the mid-ocean ridges. Assigning times to those reversals, we then had a map of the ages of the ocean floor—not the sediments, but the actual lithosphere. Today, we have enough data to produce truly remarkable maps of the ages of the ocean floor (**FIGURE 3.20**), which is not only rich in detail, but, most importantly, reveals that there are no areas of the ocean floor with crustal rocks that are much older than 180–200 million years! Knowing the ages of the seafloor, we could now calculate the actual *spreading rates* at the mid-ocean ridge. It was discovered that the speeds of plate motions are much faster in the Pacific Ocean and part of the Indian Ocean than in the Atlantic, and because of the faster spreading rates, there has been less time for sediments to accumulate on the seafloor in the Pacific Ocean (see Figure 3.12). But we're getting ahead of ourselves; let's return to our historical journey of discovery.

By the mid-1960s, after what seemed to be irrefutable evidence of seafloor spreading came as a result of paleomagnetic studies, there were still nagging suspicions among many geologists. One fear was that scientists might be picking and choosing data to fit their theories—as Wegener had been accused of doing a half-century earlier. For example: Does Africa really fit nicely against South America? Or is this an example of subjectively electing to see a pattern that fits a preconceived notion, a theory of continental drift? That fear was laid to rest in 1965, when **Sir Edward Bullard** (1907–1980) of Cambridge University published a paper with colleagues J. E. Everett and A. Gilbert Smith, in which they used a computer to find the best statistical fit

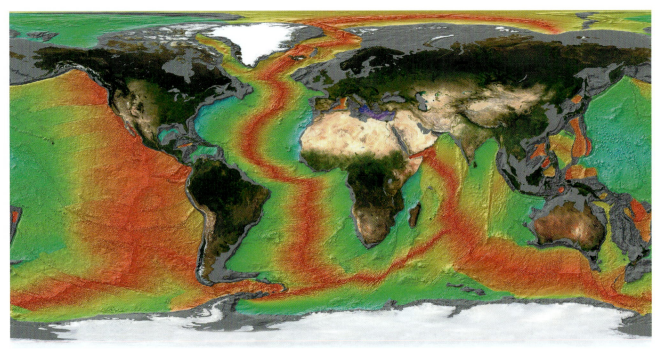

Million years

| 0 | 20 | 40 | 60 | 80 | 100 | 120 | 140 | 160 | 180 | 200 | 220 | 240 | 260 | 280 |

FIGURE 3.20 Map of the oceanic lithosphere, based on data from several sources, including paleomagnetic surveys. The relative age of the rock is presented in color contours over a high-resolution bathymetric relief map. Colors on land masses indicate areas of rich vegetation (green) and deserts (tan).

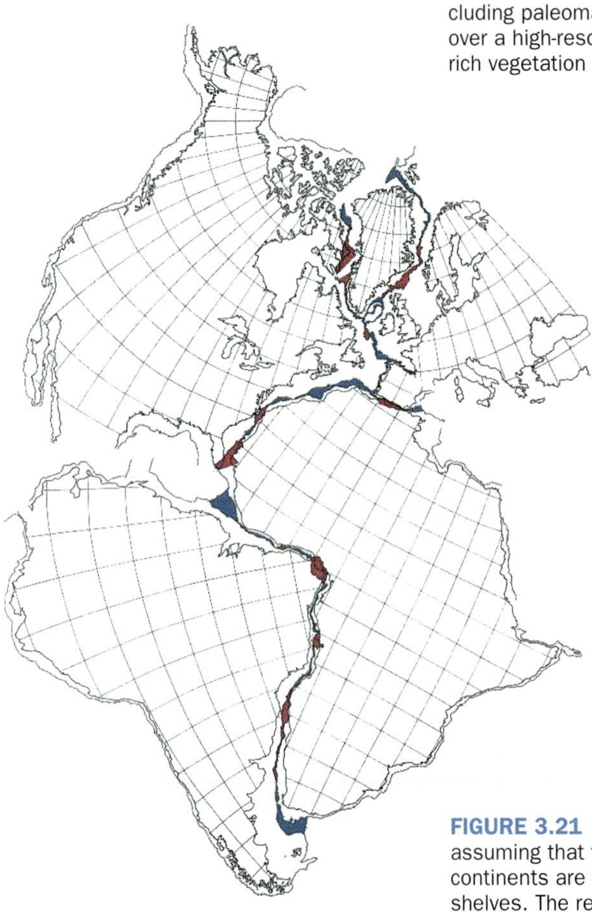

of the continents, not one that was just pleasing to the eye. Remember, computers were still quite new then, and this approach was a novel one. Rather than using the shorelines, which delineate the continents on an ordinary map, Bullard and his colleagues used the edges of the continental shelves, which they defined as the "500-fathom depth" contour (ca. 1000 m; **FIGURE 3.21**). Their result confirmed that the "fit" between Africa and South America first noticed hundreds of years earlier was in fact even better than it had appeared to be.

A unifying theory of plate tectonics

The same year that Bullard's paper appeared, in 1965, **J. Tuzo Wilson** (1908–1993) of the University of Toronto, published a paper that described how the surface of the Earth is made up of a system of "several large rigid plates" that are indeed in motion. However, he explained, their motion on the spherical Earth necessitates that there be a new category of geological fault—the **transform fault** (**FIGURE 3.22**). Wilson proposed that the surface of the Earth is covered by a network of mobile

FIGURE 3.21 The result of a computer analysis of the best fit of today's continents, assuming that they were once joined together in a supercontinent. The shapes of the continents are based on the 500-fathom depth contours off the edge of the continental shelves. The red indicates overlap; the blue, gaps.

FIGURE 3.22 (A) A diagram of a transform fault running perpendicular across a mid-ocean ridge system (or a spreading center) where new ocean floor is being formed and is spreading laterally away. Arrows indicate the directions of the oceanic plate motions. (B) Schematic diagram of the same fault, indicating the relative motions on either side of the fault; the dashed circle represents the shear zone, where the motions are opposed to one another. Seismic activity would be expected within this area, but not outside it, where the plate motion is not sheared. This is in contrast to a "regular" strike-slip fault, where there is shear (opposing motion) on both sides running the length of the fault line.

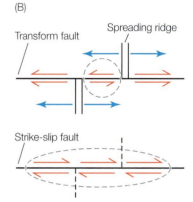

lithospheric plates, with edges that are delineated by any one of three distinct types of boundaries (**FIGURE 3.23**). One is a **divergent plate boundary**, or a spreading center, like the mid-ocean ridges where new ocean floor is formed. (From these centers, two plates diverge away from one another.) The second is a **convergent plate boundary**, where two plates collide, forming mountains or island arcs and deep sea trenches. The third type is the transform fault. With his new **Theory of Plate Tectonics** as it came to be known some time later, he was able to show that, undeniably, the continents are adrift.[22]

Transform faults run perpendicular to the spreading centers at the mid-ocean ridges, and had been known to exist for some time; but without any clear ideas about seafloor spreading, geologists were at a loss to explain how they were formed. If the mid-ocean ridges constitute a line on a spherical globe where a new surface, a lithospheric plate, is being created on either side, then

[22] Wilson, J. T., 1965. A new class of faults and their bearing on continental drift. *Nature* 207: 343–347.

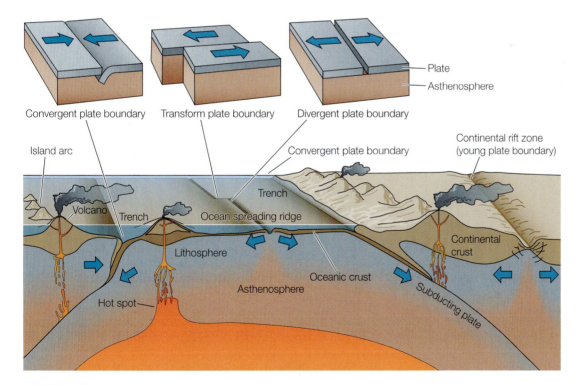

FIGURE 3.23 A summary diagram of the key processes operating in seafloor spreading and plate tectonics, with inset drawings illustrating the three types of plate boundaries (convergent, transform, and divergent).

(A)

(B)

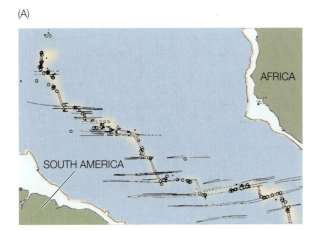
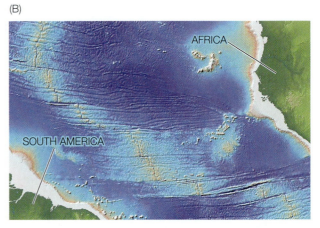

FIGURE 3.24 (A) Earthquake activity between 1955 and 1965 along the Mid-Atlantic Ridge between South America and Africa from Sykes's 1967 paper showing that, indeed, seismic activity along the transform faults is confined to the region between the ridges, as predicted by J. Tuzo Wilson in 1965. (B) A modern bathymetric map of the same region, showing the features more clearly.

the action of spreading away from that line will develop tears in the rigid plates as they move across the curved surface. Wilson explained that these lines are the mid-ocean ridges, and the tears are the transform faults. Wilson concluded his landmark paper with a simple challenge to his colleagues to investigate the validity of his concept, "because transform faults can only exist if there is crustal displacement and proof of their existence would go far towards establishing the reality of continental drift and showing the nature of the displacements involved" (p. 347).[22] Testing Wilson's idea would be a *real-time* test—whether the plates are moving *today*, versus all the evidence accumulated thus far, which were paleontological lines of evidence that the continents and seafloors had moved millions of years ago. We needed a test of whether they are moving today.

Well, the idea was tested, and in 1967 Wilson was shown to be correct. Professor **Lynn Sykes** of Columbia University's Lamont Geological Observatory published a paper showing that Wilson's prediction held up: in his study of the Mid-Atlantic Ridge, he found seismic activity on the transform faults that was restricted to the sections between the spreading ridges, and not outside that zone as would be the case with "normal" strike-slip, or "transcurrent" faults (**FIGURE 3.24**).[23]

With this new Theory of Plate Tectonics, Wegener's continental drift was to drop out of scientific lexicon, and the years that followed Wilson's 1965 paper would witness even more supportive evidence for, and refinements in, the theory. Today we know that the surface of the Earth is made up of a system of about a dozen major **lithospheric plates**, on the order of 70–100 km thick, and many other smaller plates that all move around atop the asthenosphere (**FIGURE 3.25A**). And because the plates interact with one another at the junctions between them, even today, those junctions are sites of seismic activity. In fact, they are where almost all of the Earth's approximately one million earthquakes and volcanoes occur each year (**FIGURE 3.25B**). Notice how tight the correspondence is between the plate boundaries and the locations of earthquakes. Notice also in Figure 3.25B that the rim of the Atlantic Ocean is a "quiet" one, without many earthquakes—unlike the Ring of Fire in

[23] Sykes, L. R., 1967. Mechanisms of earthquakes and nature of faulting on the mid-oceanic ridges. *Journal of Geophysical Research* 72: 2131–2153.

(A)

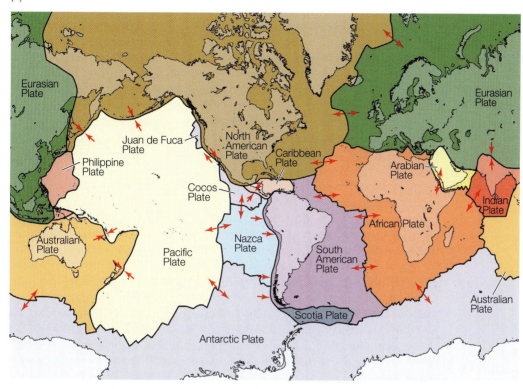

(B)

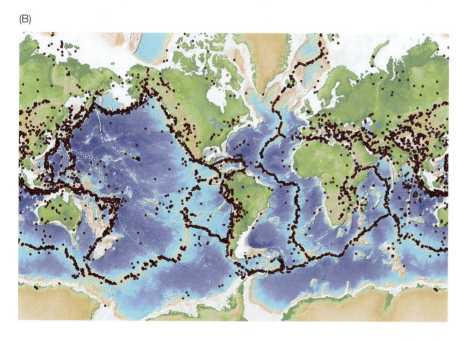

FIGURE 3.25 (A) Map of the major lithospheric plates showing their plate boundaries and the direction of plate movements (arrows). Divergent, or spreading, boundaries at the mid-ocean ridges, convergent boundaries between two plates, and transform boundaries are shown by the black lines. (B) Locations of earthquakes of magnitude 5.0 or greater between 1973 and 2009.

the Pacific. This is because the Atlantic Ocean floor west of the Mid-Atlantic Ridge and the North American continent are both part of the same North American Plate, and they move together as one. Likewise, the Atlantic Ocean floor east of the ridge is part of either the African Plate, or the Eurasian Plate (see Figure 3.25A). For this reason, the Atlantic Ocean coastlines are called **passive continental margins**, in contrast to the **active continental margins** of the Pacific Ocean.

Filling in the details

Some authors have written that the issue of continental drift was resolved once the paleomagnetic patterns in the seafloor on either side of spreading ridges were revealed, while others claim that it wasn't resolved until after J. Tuzo Wilson's proposed pattern of seismic activity in the vicinity of transform faults was confirmed. But almost everyone would agree that by the late 1960s and early 1970s, the issue had been resolved: Wegener's idea, while not correct in all its details, was in essence, a valid one. The surface of the Earth is without doubt made up of lithospheric plates, delineated by boundaries that are either zones of plate divergence, convergence, or slippage, in the form of transform faults; and not only have those plates moved about in the past, they continue to move today.

What would appear to be irrefutable evidence of the movements of lithospheric plates is being generated today by observations being made from space using one of our latest technological marvels, the **Global Positioning System**, or GPS. The GPS system is an assemblage of 24 Earth-orbiting satellites launched by the United States government between 1989 and 1994. It is not just useful for finding your way around in a car; it is also being used to measure changes in the positions of Earth's crust over time. The latest results are confirming the directions and speeds of crustal movements—even tiny movements on the order of centimeters per year—at a number of specific locations around the globe. The speeds of the North American Plate are shown to be substantially slower than those for the Pacific Ocean plates, in agreement with the estimates based on the ages of the ocean floor. Based on GPS measurements, the Big Island of Hawaii (that is, Hawaii itself), for example, is moving toward the northwest at more than 5 cm per year. And over the past eight years, it has moved north some 35 cm, and west about 60 cm. It was these kinds of direct measurements that Wegener had hoped to see one day; the evidence today certainly is overwhelming—the continents *are* adrift.

But there is irony here. Despite all the scientific results accumulated through the 1960s, and even the GPS results that continue to churn out data today in support of moving continents, which together seemed to have convinced most of the skeptics, the original criticism leveled against Wegener almost a hundred years ago—that he had not explained or identified the driving force, the mechanism, by which continents move—really hadn't changed. After the paleomagnetic seafloor mapping results were in, and after J. Tuzo Wilson's model of plate tectonics held up to closer scrutiny, we still had only an untested hypothesis as to the force driving plate motion: the idea that convection currents in the mantle were responsible for the motion of plates. Remember, this same mechanism of mantle convection had been suggested by Wegener in 1915, Holmes in 1944, and Hess in 1960. Still, as late as the early 1970s, when most scientists accepted the basic tenet of continental drift, there were still no direct observations of deep mantle convection. All we had done, really, was accumulate more evidence that the plates have moved in the past and that they continue to move today. It would be a few more decades before we had actual evidence of what makes the plates move.

Convection currents in the mantle: Further evidence

In the 1990s, seismologists were developing new ways to process the enormous volumes of data coming in from seismograph stations around the world on the thousands of individual earthquakes each month. Using computer techniques similar to those used today in medicine (e.g., CAT Scan imaging), scientists were able to discern patterns of seismic wave velocities in the mantle and to confirm what Holmes and Hess had hypothesized: there was

(A)

(B)

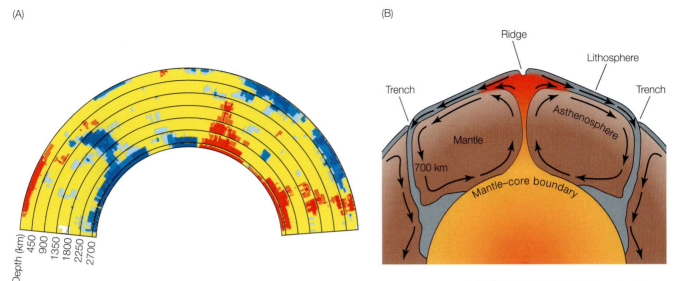

Depth (km)
450 900 1350 1800 2250 2700

FIGURE 3.26 (A) Seismic image of the Earth down to the mantle, revealing a downward-sinking slab beneath the Caribbean Sea joining other slabs near the outer core (blue) and rising mantle material (red). (B) An interpretation of the process.

clear evidence of convection cells deep below the lithosphere that extend all the way to the mantle–core boundary, some 2900 km below the surface of the Earth. **FIGURE 3.26** shows an example of such studies, where cool oceanic crust (blue) is apparently descending all the way to the mantle–core boundary where it joins a collection of other slabs of ancient oceanic crust. Likewise, the image shows rising plumes of warmer material (red), also from the mantle–core boundary.

The source of energy to move the lithospheric plates is therefore the convection currents that result from the uneven distribution of heat inside the mantle. But just how those mantle currents translate their energy into moving lithospheric plates is not all that straightforward. Arthur Holmes and Harry Hess held that the plates were simply riding on top of the flowing mantle currents, but, geologists noted, more force is required than can be explained by this "along for the ride" model. More recently, the **ridge-push**, **slab-pull model** was proposed. In this model, newly formed crust at the elevated mid-ocean ridges simply slides downhill, effectively pushing the plates away from the ridges. Likewise, as the newly formed oceanic plate cools and moves away from the ridge, its density increases, while at the same time the plate accumulates thicker and thicker layers of sediment, meaning that the plate's mass is greatest at the point farthest away, where it is being subducted. Gravity pulls it downward into the mantle, and the rest of the plate is pulled behind it. So the plate is being pulled at one end, and pushed at the other. But not all scientists buy into this idea. For one thing, the North American Plate is not being subducted—therefore, the drift of both the western Atlantic Ocean floor and all of North America would have to be due strictly to ridge-push, with no pull at a subduction zone, which some scientists have difficulty accepting. There is still a lot research being directed at just how the plates move, whether they are riding passively on the convection current, or being pushed and pulled at opposite ends.

The general acceptance of the theory of continental drift, or as scientists more commonly refer to it, Plate Tectonics, means that all the earlier observations of (and speculations about) various geological processes operating in the mantle and in the lithosphere—all those bits and pieces of the puzzle—now had a solid scientific underpinning. It would appear that such dramatic and often tragic events as earthquakes and volcanoes, though still very much unpredictable, should now be better understood. The Theory of Plate Tectonics explains the different kinds of plate motions and the interactions between them at plate boundaries (with the exception of details of continental plate collision, which we discuss below). Figure 3.23, along with the principle of mantle convection cells, encapsulates the major processes involved in plate tectonics. So, let's go over them here.

Tectonic plate boundaries

DIVERGENT PLATE BOUNDARIES   The most logical starting point for a discussion of divergent plate boundaries is the initial breakup of the supercontinent Pangaea some 225 million years ago (**FIGURE 3.27**). That event most likely began when a plume of mantle material—molten magma—rose beneath the giant continent, such that the continent split open, creating a rift valley, the floor of which was filled with basaltic oceanic crust. Thus a new ocean basin, with a new seafloor, was created. That process continued, with enormous spans of new seafloor spreading laterally away from the new mid-ocean ridge and rift system; this separated Pangaea into the ancient continents Laurasia and Gondwanaland.

This action of splitting a continent open and creating a new ocean basin has happened numerous times in the past, well before Pangaea, and is happening today. The East Africa Rift Valley is opening up now (**FIGURE 3.28**).[24] Extending across East Africa and into the Red Sea where it is flooded by the ocean, it continues north into Israel and Jordan as the Dead Sea Rift Valley. These processes of seafloor spreading at the mid-ocean rifts and in East Africa continue to produce some 4.5 cubic miles of new ocean crust each year.

CONVERGENT PLATE BOUNDARIES   Of course, the tectonic plate created by the formation and spreading of new seafloor must eventually impinge on another plate, somewhere, if the Earth is not expanding. This means that for there to be spreading centers, or divergent plate boundaries, then there have to be zones of plate convergences. These are the regions of frequent and sometimes violent earthquakes and volcanoes that result when massive lithospheric plates come together in one of three possible ways.

---

[24] The supercontinent Rodinia existed between 1100 and 750 million years ago.

FIGURE 3.27 Pangaea likely began to break up when (A) a hot mass of mantle material rose and (B) split open the continental plate above it, (C) creating a mid-ocean ridge system where (D) new ocean floor continues to be created, riding on the slowly flowing mantle material as it moves in a huge convection cell.

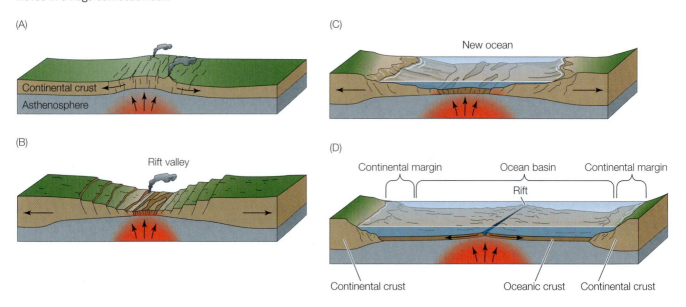

(A)

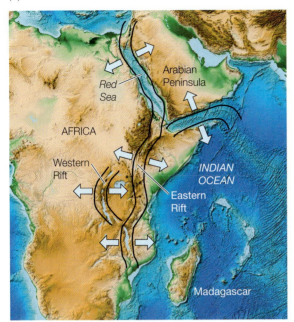

(B)

FIGURE 3.28 (A) The East African Rift Valley extends from Southeast Africa to Lebanon, and includes the Red Sea and the Dead Sea in Israel. (B) A section of the Eastern Rift.

The first possibility is *two oceanic plates converging*. In this case, one of the two plates will usually be older, and hence cooler and denser. Because of its greater density, it will slip beneath the other—it will be subducted—producing a deep sea trench (**FIGURE 3.29**). For example, we can imagine floating a sheet of plywood in a swimming pool and pushing it against a sheet of much lighter Styrofoam; the denser plywood will (usually) be subducted beneath the Styrofoam. In the case of sinking oceanic plates, that descending material will become heated, and some of it will melt and rise to the surface through cracks, thus creating volcanoes. The subduction zone where the two plates meet, and the surface expression of volcanoes usually will make a curved or arching pattern on the Earth's surface, and if the volcanic material emerges above the sea surface, it can give rise to chains of islands we know as island arcs. Examples of island arcs include

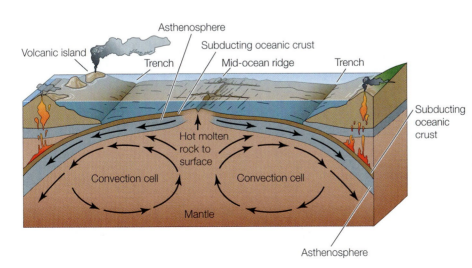

FIGURE 3.29 Diagram of convergent plate boundaries and a divergent ocean plate boundary. (Left) An oceanic plate collides with another oceanic plate and is subducted beneath it, forming a deep sea trench and an island arc. (Right) An oceanic plate collides with a continent, and is also subducted, forming a deep sea trench and a mountain range on the continent.

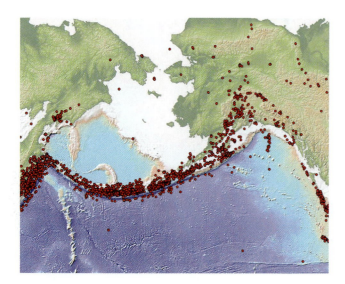

FIGURE 3.30 Locations of all earthquakes greater than magnitude 5.0 in the area of the Aleutian Islands from 1973 to 2009.

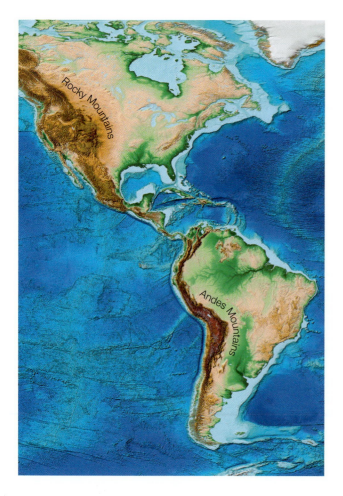

those in the northern and western Pacific Ocean, such as the Aleutian Islands that stretch out from Alaska. **FIGURE 3.30** shows recent earthquake activity resulting from the subduction of the Pacific Plate along the edge of Aleutian Islands.

In addition to the volcanic islands that result from the subduction of one oceanic plate beneath another, islands that were already formed on that plate (discussed in a minute) will move with it (because they are too light to be subducted) and will converge along that same island arc. Combined with the volcanic magma of melted oceanic crust and marine sediments, these materials form more continental crust, which accumulates into bigger and bigger islands. Some geologists think that most or all continental crust was originally formed by such an accumulation of material in island arc systems, and that over long periods of time, these island arcs become welded together into larger and larger masses of crust to form continents.

The second possibility is *an oceanic plate and a continental plate colliding*. In this case, because the oceanic plate is denser, it will be subducted beneath the continental plate (see Figure 3.29, right side). The resulting volcanoes form mountain ranges along the edges of the overriding continental plate (see Figure 3.23). This phenomenon created the great mountain ranges that run the length of western North America, Central America, and South America (**FIGURE 3.31**), and their volcanoes, such as Mount St. Helens in Washington State.

The Rocky Mountains of North America, which extend far beyond the coastal edge of the North American Plate, are a special case of mountain building. Portions of the Pacific Plate were subducted under the North American Plate at a shallow angle some 35 to 80 million years ago, wedging upward much of western North America. Because of the shallow angle of that subduction, the subducting plate was not melted to the same extent as the oceanic plate subducted under the South American Plate, under the Andes Mountains. As a result, volcanoes in the Rockies are less common than in the Andes. Subsequent erosion of the uplifted continental plate left behind the wide mountain range we see today.

And the third possibility is *two continental plates converging*. In this final case, neither is likely to be much denser than the other, and so there is no subduction. Instead, they will tend to pile upward into new mountain ranges, while at the same time their converged base will sink into the asthenosphere in order to achieve isostatic equilibrium. Often the leading edge of a continent is preceded by an

FIGURE 3.31 The Coastal and Cascade Ranges on the western edges of the Rocky Mountains of North America and the Andes Mountains in South America are on the leading edges of the North and South American Plates.

oceanic plate, which will subduct, and thus create volcanoes (**FIGURE 3.32**).

An example of colliding continental lithospheric plates occurred when the free-floating continent of India drifted north and met Asia (see Figure 3.8), crushing together with it. In the process of this collision, the highest mountains in the world, the Himalayan Mountains, were formed, at the same time that some material was forced down into the asthenosphere (remember, this is where the crustal plates are the thickest). Often ancient marine sediments that have lithified to form sedimentary rocks are uplifted in this process, thus explaining the occurrence of fossils of marine organisms high up on mountains. There are, in fact, marine fossils near the summit of Mt. Everest, the world's highest mountain.

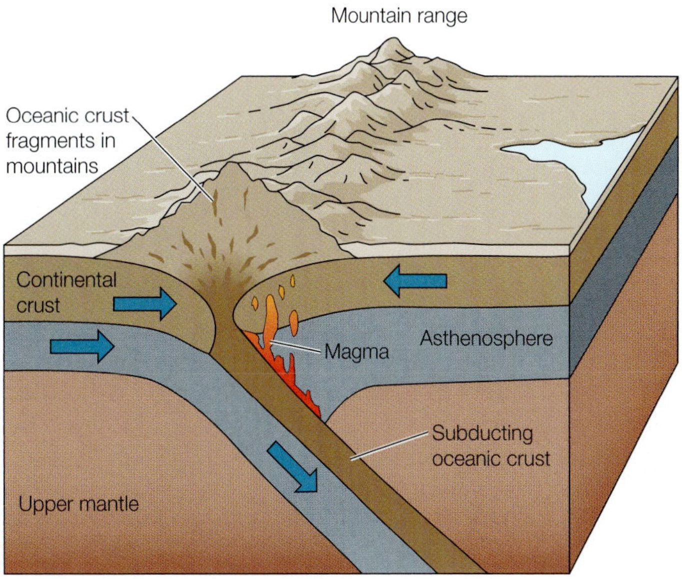

TRANSFORM PLATE BOUNDARIES At transform plate boundaries, unlike at divergent and convergent plate boundaries, crust is neither created nor destroyed. Two plates in motion slide by each other without subduction or convergence. The transform faults described by J. Tuzo Wilson are prominent features on the ocean floor, distributed along the entire length of the 37,000-mile mid-ocean ridge system. As Wilson explained, only the portions of the faults that are situated between spreading centers, where two tectonic plates shear past one another, will have seismic activity, or earthquakes. This is illustrated in **FIGURE 3.33**, which shows that the earthquake locations in the central Atlantic Ocean are confined almost exclusively to the spreading centers, or rift valleys, on the Mid-Atlantic Ridge, and those portions of transform faults that lay between the displaced spreading centers. It is only in those specific zones along transform faults that the African Plate is moving to the east (on the north side of the transform fault) and the North American Plate is moving west (on the south side of the fault), with the two

FIGURE 3.32 When a continental plate, preceded by an oceanic plate, converges with another continental plate, the leading oceanic plate will be subducted, forming volcanoes, and the two continental masses will build mountain ranges.

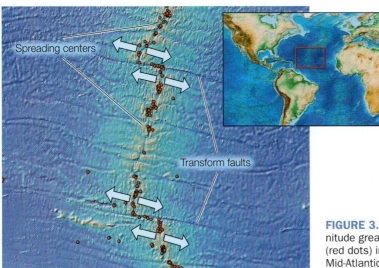

FIGURE 3.33 Locations of earthquakes of magnitude greater than 5.0 between 1973 and 2009 (red dots) in relation to transform faults along the Mid-Atlantic Ridge. The relative plate motions are indicated by arrows.

(A)

(B)

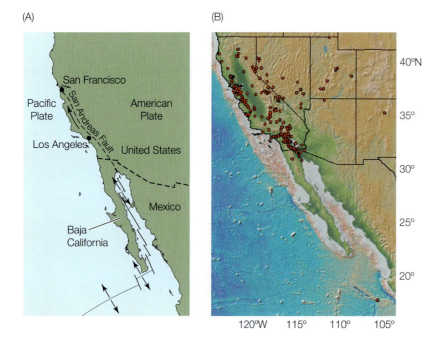

FIGURE 3.34 (A) The San Andreas fault in California is an elongated transform fault. Unlike transform faults on the ocean floor, the San Andreas also runs through continental crust. It separates the Pacific Plate on the west, which includes the city of Los Angeles, from the North American Plate to the east, which includes San Francisco. (B) A map of the locations of earthquakes greater than magnitude 1.0 recorded over a seven-day period in October, 2011; a total of 413 events are plotted here as red dots (with many dots hidden beneath others).

plates shearing past one another along that portion of the transform fault that lies between the spreading ridges.

The San Andreas Fault in California is a transform fault, but it is located on the North American continent, not on the seafloor (**FIGURE 3.34**). The San Andreas is greatly stretched in the north–south direction; on one side of the fault, a narrow slice of western California, including Los Angeles, is sliding north as part of Pacific Plate, while just on the other side, the North American Plate is sliding south. As the plates move, they store enormous amounts of elastic energy along the fault line that separates them, but only up to a point; when they finally have enough energy to overcome the friction between them, they slip. This produces earthquakes. For this reason, the west coast of the United States is tectonically active, and the frequency of earthquakes is very high (see Figure 3.34B); it is not uncommon for there to be more than 1000 measureable earthquakes in California of magnitude 1.0 or greater in a given week, with more than 50 greater than magnitude 3.0.

Other marine geological features

Plate tectonics is responsible for a number of other geological features in the oceans, such as sea mounts and guyots, coral atolls, and island chains in the middle of the ocean, all of which owe their formation to a "hot spot" in the mantle.

HOT SPOTS Although there is some debate among geologists about the nature of such features, it is generally assumed that **hot spots** in the Earth's crust are just that—stationary sources of heat in the upper mantle. As a lithospheric plate moves over a hot spot, magma may break through the surface of the plate through cracks and fissures, forming a volcano. After the plate carries the volcano past the hot spot, the volcano becomes inactive, and another new volcano may form behind it, directly above the same hot spot. The interaction of stationary hot spots and moving lithospheric plates can create island chains, such as the Hawaiian Islands (**FIGURE 3.35**). Presently, the Big Island of Hawaii is over the hot spot that is feeding the island's two active volcanoes, Mauna Loa and Kilauea, but as the plate continues to move over and away from the hot spot, these volcanoes will eventually become inactive, as have all the other Hawaiian Islands to the north of the Big Island.

A new seamount (see page 96) already named Loihi is presently forming on the south side of the Big Island (Figure 3.35B), fed by magma from the hot spot. Since 1970, the frequency and magnitude of earthquakes in this area offshore from the Big Island has been increasing. During the summer of 1996, a large number of earthquakes occurred for two months at the Loihi seamount. It began in July 1996 and by the end of August, a total of over 4000 earthquakes had been recorded, with more than 40 of them greater than magnitude 4.0. Loihi will someday (probably) become the next Hawaiian island in the chain,

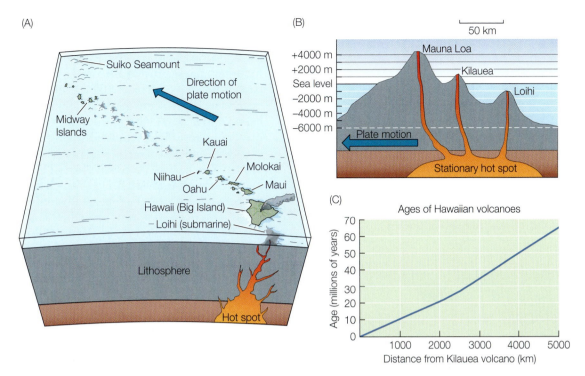

FIGURE 3.35 The Hawaiian Islands have formed as the Pacific Plate passes over a stationary hot spot in the mantle. (A) Formation of a volcanic island chain over the hot spot, and subsidence of the inactive volcanoes and islands. (B) Schematic diagram of the process: the Big Island of Hawaii is about to move out from above the hot spot, which feeds the two active volcanoes, Mauna Loa and Kiluaea; Loihi Seamount, which is now over the hot spot, may become the next Hawaiian island. (C) The ages of Hawaii's volcanoes, including extinct volcanoes that are now distant seamounts, increase with distance away from the Big Island and the stationary hot spot.

but it isn't expected to emerge for another 30,000 years.

Much earlier in Earth's geological history—that is, well before the formation of the Hawaiian Islands—that same hot spot created the Emperor Seamounts in the Pacific Ocean farther up along the chain to the northwest (**FIGURE 3.36**). Some of these may have been islands once, but have long ago sunk below the surface of the ocean, while others are undersea volcanoes that never reached the surface.

Hot spots are found not only beneath oceanic crust, but also beneath continental crust, such as in Yellowstone National Park; in fact, most hot spots are beneath continents. It is the close proximity of a hot spot in the upper mantle that contributes to the famous geysers and hot springs in the park.

SEAMOUNTS, GUYOTS, AND ATOLLS Seamounts and guyots (singular, *guyot*, pronounced "gee-oh") are extinct volcanoes on the seafloor that do not break the surface of

FIGURE 3.36 Between 30 and 75 million years ago, the Pacific Plate was moving almost directly north, but since then, it has been moving more northwest. This movement is reflected in the pattern of the seamounts, which extend to the northwest of the Hawaiian Islands.

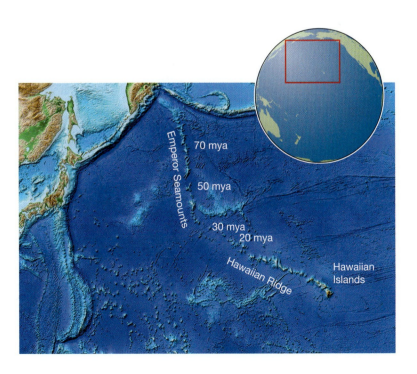

(A)

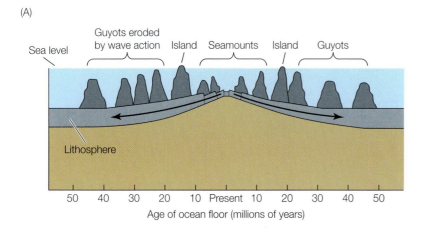

(B)

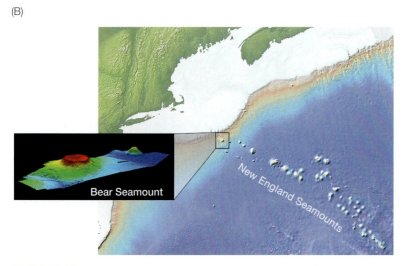

FIGURE 3.37 (A) Formation of seamounts and guyots, in this case, in the vicinity of a spreading center. They are also formed over hot spots. These extinct underwater volcanoes have subsided into the asthenosphere, but they may have once been islands; the flat-topped guyots were once islands whose volcanic peaks eroded away prior to their subsidence. Those that never reached the surface retain their peaks. (B) Bear Seamount (which is actually a guyot) in the Northwest Atlantic is one of many in the New England Seamount chain. Bear Seamount is about 200 miles off the coast of Massachusetts. The inset is looking to the northeast, with Physalia Seamount in the distance. Bear Seamount rises approximately 2000 m from the surrounding ocean floor to a depth of 1100 m from the surface.

the water. **Seamounts** are ancient volcanoes that are thought never to have emerged above the sea surface and they have retained their mountain peaks. **Guyots** are former volcanic islands that have subsided; their peaks have been shaved flat by erosion. Seamounts have been known for a long time, but flat-topped guyots were first discovered (using echo sounders) and described by Harry Hess in the 1940s, when he was a U.S. Navy officer in the Pacific during World War II. Hess named these features *guyots*, after the nineteenth-century Swiss-American geologist, Arnold H. Guyot (1807–1884). Common in the Pacific Ocean, guyots are found in all oceans (**FIGURE 3.37**).

Atolls are ring-shaped islands of coral reefs that have formed over inactive volcanoes, but how these curiously-shaped islands come into being has not always been understood. The question was the subject of a great debate between two giants of science in the late 1800s—a debate that wouldn't be settled until 1950.

Charles Darwin was one of the first to suggest a mechanism for atoll formation, as a result of observations he made on his five-year voyage on the HMS *Beagle* (1831–1836). The idea occurred to him after he experienced first-hand a large earthquake while the *Beagle* was in port at Concepcion, Chile. This earthquake was accompanied by a dramatic uplifting of the shoreline (by more than 3 meters in places), which left marine life such as seaweeds and barnacles stranded above the high-tide mark. This was a "eureka" moment: Darwin realized right away that in this uplift was a clue to the presence of marine fossils high up in the Andes Mountains. And, he reasoned, if the Earth's crust can be uplifted, it must also subside in some areas. He extrapolated that reasoning to coral atolls, and hypothesized that they are the result of volcanoes settling.

This was Darwin's first major scientific contribution, which came when he was 27 years old. He published details of his idea of how coral atolls form in 1836, in his *Voyage of the Beagle* memoir, and then in book form in 1842. He wrote that these ring-like, coral reef islands are simply ancient volcanoes that, after they become inactive, begin to sink back down into what we now know as the asthenosphere, just as all mountains do, in keeping with the principle we know as isostasy. Corals are tropical sea-anemone-like animals that secrete a rock-hard calcium carbonate structure around themselves, which we know as coral reefs. The reefs grow approximately 1 cm in height per year, which means that if the volcano sinks at a slow enough rate, the corals can keep up, and the top of the reef remains visible above sea level (**FIGURE 3.38**).

While Darwin's mechanistic interpretation of how atolls formed was well received by the scientific community, it would be challenged some 40 years

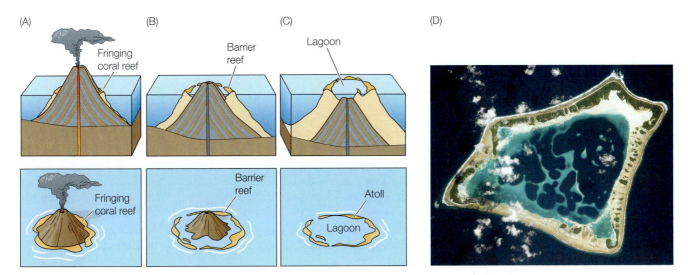

(A)　　　　　(B)　　　　　(C)　　　　　(D)

Fringing coral reef

Barrier reef

Lagoon

Fringing coral reef

Barrier reef

Atoll

Lagoon

FIGURE 3.38 (A) Undersea volcanoes that emerge above the sea surface may have coral reefs growing around their shores. If the rate of subsidence of the volcanic island into the asthenosphere is sufficiently slow, and if the rate of upward reef-building (B) by the coral animals can keep up, the volcanic peak will drop beneath the sea surface, leaving the fringing coral reef surrounding a lagoon (C). (D) Atafu Atoll in the equatorial South Pacific Ocean, as seen from space. This coral atoll is only 2 m above sea level. It was formed by the process in (A–C), a process first proposed by Charles Darwin in 1836.

later following the *Challenger* expedition (1872–1876).[25] One of the *Challenger's* major discoveries, you will recall, was the widespread occurrence of deep-sea sediments—thick layers of the skeletal remains of tiny organisms that once lived in the surface waters—covering the seafloor. This finding prompted **John Murray** (1841–1914), the second in command on that expedition, to propose in the late 1870s and early 1880s a much simpler explanation for coral atolls than Darwin's. He argued that marine sediments that sink out of the surface layers of the ocean would, over sufficiently long periods of time, form very thick layers, even thicker than had been suggested by the *Challenger* expedition a few years earlier. But, if that debris sank to great depths, where the ocean waters become caustic (e.g., acidic, for reasons we'll discuss later in this book), then that material would dissolve and not accumulate on the bottom. Murray maintained that only on relatively shallow submarine mountains (seamounts) could that debris persist, because shallow waters are less acidic. And over time, Murray reasoned, the sediment would increase in thickness until it rose to within a few hundred meters of the surface—and it was on top of these sediments, he argued, that corals grow. And, because corals grow best at the outer edges of reefs, an atoll forms around the edges of these shallow features which eventually extend to the surface as a ring-shaped island. This was a very different explanation from that offered by Darwin some 40 years earlier.

Murray shared his idea with **Alexander Agassiz** (1835–1910), son of the famous geologist **Louis Agassiz** (1807–1873). Louis, a distinguished Harvard professor, was a staunch creationist and a vocal opponent of Darwin's theory of evolution. In their day, the two great men waged a verbal battle on the issue, which lasted for decades after *The Origin of Species* was first published in 1859. Over time, however, the general consensus eventually shifted to Darwin's views, leaving Louis somewhat isolated from mainstream science. Alexander, while generally in agreement with Darwin's theory of evolution, nonetheless extolled Murray's conflicting idea about coral atolls, arguing that Darwin had it all wrong. Determined to find proof of Murray's theory, Alexander undertook numerous voyages to the tropics to study coral reefs firsthand. He never conceded defeat, and he continued to write and lecture on the topic long after Darwin's death in 1882.

It wasn't until 1950 that the matter was finally settled. In order to prepare for nuclear bomb tests, the U. S. Navy commissioned the drilling of test wells

[25]David Dobbs, 2005. *Reef Madness: Charles Darwin, Alexander Agassiz, and the Meaning of Coral* (New York: Pantheon Press).

on Eniwetok Atoll in the Marshall Islands of the Pacific. The drills cut through some 1300 meters of ancient coral, more than three-quarters of a mile, before they hit volcanic basalt. Clearly, it was coral all the way down; there were no layers of sediment, as Murray would have predicted. Darwin's explanation was the correct one.

TERRANES Further evidence of, and features that result from, plate tectonics are **terranes**. These are complicated continental land forms created when fragments of continental crust (granite), island arcs, and ancient marine sediments collide with continental plates. The fragments ride on top of the "drifting" oceanic crust, but they are too light to be subducted. As a result, they crinkle up against the continental plate they eventually collide with, forming strangely-shaped land masses. But they don't form great mountain ranges. Examples of terranes can be found in Alaska and New England as exposed masses of both granite and sedimentary rock layers occurring together in the same general location.

FIGURE 3.39 (A) The Canadian Rocky Mountains, and (B) the Grand Canyon. Each clearly shows layers of sedimentary rocks that formed at the bottom of an ancient ocean.

(A)

(B)

Marine Sediments

Pick up a rock—any rock, anywhere on the surface of the Earth—and the odds are better than 1 in 2 that it was once marine sediment. That is, more than half of all the rocks on the surface of the Earth are *sedimentary rocks*—"lithified" marine sediments. The photographs in **FIGURE 3.39** of the Canadian Rocky Mountains and the Grand Canyon aren't just pretty pictures; they show strata of sedimentary rocks that were once layers of marine sediment on the bottom of an ocean. Once uplifted, those rocks were subjected to weathering which exposed the many layers of the former marine sediments. The sedimentary rocks of the Grand Canyon, which were formed on the bottom of an ancient shallow sea sometime between 200 million and 2 billion years ago, were vertically uplifted as much as 3000 m. The Colorado River then cut through these sedimentary rock layers to reveal their strata.

What happened millions of years ago and continues to happen today can be described in a simple diagram that illustrates the Rock Cycle (**FIGURE 3.40**). Starting with mountains (we could start anywhere in the loop here), rocks are "weathered," both chemically and mechanically, into particles of various sizes. Erosion occurs as rivers and streams pick up the smallest particles and carry them in suspension to the ocean, where they are deposited. Over millennia, these deposits build up in thickness. Chemical cementing, and heat that results from the pressures associated with the thick layers of sediment, cause the accumulated sediment to become lithified—forming layers of sedimentary rocks. Those rocks can be further modified under heat and pressure to form metamorphic rocks, or they can melt completely and be extruded as volcanic material or be uplifted as new continental crust, thus starting the cycle over again.

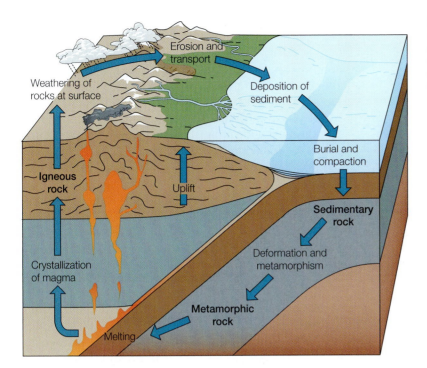

FIGURE 3.40 The Rock Cycle. Mountains erode, forming terrigenous sediments that become sedimentary rocks that become subducted. Subjected to high heat and pressure, the sedimentary rock becomes metamorphic rock, which can be further metamorphosed to igneous rocks to eventually be uplifted again into a mountain range.

Marine sediments can be defined as particles of organic (biological) or inorganic matter that collect on the bottom of the ocean in a loose, unconsolidated manner. Sediments are quite common, and we see examples of them all the time; for example, you can see suspended sediment in a gallon of apple cider. Buy a jug sometime and notice that in a few days, particles of "suspended sediment" settle to the bottom of the container and make a layer of "apple mud." In the same way, particles suspended in rivers (which makes some rivers appear muddy) settle out once those waters leave the turbulent rivers and encounter the less turbulent coastal ocean. Over time, that sediment from rivers builds up on the bottom of the ocean, sort of like a layer of dust builds up on furniture over time. If you've ever waded into marine shorelines, you've no doubt slogged through mud or beach sand, both of which are marine sediments that have accumulated over time.

A major source of marine sediment particles is from the weathering (chemical and mechanical breakup) and erosion (movement) of terrestrial rocks. But marine sediments are made up of a lot more than just pieces of rocks. Other components include dead marine organisms and their skeletons and shells; volcanic ash; chemical precipitates; and particles from space. We'll discuss each of these and their dynamics below.

Sediment sizes and dynamics

We have already seen earlier in this chapter that the thicknesses of bottom sediments vary widely in different regions of the oceans, from a "dusting" to thicknesses of several tens of kilometers. In addition, the sizes of sediment particles themselves span a wide range, from what we might call boulders (basketball-sized rocks >250 millimeters [mm] diameter), to tiny, microscopic flakes of clay. But it is principally the smaller sizes that we need to concern ourselves with here, especially particles of clay (<0.004 mm), silt (0.004–0.062 mm), and sand (0.062–2.0 mm). Beach sand, by the way, is an intermediate-size sand grain between 0.25 and 0.5 mm, about the same size as a period at the end of this sentence.

FIGURE 3.41 A thought experiment: imagine tossing a handful of dirt—with particles of various sizes—into a glass aquarium filled with water. Those dirt particles will sink to the bottom after a short while. Then, use a turkey baster to squirt water into that collected material sitting on the bottom, and you'll notice three things: (1) some particles are lifted off the bottom and only settle back down some distance away; (2) some particles roll away some distance, where they stop; and (3) some particles don't do anything, they just sit there.

The size of sediment particles is important in their **settling velocities**—how fast a particle sinks in the ocean (and in freshwaters). For example, it takes sand particles about a day to sink to a depth of 4000 m, which is the average depth of the oceans. Silt particles take about six months, and clay particles take decades. For a size range from 0.004 to 2.0 mm, a difference of just 1.996 mm, this is an extremely wide range in sinking speeds. Bottom sediments are thickest adjacent to the continents because as rivers empty into the sea from the continents, they pour out not only their water but also their sediments. The larger particles can't go very far from their source, but because the smaller particles sink more slowly, they can be transported great distances in ocean currents before they finally settle to the bottom. Only the very smallest particles are carried far enough out in the oceans to reach bottom in the deepest depths.

Sediment particles that have arrived on the bottom of the ocean do not always stay where they settle; they may still be subject to reworking; they may still be moved about. Here comes another thought experiment: imagine a handful of dirt of various size particles, from very fine silt and clays to very coarse sand, on the bottom of an aquarium, with the water around it not moving at all. All the particles just sit still on the bottom. But if we squirt some water onto the sediment-covered bottom of the aquarium using a turkey baster (as shown in **FIGURE 3.41**), we will see three things happen: (1) some particles will be picked up and become suspended in the swirling water currents—these are the smallest particles; (2) some particles will roll away along the bottom—these are the intermediate sized particles; and (3) some particles—the biggest grains—will just stay put and not move at all. The amount of displacement depends on the grain size and how hard we squirt (e.g., the speed of the current of water). This simple thought experiment and the three observations just listed illustrate three important processes: **sediment erosion**, which refers to particles being picked up in suspension; **sediment bed transport**, which refers to particles getting rolled around on the bottom; and **sediment deposition**, which refers to particles staying put or falling to the bottom some distance away, when the water currents become still again. All three phenomena depend on both particle size and water current speed, and all three are important determinants, though not the only determinants, of the distributions of marine sediments in the world's oceans (**BOX 3E**).

Types and origins of sediments

TERRIGENOUS SEDIMENT   Bits and pieces of terrestrial rocks produced by chemical and physical weathering processes are **terrigenous sediments** (from the Latin *terra*, earth). They are one of the two most abundant sediment types (the other being biogenous sediments). Terrigenous sediments are primarily particles of quartz (sand) and clay that arrive in the oceans from land by way of wind, rivers, and streams, and are also called alluvial sediments. We can see dramatic examples of this process (terrigenous sediments being transported to the ocean) going on today: satellite images of coastal waters often reveal visibly turbid waters, waters with a lot of suspended sediment particles, especially where there is a river discharging suspended sediments

BOX 3E The Hjulstrom Diagram

Once particles arrive on the sea-floor they can be redistributed by currents to an extent determined by the water current speed and the diameter of the sediment particle. A quantitative plot of that relationship, as it applies to *erosion*, *transport*, and *deposition* is given here and is called a *Hjulstrom diagram*. (Those processes are also illustrated qualitatively in our thought experiment in Figure 3.41.) The principles we've been discussing are represented in this diagram, but they apply only to particles already on the bottom—not to particles in suspension.

Notice that the Hjulstrom diagram has log scales of particle size and water current velocity, and that in general, there are two odd phenomena illustrated. First, the sediment size that is eroded (picked up and moved) most easily is sand. Second, particles smaller than sand—once they have been deposited on the bottom—are, paradoxically, not eroded easily. They require higher water velocities to erode them, because, unlike larger particles, they tend to stick together as a result of cohesive forces that act on very small particles, like clay, which are flat and have electrostatic charges that hold them to one another. (This is a lot like static electricity, which helps lint collect on your clothes.) Once the very small particles such as clay have been eroded, they will stay in suspension for a long time before they "settle out." Particles larger than sand, on the other hand, are just too heavy to be moved easily.

The phenomena illustrated in the Hjulstrom diagram explain why we have sandy beaches. Sandy beaches are made of medium-size sand (0.25–0.5 mm), the size that is most easily eroded. Once it is eroded, it settles out fairly fast compared with even finer sediment particles which stay in suspension and get carried out to deeper water offshore. This is what happens in the swash zone on a beach: as a wave breaks on a beach, the turbulence suspends sand particles that get carried with the wave farther up the beach slope, until the wave has lost its momentum and comes to a standstill. The sand particles then drop out of suspension. As the wave washes back down the beach face, it may not flow fast enough to pick up those same sand grains again. So, over time, the beach effectively accumulates these intermediate-sized sand grains. Notice in the Hjulstrom diagram that the water current velocity to erode sand-sized particles is less than that required for any other sized particles. This means that the size group is both easily picked up and is quick to fall out again.

Therefore, based on this diagram, big particles are hard to erode (which should be obvious), and so are the very small particles (which is not so obvious). This is why we have mostly very small, fine-grained sediment on the bottom way out in the open ocean basins. It is only they—the smallest particles—that can stay suspended in the water long enough to be carried great distances (hundreds of miles) from the edges of the continents where they were initially eroded, out to the centers of the oceans. And this is why big boulders stay where they are.

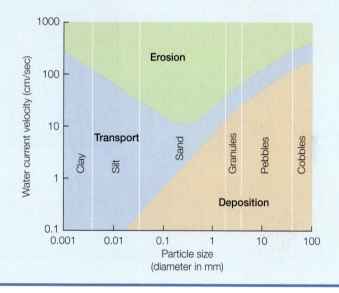

into the oceans. Examples in **FIGURE 3.42** show the Mississippi River plume in the Gulf of Mexico, where the largest river in North America dumps its high load of suspended sediments that give the river its famous "muddy waters." Also shown are the coastal waters off Cape Hatteras, North Carolina, on the East Coast of the United States, after the heavy rains from Hurricane Floyd in September of 2003.

Of the terrigenous sediment particles carried to the ocean, only the smallest particles stay in suspension long enough to be carried by currents far out to sea where they eventually settle to the bottom. Quartz particles (especially sand) do not get carried far before settling out, but clay particles (diameter <2 micrometers) do. Clay particles, also from land, can be brought out to sea by the wind, as

(A)

(B)

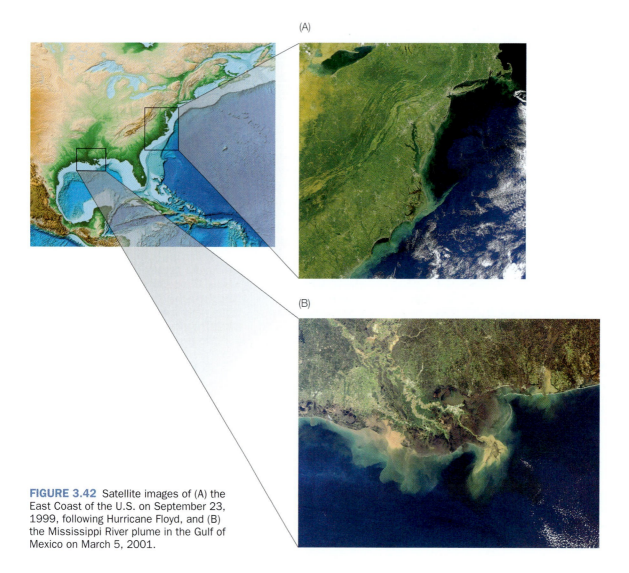

FIGURE 3.42 Satellite images of (A) the East Coast of the U.S. on September 23, 1999, following Hurricane Floyd, and (B) the Mississippi River plume in the Gulf of Mexico on March 5, 2001.

shown in **FIGURE 3.43**, which shows a satellite image of a dust storm carrying a huge load off the Sahara Desert to the Atlantic Ocean. These fine-grained clays make up as much as 38% of deep-sea sediments. They come from wind-borne dust, volcanic ash, and the like, which slowly settle to the bottom to form the fine-grained red, olive, or brown clays. The **sedimentation rate** of clays—the rate at which sediment on the ocean bottom increases in thickness—is very slow in the deep sea, on the order of less than 1 mm per 1000 years. This is only 1/1000 the sedimentation rate of larger-grained sediments closer to the coast on the continental shelves, which is about 1 mm per year, or 1 meter per 1000 years.

Terrigenous sediments include **turbidites**, layers of coarse-grained

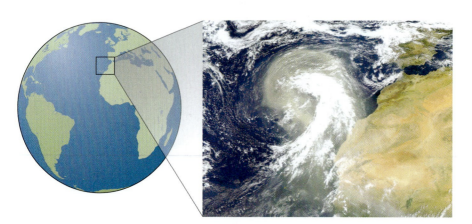

FIGURE 3.43 Dust from the Sahara Desert extending out over the North Atlantic Ocean, as recorded by satellite.

sediments interleaved with the finer sediments more typical of the deep sea. These are produced when **turbidity currents**, created by earthquakes, produce an undersea "landslide" that carries terrigenous sediments of various particle sizes farther away from the coast than would otherwise be the case. Turbidites can be deposited well beyond the edges of the shallow (<200 m) continental shelves. **Glacial** sediments are also coarse-grained particles, generally found adjacent to the coasts; they were carried by glaciers to the sea where they were deposited upon melting.

This process of terrigenous sediment delivery to the ocean—erosion and transport to the sea—must be offset, or balanced, by tectonic formation of new continental land via volcanic eruption or crustal uplift, as illustrated in the Rock Cycle above.[26] This has to be to the tune of some 100 million metric tons annually for the Rock Cycle to be in approximate steady state.[27]

BIOGENOUS SEDIMENT   The second most abundant marine sediment type is **biogenous sediment**. Most are one of two kinds: **siliceous** sediments (made of *silica*) or **calcareous** (or **carbonate**) sediments (made of *calcium carbonate*, $CaCO_3$). The particles of silica and calcium carbonate were once *in solution*, or dissolved, in seawater, but they have been "taken up" biologically by plankton and other organisms to form their various hard body parts, such as skeletons and protective shells (**FIGURE 3.44**). **Plankton** are very small living plants and animals that drift with the ocean currents; they have no, or very limited, swimming capabilities, and are usually very small, often microscopic, in size. Some are quickly consumed by other organisms and thus become part of the marine food web, but others simply die and sink before they can be eaten by something else. As those dead organisms begin sinking to the bottom, some redissolve back into solution in the seawater on the way down, and others make it all the way to the bottom, where they accumulate. Even those organisms that are consumed can still settle to the bottom: they are often "passed

---

[26] When they do not balance, we will get changes in sea level.

[27] A metric ton (MT) is 1000 kilograms (kg); one kilogram is 2.2 pounds (lb), so a MT is equal to 2200 lb. A "standard ton" is 2000 lb.

(A)

(B)
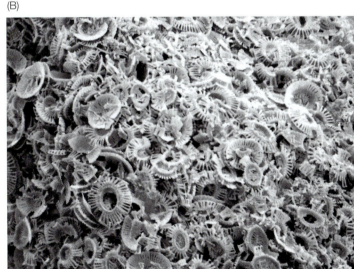

FIGURE 3.44 Origins of some calcareous sediments. (A) Scanning electron micrograph of a coccolithophore (*Emiliania huxleyi*). (B) Zooplankton fecal pellets packed with coccolith plates from *E. huxleyi* cells and collected in a sediment trap.

FIGURE 3.45 The White Cliffs of Dover on the southeast coast of England show the uplifted deposits of calcium carbonate sediments.

through" their consumers and repackaged into fecal pellets, which, because of the larger repackaged particle size, sink even faster to the bottom (see Figure 3.44B). So biogenous sediments are just what is left of once-living marine organisms. Together, terrigenous and biogenious sediments make up more than 99% of all ocean sediments.

The proportion of biogenous sediment to terrigenous sediment is generally higher in the deep sea than in coastal areas. (Out in the deep sea, far away from the coast, it only takes a little bit of biogenic sediment to completely overshadow the contribution of terrigenous sediment.) When biogenic sediments make up more than 30% of the total, those bottom sediments are called **oozes**. There are two main varieties of oozes, named for the dominant organism of which they are composed.

The **calcareous oozes** are mostly made up of the shells of a number of different planktonic species, including: (1) **foraminifera**, single-celled animals (protozoa) with snail-like shells made of calcium carbonate; (2) **pteropods**, multicellular animals (molluscs) that, like foraminifera, have calcium carbonate shells; and (3) **coccolithophores** (see Figures 3.44 and 8.14), single-celled planktonic algae, that secrete elaborate calcium carbonate plates.

Ancient calcareous oozes that have been lithified into "chalk" deposits have been uplifted by tectonic forces in various locations around the world, forming such landmarks as the famous White Cliffs of Dover (in England; **FIGURE 3.45**). Those deposits may reflect ancient oceanographic conditions that favored "blooms," or dense populations, of foraminifera, pteropods, or coccolithophores. A modern-day example of a coccolithophore bloom off the northeast coast of the United States is shown in **FIGURE 3.46**. In this instance, the single-celled coccolithophores have grown to such tremendous numbers, and become so densely packed in the surface waters of the ocean, that they (and their calcium carbonate coccoliths, or plates, that have been shed) have colored the water a milky white, making the bloom visible not only from a ship at sea, but also from space.

Calcareous oozes are not found everywhere in the deep sea; there are depths where the waters are too acidic, and the delicate calcium carbonate skeletons dissolve before they can accumulate into thick layers of sediment. For example, put a drop of acid such as strong vinegar on a seashell and watch the bubbles of CO_2 (carbon dioxide) form as the carbonate dissolves in the acid. The depth below which the waters become too caustic for calcarious sediments is called the **calcium carbonate compensation depth**, or CCD. The acidity of seawater increases with depth in the oceans (we'll discuss why later), so settling particles of calcium carbonate, or carbonate for short, are subject to dissolution as they fall slowly through the deeper depths on their way to the bottom. At about 4500 m

FIGURE 3.46 A bloom of coccolithophores (*Emiliania huxleyi*, a single-celled alga) as seen from space in the Gulf of Maine off the east coast of the United States. Image is of reflected light caused by the milky color of the surface waters as a result of the huge numbers of microscopic coccoliths shed by *E. huxleyi* into the water.

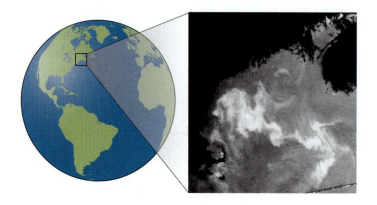

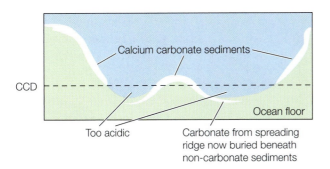

FIGURE 3.47 Distributions of calcareous sediments on either side of a spreading ridge in relation to the carbonate compensation depth.

below sea level, waters become acidic and carbonate begins to dissolve—this is the CCD. Therefore, bottom sediments below depths of about 4500 m have little calcareous ooze. The actual depths where this change occurs are deeper in the Atlantic than the Pacific because the deep waters of the Pacific Ocean are more acidic than the Atlantic. Thus the North Pacific is especially low in $CaCO_3$ ooze.

Below the CCD, the bottom is more likely to be dominated by siliceous oozes. However, the carbonate sediments deposited at the mid-ocean ridge, which is shallower than the CCD, spread laterally to deeper depths with the tectonic plates on either side of the spreading ridge. As they spread, the plates may descend to deeper depths where they will be buried beneath non-carbonate sediments that settle on top of them and as a result the calcium carbonate sediments are preserved (**FIGURE 3.47**). And as we will discuss later, the oceans are becoming more acidic as a result of increased atmospheric CO_2 additions.

Siliceous oozes make up some 14% of deep ocean sediments, and are composed of the remains of plankton with silica in their skeletons or shells. These organisms include **radiolarians**, single-celled animals, or protozoa; **diatoms**, single-celled algae; and the less abundant silicoflagellates (also single-celled algae) (**FIGURE 3.48**). (For more on radiolarians, see Chapter 9; silicoflagellates, see Chapter 8.) Like the carbonate-shelled organisms, these siliceous organisms also dissolve in seawater. Siliceous oozes are of two main types: **diatom oozes** and **radiolarian oozes**. Diatom oozes are common in high latitudes, especially in the waters around Antarctica where diatom production is high. Radiolarian ooze is more common in low latitudes, or tropical and subtropical regions.

HYDROGENOUS SEDIMENTS The **hydrogenous sediments** are mineral particles that are chemically precipitated out of the water. These are sometimes called *authigenic sediment* (from the Greek *authi*, "there") because they are produced *in situ* on the sea bottom; they are not brought there from somewhere else. Because they chemically precipitate out of seawater on the bottom of the ocean, hydrogenous sediments grow very slowly, and they constitute only a small fraction of the total sediments on the seafloor. The most prominent of these sediment particles are **manganese nodules**, which cover the bottom of many oceans of the world, and **phosphorite nodules**, which are found along some continental margins (**FIGURE 3.49**). Manganese nodules, discovered by the *Challenger* expedition, are made mostly of manganese and iron hydroxides. They grow very slowly, only 1 to 10 mm in diameter per million years. Despite their great ages, they tend not to get buried by other sediment particles settling on top of them, which should happen over such long periods of time—but it isn't clear just why that is. Other hydrogenous sediments include **evaporites**, mineral

FIGURE 3.48 Photomicrographs of live plankton that are the origins of siliceous sediments. (A) A radiolarian, a single-celled protozoan; (B) various species of diatoms, which are single-celled algae; and (C) a silicoflagellate, a single-celled alga.

(A)

(B)

(C)
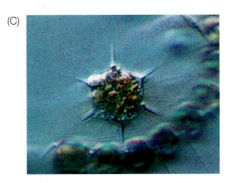

(A)

(B)

FIGURE 3.49 (A) A skate swimming among a bed of manganese nodules on the seafloor off the East Coast of the United States at a depth of about 700 m. (B) A close-up of two nodules from the Pacific Ocean.

deposits that form as seawater evaporates. These include carbonates which precipitate out of seawater first (making limestone) followed by table salt (sodium chloride). You can see examples of evaporates forming around old pipe fittings in houses where the water is hard, and around the edges of the Great Salt Lake in Utah.

COSMOGENOUS SEDIMENT **Cosmogenous sediments** are of extraterrestrial origin. They are by far the least abundant type of bottom sediment, but they do occur. For example, we only have to look up at our own Moon and notice all its impact craters to know that a lot of material has hit the Moon. Because the Moon lacks an atmosphere, the evidence of those impacts has been preserved much better than that of impacts here on Earth; but the Earth, too, has been hit with a lot of meteors (meteorites). When meteors hit the Earth's surface, they tend to melt and send upward lots of granitic materials that eventually fall back to Earth as rain-like particles, called **tektites** (**FIGURE 3.50**). Those particles are mostly just fine bits of iron, which tend to dissolve in the ocean, but they also include larger particles that don't dissolve completely, as well as glassy beads. The extremely low water content of such particles suggests that they may be flung beyond Earth's atmosphere and into space before they fall back to Earth.

FIGURE 3.50 Tektites, or cosmogenic sediments, that are formed by the impact of a large meterorite, comet, or asteroid hitting Earth. The impact flings fragments of melted terrestrial rock into the atmosphere, where they solidify into these curious shapes before falling back to Earth.

FIGURE 3.51 A range of three energy levels that influence the sizes of marine sediments. (A) A low-energy, protected tidal marsh with very small waves, or none at all and fine-grained muddy sediments. (B) A sandy beach, with small to moderate-sized waves and intermediate-sized sand grains. (C) A rocky shoreline with periodically large storm waves, with sediment present only as large cobbles and boulders. The wave sizes influence the water velocities at the water–sediment interface, which, along with sediment sizes, determine the extent to which those intertidal and near-shore bottom sediments are redistributed.

Distribution of marine sediments

In general, sediments nearest the continents are thicker than those in the deep sea, and they are of different composition. Coastal marine areas are richer in terrigenous than in biogenous sediments, simply because they are closer to the source (land). And because of the sediment redistribution processes we've just discussed, we tend to see mostly coarser-grained sediments near shore, with some exceptions. The energy levels of water acting on the different sizes of particles (especially the wave action) in different near-shore areas results in the redistribution patterns described by the Hjulstrom Diagram (**FIGURE 3.51**; see Box 3E).

Sediments on the continental shelf (at depths less than about 200 m) are called **neritic sediments**, and those farther offshore are called **pelagic sediments**. The bulk of neritic sediments is terrigenous. Relatively large grains are found near the coast, where they fall out due to their larger size, and smaller grains are found farther away, in keeping with the wave and current energy available to redistribute them.

Neritic sediments often include glacial deposits, or **glacial marine sediments**, a mix of highly variable sediment sizes which are poorly sorted at the time they are deposited. Over time, however, the water currents eventually sort the sediment grain sizes. Some sediments are **ice-rafted** (transported by icebergs), and can be deposited almost anywhere; these, too, are poorly sorted. **Turbidity currents**, which are created during an earthquake as it loosens sediment that flows downhill as a sediment/water mixture, like an undersea landslide, are one way that large sediment particles, including boulders, can be transported farther offshore than would normally be expected.

Neritic sediments have the highest sedimentation rates. As mentioned earlier, the sedimentation rate in neritic waters can be as much as one millimeter of sediment per year, whereas deep ocean, or pelagic, sedimentation rates are on the order of 1 mm per 1000 years. These aren't just terrigenous sediments: the delivery of *biogenic* sediments to the bottom is also greatest in the coastal and shelf waters as compared with the open ocean (we will learn why later), but this biogenic material is nonetheless very much diluted by the nonbiogenous, terrigenous sediments. The global pattern of these distributions is given in **FIGURE 3.52**.

The average thickness of sediments throughout the ocean, including the neritic waters, amounts to about 1 km in the Atlantic Ocean, and 0.5 km in the Pacific; that is, sediments are on average twice as thick in the Atlantic. This is because more major rivers empty into the Atlantic (the sediment supply is greater), and the Atlantic is a smaller ocean, with a smaller area over which that sediment is deposited (therefore it gets thicker faster).

(A)

(B)

(C)

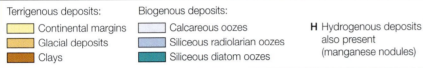

Terrigenous deposits:	Biogenous deposits:	
▢ Continental margins	▢ Calcareous oozes	**H** Hydrogenous deposits also present (manganese nodules)
▢ Glacial deposits	▢ Siliceous radiolarian oozes	
▢ Clays	▢ Siliceous diatom oozes	

FIGURE 3.52 Global distributions of dominant bottom sediment types.

The study of marine sediments is very important to the energy industry and to scientists studying climate change (**BOX 3F**). Sediments can be analyzed to reveal hints about the distributions of various resources, including petroleum reserves, and even Earth's past climates. This is a fascinating area of research, to which this introductory textbook simply cannot do justice. However, we will be revisiting marine sediments in the context of various topics that we will be covering in the chapters that follow.

BOX 3F   Methods of Studying Marine Sediments

Marine sediments on the bottom of the ocean are studied both directly and indirectly. Indirect methods include the echo sounders we've already discussed, which have proved valuable in mapping the thicknesses of sediments overlying the basaltic rock base of the ocean floor. There are several ways to study sediments directly, that is, by bringing an actual sample of bottom sediment to the surface for examination and analysis. One of the simplest, and oldest, is with a bottom "grab" (**FIGURE A**) to collect a sample of surficial sediments—the sediments of the uppermost layers.

Samples from deeper in the bottom sediments are collected with cores and are used to study sediment stratigraphy—the layering of sediments. A piston core can take core samples as long as 25 m or so (**FIGURE B**). The piston creates a suction that allows the core to penetrate downward while the core tube fills with sediment it passes through.

The deepest sampling devices, which penetrate several thousand meters into the bottom—and into the underlying crust—are drilling cores, which can collect samples not just of sedi-ment, but also of the underlying crustal rock (**FIGURE C**). The *Glomar Challenger*, a specially designed drill ship built by the National Science Foundation, was launched in 1968; the ages of crustal rock collected by the *Glomar Challenger* into the early 1970s further supported the theory of seafloor spreading. Its replacement, the *Joides Resolution*, a refitted oil exploration vessel that went into service in 1985, was joined by the Japanese drill ship *Chikyu* in the international Integrated Ocean Drilling Program, or IODP, which continues today.

Cores are kept in core libraries and are used to study details of plate tectonics, and also past climates over the last several 100,000 years or so. The deeper the core, the older the material, and by examining the unconsolidated sediments, as well as the consolidated material deeper in the core, we can learn about past ocean conditions—thousands to millions of years in the past. Scientists studying changes in the species composition of biogenic oozes, for example, draw inferences about how productive the oceans were in the past, and relate these to prob-able past climates.

FIGURE A A benthic grab sampler being retrieved aboard a research vessel. The two halves of the grab descend to the bot-tom in the open position; upon arriving at the bottom, the halves are released and close around a sample of benthic material, sealing the sample against being washed away on the way back. These samplers are used to collect the very top layers of sediments, including the marine organisms living in and on that sediment.

BOX 3F   Methods of Studying Marine Sediments *(continued)*

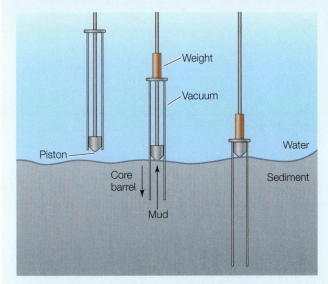

FIGURE B A piston core works by sealing the top of the core tube with a plunger (a piston) which is held in place relative to the bottom while the tube itself penetrates downward into the sediment. The vacuum seal created by the piston allows the core to penetrate and collect a sample much deeper than would otherwise be possible. (Without a piston, the friction inside the core tube would build until the core just pushed sediment in front of it as it penetrated.)

FIGURE C (Left) The *Glomar Challenger*. (Center) The *JOIDES Resolution*. (Right) Scientists aboard the *JOIDES Resolution* process a core.

Chapter Summary

- One of the eight scientific revolutions since Copernicus in 1543 is the acceptance of the theory of **continental drift**. The early evolution of various ideas involved in the development of the theory is instructive in understanding fundamental concepts of marine geology.

- New world maps showing the apparent fit between the coastlines of Africa and South America began to appear about 50 years after Columbus; these stimulated the first ideas about continental drift.

- In the nineteenth and early twentieth centuries, geologist **Eduard Suess** promoted ideas related to

continental drift, including reports of identical fossils on three widely separated continents.

- **Alfred Wegener** promoted the theory of continental drift in his 1915 book, *The Origins of Continents and Oceans*. He analyzed disparate lines of evidence, including Suess's report of fossil distributions, matching geological features on opposite sides of the Atlantic Ocean, the orientation of glacial scratch marks on rocks on different continents, and the discovery of coal in Antarctica, among his other detailed analyses.

- The sharpest criticism of Wegener's theory was that he did not provide a convincing explanation for the mechanism driving continental drift. He died in 1930, before the acceptance of his theory.

- In the late 1960s, continental drift was finally accepted based on direct lines of evidence, including the **Pacific Ring of Fire**, the locations of **mid-ocean ridge** systems, the thicknesses of bottom sediments, the **paleomagnetic properties** of the ocean floor, and seismic activity between spreading ridges along **transform faults**, to list but a few.

- The Pacific Ring of Fire is the pattern of volcanoes and earthquakes around the rim of the Pacific Ocean that results from interactions between tectonic plates.

- Mid-ocean ridge systems, revealed by echo sounders after 1925, are features where new ocean floor is formed as volcanic magma rises to the surface and solidifies into basaltic rock.

- Bottom sediments are both thicker and older away from the mid-ocean ridge systems.

- The paleomagnetic properties of the ocean floor record the Earth's periodic magnetic pole reversals, which provides strong evidence that new ocean floor was being created in the past and probably continues today along the mid-ocean ridges.

- Seismic activity between spreading ridges along transform faults are a direct result of seafloor spreading that is happening today.

- Geological features such as **seamounts** and **guyots** are the result of ancient volcanoes that settled back into the **asthenosphere**; similarly, coral **atolls** form around subsiding volcanoes. **Hot spots** in the mantle allow magma to make its way toward the surface through cracks and fissures in the crust, in some cases producing volcanic islands (most hot spots, however, are beneath continental plates).

- Both **Arthur Holmes** in the 1940s and **Harry Hess** in the early 1960s theorized that the continents might be moving atop convection currents in a plastic-like and flowing mantle. Today, we know that there are a dozen or so major **lithospheric plates**, with many smaller plates, that drift about on the surface of the Earth. The force that drives that motion is related to convection currents in the mantle, as well as the push-pull effects that result from plate formation at the mid-ocean ridges and plate destruction in **subduction zones**.

- Carried with these moving plates are the **marine sediments**, which are loose unconsolidated particles of organic and inorganic material. Terrigenous and biogenic sediments make up more than 99% of marine sediments. Terrigenous sediments are products of weathering of the continents, delivered to the sea by rivers, streams, and wind. Biogenic sediments are skeletal remains of dead marine organisms.

- Sediments are the timekeepers of the oceans. They record and store important clues to locations of energy reserves as well as to changes in the world climate in times past.

Discussion Questions

1. What were some of the early ideas of continental drift based on? Who were some of the key players, and what were their ideas? What evidence did these early characters have to support their ideas?

2. What was some of the evidence put forth by Alfred Wegener in support of continental drift? What were some of the holes in his theory? Why were his ideas unacceptable to many earth scientists of his time?

3. Why was the Pacific Ring of Fire important in the development of ideas about plate tectonics and continental drift? Why are there more earthquakes on the west coast of North America than on the east coast? Why are there transform faults? How are they formed?

4. What is paleomagnetism? What does it have to do with seafloor spreading?

5. How old are the oldest rocks on Earth? Where are they found? How old is the oldest ocean floor (the basaltic rock that it is made of)?

6. Where are marine sediments thickest? Why?

7. Why was it odd to have discovered fossils of the same species on opposite sides of the Atlantic Ocean and on so many different continents? How can that discovery be reconciled with Darwin's theory of evolution?

8. What happens when two oceanic plates collide? What happens when two continental plates collide? What happens when a continental plate and an oceanic plate collide?

9. Why are there major coastal mountain ranges on the western edges of North and South America?

10. Why was Wegener's failure to explain the mechanism behind continental drift, the forces that push and pull the continents, given such importance by his critics? Do you think that their criticism was valid? Can you think of examples of other scientific ideas and theories that have been similarly criticized in the past? Can you think of any current scientific ideas and theories that are similarly criticized?

Further Reading

Dobbs, David, 2005. *Reef Madness: Charles Darwin, Alexander Agassiz, and the Meaning of Coral*. New York: Pantheon Press.

Oreskes, Naomi, 1999. *The Rejection of Continental Drift: Theory and Method in American Earth Science*. New York: Oxford University Press.

Stewart, John A., 1990. *Drifting Continents and Colliding Paradigms: Perspectives on the Geoscience Revolution*. Bloomington: Indiana University Press.

Sverdrup, Harold U., Johnson, Martin W., and Fleming, Richard H., 1942. *The Oceans: Their Physics, Chemistry and General Biology*. New York: Prentice-Hall.

Wegener, Alfred, 1927. *The Origins of Continents and Oceans*, 4th ed., trans. John Biram, repr. 1966. New York: Dover Publications.

CHAPTER **4**

Contents

These icebergs and loose ice floes in the Southern Ocean around Antarctica illustrate several important properties of water: without these properties ice would not float, water would not freeze, and this photograph would never have been shot.

Water: Its Chemical and Physical Properties

The single most important aspect of our study of the oceans and the life forms they support is the nature of the medium in which it all happens: water. Having reviewed a little history of oceanography and marine biology and the early formation of the Earth and its oceans, we now move on to a discussion of water itself. But in order to appreciate just how important and unique the chemical and physical properties of water are—properties that underlie some of the peculiar phenomena you see every day but which you probably never thought much about—we first need to understand and appreciate the nature of chemical bonds that involve the water molecule. To do that, we need to review a little basic chemistry with respect to *atoms, molecules*, and *chemical bonds*.

Basic Chemistry: Chemical Bonds

Standard dictionaries define an en **element** as a substance that cannot be broken apart into different substances by ordinary chemical methods; the simplest form—smallest piece—of an element that still retains the properties of that element is an **atom**. A **molecule** is two or more atoms of one element or of different elements combined to form another substance. Atoms are composed of even smaller subatomic particles: a **nucleus** that contains **protons** and (usually, but not always) **neutrons**, and **electrons** that move about in the space surrounding the nucleus. The protons and electrons have an **electric charge**, but the neutron does not, hence the name; protons have a positive (+) charge and electrons have a negative (–) charge.

Atoms can combine with one another by forming chemical bonds to create even bigger and heavier molecules, which in turn take on chemical and physical properties different from their component elements. Of the three kinds of chemical bonds we need to be familiar with—covalent, ionic, and hydrogen bonds—the first two are the result of a gain, loss, or sharing of electron(s) between and among atoms. These gains, losses, and sharings are based on the "need" for charge neutrality in an atom, and for a specific arrangement of the atom's electrons.

One way to depict atoms is with the pre-1920s Bohr model. Normally, we do not use this pre-quantum mechanical model of the atom anymore because it shows electrons orbiting the nucleus the way planets orbit the sun. Instead, we now know that the electrons are moving all about the nucleus, with a probability of being located anywhere within an electron cloud (see Figure 4.3, left). Nonetheless, the older model remains useful in illustrating different arrangements of subatomic particles among the various elements. One of the simplest atoms, *helium*, is illustrated here using the Bohr model (**FIGURE 4.1**). It has two electrons, two protons, and two neutrons, shown here with its two electrons depicted as orbiting in well-defined circles around the nucleus. To put the dimensions of an atom and its subatomic particles in perspective, if

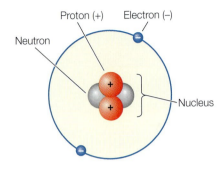

Neutron

Proton (+)   Electron (−)

Nucleus

FIGURE 4.1 The "old" Bohr model of a helium atom with electrons in orbit around the nucleus; the space occupied by the nucleus is greatly exaggerated.

the nucleus were the size of a penny, the electrons would be about 100–200 *meters* away.

Before we discuss details of chemical bonds, we need to step back a bit and consider more closely the element *hydrogen*. Hydrogen has one proton and one electron flying around the nucleus, but it does not usually have a neutron.[1] Hydrogen is the first in a series of all the known elements, a subsample of which is indicated in **FIGURE 4.2** (using the older Bohr model of the atom for convenience). As the number of electrons, protons, and neutrons, increases, the atoms get bigger and heavier.[2]

Notice in Figure 4.2 that the positive (+) or negative (−) charges of the nucleus and electrons tend to balance, or cancel each other. For example, in hydrogen, one proton (+1) and one electron (−1) balance: $(+1) + (-1) = 0$. Also, notice that atoms tend to fill what are known as their orbital shells with electrons in an orderly, prescribed fashion that is a law of nature, which dictates that to build a stable atom there should be two electrons in the first orbital shell, eight electrons in the second, eight in the third, and so on (except that the electron configurations become more complicated for the larger elements). In compliance with this condition, the hydrogen atom, which has only one electron, is short one; it "wants" one more electron, even though acquiring another would make it unbalanced with respect to charge (**BOX 4A**). It can accommodate this

---

[1] When hydrogen does have both a proton and one neutron in its nucleus, it is called *deuterium* (and when the hydrogen atoms in water are in the isotopic form of deuterium, it is called "heavy water"). If the hydrogen atom has one proton and two neutrons, it is called *tritium*. Occasionally you will see ordinary hydrogen called *protium*.

[2] An element's *atomic number* is equal to the number of protons in its nucleus; its *atomic weight* is equal to the average weight of all that element's isotopes found in nature, but it is approximately equal to the number of protons and neutrons.

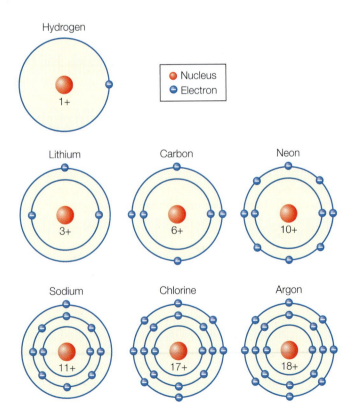

FIGURE 4.2 The Bohr atomic models of selected atoms, illustrating the progression in the numbers of electrons, protons, and neutrons.

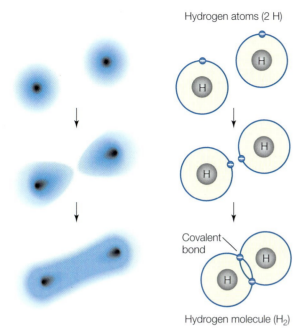

Hydrogen atoms (2 H)

Covalent bond

Hydrogen molecule (H$_2$)

FIGURE 4.3 A covalent bond between two hydrogen atoms results in the formation of hydrogen gas. The modern atomic view with electron clouds is on the left, and the diagrammatic Bohr model is on the right. Each atom shares its sole electron with the other, so that each nucleus has the requisite two electrons in its inner electron shell.

need by *sharing* an electron with another hydrogen atom, and in so doing it makes a molecule, H$_2$ or *hydrogen gas*. This sharing of an electron by the nuclei of two atoms to make a molecule is a very strong chemical bond—a **covalent bond** (**FIGURE 4.3**).

Another important example of a molecule held together with covalent bonds is the water molecule (H$_2$O), in which two hydrogen atoms and one oxygen atom are bound together. The oxygen atom is more complicated and quite a bit bigger than the hydrogen atom. The nucleus of oxygen has eight protons and eight neutrons, and it has eight electrons swirling around it (**FIGURE 4.4**). While the oxygen atom is balanced with respect to its electrical charge (eight protons and eight electrons), it is nonetheless short two electrons in its second orbital shell. One way that oxygen can get the two additional electrons is by

FIGURE 4.4 The Bohr model of the oxygen atom, with two electrons in the inner shell and six in the second shell. Though its electrical charge is balanced, the outer shell has only six electrons, making it relatively eager to form bonds with other molecules.

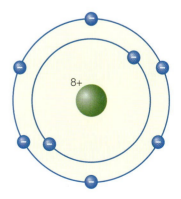

BOX 4A   Atoms and Common Sense

You may be wondering: Why do atoms effectively "want" two electrons in the first orbital shell, and eight in the second, and so on?

The short answer is that the explanation is rooted in *the theory of quantum mechanics*, which is far beyond the scope of this book. But, interestingly, the late Richard Feynman of Caltech, Nobel Prize winner in Physics and one of the key scientists behind refinements in the theory, admitted that even he did not really understand it. He wrote: "One [has] to lose one's common sense in order to perceive what [is] happening at the atomic level," and he called quantum me-

chanics an "uncommon-sensy theory" to explain the "cockeyed" behavior of electrons and matter. He told his students: "No, you're not going to understand it. ...That's because *I* don't understand it. Nobody does."

There is some comfort in Feynman's humble admission, isn't there?

8+

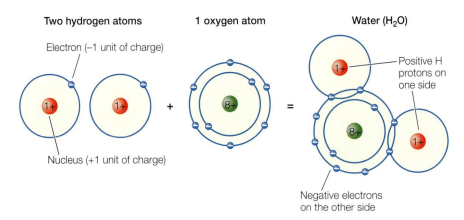

Two hydrogen atoms   1 oxygen atom   Water (H_2O)

Electron (–1 unit of charge)

Nucleus (+1 unit of charge)

Positive H protons on one side

Negative electrons on the other side

FIGURE 4.5 The water molecule is made up of two hydrogen atoms, each in need of one more electron in its first shell, and an oxygen atom, which is in need of two more electrons in its second shell. These atoms satisfy their needs by forming two covalent bonds between the oxygen and hydrogen atoms. In doing so, the two hydrogen atoms align themselves with oxygen such that the three atoms form an angle of approximately 105°. This creates a dipole—a separation of positive and negative charges.

sharing electrons with two hydrogen atoms, forming covalent bonds with each, thus creating a **water molecule**, or **H_2O** (**FIGURE 4.5**).

Some elements do more than share electrons; they lose them or acquire them. When this happens, we have an **ionic bond**, such as that between sodium (Na) and chlorine (Cl). Notice in Figure 4.2 that both chlorine and sodium have balanced charges, but their electron orbital configurations are off. Chlorine has only seven electrons in its outer shell; it is short one. Sodium, on the other hand, has a single electron in its outer shell, which is unstable; it "wants" to either gain seven more electrons or lose the one it has. When the two elements are allowed to react with one another, the chlorine acquires the "extra" electron from sodium, and both atoms become charged ions (Cl^- and Na^+). The oppositely charged ions will be attracted to one another and will form a crystalline lattice of atoms held together by ionic bonds, in a one-to-one ratio of chlorine and sodium. This crystal is ordinary table salt, NaCl. When salt is placed in water, though, the ionic bonds are broken (i.e., the salt dissolves) because of an unusual characteristic of the water molecule. So, back to the water molecule for a moment.

The sharing of electrons between hydrogen atoms and an oxygen atom in covalent bonds is a bit more complicated than Figure 4.5 would suggest. Since the water molecule has met its need for a balanced electrical charge and it has properly filled electron shells, it would seem to be a balanced and stable molecule; but it turns out that the two hydrogen atoms attach themselves asymmetrically to the oxygen. The Bohr model of the water molecule in **FIGURE 4.6** clearly shows this orientation of the two hydrogen atoms with respect to the oxygen. There are two important points to notice: (1) the two hydrogen atoms are not balanced on opposite sides of the oxygen atom; and (2) the angle between them is about 105° (the exact angle is nearer to 104.5°).

It is important to emphasize here that it is because of this angle that water is such a miraculous liquid compound, with many unique properties. That angle is responsible for another type of chemical bond, the hydrogen bond.

The hydrogen bond involves atoms that have had the position of their electrons altered. In the water molecule, both of the hydrogen

FIGURE 4.6 Two ways to diagram the water molecule, the first based on the Bohr model and the second on the space-filling model. The angle between the oxygen and hydrogen atoms is approximately 105°.

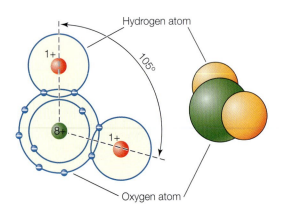

Hydrogen atom

105°

Oxygen atom

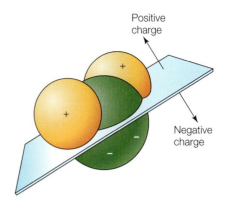

FIGURE 4.7 The water molecule's dipole moment is created by its charge imbalance, with hydrogen atoms exposing their positively charged protons and the oxygen atom, which is surrounded by its own six electrons plus the two hydrogen atoms' electrons, presenting a negative charge.

atoms' electrons are pulled toward the oxygen atom, so their protons are "exposed," effectively sticking outward (**FIGURE 4.7**). Thus, each of the hydrogen atoms has a net positive (+) charge. The oxygen atom is surrounded by its six electrons plus the two hydrogen electrons, which gives it a net negative (−) charge. As a result, the water molecule as a whole has a positive charge on the side with the two hydrogen atoms, and a negative charge on the other. This makes it a **polar molecule** (the molecule is said to possess a *dipole moment*). And because *opposites attract*, in the case of electrical charges, the charged ends of each water molecule will attract, and be attracted to, the oppositely charged ends of other water molecules. This electrostatic attraction of water molecules to one another is known as a **hydrogen bond** (**FIGURE 4.8**).

The hydrogen bond is very weak compared with the covalent bond, which is about ten times as strong, but it is nonetheless a very significant bond that imparts to water a whole host of very important physical properties (which we'll discuss shortly). Because of their weak nature, hydrogen bonds have a tendency to break apart and re-form all the time, to an extent controlled by temperature; this means that there are more hydrogen bonds between water molecules in cold water than there are in warm water. This intermittent nature, so to speak, of hydrogen bonds creates interesting molecular structures.

Hydrogen bonds between water molecules can form elongated strings of molecules and **tetrahedrons**, configurations of five molecules bound together by four hydrogen bonds (**FIGURE 4.9A**). When these strings and tetrahedrons wrap around each other, they may form hexagonal **clusters** (**FIGURE 4.9B**). Notice how the 105° angle controls the 3-dimensional geometry of the clusters, which begin to form a hexagonal, crystal-like structure. When conditions are right (when the temperature is cold enough), *all* the molecules are clustered together, and water becomes a solid, ice. This phenomenon whereby hydrogen bonds hold water molecules together—loosely in warm water, more tightly in cold water, and completely in ice—is called **cohesion**. It is precisely because

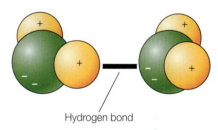

Hydrogen bond

FIGURE 4.8 The hydrogen bond between two water molecules is the result of an electrostatic attraction between the oppositely charged ends of the two molecules.

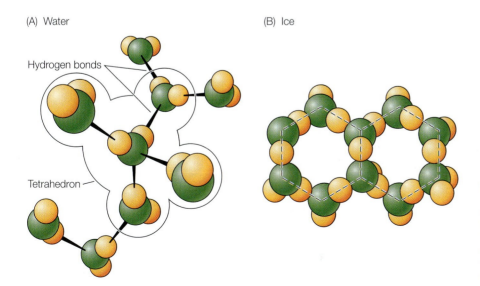

(A) Water

Hydrogen bonds

Tetrahedron

(B) Ice

FIGURE 4.9 Hydrogen bonds can form strings of water molecules. Tetrahedral bonding occurs when four hydrogen bonds form among five water molecules. When the hydrogen bonding is complete (i.e., all molecules are bonded), as in ice, they create complete hexagonal clusters of water molecules.

of this property of cohesion, and how water varies with temperature (and, as we'll see later, with the addition of salt), that water has so many unique physical and chemical properties.

Physical-Chemical Properties of Water

Water's high surface tension

Beneath the surface of a volume of water, that is, beneath the air–water interface, an individual molecule of water is surrounded by other water molecules with which to form hydrogen bonds. But the water molecules directly at the air–water interface have nothing above them in the air to bond with, so they can be thought of as stretching back onto one another and with others beneath. The tension created among those molecules at the air–water interface produces a skin-like surface, which is water's **surface tension** (**FIGURE 4.10**). It is this surface "skin" holding water molecules together that accounts for phenomena we see every day: we can "round over" a cup of water, filling the cup above the rim without spilling, and a drop of water can hang precariously below a faucet nozzle (**FIGURE 4.11**). In each case, a volume of water is held together by surface tension. It is also surface tension, pulling the surface of the water together, that produces the shiny surface appearance we often see on the ocean (or a lake or pond) when the wind is calm. And it is surface tension that allows water striders literally to walk on water. None of these things can happen with nonpolar molecules. Adding *surfactants*, such as soap or detergent, to water will destroy the surface tension by disrupting the hydrogen bonds that held the water molecules tightly together before.

> Water's surface tension lets you know when fake gasoline is being used in movies and on television: sometimes it just doesn't look right. That's because they're using water (which, of course, is a lot safer). When sprayed or splashed, water forms relatively large drops because of its high surface tension. But gasoline has very low surface tension, and instead of forming drops, it tends to fly apart in a mist-like, potentially explosive spray.

Water's great dissolving power

Water's polar molecule also enables it to interact and interfere with other molecules, especially those that have ionic bonds, because ionic molecules are similarly constructed—with offset electrical charges. This is why water is said to have great dissolving power. The ex-

FIGURE 4.10 (A) Molecules at the air–water interface have nothing above them to bond with and therefore can bond only with molecules beside them at the surface and molecules beneath them. (B) The red arrows represent hydrogen bonds between water molecules below the surface; these bonds "pull" the molecules inward toward one another. The black arrows represent water molecules at the surface; their net pull is be downward on the surface of the air–water interface. It is this downward pull that is surface tension.

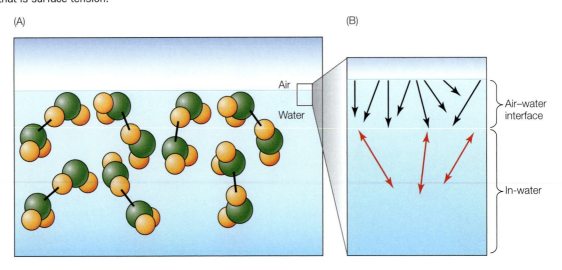

(A)

(B)

(C)

(D)

FIGURE 4.11 Examples of water's surface tension. (A) The "rounding over" of a cup of water that has been filled above the rim. (B) Surface tension holding a drop of water below the nozzle of a faucet. (C) Water striders, which can literally "walk on water." (D) The shiny, mirror-like effect of surface tension at the air–water interface.

ample already alluded to is table salt. When salt crystals are added to water, the individual sodium (Na⁺) and chloride (Cl⁻) ions in the NaCl crystalline lattice become surrounded by water molecules; water's oppositely charged ends creep in and around and surround the Na⁺ and Cl⁻ ions. The positive sodium ion (Na⁺) is attracted to the negative end of the water molecule, and chloride (Cl⁻) to the positive end, freeing them from the crystalline lattice; the salt is now "*in solution*" (**FIGURE 4.12**). This dissolving effect of water works only on polar (dipole) molecules, such as salt; it does not work on nonpolar molecules, such as oils. For this reason, oil and water don't mix. (Other compounds soluble in water, such as sugars, are molecular compounds that dissolve without dissociating into ions.)

Besides having great dissolving power and high surface tension, water is one of very few chemical compounds that is a liquid at room temperature.

Water's liquid form

A simple fact that most of us take for granted is that water is a liquid at temperatures normally encountered on Earth. But this is not a common property of hydrogen compounds. Among the hydrogen compounds listed in **TABLE 4.1**, only hydrogen fluoride (HF) is a liquid at room temperature (or almost room temperature—it boils at 19°C, or about 68°F).

TABLE 4.1	Melting and boiling temperatures for several common hydrogen compounds, including water				
	Methane (CH₄)	Ammonia (NH₃)	Water (H₂O)	Hydrogen fluoride (HF)	Hydrogen sulfide (H₂S)
Melting point (°C)	–183	–78	0	–83	–83
Boiling point (°C)	–161	–33	100	19	–60

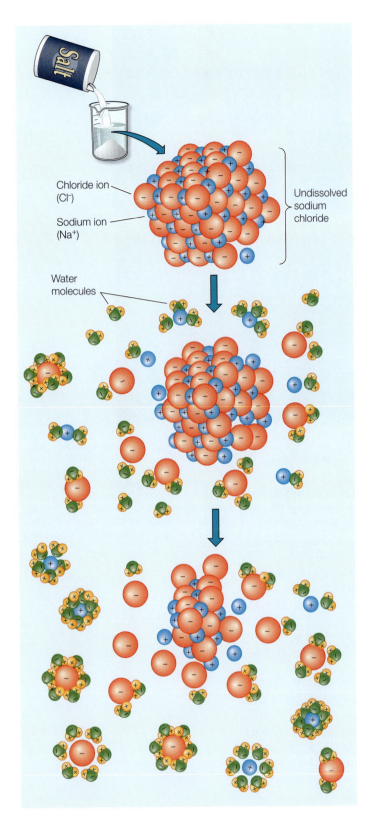

Chloride ion (Cl⁻)

Sodium ion (Na⁺)

Undissolved sodium chloride

Water molecules

FIGURE 4.12 A sequence of salt dissolving in water. When salt is mixed with water, the Na⁺ and Cl⁻ ions become enveloped by water molecules, with anywhere from one to several water molecules oriented such that the negatively charged oxygen of the water molecule (green) is adjacent to the positively charged Na⁺ ion (blue), and the positively charged hydrogen atoms (yellow) on the water molecules are oriented adjacent to the negatively charged Cl⁻ ions (red). The diagram is a snapshot of what might be seen if we could take such a picture, which is actually impossible. The molecules are all moving about at extremely high speeds and are in contact with the Na⁺ and Cl⁻ ions only for a tiny fraction of a second (on the order of one trillionth of a second).

Water's high heat capacity

Water's high **heat capacity** (or **specific heat**) is one of the most important properties that result from its hydrogen bonds. Having high heat capacity means, for example, that for a given volume of water, say a teakettle-full, you need to keep it on a source of heat, such as a stove, for a relatively long period of time to increase its temperature from room temperature to a point hot enough to brew a cup of tea. Substances with lower heat capacities would heat up faster. We can demonstrate this with a simple thought experiment, as shown in **FIGURE 4.13**, where we apply heat to different liquids and measure the change in temperature; we find that liquids such as gasoline, alcohol, vegetable oil, and turpentine, to list but a few, all heat up approximately twice as fast as water does.[3] This thought experiment brings up an important point: we need to distinguish between the terms **heat** and **temperature**. To do so we need to remember that atoms and molecules vibrate. This vibration is least energetic for solids, more energetic for liquids, and most energetic for gases. For any one of these phases (solid, liquid, gas), the vibrations increase with increasing temperature; therefore, temperature is a measure of the *speed* of motion of atoms and molecules. (The actual speeds, by the way, are quite impressive for such tiny objects—they range from about 1000 to 2000 miles per hour!)

We routinely measure temperature in degrees (either °C or °F). To increase the temperature of something, we apply heat, of course. But, heat, unlike temperature, is a measure of energy; thus we can't just measure temperature—we need a measure that includes an accounting of *how many* molecules are moving within a given volume. For example: a liter of hot air does not have as much heat as a liter of water at the same temperature; there are more molecules in a liter of water than there are in a liter of air. Also,

---

[3] Actually performing this experiment—placing highly flammable chemicals on a stove—is, of course, not advisable. Such is the value of thought experiments.

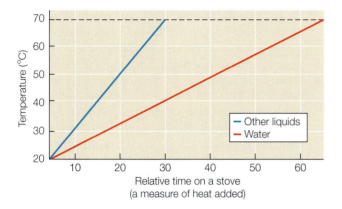

FIGURE 4.13 Results of a thought experiment in which we place a pan of water on a stove burner and measure the amount of time it takes the water to rise in temperature from 20°C to 70°C. Then we place equal volumes of other liquids in the pan and repeat the measurements. The other liquids reach 70° much faster than water does. This is because water has high heat capacity; it needs to absorb more heat in order to reach a given temperature than other liquids.

according to the same reasoning, the bottle of air would change temperature faster than the one with water. We can measure the change in temperature of the heated substance, but to measure the *amount of heat added*, we need to apply a standard, one in terms of the amount of energy required to heat a volume of water to a certain temperature. The unit of measure that we normally use for heat is the **calorie**.

The thought experiment with the two liquids on a stove demonstrates the greater heat capacity of water compared with the other liquids. In fact, water has the highest heat capacity of all but one other compound (ammonia, NH_3). Our climate would be very different if the oceans—water—did not have this capacity to absorb and hold heat from the sun. So before continuing with the basic physical properties of water, it is instructive at this point to digress for a moment and discuss in some detail the importance of water's high heat capacity on weather and climate, setting the stage for our later discussion of atmospheric circulation and ocean currents.

- The heat capacity (or specific heat) of a substance is a measure of the amount of heat (energy) required to raise the temperature of 1 gram of that substance by 1°C.
- The energy required to raise the temperature of 1 gram of water by 1°C is 1 calorie.
- Therefore, the specific heat of water is 1.0 (calorie per gram per °C).

WATER'S HIGH HEAT CAPACITY AFFECTS WEATHER AND CLIMATE   Water's high heat capacity is extremely important in the oceans' regulation of Earth's weather and climate, as we can illustrate with our earlier thought experiment. First, let's put the water back on the stove, record what happens, and then place a rock on the stove. We will see that the rock gets warmer significantly faster than does the water. In the case of granite, it will warm up five times as fast as water (the specific heat of granite is 0.2, while that of water is 1.0). And when we take the rock off the stove, it will cool off faster, as well. Quite simply, water has significantly greater heat capacity than rock, and it is precisely because water has the capacity to hold a lot of heat that the oceans are extremely important in regulating, moderating, and driving our weather and climate. Because the heat capacity of the oceans (water) is much greater than that of the continents (rocks), the oceans take longer to heat up under the influence of the sun's rays than does land, and the oceans retain that heat; they "hold heat" longer than land.

Satellite measurements of day–night temperature differences at the Earth's surface over a 24-hour period show that the day–night difference is much greater for land than for the oceans. While the oceans barely change temperature at all over the 24-hour period, the land changes by as much as 30°C over desert regions. The desert sands of Africa, for example, and the air above them may get very warm during the day, but during the night, they cool off considerably.

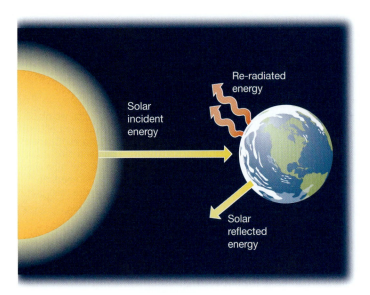

FIGURE 4.14 The components of Earth's "radiative heat budget" shows incident solar energy (sunlight) striking the surface of the Earth. Some of that solar energy is reflected back into space, especially by clouds, and some is absorbed by the Earth. Of the energy that is absorbed, some is re-radiated back into space as longer wavelength (infrared) light. The incoming and outgoing radiation are, averaged over the entire planet, equal; otherwise the Earth would either get progressively colder or warmer over time.

On the other hand, the oceans' surface waters take longer to heat up, so the daytime is not long enough for them to get very warm; by the same logic, they do not cool off much at night. This important difference in the heat capacities of land and water makes the ocean–land system a huge heat engine (more about this later), and it raises an interesting question: If the continents gain heat by day (and their temperatures rise), but they cool off again each night, where does all that heat go? To answer this we need to consider the Earth's heat budget.

EARTH'S HEAT BUDGET   Of the sun's energy (sunlight) that arrives at the Earth's surface, some is reflected off the atmosphere, off clouds, and off the surface of the Earth itself back into space, and some is absorbed as heat, making the Earth warmer. But when the surface of the Earth and its atmosphere absorb the sun's rays and become warmer, they emit heat as well. In fact, anything warmer than absolute zero (0 on the Kelvin scale, or –273°C) emits heat, or thermal energy. That thermal energy re-radiates from Earth back to space. Therefore, thermal energy is both reflected and re-radiated from Earth back to space, but the two forms of energy have different wavelengths (**FIGURE 4.14**). Reflected light energy is **shortwave** radiation, which is visible to our eyes (this, of course, is how we can look back at Earth from space and see it); on the other hand, re-radiated light energy is infrared (IR, also called **longwave** radiation), which is invisible to us. But that invisible longwave radiation can be sensed and measured by special IR-sensing cameras, such as those used by fire departments to "see through smoke." You can even sense this longwave radiation yourself, even though you can't see it; that is, you can feel your warm hand radiate heat if you hold it close to your face.

If the Earth's radiation budget is in balance—and it is—then on average each year the surface of the Earth should be getting neither warmer nor colder (we'll consider an exception, the evidence for global warming, later). This balance can be seen and measured with satellites, showing the intensity of shortwave radiation reflected back to space and longwave re-radiation. Notice that most of the reflection is off the continents (**FIGURE 4.15A**), which is direct evidence that they do not absorb as much of the sun's radiant energy as the oceans do. And, of course, because the oceans absorb more solar radiation than land does, most of the heat that is re-radiated back to space from planet Earth is from the

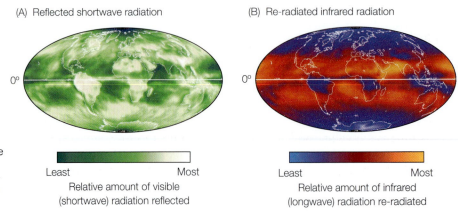

FIGURE 4.15 Satellite images of radiation from Earth for the month of March, 2000. (A) Shortwave radiation (i.e., visible light) reflected off the Earth's surface and back to space. Dark areas indicate absorption. (B) Longwave radiation (infrared light) re-radiating from the surface of the Earth.

(A)

(B)

oceans. Interestingly, much of that re-radiated longwave radiation emits from two regions, one above and one below the Equator (**FIGURE 4.15B**). These two regions would otherwise be a single large tropical region of high re-radiation if it were not for a persistent band of cloud cover, generally located on and just north of the Equator, which blocks some of the re-radiation there. This Equatorial cloud band is clearly visible just north of the Equator in satellite images (**FIGURE 4.16**; we will discuss these clouds in more detail later).

The net result of this heat absorption and re-radiation back to space is an annual balance for the globe as a whole. But within a given year, there are important differences between the heat fluxes of the land and the oceans that result in seasonal patterns. Because of water's high heat capacity, the oceans retain heat more effectively than land; they absorb and store heat energy over the summer, holding that heat longer than the land, and then re-radiate it during the winter. It is because of this phenomenon that the oceans are important in moderating the climate of coastal regions. That is, coastal areas are cooler in summer because they benefit from the ocean's heat absorption; in summer, the ocean absorbs heat from the air, making the air above the oceans cooler, which helps to cool the adjacent land. In winter, the opposite happens: re-radiation of heat from the ocean warms the air above it which also helps warm the adjacent land. This is all because of water's high heat capacity, which in turn is a result of its hydrogen bonds. As it turns out, water is even more interesting when we consider what happens when it changes phase.

FIGURE 4.16 Satellite images of visible light received by GOES (Geostationary Operational Environmental Satellites) satellites, showing clearly the cloud patterns on November 28, 2011. (A) GOES West is over the Pacific Ocean; (B) GOES East is over South America.

Water's phases: Liquid, solid, and gas

Matter generally exists in one of three possible states, or phases: solid, liquid, or gas. For most substances, phase transitions are largely a function of temperature, but the situation is less straightforward for water. Because the hydrogen bonds that hold water molecules together are affected by temperature, which controls the speed of their molecular vibration, strings and clusters of water molecules are forming and breaking all the time. At temperatures above the boiling point (>100°C), the molecules are freely flying around, with virtually all hydrogen bonds broken. At this point, water is in the form of water vapor, a gas. It is important to emphasize here that water vapor is invisible—and is not the same thing as steam or fog, which both consist of tiny droplets of liquid water suspended in

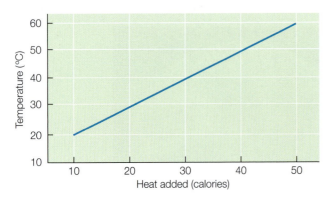

FIGURE 4.17 Results of a thought experiment in which we place one gram of water on a heat source and measure its change in temperature for each calorie of heat added. If we start with water at 20°C, we will see the temperature rise exactly one degree for every calorie of heat added.

air. At temperatures between the freezing and boiling points (0–100°C), water is liquid, with some hydrogen bonds holding the molecules together. At temperatures below the freezing point (<0°C), hydrogen bonding among all water molecules is complete, holding the water molecules tightly and making a solid (ice). But the transitions between these phase changes (from solid to liquid to gas) are not quite straightforward. In fact, it is because of the peculiarities of the phase changes in water that we use ice to cool our drinks and we sweat to cool our bodies. These peculiarities are due to water's high latent heat of melting and evaporation.

Water's latent heat of melting and evaporation

We already discussed water's high heat capacity, with reference to the amount of heat required to raise its temperature by some amount. But water is even more remarkable because of the heat involved in its phase changes. To explain, we need to recognize that there are two kinds of heat when it comes to water: **sensible heat** and **latent heat**. Sensible heat is just that—it can be sensed with a thermometer. It can be measured as a change in temperature. We demonstrated sensible heat in the simple thought experiment in which we heated two liquids (see Figure 4.13). Looking at that experiment again, this time focusing on just the water, and examining it a bit more closely (**FIGURE 4.17**), we see that to raise the temperature of 1 gram (g) of water from 20°C to 60°C (a 40°C change) requires an input of 40 calories. This relationship—1 degree per calorie—is simple and straightforward between the temperatures of 0° and 100°C. But the relationship falls apart when water changes state from a liquid to a gas (and vice versa) or from a liquid to a solid (and vice versa). The reason is water's latent heat.

The latent heat of water is not easily measured (the word *latent* means "hidden") because of the nature of water's hydrogen bonds as it changes phase. What happens is shown in **FIGURE 4.18**. Starting with ice and adding heat, we see that the temperature of the ice will increase as heat is added. For example, on a cold morning in some parts of North America, when the air temperature outdoors is well below 0°C, the ice outside will be about the same temperature as the air (e.g., well below 0°C). As the air warms up during the day, the ice will warm up as well, following the rule of 1°C for each calorie of heat added per gram of ice, but only up to a point. When the ice reaches the freezing point (0°C), it stops warming up—the temperature stops increasing even though heat is still being added. The ice holds steady at 0°C until enough heat has been added to break the hydrogen bonds holding the ice together. How much heat is required to break those bonds and convert ice to water? Close examination of the plot in Figure 4.18 reveals that at first, as heat is added, the temperature (the sensible heat of the ice) goes up. Next, the temperature stays at 0°C even though we are continuing to add heat, until *80 calories* per gram of ice have been added. This 80 calories is the **latent heat of melting**.

Once the ice melts, the liquid water will increase in temperature, from 0°C to 100°C, at the rate of 1°C per gram of water for every calorie of heat added. But, when the temperature gets to 100°C, the boiling point, where water changes state from a liquid to a gas, another

FIGURE 4.18 Results of an experiment in which we progressively add heat to a gram of ice, initially at –20°C, until we first melt the ice and then boil away the water. The curious plot of temperature versus heat input reveals the influence of the hydrogen bonds among the water molecules, which gives water its latent heat of melting and vaporization.

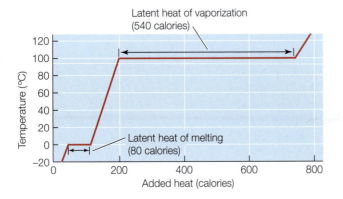

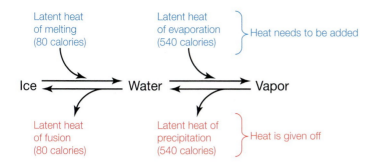

FIGURE 4.19 When water transitions from ice to water (liquid) and then to gas (water vapor), it *absorbs* latent heat at each step: 80 calories per gram of ice melted, and then 540 calories per gram of water evaporated. As water transitions in the opposite direction, from water vapor to liquid water and then to ice, it *releases* corresponding amounts of heat energy.

delay happens. Heat can be continually added once 100°C is reached, but the temperature does not increase. That energy is being used, not to increase the vibrations of the molecules, but to break hydrogen bonds. When those bonds are all broken, the water vaporizes. Notice that for this second phase change, from liquid to gas, the **latent heat of vaporization** is 540 calories per gram—much higher than the latent heat of melting, 80 calories per gram. This is why "a watched pot never boils"; when the water temperature reaches the boiling point, we don't see bubbling—the bubbles of water vapor that result when water changes phase from a liquid to a gas—right away. The pot will eventually boil, or course, but it does seem to take a long time, doesn't it?

Each of these phase changes can go either way, of course. That is, we can freeze (fuse) or melt water, and we can evaporate (vaporize) or precipitate (condense) water vapor. In the cases of precipitation and freezing, the latent heat is *released* by the water molecules to the surrounding atmosphere as the hydrogen bonds re-form (**FIGURE 4.19**).

It is difficult to overstate the importance of water's latent heat. The absorption and release of energy due to the breaking and re-forming of hydrogen bonds in water is happening all the time, and is not confined to water temperatures at 0°C and 100°C. This point becomes important when we explore some of the ramifications of water's latent heat, for which we can list a few examples:

- The latent heat of melting is why ice is so good at cooling drinks. If we just popped into our drinks a lump of plastic from the freezer that was at −18°C (0°F), it would quickly warm up and not absorb as much of the drink's heat. But water ice will absorb 80 calories per gram of ice that melts, as shown by the lower flat region in Figure 4.18.

- The latent heat of vaporization is why we sweat. Our sweat, which is water, absorbs the body's heat as it evaporates. Thus we lose body heat and we cool down. But we don't "boil" that water off our bodies, we lose water molecules almost one at a time, and as each one evaporates, it absorbs latent heat and cools our bodies. We lose 540 calories of body heat for every gram of water (sweat) that evaporates.

- Just as *evaporation* requires the input of heat, the reverse is also true: *precipitation* (or *condensation*) gives off heat. Thus, the atmosphere actually gets warmer when it rains, which is counter intuitive, because when it rains it certainly seems to get colder. One reason for this is because our bodies cool down when we stand out in the rain (precisely because of the point just made about sweating: we lose heat and get colder when we are wet because the rain water absorbs our body heat). Also, as the rain falls through the air, some of it evaporates before hitting the ground which absorbs heat, making the air temperature at ground level (as well as our wet bodies) colder. But in the rain clouds where rain is condensing from water vapor to liquid water, a lot of heat is released—540 calories for every gram of liquid water formed—which warms the atmosphere in those clouds.

■ Using the principle of latent heat, we can make a simple air conditioner by placing a wet cloth in front of an electric fan. As the water evaporates, it absorbs a lot of heat from the room air blown onto it, and the room becomes cooler. Again, 540 calories of room heat is absorbed for every gram of moisture evaporated from the wet cloth. (In fact, this is sometimes done in hot, dry climates. Of course, it would increase the humidity as the room is cooled. And there's the heat generated by the electric fan to consider as well.)

For all these reasons, water, with its ever-changing hydrogen bonds and its resulting latent heat and high heat capacity, has an enormous influence on our weather and climate. And as we'll see later, it is the release of latent heat that causes storms to intensify and is the source of energy that drives hurricanes.

Water's transparency to light

One of the most obvious, but nonetheless important properties of water, is that it is colorless. It is transparent to light in the visible range (at least, to *much* of the light in that range), which is the same range used by plants for photosynthesis. (We'll discuss more on light in the ocean later.) In short, we are lucky to have water on our planet; without it, plants could not convert solar energy into chemical energy, and life could not exist. And if it were to have even slightly different physical properties, our world would be a much different place.

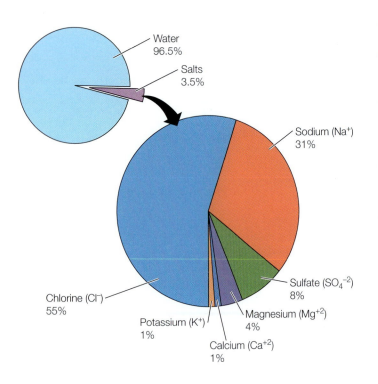

FIGURE 4.20 The six most common ions dissolved in seawater, the salts of which constitute more than 99.6% of the ocean's salinity.

Water
96.5%

Salts
3.5%

Sodium (Na⁺)
31%

Chlorine (Cl⁻)
55%

Potassium (K⁺)
1%

Calcium (Ca⁺²)
1%

Magnesium (Mg⁺²)
4%

Sulfate (SO₄⁻²)
8%

Salinity

One of water's important physical properties is its great dissolving power—it can dissolve a lot of materials, especially **salts**—chemical compounds in which a metal is bound with a nonmetal. This is, of course, why the sea is salty. We refer to the sea's saltiness as its **salinity**, which is defined as the total quantity, by weight, of dissolved inorganic solids in seawater. **FIGURE 4.20** gives a breakdown of the major salts dissolved in seawater. The salts of just six major ions make up 99.6% of the ocean's salinity; when not in solution in water, these compounds are normally found in solid, crystalline form.

So, just how salty is the sea? On average, the salinity of the world's oceans is about 3.5%. This means that these dissolved solids (salts, especially) make up about 1/35 the weight of seawater itself. This is a lot of salt! For example: if we wanted to make salt water in a tub of fresh water the size of an office desk (about 1 m × 1 m × 2 m, or 2 cubic meters [m³]), we would need to add 3.5% salt by weight. The volume of water in the tub, 2 m³, equals 2000 liters (L), and since one liter of water weighs 1000 g, the 2 m³ of water weighs 2000 kilograms (kg). To bring the salinity of that much water up to 3.5%, we would have to add 70 kg salt—that is, 154 pounds (lb), or 94 Morton's salt containers!

The salinity of seawater is nearly always between about 3.2% and 3.7%, but in oceanography, it is traditionally expressed not as a percentage (parts per hundred, or %), but in terms of parts per thousand (ppt, or ‰), or grams per kilogram (g/kg). So, 3.7% becomes 37‰, or 37 ppt.[4]

While the salts of only six major ions make up 99.6% of all the "dissolved solids" that give seawater its salinity, there are lots of other elements besides salts dissolved in seawater. **TABLE 4.2** gives you an idea of those elements in the remaining 0.4% (which includes gold, but there's not much).

TABLE 4.2	A partial list of the elements dissolved in seawater and their concentrations in milligrams per kilogram (mg/kg) of seawater		
Element	**mg/kg**	**Element**	**mg/kg**
Chlorine	18980	Copper	0.01
Sodium	10561	Zinc	0.005
Magnesium	1272	Lead	0.004
Sulpher	884	Selenium	0.004
Calcium	400	Cesium	0.002
Potassium	380	Uranium	0.0015
Bromine	65	Molybdenum	0.0005
Carbon	28	Thorium	0.0005
Strontium	13	Cerium	0.0004
Boron	4.6	Silver	0.0003
Silicon	4	Vanadium	0.0003
Fluorine	1.4	Lanthanum	0.0003
Nitrogen	0.7	Yttrium	0.0003
Aluminum	0.5	Nickel	0.0001
Rubidium	0.2	Scandium	0.00004
Lithium	0.1	Mercury	0.00003
Phosphorus	0.1	Gold	0.000006
Barium	0.05	Radium	<0.0000000001
Iodine	0.05	Cadmium	<0.0000000002
Arsenic	0.02	Chromium	<0.0000000003
Iron	0.02	Cobalt	<0.0000000004
Manganese	0.01	Tin	<0.0000000005

[4] From 1978 to 2011, the use of ppt, or ‰, as the unit of salinity fell from usage. During that time, the standard technique for measuring salinity of seawater was based on its electrical conductivity, which used a formula that left salinity unitless. Accurate to the third decimal place of salinity, electrical conductivity measures only charged forms (ions); it does not account for small concentrations of uncharged dissolved substances. More accurate measurements that account for all dissolved solids, charged and uncharged, are now required by scientists for a number of reasons, not least of which is a more accurate determination of ocean heat content. Today we refer to *absolute salinity*, which takes account of uncharged constituents and which is expressed in units of grams per kilogram, or parts per thousand. So, we are back where we started before 1978.

FIGURE 4.21 Diagram showing how river water is quickly diluted by mixing with seawater a short distance away from the point of entry.

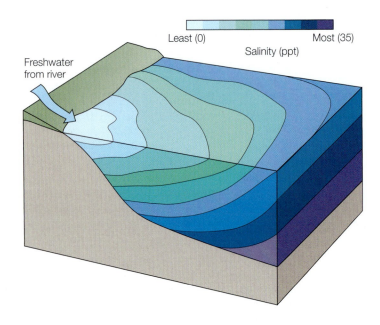

FIGURE 4.22 Contour plot of the average surface salinity off the coast of Brazil since 1950, showing the influence of the Amazon River. Notice that the Amazon's influence extends off the coast as far as 200 miles.

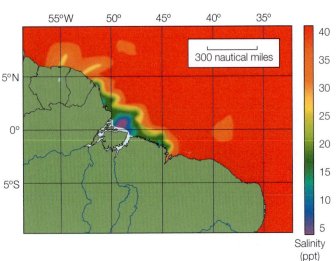

Distribution of salt in the sea

Salinity, the concentration of dissolved solids in seawater, is not the same everywhere in the oceans; it varies with location. Salinities range from 0 ppt (effectively no salt) in freshwater in some coastal locations, especially at the mouths of the world's great rivers, like the Amazon and the Mississippi, to 31–32 ppt in coastal waters (within a few tens of miles of the coast, where rivers have some influence) to as high as 37–41 ppt in some parts of the world ocean. But in the vast majority of the open oceans, the average salinity is about 35 ppt.

Salinity varies with location for a number of reasons. First, the addition of freshwater dilutes seawater and lowers its salinity. Sources of freshwater include: rain, which falls on the surface of the ocean, with regional precipitation differences around the globe; rivers, which inject freshwater at specific points (mouths of rivers); and groundwater. However, river waters—even from the world's greatest rivers, which discharge huge volumes of freshwater into the ocean—are quickly diluted by mixing with seawater. Only a few miles away from the coast, we usually see salinities that approach values similar to that of the open ocean, as diagrammed schematically in **FIGURE 4.21**. Even the huge volume of freshwater discharged from the world's largest river, the Amazon, is diluted relatively quickly, but its influence on the salinity of surface waters of the Western Equatorial Atlantic off Brazil extends as far as 200 miles offshore, where salinities are slightly lower than the oceans' average (**FIGURE 4.22**).

Thus, along the coast, salinity is controlled largely by the proximity of rivers. But in the open ocean, it is controlled by differences in evaporation and precipitation. Because evaporation removes only pure water, leaving the salt behind and increasing salinity, we might expect to find the highest salinities in the tropics and subtropics, right? Because these areas are the warmest on Earth, we would expect evaporation rates to be the highest there. This is partially correct: evaporation is highest in the *subtropics*, but it is quite a bit less in the tropics directly on the Equator. At the Equator, evaporated moisture in the atmosphere does not travel very far north or south; it falls back to the ocean as rain in roughly the same latitude from where it evaporated. In addition, low-level clouds formed by high rates of evaporation on either side of the Equator (see Figure 4.15B) tend to move toward the Equator (we'll see why in a later chapter), which means that it rains a lot at the Equator. It also means that because of all those clouds formed on the Equator, there is less direct sunlight, which leads to less evaporation. So, lots of rain and lower net evaporation rates are the rule directly on the Equator.

Most evaporation occurs at subtropical latitudes (around the Topics of Cancer and Capricorn, latitudes 23.5°N and 23.5°S), whereas there is a lot of precipitation in temperate latitudes. These patterns are illustrated in **FIGURE 4.23**. Some of the evaporated moisture from the subtropics moves to the Equator where it falls as rain, as just mentioned, but most of it falls as rain in temperate latitudes. So evaporation and precipitation vary with latitude and correspond to the patterns shown. Therefore, we can easily imagine that these global trends in evaporation/precipitation should affect the ocean's surface salinity in a predictable way—and they do. This pattern can be clearly seen in the map of annual average salinity in the surface waters of the world's oceans in **FIGURE 4.24**. Notice that high rates of evaporation produce high salinities in some locations, as just discussed. Extreme examples of salinity resulting from high rates of evaporation are found in the Eastern Mediterranean (>39 ppt) and the Red Sea (>41 ppt). Apart from the north–south salinity patterns, notice also in Figure 4.24 that the Atlantic is saltier than the Pacific (**BOX 4B**).

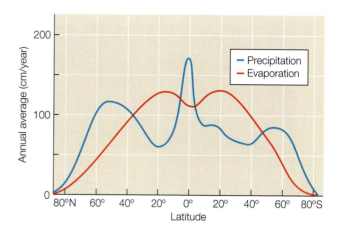

FIGURE 4.23 Plots of the annual average precipitation (rain and snow) and evaporation as functions of latitude. At the equator and in the temperate latitudes, salinity is lower because precipitation exceeds evaporation. Conversely, in the subtropics, evaporation exceeds precipitation, so salinity is higher.

Clearly, there is a lot of salt in the sea; so, where did it all come from? We've already shown that water is an excellent solvent—it can dissolve a lot of materials. So a fair question to ask is: Did all the salt in the sea come from material on land which became dissolved in freshwater and then got carried to the sea by rivers? The short answer is no; most of it did, but not all. If all the salt came only from rivers, we would expect to see in river water dilute amounts of the same elements, in the same relative proportions to one another, that we see in seawater, but that is not the case. In fact, some of the materials dissolved in seawater come from the Earth's deep interior—from the mid-ocean ridges as well as from undersea and terrestrial volcanoes. The undersea mid-ocean ridges are both a source as well as an ultimate repository for relatively small but significant amounts of dissolved solids; they are a source of manganese, iron, zinc, lithium, and cesium, and they are a sink for sulfate and magnesium. However, it is true that the majority of salt in the sea comes from weathering of the continents.

It is also fair to ask: If materials are being brought to the sea by rivers, groundwater, hydrothermal vents, and volcanos, then is the sea continuing to get saltier and saltier with time? Again, the answer is no. This is because the oceans have reached a **steady state** with regard to salinity. The amount of dissolved materials continually being added to the oceans has reached an equilibrium with the amount being lost from the oceans by way of chemical and biological precipitation processes and subsequent burial in sediments. Over the past several hundred million years the oceans have been at their current salinity, varying only slightly.

Determining salinity

Measuring the salinity of a sample of seawater is not as straightforward as it might seem. We can't simply allow a 1000-gram sample of seawater to dry up and then weigh the solid material left behind to get grams of dissolved solids

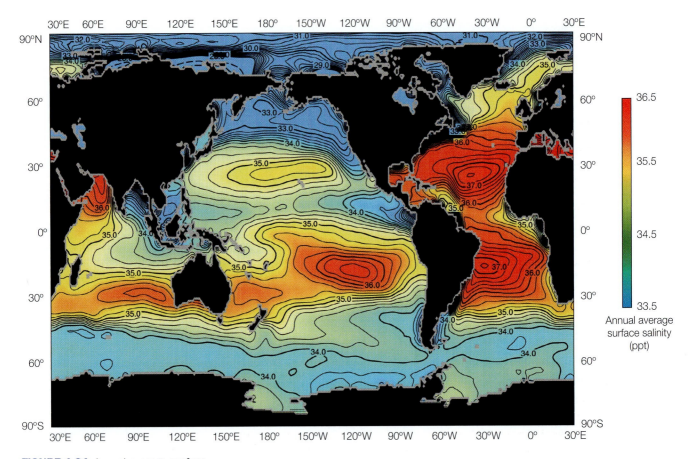

FIGURE 4.24 Annual average surface salinity of the world ocean.

per kilogram of seawater. That's because some dissolved solids in seawater are bound tightly to water, and upon drying, the seawater does not release all of its water molecules; rather, a thick, syrup-like solution—an azeotrope—is left behind. This is why, for example, it is difficult to dry out your bathing suit after swimming in the ocean; it dries much better if you first rinse it thoroughly in fresh water. Heating the seawater to assist in drying doesn't work either, because it tends to drive off gases of the materials we want to measure. So early measurements took advantage of what has been called the **Law of Constant Proportions**. This principle was based on the observation that while the total salinity of seawater may vary between about 32 and 37 ppt around the world, the proportions of the major elements in it stay the same. This finding was first published as early as 1819,[5] but it was not reliably confirmed until 1884, by William Dittmar. Dittmar analyzed the ionic composition of 77 water samples collected from around the world during the *Challenger* expedition. He

---

[5] Alexander Marcet wrote in 1819: "All the specimens of seawater which I have examined, however different in their strength, contain the same ingredients all over the world, these bearing very nearly the same proportions to each other; so that they differ only as to the total amount of their saline contents" ("On the Specific Gravity and Temperature of Sea Waters, in Different Parts of the Ocean, and in Particular Seas; With Some Account of Their Saline Contents," *Philosophical Transactions of the Royal Society of London*, 109: 161–208. p. 194). In 1864, Johan Georg Forchhammer, in his paper "On the Composition of Sea-Water in the Different Parts of the Ocean," published in the *Philosophical Transactions of the Royal Society of London* (vol. 155, pp. 203–262), analyzed hundreds of water samples and found that, if confined to only open ocean waters, the ratios of the major constituents of seawater are constant. This principle of constant proportions did not gain widespread support until after William Dittmar's detailed analyses of 77 water samples collected by the *Challenger* expedition, which he published in 1884 (*Report of Researches into the Composition of Ocean-Water, Collected by HMS Challenger, During the Years 1873–1876*, Report of the Scientific Results of the Voyage of HMS *Challenger* vol. 1, pp. 1–139).

BOX 4B Different Salinities in Different Oceans

So why is the Atlantic Ocean saltier than the Pacific? This might seem at first to be a paradox: most of the world's major rivers empty into the Atlantic, and the Atlantic Ocean is smaller than the Pacific. These two facts *alone* would lead one to expect the Atlantic to have lower surface water salinities. But it doesn't.

Reason: Because the Atlantic is smaller than the Pacific, meridional (east–west) atmospheric circulation patterns result in drier air over the Atlantic; it receives dry air masses that have been over the continents, and so there is significant evaporation. Also, much of the moisture lost via evaporation to the atmosphere over the Atlantic is carried away. In the Northern Hemisphere, some of the evaporated moisture from the North Atlantic falls as rain in the Pacific, being delivered westward across the narrow Isthmus of Panama. In the Southern Hemisphere, moisture evaporated from the South Atlantic falls as rain in the circumpolar West Wind Current around Antarctica. (We will discuss details of ocean currents in a later chapter).

reported results for the six most abundant ions (Cl^{-1}, SO_4^{-2}, Ca^{+2}, Mg^{+2}, K^{+1}, and Na^{+1}), the salts of which represent more than 99.6% of the total dissolved solids in seawater (see Figure 4.20). Because of this constancy of proportions, the ratio of any two of the major elements would be the same anywhere in the world's oceans. That is, we need to measure the actual concentration of only one of them in order to calculate all of them, and from that we can determine overall salinity relatively easily.

For example, the ratio of the two most common dissolved materials in seawater, chloride (Cl^-) and sodium (Na^+) is always 1.73 by weight (g per kg seawater, or ppt). This ratio is the same everywhere in the oceans—regardless of the total salinity! And because the easiest ion to measure of the six major ions is Cl^-, which also happens to be the most abundant ion, salinity can be calculated[6] fairly easily, as:

$$\text{Salinity (ppt)} = 0.30 + 1.8050 \times \text{chloride concentration (ppt)}$$

This became the standard method used by early oceanographers, who, as we mentioned in the first chapter, would deploy cleverly designed samplers to collect uncontaminated water samples at specific depths and then determine the salinity of those samples by way of chemical determinations of the chloride concentration. Then, about 1930, measurements of electrical conductivity began to be used, a technique based on the fact that an ionic solution like seawater conducts electricity, and the degree of electrical conductivity is a function of the salinity and temperature. Unfortunately, the earliest conductivity instruments were not very reliable and the technique was used only rarely; instead, traditional "wet chemical techniques" to determine the chloride ion concentration remained the standard method employed for quite some time. Eventually, technology developed that allowed more reliable conductivity measurements to be made, and *salinometers* began to appear in the 1960s. Today, electrical conductivity of seawater can be measured with very high precision (the equation that describes salinity as a function of electrical conductivity and temperature is a ninth-order polynomial). These days we

---

[6] Established in 1901 by the Danish scientist Martin H. C. Knudsen, in his *Hydrological Tables* (Copenhagen-London, 1901).

measure salinity almost instantly aboard ships by simply lowering an instrument called a CTD down through the water. The CTD measures electrical *c*onductivity, *t*emperature, and *d*epth (based on pressure), and sends these signals electronically up the wire to a computer on board—at a sample rate of some 40 hz, or 40 measurements per second! These instruments are not only extremely accurate and precise, but they have made it possible to collect more data points in a few hours than all the chlorinity titrations ever done before or after their invention! We are now awash in data. The remarkable contrast between what is possible today compared with the technology available to early oceanographers helps us to appreciate even more how prescient they were in developing so many highly significant findings in oceanography with so few hard-earned data points!

Density

Determining the temperature and salinity of a sample of seawater is crucial in determining the density of that water, and knowing the density of seawater is at the heart of understanding many of the processes operating in the oceans. Thus a fundamental understanding of the density of water (especially seawater) and the various ways density is influenced are among the most important concepts for us to grasp. We discussed density earlier with respect to continental and oceanic crust—and now we discuss it with respect to water.

Density, again, is mass per unit volume (the symbol for density is the Greek letter rho, ρ). Mass, in a gravitational field such as that here on Earth, is measured as weight, so density is expressed as grams per unit volume, where volume is usually given in milliliters (ml) or cubic centimeters (cm^3). By definition, 1 gram is equal to the density of 1 cubic centimeter of pure water at 4°C. To put this a different way, the density of water (at 4°C) is 1 g/cm^3. By comparison, the density of granite rock is about 2.75 g/cm^3, while the density of air is 0.0012 g/cm^3.

It is important to realize that the density of water is not constant; it varies with temperature and salinity. But the relationships are not what we might normally expect, so let's take a closer look.

$$\text{Density} \, (\rho) = \frac{\text{Mass}}{\text{Volume}}$$

By definition, 1 gram is the mass of a volume of pure water equal to 1.0 cubic centimeter (cm^3) at 4°C, and 1.0 cm^3 = 1.0 milliliter (ml).

$$\rho_{\text{water}} = \frac{\text{Mass}}{\text{Volume}} = \frac{1 \, \text{gram}}{1 \, \text{ml}} = 1.0 \, \text{g/cm}^3$$

Temperature effects on density

We begin this discussion considering only fresh water (pure water, that is; not salt water or seawater, yet). The **normal temperature effect** on the density of fresh water is to increase it at colder temperatures, and thus we might expect the density of water to conform to the relationship plotted in **FIGURE 4.25**. That is, most substances get denser as they become colder, and less dense as they become warmer. This is because temperature is a measure of how rapidly molecules (and atoms) are vibrating and moving around. Molecules of a warmed substance will "bump into one another" at a faster rate, forcing the molecules farther apart and increasing its volume. But its weight will not change—there

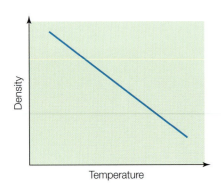

FIGURE 4.25 The relationship between the density of a substance and its temperature, showing the normal temperature effect. When a substance is warmed, its atoms (or molecules) vibrate faster, which causes them to occupy a larger volume, and so the density (mass/volume) of the warmed substance decreases. Normally that relationship is linear, as shown here.

are still only so many molecules. So if its volume increases but its mass does not change, then the denominator in the formula $\rho = $ mass/volume increases, and the density decreases.

We can perform a simple thought experiment to illustrate what happens with water. Our experiment uses a 1-liter volumetric flask (**FIGURE 4.26A**), one that allows us to measure small changes in volume of a liter of water at different temperatures. We subject the flask to different temperatures and at each one we note the volume, determine its weight using a laboratory balance, and record the results (**FIGURE 4.26B,C**). Notice that over the four experimental temperatures, in this case interpreted right to left from 20° to 0°C, the volume of water in the flask changes, but the weight does not; thus, the density changes slightly over that range of temperatures. Most important, notice that at cooler temperatures the change in density is not as pronounced as at higher temperatures—the fitted line in Figure 4.26C is steepest at 20°C. In fact, if we follow the change in density as the water sample cools from 20°C to 0°C, we see an interesting and important phenomenon: at first the density increases rapidly with decreasing temperature below 20°C in a steep, near-straight line fashion. This is the normal temperature response we predicted in Figure 4.25; that is, as the water gets colder, the speeds of molecular vibrations become slower. The molecules occupy less space, thus reducing the volume and increasing the density of the sample. But, as the water continues to get colder, our predicted response breaks down; the line begins to bend more and more because at lower temperatures there are more hydrogen bonds holding water molecules to one another. These bonds can form strings of molecules and, especially, tetrahedrons, which tend to form

FIGURE 4.26 (A) A 1-liter laboratory volumetric flask in which we determine changes in the volume occupied by an initial volume of one liter of water at a temperature of 4.0°C, recorded in (B). When we change the temperature of the water, its volume also changes. (C) A plot of the data, along with a fitted line, shows that water has a density maximum at about 4.0°C. (Actually, it's closer to 3.98°C, but we'll call it 4.0°C.)

(A)

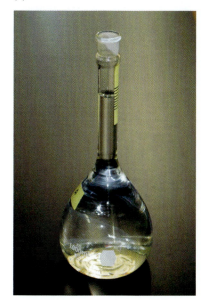

(B)

Data point	Temperature (°C)	Volume (ml or cm³)	Weight (g)	Density (g/cm³)
1	20	1001.9	1000	0.9981
2	10	1000.3	1000	0.9997
3	4	1000.0	1000	1.0
4	0	1000.1	1000	0.9999

(C)

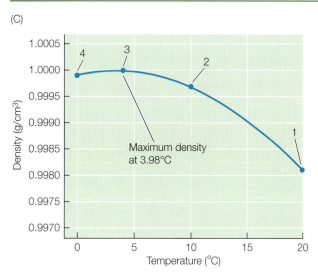

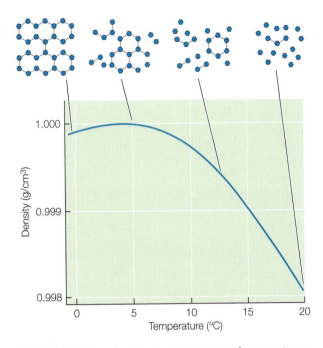

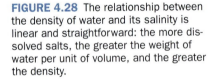

FIGURE 4.27 Density of water over a range of temperatures, indicating the relative importance of tetrahedrons of hydrogen bonds in the control of density. The more tetrahedrons there are, the greater the volume of a given quantity of water molecules. More tetrahedrons form at lower temperatures.

FIGURE 4.28 The relationship between the density of water and its salinity is linear and straightforward: the more dissolved salts, the greater the weight of water per unit of volume, and the greater the density.

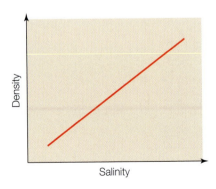

clusters of water molecules. When these tetrahedrons form, the angle between the oxygen atom and the two hydrogen atoms increases from 105° to 109°, which increases the volume and therefore reduces the density. When the water temperature cools to 3.98°C, this **cluster effect** from the formation of tetrahedrons now dominates the change in density, so that rather than continuing to get denser with lower temperatures, the water now becomes *less dense*! We have passed the point of water's maximum density; ρ_{max} is at 3.98°C.

The curved relationship between the temperature and density of water represents two opposing molecular effects being expressed at the same time. The *normal temperature effect*, whereby lowering the temperature reduces the speed of molecular vibrations and reduces the volume the molecules occupy, which increases the density; this process dominates until the temperature drops to 3.98°C. The *cluster effect*, which increases the volume and decreases the density, works against the normal temperature effect; it becomes more important—more tetrahedrons and clusters of tetrahedrons form—at lower temperatures. The normal temperature effect dominates at temperatures above 3.98°C, but below that, the cluster effect takes over and dominates such that as temperatures drop from 3.98° to 0°C, the density actually *decreases* (**FIGURE 4.27**). Eventually, ice forms, which is a complete lattice structure of hexagonal clusters. We'll return to ice after we discuss the influence of salinity on this intriguing relationship between temperature and density.

Salinity effects on density

Dissolving salt in water to increase its salinity will also increase the density. The reason is simply that we are packing more material—the dissolved salt—into nearly the same volume of water. The water molecules pack closely around these charged particles and therefore they occupy a volume that is less than their sums. That is, the mass increases, but the volume does not (in fact, it decreases slightly because of this tight packing of charged particles). So, if we increase mass, but not volume, then density increases. Therefore, the relationship between salinity and density is straightforward; with increasing salinity, density increases in a linear fashion (**FIGURE 4.28**).

Ice

The anomalous property of water we've been discussing—whereby upon cooling it reaches its maximum density before freezing, and then becomes less dense as it continues to cool to the freezing point—explains why the Earth is not covered with more ice than it is. For example, consider what happens to a freshwater lake as winter sets in (**FIGURE 4.29**), keeping in mind that fresh water and even slightly saline seawater have their maximum densities at temperatures above the freezing point. In summer, we might have a temperature profile that looks like profile 1 in Figure 4.29, where the surface is >16°C, and there is a **thermocline**, which is a depth interval where the temperature changes rapidly, beginning at a depth of about 10 m; temperatures decrease rapidly with depth to about 20 m, below which the temperature is fairly constant to the bottom, at about 8°C. As summer turns to fall, the warm surface layer of the lake cools as it loses heat to the atmosphere, to a point where it becomes denser than the warmer water layer just beneath it, and sinks. This surface water is replaced, of course, with warmer and lighter water from beneath. This, too, becomes colder when it loses heat to the cold air above, and as it becomes denser than the water beneath it, it too sinks, mixing with waters in the thermocline, and even making that water slightly warmer as a result (profile 2). This process continues until the entire water column, from top to bottom, is **isothermal**—the same temperature (profile 4). The lake will continue to lose heat at the surface, making denser water, which sinks, mixing with warmer water beneath, and continuing to cool the lake (profile 5). This process of surface water sinking is called **fall overturn**, and it continues throughout the fall and early winter. Eventually, the entire water column, from the top to the bottom, reaches its maximum density at a temperature of 3.98°C (profile 6). After that, any further heat loss from the surface water will make that water less dense than the layer beneath it (profile 7). Eventually, a cold calm night extracts more heat, making 0°C water at the surface, and ice can now form (profile 8). But if for any reason the lake does not become vertically isothermal at 3.98°C, it will not freeze; instead, fall overturn continues, which does not allow the surface to cool to 0°C. Such is the case for some deep alpine lakes—they are just too deep and the winter isn't long enough to make 3.98°C water from top to bottom. Now, if water were not such a wondrous liquid, if its solid form (ice) were also its densest, as is the case for nearly all other substances, our high-latitude lakes (which freeze in winter) would fill with ice from the bottom up.

Not only is the process of water freezing interesting (and important), but ice itself is also very interesting. Upon reaching the freezing point of 0°C, fresh water will begin to form hexagonal clusters of molecules, releasing latent heat in the process, until it has given up all its latent heat of fusion (80 calories per gram), and freezes into the solid crystalline structure we know as ice. And remember: the clusters are made of tetrahedrons of water molecules, which stretch the angle between the oxygen atom and its two hydrogen atoms, from 105° to 109°. This greatly expands the volume. In fact, the space occupied by *27 water molecules of liquid water now can fit only 24 water molecules* (**FIGURE 4.30**). The volume of the water has expanded 8.3%, and its density has decreased correspondingly. Thus the ice floats easily at any temperature, even though as it continues to cool down, its density increases somewhat.

This expansion of ice is why freezing water can be so destructive, such as when you allow the cooling water in your car engine to freeze. It expands (by 8.3%), and if it doesn't have anywhere to expand to (which it doesn't in your engine block), it will crack the steel engine block; the strength of these "weak" hydrogen bonds is quite impressive! This is also how rocks get cracked into smaller pieces in regions

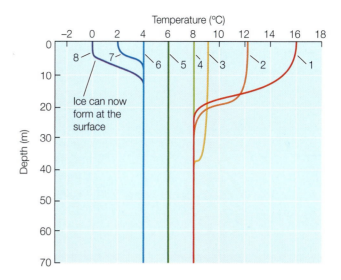

FIGURE 4.29 An example of how the temperature profile in a freshwater lake in North America would likely change from summer to winter. Profile 1 shows a surface-water temperature of 16°C, and a thermocline between about 10 and 20 m depth. When the air temperature gets colder than the surface-water temperature as fall and winter set in, the lake water loses heat to the atmosphere. The surface water becomes colder and denser than the layer just beneath it, so it sinks, as illustrated here for profiles 1–3. (The reason profiles 2 and 3 show the deeper water actually getting warmer is that winds create surface waves that mix warm water downward, making a slightly warmer temperature there.) This continues (profiles 3 and 4), and eventually the entire water column is vertically isothermal (profile 5). Continued heat loss and sinking of surface water continues until the entire water column is 3.98°C (profile 6), after which further heat loss at the surface creates less dense water that floats and eventually freezes (profiles 7 and 8).

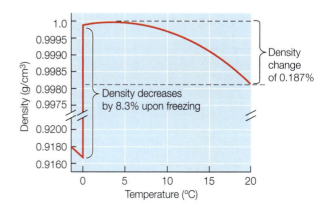

FIGURE 4.30 When water freezes, all its molecules form tetrahedrons of hydrogen bonds, expanding the volume and decreasing the density. Because the density of ice is less than that of liquid water, it floats. Note the break and change in the *y*-axis.

with cold winters, and why roads seem to fall apart as water fills in the tiny asphalt cracks and freezes, making potholes (and why it is never a good idea to put cans of beer in a freezer to chill them quickly).

This dramatic expansion of ice is also at least partly the reason why an ice skate works. That is, the immense pressure directly under the steel skate blade (on the order of 1500 lbs/in²) compresses the ice, breaking hydrogen bonds, and forcing it back into liquid form—which makes it slippery; therefore, skating is actually done on a thin layer of water, not ice. But this lasts for only a split second while the skate is there; after the skate has released the pressure, the ice immediately freezes again. And it works only if the temperature of the ice is within a degree or so of the melting point to begin with (which happens even if the ice is much colder because of heat generated by the friction between the moving skate and the ice). But it isn't just this pressure-forced phase transition that makes ice slippery. A phenomenon called the *surface melting effect* plays a role, as well. For example, we know that very light objects easily slide across an ice surface, clearly without having to melt the ice under heavy pressure. This is because the molecules in ice at the ice–air interface have nothing to bond with, and there tends to be a monolayer of liquid water molecules on the surface of ice.

Interestingly, you can actually feel (and hear!) this pressure-induced phase change effect by slowly biting down on an ice cube; but to do this experiment, be sure to let the ice cube warm up in your mouth to a temperature of about 0°C (and be careful not to break a tooth!).

Properties of Salt Water

The dissolved solids that give seawater its salinity interfere with the formation of hydrogen bonds, and therefore they interfere with the interesting properties we just discussed for fresh water. Adding salt significantly influences four physical properties of water, as described below.

Increasing salinity decreases heat capacity

Addition of enough salt to make full-strength seawater (ca. 35 ppt) lowers the heat capacity of the water—but only by about 4%. This is interesting, but not nearly as important in oceanography as the next two phenomena.

Increasing salinity increases density

Adding salt to fresh water significantly increases its density, as we showed conceptually in Figure 4.28. An actual plot of the density of water over the range of salinities normally found on Earth, from freshwater (0 ppt) to about 40 ppt, is shown in **FIGURE 4.31** (for water held at 4°C). The effect is linear. Adding enough salt to make full-strength seawater (ca. 35‰) will increase density by about 2.8%. For example: the density of 4°C freshwater (0‰) is 1.000 g/cm³, and the density of 4°C seawater (35‰) is 1.028 g/cm³.

Increasing salinity lowers the temperature of maximum density

When salt is added to fresh water, the temperature of maximum density is no longer 3.98°C; it becomes significantly lower, depending on how much salt is added. This is because salt interacts with the charged ends of water molecules and thus interferes with the formation of hydrogen bonds and

FIGURE 4.31 The density of 4°C water of varying salinity, from fresh water (0 ppt) to 40 ppt salt water. Fresh water has a density of 1.0 g/cm³, while average seawater has a density of 1.028 g/cm³.

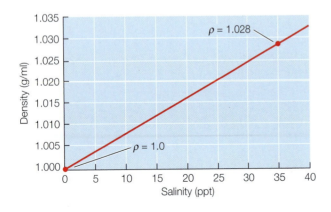

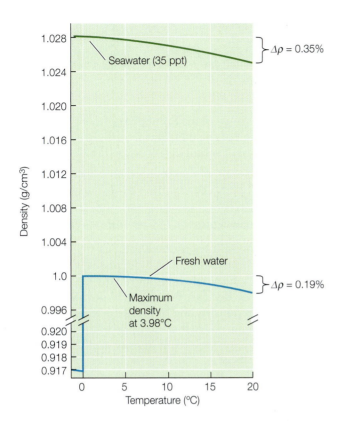

FIGURE 4.32 Plots of both fresh water and seawater densities as a function of temperature. While fresh water has a maximum density at about 4°C, seawater reaches its maximum density well below 0°C, not shown.

impedes the formation of tetrahedrons and clusters. Thus, the importance of clusters in decreasing the density of water is suppressed in seawater. Therefore, the relationship is not as curved downward as it is for fresh water; the normal temperature effect has more influence on density than does the cluster effect. As salt water cools, the normal temperature effect increases density faster than it does in fresh water over the same temperature range. **FIGURE 4.32** shows how the density of fresh water and salt water responds to changes in temperature. It illustrates two important points: first, in full-strength seawater, salt interferes with hydrogen bonding. So the density of seawater increases faster as the temperature drops, giving a steeper curve. In seawater, density changes some 0.35% over a 20°C change in temperature, whereas it changes only 0.19% in fresh water over the same temperature range. Second, the cluster effect never does result in density reversing as temperatures drop, the way it does in fresh water. Seawater just keeps getting denser and denser with decreasing temperature. In fact, not only is the temperature of maximum density of seawater depressed to below 3.98°C, but it does not reach its densest point until −1.9°C, the freezing point for 35 ppt salt water!

The temperature of maximum density as a function of salinity is plotted in **FIGURE 4.33**. The temperature of ρ_{max} ranges from 3.98°C for fresh water, of course, to less than −1.0°C for seawater. Notice that we stopped the plot well before reaching the average ocean salinity of about 35‰. The reason we did this becomes clear once we consider the effect of salt on depressing the freezing point of water.

Increasing salinity depresses the freezing point

Most of us already know that adding salt to water lowers the freezing point, which is why salt is spread on roadways to prevent freezing and melt ice in some parts of North America. The freezing point of seawater can be depressed to well below 0°C, depending on how much salt is in the water. Again, the

FIGURE 4.33 The relationship between the temperature of maximum density and the water's salinity. The plot shows that fresh water (0 ppt) has its maximum density at about 4°C (circled in red), while water of 23 ppt salinity, for example, has its maximum density at a temperature of −1.0°C (circled in green).

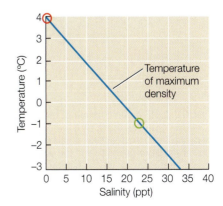

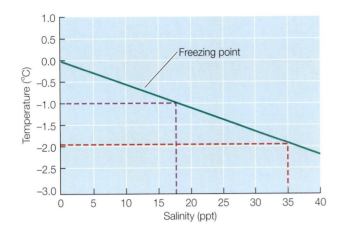

FIGURE 4.34 The freezing point of water is depressed, or lowered, as the salinity increases. For example, at a salinity of 18 ppt (purple dashed line), the freezing point is depressed from 0°C to –1°C, and at 35 ppt (red dashed line), the average salinity of the oceans, the freezing point is depressed to –1.91°C.

effect of salt is to impede the formation of hydrogen bonds and the formation of clusters. This freezing-point depression as a function of salinity is illustrated in **FIGURE 4.34**, which shows that the effect can be as great as –1.0°C at 18‰, and –1.91 at 35‰. The freezing-point depression can be approximated as

$$\text{Freezing-point depression } (-°C) \approx -0.054 \times \text{salinity } (‰)$$

For example, the salinity in the Dead Sea (in Israel and Jordan) ranges from 200‰, which depresses the freezing point to –10°C, to as much as 400‰, which depresses it to about –20° (**BOX 4C**).

Thus, adding salt to fresh water will decrease heat capacity, increase density, lower the temperature of maximum density, and depress the freezing point. With respect to the last two phenomena, notice that if we plot both on the same graph (**FIGURE 4.35**), the temperature of maximum density and the freezing point are the same (–1.33°) at a salinity of 24.7‰. At all salinities greater than 24.7‰, the temperature of maximum density is the same as the freezing point, which means that, unlike fresh water, salt water reaches its maximum density before it freezes. Moreover, this means that the dashed portion of the maximum density line in Figure 4.35 represents temperature-salinity characteristics of water that *do not exist in nature*.

All four of these properties associated with the addition of salt (decreased heat capacity, increased density, lower temperature of maximum density, and depressed freezing point) are important in interpreting how the world ocean works, especially at high latitudes (in cold climates, where the water is cold) and where fresh water meets the sea. Let's compare the phenomenon of fall overturn in freshwater lakes with the same process in seawater (**FIGURE 4.36**). We begin the same as we did for the lake: a summer surface temperature of about 16°C (profile 1) with a thermocline between 10 and 20 m. As fall approaches, the surface water layer loses heat to the atmosphere, becomes colder and denser, and sinks, replacing and mixing with the layers beneath (profiles 2 and 3). Cooling at the surface continues until we again have vertically isothermal water, top to bottom, of first 8°C, then 6°C, then 3.98°C, as well as all temperatures in between, of course. But here is where

FIGURE 4.35 A combination of the plots in Figures 4.33 and 4.34; the plotlines cross at a temperature of –1.33°C and a salinity of 24.7 ppt. At salinities greater than 24.7 ppt, the temperature of maximum density and the freezing points are one and the same. That is, lowering the temperature of a volume of water that has a salinity greater than 24.7 ppt will cause it to freeze before it reaches the "predicted" temperature of maximum density.

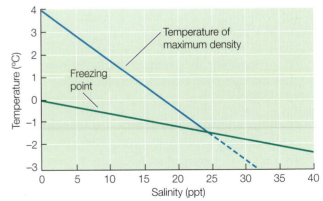

BOX 4C   Salt and the Surface Melting Effect

We should point out that road salt, which is used to melt ice on roads in regions of colder climates, works a little differently. That is, when we throw grains of rock salt onto solid ice, the ice melts. The reason is tied to the surface melting effect discussed earlier; that is, the top molecular layer of water molecules in the ice, the layer in direct contact with the salt crystal, is in a semiliquid state and will begin to dissolve the salt. This creates very high salt concentrations (high salinities) right next to the rock salt, which can further melt more and more ice. Often, when rock salt is applied to a relatively thick sheet of ice, the salt grain will drill downward into the ice as it melts the ice immediate surrounding it, leaving behind a conspicuous hole in the ice.

the ocean differs from lakes. Even after a temperature of 3.98°C is reached from top to bottom, the continued loss of heat at the surface produces colder seawater that *continues to sink*—and with continued heat loss, we simply get a colder and colder vertically isothermal water column. Assuming that this is 35‰ seawater, this process continues until the entire water column has reached –1.91°C, the temperature of maximum density, which is also the freezing point. With continued heat loss, the surface can now freeze over.

This cooling of the entire water column to temperatures much lower than 3.98°C requires much more heat loss than that experienced by a freshwater lake; the entire water column must be cooled another 5.9°C! That additional heat loss requirement, represented by the area shaded in purple in Figure 4.36, is why the ocean seldom freezes—except at the poles or in very shallow water. Even in regions that experience cold winters, such as in New England, we see seawater freezing only in very shallow areas or in exceptionally cold winters.

Once seawater does freeze, it is important to note that sea ice is a bit different from freshwater ice in that it is very brittle. This is because as the freezing point is reached and the seawater freezes, the dissolved salt in that water is excluded from the crystalline lattice. This leaves behind a salty brine as more and more of the ice forms out of pure water; that brine sinks, making

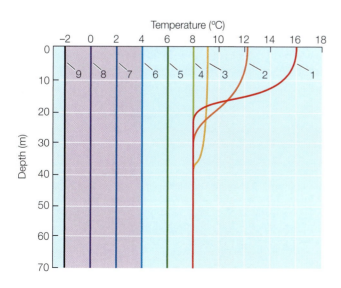

FIGURE 4.36 An example of how the temperature profile in the ocean, at 35 ppt, would likely change over time from summer to winter (in a cold region of the ocean). As in Figure 4.29 for a freshwater lake, temperature profile 1 shows a surface-water temperature of about 16°C, and a thermocline between about 10 and 20 m depth. Also as in Figure 4.29, when the air temperature gets colder than the surface-water temperature, the ocean water loses heat to the atmosphere, making the surface water colder, and therefore denser, than the layer just beneath it, and it sinks. This process continues just as in the freshwater lake example, except that the ocean continues to become denser even after the water reaches 4°C. It continues to lose heat to the atmosphere until the water column reaches the temperature of maximum density, which, for this example, is –1.9°C, the freezing point.

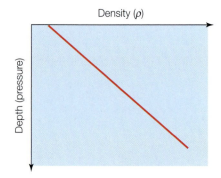

Density (ρ)

Depth (pressure)

FIGURE 4.37 The relationship between the density of water and pressure. It is linear and straightforward: the greater the depth, the greater the pressure, and the greater the density of water—freshwater or seawater. The pressure reduces the volume slightly, forcing the water molecules closer together, and therefore increasing the density.

holes as it drills down into the base of the layer of ice, leaving behind air pockets. Consequently, sea ice is not very sturdy compared with freshwater ice (caution: don't walk on it). But the actual ice itself, after the brine seeps out, is pure water. The fact that sea ice is pure water, plus the fact that a lot of the ice in icebergs is from packed snow, is why some enterprising groups have proposed hauling Antarctic icebergs to desert regions as a source of fresh water.

Increasing pressure increases density

Like the effect of salinity, the effect of pressure on the density of seawater is straightforward. As pressure increases (which happens with increasing depth in the ocean) the density increases, simply because the volume in the density equation has been reduced by compression under pressure (**FIGURE 4.37**). While it is commonly stated that "water is incompressible," that statement is wrong. Water is not nearly as compressible as air, for example, but it is nonetheless compressible. It certainly takes a lot more pressure to compress it than to compress air—but that condition is easily met in the depths of the ocean.

Because pressure is a straightforward, linear function of depth in the ocean, changes in density that result solely from changes in pressure are easily computed. That is, while the density of seawater is a function of the temperature, salinity, and pressure, the *change* in density of a sample of water that changes its depth (as it is brought to the surface, for example) will be due solely to the change in pressure. It will become less dense in a straightforward manner. Therefore, oceanographers are more concerned with the density of water that results from its temperature and salinity only.

Combined effects of temperature and salinity on density

As we have just seen, both temperature and salinity affect the density of salt water. As temperature increases, density decreases; as salinity increases, density increases. This means that water of different temperatures and salinities can have the same density. For example, notice how the different temperature and salinity data pairs in **FIGURE 4.38** all fall on a curved line—a line of constant density. Such a plot for a range of temperatures and salinities produces what is known as a **temperature-salinity diagram**, or **T-S diagram** (**FIGURE 4.39**), which is a plot of lines of constant density over a range of temperatures and salinities. The T-S diagram in Figure 4.39 shows, for example, how the density of water held at 15°C increases as its salinity increases (red arrow), and how 34‰ water that gets warmer becomes less dense (thin blue arrow). This graph is quite intuitive if you think about it, except for the curved lines of constant density.

(A)

Temperature (°C)	Salinity (ppt)	Density (g/cm³)
3.5	33.8	1.027
6.0	34.2	1.027
10.0	35.0	1.027
14.0	36.0	1.027
18.0	37.2	1.027

(B)

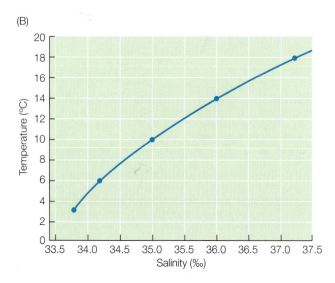

FIGURE 4.38 (A) Five samples of seawater representing a range of temperatures and salinities. All have the same density of 1.027 g/cm³. Note that this is possible only because the effect of higher temperatures on water density is effectively canceled out by the effect of higher salinity, and vice versa. (B) The density of the five samples plotted as a function of both temperature and salinity. Notice that those points generate not a straight line, but a curve.

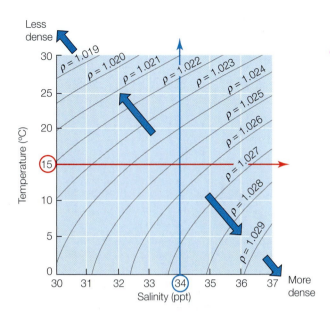

The T-S diagram in **FIGURE 4.40** reveals an interesting aspect of the density of seawater and how both temperature and salinity exert controls. For example, if we took two samples of water from the ocean, with temperatures and salinities as given in the table in Figure 4.40A, and plotted them (Figure 4.40B) we would see that they had the same densities, sigma-t (σ_t) = 25.0 (**BOX 4D**). However, if we mixed the two water samples, we would get a mixture that has a temperature and a salinity that is the average of the two original samples, new values of temperature and salinity that are halfway between the original values. But the resulting density is not the halfway point between the two starting values, which are equal to one another; instead, the density has changed—the mixture has become *denser* than either of its two constituent samples! We mixed two volumes of water of the same density, and made denser water. This means that out in the ocean, if two water masses of the same density, but different temperature and salinities, come together, the resulting mixture will be denser than either one of the original water masses. So, the new mixture must sink.

(A)

	Point 1 (•)	Point 2 (•)	Halfway (Point 1–Point 2) (•)
Temperature (°C)	2.6	15.0	9.7
Salinity (ppt)	31.3	33.5	32.4
Sigma-t (σ_t)	25.0	25.0	25.2

(B)

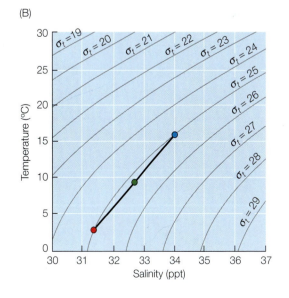

FIGURE 4.40 (A) Data for two water masses of the same density, but with different temperatures and salinities (red and blue points). (B) What happens when the two mix? They make a water mass with the average temperature and salinity of the two sources, but with a greater density than either one (green point).

Notice in the T-S diagram (see Figure 4.39) that the lines of constant density of seawater range from $\rho = 1.019$ to 1.029 g/cm^3. Over the normal range of temperatures and salinities in the oceans, it is only the last two decimal places of density that change. So it is a lot easier if we simply drop the "1.0" and consider the "density anomaly from 1.0"—which is how much the density of a sample differs from that of fresh water.

For example, the density of a sample of seawater at the surface of the ocean (at 0 m, where pressure, p, is 1 atmosphere) with a temperature of 10°C and salinity of about 33.2‰ would be $\rho_{s,t,p} = \rho_{33.2,10,0} = 1.025$ g/cm^3, and the density anomaly $\sigma_{s,t,p}$ would be computed as:

$$\sigma_{s,t,p} = (\rho_{s,t,p} - 1) \times 1000 = 25.0$$

We can simplify $\sigma_{s,t,p}$ even further by ignoring pressure, which increases as a direct function of depth. This gives us sigma-st ($\sigma_{s,t}$), or for simplicity, just sigma-t (σ_t)

$$\text{sigma-t} = \sigma_t = \sigma_{s,t,0} = (\rho_{s,t,0} - 1) \times 1000 = 25.0$$

Vertical Structure in the Ocean

We have been dwelling on the effects of temperature and salinity on the density of seawater because they are important in the vertical structure of the ocean. Temperature and salinity, and therefore density, vary with depth, which is extremely important not just to the physical structure of the ocean, but to processes that govern the biology of the oceans.

Differences in water density result in layering; a less-dense layer will be buoyant and thus float upon a denser layer beneath. Buoyancy is determined by relative density differences, which are controlled by both the temperature and salinity. For example, if we were to measure a water column profile of both temperature and salinity with depth somewhere in the ocean, using a conductivity, temperature, and depth (CTD) system such as that in **FIGURE 4.41**, we might expect to see several things. First, we might expect the shallower waters to be warmer than those at deeper depths; it is the surface of the ocean, after all, that receives direct sunlight and which we would expect to become warmer, and therefore, less dense (**FIGURE 4.42A**). Likewise, if this were a coastal location, we might expect it to be influenced by nearby rivers, with a low level of salinity and therefore, low density. What we will find is that the warmer, low-salinity water is sitting on top of colder, high-salinity water beneath (**FIGURE 4.42B**). The depth interval separating the two layers is, in this case, both a thermocline (defined earlier as a zone of rapidly changing temperature with increasing depth) and a **halocline** (from the Greek *hals*, salt), a zone of rapidly changing salinity with increasing depth. A thermocline is obvious to anyone who has ever done a surface dive in a lake; sometimes even your feet can feel it while treading water, depending on how tall you are and how deep

FIGURE 4.41 A modern CTD (conductivity, temperature, and depth) system being deployed aboard a research ship. The CTD is an electronic sensing system that measures temperature and salinity as it is lowered through the water, and sends those data back to a computer on the ship. The instrument package also has water sampling bottles that can be closed at a desired depth of interest to bring back samples for chemical or biological analysis.

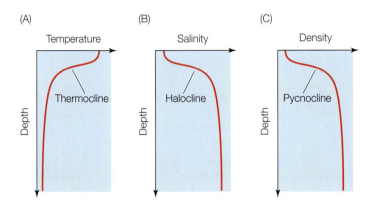

FIGURE 4.42 Examples of the profiles that are produced with data collected by the CTD: Vertical profiles of (A) temperature, (B) salinity, and (C) density.

the thermocline is. Sharp haloclines are often found near the coast, especially near the mouths of estuaries, where fresh water empties into the ocean. The fresh water floats in a relatively thin layer at the surface. The corresponding density profile that would result from these temperature and salinity profiles resembles that of salinity, and is the mirror image of temperature, which should make intuitive sense (**FIGURE 4.42C**). Warmer water is less dense than colder water, and low-salinity water is less dense than higher-salinity water. Note that the **pycnocline** (from the Greek *pyknos*, dense), a zone of rapidly changing density with increasing depth, coincides with the depths of the thermocline and halocline. In this case, the vertical density structure is the result of both the temperature and the salinity distributions, but this isn't always the case.

You should also be aware that under unusual circumstances, vertical profiles of temperature and salinity can look quite odd. For example, **FIGURE 4.43** shows a hypothetical temperature and salinity profile of seawater somewhere in the tropics, where the air temperature is generally warm and rates of evaporation are high. The salinity profile, which might otherwise resemble Figure 4.42B with fresher water nearest the surface, is saltier right at the surface. This reflects high evaporation at the surface, but despite the higher salinity, that surface layer does not become dense enough to sink. This is because in this case, the relatively high surface temperature has a greater influence on the density of that top layer of water, making it less dense than the higher salinity does (which would make it denser); the increased surface salinity is insufficient to promote its sinking. The T-S diagram is helpful in evaluating these phenomena with actual examples.

The vertical structure of the ocean and the propagation of sound waves

We have already discussed how seismic waves refract (see Chapter 3). The same principles apply whether we are talking about light waves, ocean waves, or sound waves: they all tend to bend toward the medium in which wave speed is reduced. In the case of sound in the sea, this phenomenon becomes important. The speed of sound in water is very fast, about 1450 meters per second (m/sec), which is about four times as fast as the speed of sound in air. It is because of this high speed that, when swimming underwater, you cannot determine which direction a sound comes from. You lose your "stereoscopic" hearing because the sound arrives at both ears at almost the same time.

FIGURE 4.43 (A) An example of an "odd" pattern of temperature and salinity that nonetheless results in (B) the same density profile as that in Figure 4.42C. The salinity at the surface is higher than would normally be expected (e.g., compare with Figure 4.42B) because of increased evaporation. The water is sufficiently warm that, despite the higher salinity, it remains less dense and does not sink.

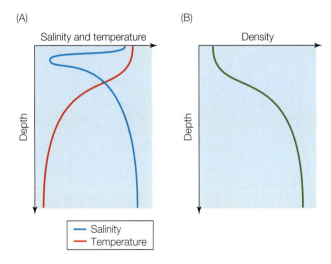

In water, the speed of sound responds primarily (but not exclusively) to two factors: temperature and pressure. Of the two, temperature has the greater effect; and as these two variables change with depth in the water column, the path of sound waves will change as well, as they refract toward the medium of slower speed.

The refraction path of sound waves is analogous to the path of a car that hits soft dirt on the shoulder of a road: the car can't go as fast with soft dirt under its tires, and so both tires on the side of a car in dirt will tend to "pull," or bend, the path of the car to that side. Sound waves in water travel faster as temperature increases *and* as pressure increases, which can lead to refraction if the water column has vertical structure. To see what happens, we'll separate the two variables. First, let's consider only pressure, keeping temperature constant. We know that pressure increases at a constant rate with depth. Therefore, if we emit a sound at the surface of the ocean and trace the path of the resulting sound waves, the sound waves will refract upward toward the surface where the speed of sound is slower because pressure decreases toward the surface (**FIGURE 4.44A**). Now let's look at the temperature effect, keeping pressure constant. Assuming that temperature is decreasing rapidly with depth, when we emit a sound at the surface, we might see a refraction pattern like the one shown in **FIGURE 4.44B**: the sound waves will refract downward toward the cold water where the speed of sound is slower. Combining both effects, temperature and pressure, and remembering that temperature has a greater effect than pressure, we would see a vertical structure in the relative speed of sound waves in the sea. In the thermocline, which extends from the surface to about halfway down in the plot of temperature and pressure in **FIGURE 4.44C**, the temperature drops rapidly (which would slow the speed of sound waves); the effect of this temperature drop is to overwhelm the increasing effect of pressure, causing a reduction in speed from the surface to the base of the thermocline. In the lower water layer, the temperature is isothermal again—it does not change the rest of the way to the bottom—and so the change in pressure dominates, and the speed of sound increases. The main point in this example pattern of sound velocities with depth in the ocean is the creation of a *depth of minimum sound velocity*, toward which sound waves will refract. In this deep layer, sound waves that are generated there are trapped there, sort of like a fiber optic cable where light waves are "trapped." This is known as the **deep sound channel**, where the velocity of sound waves may be reduced from about 1450 m/s to 1425 m/s—but within that layer, sound transmission is very efficient, like light inside a fiber optic cable (**FIGURE 4.45**). The sound energy is not lost quickly—it keeps bending back toward the depth of the sound channel.

This phenomenon of a deep sound channel in the ocean makes it theoretically possible to transmit sound waves around the world ocean. The U.S. Navy has done just that, from the Indian Ocean to the coast of Oregon in the Pacific Ocean. Also, because sound velocity in water is proportional to temperature, it is possible to detect minute changes in water temperatures averaged across an entire ocean basin by measuring transit times for sound to travel across an ocean in the deep sound channel. A change in water temperature as small as 0.001°C can be detected with this technique; thus it might be possible to monitor future effects of global warming. But these kinds of acoustic monitoring research programs are not without environmental concerns. We know that the great whales also use the deep sound channel, and similar sound frequencies, to communicate with one another across great distances, and we cannot predict what effect our artificial sound productions might have on them.

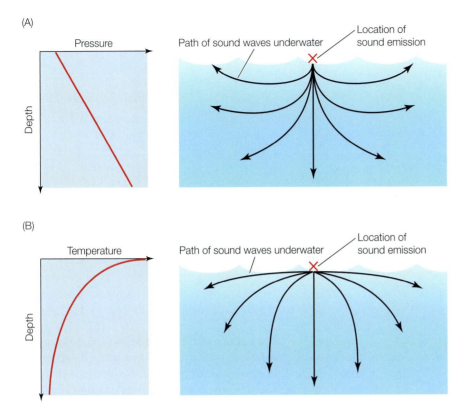

(A)

Pressure

Depth

Path of sound waves underwater

Location of sound emission

(B)

Temperature

Depth

Path of sound waves underwater

Location of sound emission

(C)

Temperature and pressure

Depth

— Pressure
— Temperature

Speed of sound

Depth

FIGURE 4.44 The behavior of sound with depth in the ocean depends on both temperature and pressure, which, of course, change with depth. If we examine how sound behaves with respect to each variable separately, we observe the following phenomena. (A) Because sound travels faster in water that is under greater pressure, sound waves emitted at the surface tend to refract back toward the surface, where the speed of sound is slower. (B) Sound travels faster in warmer water, so sound waves emitted at the surface of a water column that exhibits a decrease in temperature with depth tend to refract downward. (C) The combined effects of both temperature and pressure are considered, with the understanding that temperature changes tend to dominate changes in pressure. The result is a layer beneath the thermocline where the speed of sound is slowest, but it increases again with increasing depth. Therefore, sound waves tend to refract toward that depth layer, producing what is generally known as a deep sound channel (see Figure 4.45).

In addition to the deep sound channel created by a depth stratum where the velocity of sound is at a minimum, often there is also nearer the surface a *maximum velocity depth interval* known as a **shadow zone**, which makes it possible for submarines, or anything else, to hide from sonar signals emitted by surface ships. The maximum sound velocity stratum exists because the vertical structure of temperature in the upper ocean often exhibits what we call an **upper mixed layer**. This is a layer near the surface of relatively uniform temperatures, having been mixed by surface winds and waves. Just beneath that layer, however,

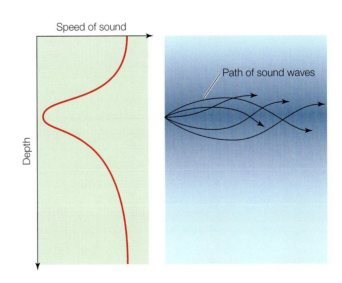

Speed of sound

Depth

Path of sound waves

FIGURE 4.45 The deep sound channel is a subsurface depth layer where the speed of sound is minimal compared with depths above and below it. In this channel, a sound emitted by an animal or a machine will tend to be "trapped" in that layer.

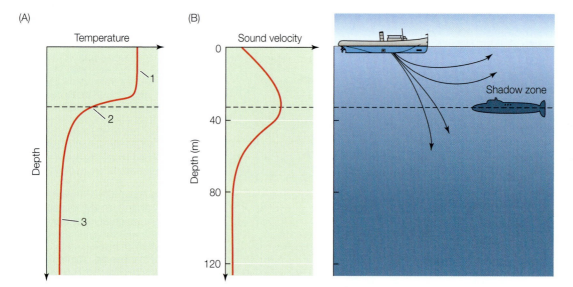

FIGURE 4.46 (A) Sound velocity with depth graphed as a function of water temperature. Because the shallowest depths often show little or no change in temperature (this is called a surface, or upper, mixed layer), such as that shown at area 1, pressure changes will dominate, resulting in an increase in sound velocity with depth—but only until the temperature begins to drop in the thermocline (area 2), below which the water temperatures remain relatively cold (area 3). (B) This temperature profile will result in a zone of higher sound velocity at the top of the thermocline, which is just above the deep sound channel of slower sound velocity. (B, right) The "shadow zone" which results from the pattern of sound wave velocities near the surface.

is a relatively sharp thermocline (**FIGURE 4.46A**), where the temperature gets colder with increasing depth. The speed of sound in the upper mixed layer will be controlled primarily by pressure change, since temperature does not change over this shallow depth interval (area 1 in Figure 4.46A). Thus, the speed of sound starts to increase with depth, reaching a maximum at the base of the upper mixed layer (approximately the depth at location 2 in Figure 4.46A), before it decreases with depth in the thermocline where the temperature gets colder (area 3 in Figure 4.46A). This gives a high-velocity depth stratum; sound waves will refract away from this shallow area, and propagate back toward the surface, as well as toward the bottom, thus creating the shadow zone, in which submariners learned to "hide" during World War II (**FIGURE 4.46B**).

Chapter Summary

- **Atoms** are made up of subatomic particles. They form chemical bonds with one or more other atoms to form **molecules**.

- Three types of chemical bonds were discussed: covalent, hydrogen, and ionic bonds.

- In a **covalent bond**, two atoms share one or more electrons, as in the **water molecule**, in which one oxygen atom and two hydrogen atoms share two electrons among them.

- When the negatively charged oxygen atom on the side of a water molecule is co-attracted to the oppositely charged hydrogen atoms of another water molecule, we get a **hydrogen bond**, a relatively weak but extremely important chemical bond.

- An **ionic bond** is formed between two atoms when one of them completely acquires an electron from the other atom. This creates two charged **ions** that co-attract one another and form a solid, crystalline lattice of atoms. Ordinary table salt, sodium chloride (NaCl), is an example of a molecule held together by ionic bonds.

- Water is a **dipole**, or **polar**, molecule with oppositely charged sides. This results because the three atoms form an angle of approximately 105° and the electrons in the two hydrogen atoms are pulled toward the oxygen atom, leaving their positively charged hydrogen protons sticking out on one side of the molecule, while on the other side of the molecule, oxygen's electrons stick out.

- Because of its polar characteristics and tendency to form hydrogen bonds, water has a number of unusual chemical and physical properties, including high surface tension, high dissolving power, and high heat capacity, among others.

- **Surface tension** results because molecules directly at the air–water interface have nothing above them in the air to form hydrogen bonds with, so they stretch back onto one another and with others beneath, creating a skin-like surface.

- Water's **high dissolving power** allows it to dissolve a large number of materials, especially **salts**.

- Water has a **high heat capacity**, requiring a greater input of heat energy to warm it than to warm equal volumes of other liquids or most solids.

- Because of water's hydrogen bonds, phase transitions in water require (and release) additional heat, known as the **latent heat of melting** and the **latent heat of evaporation**.

- Seawater is characterized by its **salinity**, the concentration by the weight of dissolved inorganic solids in seawater. Salt in the sea comes mostly from terrestrial weathering, but also, in small quantities, from the Earth's deep interior. Salinity varies with location throughout the world ocean.

- The density of water varies with temperature and salinity. Temperature affects water density through two opposing molecular effects: the **normal temperature effect** and the **cluster effect**.

- Salt water is denser than fresh water in proportion to its salinity. Salt water exhibits a less pronounced cluster effect on density with changes in temperature than fresh water because the dissolved particles in salt water interfere with the hydrogen bonds among water molecules.

- Fresh water is densest at 3.98°C, while salt water is densest at colder temperatures, depending on the salinity. In waters of salinities greater than 24.7 ppt, the temperature of maximum density and the freezing point are the same.

- Ice is the least dense form of water. It forms when all water molecules have completely formed **tetrahedrons** and hexagonal clusters, thus forming a solid. Because of changes in the molecular structure when tetrahedron bonds form, ice expands, making it less dense than liquid water.

- The ocean's vertical structure can be described in terms of vertical changes in temperature, salinity, and density. The depth intervals where each of these changes rapidly are called a **thermocline**, **halocline**, and **pycnocline**, respectively. The vertical structure affects the movement of sound waves through layers of water.

Discussion Questions

1. What is an element? A molecule? An atom? What are the three subatomic particles and their characteristics that we discussed? What two criteria must be satisfied for an atom or molecule to be balanced and stable?

2. Discuss the nature of covalent, ionic, and hydrogen chemical bonds.

3. What is a salt? How do salts form crystals? Why does salt dissolve so readily in water?

4. What are the most common constituents that give seawater its salinity? Why do the relative proportions of those major constituents not vary with location or depth in the world ocean, but the total salinity of seawater does? What processes help maintain that constancy?

5. How much table salt would you need to add to a 1 cubic meter tub of freshwater in order to make a final salinity of 34 ppt?

6. Where did the salt in the sea come from? Is it still being added to the oceans? If so, are the oceans getting saltier? Why, or why not?

7. What are some examples of surface tension? How might changes in temperature and salinity, individually and together, affect the surface tension of water? For example, how might changes in temperature and salinity alter the size of a drop of water falling as rain, or drops of water splashing over the side of your ship in a storm?

8. Explain why coastal areas in summer are generally cooler than inland areas. What are the differences in winter?

9. Explain the probable origin of the expression: "A watched pot never boils."

10. Why does a 10-pound bag of ice at an initial temperature of 0°F (−18°C) chill a six-pack of soft-drink cans better than the same weight bag of 0°F iron, rocks, wood, or air?

11. Explain why alpine lakes in mountainous regions of Switzerland never freeze, even though their winters are quite cold.

Further Reading

Feynman, Richard P., 1985. *QED: The Strange Theory of Light and Matter*. Princeton, NJ: Princeton University Press. In his inimitable style, Feynman explains that quantum mechanics is virtually impossible to understand, but that the theory works anyway.

Gabianelli, V. I., 1970. Water—The fluid of life. *Sea Frontiers* 16(5): 258–270.

Gribbin, John, 1984. *In Search of Schrödinger's Cat*. New York: Bantam. A great read; written for the lay person on the fascinating, weird, even magic-like, world of quantum physics.

Kurlansky, Mark, 2003. *Salt: A World History*. New York: Penguin Books.

MacIntyre, F., 1970. Why the sea is salt. *Scientific American* 223(5): 104–115.

Pilson, Michael E. Q., 1998. *An Introduction to the Chemistry of the Sea*. Upper Saddle River, NJ: Prentice-Hall.

Riley, J. P., and Chester, R., 1971. *Introduction to Marine Chemistry*. London: Academic Press.

CHAPTER **5**

Contents

Hurricane Frances, seen here in a spectacular satellite image taken just before making landfall in Florida in early September of 2004, was born over the warm surface waters of the eastern subtropical Atlantic Ocean in late August that same year. It is only over warm ocean waters that hurricanes can form, as they derive energy from the latent heat released by rising masses of warm, humid air.

Atmospheric Circulation and Ocean Currents

Thus far our discussion of water has focused on the important roles played by hydrogen bonds in the water molecule and how they are responsible for water's rather unusual physical and chemical properties. One of these properties is its high heat capacity which enables water—and the oceans—to store heat; this is in sharp contrast to the continents, which have a much lower heat capacity. It is water's capacity to store heat that drives our weather and climate, which in turn are the driving forces behind our major ocean currents. In this chapter we focus on those processes that convert the Sun's energy into our planet's ocean and atmospheric circulation patterns—why we have winds and how those winds force ocean currents. We will start by laying a foundation of some very basic principles.

In addition to water's high heat capacity, there are two other seemingly obvious characteristics of water that enable the oceans to store a lot of heat—much more heat than can be stored in land (**FIGURE 5.1**). First, water is **transparent**; land is not. The Sun's rays can penetrate water and spread heat energy deeper, allowing the oceans to store the heat in a greater volume. Second, the oceans can **mix vertically**; except on time scales of hundreds of millions of years, land cannot. The vertical mixing of surface waters is accomplished by various processes, including winter convection, which we discussed in the last chapter, as well as by the wind and resulting ocean waves. Water, therefore, has at least three general characteristics that enable it to store great quantities of heat: (1) it has high heat capacity; (2) it is transparent to light; and (3) it can mix vertically. We discussed water's heat capacity in the last chapter, and so we begin here with a discussion of light.

Light in the Sea

As we discussed in Chapter 1, Edward Forbes suggested that there is no life in the sea below 550 meters depth because, for one thing, there is no light down there. He was wrong; there is life even at the ocean's greatest depths, but Forbes nonetheless made a good point about the absence of light at depth in the ocean. It does get dark—quickly—as we venture deep beneath the surface.

While the oceans are indeed transparent to the passage of light, and the Sun's rays do penetrate down through the upper layers of the ocean, not all of the sunlight that passes through the atmosphere to the sea surface penetrates very deep. The degree to which light penetrates the ocean depths is very important to marine life—not only for animals to see, but also for photosynthesis by marine algae. Of the sunlight that reaches Earth, some is absorbed and reflected by clouds and the atmosphere before it even reaches the ocean surface, and of the sunlight that strikes the surface of the ocean, some of that is also reflected back into space. Light that is not reflected penetrates downward where it is **attenuated** (lost), by a combination of **absorption** and **scattering**.

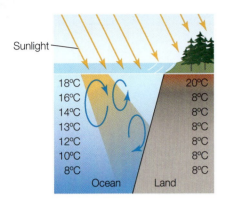

FIGURE 5.1 Diagram comparing the relative penetration of the Sun's rays, and therefore heat from the Sun, into shallow depths of the ocean and land. Generally speaking, the Sun's rays warm only the very surface of the land, as shown by the relatively colder ground temperatures immediately beneath the surface. In the oceans, heating extends to greater depths.

Light energy is absorbed by several components of water; water molecules themselves absorb some of the light energy, but much more is absorbed by the various particles suspended in the water and dissolved materials, especially **dissolved organic matter** (**FIGURE 5.2**). You have no doubt seen examples of this dissolved material, probably without thinking much about it. For example, ponds, lakes, and rivers often have a yellow-brown color, which results from once-living things—especially plants—that have died, decayed, and eventually dissolved in the water. In such cases, that colored water is a lot like a cup of tea, which is the result of immersing in hot water a porous paper envelope with dried leaves inside. Some of the organic materials in those leaves dissolve and give color and flavor to the hot water. When you allow the tea to steep, to make stronger tea, it colors the water more and more, and, in the process, less light penetrates to the bottom of your cup; the light is absorbed. Light that is absorbed is effectively lost, having been converted to another form of energy—heat energy; the light energy has therefore been captured by whatever absorbed it.

In addition to being absorbed, a lot of light in the sea is scattered: the light reflects off particles, including air bubbles and dissolved materials, as well as off water molecules. Light scattering is an important process because it effectively increases the path length that a light ray travels, as illustrated in Figure 5.2. Increasing the path length will therefore result in even more light absorption.

The attenuation of light with depth in the sea is rapid, as Figure 5.2B reveals for light measurements made in coastal ocean waters. Figure 5.2C gives the percent of total sunlight arriving at the surface of the sea that penetrates to a particular depth; the light disappears quickly with increasing depth, which is characteristic of an **exponential decay**. Light values (I) attenuate exponentially with depth according to a simple equation known as **Beer's Law** (**BOX 5A**),

(A)

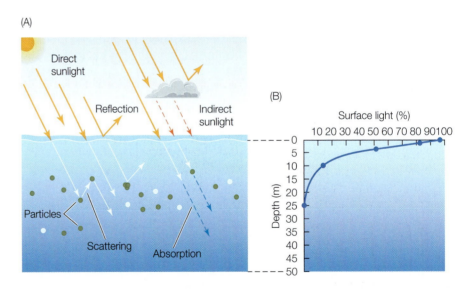

(B)

FIGURE 5.2 (A) Some of the sunlight that strikes the sea surface is reflected; some is absorbed by dissolved and particulate materials, and some is scattered by particles, including bubbles. Scattering increases the path length of a ray of light, which allows more light to be absorbed. This attenuation, or loss, of light with depth is rapid—it is exponential—as shown by (B) the plot and (C) the data table.

(C)

Depth below surface	0	1 cm	10 cm	1 m	3.4 m	10 m	25 m
Surface light (%)	100	99.8	98.0	81.8	50.6	13.5	0.67

BOX 5A   Beer's Law $I_z = I_o e^{-kz}$

The rate of light attenuation—the rate at which light, abbreviated "I," is lost as it penetrates deeper into the ocean—is constant with respect to depth. This loss of light as it penetrates downward can be expressed by a simple differential equation, which states that the change in light (dI) as depth changes (dz) is a constant (k) with respect to depth (z). That is:

$$\frac{dI}{dz} = -kz$$

The solution of this equation is known as Beer's Law:

$$I_z = I_o e^{-kz}$$

where z = depth in meters; I_o = amount of light received at surface (e.g., 100%); I_z = amount of light that penetrates to depth, z, in meters and k, the diffuse attenuation coefficient, is a constant with units of m^{-1} (we use

negative k to indicate loss of light).

Let's work through an example. For an attenuation coefficient of 0.17, what is the depth of the 1.0% surface light (I_o)?

$$I_z = I_o e^{-kz}$$
$$k = 0.17$$
$$I_z / I_o = 0.01 = e^{-kz}$$
$$ln\,(I_z / I_o) = -kz = -4.065$$
$$z = -4.065 / -0.17$$
$$z = 27.08 \text{ m}$$

Examples of how light is attenuated with depth for two values of k are shown in the graph here. Notice how light attenuates faster with depth for water with a greater value of k. At 10 m depth, the water sample with $k = 0.20$ has only 13% of the surface light remaining, whereas the water with $k = 0.10$ has 30% remaining at 10 m.

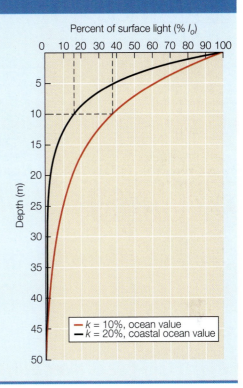

which is $I_z = I_o e^{-kz}$. It holds that the *rate* of light attenuation (the rate at which light is lost as it penetrates deeper) is *constant*. That is, at each depth interval through which that light passes—every meter, for instance—a certain percentage of that light is lost. For example, if there is a loss of 50% of the light for every meter light penetrates, then we would have one half of the surface light at 1 m depth; at 2 m, we would have half that, or 25% of the surface light; at 3 m, we would have half that, or 12.5% of the surface light; etc. The coefficient used here, 50% or simply 0.5, is our "constant rate" and is referred to as an **extinction coefficient**. It can range from 0 to 1.0. An extinction coefficient of 0 means that no light is lost, and 1.0 means that all the light is lost.

In oceanography we refer to the extinction coefficient as the **diffuse attenuation coefficient**, which has the symbol k in Beer's Law. Because it determines the amount of light lost to absorption and scattering, k is a function of the degree to which the water is clouded with dirt and other foreign materials. If there is a lot of dissolved material coloring the water, as there is in a cup of tea, for example, then k is relatively large; this is often the case in coastal waters. Also, if there are lots of particles suspended in the water, as there is in "muddy" river water, then k will likewise be relatively large (but still between 0.0 and 1.0; **BOX 5B**). The larger the value of k, the faster the light disappears with increasing depth. The figure in Box 5A compares how light penetrates at two values of k: 0.1 or 10%, which is a typical open ocean value, and 0.2 or 20%, which is a typical coastal ocean value. It shows, for example, that in open ocean waters, about 38% of surface light reaches a depth of 10 m, but in coastal water, only 13% of surface light reaches that depth.

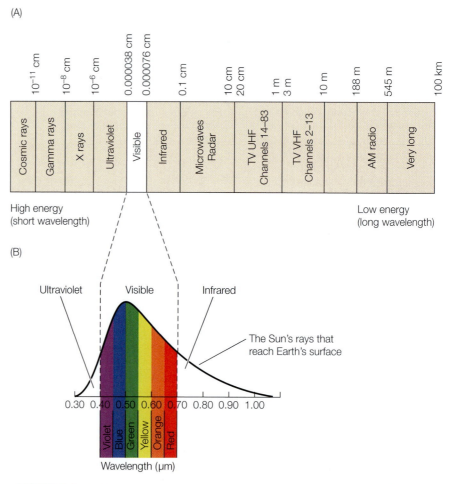

FIGURE 5.3 (A) The spectrum of electromagnetic radiation, which is energy that is propagated as waves, even in a vacuum. It includes radiation that ranges from very high-energy, short wavelengths to very low-energy, long wavelengths. The spectrum includes visible light—waves of energy that we can see with our eyes—but this is only a very small part of the spectrum. (B) The visible band of wavelengths includes all the colors from violet, the shortest visible wavelength at about 0.4 micrometers (μm), to red, the longest visible wavelength, at about 0.7 μm. Notice that not all the visible wavelengths of light arrive at the Earth's surface with the same intensity; differences in intensity are indicated by the height of the color bands in the bar graph.

Not only does the *quantity*, or total amount, of light change with depth, the *quality* changes as well; in this case, quality refers to wavelength, which is the color. To explain, we need to review a little more about what exactly light is (**FIGURE 5.3**). Light is a form of electromagnetic radiation. It occupies a portion of the **electromagnetic spectrum**, which spans the range of wavelengths from cosmic rays (very high-energy, short wavelength waves) to radio waves (low energy, long wavelength waves). The electromagnetic spectrum includes *visible light*, the part of the spectrum that is visible to human eyes, between the wavelengths of about 400 to 700 nanometers (nm), or 0.4 to 0.7 micrometers (μm).[1] And, not coincidentally, notice

- A nanometer (nm) is 10^{-9} meters
- A micrometer (μm) is 10^{-6} meters
- A millimeter (mm) is 10^{-3} meters

So: 400 nm = 0.4 μm = 0.0004 mm = 0.0000004 m.

[1] It is because of these visible wavelengths that light microscopes are limited in their resolving power. They are capable of magnifying only so much, and they cannot focus on anything smaller than the wavelengths of visible light. For example, a marine bacterium, which is about 0.4 μm in diameter, is just a blurry spot—no matter how expensive your microscope is.

BOX 5B   How to Measure Light Attenuation

The value of k can be estimated in a number of ways. The best way is to use an underwater light meter. Another older, more primitive, but nonetheless surprisingly simple way is to use a Secchi Disc.

A Secchi Disc is just a steel disk, about 18 inches in diameter, usually painted with alternating black and white quadrants, which is lowered into the water until you can't see it anymore. The depth in meters where it disappears from view is the Secchi Depth (in meters). It doesn't matter how big the disk is, it is just a matter of how deep you can lower it and still see it. This is sort of like watching the bottom disappear from sight as you row a boat into deeper water. The depth where you can no longer see the bottom is the same as the Secchi Depth. So, k is estimated as:

$$k = \frac{1.7}{\text{Secchi Depth (m)}}$$

Also, the Secchi Depth multiplied by 2.5 is nearly always equal to the 1% light depth, which is the depth where all but 1% of the sunlight has disappeared. The 1% light depth is an interesting metric because usually it is the maximum depth where there will be enough sunlight for photosynthesis (generally speaking, that is; there are exceptions). Thus we call this the depth of *the euphotic zone* ($\approx$ 1% surface light [I_0]).

FIGURE 5.4 There is selective attenuation of light with depth in the sea. That is, not all colors, or wavelengths, of light penetrate equally down through the water column. The light that is attenuated least is in the blue and blue-green range, while red light is attenuated quickly.

in Figure 5.3B that most of the light from the Sun that penetrates the atmosphere is in the *visible range*, which makes sense in evolutionary terms, in that we have evolved eyes to be sensitive to this range.

Not all wavelengths of visible light penetrate water equally. We can see in **FIGURE 5.4** that there is **selective attenuation** of longer wavelengths of light as they penetrate deeper into the ocean; that is, red light is absorbed most—it is the first color to disappear. Because the longer wavelengths, especially red, are absorbed before they penetrate very deep, scuba divers see mostly gray colors at depth underwater, rather than bright red or orange colors. For this reason, underwater color photography in relatively deep waters requires a flash in order to see the true colors.

Also, this same phenomenon—selective attenuation—is why the sea is blue to blue-green in color. It is precisely because the blue wavelengths of light are attenuated least that they are more likely to be scattered back out of the water to our eyes. (The same phenomenon holds for the atmosphere—sort of, anyway: blue light is both absorbed and scattered significantly by air molecules, while the other wavelengths pass though; thus the strongest wavelengths of light making it to our eyes are blue, coming from all angles overhead.) We turn our attention now to how this light energy from the Sun drives atmospheric circulation and our ocean currents.

Atmospheric Circulation

In this section we will start to pull together the seemingly unrelated topics we've been discussing so far, such as: the water molecule and the various physical properties of water; Earth's radiative heat budget; the behavior of light in the sea; even latitude, as it constrains solar radiation and controls the seasons. But first, in order to fully appreciate the dynamics of ocean currents, the effects the oceans have on climate and weather, and how seasonal differences in weather, winds, and ocean currents respond to seasonal patterns of wind and weather, we need to take a closer look at air itself.

Air

Chemically, air is made up of about 78% nitrogen (N_2) and 21% oxygen (O_2) by volume, with the remaining 1% made up of other trace gases, which include carbon dioxide (CO_2). But water vapor can be as much as 4% of the volume of air. And when air has more water vapor, the percentages of oxygen and nitrogen become less (we'll come back to this later).

Notice that we are using volume percentages to refer to the elemental composition of air, and not the weights of those elemental components, as we did when we discussed the various materials dissolved in seawater. The reason for this goes back to the Italian scholar and scientist, **Amadeo Avogadro** (1776–1856) who in 1811 determined that equal volumes of gases contain the same number of molecules, regardless how tiny or how massive those molecules are (**BOX 5C**). Based on this simple but profound discovery, we can immediately see that a liter of nitrogen gas and a liter of oxygen gas have the same number of molecules but different weights because oxygen is heavier than nitrogen. Therefore, their densities will be different. An important point we want to get across here is that air has mass—not a lot compared with water, of course, but it has mass, nonetheless—and therefore air has density.

Density of air and atmospheric pressure

Air may not be very heavy, but its mass becomes impressive nonetheless when we add it up. For example, a square meter column of air (1.0 m^2) that extends from the ground to the top of the atmosphere (some 40–50 km high) weighs about 10,000 kg (**FIGURE 5.5**). The weight of that air on the surface of the Earth exerts a pressure on everything around it, called **atmospheric pressure**. At sea level on the Earth's surface, the atmospheric pressure is 10,000 kg/m^2. Put in the English foot-pound system, a column of air one inch square (1.0 in^2) to top of the atmosphere weighs 14.7 pounds, giving a pressure from that weight of 14.7 lb/in^2 (commonly given as psi, or pounds per square inch).

This is a lot of pressure, but we don't normally feel it on our bodies; in fact, we are normally completely unaware of it. But consider this: the average human being has a body surface area of between 1.5 and 2.0 square meters, which means, for example, that you and I experience between 15,000 and 20,000 kg of pressure on our bodies—just from the weight of the air in the atmosphere. Yet we are oblivious to it.[2] However, we can sense very small *changes* in that pressure when we ride an elevator up or down a tall building. As the elevator ascends, we experience less and less atmospheric pressure, because

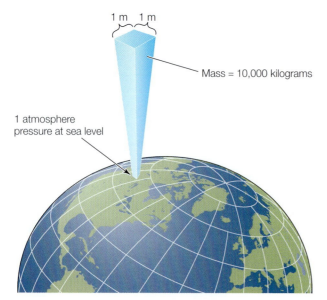

FIGURE 5.5 A column of air, 1 meter by 1 meter square, extending all the way to the top of the atmosphere, about 50 km high, weighs 10,000 kg. That weight means that anything on the surface of the Earth will experience atmospheric pressure of 10,000 kg/m^2, or about 14.7 lb/in^2. Relationship to the Earth is exaggerated.

1 m   1 m

Mass = 10,000 kilograms

1 atmosphere pressure at sea level

---

[2] We are unaware of the great pressures on our bodies because, with the exception of body cavities such as our inner ears and sinuses, we are made up mostly of fluids, which, unlike air cavities, are nearly incompressible. But we can gain an appreciation of this phenomenon with respect to pressure underwater with a simple kitchen experiment (water is much heavier than air, so it helps us in this simple experiment). If you were to put a loose-fitting rubber glove on one of your hands (to create a kind of air cavity between the glove and your hand) and just hold your hands in the air, you'd notice nothing special; you just have a glove on one hand. But if you stick your gloved hand into a pail of water, you will immediately notice the pressure exerted by the water on the glove as it is pressed against your skin. Stick your ungloved hand in the water and you'll notice the difference: that your ungloved hand doesn't feel the same effect of such a slight increase in pressure in the water.

BOX 5C   Avogadro's Number

Amadeo Avogadro determined in 1811 that there is a finite number of molecules (or atoms) of a gas in a specific volume. This is a law of nature: **Avogadro's Number** says there are 6.02×10^{23} molecules in a mole, and this mole of gas molecules (or atoms) will occupy a volume of exactly 22.4 liters no matter what the gas is—water vapor, or H_2, or O_2, etc.

A **mole** is defined as Avogadro's number of atoms or molecules of any substance. Note that because a *mole* is defined as Avogadro's number of particles (atoms or molecules), the *atomic weight* of an element is the weight in grams of one mole of atoms of that element (and molecular weight is the weight in grams of one mole of molecules).

For example, a mole of hydrogen (with 1 proton and 0 neutrons) weighs 1.0 gram; a mole of oxygen (with 8 protons and 8 neutrons) weighs 16.0 grams. And a mole of any other element has the same number of atoms (6.02×10^{23}) as a mole of any other element.

with each passing floor there is less atmosphere above us. We can sense that change in pressure in our air cavities, such as our ears, as the air pressure outside our eardrums becomes lower than that in our inner ears; we also can feel this sensation in our sinus cavities if we have a head cold (which makes it difficult to adjust to those changes in pressure).

Atmospheric pressure is normally expressed as an average, because the actual pressure at any given time and location on the Earth's surface can and does change as a result of changes in the density of air. The density of air changes with temperature, with the pressure that the air is under, and with changes in humidity, (the amount of gaseous water vapor the air contains). We'll discuss each of these factors next.

TEMPERATURE EFFECTS   The effect of temperature on the density of air is straightforward: the warmer an air mass, the less dense it is.[3] Warming an air mass (adding heat) will cause it to expand, and since the density of the air mass is equal to its mass divided by its volume ($\rho = \text{mass}/\text{volume}$), its density will decrease. This will cause that less dense air mass to rise to an altitude where it is surrounded by air of the same density. Cooling an air mass (extracting heat) will cause it to contract, so its density will increase, causing it to descend. It will sink through the atmosphere, stopping only when it is surrounded by air of the same density or when it arrives at the Earth's surface.

PRESSURE EFFECTS   Unlike water, air is very compressible, so pressure effects are very important in the atmosphere. If we compress air (e.g., increase pressure) or if we reduce pressure, the volume will change accordingly. High pressure "squeezes" a volume of air, making its volume smaller; lower pressure allows it to expand. And because density (ρ) = mass/volume, the density of the air mass will also change in response to pressure.

Air pressure changes dramatically with altitude and is determined by how much air there is above a given altitude and the weight of the air above that level. Therefore the pressure acting on an air mass is proportional to its eleva-

---

[3] An air mass is a large body of air with relatively uniform temperature and humidity, and hence uniform density.

FIGURE 5.6 (A) Atmospheric pressure in millibars with altitude above sea level. Notice that the relationship is greatly curved and unlike that for pressure with depth in water, which is linear (B). Notice that the pressure scale for depth in the ocean is decibars, or tens of atmospheres, versus that for altitude, which is in millibars, or thousandths of an atmosphere. Relationship to the Earth is exaggerated.

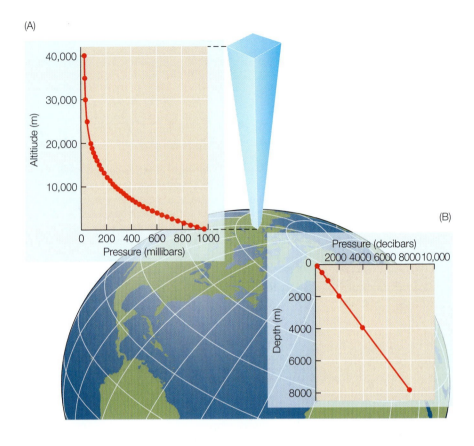

tion, or altitude, above sea level.[4] This is illustrated in **FIGURE 5.6A**, where the pressure is plotted from sea level to 40,000 m (40 km) altitude, above which there is very little atmosphere left; at 40–50 km, we are at the edge of space. Because of the compressibility of air, the change in atmospheric pressure with altitude is curvilinear, being greatest nearest the surface where the air is compressed and therefore densest. This results in an initially rapid drop in pressure with increasing altitude, but above about 10,000 m, for instance, the pressure does not decrease nearly as fast. This is unlike the change in pressure with depth in the ocean, where, because water is far less compressible than air, pressure increases linearly from the surface to the seafloor (**FIGURE 5.6B**).[5] Regardless of the shape of the curve, we can see from Figure 5.6 that at higher altitudes, the atmospheric pressure is lower. For example, on the top of a very tall mountain there is not much air—the atmosphere gets thinner and thinner with increasing elevation—so the air pressure up there is low.

A change in pressure affects not only the volume of an air mass, but also the temperature of the air mass undergoing that pressure change; this is because compressing an object will cause it to get warmer. Therefore, when a volume of air descends and the pressure on it increases, it becomes compressed and gets warmer, and vice versa. (In Chapter 2, we discussed this basic principle with respect to the Earth's deep interior, which we compared with a copper penny squeezed in a bench vise; compression heats it up.) For example, we

---

[4]We use the term *elevation* to refer to the height of *land* above sea level, as in the elevation of a mountain; we use *altitude* to refer to height above sea level in *air*, as in the altitude of a cloud or an airplane.

[5]The reason the change in air pressure is curvilinear is that air is compressible; it compresses under its own weight, and so its density changes in relation to altitude, increasing from the top of the atmosphere to the surface of the Earth. Water, on the other hand, is far less compressible. And while its density, and hence its weight, also change (with temperature, salinity, and pressure), that relative change is far smaller than what we see for air.

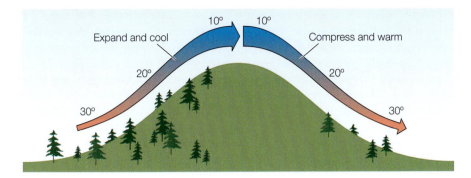

FIGURE 5.7 Illustration of what happens to the temperature and volume of a mass of air that is forced to ascend or descend in altitude. Ascending air becomes surrounded by lower pressure as it rises, which allows that air to expand and become cooler. Descending air encounters greater pressure, becomes compacted into a smaller volume, and is heated in the process.

compress air when we fill an air cylinder, such as a scuba tank; as air is forced into the tank, the air inside becomes not only denser, with more molecules being packed into the confined volume of the tank, but also warmer. In fact, the air and the tank holding it get very warm (too hot to hold!), which is why a scuba tank is placed in a container of water when it is being filled.

Conversely, reducing the pressure on an air mass will allow it to expand, which is what happens to an ascending air mass; as it rises, it expands and cools. Referring back to the scuba tank analogy, we can reverse the situation: when we open the valve on the tank and allow the tank to empty, the escaping air will greatly expand and cool, condensing and freezing any water vapor in the escaping air immediately as it exits the tank, forming ice on the valve.

These phenomena are illustrated in **FIGURE 5.7** for a volume of air that changes its position vertically, such as often happens when a wind blows up the slope of a hill or mountain. The atmospheric pressure surrounding that air mass decreases as it increases in elevation, which allows it to expand and therefore become cooler. And on the other side of that mountain, should the wind continue to blow, that same air mass will descend, contracting into a smaller volume and therefore becoming denser again, and as it does so, it will get warmer.

It is important to note here that changes in altitude or elevation are not the only factors that change pressure—atmospheric pressure at sea level can change without such vertical motions (more on this later).

HUMIDITY EFFECTS   Humidity is a measure of how much water vapor is contained in a volume of air. This is usually given as the **dew point** temperature,[6] which is the temperature at which the air has become saturated with water vapor and cannot hold any more; should the air temperature drop, the water vapor will condense into liquid form. Warmer air can hold more water vapor than cooler air. Recall that water vapor can be as much as 4% of the volume of an air mass; humidity is variable. But as the humidity changes over the range from 0 to 4%, the density of that air also changes. This is because, as Avogadro's law states, for every molecule of water vapor that creeps into a volume of air, a molecule of either oxygen gas (O_2) and/or nitrogen gas (N_2) gas (which together make up 99% of air) is displaced. Therefore, because of the differences in the molecular weights of O_2, N_2, and water vapor (H_2O), changes in humidity change the density of the air mass.[7] The end result is that air that becomes more humid becomes less dense and vice versa.

---

[6] Humidity is also expressed as *relative humidity*, the ratio of the water vapor content of an air mass at a given temperature to the saturation value for that same temperature. Normally the ratio is multiplied by 100 and expressed as a percentage.

[7] The molecular weight of O_2 is 32 g/mole, which is greater than that of N_2 (28 g/mole) which is, in turn, greater than that of H_2O (18 g/mole). Therefore, humid air is lighter than drier air because H_2O (a light molecule) has taken up a volume of air previously occupied by a molecule of either nitrogen gas (N_2) or oxygen gas (O_2), each of which is heavier than water vapor.

Many people assume the opposite—that humid air is denser than drier air. Sure, humid air does feel heavy and thick, especially if it is a warm day, but this effect is an illusion: we feel the "heaviness" of the air because we are less able to evaporate sweat from our bodies when the atmosphere around us is already near the dew point—the air cannot absorb much more water vapor and so evaporation off our bodies is impeded. It is interesting to note that this mistake is commonly made with reference to the flight of a potential home run in baseball: it is often said that the ball is "held up" in the "dense humid air," but this is wrong; the exact opposite is true. There is greater resistance to the flight of a baseball in cool, low-humidity air, which is denser. This general misconception is especially pronounced, for some reason, with Major League baseball announcers.

Pressure also influences the amount of water vapor that an air mass can hold, but the relationship is counter-intuitive. For example, we might expect that increasing pressure would be like squeezing a sponge, and that the water would precipitate out of a high-pressure air mass, but in fact, the atmosphere works just the opposite way. "Squeezing air," increasing the atmospheric pressure, allows an air mass to hold more water vapor, not less. Decreasing the atmospheric pressure, on the other hand, results in the air losing its water vapor—the water vapor will condense from a gas to a liquid. Sometimes you can see this phenomenon when you quickly pop the cap off a bottle of Coke; releasing the pressure quickly can actually produce a small cloud of tiny water droplets in the neck of the bottle (this cloud, or fog, is liquid water). As we'll see later, it is precisely this low atmospheric pressure that is the basis for rainstorms and other severe weather phenomena.

Solar heating of the Earth and atmosphere

Coming back to sunlight and how it heats the Earth, it is important to realize that the Sun's rays are not evenly distributed on our spherical Earth. The intensity of solar radiation varies with latitude and with season.

UNEVEN HEATING WITH LATITUDE   The Earth is a sphere, and therefore the Sun's rays strike the Earth's surface at angles that depend on the latitude and the particular time of year; as a result there is differential heating with latitude. For example, as shown in **FIGURE 5.8**, which illustrates the Northern Hemisphere summer, the Sun is directly over the Tropic of Cancer (23.5°N), so its rays strike the surface of the Earth more directly there. At latitudes farther north and or south of the Tropic of Cancer, the Sun's rays strike the Earth's surface at a greater angle, thus spreading the Sun's energy over a larger area

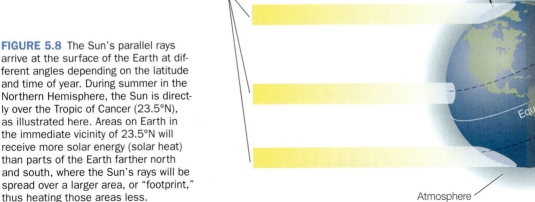

FIGURE 5.8 The Sun's parallel rays arrive at the surface of the Earth at different angles depending on the latitude and time of year. During summer in the Northern Hemisphere, the Sun is directly over the Tropic of Cancer (23.5°N), as illustrated here. Areas on Earth in the immediate vicinity of 23.5°N will receive more solar energy (solar heat) than parts of the Earth farther north and south, where the Sun's rays will be spread over a larger area, or "footprint," thus heating those areas less.

(a larger "footprint"). In addition, because of their greater angle of incidence, the Sun's rays will be more likely to reflect into space, effectively skipping off the inclined surface of the Earth.

Because of the latitudinal differences in the Sun's footprint and reflection, it stands to reason that the tropics and the subtropics (those areas that fall between the Tropics of Cancer and Capricorn, 23.5°N and 23.5°S) receive more of the Sun's heat over the course of a year than latitudes farther north or south, which is in fact the case (see Figure 1.6).

Data plotted in **FIGURE 5.9** show that the amount of solar radiation that the Earth absorbs over the course of a year is greatest in the tropics and subtropics, and least at the higher latitudes. This is what we would predict based on what we've been discussing (except for the slight dip in the plot near the Equator, which, we noted in Chapter 4, is due to the band of clouds encircling the globe there). Using satellites, it is possible to measure the amount of radiation that *leaves* the surface of the Earth as well, which Figure 5.9 reveals is not the same as that being *absorbed*—there is an imbalance. In the tropics and subtropics, more radiation is received and absorbed by the Earth than is emitted, but the opposite is true for the higher latitudes, where the Earth's surface is—somehow—emitting more radiation than it is receiving. Since the tropics and subtropics are gaining more heat than they emit, it would seem that they should be getting warmer with each passing year. And since the higher latitudes emit more than they receive, it would seem that they should be getting colder. But neither is actually happening, which means that something else is going on.

In order to balance this apparent inequality in heating of the Earth's surface, and to explain the measured excess heat radiating out to space at higher latitudes, there needs to be some form of net energy transport mechanism at work. That transport is accomplished by two processes: (1) the atmospheric **latitudinal heat pump**, and (2) the major ocean currents, which we will get to shortly.

In Chapter 4, we discussed the latitudinal differences in global patterns of evaporation and precipitation as they influence the oceans' surface salinity. Those same patterns also are behind a net energy transport system that is driven by water's latent heat of evaporation and precipitation. Remember, most of the evaporation in the oceans occurs in the subtropics, on either side of the Equator near the Topics of Cancer and Capricorn. The heat that drives this evaporation is extracted from the oceans and lower atmosphere there, thus cooling the oceans and atmosphere and helping to offset the excess incoming

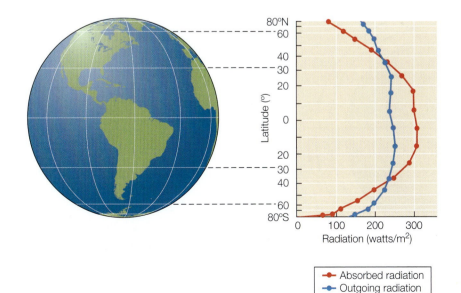

FIGURE 5.9 Pole-to-pole plots of the amount of incoming solar radiation absorbed by the Earth's surface and the amount of outgoing radiation leaving the Earth's surface over the course of a year. The slight dips in each of these plots are the result of clouds along the meteorological equator, at about 8°N.

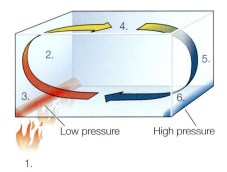

FIGURE 5.10 A simple model of an atmospheric convection cell. (1) If we were to heat one side of a closed box of air, (2) the air on that side of the box would expand as it warms, become less dense, and rise. (3) Atmospheric pressure on the floor on that side of the box would now be relatively low because of the reduced weight of the low density rising air there. (4) Upon reaching the top of the box that air would have nowhere to go except across the ceiling to the other side, where, having lost heat along the way, (5) it becomes dense enough to fall. (6) Atmospheric pressure on the floor of the box on that side is now high relative to that on the opposite side.

FIGURE 5.11 (A) Schematic and (B) three-dimensional diagrams of the theoretical convection cells, one in each hemisphere, that result from unequal heating of the surface of a nonrotating Earth.

solar radiation there. When that same water vapor (the humid air mass) moves to higher latitudes, it condenses and falls as rain, releasing its store of latent heat and warming the atmosphere there. It is this movement of water vapor that transports heat to higher latitudes.

ATMOSPHERIC CONVECTION CELLS The preceding discussion raises an important question: How and why do those air masses move to higher latitudes? The answer is related to differences in air density. Air masses acquire the properties (temperature, humidity) of the Earth's surface over which they reside, whether it be cold, dry land or a warm ocean, which means different areas on the surface of the Earth have air masses of different densities. If a certain geographic area gains or loses more heat than a neighboring area, the resulting differences in air density between the two areas will produce atmospheric **convection cells**.

A simple model of an atmospheric convection cell, represented as a closed box, is given in **FIGURE 5.10**. If the air on one side of the box is heated, causing it to expand and rise, it will create an area of low atmospheric pressure on the bottom of the heated side of the box; the column of air on that side, having been heated and expanded, is less dense, and therefore the column of air from the bottom to the top of the box weighs less. That air rises and is replaced by air that moves across the floor of the box. The rising air encounters the roof of the box and flows laterally across the top of the box to the other side, losing heat as it does so. Upon cooling and becoming denser, it falls back down on the opposite side. This relatively dense, falling air creates beneath it an area of high atmospheric pressure on the floor of the box. The continued heating of air on the warm side of the box, and the continued loss of heat on the opposite side, sets up a circulation pattern of both vertical and horizontal air motions that is a convection cell.

We can apply this simple convection cell model to explain the large-scale atmospheric circulation on the Earth. But before we do this, we first need to make a big assumption: that the Earth is *not* rotating.

On a nonrotating Earth, we would see the general pattern of global air motions diagrammed in **FIGURE 5.11**. Earth would have two huge, three-dimensional atmospheric convection cells, one in each hemisphere, as a result of the differential

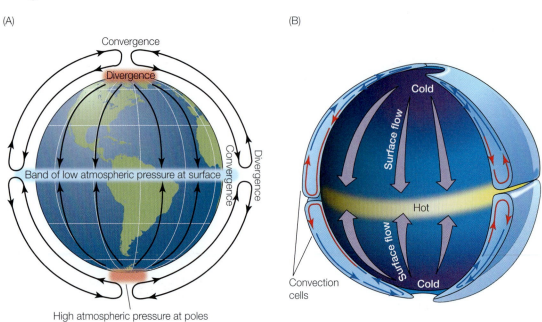

heating of the Earth's surface. The area of greatest warming, in the tropics and subtropics, would create a band of warm, less dense air encircling the Earth. That band of warm, humid, low-density air—a band of low atmospheric pressure—would tend to rise to the top of the **troposphere**, the lower atmosphere. (The troposphere extends to about 20 km altitude at the Equator, and includes nearly all of the air involved in weather phenomena.) Upon reaching the top of the troposphere, the rising air band would divide, and move equally to the north and to the south; this division would constitute an area of high-altitude **divergence**. Beneath the Equatorial band of rising air we would see air rushing in at the surface to replace it, from both the north and the south, which would come together in a **convergence**. Meanwhile, the high-altitude air would move away to both the north and the south until it loses its heat and humidity, thus becoming denser. Eventually it would fall back to the Earth, but (in our model) not before it had made its way to the poles. That falling mass of cold and dry air would create a surface area of high atmospheric pressure at the poles. The falling air in turn would move away from the poles and back toward the tropics, thus completing the single convection cycle in each hemisphere.

This simple model of the global atmospheric circulation pattern produces surface winds that, in the Northern Hemisphere, flow *only* from north to south, and in the Southern Hemisphere, form south to north. The returning high-altitude winds flow in exactly the opposite directions from the surface winds beneath them. The result is a single band of low atmospheric pressure around the Equator, and two areas of high atmospheric pressure, one at each pole.

Based on only the simplest of principles—that the density of air changes with temperature and humidity, and that there is an effective top of the atmosphere, similar to the theoretical convection cell in the box in Figure 5.10—we have "modeled" the Earth's global atmospheric circulation. There is only one problem with this simple but elegant model, and that is: the Earth *is* rotating—which makes things a bit more complicated.

The Coriolis effect

Our theoretical model of global atmospheric circulation cells is made more complicated when we consider the fact that the Earth rotates on its axis once each day. This rotation influences the large-scale motions of air masses in a way not predicted by our simple model. In this section we are going to demonstrate that moving objects in the Northern Hemisphere experience a tug—a *deflection*—to the *right* of their line of direction, and in the Southern Hemisphere, there is an identical, but opposite deflection to the *left*. This phenomenon is known as the **Coriolis effect**, or **Coriolis force**,[8] named for **Gaspard-Gustave de Coriolis** (1792–1843), who worked out the physics behind the deflections on rotating spheres. To explain how this important phenomenon operates, we are going to use simplified graphical models and thought experiments.

Imagine an ordinary phonograph turntable (turntables are becoming obsolete, but surely we are all familiar with them). We have diagrammed one in **FIGURE 5.12**, and have made the record spin in the direction opposite to what

FIGURE 5.12 A thought experiment demonstrating the Coriolis effect using an ordinary phonograph turntable and record, rotating counterclockwise, upon which we place a glass marble that has been dipped in red paint. (A) While the paint is still wet, we flick the marble toward a target across the room. (B) Then we stop the record and notice the path of the marble left by the paint on the record. While we thought we had shot the marble in a straight line, an observer sitting on the record would disagree; to that person, the path of the marble was clearly curved. The evidence is the curved line of paint.

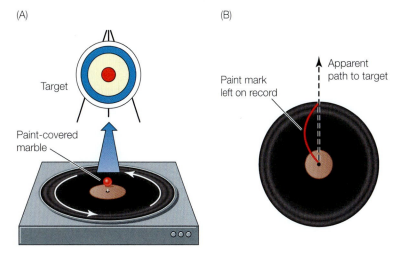

---

[8] This important phenomenon is actually an *acceleration* that produces an *apparent force*, and you will see references to both the *Coriolis effect* and the *Coriolis force*. The terms are used interchangeably.

it normally does (phonograph turntables normally spin clockwise; the reason we've reversed it will be apparent in a minute). If we start the phonograph, place a glass marble covered with wet paint near the center of the record, and then "shoot" the marble—by giving it a quick flick with our finger—across the record to a target some distance away, we will see an interesting phenomenon. If we are a good shot, we will hit the target with the marble and it will seem to us that the marble traveled to it in a straight line. But if we stop the record and examine the path of the marble left by the wet paint, the track on the record will show that the marble curved to the right, rather than traveling in a straight line. Our experiment gives two conflicting results: someone standing beside the turntable watching this experiment will see the marble travel in a straight line to the target; but an observer sitting on the spinning record would see the marble curve to the right. So, which result is correct? The answer is: both are. The trajectory we see depends on our frame of reference.

Here is another thought experiment, similar to the phonograph one, but on a larger scale: imagine that you are out in space looking down on the Earth from a point directly above the North Pole. You are watching a cannonball being fired from the North Pole to a target in Southeast Asia (**FIGURE 5.13A**). It will take several hours for the cannonball to reach its target, some six thousand miles away, and while it is in flight—traveling in a straight line toward the target—the Earth rotates beneath it, effectively moving the target to the east, and moving the Indian Ocean into position to receive the cannonball. To an observer on the surface of the Earth, that cannonball would appear to travel in a graceful arc (**FIGURE 5.13B**), curving to its right; but out in space, you would see quite clearly that the cannonball is flying in a straight line.

These are examples of the Coriolis effect, which causes moving objects in the Northern Hemisphere to be deflected to the *right* of the direction they are headed and moving objects in the Southern Hemisphere to be deflected to the *left* of the direction they are headed. To imagine this opposite deflection in the Southern Hemisphere, just repeat the thought experiment in Figure 5.13, observing from a point above the South Pole. Viewed from there, the Earth would be rotating counterclockwise, right? (Just imagine that you can look behind this page and stare back at the two globes in Figure 5.13; which way would they be spinning?) In the Southern Hemisphere, the cannonball would curve to the left.

This effect, whereby moving objects in the Northern Hemisphere experience a deflection to their right, and vice versa in the Southern Hemisphere, is extremely important in understanding how air masses and ocean currents move

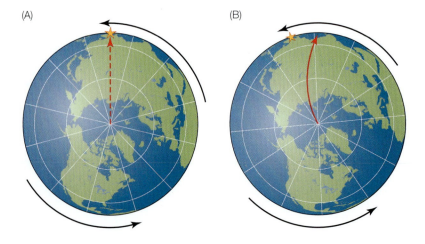

(A)   (B)

FIGURE 5.13 The Earth, viewed from space over the North Pole, spins counterclockwise such that if we repeated our phonograph turntable experiment, we'd see the same thing. A cannonball, shot in the straight line (A) from the North Pole toward Southeast Asia at a speed of about 1500 miles per hour would curve to the right and miss its target. (B) It would instead arrive in the Indian Ocean about 4 hours later.

(not just paint-covered marbles and cannonballs). But the actual deflection, or how much the moving object curves, will depend on how fast the object is moving, how far the object travels over the moving reference frame (the rotating Earth), and the latitude at which the motion happens.

To illustrate the importance of speed, for example, we can imagine using, instead of a cannon, a spacecraft traveling at nearly the speed of light toward Southeast Asia. At that speed, there would be no noticeable Coriolis effect because the object is moving so much faster than the Earth spinning beneath it. Motions over very small distances are also unlikely to exhibit a noticeable Coriolis effect. While such deflections may be unnoticeable, or insignificant, the Coriolis effect nonetheless does operate.

You may have heard that water flushed down a toilet bowl in the Northern Hemisphere will develop a whirlpool that spins counterclockwise, whereas in the Southern Hemisphere it spins clockwise. This anecdote has some truth to it—there will be a Coriolis effect—but there is little chance you will ever witness it. The reason is that frictional effects and the imperfect shape of the average toilet bowl will completely overwhelm any effect of the Earth's rotation, and the direction of the whirlpool will be more a function of the shape of the bowl itself and, especially, the orientation of the openings that allow water to enter and leave. However, it is possible to build the perfect toilet, one that overcomes the shortcomings in your and my bathrooms, and with such bowls, the Coriolis effect has been demonstrated in the laboratory.

As we have just shown, fast-moving objects and motions across small distances will experience an insignificant Coriolis effect. But what about objects such as air masses that are not moving too fast and cover a sufficiently long distance? At those scales the Coriolis effect is indeed important. The deflection will vary with latitude. Let's explore what happens.

The magnitude of the Coriolis effect increases from zero degrees latitude (on the Equator), where there is no Coriolis effect, to a maximum effect at the poles, which means that moving objects at higher latitudes will experience a greater deflection. This is because anything on the surface of the Earth is actually "flying through space" to the east as Earth rotates, at a speed that is a function of latitude (**FIGURE 5.14**), where $\emptyset$ = latitude, and:

$$\text{Speed (statute mph)} = 1000\,[\cos(\emptyset)]$$

If we were standing on the Equator (0° latitude), our speed through space would be about 1000 mph, because the Earth rotates once every 24 hours, and it is about 24,000 statute miles around the world. (Actually, Earth's circumference is about 24,870 statute miles at the Equator, which gives a speed of 1036 mph. For simplicity, we round that off to 1000 mph.) At 60°N we would be moving about 500 mph to the East through space, and at 30°N we would be moving about 866 mph. Because of the dependence on latitude, motions on Earth at higher latitudes will experience a greater Coriolis effect than objects at lower latitudes.[9] This is illustrated in Figure 5.14, which shows the deflections of two moving objects, one

FIGURE 5.14 The speed at which any object on the surface of the Earth is flying through space as Earth rotates on its axis once each day is a function of the cosine of its latitude. At the Equator (0°) that speed is approximately 1000 miles per hour; at 30° (north or south) it is 866 mph; at 60° it is 500 mph. Therefore, an object (red) starting at the position shown that moves to the south will be traveling over an Earth surface that is moving to the east at a progressively faster speed than the object is; rather than flying straight south, that object will be deflected to its right as shown. Furthermore, because it is at higher latitudes, it will experience a greater Coriolis effect than an object (black) at lower latitudes; notice that the northward-moving object starting from South America does not curve to the right as much.

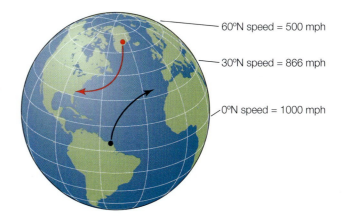

60°N speed = 500 mph

30°N speed = 866 mph

0°N speed = 1000 mph

---

[9] For example, an object at rest on the surface of the Earth at 70°N, where its speed through space is 1000 × cosine(30°), or 343 mph, which moves directly south to 60°N—a distance of 10° latitude, or 600 nautical miles—will find itself over an area of the Earth's surface that is moving beneath it at 500 mph, which is a difference of 158 mph. But an object that moves south from 30°N, where it is moving at 866 mph through space, to 20°N (also a distance of 600 nautical miles) will find itself over an area of Earth's surface that is moving 939 mph, or only 73 mph faster. The curve to the right is thus more pronounced at the higher latitudes.

that starts out just north of the Equator, at about 5°N, and another that starts out farther north, at about 65°N; the trajectories of both objects curve to the right, but the degree of curvature is greater for the object at the higher latitude.

In the two examples just discussed by way of thought experiments (see Figures 5.12 and 5.13), we can envision quite easily the curving motion of an object relative to a moving frame of reference—the rotating Earth—as long as the object's motion is along the north–south axis. That is, we have only considered motions in the north–south directions, which raises the question: does the Coriolis effect also work on objects moving in the east-west directions? The answer is yes. For example, in our thought experiments, notice that as the objects begin to curve to the right (we continue this discussion for the Northern Hemisphere only) they no longer move only along the north–south axis, but they aquire an east–west component to their motion as well, and yet the objects continue to curve. However, our thought experiments really only accounted for the conservation of angular velocity of moving objects along the north–south axis (across a phonograph record, and across a rotating Earth viewed from above), which is an excellent way to visualize moving objects curving to the right. But the Coriolis effect causes moving objects to be deflected regardless of the direction of their motion. In fact, the continued Coriolis deflection would, over time, result in the moving objects tracing out a circular path, bringing the object back to a position very near its original starting point. Called **inertial oscillation**, the curvature (or circumference) of that near-circle will be larger at lower latitudes and smaller at higher latitudes because of the latitudinal change in the Coriolis just discussed, and will depend on the object's speed. But visualizing that continued curving of the object's path is no longer straightforward if we consider only its path relative to the moving reference frame.

As the moving objects in our thought experiments start to move in a north–south direction, they will be deflected, as we just saw demostrated, and therefore they will aquire an east–west component to their motion; as a result they will experience changes in the relative strength of **centrifugal force** acting on them. **FIGURE 5.15A** shows centrifugal forces as vectors (yellow arrows) relative to latitude on the rotating Earth. These are the *apparent* forces that tend to pull a rotating object outward (**centripetal force** is the opposite, and pulls inward to balance centrifugal force). The example most of us are familiar with (and which we use to explain the forces acting on Earth-orbiting satellites; see Appendix A) is that of swinging a bucket of water over our heads in a circular pattern; if we swing the bucket fast enough, the water stays in the bucket even when it's upside-down. Centrifugal force holds the water in, while our arm exerts an equal but opposite centripetal force. On the surface of the rotating Earth, centrifugal force is greatest at the Equator, which is where Earth's rotational speed is greatest, and it decreases to zero at the poles. This is shown by the relative lengths of the yellow arrows in the illustration; they are longest at the Equator.

Gravity, the centripetal force on Earth, pulls us toward the Earth's center of mass, as shown by the dashed red arrow in **FIGURE 5.15B**, which represents the forces acting on an object that is not moving on the surface of the Earth. Here, the centrifugal force and gravitational force vectors do not point in opposite directions the way the forces in our bucket-over-the-head example did (see Appendix A, Figure A)—they do not balance or cancel each other. There would be a resulting net force, vector 1, that would tend to pull objects on the surface of the Earth toward the Equator. This doesn't happen because the Earth is not a perfect sphere—it is significantly fatter at the Equator, which means that sliding along the Earth's surface toward the equator would actually be a slight *uphill* motion. This shape of the Earth resulted back when the early Earth was melted; rotation distorted its shape then, and it remains distorted. Therefore, the Equator-ward centrifugal force (vector 1) and the downhill, gravitational force pulling the object toward the North Pole (vector 2) are exactly equal in magnitude but opposite in direction; they balance one another. The result is that stationary objects on the surface of the Earth do not experience any net horizontal forces, and therefore they do not move. So far so good.

Once an object begins to move—whether north–south or east–west—the forces become unbalanced. The object will experience a change in the magnitude of the centrifugal force vector (yellow in Figure 5.15) which will change the balance between the downhill, pole-ward gravitational force and the uphill, Equator-ward centrifugal force. Therefore, objects moving east, or into the page as we look at **FIGURE 5.15C**, or objects that begin moving north and are deflected to the right, which is to the east, will be moving east faster than the Earth beneath them, and therefore they will experience a greater centrifugal force than when they are at rest, represented by vector 3. The result is a *net* Equator-ward force, shown by vector 4. Consequently, they will be pulled south, which is to the right of the direction of movement. Objects moving west (**FIGURE 5.15D**), or out of the page, will be moving more slowly than the Earth beneath them, and they will therefore experience less centrifugal force than they do at rest by an amount equal to vector 5. The result is a *net* pole-ward force shown by vector 6. Objects influenced by this force will be deflected to their right, which is toward the north. Each motion, east or west, produces a deflection to the right.

The important point in this discussion is that the moment an object begins to move on the rotating Earth, that motion alters the balance between gravity and centrifugal force, regardless of the direction of movement. Therefore, the Coriolis effect deflects moving objects to the right in the Northern Hemisphere to a degree determined by the speed of the object, its latitude, and the period of time it is in motion. This is shown mathematically in **BOX 5D**, along with an example calculation of what happens to a baseball when it is thrown.

If we take the Coriolis effect into consideration, our simple model of atmospheric circulation (convection) cells on a nonrotating globe will change. Now, instead of a single convection cell in each hemisphere, we find that we have three.

Global atmospheric circulation and the effect of Coriolis

If we compare the theoretical atmospheric convection patterns we would predict on a

FIGURE 5.15 (A) Relative centrifugal force on an object at rest on the surface of the Earth that results from the Earth's rotation. (B) Gravity (dashed red line) pulls an object at rest inward, toward the Earth's center of mass, while centrifugal force (yellow) tends to throw the object outward at right angles to the axis of rotation. The result of these two forces would normally pull the object toward the Equator (vector 1), but because the Earth is actually fatter at the Equator, that force is exactly balanced by the downhill gravitational force toward the North (vector 2), and so a stationary object does not experience any net force. (C) An object moving toward the East, or into the page, increases its centrifugal force by an amount represented by vector 3, which results in a net combined force toward the Equator (vector 4)—thus, deflecting its eastward motion to the right. (D) An object moving toward the west, or out of the page, would decrease centrifugal force by an amount represented by vector 5, thus resulting in net combined force to the North (vector 6), which would pull the object toward the pole, deflecting its motion to the right as well.

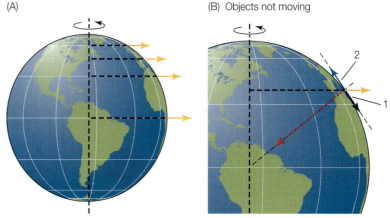

(A)

(B) Objects not moving

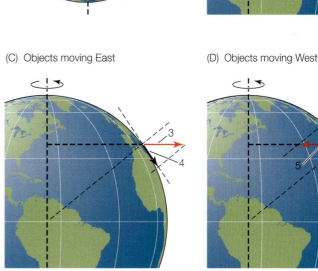

(C) Objects moving East

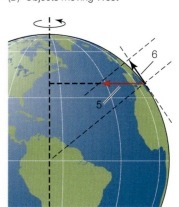

(D) Objects moving West

BOX 5D Mathematical Derivation of the Coriolis Effect on a Baseball

The Coriolis effect, the deflection of moving objects to the right in the Northern Hemisphere and to the left in the Southern Hemisphere, is an acceleration, which can be described mathematically as follows:

$$\text{Coriolis acceleration} = \frac{d^2x}{dt^2} = fv$$

This acceleration is expressed as a change in position x over time t, directed 90° to the right of an object's motion at velocity v. The Coriolis parameter, f, is a function of latitude and the Earth's rate of rotation:

$$f = 2\Omega \sin\theta$$

Ω = Earth's rotational velocity = $7.29 \times 10^{-5} \sec^{-1}$ (radians/second)

θ = Latitude

The solution, giving the change in an object's position, is:

$$x = \frac{1}{2}(fv_0)t^2 + u_0 t + x_0$$

where v_0 = the velocity (m/sec) of the moving body in question, u_0 = the velocity to the right resulting from the Coriolis effect, x = the distance the object is deflected to the right, and t = time (seconds).

For example, at 45°N (where I live), $f = 1.03 \times 10^{-4}$. Setting $x_0 = 0$ (initial position of the object in motion) and $u_0 = 0$ (initial velocity of the object's motion to its right by

Coriolis), the distance that an object will be deflected right by Coriolis (x) at 45°N is:

$$x = \frac{1}{2}(fv)t^2 = 0.515 \times 10^{-4}vt^2$$

Since we have already used a baseball example when we discussed humidity and the flight of a homerun ball, let's see what the Coriolis effect does to a baseball thrown in from the outfield to home plate. We will assume that the distance from the outfield to home plate is 99 meters and that the outfielder throws the ball at 33m/sec (v) with a top-spin only (no curveballs or screwballs). It will take 3 seconds to get there (t), which means, when we plug these values in, that the ball will be deflected to the right by Coriolis approximately 1.5 cm (x)—a tiny deflection, and one that

most certainly would not be noticed by the outfielder throwing the ball or by the catcher at the plate. The ball got there too fast to be deflected very much.

So, now we'll assume the ball is weightless, so it can be thrown slower and still reach home plate without bouncing and messing up our calculations. If the ball is thrown at the ridiculously slow speed of 0.5 m/sec, which is a slow walking speed, it will take 200 seconds to reach home, and will be deflected 103 cm—more than a meter! This, the catcher and the outfielder will notice! This is the speed of a fast ocean current (0.5 m/sec = 1 kt) such as the Gulf Stream. The main point here: slow but distant motions over long distances (like those of ocean currents) are affected most by the Coriolis force.

nonrotating globe (Figure 5.11) with those on a rotating Earth (**FIGURE 5.16**), the first thing we notice is that the situation on a rotating Earth it is much more complicated; but if we walk through it slowly, you will see that the fundamental principles we've been discussing are still preserved. We've just made the winds curve as a result of Coriolis.

In reality, as in our earlier model, the air on the surface of the Earth at the Equator is warmed, on average, more than surface air at higher latitudes. It therefore becomes less dense and rises. As in the nonrotating model, this rising air at the Equator also forms an area of low atmospheric pressure that encircles the Earth. But this is where the similarity ends. On a nonrotating globe, as the air at the Equator rises to higher altitudes, air at the surface rushes in to replace it, but on our rotating Earth those air motions are acted upon by Coriolis; they are deflected to the right in the Northern Hemisphere and to the left in the Southern Hemisphere. Therefore, the surface winds between 30°N and the Equator will be unable to blow straight south to the Equator; they will be deflected to the right. For this reason, the winds in this band blow from the northeast in the Northern

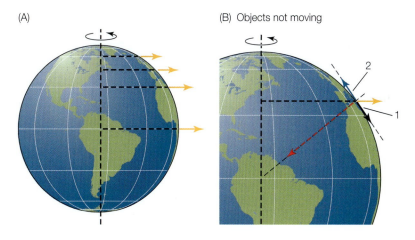

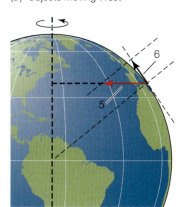

(A)

(B)

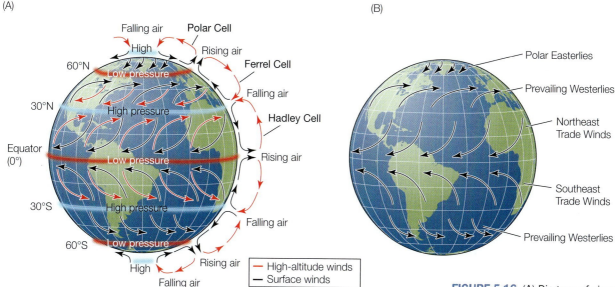

FIGURE 5.16 (A) Diagram of air motions on a rotating Earth under the influence of the Coriolis effect. Three atmospheric convection cells in each hemisphere result: the Hadley Cell, between 0° and 30°N or S; the Ferrel Cell between 30° and 60°N or S; and the Polar Cell between 60° and 90°N or S. (For simplicity, only the two Hadley Cells and the Northern Hemisphere Ferrel Cell have both the surface and high-altitude winds indicated.) (B) The pattern of winds that results on the surface of the Earth from the three convection cells in each hemisphere.

Hemisphere; these are the **Northeast Trade Winds**. The same (but flip-flopped) phenomenon happens south of the Equator, making the **Southeast Trade Winds** between 30°S and the Equator. Directly on the Equator, the winds are light and variable. This is because, for the most part, once those winds from the north and the south approach the Equator, they begin to rise vertically, retaining little or no horizontal motion—this area is known by mariners as the **Doldrums** because of the poor sailing conditions. This rising air is also responsible for the clouds we see hugging the Equator over oceans. The rising humid air expands, causing water vapor to condense, forming tiny droplets of liquid water, which form clouds and rain. The air that rises at the Equator can rise only so high before it reaches the top of the troposphere. And because air from beneath continues to rise, that high-altitude air is forced to move laterally away from the Equator as high-altitude winds—some air goes south, and some goes north. Again, the Coriolis force causes these high-altitude air motions to be deflected to the right in the Northern Hemisphere, and to the left in the Southern Hemisphere, as illustrated in Figure 5.16. Between the Equator and 30°N, directly above the Northeast Trade Winds, are high-altitude winds that blow from the southwest—in just the opposite direction to the Trade Winds. These high-altitude winds, at about 30°N and S, create the **Subtropical Jet Stream** in each hemisphere, which are ribbon-like winds that blow directly east (in each hemisphere) at very high speeds (on the order of 100 mph; see Box 5E).

Also at about these latitudes (30°N and 30°S), the high-altitude air has become dense enough, having cooled and lost humidity, that it sinks back toward the surface. This cold and dry sinking air results in high atmospheric pressure at the surface at 30°N and 30°S. Because this sinking air is low in humidity, there is very little cloudiness or precipitation at these latitudes. Over land, these are generally the latitudes of the world's major deserts; on the oceans, they are called the **Horse Latitudes**, where surface winds are weak.

Looking now only at the Northern Hemisphere, in order to simplify our explanation, we can see in Figure 5.16 that the high-altitude air descends in a band encircling the globe at 30°N, reaching the Earth's surface where it again moves laterally, both to the south and to the north. The air that moves south is acted upon by the Coriolis force, which deflects those winds to the right, thus reinforcing the Northeast Trade Winds, while the air that moves north is also deflected to the right, making the winds we know as the **Prevailing Westerlies** between 30°N and 60°N.[10] It is because of these Prevailing Westerlies that much of the continental United States north of 30° experiences weather systems that cross the continent from the west. In fact, ocean storms that develop in the Pacific Ocean often move across the country to the east coast.

It is at about 60°N that we find the **Polar Jet Stream**, which, like the Subtropical Jet Stream, also flows from west to east (**BOX 5E**); the Polar Jet is created as the Westerlies encounter cold air from the polar region and rise over it, all the while being deflected to the right as that air ascends, creating another ribbon-like, high-altitude, high-velocity wind. The air directly over the North Pole (90°N) is cold and dry. It sinks there, creating high pressure, and spreads to the south (the only direction there is at the North Pole, of course). Those winds also respond to the Coriolis force and are turned to their right, forming the **Polar Easterlies** between 60°N and the North Pole. As the Polar Easterlies approach 60°N, some of their mass has become warmed sufficiently that some air rises, but most of that still-cold Arctic air cuts beneath the air flowing North, thus lifting that air to form the Polar Jet Stream.

From this more elaborate, but still greatly simplified theoretical model of atmospheric convection, we would predict that there is not just one convection cell in each hemisphere, as in the model for a nonrotating globe (as in Figure 5.11), but rather, three cells in each hemisphere. These three cells are called the **Hadley**, **Ferrel**, and **Polar Cells**. They would produce on the surface of the Earth latitudinal belts of high and low atmospheric pressure that extend around the globe (for an ideal Earth, that is), and produce the Earth's major global wind patterns.

Our discussion thus far, as we have been emphasizing, has been theoretical, and you may be wondering: Do we see any evidence of these bands of high and low atmospheric pressure? The answer is yes. Actual measurements show the general pattern quite well, with some important differences. **FIGURE 5.17** shows average surface atmospheric pressure in winter and summer for the globe. Notice the prevalence of generally high pressure at about 30°N and 30°S, as we predicted. But the high and low pressure belts are not continuous; they are broken up into cells by the contrasts (particularly, their different heat capacities) between the oceans and the continents. Notice also in Figure 5.17 that there is a seasonal variance, with the high and low pressure cells moving south in winter. Patterns of atmospheric pressure over the oceans and the continents correspond with the properties of the Earth surface over which they occur. Cold, dry land makes a cold, dry air mass that is relatively dense, and a surface area of high atmospheric pressure. This is an area of sinking air. Warm water, on the other hand, makes a warm and humid air mass, and a

FIGURE 5.17 Average Northern Hemisphere summer (A) and winter (B) atmospheric pressure at sea level in millibars. Notice how the high pressure and low pressure areas correspond with the bands predicted at the Equator and 30°N and S, but are broken into discrete cells.

(A) Summer

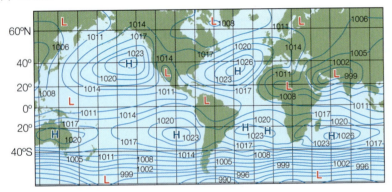

(B) Winter

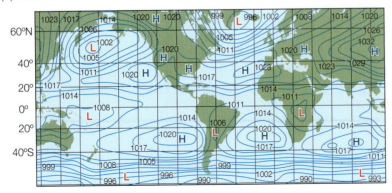

---

[10] Winds are named for the direction from which they blow; a north wind, for example, comes out of the north and blows toward the south. Unlike winds, ocean currents are named for the direction in which they are headed. For example, a westerly current flows toward the west.

BOX 5E   The Jet Streams

At about 30°N and 30°S, the high-altitude winds shown in Figure 5.16 are moving their fastest—those winds have been deflected to their right and are now moving more east than north (**FIGURE A**). These latitudes are where, on average, we find the high-altitude *Subtropical Jet Streams*, which flow west-to-east, and are "ribbons" of fast moving air (>100 mph) at the top of the troposphere (the lower layer of the atmosphere where most of our weather occurs).

At 60°N and 60°S are the latitudes of the *Polar Jet Streams*. These develop similarly to the Subtropical Jet Streams except that they are the result of ascending air currents. As the surface winds of the Ferrel Cell blow initially from 30°N to 60°N, for example, they are deflected to their right by Coriolis, forming the Prevailing Westerlies. But they also ascend as they ride up and over the dense air flowing south from the Polar Easterlies. That rising air also increases in speed as it is deflected to its right (**FIGURE B**).

The Jet Streams, both the Subtropical Jet and Polar Jet, in both hemispheres (four of them in all) flow west to east. **FIGURE C** shows a cross section of the Hadley, Ferrel and Polar Cells, and the Jet Streams between them. Often the Tropical and Subtropical Jet Streams merge into a single "Jet Stream."

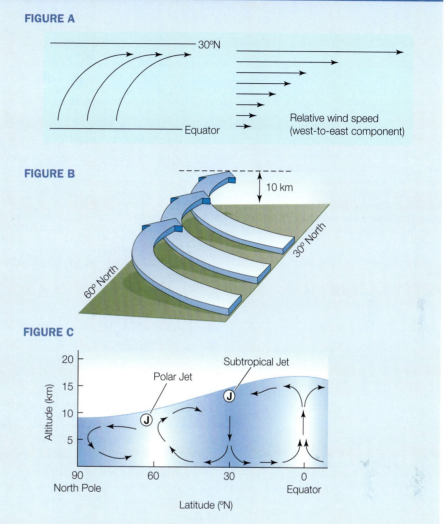

FIGURE A

30°N

Equator

Relative wind speed (west-to-east component)

FIGURE B

10 km

30° North

60° North

FIGURE C

Polar Jet

Subtropical Jet

Altitude (km)

90 North Pole

60

30

0 Equator

Latitude (°N)

---

surface area of low atmospheric pressure. The air here is less dense and will tend to rise. These phenomena produce a seasonal pattern between the oceans and the continents; in summer in the Northern Hemisphere, the land is warmer than the ocean. The warm land heats the air above it, which rises and creates low atmospheric pressure at the surface over land (see Figure 5.17A). On the other hand, the air over the ocean is relatively cool; it sinks and creates high atmospheric pressure at the surface over the ocean. In winter, the reverse is true. Because of the high heat capacity of water, the oceans remain warmer than land, while the land loses its heat more easily. Thus in winter we have high pressure over land (see Figure 5.17B) with descending air falling onto it.

The differences in heat capacities between land and the oceans, then, breaks up the theoretical *belts* of highs and lows into discrete *cells* of high and low pressure. And these high and low pressure cells give rise to various weather phenomena.

Smaller-scale patterns (weather)

In addition to the global patterns of high and low pressure, and their shifts to the north and south with the changing seasons, air masses interact with

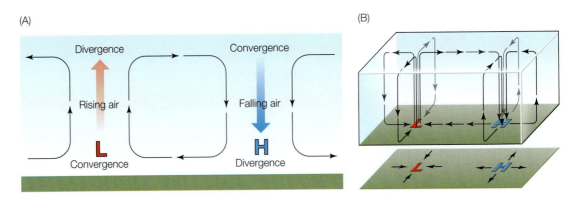

(A)

Divergence

Convergence

Rising air

Falling air

L

H

Convergence

Divergence

(B)

FIGURE 5.18 (A) Schematic cross section through an atmospheric convection cell. Rising air, over a low-pressure area on the surface of the Earth, is coupled with descending air over a high-pressure area. (B) Three-dimensional depiction—without the influence of Coriolis considered.

one another over much shorter periods of time, and over much smaller areas of the Earth's surface, in ways that can result in storms. One kind of storm happens when the boundary—or **front**—separating major air masses becomes enhanced as the density gradient between the two becomes sharper. This is most common in winter, when air temperatures and density differences on either side are most pronounced.

High and low pressure cells on the surface of the Earth are often coupled with one another, as shown in **FIGURE 5.18**. The areas of high and low pressure result from surface (ground-level) areas of convergence (low pressure) that accompany warm and/or humid rising air, and areas of divergence (high pressure) of cold, dry, descending air. Once a localized area of low pressure forms, for example, the air rushing in to replace rising air will begin to encircle the low pressure, because of the Coriolis effect, creating a counterclockwise rotation of winds around the low (**FIGURE 5.19**). The opposite happens around an area of high pressure, where relatively dense air (dry and cold) falls and spreads laterally at the Earth's surface, giving a clockwise rotation around the high. What happens is this: as air rushes toward a low-pressure area on the surface of the Earth in order to replace the air that is ascending, it experiences a deflection to the right by the Coriolis effect. This causes the rushing air to curve to the right before it actually reaches the center of the low, thus producing a counterclockwise rotation around the low. Just the opposite happens in an area of high pressure. As the falling air rushes away from the center of the high, it too is acted upon by the Coriolis force, causing the winds to curve to their right, and thus forming a clockwise rotation around the area of high atmospheric pressure. Sometimes, these areas of low pressure and their associated air motion can become significant, producing full-fledged storms.

EXTRA-TROPICAL CYCLONES Powerful winter storms often develop along the boundary between the Polar Cell and the Ferrel Cell (again, limiting our discussion to the Northern Hemisphere). Already a band of low atmospheric pressure, this boundary is also an area of horizontal shear, where winds of opposite directions—the Polar Easter-

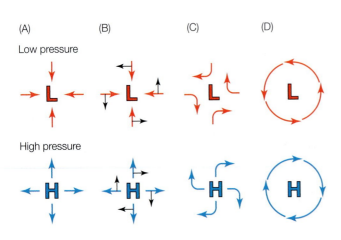

(A)   (B)   (C)   (D)

Low pressure

High pressure

FIGURE 5.19 The development of rotation of winds around areas of low and high atmospheric pressure. As air rushes toward a low pressure area (A) to replace ascending air, it is deflected to the right (B,C) by the Coriolis Effect (black arrows), thus producing a counter clockwise rotation of winds around the center of that low pressure area (D). The opposite occurs for sinking air in a high pressure system. As that sinking air spreads laterally (A), it too is deflected to the right (B,C) by the Coriolis effect, creating a clockwise rotation around the high (D).

lies and the Prevailing Westerlies—meet one another (**FIGURE 5.20**). If a ripple, or meander, develops along that front between the two oppositely moving air masses, then the wind on the north side of that front will be forced to move faster than the wind to the south because of the increased diameter of the curved path on the north side (Figure 5.20C). This is a lot like running around a track in the outside lane, but keeping pace with a runner on the inside lane—you will have to run faster than your opponent. This is important because air that is moving, that is, a *wind*, has lower atmospheric pressure than air that is not moving, or air that is not moving *as fast*. The result is a growing "wave" of low atmospheric pressure in the front, which tends to propagate—to move to the east—along the front, and which may even eventually begin to "pinch off" to form a more intense low pressure cell. That cell is called an **extratropical cyclone** (also called a mid-latitude cyclone, or a *nor'easter*) because it rotates *cyclonically* (around its low pressure center, which is counterclockwise in the Northern Hemisphere). In theory, the frontal boundary where such storms

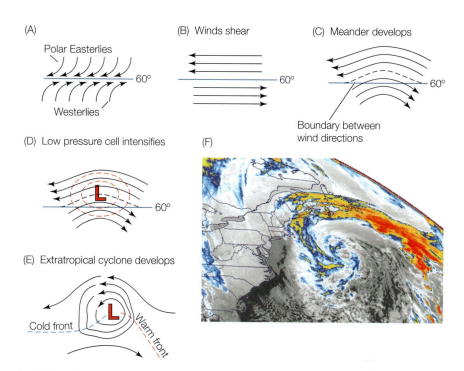

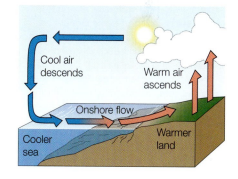

FIGURE 5.20 As an extratropical cyclone, or nor'easter, develops, the frontal boundary between the Ferrel and Hadley Cells, where the Polar Easterlies and Prevailing Westerlies meet (A), is a shear zone, where winds are moving in opposite directions (B). A meander develops on what is already a band of low atmospheric pressure (C), which can intensify (D) into a storm. (E) Colder air from the north circulates around the low and forms a cold front, or leading edge; likewise, warmer air from the south circulates around the low forming a warm front at the leading edge. (F) The "perfect storm" was an example of a particularly powerful extratropical cyclone. This image is a color-enhanced satellite image. The counterclockwise circulation is clearly evident.

develop is located at about 60°N, but it is often displaced farther south in the Northern Hemisphere winter, when the contrasts in air temperatures across that front are greatest, bringing interesting winter weather to southern Canada and the United States.

As storms develop (that is, as the low-pressure cell intensifies, atmospheric pressure drops lower and lower, and more and more air rises), the rising air mass releases its water vapor content (its humidity) forming clouds and rain, which releases even more energy to the storm because of the latent heat of condensation. That heat causes the rising air, which is already expanding as it rises, to expand even more. Thus, under the right conditions (e.g., over relatively warm water), these storms can grow into monsters, an extreme example of which was the "perfect storm" off the U.S. East Coast in October of 1991 (see Figure 5.20F).

SEA BREEZES, LAND BREEZES, AND MICROCLIMATES On a smaller spatial scale at the seashore, we often see a daily pattern of sea breezes and land breezes (**FIGURE 5.21**), particularly at times of the year when there are large differences in the heating and cooling between the land and the ocean. During the daytime, solar heating of the land makes the air directly

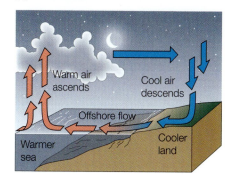

FIGURE 5.21 (A) Sea breezes result during the daytime when air over land is heated more than that over the sea, causing that air to rise and be replaced by the cooler air over water. (B) Just the opposite happens at night, giving a land breeze.

over it less dense, and it rises. This creates low pressure at the surface, and therefore the cooler, denser air from over the water spills in over the land to replace it. The reverse happens at night; the land gives up its heat, becoming an area of high atmospheric pressure relative to the warmer water. Air sinks over the land and flows out over the relatively warmer, low-pressure water.

The rising air over land during the onshore sea breeze also produces clouds. The reason is that the moist air from over the water is warmed by the land, and rises; cooling and lowered pressure as it rises means that it cannot hold as much water vapor. The result is the formation of clouds, collections of water droplets that are too small to fall as rain. Often you'll see this phenomenon while sitting at the beach. That is, there will be clouds above and behind you as you face the ocean, but clear blue skies out over the water. And you can watch those clouds move slowly out over the water, where they soon begin to dissipate. (So, sunny skies over the ocean do not always mean sunny skies right on the beach, which can be quite frustrating to beachgoers.)

A similar phenomenon, but on a somewhat larger scale, occurs when air masses move from over the ocean to land and then ascend over mountains.

BOX 5F Santa Ana Winds

During the fall and early winter we're usually hearing about Santa Ana winds and wildfires in Southern California. What happens is this: High atmospheric pressure builds over the Continental U.S., because the land loses heat more rapidly during the fall than the ocean does. This is especially pronounced over the desert regions of the American West. The winds around the high pressure blow clockwise, and spill over the coastal mountains to the Pacific Ocean, as shown in the figure here. The descending air, which is already dry, becomes compressed as it flows to sea level. This compression heats the air, reduces the relative humidity making the air even dryer than it was over the desert, and accelerates the winds, especially though mountain passes.

The result is often extreme drying conditions for the vegetation on the western side of the coastal mountains, where wildfires become epidemic. The smoke from these extensive wildfires is sometimes visible form space, as this satellite image shows.

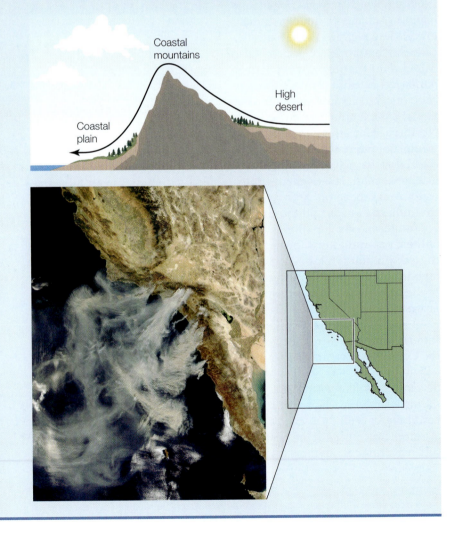

This happens, for example, on the west coasts of North and South America. As the coastal air from the Pacific Ocean rises over the mountains it expands under the lower atmospheric pressure, and produces clouds and rain on the western sides. The descending air on the eastern side of the mountains is therefore dry, having been stripped of its moisture, and the lands to the east of the mountains are apt to be desert; the Mohave Desert in California is an example. These processes are also important in producing the microclimates in such areas as the mountainous regions of the Hawaiian Islands, where it commonly rains far more often on the windward side of the mountainous islands. An extreme example of ascending and descending air masses creates the hot, dry winds in California known as Santa Ana Winds (**BOX 5F**).

HURRICANES   The development of low-pressure cells along a front separating major air masses, such as between the Polar and Ferrel Cells, is not the only way that storms can form; they can develop without a front as well. An extreme example of this is a hurricane. **Hurricanes** are basically thunderstorms that just keep growing (**FIGURE 5.22**). Their initial development is confined to warm ocean waters in the tropics, in regions that have surface-water temperatures warmer than 26°C. Unlike thunderstorms over land, they continue to build strength as long as their source of energy, the warm ocean water (providing warm, humid air), is still beneath them. Initially, the thunderstorm develops when that warm, humid air rises and expands. This expansion causes water vapor to condense into water droplets (that form clouds and rain), releasing latent heat. That latent heat warms the cooler air around it inside the thunderstorm, which causes that air to expand in turn, which becomes less dense, and it rises as well. All this rising air is replaced by more warm, humid air that flows up from the surface in contact with the warm ocean water. And because all that rising air expands with altitude, the storm gets even bigger as that rising air spreads laterally out from the top of the storm. As this continues, the weight of the air above the ocean surface (atmospheric pressure) is greatly reduced. This reduction of surface atmospheric pressure is what is meant by the storm *intensifying*. As these processes reinforce one another, the brewing thunderstorm grows and becomes more and more intense.

As air rushes into the center of the area of low atmospheric pressure, replacing the rising air, it responds to the Coriolis effect; that is, as surface winds blow toward and around the intensifying low pressure at sea level, they develop a counterclockwise rotation.[11] Eventually, the intensifying winds that circle the low-pressure center at sea level become sufficiently large and intense that an "eye" develops in the center of the winds, where cold dry air descends down

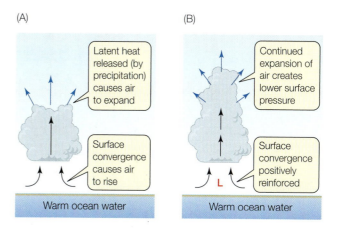

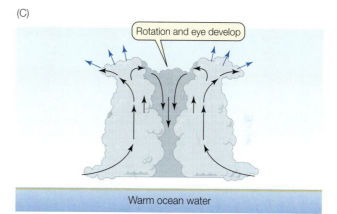

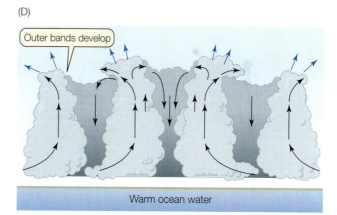

FIGURE 5.22 Cutaway views of initial stages in the development of a hurricane. First, a thunderstorm develops over warm ocean waters (A). As long as the storm resides over the warm ocean, it continues to build, with warm, humid air rising (black arrows), forming clouds and rain that release latent heat, thus adding to the continued expansion of the storm (blue arrows), which in turn draws in more warm humid air from the ocean surface (B). The process continues, and at some point rotation develops around the surface area of low atmospheric pressure, eventually forming an eye (C). With continued strengthening, outer bands develop in the rotating storm (D).

---

[11] At high altitude, however, the clouds may rotate in the opposite direction from that at sea level; the rising air exits the storm at the top, spreads laterally, and rotates in a clockwise direction in response to the Coriolis effect. This produces the spiral-like pattern in upper level clouds that we often see in satellite images of hurricanes.

BOX 5G From Tropical Disturbance to Hurricane

Hurricanes go through a number of stages in their development. They begin as a *tropical disturbance*, which is essentially a very large thunderstorm that lasts for more than a day, and which may or may not have high wind speeds associated with it. These disturbances generally move as wave-like phenomena across the tropical oceans from east to west. They may develop next into a *tropical depression*, when the storm develops rotation and has wind speeds up to 33 knots (kts, nautical miles per hour). The growing storm is termed a *tropical storm* when wind speeds exceed 34 kts, but are less than 63 kts. It becomes a *hurricane* when the wind speed equals or exceeds 64 kts. The intensity, or strength, of a hurricanes is classified on the basis of its wind speeds, from a category 1 hurricane (64–82 kts) to a category 5 (>135 kts); this is known as the Saffir–Simpson Scale.

the center of the storm. The counterclockwise rotation at sea level continues to develop, and outer bands of low pressure form as a result of all the rising and descending air, and the thunderstorm grows into a hurricane (**BOX 5G**).

In order for a hurricane to develop, there are several necessary conditions: (1) Warm waters, of at least 26°C (80°F), must extend from the surface to a depth of about 80 m. If they do not extend to such a depth, the developing storm will mix the waters such that cold deep waters can mix upward, denying the storm of its energy source (the warm water), effectively killing the storm. (2) At the surface, there must be a localized area of low-pressure as the initial thunderstorm grows; this low-pressure cell is needed to generate rotating winds from the Coriolis effect. (3) Preexisting winds over the area of storm formation must come from the same direction at similar speeds at all altitudes. That is, there cannot be a vertical wind shear, which would effectively slice off the top of the developing thunderstorm, thus preventing the storm from growing.

We would expect to find that most hurricanes develop on the Equator because this is where we would expect to find the warmest ocean water, but neither is the case. Although the waters on the Equator are indeed quite warm, the warmest waters in the Atlantic Ocean are just north of the Equator (**FIGURE 5.23**). Surface waters at the Equator might hold enough heat to support hurricanes, but there is no Coriolis effect directly on the Equator. Therefore, hurricanes cannot develop there. South of the Equator the Coriolis effect becomes important again, but the waters there are not warm enough to support the development of hurricanes.

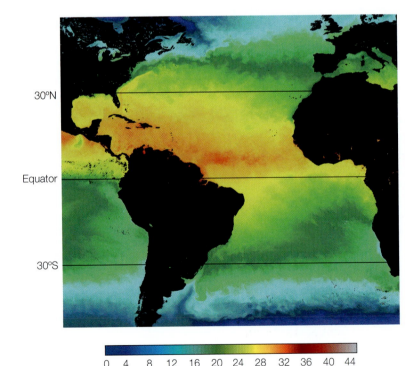

Sea surface temperature (°C)

FIGURE 5.23 Composite satellite image of sea-surface temperature in the Atlantic Ocean for early September 2011; warmer colors correspond with warmer surface water temperatures. Notice that the warmest surface temperatures are north of the Equator.

(A)

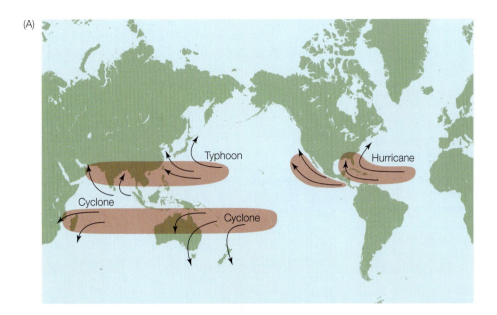

(B)

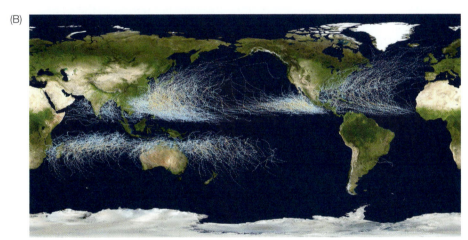

FIGURE 5.24 (A) "Spawning grounds" or areas of the world ocean where hurricanes, cyclones, and typhoons develop (pink), and the average storm trajectories. (B) Storm tracks for all hurricanes, typhoons, and cyclones from 1950 to 2005.

Hurricanes are not confined to the Atlantic Ocean; they also occur in the Pacific and Indian Oceans, where they have different names (**FIGURE 5.24A**). In the western Pacific in the Northern Hemisphere, they are called **typhoons**, and in the Indian Ocean and in the Southern Hemisphere of the western Pacific, they are called tropical **cyclones**. **FIGURE 5.24B** shows the storm tracks, or the paths followed, by all hurricanes, typhoons, and cyclones between 1950 and 2005. Notice that in the Northern Hemisphere they tend to track east-to-west before changing direction and curving to the north, which is to the right of where they are initially heading; they also track east-to-west in the Southern Hemisphere, but there they curve to the left of their track line. In both cases, the storm tracks are steered initially by the prevailing winds—the Northeast and Southeast Trade Winds—and by the Coriolis effect. In the case of Atlantic Ocean hurricanes, once they turn to the north and reach about 30° latitude, they often track "harmlessly out to sea." But the actual tracks are highly variable, and the storms can strike almost any part of the Central and North American coastline.

The particular track a hurricane takes is of obvious importance to those of us who live in hurricane-prone areas. We need to pay attention not only to the

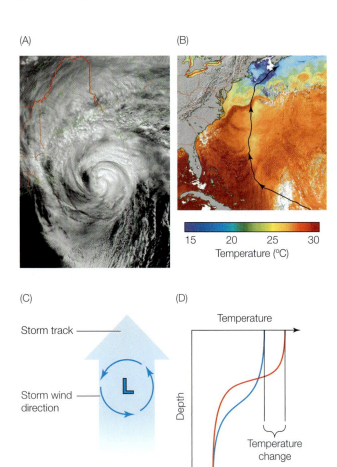

FIGURE 5.25 (A) Satellite image showing cloud pattern of Hurricane Edouard as it tracked up the east coast of the United States in September 1996. (B) Satellite image of sea-surface temperatures with the storm track superimposed. Notice that the waters on the right side of the storm track are cooler. This is because the wind speed on the right side of a hurricane, relative to its storm track, is greater, as illustrated in (C); the direction of the storm track and the storm's wind direction are the same on the right side, so the net wind speed is the sum of the two. On the left side, the wind direction is opposite to the storm track, and the net wind speed is the difference between the two. The high wind speeds on the right side of the storm have mixed warm surface waters downward and cooler waters upward, more so than on the other side of the storm, changing the temperature profile as illustrated in (D). Warm surface waters and a relatively shallow thermocline (red) become redistributed to produce cooler surface water temperatures (blue).

path of the center of the storm when it makes landfall to determine how it may impact populated areas, but also to the winds around the storm's periphery. These are also important because hurricanes are most destructive on their right side relative to the direction they are tracking. This is because on the east side of a storm tracking to the north, the speed at which the storm is moving along its track is *added* to the speed of the winds circulating around the eye, but on the storm's west side, the storm's tracking speed is *subtracted* from its circulating wind speeds. For example, a hurricane in the Gulf of Mexico that is tracking to the north at 20 miles per hour (mph) and which has winds around the central eye of 100 mph, will have winds of 120 mph on the east side, but the winds on the west side will be "only" 80 mph. Winds this strong are extremely destructive, and, for this reason, it is important to have accurate forecasts of storm tracks. One effect of this difference in wind speeds on either side of a northward-tracking hurricane is shown in **FIGURE 5.25** for Hurricane Edouard in 1996. The stronger winds on the right side of Edouard's northerly storm track can be seen to have mixed the surface waters of the ocean, making them cooler on the right side in the storm's wake.

If a hurricane should stall or slow down as it tracks across warm surface waters of the ocean for a period of time, or if those warm ocean waters do not extend deep enough, the storm's wind mixing of the upper ocean could lower the surface water temperature to the point that its source of energy is cut off, and in effect, the storm would kill itself.

When hurricanes make landfall, they not only bring torrential winds and rain, but they can also pile up ocean water ahead of them—a phenomenon called **storm surge**. The winds on the right side of the storm relative to its track direction are not only the strongest winds, but they are also blowing toward the shoreline the storm is tracking toward, while on the left side the winds are blowing offshore. On the right side, then, the winds will push surface water toward the coast, effectively piling up a wall of water in front of it as the storm comes ashore, much the way a snow plow piles snow in front of it.[12] The storm surge from Hurricane Katrina in 2005, for example, was estimated at about 7.6 m (25 ft) above sea level along the coast of Mississippi.

Ocean Circulation

The reason we have been spending time on atmospheric circulation and winds associated with storms is that wind is the main force driving ocean currents. It is wind that translates the energy of the Sun to the surface of the ocean,

---

[12] A second, much less important component of a storm surge is that associated with the low atmospheric pressure, which tends to have an elevated sea level directly beneath it because of the much reduced weight of the atmosphere above the sea surface.

making surface currents. Quite simply, a wind blowing across the surface of the ocean will transfer momentum to the ocean surface, thus driving a surface current. For example, if a wind blows over the surface of water for at least 10 hours, the water beneath that wind will flow downwind at about 2% of the wind speed. The wind's energy is transferred to the water via friction, producing waves, which themselves help to catch the wind energy—like sails. We are going to take this simple principle a step further and see how the global pattern of zonal winds on the surface of the Earth (see Figure 5.16) produces the world ocean's wind-driven circulation. But, again, to do so we need to consider the Coriolis effect.

Ekman currents

When we include the effects of the Earth's rotation, our simple image of a wind blowing across the surface of the ocean becomes a little more complicated. When the wind begins to blow across the surface of some body of water, it will indeed produce a surface current that will flow directly downwind, but at a speed significantly less than the wind speed. At least that's what happens initially. But after some hours of adjustment—a significant portion of a day—the surface current will respond to the Coriolis force and will flow at an angle to the right of the wind that is pushing it (in the Northern Hemisphere). So, right at the surface, where the energy from the wind is imparted to the surface of the ocean, we will have a balance between the wind force pushing the water downwind and the Coriolis force pushing the water to the right; the net result is a thin surface layer of water that flows exactly 45° to the right of the wind (**FIGURE 5.26**).

Below that thin surface layer, the current will do something curious. The now-moving surface layer is itself transferring momentum to the layer of water just beneath it—just as the wind transferred momentum to it. Initially, the deeper water layer moves in the same direction as the water that pushes it, but with time, it also adjusts to the Coriolis force, and thus it is also deflected a few *more* degrees to the right, while losing some energy. Now the average flow of both layers will be deflected slightly more than 45° to the right of the wind.[13]

This process, in which one layer of moving water transfers momentum to the layer beneath it, which responds to the Coriolis force and in turn transfers momentum to the layer beneath it, continues downward through the water column, with each of the deeper layers deflected farther to the right by the Coriolis force while also losing energy. This continuous downward translation of momentum and twisting is known as the **Ekman spiral** (**FIGURE 5.27**).

The actual depth over which these twisting currents develop depends on latitude and wind speed, but is usually several tens of meters. That is, the spiral extends only so deep—to a depth called the **Ekman depth** (or, the depth of the **Ekman layer**).[14] This Ekman layer is achieved because at some point a depth is reached where the *depth-averaged*

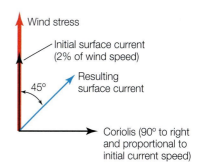

FIGURE 5.26 A wind blowing directly north in the Northern Hemisphere will force a surface current toward the north initially, but after a few hours, that current will experience a deflection to its right by Coriolis that is proportional to its speed. The result is a surface current that flows 45° to the right of the wind direction.

FIGURE 5.27 At the very surface beneath a wind, there will be a current flowing 45° to the right of the wind. Beneath that is a current that has been deflected farther to the right and which is not moving as fast, and so on down through the depths, as shown, making the Ekman Spiral. Eventually we reach a depth where the current is flowing directly opposite to the wind (e.g., 180° from the wind), and at depths deeper than that, there are currents that have been rotated more than 180°. Eventually, we reach a depth where the sum total of the currents flow exactly 90° to the right of the wind, as shown by the broad arrow; this is the Ekman current. Over that depth interval, then, Coriolis is acting 90° to the right of that integrated current. This means that Coriolis and the wind are in exact opposition, oriented 180° apart from one another, and therefore the downward spiral of currents is as deep as it will get.

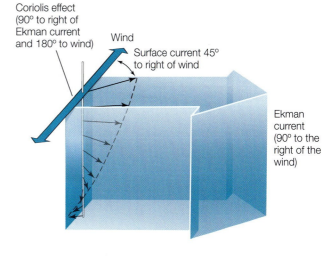

---

[13] Although we are describing discrete layers of water in this explanation, the ocean actually behaves in a vertically continuous manner. Talking about layers just helps us to visualize the process.

[14] Theory predicts that the Ekman depth can extend to some 300 m, depending on the wind speed and latitude. But direct observations show that, except in winter, about 90% of the time the Ekman depth is less than 25 m. Price, J. F., Weller, R. A., and Schudlich, R. R., 1987. Wind-driven ocean currents and Ekman transport. *Science* 238: 1534–1538.

current flows exactly 90° to the right of the wind stress. Even though the current at that depth may be rotated more than 180° from that of the wind that is forcing it (see Figure 5.27), the sum total of those currents—the average flow—will be exactly 90° to the right of the wind direction, and the Coriolis force will therefore be directed another 90° to the right of that. This means that the two forces, the wind and Coriolis are working 180° from one another: they cancel each other out. This current in the top few tens of meters, which flows 90° to the right of the wind, is known as the **Ekman current**.[15]

The preceding explanation is a rather complicated way of stating the following: under the influence of a wind, there will be a near-surface Ekman current in the top few tens of meters that flows exactly 90° to the right of the wind direction. Therefore, in the Northern Hemisphere, if a wind is blowing toward the North across the surface of the ocean, there will be an Ekman current that flows East.

The major ocean gyres

It is because of these shallow Ekman currents that we have the major ocean **gyres**, which are made up of ocean currents that extend to depths of some 800–1000 meters. To explain how the Ekman currents—currents in only the top few tens of meters—can have such a major influence on the circulation of the world's oceans over much greater depths, we need to refer back to the pattern of global winds we discussed earlier. To simplify our discussion, we will focus on only one of the major ocean basins, the North Atlantic.

FIGURE 5.28A shows a simplified schematic of the North Atlantic basin and its major surface wind fields: the Northeast Trade Winds between the Equator and about 30°N and the Prevailing Westerlies between 30°N and 60°N, and how the North Atlantic surface waters respond—by producing, literally, a "pile of surface water" in the center of the ocean basin. The winds produce shallow surface Ekman currents that are directed into the center of the basin, and those currents continue to direct surface water into the center of the basin as long as the winds continue to blow. But the pile of water can get only so high, because the wind is pushing water into a hill against the opposing force of gravity. Something has to give. What happens is that the elevated surface water in the center continuously sinks and is replaced by the surface Ekman currents; then, once those sinking waters are below the surface Ekman layer (and beyond the influence of the winds), they spread laterally. That spreading mass of water beneath the thin Ekman layer, extending to the depth of the permanent thermocline, between about 800 and 1000 m,[16] also responds to the Coriolis force, which causes it to flow as a clockwise gyre in the North Atlantic Ocean, giving us what are generally known as **geostrophic currents**.[17] The way this happens is illustrated in **FIGURE 5.28B**: the mass of water that spreads laterally to the west (on the left of the diagram) is deflected by the Coriolis effect to its right (into the page), while

---

[15] It was Fridtjof Nansen who first called attention to this phenomenon. He noticed that the *Fram*, which was locked in the ice in the Arctic, seemed to drift at an angle to the right of the wind—between 20–40° to the right of the wind direction. Nansen realized that this was the result of the Earth's rotation, but he was unable to formulate a mathematical description of the drift. Upon his return from the Arctic, he gave his data to V. W. Ekman, a physicist, who worked out the math and published his results in 1905. The current might have been called the Nansen current, if only he could have worked out the math himself (there is a lesson here: learn the math).

[16] The permanent thermocline is where the water temperatures get colder with depth; this is where waters that came from the poles may be found. The permanent thermocline is to be distinguished from the seasonal thermocline, which, at higher latitudes, develops each spring and summer and is destroyed each fall and winter.

[17] Geostrophic currents are ocean currents that result from a balance between the Coriolis force and the force from the gradient in pressure produced by the elevated sea level.

(A)

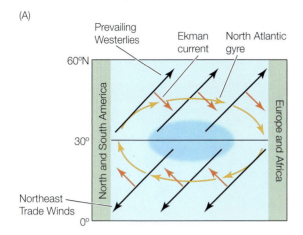

FIGURE 5.28 (A) Schematic representation of the Northeast Trade Winds and the Prevailing Westerlies that blow in the North Atlantic Ocean (see Figure 5.16). The resulting Ekman currents (orange arrows) flow to the right of the wind and force surface water into the center of the North Atlantic Ocean, essentially piling it up to produce slightly elevated sea level there (dark shading). (B) The pile of surface water has nowhere to go except downward, depressing the depth of the permanent thermocline, and spreads laterally, creating currents between the surface and the permanent thermocline that respond to Coriolis, which deflect them to the right in the Northern Hemisphere. The result is the North Atlantic gyre—the yellow arrows in (A)—a large-scale current system that flows clockwise around the entire ocean basin.

(B)

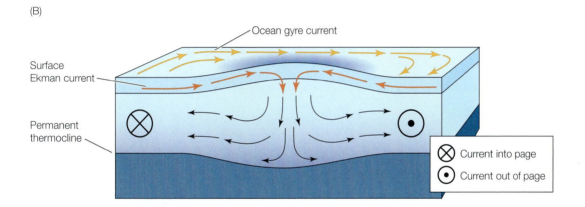

water that spreads east is also deflected to the right (out of the page). In the Northern Hemisphere, these motions produce a giant clockwise pattern of currents in the upper ocean, between the surface and about 800–1000 m, which we refer to as a gyre.

The pile of water in the "centers" of the ocean basins, created by the surface Ekman currents, is real and not just an abstract concept: it can be mapped by satellite altimeters that are capable of measuring sea-surface elevations to within a centimeter. And it can be computed from the vertical density changes in the ocean, which in turn are based on measurements of temperature and salinity, as we discussed in Chapter 4. Sea level is elevated as much as 1.5 m in the North Atlantic (between North America and Bermuda), for example. However, for reasons similar to those that result in the western intensification of boundary currents (discussed next), these hills are not directly in the centers of the ocean basins, but are displaced to the west of center.

The same processes that create the ocean gyre currents in the North Atlantic are also at work in the other ocean basins: The South Atlantic, the North and South Pacific, and the Indian Ocean. **FIGURE 5.29A** is a schematic diagram of those gyres as they should operate, based on the principles described above. We would predict the presence of ocean currents that produce a clockwise rotating gyre system in both the North Atlantic and North Pacific Oceans, and counterclockwise gyres in the South Pacific, South Atlantic, and Indian Oceans. Indeed, those predicted current systems agree well with actual current systems in those oceans (**FIGURE 5.29B**).

(A)

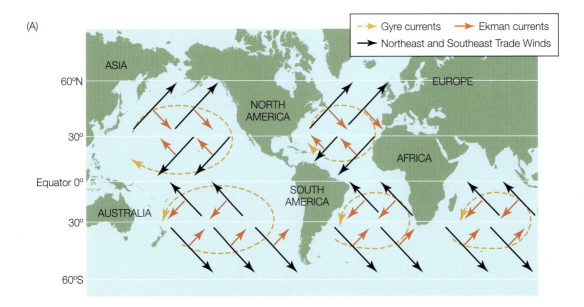

(B)

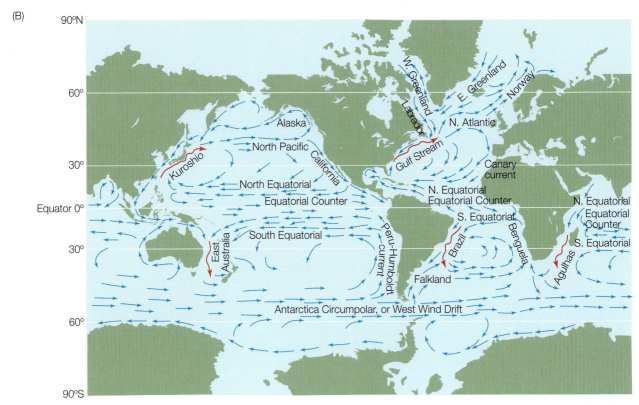

FIGURE 5.29 (A) Schematic diagram showing what we would predict for the gyre current circulation (yellow) in the North and South Pacific Oceans, the North and South Atlantic Oceans, and the Indian Ocean based on the principles in Figure 5.28. The global winds are shown as black arrows, and the resulting Ekman currents are shown as orange arrows. (B) The actual major ocean currents. Two important features shown here are the Equatorial Counter Currents and the intensification of the Western Boundary Currents (the Kurioshio, Gulf Stream, East Australia, and Brazil Currents). Also shown is the Antarctic Circumpolar Current (or West Wind Drift).

Western Boundary Currents

Our theoretical diagram has missed some important features of the world's ocean currents (indicated in Figure 5.29B). For instance, we did not discuss the Antarctic Circumpolar Current, which is an expression of the southern limbs of the ocean gyres in the South Atlantic, South Pacific, and Indian Oceans. This current results from the prevailing westerly winds in the Southern Hemisphere blowing, unobstructed by land masses, literally around the world. Nor did we discuss details of the smaller-scale currents such as those in polar regions, or in the Mediterranean Sea. But two other discrepancies are important and need to be considered: the intensification of Western Boundary Currents and equatorial currents.

Notice that the southward-flowing Canary Current in the North Atlantic is very broad compared with its counterpart on the other side of the ocean, the northward flowing Gulf Stream, which is very narrow (**BOX 5H**). The Gulf Stream is a much thinner and swifter-flowing current, and is the result of a phenomenon known as the **intensification of Western Boundary Currents** and is related to changes in the strength of the Coriolis effect with latitude. Similar intensified Western Boundary Currents (red arrows in Figure 5.29B) are present in each ocean basin; they are the Kuroshio Current in the North Pacific, the East Australia Current in the South Pacific, the Brazil Current in the South Atlantic, and the Agulhas Current in the Indian Ocean. The details of this phenomenon are beyond the scope of this book, but suffice it to say that without the Coriolis effect—if the Earth were not rotating—and with all else being the same, we would see identical current systems on both sides of the major ocean basins. Also shown in Figure 5.29, but which is not readily evident from our discussion thus far, is the presence of Equatorial Counter Currents in the Atlantic, Pacific, and Indian Oceans.

Equatorial currents

Our predicted pattern of major ocean currents did not include Equatorial Counter Currents because it didn't take into account the fact that the geographic Equa-

BOX 5H The Gulf Stream

The Gulf Stream is one of the Western Boundary Currents. It was recognized as an important ocean current even before 1769, when Benjamin Franklin published his famous map (**FIGURE A**). Franklin made a lot of trips to Europe, and he became interested in how ocean currents helped and hindered those voyages across the Atlantic. Matthew Fontaine Maury's later map (**FIGURE B**), published in 1855, was more detailed than Franklin's, but both showed a coherent pattern to the Gulf Stream, which mariners had long ago known was not the case—because the Gulf Stream is actually highly variable.

Much progress was made in our understanding of the Gulf Stream in the twentieth century, especially the 1950s and 1960s. We now know that as a Western Boundary Current, the Gulf Stream is "intensified," as a result of changes in the Coriolis effect with latitude. If there were no Coriolis effect, then the North Atlantic gyre, for example, would be symmetrical, as shown in **FIGURE C**, with equal currents

on both sides of the Atlantic. But with Coriolis, we have swift, narrow currents on the western side of the ocean gyres, such as the Gulf Stream.

It wasn't until satellites were used that we gained a full appreciation of the Gulf Stream's variability. Satellites revealed a very complicated picture indeed, and showed that the Gulf Stream is made up of rich detail, with many contributing currents of varying sizes and velocities, as well as rings (or eddies) and meanders being quite common, as seen in this 7-day composite

image of sea surface temperature, showing clearly the warm waters of the Gulf Stream flowing north (**FIGURE D**). The Gulf Stream, and Western Boundary Currents in general, tend to develop "waves" or "meanders" along their length, which are visible in Figure D (top). These meanders tend to grow larger and some may even break off from the main stem of the current, creating spinning rings of water, called *Gulf Stream rings*.

"Warm core" Gulf Stream rings form on the landward side of the core of the

(continued)

FIGURE A

FIGURE B

BOX 5H   The Gulf Stream *(continued)*

Gulf Stream, thus capturing warmer water from the seaward side. This spinning bolus of warmer water is now surrounded by the cooler water that resides over the slope and shelf. "Cold core" rings do just the opposite, enclosing cooler coastal (slope and shelf) water on the seaward side of the Gulf Stream. These rings can last for many months (sometimes more than a year), especially the cold core offshore rings; the warm core rings on the inside of the Gulf Stream tend to erode away more quickly as they rub up against the shallow continental shelf.

The Gulf Stream becomes the North Atlantic Drift as it continues its path across the Atlantic, bringing its warm waters, and heat, to northern Europe and as such it is important in the latitudinal heat pump. Future changes in the Gulf Stream

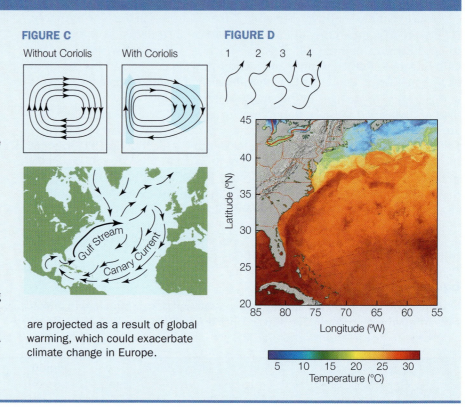

FIGURE C

Without Coriolis   With Coriolis

FIGURE D

are projected as a result of global warming, which could exacerbate climate change in Europe.

tor is not, in fact, where the Southern Hemisphere and Northern Hemisphere Trade Winds meet; they actually meet a few degrees north of the Equator. This **Intertropical Convergence Zone**, or ITCZ, where the two Hadley Cells meet, is also called the **meteorological equator**. It varies in position seasonally, but generally is north of the geographic Equator as a result of the inequality in the distributions of the continents and oceans in the Northern and Southern Hemispheres. (There is more continental land area in the Northern Hemisphere and more ocean area in the Southern Hemisphere.) Because the ITCZ is north of the Equator, the Southeast Trade Winds blow across the Equator (**FIGURE 5.30A**). These winds are acted on by a Coriolis effect that reverses direction on either side of the Equator: in the Southern Hemisphere, the Southeast Trades experience a deflection to the left, but once they cross the Equator, they are deflected to their right. So, between the geographic Equator and the ITCZ we have an eastward wind component (i.e., a westerly wind) that drives an eastward-flowing surface current, the Equatorial Counter Current (**FIGURE 5.30B**). Equatorial Counter Currents are present in each ocean, but they are best developed in the Pacific Ocean. So we have three major Pacific Equatorial surface currents—the North Pacific Equatorial Current, which is part of the North Pacific gyre, and the South Pacific Equatorial Current, which is part of the South Pacific gyre (both currents flow east-to-west), and between them, the Pacific Equatorial Counter Current, flowing west-to-east. Changes in these currents are important in the weather phenomenon El Niño (see Appendix B).

There is also a fourth Equatorial Current that is well developed in the Pacific Ocean, but it isn't a surface current, it is a subsurface feature. It exists because: (1) the North and especially the South Equatorial Currents push a lot

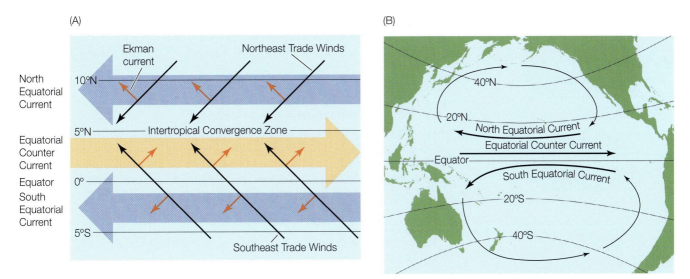

of surface water to the west and thus pile up water on the western side of the Pacific Ocean, and (2) because there is no Coriolis effect exactly on the Equator. To balance the flow of surface water across the Equatorial Pacific—to get water back across the Pacific Ocean to the eastern side—there is the **Equatorial Undercurrent**, or Townsend Cromwell Current, which is a thin, ribbon-like subsurface counter current that is positioned directly at the Equator, and runs beneath the South Equatorial Current (**FIGURE 5.31**). The existence of this unusual current was predicted based on theory, and makes sense if we think about it. If the North and South Equatorial Currents are piling water up on the western side of the Pacific, and are primarily surface currents, then how does this huge flow of water balance out? That is, how does water get back to the east? Does it all circle back as part of the great ocean gyres? Some does, but a lot of it is returned in an interesting and unique way: because there is no Coriolis effect directly on the Equator, then it is possible that a current directly on the Equator can flow toward the east, (that is, it flows "backward"). But because the South Equatorial Current is flowing at the surface and straddles the Equator, this backward-flowing current must lie beneath it, at a depth deeper than the Ekman layer, where there is no wind influence. It can flow "backward" (from west to east) because Coriolis, in essence, "pulls" it back to the Equator, keeping it there any time it veers to the north or the south. It is stuck on the Equator. Interestingly, such a current could not flow east-to-west. An east-to-west current would be incapable of staying on the Equator, because any deviation or meander to the north or south, as well as any part of the current's flanks that extend across the Equator, would expose it to Coriolis, which would pull it away from the Equator (see Figure 5.31).

Density-driven thermohaline circulation

So far we have learned that one force behind ocean currents is the wind blowing across the surface of the ocean, which creates motions that respond to the Earth's rotation, giving us the system of major ocean currents, boundary currents, and equatorial currents at the surface and subsurface. These all fall under the general category of the **wind-driven circulation** of the world ocean. They are motions created in the Ekman layer (top few tens of meters) which influence water motions down to depths of 800–1000 m. But, the ocean averages 4000 m. What about currents down there? Do those waters flow, or are they stagnant? The deep waters of the world's oceans between 1000 m and the bottom do move, but their circulation is density-driven rather than

FIGURE 5.30 (A) The Southeast Trade Winds cross the geographic Equator to the ITCZ, where the two Hadley Cells meet. There, they experience a Coriolis force that reverses direction on either side of the Equator, producing first an east-to-west current (the South Equatorial Current), which is not just south of the Equator, but also straddles it, and second, just north of the Equator, a west-to-east wind. The Equatorial Counter Current lies between two major ocean gyres (B).

FIGURE 5.31 An eastward-flowing current on the Equator that veers either north or south will be pulled back to the Equator by Coriolis. If it ventures north, it experiences a pull to its right, redirecting it south back toward the Equator. If it ventures south, it experiences a pull to its left (in the Southern hemisphere), redirecting it north and back toward the Equator. But a current flowing in the opposite direction, from east-to-west, would be pulled apart by Coriolis. It is this principle that allows there to be a subsurface current, flowing west-to-east, directly on the Equator, beneath the South Equatorial Current.

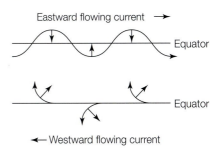

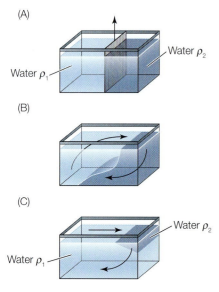

(A)

Water ρ_1

Water ρ_2

(B)

(C)

Water ρ_2

Water ρ_1

FIGURE 5.32 Imagine an aquarium with a moveable divider in the center (A), which has water on one side that is saltier, and hence denser, than the other. If we add food coloring to the denser water and then remove the divider (B), we will see the dense water slip slowly beneath the fresher, less dense water. This is an example of a density-driven current, which, in this case, flows along the bottom of the aquarium from one side to the other. (C) Now imagine the same aquarium without a divider, but which has heat extracted from the surface of one side. That colder, and hence denser, surface water, a wedge shape here, would also slip beneath the warmer, less dense water, creating a similar density-driven current.

wind-driven. These currents constitute the **thermohaline circulation** (from the Greek *therme*, heat, and *hals*, salt) of the world ocean. The principles we discussed in Chapter 4—the density of seawater, the roles that salinity and temperature play in controlling density, and the vertical water column stability that results from vertical variations in density—all form the basis for this discussion of the deep ocean circulation, which can be visualized with the aid of a couple thought experiments.

Imagine an aquarium full of water with a moveable partition dividing it in half, as sketched in **FIGURE 5.32A**. If we were to make the water on one side denser than that on the other, by adding salt, for instance, and color the water on the denser side with food coloring, and then pull that partition out, we would see the "density-driven" currents as sketched. The denser water would sink and slide beneath the less dense water, which in turn would spread over the dense water. In a second experiment (**FIGURE 5.32C**), imagine that we are extracting heat just at the surface of one side of an aquarium (no partition this time) to make denser water. That surface layer of cold, dense water would sink, like the salty water in our first experiment. The difference here is that the denser water, if continually being created at the surface, will continue to sink and spread throughout the aquarium at depth (but only for a while, of course, because the aquarium is pretty small; eventually, it will reach a nearly uniform temperature throughout its volume and freeze).

Out in the deep waters of the real ocean, such density-driven currents are flowing all the time and constitute the deep thermohaline circulation of the world ocean. Being slow and down deep, those currents are invisible, so to speak, to observers on ships at the surface, which has made them difficult to study. But the oceans' deep and bottom waters have been tracked based on their temperature and salinity characteristics, which tend to be preserved, much like the water masses we sketched in the two aquarium thought experiments.

Like air masses, which have temperature and humidity signatures, water masses have temperature and salinity (T-S) signatures. Because T-S characteristics are a lot like fingerprints and can be used to identify and follow water masses as they flow in the deep layers around the world ocean (**BOX 5I**), we can use T-S analyses to study thermohaline circulation patterns. We have already discussed how, under the right conditions (e.g., cold weather, such as at the poles) we can have convective sinking of the ocean's surface waters; this happens only if the waters become cooled enough, such that in combination with sufficiently high salinities, they become dense enough that they produce an unstable water column; the surface water becomes denser than the water beneath it, and it sinks. The deep and bottom waters of the oceans are formed in just such a manner.

FIGURE 5.33 shows a cross section of the Atlantic Ocean, from pole to pole, and how salty waters at the surface in the North Atlantic Ocean sink (upon being cooled sufficiently) and spread toward the Equator. Along the way, deep and bottom waters from the North Atlantic ride over bottom water that was formed in a similar manner, but in the Southern Ocean, off Antarctica. Using T-S characteristics, oceanographers have been able to piece together this cross section of water masses. Similar deep water masses are not formed in the North Pacific Ocean; the high latitude North Pacific is too shallow and its surface waters are less saline.

Some of the water masses in Figure 5.33 have very interesting and unique properties. One such water mass is the Mediterranean Sea that spills out of the Strait of Gibraltar and into the Atlantic Ocean between Africa and Europe. The internal circulation in the Mediterranean Sea is driven by intense evaporation at the surface; this area is extremely arid, and there is very little precipitation. So, the surface waters, although they get very warm, also get very salty, and eventually become dense enough to sink and flow out of the Mediterranean and into the

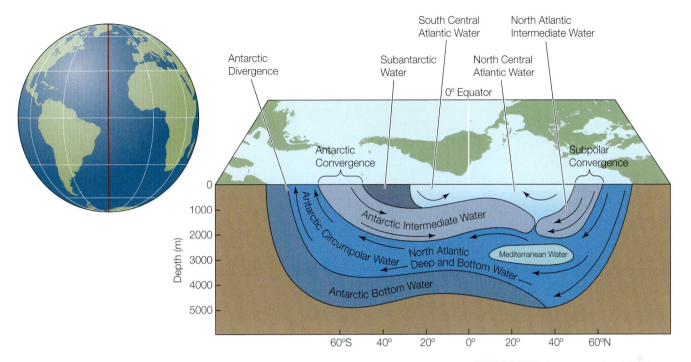

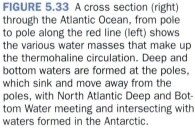

FIGURE 5.33 A cross section (right) through the Atlantic Ocean, from pole to pole along the red line (left) shows the various water masses that make up the thermohaline circulation. Deep and bottom waters are formed at the poles, which sink and move away from the poles, with North Atlantic Deep and Bottom Water meeting and intersecting with waters formed in the Antarctic.

Atlantic Ocean at relatively deep depths (ca. 1000 m), while surface water from the Atlantic Ocean flows in at the surface to keep a balance. The result is an easily identifiable water mass of warm temperature and high salinity—but which is now established at a depth where its surrounding waters are of equal density.

Unlike the wind-driven surface circulation, these processes of thermohaline circulation produce currents that are very sluggish—they take decades to centuries to move across an ocean basin. But they are very important and have been closely studied for a number of reasons. For one, deep ocean circulation is important in climate; and for another, that deep flow is important to the deep water transport of dissolved nutrients, which are important to the ocean's biology, as we will see in Chapter 7. This interconnectedness of the deep thermohaline circulation effectively produces a **great ocean conveyor belt** (**FIGURE 5.34**).

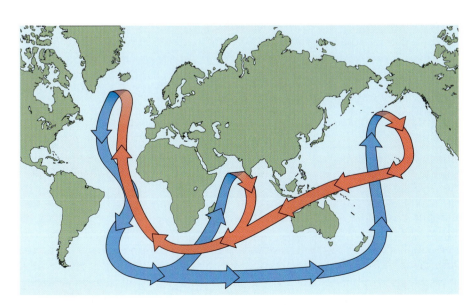

FIGURE 5.34 The "great ocean conveyor belt," a term coined by Wallace Broecker of Columbia University, interconnects deep and bottom water flows in the world ocean (blue path) with surface circulation (red path). Deep and bottom waters formed in the North Atlantic spread throughout the oceans, combining with additional deep and bottom waters from the Antarctic. Those waters eventually upwell to the surface again and are brought back to the deep water formation sites.

BOX 5I Measuring Ocean Currents

Ocean currents are measured using both direct and indirect methods. Direct methods include current meters, which are usually deployed from buoys. Signals are sent up the wire to the surface where they may be transmitted via satellite to a shore-based facility. The use of current meters to measure the flow of water past a fixed position in the sea is called the *Lagrangian method*. A buoy may have various instruments in addition to current meters attached to a cable running between it and an anchor; such a buoy is being deployed here by University of Maine scientists (**FIGURE A**).

Another direct method is the use of drifters, objects fitted with Global Positioning System (GPS) devices that are released at sea and are carried with the ocean currents. They are tracked as they send their GPS positions to a receiving station. Connecting the positions over time will produce a path on a map revealing where a given parcel of water has been and when. This is called the *Eulerian method*. A surface "window shade" drifter is being deployed in **FIGURE B** by Woods Hole Oceanographic Institution scientists. It floats just beneath the surface and moves with the surface current, reporting its position every 15 minutes.

An indirect method of measuring ocean currents, known as the *geostrophic method*, is to compute the density field using measurements of temperature and salinity. Differences in the average density from the surface to some reference depth (the bottom, for instance) will indicate variations in sea level.

For example, a coastal location under the influence of freshwater discharging from a river would have less dense water near shore and therefore an elevated sea surface near shore, as shown in **FIGURE C**. That near-shore water will initially begin to flow downhill toward the area of lower sea level offshore, but it adjusts to the Coriolis force and in the Northern Hemisphere that water will flow parallel to the sea surface slope, giving a geostrophic current.

FIGURE A

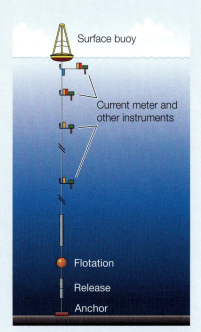

Surface buoy

Current meter and other instruments

Flotation

Release

Anchor

FIGURE B

Flotation devices

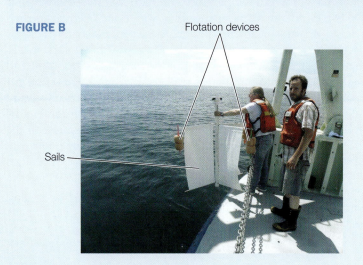

Sails

FIGURE C

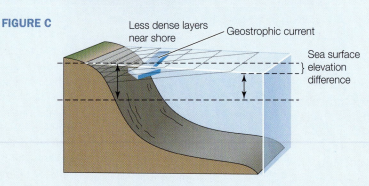

Less dense layers near shore

Geostrophic current

Sea surface elevation difference

Upon sinking in the North Atlantic Ocean, the deep water flows away from the area of origin and enters a worldwide circulation pattern. As the flow of cold North Atlantic Deep Water (NADW) moves south, it too experiences a western intensification because of the change in Coriolis with latitude. The result is a *deep* western boundary current that flows south along the edge of the outer continental shelf of North America and South America. Using T-S analyses, we can track this water mass all the way to the Southern Ocean. From there, the cold dense waters are joined by additional deep waters formed in the Antarctic, and they flow to the east and enter the deep basin of the Indian Ocean. Finally, the flow enters the Pacific Ocean where it can be traced north to the east coast of Japan and the Kamchatka Peninsula in northern Russia.

This deep-water circulation of the world ocean is extremely important in controlling the Earth's climate and weather. For example, the cooling and sinking of surface water in the North Atlantic results in a net heat flux to the atmosphere there. In fact, variations in intensity of North Atlantic Deep Water formation have been implicated in the history of Earth's glacial and interglacial periods.

The deep flow is important in biological oceanography because it accumulates materials that "rain" down (sink) from the overlying surface waters, and thus the deeper waters become enriched in dissolved substances such as the inorganic nutrients important for plankton growth. The deep water thermohaline circulation determines the distributions of inorganic nutrients throughout the deep ocean basins. The conveyor belt process thus results in a nutrient enrichment of those waters farthest removed from the deep water source (the North Atlantic). So, deep waters of the Pacific Ocean, being older than deep Atlantic or Indian Ocean waters, have the highest concentrations of dissolved nutrients.

Chapter Summary

■ The creation of winds, storms, and how winds drive the major ocean currents are fundamental processes in atmosphere circulation.

■ The Sun is the ultimate source of energy driving atmospheric and ocean circulation. The oceans absorb and store great quantities of heat from solar radiation—far more than the continents do—because of water's **transparency**, which allows sunlight to penetrate and heat to deeper depths, and especially because of water's high heat capacity. The oceans **mix vertically**, mixing that heat throughout the upper layers of the ocean.

■ Penetration of sunlight in the oceans is important not only to heat storage, but to biology as well. Light that penetrates the upper layers of the ocean is **attenuated** quickly (exponentially) with depth because of **scattering** and **absorption** due to dissolved and particulate materials in the water, as well as by the water molecules themselves. Not all wavelengths of light penetrate equally; red wavelengths are attenuated quickest, and blue and blue-green wavelengths penetrate deepest.

■ The density of air is affected by temperature, pressure, and humidity.

■ Solar heating of the Earth is uneven because the angle of incidence at which sunlight arrives varies with season, changing the surface area over which the Sun's energy is spread seasonally as well.

■ In theory, the Earth has three atmospheric **convection cells** in each hemisphere due to the Coriolis effect.

■ The **Coriolis effect** (or force) imparts a deflection, causing a curving motion to any moving object on the rotating Earth—to the right in the Northern Hemisphere and to the left in the Southern Hemisphere. The magnitude of the Coriolis depends on the speed of the moving object and on the object's latitude. The Coriolis effect is zero (nonexistent) directly on the Equator, and is maximum at the poles.

- The Coriolis force directs the major surface wind fields: major surface winds are the **Northeast** and **Southeast Trade Winds**, the **Prevailing Westerlies**, and the **Polar Easterlies**. Those winds result from areas of high and low atmospheric pressure on the surface and produce the jet streams at high altitudes. At times, low-pressure areas grow into storms, including hurricanes.

- The patterns of surface wind fields create the large scale ocean circulation patterns, the wind-driven currents.

- **Ekman currents** are generated in the top several tens of meters of the ocean when the wind blows across the ocean surface. Ekman currents move at exactly 90° to the right of the wind direction in the Northern Hemisphere.

- Ekman currents produce a surface flow that pushes water into the centers of the major ocean basins, elevating sea level there. That surface water sinks to as deep as 800–1000 m, and then spreads laterally. The spreading waters are deflected to their right by Coriolis, producing the large-scale gyre current system in the North Atlantic. Because of the Coriolis force, the elevated sea surface is not in the exact centers of the ocean basins, but is displaced to the west. Because the Coriolis effect increases as currents flow toward the poles and decreases as they flow toward the Equator, there is an intensification of boundary currents on the western edges of ocean basins. The Gulf Stream is an example.

- In addition to the **wind-driven circulation** of the world ocean, there is a also a density-driven circulation (also known as **thermohaline circulation**). This circulation occurs between the bottom and depths of about 1000 m. It is the result of high-salinity waters losing heat at high latitudes and becoming dense enough to sink to great depths. These deep, bottom waters flow slowly throughout the world ocean.

Discussion Questions

1. Why do the oceans store more heat than the continents? How does heat input from solar radiation vary with latitude and season? Why?

2. Explain what happens to light from the Sun as it first arrives at the top of the atmosphere, penetrates the atmosphere, and then arrives at the sea surface. Explain how the light quantity and quality (wavelength) change with depth in the ocean.

3. How do changes in temperature, pressure, and humidity affect the density of air? How might changes in pressure occur? What are surface areas of high and low atmospheric pressure? Explain how changes in surface pressure can produce convection cells.

4. Does all of the Earth receive the same amount of solar heating throughout the year? Explain. If heat is constantly being input to the surface of the Earth from solar radiation, is the Earth getting warmer with time? Explain.

5. Explain the processes that lead to the creation of two huge atmospheric circulation (convection) cells on a globe that is not rotating. Why does the rotation of Earth result in a number of such atmospheric circulation cells?

6. What is the Coriolis effect? Explain what happens to a moving parcel of air at 30°N as a result of Coriolis. What happens at 75°S? Is there a difference between the curvature of the trajectories at these different latitudes? Why?

7. In what direction is the wind blowing, on average, at an altitude of 10,000 m at 45°N? What about at the surface (sea level)?

8. Why do hurricanes require warm water in order to form? What happens when a hurricane makes landfall? Why are there few if any Atlantic hurricanes south of the Equator?

9. What is an Ekman current?

10. What drives currents in the deep sea, below the surface wind-driven currents?

Further Reading

Broecker, Wallace S., 2010. *The Great Ocean Conveyor, Discovering the Trigger for Abrupt Climate Change*. Princeton, NJ: Princeton University Press.

Persson, A. O., 1998. How do we understand the Coriolis force? *Bulletin of the American Meteorological Society* 79(7): 1373–1385.

Persson, A. O., 2005. The Coriolis Effect: Four centuries of conflict between common sense and mathematics. *History of Meteorology* 2(815): 1–24.

Price, J. F., Weller, R. A., and Schudlich, R. R., 1987. Wind-driven ocean currents and Ekman transport. *Science* 238: 1534–1538.

Stommel, Henry, 1965. *The Gulf Stream, A Physical and Dynamical Description*. Berkeley, CA: University of California Press.

Stommel, Henry M., and Moore, Dennis W., 1989. *An Introduction to the Coriolis Force*. New York: Columbia University Press.

Contents

Formed by strong winds in ocean storms, long, graceful swells can travel hundreds of miles across the open ocean—leaving the storm that created them far behind. But, upon encountering coastlines and shoaling depths, they come ashore and die as surf, such as this spectacular plunging breaker.

Waves and Tides

Standing at the rail of a ship, or just sitting on a beach, people will stare off into the distance, almost hypnotically, while watching waves. Their rhythmic motions affect us almost as if they were a form of visual music—for some reason we just seem to enjoy watching them. This may be part of the reason that many of us find it so relaxing to live on or near the ocean, or to visit the coast every so often. But when we look closely at waves, we find that their motions reflect a lot of interesting physical phenomena. In this chapter we will explore some of those phenomena—what ocean waves are and how they behave, from the smallest surface waves, less than 2 centimeters (cm) in length, to the largest waves, the ocean tides.

The Basics of Ocean Waves

To begin our discussion, we need to master some of the jargon (the terminology) for the various parts of a wave (**FIGURE 6.1**). Surface waves on the ocean are displacements of the sea surface that oscillate up and down, and they usually propagate—or move—along the surface. Each wave has a **crest** and a **trough**. The horizontal distance between one crest and the next, or between one trough and the next, is a **wavelength**. The **period** is the time it takes for one complete wave cycle; this is usually measured as the time between two successive crests or troughs. The displacement of the sea surface can be measured in two ways: (1) as the **wave height**, which is the vertical distance from the top of a crest to the bottom of a trough, or (2) as the wave **amplitude**, the vertical distance that the sea surface is displaced from a position at rest. The wave's amplitude is equal to one-half the wave height.

Waves on the surface of the ocean or any other body of water are the result of a disturbance of an otherwise flat surface. Therefore we can break down the formation of a wave as being the result of two opposing forces: The first is the **disturbing force**, that which causes the sea surface (the air–sea interface) to change its elevation by some amount, at some location. The disturbing force can be any number of potential forces: the wind blowing across the surface of the ocean, an earthquake that moves the ocean floor, or simply a rock thrown into the water. The second is the **restoring force**, which acts to bring the elevated sea surface back to its position at rest. The effects of these two forces, the disturbing and restoring forces, don't end immediately after they act, the way they would if a rock were thrown into a pile of sand. Instead, as the sea surface is restored, its momentum causes the sea surface to overshoot its original spot, and the surface oscillates, thus making a wave.

The waves shown in **FIGURE 6.2** were all formed by wind blowing on the surface of the ocean, either in the immediate area where the photo was taken, or much farther away. Because wind is the chief disturbing force that creates waves on the sea surface, the commonest waves in the world ocean are **wind waves**. And as the wind acts on the sea surface to create these waves, the waves undergo an interesting transition.

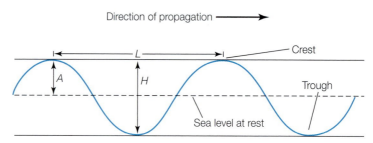

FIGURE 6.1 A rock tossed into a calm body of water generates surface gravity waves that propagate outward in all directions. The parts of such waves include the crest, or top, of the wave; the trough, the lowest point; the wavelength (*L*), the length from crest to crest or trough to trough; the wave height (*H*), the vertical distance from the wave crest to the trough; the wave amplitude (*A*), one half the wave height.

Capillary waves

The tiniest waves on the surface of the ocean are **capillary waves**. These are the first waves to form when the wind blows, as energy is transferred from the moving air to the water. What happens is best visualized this way: if you were to blow across the surface of a cup of coffee or tea in order to cool it down a bit, you would see small ripples develop as the energy from your blowing is transferred to the liquid. This transfer of energy happens because even though the air you are blowing is directed parallel to the surface of the liquid, air that is moving has lower pressure than non-moving air (the air above the liquid just before you started to blow). Therefore, right above the air–water interface,

FIGURE 6.2 Ocean waves can take many forms, as these examples show. (A) Waves during a relatively calm period with only light winds. Smaller waves can be seen riding on top of larger waves, creating a confusion of surface wave patterns and making it difficult or impossible to determine the direction any are coming from. (B) Closer to shore, wave patterns are different; they usually arrive from offshore, such as these surf waves breaking on a rocky coastline. (C) An almost perfectly calm sea with little or no wind blowing except near the horizon, where a slight disturbance can be seen. Only a gentle swell is obvious on the sea surface. (D) A storm wave sending spray over the bow of a 176-foot-long research ship.

slight differences in wind speed and the resulting differences in lowered air pressure lifts some of the water upward into ripples. These ripples that form initially are capillary waves (also called cat's-paws); their wavelength is less than 1.73 cm, which is just over ¾ inch. It is these tiny capillary waves that mark the beginning of a process that, under the influence of a wind, eventually builds larger waves such as those we are more accustomed to seeing on the surface of the ocean. **FIGURE 6.3** shows these little capillary waves beside our research ship one day when a gentle breeze was blowing.

For these very small capillary waves, the restoring force is dominated by surface tension, which results from the pull of hydrogen bonds at the air–water interface (see Figure 4.10). Gravity also pulls the wave downward, but at these very small wave sizes, surface tension pulls harder. Gravity becomes important when the waves get bigger. When the wave's initial upward bulge is formed, the water's skin-like surface is stretched, but returns to its original, flat position as the hydrogen bonds pull the surface back. The alternating dominance of those two forces creates an oscillation as the tiny wave is first formed. Such a forced displacement of a surface and its subsequent return to its original position is analogous to what happens when we beat a drum. The drumskin is depressed by the drumstick, and then the drumskin's tension restores the drum surface to its original flat shape, but then it overshoots the level of the original surface and oscillates or vibrates, creating sound waves. The same thing happens when we pluck a guitar string; the string's tension pulls it back to where it started, and then it overruns that position, oscillating back and forth in a vibration that creates a musical tone. Both of these examples using musical instruments represent the action of stationary or **standing waves**, but they illustrate nicely the surface-tension-restoring force that creates capillary waves in water. While there can also be standing waves in water (called seiches; see page 207), we are more accustomed to seeing waves that move across the surface of the ocean: **progressive waves**. So, back to our discussion of how wind waves are formed on the ocean.

Surface gravity waves

If the wind continues to blow across the surface of the ocean, the small capillary waves, which now have an inclined surface for the wind to blow against (**FIGURE 6.4**), will continue to acquire energy from the wind, which results in the waves growing bigger. The wave will increase both in height and wavelength, and once the wavelength exceeds 1.73 cm, the restoring force is no longer dominated by surface tension, but by gravity. Beyond 1.73 cm, the extremely weak surface tension of water is no longer capable of pulling the surface back to a flat position. The wave transitions from a capillary wave to a surface **gravity wave**. Gravity causes the crest of a surface gravity wave to, quite simply, fall downward. As with the capillary wave, the gravity wave falls, its momentum causes it to overshoot its original position, producing an oscillating wave. Under the influence of a wind blowing that wave horizontally across the surface of the ocean, we have a progressive wave which moves, or propagates, away from its point of origin.

It is important to realize that progressive waves are moving energy, not water. It is the wave form itself (the shape of the wave) and the energy the wave carries with it that are propagated. This phenomenon is illustrated in Figure 2.23, which shows how we can force a wave down the length of a rope

FIGURE 6.3 Capillary waves are very short-wavelength waves that can eventually transition to surface gravity waves.

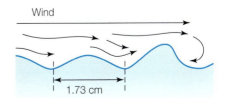

FIGURE 6.4 A capillary wave on the surface of the ocean provides a face for the wind to blow against, making for a more efficient transfer of wind energy to the ocean.

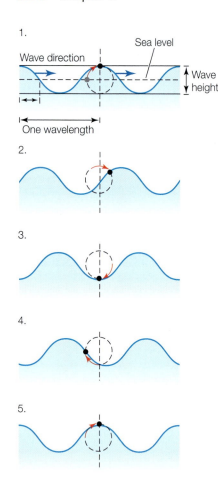

FIGURE 6.5 The orbital path, equal to the wave height, traced by a particle of water on the surface of the ocean as a wave passes from left to right. (1) A wave crest approaches from the left, carrying the water particle up and forward to the right. One-quarter wavelength later (2) the particle has continued to move forward to the right, but also downward. (3) As the next quarter-wavelength passes, the particle continues to fall downward, but now is being pulled backward into the wave trough. (4) As the next crest approaches, the particle is lifted upward, while continuing to be pulled backward. (5) As the wave crest approaches, the particle is lifted upward and forward. When a full wavelength has passed, from crest to crest, the particle has completed its circular wave orbit. The wave itself has moved horizontally from left to right, but the water, as shown relative to a stationary dashed line in the background, has not. The water particle did not move any net distance horizontally or vertically.

attached to a wall. The wave form moves along the length of rope, but the rope itself remains in your hand at one end and attached to the wall at the other.

If we were to examine closely the particles of water in a surface gravity wave both at the sea surface and immediately beneath, we would see that water particles don't just move up and down as waves pass by; they actually trace out a nearly perfect circle called a **wave orbit** (**FIGURE 6.5**). To explain what happens, let's start by considering just the surface of the ocean; we'll consider subsurface waters later. Figure 6.5 shows that in the crest of the wave, the water particles are moving forward in the direction of propagation with the wave itself, but in the trough, they are moving backward (!), and they complete the trace of a circle as the wave passes. With each passing wave, the water particles return almost exactly to their starting position—the wave moves, but the water does not. While waves in general behave according to this principle, there is an exception in the case of these surface gravity waves in water, whereby there is a very small but significant net movement of the near-surface water itself in the direction the wave is moving. This phenomenon is known as **Stokes Drift** (**FIGURE 6.6**).

(A)

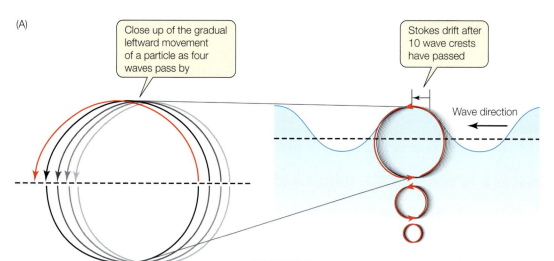

Close up of the gradual leftward movement of a particle as four waves pass by

Stokes drift after 10 wave crests have passed

Wave direction

FIGURE 6.6 Have you ever wondered why floating debris, such as seaweed, driftwood, and trash accumulate on beaches rather than being washed out to sea? Two phenomena are responsible: the refraction of waves toward beaches and Stokes Drift. Even though we know that waves transport energy and not materials, that isn't strictly true when it comes to matter in the near-surface waters. Because wave orbits decrease in size beneath the surface, an individual wave orbit will have a slightly smaller arc at its base, in the trough, than when it is at the crest. This produces a corkscrew action and very slowly transports water and debris in near-surface waters in the direction of wave propagation (A). These diagrams illustrate the principle, with the most recent orbit in red and the earlier orbits in gray. The inset shows a close up and movement of a particle (the scale is exaggerated; actual movement is slight). In near shore waters, this direction is toward the shoreline, where various materials can collect, such as the seaweed in a strand line on this beach (B).

(B)

Strand line

This phenomenon of wave orbits is not confined to the surface of the ocean; as you might logically expect, water particles beneath the surface also travel in orbits. But, unlike the surface-wave orbits, in which the diameter of the orbit is equal to the wave height, wave orbits beneath the surface have diameters that are greatly diminished with increasing depth (**FIGURE 6.7**). At a depth of only one-half the wavelength, the orbital diameters are reduced to a mere 4% of what they are at the surface.

Scuba divers are well aware of this phenomenon of wave orbits. In a rough sea, or with swells passing (more on swells below), divers at or near the surface will be jostled about with these wave orbits, but upon diving deeper, they quickly find themselves in relatively calm waters. They do not have to go very deep to experience this diminished wave action.

Except for those of us who have experienced this phenomenon firsthand, the concept of wave orbits is somewhat difficult is to imagine. Figures 6.5 and 6.7 help to visualize the phenomenon, but they aren't nearly as effective as direct observation. Apart from scuba diving, one of the best ways to observe these orbital motions is to stick your head over the edge of a dock with your face just a few inches above the water and look at the waves passing by. Assuming that there are waves present, watch how objects on and just beneath the surface move about. Look at the little air bubbles just below the surface of the water as well as the tiny flecks of particulate material suspended in the water, and you'll see the wave orbits. Better still: go to the beach and tread water just beyond the breakers—if you can swim, that is. You can feel yourself traveling in these orbital motions as waves approach and pass by you. Or, stand in waist-deep water. As the trough in front of a swell approaches you, and before the crest arrives and then breaks, you will feel yourself being pulled toward the oncoming wave as water rushes by you toward the oncoming crest. And after the crest passes you will feel a momentary pull toward shore as water rushes past you in the opposite direction. The main point here, and in Figures 6.5 and 6.7 is that water particles return to very near their original positions after a wave passes by. Again, this is because, in general, the wave transports energy, not material (with Stokes drift being the exception).

Because wave orbits extend beneath the surface, it follows that a surface gravity wave requires some minimum depth of water in order to exist. Contact with the bottom generally interferes with the wave orbits, resulting in the wave losing energy due to friction, and slowing down. Waves in deep water do not encounter this problem. For this reason, we categorize surface gravity waves according to the relationship of their wavelength to the water depth.

DEEP WATER GRAVITY WAVES   As the name implies, **deep water gravity waves** are surface waves that occur in water that is deeper than one-half their wavelength; the bottom does not interfere with the wave orbits. The speed, or velocity, of propagation of a deep water wave is 1.25 times the square root of the wavelength; that is, $V = 1.25\sqrt{L}$ (**BOX 6A**). Notice that depth is not in the equation. When a deep water wave propagates to where the water's depth becomes less than one-half its wavelength, it will begin to be impeded by friction, and its speed will diminish.

SHALLOW WATER GRAVITY WAVES   In shallow water, the wave orbits of a deep water wave will be deformed and flattened into ellipses (**FIGURE 6.8**). This frictional encounter with the bottom causes

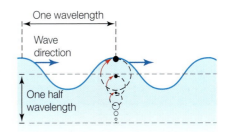

FIGURE 6.7 Wave orbits continue with depth beneath a surface wave, but their diameters quickly diminish. At the surface, the diameter of a wave orbit is equal to the wave height. But the diameter of wave orbits at a depth equal to one half the wavelength is only about 4% the diameter of those at the surface.

FIGURE 6.8 In a shallow water wave, the bottom causes the wave orbits to flatten. This happens when the bottom depth is on the order of 1/20 that of the wavelength. Just off the bottom, the water motions flow back and forth. Note: This diagram is distorted vertically in order to show more clearly the flattened orbits. In nature, the orbits would be even flatter than shown here, where we have drawn the bottom depth to be nearer one-third the wavelength.

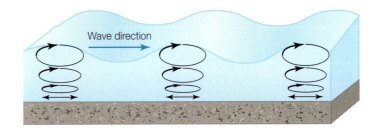

BOX 6A   The Speed of Ocean Waves

Wave velocity (V, in meters per second, m/s) can be described as a function of wavelength (L, in meters, m) and period (T, in seconds, s). These are related as:

$$\text{Velocity } (V) = \frac{\text{Wavelength}(L)}{\text{Period}(T)}$$

We can rearrange this equation to solve for T and L, as:

$$\text{Period}(T) = \frac{L}{V}$$

$$\text{Wavelength}(L) = VT$$

In terms of its wavelength (L), the full wave equation for velocity (V) is:

$$V = \sqrt{\frac{gL}{2\pi} tanh \frac{2\pi h}{L}}$$

Here, g is the acceleration of gravity, 9.8 m/sec^2; L is the wavelength; π is 3.14; h is the water depth; and $tanh$ is the hyperbolic tangent.

If the water is very deep, or greater than 1/2 the wavelength—that is, if we are dealing with a deep water wave—then the quantity $tanh 2\pi h/L$ is approximately equal to 1.0, and so we can drop that part from our equation. Thus, the speed of a deep water wave will depend only on its wavelength, and not on the water depth, as:

$$V = \sqrt{\left(\frac{gL}{2\pi}\right)} = \sqrt{\left(\frac{9.8 \times L}{6.28}\right)} = 1.25\sqrt{L}$$

In very shallow water, where the depth is small compared with the wavelength, that is, when the quantity $2\pi h/L$ is small, then $tanh 2\pi h/L$ will be approximately equal to $2\pi h/L$. Therefore, the wave equation above reduces to:

$$V = \sqrt{gh}$$

This is the equation for the speed of shallow water waves, where the water depth is shallower than 1/20, or 0.05×, the wavelength (that is, where $z = 0.05L$).

For intermediate water waves, where the depth is between 1/2 and 1/20 the wavelength, the full equation must be used to solve for wave speed.

---

the wave to lose energy and diminishes the wave speed. As the water gets shallower than 1/2 the wavelength, the bottom becomes more and more important in controlling the speed; the circular orbits become flattened and the wave loses more energy due to friction with the bottom. Eventually, when the water shoals (becomes shallow) to a point where the depth is less than 1/20 the wavelength, the waves become **shallow water gravity waves**. The velocity of these waves is controlled entirely by the water depth.

INTERMEDIATE GRAVITY WAVES   Between the depths for deep water waves and shallow water waves—between 1/2 and 1/20 of the wavelength—we have **intermediate gravity waves**. Intermediate waves "feel" the bottom; their proximity to the bottom distorts their wave orbits, and to varying extents, this distortion affects the speed of these waves. But wavelength also is important, and so the velocity of intermediate gravity waves is determined by both the water depth and the wavelength (see Box 6A).

Seas and swells

Thus far in our discussion of wind-generated waves, we have illustrated them as smooth sinusoidal curves, but in fact, locally generated wind waves don't look like that. Those smooth sinusoidal waves are actually **swells**, which are surface waves that have a relatively long wavelength, are symmetrically shaped, and have a relatively long period, on the order of 10 seconds or so. Swells are generated remotely, some distance away from where they assume these characteristics. *Locally-generated* waves created by the wind blowing on the sea surface are known as **seas**, which are relatively short and steep waves, with an abrupt crest (**FIGURE 6.9**).

Swells begin as seas that propagate across the ocean to an area miles away from the influence of the wind that created them, where they "flatten out" into swells. So, capillary waves become seas, which can eventually become swells, but there are a few more details we need to discuss with respect to the evolution of each of these developmental stages in the formation of wind waves.

Formation and evolution of wind waves

The minimum wind speed required to generate a wave disturbance is on the order of 1–2 knots. (One knot, or 1 nautical mile per hour, is about 51 cm/s, which is a slow walking speed for an average person.) You and I will only begin to feel a breeze on our faces when the wind increases to a speed of about 1 knot. Wind blowing at this speed across the surface of a body of water will generate capillary waves. Then, as the wind continues to blow, adding energy to the wave and growing it longer than 1.73 cm, the wave will transition from a capillary wave to a gravity wave (**FIGURE 6.10A**).

Once the early gravity wave has formed, its initial velocity is only about one-third the velocity of the wind that formed it. But as the wind continues to blow at the same velocity, all the while continuing to add energy to the wave, the wave will grow in both wave height and wavelength. And as it grows, a curious phenomenon will occur: the surface gravity wave's height (H) will increase faster than its wavelength (L). This means that as the wave grows, it gets steeper. A wave's height cannot continue to increase faster than its wavelength indefinitely; if it were to do so, we would eventually have sky-high but pencil-thin slivers of waves on the surface of the ocean. Instead, the wave eventually collapses on itself—it breaks—when it reaches a **critical steepness**, which is determined by its height relative to its length (**FIGURE 6.10B**). That critical steepness is reached when the ratio of wave height to wavelength ($H{:}L$) is 1:7; at this point the wave crest will exhibit a steep profile, with a characteristic angle of about 120°. Once the critical steepness is reached, even though the wind may continue to blow at a constant velocity, the wave cannot grow any higher; it has reached its maximum height, and the energy added

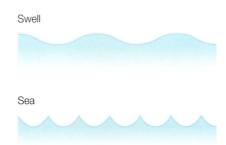

FIGURE 6.9 Shapes of swells and seas. Swells are generally longer-wavelength, gently sloping symmetrical waves, whereas seas are asymmetrical, with steeper crests than troughs.

(A)

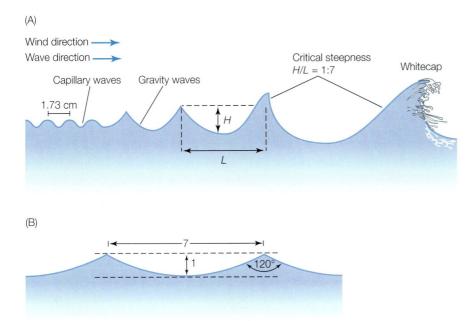

(B)

FIGURE 6.10 Stages in the development of a sea. (A) Capillary waves grow to gravity waves once their wavelength exceeds 1.73 cm. Wind continues to add energy to the gravity waves and so they grow, gaining wave height faster than wavelength (vertical scale exaggerated). Eventually, the ratio of wave height to wavelength exceeds a critical value of 1:7 and the wave breaks, forming a whitecap. (B) The critical steepness diagram to scale.

FIGURE 6.11 Photograph taken from a ship at sea where the sea has become fully developed for that wind speed, which on this particular day was about 20 knots. The seas have reached their critical steepness for that wind speed, and energy added to the sea as the wind continues to blow at 20 knots is dissipated as white caps.

by the wind after that simply causes the wave to break, producing **whitecaps**, which dissipate the wave's energy (**FIGURE 6.11**). After whitecaps form, we have an equilibrium, where the energy imparted to the sea by the wind matches the energy dissipated by the breaking whitecaps. If the winds do not increase in velocity, the waves cannot get any bigger—wave height will not increase. This sea state is called a **fully developed sea**.

The attainment of a fully developed sea *for a particular wind speed* depends on two factors: the **duration** of time the wind is blowing, and the length of ocean over which that wind blows, the **fetch** (**FIGURE 6.12**). In other words, it takes some time for the wind to build the seas, and the span of ocean over which it blows has to be big enough. A brief high-velocity wind gust will not build seas that are nearly as big as a sustained high-velocity wind will build; but even a sustained high-velocity wind will not make big seas in a swimming pool. For there to be a fully developed sea when the wind is blowing 20 knots (kt), for instance, which is quite common out on the open ocean, we see in Figure 6.12A that the wind speed must be maintained for a minimum of 10 hours, and it must blow over a fetch of nearly 100 nautical miles. We can also see in Figure 6.12A that for wind speeds higher than 20 kt, the required fetch and duration become quite large—conditions that are only rarely met in nature.

But when those conditions are met—and sometimes they are—the seas can become very large. How large? Seas that can be produced in fully developed sea at different wind speeds are shown in Figure 6.12B. A fully developed sea in a 40 kt wind, for example, will have waves that average 7 meters (m) in height—and one in ten of those waves will exceed 16 m, or about 50 feet. That is a big wave by any standard; fortunately, such waves are relatively rare. Even though 40 kt winds are not uncommon in ocean storms, we don't see 16 m waves very often because both the fetch and duration criteria for a fully developed sea are rarely met. Storms are usually on the order of 100 to 200 miles wide, which is less than the required fetch, and they usually last less than 24 hours, which is less than the required duration for a fully developed sea at this wind speed. Nevertheless, ocean waves can still get to be quite large in a storm, even if conditions for a fully developed sea are never reached.

The sizes of waves that can be expected for a given wind speed—even though conditions for a fully developed sea may not be met—are given in the **Beaufort Wind Force Scale** (**TABLE 6.1**). The Beaufort Scale is an empirical scale based on *observed* sea conditions and provides an estimate of wave sizes one might expect to see for a particular wind speed. In general, the wave height of the seas that result from the wind is related to: (1) the wind speed; (2) the duration of that wind; and, (3) the fetch. When extreme values for each of

(A)

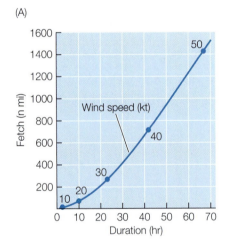

(B)

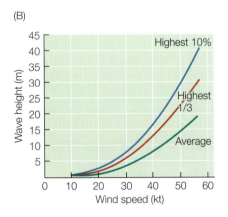

FIGURE 6.12 (A) Plot of the required fetch (in nautical miles) and duration (in hours) for there to be a fully developed sea at the indicated wind speeds (in knots). (B) Wave heights (m) for a fully developed sea at the indicated wind speeds, presented as the height of the highest 10% of waves, the highest 1/3 of waves, and the average wave heights.

these three criteria occur, waves can get to be quite large. As Table 6.1 reveals, hurricane-force winds (greater than 63 kt) can build seas with average wave heights greater than 46 feet!

In case you are wondering just how big a wave can get: the biggest wave on record was observed by the crew of the USS *Ramapo* during a storm in the Pacific Ocean in 1933 that was estimated at some 34 m (112 feet) in height. (That wave is now thought to have been a rogue wave; see page 204). The estimate was based on observations made by an alert (and obviously not seasick) watch officer on the bridge, who lined up by sight the ship's crow's nest with the horizon, while at the same time lining up the bow with the horizon (**FIGURE 6.13**); this allowed a calculation of the wave height. But, let's face it: the biggest wave that actually has occurred out on the ocean, and which was also observed by people aboard a ship at the time, probably sank the ship and with it the crew who saw it.

WAVE DISPERSION   After wind-generated surface gravity waves are formed, they propagate away across the sea at a speed that is dictated by their wavelength, which grows longer over time. And as those waves move away, a few other interesting phenomena occur.

Once waves propagate to an area where the wind is no longer adding energy to the waves—as they leave the area of the storm that generated them, for instance—the waves begin to flatten and lengthen. Their wave heights decrease in

TABLE 6.1   The Beaufort Scale of average wave heights for given wind speeds[a]

Wind speed (kt)	Miles per hour (mph)	Average wave height (ft)
<1	<1	0
1–3	1–3	0
4–6	4–7	<0.3
7–10	8–12	0.3–1.6
11–16	13–18	1.6–4
17–21	19–24	4–8
22–27	25–31	8–13
28–33	32–38	13–20
34–40	39–46	13–20
41–47	47–54	13–20
48–55	55–63	20–30
56–63	64–72	30–46
>63	>73	>46

[a]These data are based on observations, unlike those in Figure 6.12, which are based on theoretical calculations.

(A)

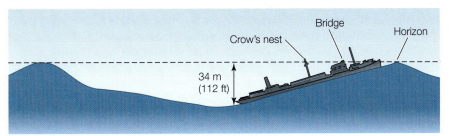

(B)

FIGURE 6.13 (A) The USS *Ramapo*, built and launched in 1919, and which remained in service until 1946, observed what is still believed to be the largest wave ever recorded. (B) The 112-foot wave occurred in 1933 in the Pacific Ocean, and was measured using the geometry shown here and as explained in the text.

(A)

(B)

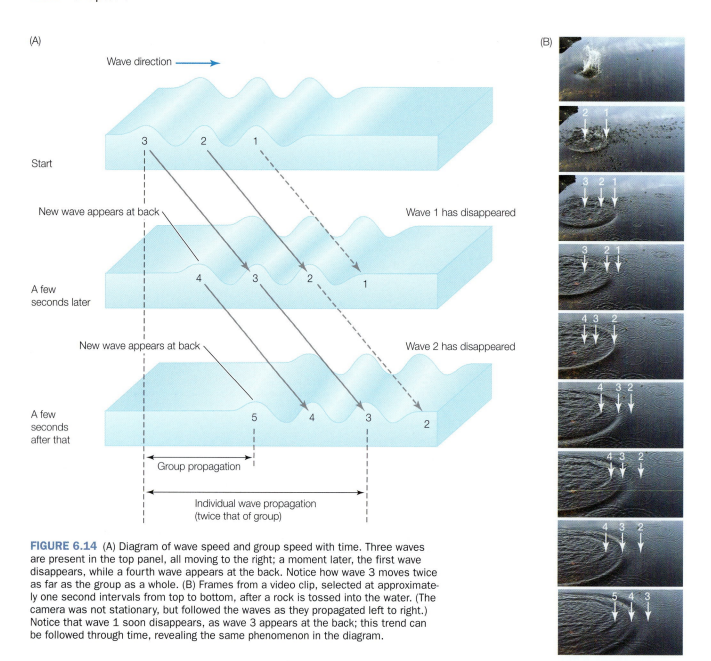

FIGURE 6.14 (A) Diagram of wave speed and group speed with time. Three waves are present in the top panel, all moving to the right; a moment later, the first wave disappears, while a fourth wave appears at the back. Notice how wave 3 moves twice as far as the group as a whole. (B) Frames from a video clip, selected at approximately one second intervals from top to bottom, after a rock is tossed into the water. (The camera was not stationary, but followed the waves as they propagated left to right.) Notice that wave 1 soon disappears, as wave 3 appears at the back; this trend can be followed through time, revealing the same phenomenon in the diagram.

response to the pull of gravity, and, in order to conserve energy, their wavelengths necessarily increase. Their period, however, does not change, and so the waves now move faster than they did at the point of generation. If the period (T) doesn't change, then increasing the wavelength (L) will increase the velocity (as explained in Box 6A, $V = L/T$).

This decreasing wave height and increasing wavelength decreases the waves' steepness and the sea transitions to a swell. So areas of the ocean away from the area of wave generation will experience long-wavelength, fast-moving swells; the smaller, slower waves will not have had time to get there, or they may have dissipated (lost their energy and just plain died) before getting there. The waves are said to have **dispersed**, making groups or trains of swells that have become separated out into groups of similar wavelengths and so they travel at similar speeds.

An important point here is that the longer-wavelength, faster-moving swells often arrive on coasts that are hundreds of miles away from the storms that created them, and they always precede the arrival of the storms that create them! This is why we see big surf on some beaches far removed from, or prior to the arrival of, a hurricane or other large ocean storm. These are the big, well-formed waves that surfers dream of.

There is another curious phenomenon at work here as well: within a wave train, the **wave speed** of individual waves is *twice as fast* as the **group speed**. This means that the waves created by a storm far out at sea take twice as long to arrive on the beach as we would expect them to based on their individual wave velocities (**FIGURE 6.14**). You can observe this phenomenon of the group speed being half the individual wave speed by simply tossing a rock into a calm body of water and watching closely what happens. As the waves spread outward, the front wave always disappears—and just before it disappears, you will notice that it is moving twice as fast as the group. At about the same time that the leading wave disappears, a new wave appears at the back of the train to replace it. The new wave seems to appear magically out of nowhere, but its appearance is simply a way of conserving energy. You would also notice that as the waves move away from the disturbance, they get longer and move faster. The important points here are: the group speed is slower, by one half, than the individual wave speeds; and, the farther away a swell propagates, the longer its wavelength and the faster its speed.

Surf

As waves approach the coast and encounter shallower water where the depth decreases to less than one-half the wavelength, they begin to feel the bottom and transition from deep water waves to intermediate water waves. These waves continue to slow down as they approach the shore, losing energy through friction with the bottom. Thus, the waves following in the wave train will "catch up" with the front waves, and they will all begin to pile up (**FIGURE 6.15**). As the waves slow down, they get steeper: their wavelengths shorten and their height increases. Then, when the water depth (Z) is approximately equal to 1.3 times the wave height (H) (or, when $H/Z = \frac{3}{4}$) the wave reaches its critical steepness, and like the white caps we just discussed for conditions offshore, the wave breaks, creating **surf**. So, if you see a wave breaking on the beach, you can estimate the water depth where the

FIGURE 6.15 As a deep water wave approaches shore and the water depth shoals to one-half the wavelength ($L/2$), the wave will begin to feel the bottom and transition to an intermediate water wave and then to a shallow water wave. This causes the wave to slow down, which in turn compresses the energy and therefore increases the wave height (H). Stated another way: as the wave speed drops, other waves behind begin to catch up with the leading waves, causing the wavelengths to shorten and the wave heights to increase. Eventually, when the bottom depth becomes critical (i.e., when the depth is approximately 1.3 × H), the wave will break as surf.

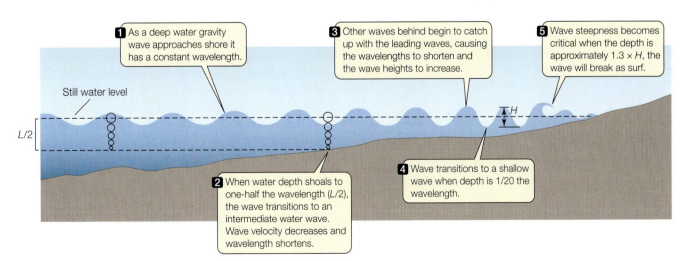

1 As a deep water gravity wave approaches shore it has a constant wavelength.

2 When water depth shoals to one-half the wavelength ($L/2$), the wave transitions to an intermediate water wave. Wave velocity decreases and wavelength shortens.

3 Other waves behind begin to catch up with the leading waves, causing the wavelengths to shorten and the wave heights to increase.

4 Wave transitions to a shallow wave when depth is 1/20 the wavelength.

5 Wave steepness becomes critical when the depth is approximately 1.3 × H, the wave will break as surf.

Still water level

$L/2$

H

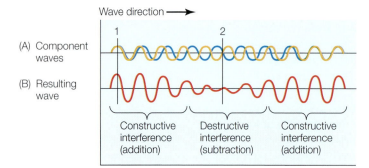

Wave direction ⟶

(A) Component waves

(B) Resulting wave

Constructive interference (addition) Destructive interference (subtraction) Constructive interference (addition)

FIGURE 6.16 (A) Diagram of how two sets of waves of equal wave heights but unequal wavelengths would interfere with one another to produce a wave that is the sum of the two original waves (B). At point 1, the two waves, each propagating to the right, are in phase with one another; the crests of each wave coincide, as do the two troughs. Thus, when they interfere, they produce a larger wave that is the sum of the two, with twice the wave height of either of the two original waves. This is constructive interference. Because the two wave types have different wavelengths, they eventually propagate to point 2, where they are now out of phase. Their crests coincide with the other's troughs, canceling out one another. This is destructive interference. Still later, the two wave sets propagate to a point where they are back in phase with one another, again producing waves that are the sum of the two component waves.

(A)

(B)

wave begins to break. For example, a 10-foot wave will break when the depth is about 13 feet. This is why in the old days, ship captains would post a lookout to spot breakers as the ship approached an unknown or uncharted shore. The size of the breakers (the surf) would give a measure of the bottom depth and thus allow the lookout to spot shoal areas and reefs.

Wave interference

Waves on the surface of the ocean interact with one another and with objects such as shorelines and sea walls in interesting and important ways. This is because waves are additive. That is, when the crest of one wave propagates past, or intersects, another wave, the two intersecting waves produce a new wave, the components of which are the sum of the two; the two crests are added to one another, and the two troughs combine to produce a single trough that is proportionally deeper.

When two waves that are **in phase** with one another—their crests and troughs coincide—intersect, the result is **constructive interference**, where a new wave is formed with a new wave height that is the sum of the two intersecting wave heights (**FIGURE 6.16**). If the waves are **out of phase**—with crests coinciding with troughs—there is **destructive interference**, which tends to cancel out the two component waves.

This constructive and destructive wave interference explains why you sometimes see waves at the beach coming in sets of several larger waves arriving for a minute or two, for example, followed by a period of relative calm, with only smaller waves coming ashore. You are seeing the effects of addition and subtraction among random groups of waves, perhaps arriving from different points of origin far offshore where they were created, and as such they may alternate in and out of phase with one another.

ROGUE WAVES Wave interaction may also explain the existence of rogue, or freak, waves. **Rogue waves** are huge waves that, although rare, are sometimes unexpectedly encountered by ships at sea (**FIGURE 6.17**). The causes of these unusual waves are complex, and we do not understand them well enough to predict their occurrence, but many are thought to be the result of constructive wave interference of multiple waves. Recent research on these phenomena is revealing that rogue waves occur in both standing wave forms, the results of interference of multiple waves, and progressive wave forms, about which even less is known.

FIGURE 6.17 (A) Photograph taken in the Bay of Biscay, off the coast of France, of a merchant ship in heavy seas as a rogue wave looms astern. (B) Photograph taken from the SS *Spray* in 1986 in the Gulf Stream off of Charleston, South Carolina.

Given the extremely large number of wave-generating phenomena (storms) out in the oceans, and the tendency for them to produce waves that can travel hundreds, even thousands of miles, there is a low, but very real statistical probability of multiple large waves coinciding with one another at the same time and place. When they do coincide, unusually large waves seem to appear out of nowhere. The biggest rogue waves are so powerful they can break the hulls of large ships, and in fact, the largest ships are especially vulnerable. Because of the unusually large wave heights and long wavelengths of rogue waves, large ships can find themselves improperly supported as they encounter such phenomena, sometimes resulting in their hulls cracking.[1] It has been reported that one to two large ships a week (that's 50–100 ships every year!) sink under mysterious circumstances, and of those, about 10% are supertankers.[2]

Some rogue waves are really quite common—the smaller ones, that is. For example, look out the window if you happen to be on airliner crossing the ocean. The broadly-spaced patches of whitewater (e.g., foam from breaking waves) you see on the ocean surface some six to seven miles beneath the plane are often the result of breaking rogue waves.

Most photographs taken of rogue waves at sea (see Figure 6.17) are of poor quality, for obvious reasons—there's not much time to think about photography when one approaches your ship! An eye-witness account of the wave hitting the SS *Spray* in Figure 6.17 follows:

> A substantial gale was moving across Long Island, sending a very long swell down our way, meeting the Gulf Stream. We saw several rogue waves during the late morning on the horizon, but thought they were whales jumping. It was actually a nice day with light breezes and no significant sea. Only the very long swell, of about 15 feet high and probably 600 to 1000 feet long. This one hit us at the change of the watch at about noon. The photographer was an engineer (name forgotten), and this was the last photo on his roll of film. We were on the wing of the bridge, with a height of eye of 56 feet, and this wave broke over our heads. This shot was taken as we were diving down off the face of the second of a set of three waves, so the ship just kept falling into the trough, which just kept opening up under us. It bent the foremast (shown) back about 20 degrees, tore the foreword firefighting station (also shown) off the deck (rails, monitor, platform and all) and threw it against the face of the house. It also bent all the catwalks back severely. Later that night, about 19:30, another wave hit the after house, hitting the stack and sending solid water down into the engine room through the forced draft blower intakes.[3]

SEA WALLS Less monstrous, but still destructive examples of wave interference occur when waves reflect off barriers such as sea walls. **Sea walls** are walls of rocks intended to protect shore-side homes from damaging waves during a storm as well as to prevent shoreline erosion from storm waves. However, they sometimes promote wave interference by reflecting incoming waves back out to sea, where they meet and combine with their sister waves coming in from offshore. When the crests of incoming and outgoing waves combine, we get large standing waves that often break and create additional turbulence that

---

[1] If a rogue wave elevates the midsection of a large ship, leaving the bow and stern nearly out of the water, the ship's hull may crack in the middle; the same result ensues if the ship encounters a deep trough such that only the bow and stern are properly supported, with the midsection nearly out of the water. Large ships are most vulnerable because they are more likely to be similar in length to the wave length, and they are simply not designed to withstand such unusual stresses.

[2] Source: British Broadcasting System.

[3] Account by Captain G. A. Chase when aboard the SS *Spray* in 1986, personal communication with author. See also Rainey, R. C. T. 2002. *17th International Workshop on Water Waves and Floating Bodies* (England: Peterhouse, Cambridge).

FIGURE 6.18 Sea walls in front of homes on an eroding beach in Southern Maine. Storm waves can reflect off the sea walls and, with constructive interference with oncoming waves, build even larger waves that erode the beach even further, undercutting and destroying the sea wall. The remains of two earlier sea walls that have been destroyed are indicated by the arrows. This photograph was taken at about half-tide; at high tide the water is up against the sea wall, leaving no beach sand exposed, and making the wall vulnerable to erosion.

resuspends beach sand, thus contributing to additional storm erosion (**FIGURE 6.18**). Sea walls will protect your house, but they may also help to wash away the beach in front of your house.

Wave refraction and diffraction, and longshore currents

Like the seismic waves we discussed in Chapter 3, ocean waves also bend, or **refract**, toward an area where the wave speed is slower. As we just discussed, surface gravity waves slow down as they near the shoreline because of friction with the shoaling bottom. Waves that initially approach the shoreline at an angle, as diagrammed in **FIGURE 6.19**, will be slowed down more on the shallower shoreward side, and thus will bend, or refract, toward that side. This is why waves always seem to arrive on beaches with their fronts nearly parallel to the shore. An example of this kind of refraction can be seen in Figure 6.19C. Waves will also **diffract** when they pass by an obstacle such as a jetty (Figure 6.19B).

Wave refraction does not always steer waves such that they arrive 90°, or perpendicular, to the shoreline (with their wave fronts parallel to the shoreline);

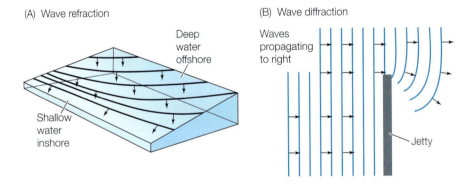

(A) Wave refraction

Deep water offshore

Shallow water inshore

(B) Wave diffraction

Waves propagating to right

Jetty

(C)

FIGURE 6.19 (A) Waves refract toward the shallower water depths where the shallow water gravity wave speed is slower; the result is that waves orient themselves such that they arrive from offshore almost parallel to the shoreline. (B) Viewed from above, waves will diffract around an obstruction, such as a jetty extending out into the water. As the wave fronts pass the object, they are effectively cut off on one end, but they continue to propagate from the edge of the jetty as circular wave fronts emanating from a point, which results in the waves effectively curling around the end of the obstruction. (C) Aerial photograph of diffraction and refraction.

(A)

FIGURE 6.20 Waves that arrive at an oblique angle on a beach create a longshore current in the swash zone. (A) Waves arriving at an angle on a beach. (B) The same beach on the same day. The arrows indicate the path of water and suspended particles washing first up the beach with an alongshore component, slowing down as the water curls to the right, before flowing back down the beach face with some complementary alongshore component.

(B)

Path of waves in swash zone along with suspended sand particles

Net longshore current

often waves arrive at an angle. In such instances, the oblique arrival of waves on a beach can produce a **longshore current**. A breaking wave that arrives from directly offshore, and not at an angle, washes straight up the beach face (in the **swash zone**), comes to a stop, and then washes back down the beach face. But waves that arrive at an angle wash up the beach face at an angle, and thus there will be an alongshore—along the beach—component to the water motion, as well as a net transport of water along the beach (**FIGURE 6.20**). This produces a net current that flows along the shore, not just in the swash zone, but also in the strip of water adjacent to the beach. You may notice this effect while swimming at a beach, as you are gradually carried along the shoreline. These longshore currents also transport sediment that is resuspended by the waves in the swash zone, often carrying significant loads of beach sand along the beach.

Seiches

A **seiche** (pronounced "saysh") is a form of standing wave that can be caused by a storm surge or simply by a steady wind blowing toward one end of an enclosed body of water such as a lake (**FIGURE 6.21**) or

FIGURE 6.21 A seiche in a lake, where the water sloshes back and forth about a node (which is where the depth remains constant), where the water levels rise at one end of the lake while dropping at the other, back and forth. Notice how the standing wave of this seiche (as all seiches) has a wavelength that is twice the length of the lake.

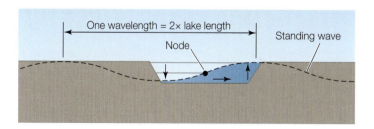

One wavelength = 2× lake length

Node

Standing wave

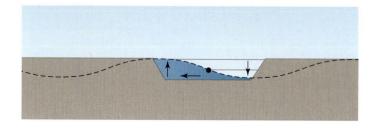

a semi-enclosed body such an a coastal bay or inlet. The wind, if it is blowing at the right speed and direction relative to the size and shape of the body of water, can pile water up at the downwind end. When the wind stops, the water sloshes back as a standing wave. The sloshing will occur at the **resonant frequency** of the body of water in question (we discuss resonance in greater detail on page 221).

We've all experienced this effect of resonance when we slosh a cup of coffee or a pan of water back and forth, setting in motion a seesaw-like wave. But a seiche doesn't have to occur an enclosed basin, like a lake. Seiches also are formed in semi-enclosed bodies of water (harbors, bays, and inlets) with the outside sea level serving as the outer boundary within which the standing wave oscillates.

FIGURE 6.22 (A) Photograph taken as the first waves from December 2004 Indian Ocean tsunami came ashore. (B) The destruction left behind in Banda Aceh, Indonesia, in early January 2005.

(A)

(B)

Tsunami (or seismic sea waves)

Incorrectly called tidal waves, **tsunami** (singular, *tsunami*) are extremely long-wavelength ocean waves, with wavelengths that can be some 200 km (125 miles), but very small wave heights, often less than 1 m. They are usually caused by an undersea earthquake or landslide that displaces a large volume of seawater, thus disturbing the ocean surface and creating a huge wave. The wave propagates across entire ocean basins at an extremely fast speed. Because the wavelength is so long, it propagates as a shallow water wave: 1/20 of a 200 km wavelength is 10,000 m, and the average depth of the oceans is about 4000 m. Thus, the speed of a tsunami is determined by the depth; as shown in Box 6A, the speed will be equal to the square root of g (gravity) times h (the depth), or $V = 3.1\sqrt{h}$. For the Pacific Ocean, which has an average depth of 4600 m, the speed of a tsunami is on the order of 210 m/s, or some 470 mph—this is the speed of a jet airliner.

Tsunami are notoriously destructive (**FIGURE 6.22**). Because of their small wave height (<1 m) and long wavelength (200 km), they can pass beneath ships at sea without being noticed; however, they wreak havoc when they come ashore. Like the waves diagrammed in Figure 6.15, their wavelengths will shorten and their wave heights will increase as the bottom shoals, creating wave heights that reach several tens of meters high as they come ashore. But just before the first tsunami wave crest arrives on shore, there will be a trough—a very dramatic trough—that will appear to drain the coastal ocean in a matter of minutes. Tragically, this phenomenon draws curious onlookers, who soon find themselves facing monster wave crests that follow each trough.

The great Alaska earthquake of 1964 (see Chapter 3) generated a tsunami that propagated some two thousand miles across the Pacific Ocean to the Hawaiian Islands in only about five hours. Along the way, the massive wave passed by ships at sea completely unnoticed, but when it arrived at Hawaii, it inundated Hilo on the Big Island with a series of huge waves that came ashore.

One of the most powerful and destructive tsunami in history occurred in the Indian Ocean on December 26, 2004. That day a massive earthquake caused by an extensive subduction event beneath the sea floor off Sumatra registered a magnitude between 9.1 and 9.3, making it one of the most powerful earthquakes ever recorded. It created a series of tsunami throughout the Indian Ocean, and was responsible for the deaths of about 250,000 people. Indonesia was hit hardest by the waves, which were estimated at greater than 30 m high; Sri Lanka, India, and Thailand also suffered extensive damage and loss of lives (**FIGURE 6.23**).

The 2004 Indian Ocean Tsunami was detected by both warning centers operated by NOAA, the Alaska Tsunami Warning System, and the Pacific Tsunami Warning Center in Hawaii. They responded almost immediately, within minutes, with a limited warning that was revised several hours later, which called for a possible six-foot wave in Hawaii. Hawaii did experience a minor tsunami, but it was only a three-foot wave, which did not cause major problems there.

The Indian Ocean tsunami was a wake-up call for the U.S. National Oceanic and Atmospheric Administration (NOAA), and for the U.S. Congress, which responded by investing in much-needed improvements to our system of monitoring for tsunami. Congress appropriated millions for NOAA to make upgrades to the warning system, which depends on seismograph stations, sea-level and tide gauges, and a system of 39 tsunami detection buoys in the Pacific and Atlantic Oceans. The buoy system has bottom pressure sensors that can detect even small changes in bottom pressure that result from the passage of tsunami.

Most recently, on March 11, 2011, a 9.0 magnitude earthquake struck the island nation of Japan. With the epicenter just 70 km off Japan's eastern coast, it is the most powerful earthquake ever recorded in Japan, and is one of the five most powerful earthquakes ever recorded anywhere. The earthquake caused extensive damage to buildings and homes, but the worst was yet to come. Minutes after the earthquake, tsunami waves estimated at some 40 m in height came ashore, inundating coastal communities and causing extensive damage to a nuclear power plant. Japan has yet to fully recover from this disaster.

FIGURE 6.23 Map of the first ten hours of propagation of the 2004 tsunami across the Indian Ocean, as simulated by a computer model; the wave continued to propagate beyond the Indian Ocean and was detected around the world.

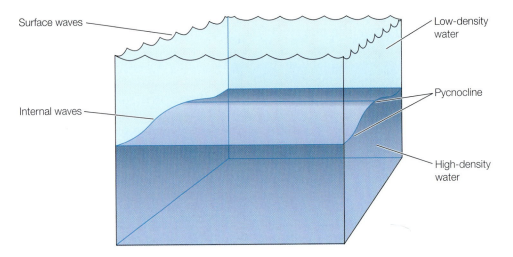

Surface waves

Internal waves

Low-density water

Pycnocline

High-density water

FIGURE 6.24 Diagram of an internal wave propagating along a pycnocline, separating lower-density water at the surface from higher-density water at depth.

Internal waves

An interesting category of ocean waves are **internal waves**, which are ubiquitous throughout the world ocean, although most often they are not obvious to an observer on the surface. Usually, these are **interfacial** waves, similar to surface gravity waves, but instead of propagating along the air–water interface, these waves propagate along a pycnocline (such as created by a thermocline; **FIGURE 6.24**). Recall from Chapter 4 that a pycnocline is a zone of rapidly changing water density below the surface. But because the density differences of water on either side of the pycnocline are extremely small compared with the difference between water and air, these waves tend to move much more slowly than surface waves, and appear to move almost in slow motion. You can witness this for yourself by making internal waves in a bottle of salad dressing. For example, if you allow a bottle of oil and vinegar (vinegar is basically acidic water) to separate into layers, with the oil sitting on top of the vinegar, and then gently jostle the bottle, you can produce slow-motion waves that are clearly visible at the interface between the layers. Also, scuba divers may have experienced internal wave phenomena without understanding what happened. If a diver is at a depth at or near the thermocline as an internal wave passes, he or she will experience a periodic, abrupt change in temperature of the water as the thermocline oscillates up and down with the passing wave.

As internal waves propagate, they sometimes leave evidence of their subsurface presence in peculiar patterns on the surface of the ocean. These patterns result from alternating vertical water motions as the waves' crests and troughs pass beneath, pushing water upward and then downward. Deeper waters rising upward to the surface tend not to support surface waves, and these features appear as long streaks of calm water, or surface slicks. When those same waters spread horizontally at the surface and eventually pass beneath adjacent water that is warmer and lighter, **foam lines** may be formed. Too buoyant to sink, the foam, as well as other floating materials, collects and accumulates at the surface along the area of convergence.

Tides

Tides are the largest of all ocean waves in terms of their wavelength, and most of the world ocean experiences them. They are large-scale water motions that result from the gravitational attractions between the Earth and its nearest celestial bodies, especially the Sun and the Moon, all of which pull on one another and influence each other's orbits in space. The gravitational pull on the oceans produces tidal waves and their associated tidal currents, which result in a massive sloshing about of the oceans as the Earth rotates. As the Earth, Moon, and Sun change positions relative to one another over the course of a day, a month, a year, and even longer periods of time, the tidal waves and currents vary in their size and velocity. These variations are the astronomical high and low tides you often hear about, especially when a tropical storm or hurricane looms in the forecast; storm surges are either added to or subtracted from the tides, and coastal flooding can be exacerbated during astronomical high tides.

The water motions that result from gravitational forces are also influenced by the shapes and sizes of oceans and seas and their coastlines, which act together to produce tides that don't just go up and down, as most people tend to think of them—they create strong horizontal tidal currents. It is the flow of those tidal currents in response to tidal waves that cause the tide to flood and ebb, or to come in and go out, thus changing local sea level over the course of hours. And those same tidal currents are very important to a lot of oceanographic processes, which in turn influence ocean biology, as we'll see later.

Around the world's ocean, tides vary greatly in their **tidal range**—the difference in local sea level between high and low tides—with some parts of the world experiencing tides that are barely noticeable, while other areas have huge tides (**FIGURE 6.25**). Tides also vary throughout the world in their periodicity: the tide may come and go once a day, twice a day, or exhibit different proportions of each pattern (**FIGURE 6.26A**). Parts of the world where tides come and go twice a day have a **lunar**, or **semidiurnal**, **tide**. Such regions are "in tune"[4] with the gravitational attraction of the Moon. In other places, there may be one high tide and one low tide each day—this is a **solar**, or **diurnal**, **tide**. These areas are more in tune with the gravitational attraction of the Sun. In parts of the world where both the Sun and Moon exert significant influence, we have mix of solar and lunar tides; these are called **mixed tides**.

Much of the west coast of North America experiences mixed tides (**FIGURE 6.26B**). Notice that Los Angeles, California, experiences approximately two

---

[4] By being in tune, we are referring to the resonant frequency of a region. This concept of resonance is explained further below in this chapter.

FIGURE 6.25 High tide (A) and low tide (B) in the Bay of Fundy, Canada, which has the greatest tidal range in the world, exceeding 15 m (50 feet).

(A)

(B)

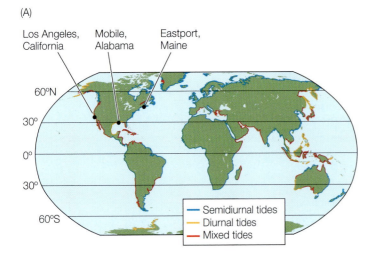

(A)

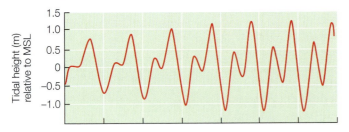

(B) Los Angeles, California

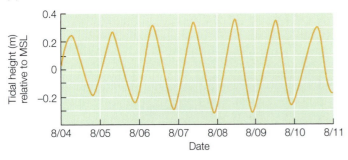

(C) Eastport, Maine

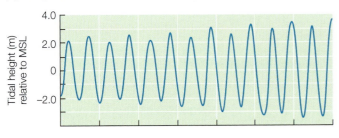

(D) Mobile, Alabama

FIGURE 6.26 (A) Distributions of the types of tides around the world, with examples of each: (B) Los Angeles, with a mixed tide; (C) Eastport, Maine, with a semidiurnal tide; and (D) Mobile, Alabama, with a diurnal tide. Observed tidal heights for the week of August 4–12, 2010, are plotted in meters relative to mean sea level (MSL).

high tides and two low tides each day (each lunar day, that is),[5] but that one of the high tides and one of the low tides is dramatically bigger than the other. Diurnal tides dominate in the Gulf of Mexico, as shown for Mobile, Alabama, where the tide comes and goes once a day. The Gulf of Maine (in Eastport, Maine; **FIGURE 6.26C**), on the other hand, has semidiurnal tides, with two high tides and two low tides each day. But notice that one of the two high tides is higher, and one of the two low tides lower, in an alternating pattern. Because of these inequalities, the Gulf of Maine actually has a mixed, semidiurnal tide, similar to that of Los Angeles, but nowhere near as extreme. Notice also in Figure 6.26 that the tidal range in Eastport is on the order of 7 m (23 feet), whereas the tidal ranges in the Gulf of Mexico (Alabama; **FIGURE 6.26D**) are only about 65 cm (just over 2 feet). This variation gives us a hint of the unequal distribution of tidal energy in different parts of the world ocean. The tides in the Gulf of Maine are not only large; they increase toward its upper reaches in the Bay of Fundy, where we find the largest tidal ranges in the world (we'll come back to the tides in Gulf of Maine and the Bay of Fundy below).

Understanding the forces at work

In order to understand and appreciate the forces that create tides, we're going to rely on yet another thought experiment. Our approach is based on what is generally known as the equilibrium theory of the tides, which was first developed by Isaac Newton in 1687 as part of his universal theory of gravity. But unlike Newton, we aren't going to derive the mathematical formulation of the tide-generating forces; instead we're going to use diagrams and logical arguments in order to gain a conceptual understanding (a more quantitative description is given in Box 6B.) In short, Newton showed that the gravitational attraction between two celestial bodies is proportional to the product of their masses divided by the square of the distance between them. That means that more massive objects, like the Sun, have a much stronger gravitational attraction than smaller bodies, such as the Moon. However, their gravitational attractions on other objects diminish with distance. Because the Sun is so much farther away from Earth than the Moon, its gravitational

---

[5] Because the Earth is rotating as the Moon orbits it, we need to introduce the *lunar day*, also called a tidal day—the time it takes for the Earth to complete one full rotation on its axis with respect to the Moon, about 24 hours, 50 minutes, 28 seconds.

effect on the tides is only about half (46%) that of the Moon.

Our thought experiment will examine what happens to the tides on Earth in response to the gravitational attraction of the Moon, ignoring for the moment the influence of the Sun. To do so, we are going to imagine a hypothetical, simplified Earth without any continents, and only a thin skin of water covering the entire surface of the planet (**FIGURE 6.27A**). The Moon, shown off in the distance, exerts a gravitational attraction, which, as Newton explained, is inversely proportional to its distance from Earth. The different lengths of the vectors in the sketch are intended to show the relative proportions of the strength of the Moon's gravitational attraction at various positions around the world. It shows that the Earth experiences—everywhere around the world—a gravitational pull by the Moon in the directions indicated, being greatest at points on Earth that are closest to the Moon. Notice that the pull of the Moon's gravity is felt all the way through to the opposite side of the world—the presence of the Earth is no shield against gravity. The key point here is that locations on the Earth closest to the moon experience the greatest gravitational pull.

Given the pull of the Moon's gravity as diagrammed in Figure 6.27A, it would logically follow that anything not tied down to the surface of the Earth, such as an ocean, would tend to slide along the Earth's surface in the directions indicated by the arrows in **FIGURE 6.27B**. Therefore, that gravitational attraction would tend to pull water such that it flows to one side of the spherical Earth; that is, it would flow toward the spot directly under the Moon and pile up, producing a bulge. Thus we would expect a high tide on the side of the Earth facing the Moon; everywhere else around the Earth, from 90° away and beyond, we would have low tides, as shown in the sketch.

We can demonstrate the phenomenon illustrated in Figure 6.27B with another brief thought experiment. Imagine a bowling ball (call it "Planet 1") that we have attached to a length of wire. Imagine also that we dip that ball into a large bucket of latex paint, and then pull it out, leaving it hanging from the wire; the thin layer of wet paint coating the ball would constitute Planet 1's ocean. We would see what was initially a uniform layer of wet paint begin to slide across the surface of the bowling ball where it would collect on the bottom of the ball in response to the gravitational attraction of Earth. That is, the paint would pile up and create a bulge—the paint would be thickest, or deepest, on the side of the ball facing the Earth; this phenomenon is analogous to our hypothetical high tide in Figure 6.27B.

Based on our brief painted bowling ball thought experiment, it would seem that we have confirmed what we expected, based on the arguments in Figure 6.27: that there must be a *single* tidal bulge on one side of our hypothetical Earth that results from the Moon's gravitational pull of water toward the point on the Earth closest to the Moon. So far so good, except that this is wrong. Here on the real Earth, there are actually *two* tidal bulges

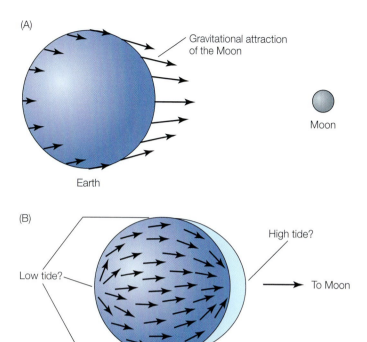

FIGURE 6.27 (A) Sketch of an imaginary Earth with all the continents removed and a single, uniform ocean covering its surface. Vectors represent the relative pull of the Moon's gravity, which is greatest at points closest to the Moon. (B) Sketch of the tidal currents on the surface of a hypothetical Earth that might be expected to result from just the gravitational attraction of the Moon as in (A). We would expect a flow of waters around the surface of the Earth to a point in line with the Moon, where they pile up, creating a high tide. But this is *not* what actually happens, as explained in the text.

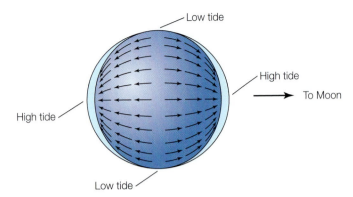

FIGURE 6.28 The directions of water motions on the surface of the real Earth under the influence of the Moon's gravitational attraction. Unlike our hypothetical Earth in Figure 6.27, there are actually two identical but opposite forces pulling water to opposite sides of the Earth, the result of which produces two high tides and two low tides.

attributable to the pull of the Moon. In reality, there is another high tide on the opposite side of the globe (**FIGURE 6.28**), due to a second set of forces on the surface of the Earth, which we explain next.

To understand why there is this second set of forces, we need to appreciate the fact that not only does the Moon orbit the Earth, but the Moon and Earth orbit each other, around a point that is the center of mass of the Earth–Moon pair. Continuing our original thought experiment: If the Earth and Moon were the same size and mass, they would orbit around one another the way a dumbbell would spin on its handle, as shown in **FIGURE 6.29A**. Because the mass of each of the two balls on the ends of the handle is the same, the balance point (the fulcrum) is exactly halfway between them. Imagine how the dumbbell might rotate around that center of gravity, thus mimicking two planets of identical mass that orbit around one another. As the two planets of equal mass orbit one another, they are held in place by their mutual gravitational attraction, which exactly balances the centrifugal force pulling them outward, indicated by vectors of equal length in Figure 6.29B (see Appendix A). But if we make one of the planets smaller (now we will call it a "moon"), the balance point—the center of gravity—moves closer to the larger ball (Figure 6.29C). Eventually, if the smaller ball (the moon) gets small enough, the balance point, the center of mass, will be located beneath the surface of the larger ball. This is actually the case for the Earth–Moon pair. This is illustrated in the sketch in Figure 6.29D, where the fulcrum point is set at the center of mass of the Earth–Moon pair, 2903 miles away from the exact center of the Earth. In other words, the balance point is about 1000 miles, or one-eighth of the Earth's diameter, beneath the Earth's surface.

This means that as the Moon orbits the Earth, the Earth orbits the Moon as well, but it isn't as noticeable because, rather than tracing out a large orbital circle as the Moon does, the Earth orbits a much tighter circle around the center

FIGURE 6.29 (A) Sketch of a dumbbell with two weights of equal mass. They would balance at a point in the middle of the handle, such that a fulcrum placed there would allow the dumbbell to spin around that point. The spinning dumbbell is analogous to an "Earth–Moon pair" in which both bodies are of the same mass. The dashed arrows show the orbital path of the two. (B) As the Earth and Moon orbit one another around their center of mass, they would each experience a gravitation attraction (g) to the other mass and a centrifugal force (CF) equal but opposite to that force, as shown here for just the Earth. (C) Reducing the mass of the Moon causes the balance point to move closer to the Earth. (D) The actual Earth–Moon pair has a center of mass that is beneath the surface of the Earth. Even though the center of mass of the Earth–Moon pair is no longer out in space between them, they each experience a centrifugal force that is equal to the gravitational attraction by the other mass, as in (B).

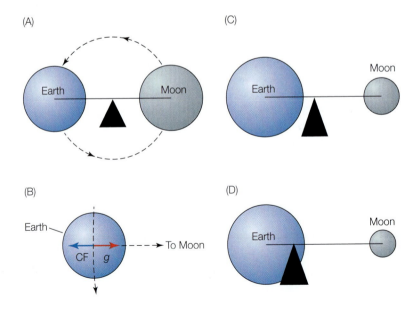

FIGURE 6.30 (A,B) As the Moon orbits around the center of mass of the Earth–Moon pair, the Earth orbits around the same point, which is beneath the surface of the Earth. Earth's orbit around this point, then, will appear as a wobble (C) that will create a centrifugal force (CF) that is equal to but opposite to the average gravitational attraction (g) of the Moon.

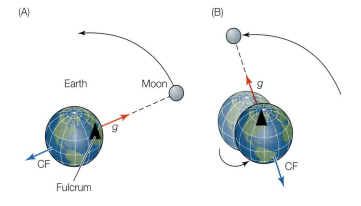

of mass deep inside itself, and *wobbles* (**FIGURE 6.30**). But the Earth still has a centrifugal force as a result of that wobble, and that centrifugal force is still exactly equal to the Moon's average gravitational attraction.[6] Therefore, the entire Earth and everything on it experiences a centrifugal force as the Earth wobbles around the center of mass of the Earth–Moon system, which is equal in magnitude, but opposite in direction, to the average force of the Moon's gravitational attraction.

These two forces, the Moon's gravitational attraction and centrifugal force, are equal to one another only along a plane that runs through the center of the Earth and as an average for the whole Earth (**FIGURE 6.31**). Everywhere else on Earth the two forces are not equal, and it is this inequality that creates the two tidal bulges. This is because while the centrifugal force on Earth is the same everywhere—directly under the Moon, on the opposite side of the Earth from the Moon, anywhere—the Moon's gravitational attraction is not (**BOX 6B**). It varies with the distance to the Moon, which, obviously, isn't the same everywhere on the surface of the Earth. On the side of the Earth nearest the Moon, the Moon's gravitational attraction is greater than the Earth's centrifugal force; on the side of the Earth opposite the Moon, the Moon's gravitational attraction is less than centrifugal force. But, again, on average for the whole planet Earth, they both are equal, which must be the case in order for them to orbit one another.

This inequality in forces on opposite sides of Earth creates the tidal flows in Figure 6.28 that form the two identical bulges—two high tides—on opposite sides of the Earth. As the Earth rotates once each day, the two bulges (the two high tides) will always be on the side of the Earth toward the Moon and on the opposite side away from the Moon. Therefore, the rotating Earth essentially slides beneath each of the two bulges once every day (every lunar day), as

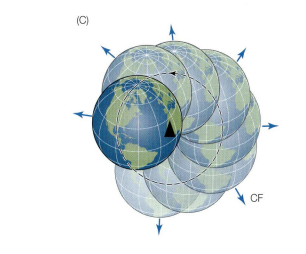

FIGURE 6.31 Diagram of the relative importance of the Moon's gravitational attraction (g, red arrows) and the Earth's centrifugal force (CF, blue arrows). CF is the same everywhere on the surface of the Earth, as indicated by the blue arrows of equal length, and in the center of the Earth, CF and g are equal. But, because locations on Earth on the side of the Moon are closer to the Moon, they experience a greater g, and so g exceeds CF there. On the opposite side from the Moon, the greater distance from the Moon results in a smaller g, and so CF is greater than g there. The net forces that result are indicated by the black arrows. These forces are explained further in Appendix A.

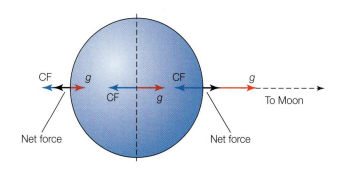

[6]The point inside the Earth that is the center of mass of the Earth–Moon system is not a stationary, or fixed, point inside the Earth; it essentially moves through the Earth as the Moon orbits such that it is always beneath the surface of the Earth that is closest to the Moon. This is an important detail because it influences the way the planet wobbles around the point: it is as if you were to hold firmly a basketball with both hands, and then move it in circles, like the "Earth" in Figure 6.31C. This is not the same as drilling a hole off-center in the basketball, inserting a stick, and then spinning the basketball around the stick.

BOX 6B The Tide-Generating Forces

Isaac Newton showed that the gravitational attraction between two bodies is directly proportional to their masses, M, (more mass means greater gravitational attraction), and inversely proportional to the square of the distance between them, d, (greater distance means less attraction). Assigning Earth's mass a value of 1, then the gravitational attraction of the Moon, g, at the center of mass is proportional to

$$\frac{M_1 M_2}{d^2} = \frac{m}{d^2}$$

This means that the net force is

$$F = G\frac{M_1 M_2}{d^2}$$

In that equation, G is the gravitational constant. Because the Earth and Moon orbit one another around their common center of mass, the Earth and the Moon experience a centrifugal force (CF), which at the center of the Earth, is equal to the Moon's gravitational attraction (g), as shown in **FIGURE A**. The variable r is the radius of the Earth.

CF is the same everywhere on the Earth, as in **FIGURE B**, but g is not everywhere the same. It varies according the distance from the center of the Moon. At Position 1, which is the closest point to the Moon, g is greater than CF. The Moon's gravitational attraction, g, is therefore proportional to:

$$\frac{m}{(d-r)^2}$$

The Earth's CF is proportional to

$$-\frac{m}{d^2}$$

The sum of the two forces, CF and g, gives the net force at Position 1 as proportional to:

$$\frac{m}{(d-r)^2} - \frac{m}{d^2} = \frac{2mr}{d^3}$$

Using the same logic, the resulting force at Position 2, on the opposite side of the Earth, is proportional to:

$$\frac{m}{(d+r)^2} - \frac{m}{d^2} = \frac{-2mr}{d^3}$$

At positions along the vertical dashed line, which runs through the Earth's center of mass, CF and g are equal and are proportional to:

$$\frac{m}{d^2}$$

However, g is directed along the line indicated toward the center of mass of the Moon, and so it has a small component directed toward the center of the Earth.

Forces at other points can be determined trigonometrically. The resulting relative forces are illustrated in **FIGURE C**, and are responsible for the tidal currents illustrated in Figure 6.28.

FIGURE A

Earth (mass=1)

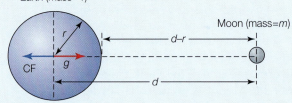

Moon (mass=m)

FIGURE B

On side away from moon, g is less than CF

At center of mass, CF and g are equal

On side toward moon, g is greater than CF

Position 1

Moon

Position 2

FIGURE C

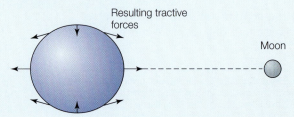

Resulting tractive forces

Moon

illustrated in the sketch in **FIGURE 6.32**. (We should note here that the actual tidal heights that result will be related to the angle between the Earth's orbital axis and the Moon's orbital plane; this is an additional complication, but it is a relatively minor one that we don't need to consider in this discussion of the basics of tides.)

Combined influences of the Sun and Moon

While the strongest force producing the Earth's tides is the Moon's gravity, which is about twice as strong as that of the Sun, the Sun nonetheless exerts a significant gravitational force. The Earth's tides are therefore the result of a combination of gravitational attraction by the Sun and Moon, such that water flows toward each of them as they and the Earth fly through space relative to one another. Once we add the gravitational pull of the Sun to our simple conceptual model of the lunar tides, and incorporate a consideration of the orbital dynamics of both the Moon and the Sun, the tides get much more complicated.

The Sun and Moon don't "go around" the Earth at same rate. The Sun "goes around" once a day, while the Moon "goes around" about once every month. (Of course, the Moon orbits the Earth and they both orbit the Sun.) This means that there are important tidal phenomena that occur at fortnightly (bi-weekly) intervals in relation to the phases of the Moon. The orbital alignments of the Sun–Earth–Moon system are lined up in a straight line twice a month—when we have a full moon and when we have a new moon (**FIGURE 6.33A**). Each results in the gravitational attractions of the Sun and Moon effectively working together. During a full Moon, for example, the Moon pulls in a direction that is 180° opposite from the pull of the Sun, thus reinforcing the two bulges on opposite sides of the Earth. Likewise, their centrifugal forces are also directed 180° from one another, also reinforcing the double bulges. During a new Moon, the effect is as if the Moon and Sun were welded together into a single, larger body. In each case, the gravitational attractions and centrifugal forces are additive, which gives us tides that are significantly larger than average during a full or a new Moon. That is, the bulges of water—the high tides—are higher, which means that the low tides must be lower. These are called **spring tides**.

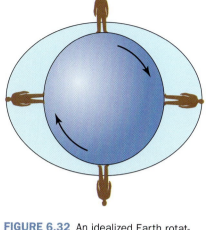

FIGURE 6.32 An idealized Earth rotating beneath an ocean without continents would have two high tides and two low tides. As the Earth rotates an imaginary person standing on it would experience two high tides and two low tides each lunar day. That is, the tides aren't moving in this conceptual model; the Earth is moving under them.

FIGURE 6.33 (A) Weekly orientations of the Sun, Earth, and Moon orbital system with phases of the Moon indicated. Twice a month, the Sun, Earth, and Moon line up, giving us a full moon and a new moon. In those instances, the gravitational attractions reinforce each other such that we see spring tides—high tides are higher and low tides are lower than average. (B) When the Sun, Earth, and Moon are oriented at right angles, their gravitation forces are perpendicular and there are no additive effects; these are when we have the first quarter and third quarter phases of the Moon, which correspond with the neap tides—the lowest high tides of the month and the highest low tides of the month.

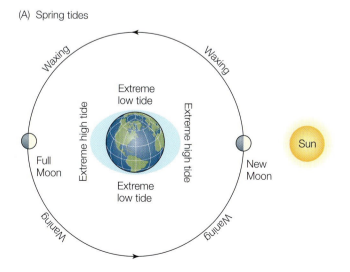

(A) Spring tides

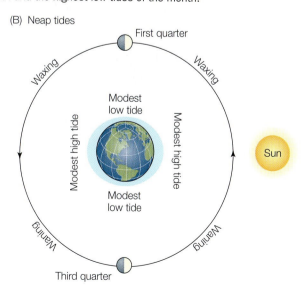

(B) Neap tides

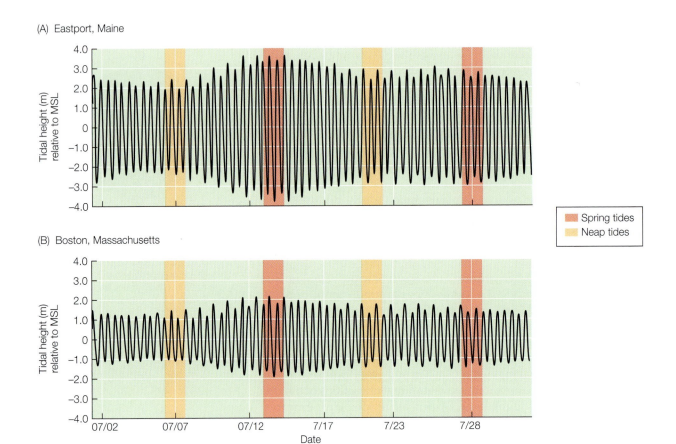

(A) Eastport, Maine

(B) Boston, Massachusetts

Spring tides
Neap tides

FIGURE 6.34 Observed tides recorded at (A) Eastport, Maine, and (B) Boston, Massachusetts, for the month of July 2010. The tidal height relative to mean sea level (MSL) is plotted; high tides are given as meters above MSL, and low tide as meters below MSL.

Using the same logic, we will also see twice a month the Sun, Moon, and Earth oriented at right angles to one another, which is when we have "half moons" (also known as the first and third quarter phases of the Moon; **FIGURE 6.33B**). When this happens, the gravitational forces are not additive; because they are at right angles, they have no effect on one another. But that orientation does make the tides a bit more complicated. Because the gravitational forces are working at right angles, we have the Sun (or Moon) pulling the tide to one side of the Earth, while the Moon (or Sun) pulls the tides to a point 90° away. Twice a month, then, we have minimal tidal ranges, when high tides are not very high and low tides are not very low. These are called **neap tides**.

An example of this spring–neap cycle over the course of a month can be seen in the sampling of tide records for Eastport, Maine, and Boston, Massachusetts, in 2010 (**FIGURE 6.34**). Notice that the fortnightly pattern of spring and neap tides is apparent for both ports, with spring tides on about July 14 and 28, and neap tides on July 5 and 19. In addition, notice how the tides are mixed semidiurnal, with slightly different levels for successive high tides and low tides. Finally, notice that the tidal ranges are markedly different from one another; the tidal ranges in Eastport are twice those in Boston. The reasons for differences in tidal ranges such as these are related to the shapes of ocean basins and their coastal areas.

Tides in ocean basins

Our thought experiment on the forces controlling the tides has produced a conceptual model of what we might expect on our planet if there were no continents and only one big ocean. Once we add the continents, the Earth's oceans are no longer just a wet surface on a smooth ball; there are now obstacles that prevent the easy flow of tidal currents. The addition of the continents and consideration of bottom friction as it impedes tidal currents produces an interesting variation in the nature of tides.

Because of the Earth's rotation, the tides respond to the Coriolis effect and do not flow in straight lines; instead they flow as waves that tend to slosh around the insides of embayments (bays), both large and small. As a flooding tide (or rotary tidal wave) enters a coastal embayment, the tide enters as a wave that tends to hug the right side of the bay, keeping the shoreline to its right as it rotates around the basin (**FIGURE 6.35**). (In the Southern Hemisphere, rotary tidal waves rotate in the opposite direction). Along the coastal edges of the bay, the changes in depth between the crest and troughs of the tidal wave are greatest, and toward the open end of the bay there is a point where there is no change in water depths—this is known as an **amphidromic point**.

This phenomenon of a rotary tidal wave is easy to visualize or to demonstrate to yourself in your kitchen (for an enclosed body of water, not a semi-enclosed bay). For example, if you hold a pan of water—an ordinary dishpan, preferably a round one for this experiment—and then swirl it around just right, you can create a very dramatic and well-formed wave that rotates around the edges of the pan (Figure 6.35B). As that wave rotates, you will see the water depths at the edges of the pan increase and decrease as the crests and troughs pass

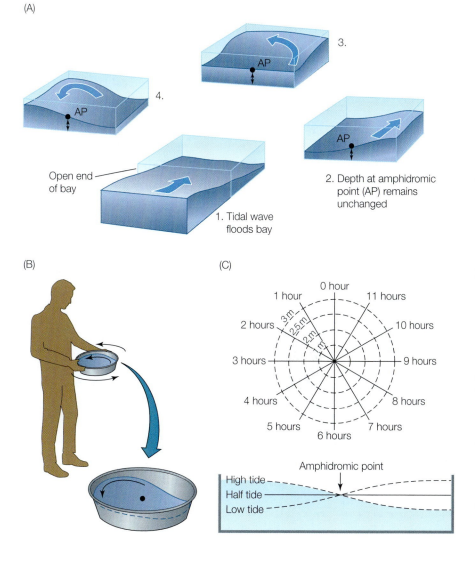

(A)

3.

AP

4.

AP

Open end of bay

2. Depth at amphidromic point (AP) remains unchanged

1. Tidal wave floods bay

AP

(B)

(C)

0 hour
1 hour   11 hours
2 hours   3m 2.5m 2m 1m   10 hours
3 hours ┼─────────────────┼ 9 hours
4 hours   8 hours
5 hours   7 hours
6 hours

Amphidromic point
High tide
Half tide
Low tide

FIGURE 6.35 (A) Diagram of a tidal wave entering a bay that is open to the ocean at one end (1). In the Northern Hemisphere the incoming wave will hug the edges of the bay (2, 3), keeping the bay to the right of the wave's progression as it rotates around an amphidromic point at the opening of the bay (4). (B) A person swirling a dishpan of water just right can make a wave that rotates around the edges of the pan. The water depth in the very center of the pan will remain unchanged, while water depth at the edges of the pan will increase and decrease as the wave's crest and trough pass by. This is analogous to a tidal wave that rotates around an ocean basin. (C) A hypothetical ocean basin diagrammed looking down from above as well as in cross section. The solid lines are co-tidal lines that mark the position of the crest of the rotary tidal wave each hour into the lunar tide. The dashed concentric circles give the tidal heights in meters (tidal height is the depth differences between high and low tide) across the basin from the center to the outer edge. There is no change in water depth in the center, which is the amphidromic point.

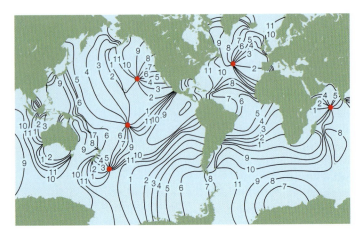

FIGURE 6.36 Amphidromic points (red points) in the world ocean, along with co-tidal phase lines, which approximate the location of the crest of the tidal wave for each hour into the 12 hour lunar tidal cycle.

by; these are analogous to high and low tides. And in the center of the pan you will notice an amphidromic point, where the water depth doesn't change. In an ocean basin (versus a dishpan) we would see the tidal wave rotate similarly as in Figure 6.35A, but in this case, for the lunar tide. Figure 6.35C shows a cross section of how the water depths change as the tidal wave completes one cycle, and also a plan view (Figure 6.35C), with locations of the crest of the tidal wave for each hour into the lunar tidal cycle; also shown in that diagram are concentric circles that represent approximate tidal wave heights, from zero at the amphidromic point to a meter or so around the edges. These rotary tidal currents and their amphidromic points are found throughout the world ocean (**FIGURE 6.36**).

It is this complication resulting from the presence of shorelines, and the relative size scales of those coastal features, that help to alter the relative importance of the gravitational forces from the Sun and Moon, thus giving us the various tide types (semi-diurnal, diurnal, and mixed tides) that are found around the world (see Figure 6.26).

Tides in the Gulf of Maine and the Bay of Fundy

The rotary tidal wave in the North Atlantic Ocean rotates counterclockwise, thus forming a tidal wave that propagates down the east coast of North America, from north to south (**FIGURE 6.37A**). When this North Atlantic tidal wave crest is sitting outside the Gulf of Maine (and other coastal bays, harbors, and estuaries on the east coast of North America) the sea level will be higher outside, farther offshore in the Atlantic Ocean, than inshore and inside the Gulf of Maine, thus creating a flow of water from high sea level

FIGURE 6.37 (A) The co-tidal phase lines (red), approximate locations of the crest of the tidal wave each hour into the lunar tidal cycle around the North American amphidromic point. The co-range lines (blue) give the tidal height in meters, and the black arrows indicate the counterclockwise movement of the tidal wave crest. (B) Currents in the Gulf of Maine and Bay of Fundy.

(A)

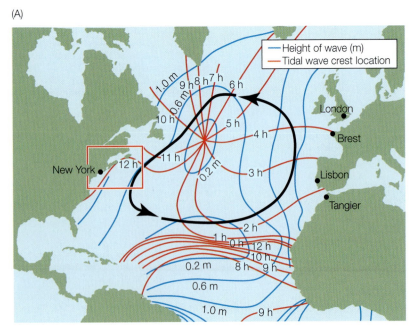

(B)

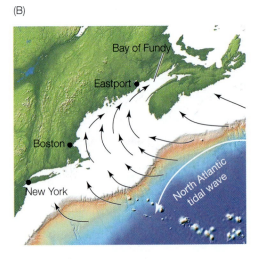

toward low sea level (downhill); this creates a tidal current that floods the Gulf (**FIGURE 6.37B**). Those tidal currents will continue to flow into the Gulf of Maine until the trough of the tidal wave moves into position outside the Gulf, at which time the Gulf drains. It turns out that this time period between successive Atlantic tidal wave crests and troughs in the North Atlantic Ocean is about the same as the resonant frequency of the Gulf of Maine—about 12.4 hours, or half a lunar day, the time between successive lunar high tides. And this is one of the factors—a major one—that produces the biggest tides in the world in the upper reaches of the Gulf of Maine, in the Bay of Fundy.

As we saw in Figure 6.34, the tides in the Gulf of Maine are quite large, with spring tides ranging from about 4 m in Boston, Massachusetts (the western Gulf), to more than 7 m in Eastport, Maine (the eastern Gulf). But the biggest tides in the world are still farther to the east in the upper reaches of the Bay of Fundy at the extreme eastern end of the Gulf of Maine (**FIGURE 6.38A**). The reasons for these exceptional tides are two-fold: first, as just mentioned, the resonant frequency of the Gulf of Maine is very close to that of the lunar tide. Second, the Bay of Fundy's location within the Gulf of Maine and its narrowing shape focus the tides, as we explain next.

We are all familiar with the principle of resonant frequencies, but most of us don't think about it very often. Examples include the tone that can be achieved when we blow air across the top of an empty bottle at just the right speed—we get a standing wave pattern in the bottle that produces a tone unique to that bottle. It resonates at the bottle's resonant frequency. We can also often get a standing wave resonating in a coffee cup—usually unintentionally, spilling the coffee (**FIGURE 6.38B**). That frequency is something close to 0.2 seconds. And, of course, who hasn't made dramatic waves when sitting in a bathtub, forcing a wave that sloshes back and forth, at a frequency of about 1.5 seconds? OK, maybe you haven't; if that's the case, then try this little experiment (but do it standing up in the tub; you'll get better results). First, bend over and move your hands forward and backward through the water, at a fairly fast rate. You'll find that all you have done is make a mess by splashing water everywhere. Then do the same but move your hands very slowly; if you do this right, all you will see is your hands dragging through the water, with no waves or splashing. But if you move

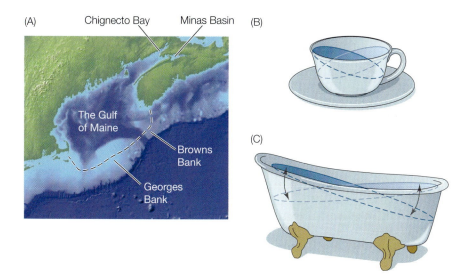

FIGURE 6.38 (A) The Gulf of Maine. The undersea features, Georges Bank and Browns Bank, make the Gulf semi-enclosed. (B) The resonant frequency of a coffee cup is about 0.2 second; it is short because the cup is so small. (C) The larger bathtub has a frequency of about 1.5 seconds; this is still quite short. The much larger Gulf of Maine (A), while not a well enclosed container like the coffee cup or bath tub, has bottom features and offshore banks that make it semi-enclosed (as indicated by the dashed line). The resonant frequency of the Gulf is estimated at about 12.4 hours—which is very close to the frequency of the North Atlantic tidal wave.

your hands back and forth at just the right frequency—about 1.5 seconds, the resonant frequency for the tub—you will get a very nice standing wave sloshing back and forth in the tub, as in **FIGURE 6.38C**, with the water rising up on one end and then the other (small children perform this experiment masterfully, by the way).

A similar phenomenon happens in the Gulf of Maine. The North Atlantic tidal wave is just about the right frequency such that no sooner has the higher sea level offshore from a passing tidal wave crest filled the Gulf of Maine, which takes some time, than the trough of the tidal wave passes by, with just enough time to drain the Gulf. These successive crests and troughs are at the right frequency that they reinforce the successive filling and draining of the Gulf, analogous to our hand motions in the bath tub experiment, thus creating fairly large tides in the Gulf of Maine.

But why are the tides in the Bay of Fundy so much bigger than tides in other places? The answer is in the bay's shape and its position inside the Gulf of Maine. For example, looking back at the sketch of the bathtub in Figure 6.38, imagine that the tub is made of soft modeling clay, such that we could pinch one end of the tub into a narrow point. This would be analogous to the shape of the upper reaches of the Bay of Fundy, with its two pointed bays, Chignecto Bay and Minas Basin. Then, when the next wave crest arrives at that end of our clay bathtub, you can imagine what will happen: the wave will focus water up the narrowing wedge, reaching a very high water level at the far end, and perhaps even shooting water out that end of the tub. This basically is what happens in the Bay of Fundy.

The importance of tides

Tidal motions on Earth represent huge amounts of energy. In fact, the energy dissipation associated with those tidal flows rubbing against the ocean bottom is slowing down the rotation of the Earth. A few hundred million years ago (380 mya), there were about 400 days in a year—the Earth rotated 400 times for each complete orbit of the Sun—and a day was only about 22 hours long. Today the Earth's rotation has slowed such that it takes 24 hours to rotate once, and it completes only 365 rotations for each orbit around the Sun. The effect the tides have on the Earth's rotation is a lot like what happens to a baseball in flight when its cover is partially torn off; the additional friction will slow the ball's rotation. Similarly, the constant friction of tides rubbing against the ocean floor is progressively slowing down the Earth's rotation.

TIDAL POWER DEVELOPMENT   As oil reserves continue to deplete and as world oil markets keep raising the price of a barrel of oil, a number of companies are looking very closely at harnessing tidal power. The kinetic energy associated with the tides, which is simply lost in slowing the Earth's rotation, is the target of their efforts. The idea is simple: by placing a turbine in the flow of tidal currents, it is possible to extract energy for use in generating electricity. There are a number of technical variations of this basic principle. For example, one major tidal power station in operation today is the 240 megawatt (MW) facility in St. Malo, France, on the Rance Estuary, which was built in 1966. There are also several experimental facilities, such as the 20 MW unit at Annapolis Royal in the Bay of Fundy, Canada, built in 1984. These two examples were built with dams that capture the tidal flow on one or both tides (flood and ebb). But dams are not the only ideas: tidal turbines are also being proposed in various configurations throughout regions where there are sufficient tidal current speeds. Areas of the world ocean where tidal power development is considered feasible require a

FIGURE 6.39 Areas of the world ocean where tidal energy is considered sufficient to make tidal power development feasible. Notice that all are in coastal and inshore areas, where tidal current speeds are greatest, as modified by shoaling depths and a concentration of those flows, in contrast to the open ocean.

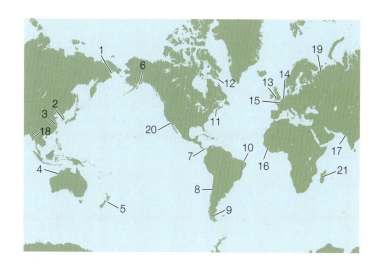

minimum 4 m tidal range which, it turns out, is quite common (**FIGURE 6.39**).

Target sites

1. Siberia	11. Bay of Fundy
2. Inchon, Korea	12. Frobisher Bay, Canada
3. Hangchow, China	13. Wales, UK
4. Hall Point, Australia	14. Antwerp, Belgium
5. New Zealand	15. Le Havre, France
6. Anchorage, Alaska	16. Guinea
7. Panama	17. Gujarat, India
8. Chile	18. Burma
9. Punta Loyola, Argentina	19. Semzha River, Russia
10. Brazil	20. Colorado River, Mexico
	21. Madagascar

TIDAL MIXING   Tidal currents are important to the biology of the ocean in a number of ways. The currents are important, of course, in transporting plankton and smaller organisms, especially in their early developmental stages. But tidal currents are also one of the main reasons why some coastal areas are very productive biologically: tidal currents create tidal mixing.

Tidal currents run from the surface to the bottom—they encompass the entire water column. Because tidal currents extend to the bottom, they rub along the bottom, and in the process, they create a turbulent mixing action. Rivers are analogous: they also flow from top to bottom, and where the river bottom is shallow enough, the river current speed swift enough, and the bottom rough enough, we often see rapids, or whitewater. Vertical mixing there extends from the bottom all the way to the surface, giving us the visible turbulence.

Likewise, where coastal waters are shallow enough, we sometimes see tidal currents that completely mix the waters from the bottom to the surface, homogenizing both the temperature and salinity (and hence, the density). The tidal mixing process is illustrated in **FIGURE 6.40**. The arrows represent turbulent mixing currents that extend upward from the bottom to a fixed distance from the bottom. That distance is determined by the tidal current speeds and the bottom roughness. For example, inshore waters that are shallow enough for mixing to extend to the surface, as in the left side of Figure 6.40, will not exhibit a thermocline; despite the Sun's rays warming of the surface waters, the waters stay well mixed. Instead, heated surface waters are simply mixed with the colder waters at depth, keeping the surface waters from becoming very warm. But farther offshore, the waters become too deep for this to happen; the tidal mixing does not extend far enough up off the bottom, and surface heating is unaffected. Here we routinely see warm surface waters separated from colder deeper waters by a thermocline. The overall pattern is one of cool surface water temperatures in inshore areas, warm water at the surface offshore (at the surface above the thermocline), and cold waters on the bottom offshore. It is often in these tidally mixed areas that deep nutrient-rich waters are mixed upward, stimulating biological productivity—the subject of our next chapter.

FIGURE 6.40 Schematic diagram of tidal mixing in coastal waters, creating a mixed zone against the coast of cooler waters. In deeper water, the surface waters offshore are warm, too far removed from the bottom to be mixed.

Cool near-shore water

Warm surface water

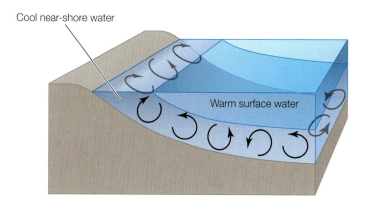

Chapter Summary

- Waves have a **crest** and a **trough**. The horizontal distance between one crest and the next, or between one trough and the next, is a **wavelength**. The **period** is the time it takes for one complete wave cycle, usually measured as the time between two successive crests or troughs.

- The displacement of the sea surface created by a wave can be measured as the **wave height**, which is the vertical distance from the top of a crest to the bottom of a trough, or the wave **amplitude**, the vertical distance that the sea surface is displaced from a position at rest. The wave's amplitude is equal to one-half the wave height.

- Waves are the result of a **disturbing force** and a **restoring force**. One disturbing force is the wind, which gives us wind waves; one restoring force is gravity.

- As wind blows and imparts energy to waves, waves grow in length and height, but they grow faster in height than in length. Eventually a **critical steepness** value of the ratio of wave height to wavelength is reached (when $H{:}L = 1{:}7$) and the wave breaks, forming **whitecaps**.

- It is generally true that surface waves move energy, not mass, but there is an exception: **Stokes Drift**, which is what brings flotsam to shore.

- **Wind waves** begin as small **capillary waves**, with a wavelength less than 1.73 cm, for which the restoring force is the water's surface tension. Once formed, and as the wind continues to blow and their wavelength increases beyond 1.73 cm, the waves transition from capillary waves to surface **gravity waves**, for which the restoring force is gravity.

- There are two categories of surface gravity waves: **shallow water waves** and **deep water waves**. These waves differ in the relative importance of water depth and wavelength in controlling the speed they move across the sea surface.

- **Wave speed**, or velocity, is determined by the wavelength for deep water waves, for which the depth is greater than 1/2 their wavelength; by the water depth for shallow water waves, for which the water depth is less than about 1/20 of their wavelength; and by both factors **for intermediate water waves**.

- **Surf** results when surface gravity waves come into shoaling water depths; the waves lose speed due to bottom friction in the shallower water, which shortens the wavelengths while increasing the wave heights. Eventually, the waves reach their critical steepness, become unstable, and break as surf, usually when the water depth is about 1.3 times the wave height.

- **Seas** are locally formed by the wind and have a steep profile, whereas **swells** are more graceful, symmetrically shaped waves. Swells develop from seas that have propagated out of the area where they were formed and have lost their steepness.

- The size of wind waves is determined by three major factors: the wind speed, the **fetch** (distance over which the wind blows), and the **duration** of that particular wind speed. Depth is also important, in that deeper waters can support larger waves than shallower waters can.

- Waves are additive. When two sets of waves are **in phase**, they produce a wave with a wave height that is the sum of the two. When they are **out of phase**, they produce a wave with a wave height that is the difference between the two. It is this additive effect that sometimes produces **rogue waves**.

- **Tsunami** are usually produced by earthquakes. They have very long-wavelengths, on the order of 200 km or so, but very small wave heights, on the order of 1 m, and propagate as shallow water waves at high speeds. When tsunami encounter shallow coastal waters, their wavelengths shorten and wave heights increase greatly, sometimes to more than 20 m.

- **Tides**, the largest ocean waves, are the result of gravitational attractions of the Sun and Moon (and other celestial bodies as well) on the Earth's oceans. The Sun and Moon account for most of the tide-generating forces, with the gravitational attraction of the Moon accounting for more than twice that of the Sun. There is a fortnightly cycle of **spring tides** and **neap tides**.

- Because of the natural **resonant frequencies** of different areas of the world ocean, some areas are more in tune with the daily cycle of the Sun's gravitation attraction than other areas. In these areas, we see **diurnal tides**, with just one high tide and one low tide per day. In areas that are more in tune with the Moon's gravitational forces, we see **semidiurnal tides**, with two high tides and two low tides per day. Still other areas exhibit a combination of both influences, producing **mixed tides**. Extreme tidal ranges, such as in the Gulf of Maine and Bay of Fundy, result when the resonant frequency of a body of water closely matches the lunar tidal frequency, and is modified by the basin shape.

■ Tides are important for many reasons. In areas of the ocean where tides are large, they may be used to generate electricity. Also, tidal currents mix coastal waters and deliver deep-water nutrients to the near-surface waters, where they stimulate biological production.

Discussion Questions

1. What are some examples of disturbing forces that can create waves in the ocean?

2. What controls the speed of propagation of a surface water wave? Why do surface water waves slow down as they move onto shallower waters?

3. Make a sketch of a deep water gravity wave that is moving as a swell and explain the changes that it undergoes as it moves into shallow water on its way to a beach.

4. Explain what happens to the sea surface as the wind begins to blow, and as the wind continues to blow at a constant wind speed.

5. What factors are important in determining the maximum size a wind wave can achieve?

6. What are wave orbits? How do they change with depth below the surface?

7. What is the difference between a sea and a swell? Why do waves from distant storms outrun the storm itself, producing large surf on coastlines far removed from the storm?

8. What are whitecaps? How and why are they formed? What is meant by a fully developed sea?

9. Explain what is meant when we say that the group velocity of waves is equal to one-half the speed of individual waves in that group.

10. Why do some regions of the world ocean experience semidiurnal tides, others diurnal tides, and still others mixed tides? What is meant by these terms?

11. What might be the environmental implications of harnessing a significant fraction of the tidal energy in an area with large tides, such as the Bay of Fundy?

Further Reading

Bascom, Willard, 1964. *Waves and Beaches: The Dynamics of the Ocean Surface*. Garden City, NY: Anchor Books, Doubleday.

Gonzalez, F., 1999. Tsunami! *Scientific American* 280(5): 56–65.

Junger, Sebastian, 1997. *The Perfect Storm*. New York: W. W. Norton. This is the real-life thriller on which the movie (2000) was later based.

Knaus, John A., 2005. *Introduction to Physical Oceanography*. Englewood Cliffs, NJ: Prentice-Hall.

Korgen, S., 1995. Seiches. *American Scientist* 83(4): 330–341.

Sverdrup, Harald U., Johnson, Martin W., and Fleming, Richard H., 1942. *The Oceans: Their Physics, Chemistry, and General Biology*. Englewood Cliffs, NJ: Prentice-Hall.

Contents

The diversity and abundance of life in the oceans is exemplified in tropical coral reef systems. The variety of forms and colors of the reefs themselves, along with the benthic animals and fishes that make these stunningly beautiful seascapes their home, is matched nowhere else on Earth.

Introduction to Life in the Sea

Life in the sea is a source of endless fascination for most of us, but, at the same time, very few of us are aware of the fundamentals that support and shape the oceans' biological wonders. Part of the reason for this, which also reinforces our fascination, is that marine ecosystems are highly *diverse*, with creatures that seem to come in all sizes, shapes, colors, and behaviors. In fact, the oceans boast tens of thousands of species, from the smallest organisms on Earth to the largest animal that has ever lived, the blue whale, and a myriad of species in between. It should not surprise us, then, that marine environments are complex—after all, unlike terrestrial environments, the oceans are a *three-dimensional* world; its plants and animals are not stuck to any one surface the way most of life on land is. While terrestrial ecosystems do extend beneath the soil surface, and of course, birds have escaped the confines of direct contact with the Earth, the oceans are very different—they are far more complicated when that third dimension, the *ocean depths*, is considered. And, of course, it is precisely because of this third dimension that so many of the oceans' wonders lie beneath the surface and well hidden from our gaze. The world of plankton, which lies at the heart of it all, can be seen only with microscopes. Even the ocean world we can see, from tide pools to shallow coral reef systems, reminds us that life in the sea is very different from the world we have come to appreciate. And that difference seems to draw us in.

Most people are familiar with flower gardens and forests and how they grow; and we all seem to understand deer and squirrels, big trees, and farms that produce our food supplies. Yet the oceans and what lives in them remain foreign to us in so many ways; we just don't have a good feel for how marine organisms fit into their watery world. So, let's begin our exploration of life in the sea by covering a few basic principles.

The Basics of Marine Biology: Photosynthesis and Respiration

As we have seen in the opening chapters, so many of the physical dynamics of the oceans are driven by the Sun's energy, and without the Sun our planet would be little more than a cold rock in dark space. But the Sun drives more than just the physics here on Earth; it is also the most important source of energy for the biological realm—in the oceans and on land. This is because of an important biochemical process, **photosynthesis**, which first appeared on Earth as much as 3.8 billion years ago. By about 2.4 billion years ago, photosynthetic organisms that produce molecular oxygen (O_2) had become firmly established, and Earth's atmosphere had, for the first time, sufficient levels of oxygen to support **aerobic** organisms, organisms that require oxygen.[1] There

[1] Buick, R., 2008. When did oxygenic photosynthesis evolve? *Philosophical Transactions of the Royal Society of London, B* 363: 2731–2743.

was no going back to the way things were—the Great Oxidation Event had forever changed our atmosphere. Today, molecular oxygen makes up some 21% of our atmosphere, thanks to photosynthesis on land and in the oceans.

Nearly all life in the sea is supported by photosynthetic organisms that absorb the Sun's energy, convert it to chemical energy, and then use that chemical energy to make organic material: carbohydrates, or "food." The carbohydrates are a form of stored chemical energy for use by the photosynthetic organisms as well as for other organisms that feed on them.

We need to take a moment here to emphasize that production of food in the sea is not limited to photosynthesis; some is produced by bacteria using *chemical* energy to synthesize organic materials, not light energy; this process is **chemosynthesis**, which, as we'll see in Chapter 12, is important in parts of the deep sea. However, chemosynthesis is not nearly as important to total biological productivity in the oceans as photosynthesis. Regardless of the process, chemosynthesis or photosynthesis, the organisms that perform such biological production are termed **autotrophs**, meaning that they make their own food; the **heterotrophs** are those organisms that cannot make their own food, so they have to eat.

Collectively, the autotrophic organisms are the **primary producers**. They are the first in line, so to speak; they produce the food at the base of the marine food chain. The *rate* at which they do so is called **primary production**. It is important that we not confuse primary production with **biomass**, which is a measure of how much food there is at any one time. So, *biomass* is a static measure of how much (i.e., how many grams), and *production* is a dynamic measure of how much is made per unit time (e.g., how many grams per day).

Most of the photosynthetic primary production in the sea is performed by the **phytoplankton**, a diverse group of single-celled organisms that includes the **microalgae**, the single-celled **algae** (nonvascular plants), and many other taxa, which we discuss in greater detail in Chapter 8.[2] (On land, in contrast, most of the primary producers are large vascular plants, such as trees and grasses.) Marine algae may be large or small; the **macroalgae** are the larger forms that include the kelps and seaweeds that most of us have seen at one time or another (**FIGURE 7.1**). We are aware of these macroalgae precisely because we can see them; they are large and they occur in relatively shallow

---

[2] Phytoplankton are the photosynthetic members of the *plankton*, a term that encompasses all the organisms in the oceans that freely drift about with currents, and are capable of only weak swimming abilities. We will be discussing the plankton in much greater detail in the next two chapters.

FIGURE 7.1 (A) A macroalgae-dominated intertidal shoreline in New England. (B) A kelp forest off the coast of Washington state.

(A)

(B)

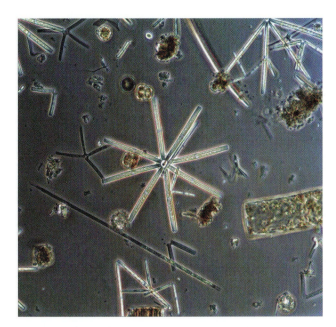

FIGURE 7.2 Photomicrograph of live phytoplankton, photographed soon after collection at sea.

water near shore, where we routinely encounter them. As impressive and as interesting as they are, the macroalgae are not nearly as important to the biological productivity of the oceans on a global scale as the phytoplankton, which we cannot easily see (not without a microscope, that is). It is the tiny, single-celled phytoplankton (**FIGURE 7.2**) that are responsible for the vast majority of the oceans' primary production.

Phytoplankton are generally small, and are measured in micrometers, or microns (μm). In their miniature word, everything these cells need for photosynthesis and growth is dissolved in the water that surrounds them and is transported into their cells across a cell membrane, while wastes are transported in the other direction.

Photosynthesis can be described as the process whereby organisms use the Sun's energy to synthesize organic matter (simple sugars) from inorganic matter (carbon dioxide). They use **photosynthetic pigments** to capture the Sun's energy and convert it into chemical energy, which is stored inside their cells in the form of a high-energy molecule, glucose.

1 micrometer, or 1 micron (1 μm)
= 10^{-6} meters (1 millionth of a meter)
= 10^{-3} millimeter (1 thousandth of a millimeter)

Of these photosynthetic pigments, **chlorophyll** is the most common in both the macroalgae and the phytoplankton. Most algae, including both the macroalgae and the microalgae, have other photosynthetic pigments, called **accessory pigments**, in addition to chlorophyll. Those pigments, which effectively mask the green color of chlorophyll, have been used to classify groups of algae (more about this in Chapter 8). The chlorophyll molecule absorbs primarily blue light (as well as some red light) and reflects green light; not coincidentally, blue wavelengths of light, remember, penetrate deepest into the sea. But even blue light attenuates quickly with depth, such that photosynthesis is possible only at relatively shallow depths, with the vast majority confined to the top 100 meters (m) or so of the ocean.

The energy required to drive photosynthesis comes from the Sun; we can summarize the process as a chemical equation:

$$CO_2 + H_2O + Energy \rightarrow CH_2O + O_2$$

Photosynthesis consumes water and carbon dioxide to chemically "fix" inorganic carbon. That is, it converts inorganic carbon dioxide, CO_2, that is dissolved in

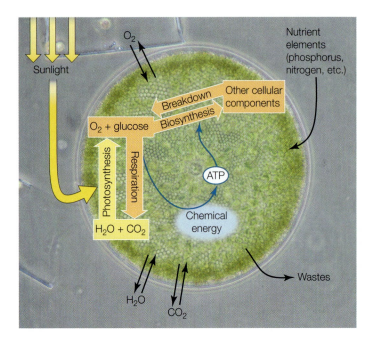

FIGURE 7.3 A schematic diagram illustrating the processes of photosynthesis, respiration, and biosynthesis (growth) in a generic phytoplankton cell, including exchanges with seawater across the cell surface.

seawater, and water, H_2O, to make organic carbon in the form of a carbohydrate, CH_2O, or $C_6H_{12}O_6$ (both formulas represent glucose, a simple sugar), which is stored in the cells (**BOX 7A**). In the process, molecular oxygen, O_2, is liberated into the surrounding seawater, thus making it available to other organisms as well as to the algae themselves (**FIGURE 7.3**).

Phytoplankton and other algae don't just perform photosynthesis, they also *respire*—continuously. In the absence of light, when they are unable to perform photosynthesis, they only respire. They all need oxygen to respire—which means that the **chemical equation of life**, as we have called it, also runs in the reverse direction. **Respiration** is simply the reverse of photosynthesis; i.e.,

$$CH_2O + O_2 \rightarrow CO_2 + H_2O + Energy$$

This is the **oxidation** of organic matter, or the *burning* of food material. In this case, that food material is the glucose created in photosynthesis (**BOX 7B**). The energy released from respiration is converted to another form of chemical energy needed for other cellular processes; those other cellular processes include **growth**.

While growth and photosynthesis are closely tied to one another, they are not synonymous. Notice that all photosynthesis does is make *sugar*; but to grow, to increase its biomass, the cell needs to make more than sugar—it must synthesize other cellular materials to build new tissues.

BOX 7A   Photosynthesis and Respiration: The Basics

Photosynthesis and respiration can be described by the following simplified chemical equation:

Energy in

$$CO_2 + H_2O \rightleftharpoons CH_2O + O_2$$

Energy out

This equation is sometimes called the *chemical equation of life*. Notice that it can run in both directions, to the right or to the left. When it runs to the right, consuming energy, carbon dioxide, and water to make carbohydrate and oxygen, we have photosynthesis; that is, the reaction is:

$$CO_2 + H_2O + Light\ energy \rightarrow CH_2O + O_2$$

When the reaction runs to the left, carbohydrate is oxidized, making carbon dioxide and water, and releasing energy. This is respiration; i.e.,

$$CH_2O + O_2 \rightarrow CO_2 + H_2O + Energy\ (heat)$$

But note well: This extremely simplified reaction is for summary purposes only. It does not include any of the intermediate biochemical steps, such as the light and dark photosynthesis reactions, the Calvin Cycle and so forth. Some authors use a more complete equation, which represents fundamentally the same thing:

$$6\ CO_2 + 6\ H_2O \rightarrow C_6H_{12}O_6 + 6\ O_2$$

We use these equations only to show the starting and ending points, which are valuable in the construction of ecosystem carbon or energy "budgets."

The **biosynthesis** of new cellular materials—which is what *growth* is—is done using the energy produced by respiration, as diagrammed for an individual phytoplankton cell in Figure 7.3. Some of the chemical energy released in respiration is stored in the form of high-energy chemical bonds in the molecule **ATP, adenosine triphosphate**. To generate ATP, the cell requires more than just the basic ingredients we've been discussing so far. That is, only three elements—carbon, hydrogen, and oxygen—are involved in the primary photosynthesis reaction; but the synthesis of the ATP molecule requires phosphorus. The cell usually gets the phosphorus atoms from **phosphate** (PO_4^{3-}), which is dissolved in seawater and is transported across the cell membrane into the phytoplankton cell. The attachment of a third atom of phosphorus to the molecule **adenosine diphosphate** (**ADP**) makes ATP, a high-energy compound. When one phosphorus atom is separated from ATP, the stored chemical energy is released and is now available to drive *biosynthesis*—the chemical reactions that synthesize more complex molecules needed for growth of new body materials.

For biosynthesis to occur, for there to be growth, the cells need other elements besides phosphorus with which to biosynthesize the various cellular tissues and other constituents. Those other elements are known collectively as **nutrients**. So what are these nutrients, and where do they come from?

Photosynthesis makes the organic carbon "fuel" that algal cells require for energy and for growth, usually in the form of a simple carbohydrate sugar, glucose. But *growth* is a biological process not synonymous with photosynthesis; it is the biosynthesis of new biomass using the products of photosynthesis. *Biosynthesis* requires other elements in addition to the carbon, hydrogen, and oxygen stored in carbohydrates; it requires nutrient elements.

BOX 7B   Important Biologically Mediated Chemical Reactions in the Oceans

In discussions of the chemical reactions involved in various biological processes, such as photosynthesis and respiration, you will often see references to two general classes of chemical reactions, acid–base and oxidation–reduction reactions. A brief explanation of what they are is warranted here. Each of these reaction types is important in the biologically mediated reactions in the oceans.

Acid–Base Reactions

As the name implies, these chemical reactions involve an acid and a base. An acid is a chemical compound that is a proton donor; that is, it donates one or more hydrogen ions, H^+, to another substance (H^+, as we learned in Chapter 4, is a proton). Common examples are hydrochloric acid, HCl, and sulfuric acid, H_2SO_4. A base is a proton acceptor. Common examples of bases include sodium hydroxide, NaOH, and potassium hydroxide, KOH; in these compounds it is the hydroxide ion, OH^-, that accepts the proton (H^+). However, there are many other chemical compounds that also behave as bases and accept protons; one of these is ammonia, NH_3. In the case of ammonia, the entire molecule accepts a proton (it combines with a hydrogen ion) to produce ammonium, NH_4^+.

An example of an acid–base reaction is that of hydrochloric acid and water:

$$HCl \rightarrow H^+ + Cl^- \text{ and } H^+ + H_2O \rightarrow H_3O^+$$

The HCL dissociates in water to give H^+ and Cl^-; the proton (H^+) combines with water to give H_3O^+. It is this H_3O^+ that is sensed by a pH meter to give us a measure of acidity; usually we see H_3O^+ expressed in a shorthand notation as simply H^+.

Oxidation–Reduction Reactions

In acid–base reactions, as we just showed, protons are transferred from a proton donor to a proton acceptor. In oxidation–reduction reactions, also called redox reactions, *electrons* are transferred from an electron donor to an electron acceptor. Atoms that lose electrons are oxidized, and atoms that gain electrons are reduced.

For example, this is what happens to iron when it rusts: electrons are transferred from iron, Fe, to oxygen, O_2, as follows:

$$4\,Fe + 3\,O_2 \rightarrow 2\,Fe_2O_3$$

In this reaction, each iron atom transfers 3 electrons, for a total of 12 electrons; those electrons are accepted by 6 oxygen atoms. Iron is oxidized and oxygen is reduced.

Nutrients and Limiting Factors

Of the various chemical elements necessary for life, the three most obvious are those used in our equation of life: oxygen, carbon, and hydrogen, all of which we find in plentiful supply in the oceans. That is, there is plenty of carbon dioxide and, of course, water available to supply the C, O, and H for the photosynthesis reaction. As for all the other elements needed by cells—i.e., the **nutrient elements** such as phosphorus and many others—they are not so abundant in seawater, and their shortage can at times set an upper limit to the amount of biological production possible. Thus, these nutrient elements are called the **limiting nutrients**.

Elements that constitute the major and minor nutrients dissolved in seawater, and which are required by all living organisms, are revealed in an analysis of the average chemical composition of a typical phytoplankton cell, such as that given in **TABLE 7.1**. For example, if we were to collect a mass of phytoplankton cells by filtering a volume of seawater, grind them up in a blender until we have what amounts to a mushy soup, and then conduct a chemical analysis of all the elements in that soup, we would see the approximate percentages of elements given in Table 7.1. The nutrient elements are those that make up less than about 5% of the cell's biomass. These minor elements are what we normally refer to as the **nutrients**, or plant nutrients (we do not usually refer to C, H, and O as nutrients). The nutrients fall into two groups, based on their percentage composition in phytoplankton and other algae: the **macronutrients** and the **micronutrients**. It is these macro- and micronutrients that can set an upper limit to biological production in the ocean; sometimes, they run out, and thus constitute a **limiting factor** that determines the amount of biological production possible. This is an important point that deserves a bit of emphasis: what we mean here is that for the cells to grow, they must obtain nutrients from the seawater around them. If they cannot, because those nutrients are either unavailable or in low concentrations, then the cells may run out of necessary materials and cease growing (**BOX 7C**).

This principle of *limiting factors* has been known for a long time, having been recognized early in the development of agriculture. The term itself is generally attributed to Justus von Liebig (1803–1873) who noted in 1840 that "growth of a plant is dependent on the minimum amount of foodstuff present."[3] By *foodstuff* he meant what we now know as nutrients. This principle is well known to gardeners: they routinely add fertilizers—nutrients—to enrich the soils and promote further growth. There are other everyday situations in which the basic concept of limiting factors holds. For example, imagine you are making a few batches of chocolate chip cookies, but you discover after baking the first batch that you have less than a cup of sugar left—this means that sugar has become your limiting factor. If you had more sugar you could make more cookies, but the amount of sugar available has set an upper limit to cookie production, which in this case ends with only a single batch.

TABLE 7.1	Approximate elemental composition of phytoplankton and other algae as a percentage of tissue weight
Chemical	**Percent Tissue Weight**
Major Constituent Elements	
Oxygen	~60%
Carbon	~20%
Hydrogen	~10%
Nutrient Elements (or just "nutrients")	
Macronutrients: (nitrogen and phosphorus and sometimes silicon)	1–5%
Micronutrients: (S, Na, Cl, B, Mn, Mg, Zn, Si, Co, I, F, Fe, Cu, and others)	<0.05%

[3] Justus von Liebig, 1840. *Organic chemistry in its applications to agriculture and physiology* (London: Taylor and Walton).

BOX 7C   The Difference between Nutrients and Primary Production on Land and in the Ocean

On land, nutrients come from the ground, which is where everything gets deposited after death. Nutrients are released from dead and dying biomass by way of bacterial degradation and become dissolved in the soil's moisture. Therefore the nutrients are retained nearby, stored where roots can take them up again.

In the ocean, it is much more complicated. Dead organisms usually sink to relatively great depths before bacterial decomposition releases the nutrients in dissolved form. Because there is insufficient light for photosynthesis at depth in the oceans—it is dark down there—the newly released dissolved nutrients cannot be taken up by algal cells that might still be alive at depth. Because the dissolved nutrients aren't taken up, except by some groups of bacteria, they tend to accumulate.

The plant nutrients often—but not always—set an upper limit to phytoplankton production in the sea. Sometimes they are either absent or in such low concentrations that phytoplankton cells are incapable of absorbing them out of the seawater; at such times, their shortage sets the upper limit to the amount of biomass that can be produced. There are parts of the oceans that simply run out of one or more of these limiting nutrients at certain times of the year. In some areas, the limiting nutrient may be one of the macronutrients, such as nitrogen or phosphorus, and in others, it may be one of the micronutrients, such as iron (Fe; see Table 7.1). Either way, those waters become nutrient-limited and can support no further plankton growth. And as we will discuss shortly, it is the growth of phytoplankton that depletes those nutrients.

In addition to nutrients, there are other factors that can limit biological production in the oceans. For example, in 1905, Blackman suggested that **light intensity** and **temperature** should be added to the list of limiting factors in agriculture.[4] We do not normally consider temperature to be a limiting factor in photosynthesis and growth by phytoplankton in the oceans, although one can easily imagine situations where it might be—at near-freezing or very high temperatures in shallow, tropical lagoons, for instance. We'll return to temperature and its role in marine biology later. But it is usually light *or* nutrients, seldom both at once, that limits biological production in the sea. Both are required for phytoplankton and other algae to carry out photosynthesis and growth: they need light for photosynthesis and they need nutrients to grow.

The product of that primary production, the result of photosynthesis and growth of new plant biomass, is the creation of organic material that serves as food for consumption by heterotrophs (animals) up the food chain. In order to understand how it all fits together, from sunlight and nutrients to animals such as fish, we need to see how light and nutrients come together in the three-dimensional ocean.

Biological Production in the Oceans

The two most important quantities required for photosynthesis and growth of phytoplankton in the sea, light and nutrients, do not occur in abundance everywhere in the oceans. In Chapter 5 we learned that light disappears rapidly

---

[4]Blackman, F. F., 1905. Optima and limiting factors. *Annals of Botany* 19: 281–295.

(A)

Relative light intensity

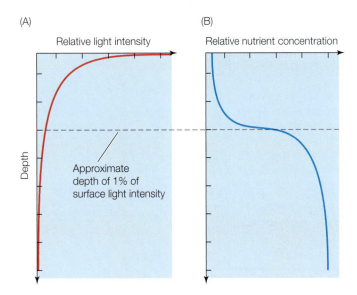

Depth

Approximate
depth of 1% of
surface light intensity

(B)

Relative nutrient concentration

FIGURE 7.4 Relative changes in light intensity and nutrient concentrations with increasing depth in the ocean. (A) The relative depth to which 1% of the surface light penetrates, in relation to (B), a nutrient profile typical of a stratified water column in the tropics or subtropics, or during the warmer months at higher latitudes.

with depth—it is attenuated exponentially (**FIGURE 7.4**). Notice in Figure 7.4, which represents a typical stratified water column where a pycnocline isolates an upper mixed layer from deeper waters beneath, that the concentrations of nutrients are low in the upper well-lit depths of the ocean, but they begin to increase at about the depth where only 1% or so of the surface light intensity remains. There is plenty of light near the surface, but only very low concentrations of nutrients. And conversely, there are high concentrations of nutrients at depth, but little or no light. It is because of this mismatch—this conundrum—that nutrients are said to limit phytoplankton production in stratified waters. That is, in the surface layers of a stratified water column, the phytoplankton have plenty of light for photosynthesis, but they often run out of dissolved nutrients. The situation depicted in Figure 7.4 is one in which the cells have absorbed the nutrients in the shallow surface waters more quickly than they can be replenished by upward mixing of nutrients from deeper waters. In contrast, on land, nutrients come from the ground, where everything gets deposited after death. Nutrients are released from dead and dying biomass by way of bacterial degradation and become dissolved in the soil's moisture, where they are stored until plant root systems take them up again.

The reason that nutrients increase in concentration with depth is related to the uptake of dissolved nutrients by phytoplankton cells in the surface waters, and the "raining" down of that particulate biomass, in the form of dead cells, to deeper depths. To understand how those particulate nutrients get recycled back into dissolved form and back to the oceans' surface waters, we first need to discuss how dissolved nutrients move back and forth—how they cycle—between seawater and the photosynthetic autotrophs.

The two most important macronutrient elements that can become limiting for marine life, especially phytoplankton, are nitrogen (N) and phosphorus (P) (see Table 7.1). (A third, silicon (Si), is important to certain groups of phytoplankton as well, but we'll discuss that in Chapter 8.) Phosphorus and nitrogen are both present in dissolved form in the oceans, but they come from very different sources. These two nutrient elements are required by all phytoplankton and other algae in the sea, and yet each may be in short supply, depending on the location, the time of year, and so forth—and each is cycled quite differently from the other.

The phosphorus cycle

Dissolved phosphorus, mostly in the form of phosphate, (PO_4^{3-}), comes from the weathering of rocks. That is, it enters the oceans from the continents by way of rivers, and from the Earth's interior by way of volcanoes, deep-sea vents, and so forth, because rocks simply dissolve over long periods of time. For example, if we were to carve a bowl out of a rock and keep it filled with water—replacing any water that evaporates—we would see that over a very long period of time, the dissolved forms of the very same materials that the rock itself is made of would accumulate in the water in the bowl.[5] In particular,

---

[5] This is especially true if we use rainwater in our experiment; rainwater is weakly acidic and dissolves materials more aggressively.

assuming we use the right type of rock, we would see an accumulation of dissolved phosphate (PO_4^{3-}) and other materials (recall the long list of chemicals in seawater, Table 4.2). The same thing happens in the oceans—they receive waters that have materials from rocks dissolved in them. Over geological time, the process arrived at a steady state in which the dissolved phosphate currently in the oceans is neither increasing nor decreasing; the inputs are exactly matched by the losses (as we discussed in Chapter 3 for salinity). In the oceans, phosphorus is recycled rapidly because it is continually passing between living cells and seawater.

Although phosphorus is often the limiting nutrient in freshwater, it is normally not limiting in surface waters of the oceans the way that nitrogen is (**BOX 7D**). There are a couple reasons for this. One is that, for the most part, phosphorus does not sink with dead organisms to the extent that nitrogen does. Instead, it is released relatively quickly in dissolved form in the surface waters. Recycling is fast for phosphorus because, unlike nitrogen, which is used in the structural composition of the cell, phosphorus is mostly used in metabolic reactions. Thus, it is released into dissolved form very soon after the cell dies—before the dead cells (**detritus**) sink too deep—and, therefore, the phosphorus is made available to be recycled into another phytoplankton cell. Another reason phosphorus is seldom the limiting nutrient in the oceans is that even though there is not a lot of dissolved inorganic phosphorus in seawater, algal cells need only about 1/16 as many phosphorus atoms as nitrogen atoms. Interestingly, and not coincidentally, the relative proportions of nitrogen and phosphorus dissolved in the oceans is virtually the same as in living cells. Both seawater and living cells have 16× as many nitrogen atoms as phosphorus atoms. Examples of the relative concentrations of dissolved nitrogen and phosphorus in seawater, conforming to the Redfield ratio of 16 nitrogen atoms to 1 phosphorus atom, are given in Figures 7.6 and 7.7. Finally, even though phosphorus is recycled quickly in the surface waters of the oceans, some nonetheless does sink in particulate form to deep depths before it is released in dissolved form, and it is in the deep waters of the oceans that we find the highest concentrations of dissolved phosphate.

Alfred C. Redfield (1890–1983) of the Woods Hole Oceanographic Institution first noticed that the proportions (the molar ratio) of carbon, nitrogen, and phosphorus in phytoplankton and seawater were virtually the same, being 106:16:1. This has become known as the Redfield ratio.

The nitrogen cycle

Nitrogen is the most important limiting nutrient element in the oceans, or at least throughout the majority of the most biologically productive parts of the

BOX 7D Nitrogen and Phosphorus Limitation in the Sea and in Freshwater

Phosphorus is often limiting in freshwater because, relatively speaking, there are more diazotrophs—nitrogen fixers—in freshwater than in the oceans. With more organisms making bioavailable nitrogen from N_2 in freshwaters, nitrogen is less likely to run out as compared with phosphorus, the supply of which depends on the weathering of rocks, which is a slow process. Therefore, in freshwater ecosystems, phosphorus is more likely to run out before nitrogen.

In the oceans, there is less nitrogen fixation than there is in freshwater—there are fewer diazotrophs—so even though there is not a lot of phosphorus present, it is nitrogen that is more likely to run out first.

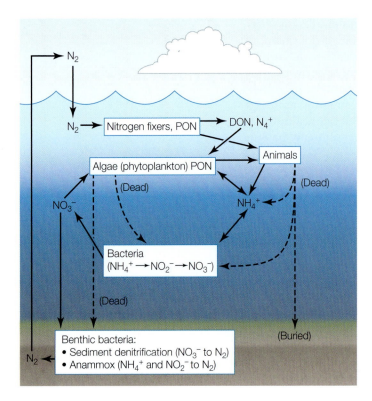

FIGURE 7.5 Flow diagram of the major processes involved in the marine nitrogen cycle. See text for details.

world ocean.[6] In near-surface waters throughout much of the year, it often becomes depleted to concentrations too low for phytoplankton to take up, and this usually happens well before phosphorus becomes depleted. Therefore, nitrogen is generally limiting before phosphorus is. To understand this dynamic, we need to examine how nitrogen cycles from its source in the atmosphere to dissolved form in seawater and back and forth through biological organisms. **FIGURE 7.5** shows two major processes: (1) the consumption and (2) the regeneration of nitrogen. Each process generally happens at different depths in the ocean: consumption (or uptake) by phytoplankton occurs at the surface, and regeneration is usually done at deeper depths. What happens is this:

Nitrogen, unlike phosphorus, comes from the atmosphere as N_2 gas (molecular nitrogen), which makes up 78% of the atmosphere. Molecular nitrogen, along with gases such as CO_2 and O_2, can freely dissolve across the air–sea boundary, so N_2 is dissolved in surface waters of the ocean in roughly the same proportion as it is found in the atmosphere. But most marine photoautotrophs cannot use molecular nitrogen because it does not easily react chemically. Before it can be used by most living cells, nitrogen must be chemically changed through a process called **nitrogen fixation**. In this process, the bond between the two nitrogen atoms in N_2 is broken, enabling the two nitrogen atoms to be incorporated into more *bioactive* forms. Nitrogen fixation is accomplished in the oceans (and on land) by **nitrogen fixers** or **diazotrophs**, specialized groups of bacteria and a few species of bacteria-like algae. Nitrogen fixation transforms dissolved N_2 into cellular components of the diazotrophs, such as amino acids and proteins, for example, thus making **particulate organic nitrogen** (**PON**). When the diazotrophs excrete nitrogen into seawater and when they die, they release a biologically useable form of nitrogen, known as **dissolved organic nitrogen** (**DON**).

Some of the fixed nitrogen is released by diazotrophs as **ammonium** (NH_4^+) as part of the cells' normal metabolic activity. When the cells die and decompose, they release most, if not all, of their nitrogen in dissolved form, either as various forms of dissolved organic nitrogen, or as ammonium once bacteria have acted upon the dead cells. Heterotrophs, too, release ammonium. However, unlike autotrophs, heterotrophs obtain their nutrient requirements directly from the food they ingest, as do most animals in general (including you and me). When marine heterotrophs consume and metabolize autotrophs, they, like the diazotrophs, release ammonium in dissolved form as a metabolic waste product into the surrounding seawater. The net result is the release of both dissolved organic and **dissolved inorganic nitrogen** (principally as NH_4^+) into the surrounding seawater. If this release of dissolved inorganic nitrogen happens in the surface waters, where there is sufficient light available for photosynthesis, it may

[6] An important exception is the micronutrient element iron, which because of its terrestrial source, becomes limiting in many open ocean environments. Often there are relatively high concentrations of nitrogen and phosphorus in surface waters of open oceans, but little if any measureable phytoplankton. We now know that phytoplankton production in these remote areas is limited by iron—in particular, by the delivery of iron-laden dust particles carried by the winds, as we discussed in Chapter 3.

then be taken up by phytoplankton and incorporated into cellular materials such as proteins (e.g., back into particulate organic nitrogen, PON). Like the diazotrophs, phytoplankton also excrete ammonium (NH_4^+) and they may also leak DON, making that nitrogen available for other cells, and thus recycling it in the surface waters of the oceans. But most of the organic matter from phytoplankton primary production (that is, the phytoplankton cells themselves) and even most of the small animals that eat them sink to deep water as particulate material before being passed too far up the food chain.

On the way to the bottom, bacteria decompose the organic matter, and in so doing they also produce ammonium, which is released into the surrounding seawater; but the nitrogen doesn't stay in that form for long. In those deep waters, the NH_4^+ so produced is acted upon by still other kinds of bacteria which oxidize it to **nitrite (NO_2^-)** and then to **nitrate (NO_3^-)**, a process called **nitrification**. Nitrification occurs only in the dark and is relatively slow, often taking weeks to months; it occurs primarily in the deep waters of the oceans. It takes weeks to months for some forms of particulate organic matter to sink several hundred meters in the ocean, and then, when it has sunk to depths of total darkness, the rate of bacterial nitrification is slow, too.[7] As a result, the end product of nitrification—nitrate (NO_3^-)—reaches its highest concentrations well below the surface, between 500 and 1000 m or so in the open ocean (**FIGURE 7.6**). Notice also in Figure 7.6 that the deep-water concentrations of both nitrate and phosphate increase from the Atlantic to the Indian to the Pacific Oceans, reflecting an accumulation over time and distance as those deep waters traverse the ocean depths as part of the global thermohaline circulation pattern, described in Chapter 5 as the Great Ocean Conveyor Belt. As those deep and bottom waters slowly flow throughout the world ocean, they accumulate detrital materials raining down from the shallow waters above, and therefore the concentrations of nutrients increase with the ages of those waters; because the oldest deep ocean waters in the world are in the Pacific Ocean, those waters have the highest nutrient concentrations. A similar phenomenon of higher nutrient concentrations at depth can also be seen on the relatively shallow continental shelves, less than 200 m or so (**FIGURE 7.7**).

Some of the sinking particulate material makes it all the way to the bottom, especially on the continental shelves but also in the deep sea, and once there, it may be buried in the sediments while it is still in particulate form. However, much of the particulate organic nitrogen in detritus is decomposed before it reaches the bottom, and therefore we see dissolved inorganic nutrients (NO_3^-, in this example) in high concentrations all the way to the bottom in the deep

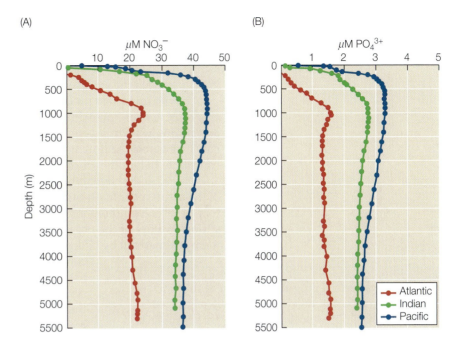

FIGURE 7.6 Vertical profiles of (A) nitrate and (B) phosphate concentrations in the Atlantic, Indian, and Pacific Oceans.

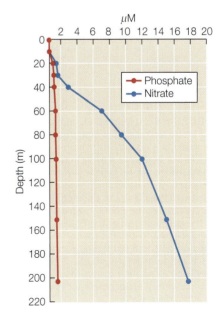

FIGURE 7.7 Vertical profile of nitrate and phosphate concentrations on the Northwest Atlantic continental shelf.

---

[7] Some forms sink much faster; fecal pellets of zooplankton sink especially fast because they are very compact particles.

sea (see Figure 7.6), concentrations that are much higher than in the surface waters. Finally, certain kinds of bacteria in bottom sediments of the deep ocean and on continental shelves are capable of converting NO_3^- back to molecular nitrogen in a process called **sediment denitrification**. That N_2 can thus make its way back to the atmosphere, completing the nitrogen cycle.

Another important process in the nitrogen cycle, called **anammox** (short for *an*aerobic *amm*onium *ox*idation), in which N_2 is produced from ammonium and nitrite, was discovered quite recently (see Figure 7.5).[8] This process occurs in **anaerobic** sediments and deep-sea vents (where O_2 is absent) and may contribute half of the world ocean's total denitrification.

It is because much of the organic material produced in the surface waters of the oceans eventually sinks that deep ocean waters are the ultimate repository of dissolved inorganic nitrogen and phosphorus, and it is also why the surface waters of the oceans are usually depleted of nutrients, having had them stripped out by biological activity. In the absence of some mechanism to bring those deep waters and their nutrient loads back to the surface sunlit layers, the surface water nutrients simply run out faster than they can be resupplied via their sources (the atmosphere for nitrogen and the weathering of rocks for phosphorus). The key, then, is for that deep-water storehouse of nutrients to be brought to the surface. Therefore, it isn't just biology and chemistry that are responsible for nutrient cycling in the sea; dissolved nutrients move with the waters they are in, and this is how they are returned to the sunlit surface layers—by way of the upwelling processes we discussed in Chapter 5.

Light at the surface but nutrients down deep in the sea presents a problem. Before there can be any growth (biosynthesis) of phytoplankton and higher-trophic-level animals that eat algae, deep-water nutrients need to be brought back to the surface of the oceans, where there is sufficient light for photosynthesis—otherwise, biological production in the oceans would eventually run down and stop altogether. This is where our understanding of physical oceanography and upwelling becomes useful.

Upwelling

Recall from Chapter 5 that the oceans have vertical currents (upwelling and downwelling) in addition to the horizontal currents we normally think of. There are two fundamental types of upwelling: wind-driven coastal Ekman upwelling, and Equatorial divergent upwelling (**FIGURE 7.8**). These vertical currents are at the heart of the oceans' biological productivity—the greatest primary production occurs in upwelling regions, where deep, nutrient-rich

---

[8] For a review, see Dalsgaard, T., Thamdrup, B., and Canfield, D. E., 2005. Anaerobic ammonium oxidation (anammox) in the marine environment. *Research in Microbiology* 156: 457–464.

(A) Coastal Ekman upwelling

(B) Equatorial upwelling

FIGURE 7.8 (A) Coastal Ekman upwelling in the Northern Hemisphere. In this example, a wind blowing from the north forces surface waters away from the coast, which are replaced by waters from beneath. (B) In equatorial upwelling, the westward-flowing South Equatorial Current straddles the Equator, which results in the portion north of the Equator being deflected to its right (or north), and the portion south of the Equator being deflected to its left (or south). This divergence along the Equator results in a compensatory replacement, or upwelling, of subsurface waters.

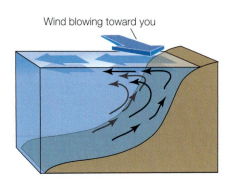

Wind blowing toward you

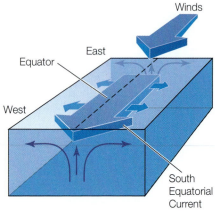

Winds

East

Equator

West

South Equatorial Current

waters are delivered to the surface sunlit layers. There, phytoplankton can multiply and produce the biomass that drives the marine food chain.

Knowing how the winds behave around the globe, we can thus predict where we might expect to see elevated rates of primary production as a result of the upwelling of deep-water nutrients. For example, the global surface wind fields for the Northern Hemisphere winter and summer are given in **FIGURE 7.9**. Based on these wind fields, we can predict the major areas of coastal Ekman upwelling in each hemisphere, which, in addition to waters on the Equator, should correspond to biologically productive regions of the world ocean. Also in Figure 7.9, notice that there is little change in the winds over the course of a year, except for the northwest portion of the Indian Ocean (red circles), which reverse with the change of season—this results from a shift in the position of the meteorological Equator (the Intertropical Convergence Zone) between winter and summer that drives the winter monsoons in this part of the world, as we discussed in Chapter 5.

One of the first compilations of the global pattern of primary production in the oceans produced the now-quite-famous illustration reproduced in **FIGURE 7.10**, which presents the world ocean's measured rates of primary

Northern Hemisphere winter

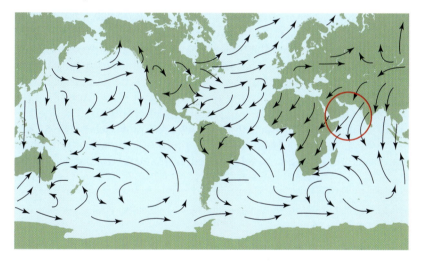

Northern Hemisphere summer

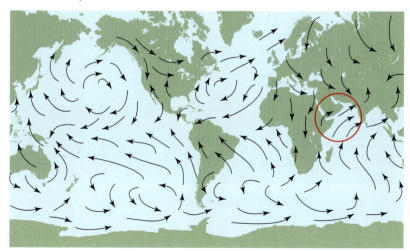

FIGURE 7.9 Global wind fields in winter and summer. Notice that in the summer, the wind direction in the northwest Indian Ocean (circled in red) off the Arabian Peninsula reverses, producing a monsoonal Ekman upwelling there.

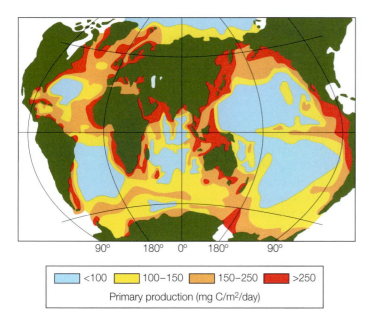

FIGURE 7.10 A map of measured rates of primary production on the world ocean.

productivity as of about 1970. That map represented some 7000 individual stations where shipboard measurements of phytoplankton primary production had been made (**BOX 7E**).[9] The correspondence between areas of high primary production and the upwelling areas we would predict is remarkable. The map shows coastal upwelling areas of high production; low production in open oceans, where there is no significant upwelling; and bands of high production along the Equator, especially in the Pacific Ocean. Notice also the high rates of primary production at high latitudes in the North Pacific and North Atlantic Oceans; that productivity is the result of another kind of upwelling: *winter convective mixing*. We'll come back to this phenomenon shortly. Also, notice in Figure 7.10 that similarly high rates of production were not recorded in the Southern Ocean off Antarctica; the reason is related to iron limitation.[6]

About a decade after this map was made, it became routine to determine patterns of phytoplankton biomass over the entire world ocean using satellite images. The satellite sensor simply records the intensity of green light leaving the ocean surface; this "greenness" of the waters corresponds to the concentration of chlorophyll in the upper layers of the ocean, which in turn, corresponds to the density of phytoplankton cells. Areas of the ocean where the satellite senses higher intensities of green wavelengths means they have relatively higher concentrations of phytoplankton chlorophyll. The satellite image in **FIGURE 7.11** shows estimated concentrations of chlorophyll plotted as contours of cooler and warmer colors (blues are low concentrations and reds are higher). The areas high in chlorophyll are the areas of greater phytoplankton biomass. Data in this particular image were collected over a

[9] Koblentz-Mischke, O. J., Volkovinsky, V. V., and Kabanova, J. G., 1970. "Plankton primary production of the world ocean," in *Scientific Exploration of the South Pacific*, ed. W. S. Wooster (Washington: National Academy of Science).

FIGURE 7.11 A composite satellite image of relative ocean chlorophyll concentrations during the Northern Hemisphere spring, produced by combining images collected from March 21 to June 20, 2006. Warmer colors correspond to higher chlorophyll concentrations. Black areas represent cloud cover where no data are available.

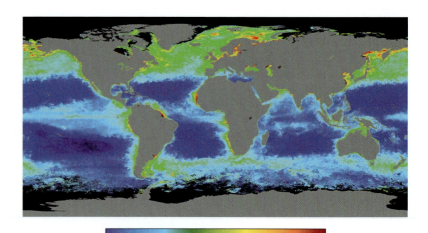

Low   High

Chlorophyll concentration

BOX 7E Measuring Primary Production

The data plotted in Figure 7.10 are based on shipboard measurements of primary production by phytoplankton. But exactly how is that done? Let's go back to the basic chemical equation that describes photosynthesis:

$$CO_2 + H_2O \rightarrow CH_2O + O_2$$

If we make a big assumption that the measured increase in organic carbon (CH_2O) is indeed a measure of new phytoplankton biomass, then we can measure primary production three ways, all based on this equation. We simply measure the change in any one of the quantities, CO_2, H_2O, CH_2O, or O_2.

First, we could place a sample of seawater with phytoplankton cells in it into a clear glass bottle (allowing sunlight to pass through to the cells) and then measure the incorporation of inorganic carbon from CO_2 into particulate carbohydrate, CH_2O, by chemically labeling the carbon atoms with radioactive carbon-14 ($^{14}CO_2$). One simply measures how much particulate carbon is produced per unit time. This was the method employed in Figure 7.10 and is the standard technique employed today. A second method involves monitoring the release of dissolved O_2 in the bottle, since it is relatively easy to measure oxygen in seawater. Third, we can simply count how many new phytoplankton cells are formed (this also gives an estimate of how much CH_2O is produced). And, of course, a fourth method would be to measure the disappearance of H_2O in the bottle—but that would be a little difficult.

3-month period (March to June) and averaged in order to fill in gaps that result from cloud cover. Notice that the locations where high rates of *primary production* were measured (see Figure 7.10) correspond very closely with areas of high phytoplankton *chlorophyll concentrations*. But take special note here: biomass is not the same thing as primary productivity, although the biomass patterns around the globe certainly are correlated with rates of primary productivity.

Winter convective mixing and seasonal vertical stratification

Both Figures 7.10 and 7.11 indicate greater biomass and productivity at high latitudes, especially in Arctic and Subarctic waters. This brings up an interesting phenomenon: deep vertical mixing in winter at high latitudes, or **winter convective mixing**.

Recall our discussion in Chapter 4 about what happens in the fall and winter at higher latitudes—latitudes where the winters get cold and the surface waters lose heat to the atmosphere, which increases density to an extent that those surface waters sink. At those latitudes the upper ocean undergoes a seasonal transition from a relatively warm, shallow, near-surface layer during the warmer months of the year, which is isolated from deeper, colder, and denser waters by a thermocline (and therefore, a pycnocline), to a situation in the fall and winter where the surface waters get cold and dense and therefore sink. That sinking *mixes* the water layers, eroding the thermocline in the process, but—and here's the important point—deep nutrient-rich waters are mixed upward to the surface as a result (**BOX 7F**). Low-nutrient surface waters are homogenized

> Primary production and biomass are related to one another, but very different. One way to see the difference is to imagine two hayfields, one with and one without sheep. The grass may grow at the same rate in both fields (the rates of primary production in each are equal), but the hayfield without grazing sheep will have significantly taller grass (the biomass is greater).

FIGURE 7.12 A typical temperature profile in winter in a temperate-to-high-latitude sea, where winter convective mixing has depressed the thermocline. Also shown is the probable path of water during the time of active convective mixing.

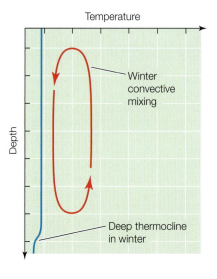

FIGURE 7.13 The seasonal development of the thermocline in a temperate-to-high-latitude sea following the winter period of deep convective mixing illustrated in Figure 7.12. The trend is one of a shoaling thermocline. The arrows depict the probable path of a parcel of water as it is vertically mixed for each month from February to May, revealing a shallow upper mixed layer in May in contrast to the deep one in winter.

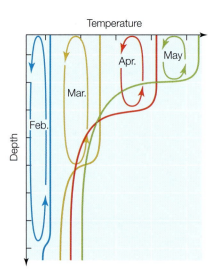

with deeper nutrient-rich waters. The net effect is a nutrient recharge, or a replenishment of nutrients at the surface, in winter in high-latitude areas that get cold enough for this convective sinking to occur. But even after deep-water nutrients are mixed into the surface layers, primary production nonetheless remains low. The reason is that in winter, especially at high latitudes, light levels are low, and the system is **light limited**. Light levels are low at high latitudes in winter because the sun angle is low, the days are short, and, most importantly, because the winter water column is not vertically stratified—there is no upper ocean thermocline or pycnocline. This means that those waters will vertically mix easily, such that a phytoplankton cell suspended in those waters will be mixed up and down throughout much of the upper water column, the net result of which is for the cell to receive very little light exposure, too little for photosynthesis (**FIGURE 7.12**). So, even though winter nutrient concentrations in the surface waters are high, primary production by phytoplankton remains low. That all changes, though, as winter transitions into spring and summer.

The spring phytoplankton bloom

As the seasons change from winter to spring in temperate and high-latitude oceans, the water column will again begin to stratify. As the days get longer and the Sun climbs higher in the sky, the oceans gain heat at the surface, and another thermocline is created. That thermocline will tend to maintain the phytoplankton cells above it, in the upper part of the water column, which we call the **upper mixed layer** (**FIGURE 7.13**). Wind and waves provide the energy to do the mixing in the upper mixed layer. We can be confident that vertical mixing in this upper mixed layer is in fact confined to the depth interval between the surface and the depth of the thermocline precisely because there *is* a thermocline. If mixing were deeper, the thermocline would be deeper as well (it is really quite intuitive, if you think about it).

At some point, as late winter and springtime light levels increase and as the depth of the seasonal thermocline shoals, a sort of compromise condition is reached, where the average light levels in the upper mixed layer eventually become high enough for photosynthesis. Because the concentrations of nutrients are already high from the previous winter's convective mixing with deep, nutrient-rich waters, conditions become right for the **spring phytoplankton bloom**. In many ways, the spring phytoplankton bloom is a lot like spring on land throughout much of the Northern Hemisphere: the days are longer and light levels are higher, and the grass and tree leaves begin to grow again; the terrestrial environment seems to come alive once more after what always seems like a long, cold winter. That is, things *bloom*.

In spring in the ocean a balance is achieved between seasonal light levels and vertical mixing such that phytoplankton can bloom. It describes what is generally referred to as the Sverdrup model of the spring bloom, after **Harald Ulrik Sverdrup** (1888–1957), a Norwegian oceanographer and meteorologist who was the first to quantify the relationship in a landmark paper published

BOX 7F   Winter Cooling and Sinking

A simple everyday example of vertical convective mixing: slowly add cold milk to a cup of hot coffee. The milk will sink to the bottom, and essentially stay there. There will be a strong thermocline separating the warm coffee at the surface from the cold milky layer at the bottom. This thermocline constitutes a pycnocline that inhibits vertical mixing between the two layers. The milk on the bottom is just like the dissolved nutrients in deep ocean waters.

If you gently float an ice cube on the surface of the coffee, you will see slugs of cold ice-melt and coffee sink through the lower milky layer, mixing with and bringing some of the milky layer toward the surface. This is what winter convective mixing does to recharge nutrients into the surface waters of the oceans every winter wherever the winter temperatures get cold enough.

in 1953.[10] The following is a detailed explanation of what Sverdrup was describing in his paper.

In **FIGURE 7.14** we have plotted how phytoplankton respiration and photosynthesis change with depth. Notice that photosynthesis is confined to the upper water column, or relatively shallow depths; in this example, the photosynthesis curve exactly matches the light curve in Figure 7.4. And just like the light curve, photosynthesis decreases rapidly with depth. Respiration by the same phytoplankton cells, however, is assumed here to remain constant with depth; the cell losses—respiration—are assumed to occur at the same rate independent of depth and independent of light levels.[11] This means that there is a depth in the ocean such that if we were to suspend a group of phytoplankton cells there, not allowing them to move vertically at all, their photosynthesis and respiration (the production of phytoplankton and their losses) would match exactly; this point is called the **compensation point**; the light intensity at that depth is called the **compensation light intensity**. That is, our chemical equation of life would be running

[10] Sverdrup, H. U., 1953. On conditions for the vernal blooming of phytoplankton. *Journal du Conseil International pour l'Exploration de la Mer* 18: 287–295.

[11] Sverdrup's reference to respiration in his 1953 paper has drawn criticism over the years. The term respiration as used by Sverdrup is often mistakenly assumed to refer explicitly to cellular respiration by the phytoplankton cells. However, Sverdrup freely substituted the words "destruction" and "loss" for respiration in his paper, and explained that he was actually describing a balance between phytoplankton production and phytoplankton losses—by cellular respiration as well as by way of consumption by animals (zooplankton). Sverdrup's term "respiration" is preserved in this book, but it is to be understood that it is intended to encompass all forms of phytoplankton losses, including losses from sinking and ingestion by zooplankton.

FIGURE 7.14 (A) The relative changes in photosynthesis and "respiration" (representing all phytoplankton losses, per Svedrup's definition of respiration) with depth in the water column. The depth where the lines cross is where the rate of photosynthesis is equal to that of respiration, which is known as the compensation point, or compensation depth. (B) Same as for that in (A), except for vertically-integrated photosynthesis and respiration. In this example, the critical depth is the depth where integrated photosynthesis, the average photosynthesis experienced by a cell as it is vertically mixed throughout the water column to a particular depth, equals the integrated respiration. In the case of deep winter convective mixing, such as that in Figure 7.12, integrated respiration would exceed integrated photosynthesis, as indicated by dashed line 2. In the case of a shallow mixed layer, such as indicated by dashed line 1, integrated photosynthesis would exceed integrated respiration.

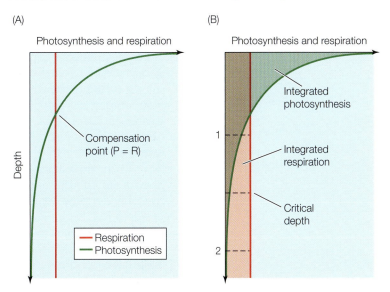

at the same rate to the right (photosynthesis) as it does to the left (respiration). The cells would neither gain biomass via photosynthetic production nor lose it. The population of cells would be just keeping even.

The compensation-point principle can be somewhat misleading in that the situation *almost* never happens in the ocean. That is, it is very unusual for there to be no vertical mixing of waters where phytoplankton cells can hover at one depth and not be mixed upward or downward. Certainly, such is not the case in winter during the period of convective mixing when the waters are actively moving back and forth between the surface and relatively great depths. This means that a phytoplankton cell in winter would be moving up and down with the mixing waters (see Figure 7.12), receiving an average of the light throughout that depth range. Because that average light would be quite low in winter, the rate of photosynthesis would be low as well.

Figure 7.14B incorporates the effects of vertical displacements of phytoplankton cells as the waters vertically mix and represents the *vertically integrated* values, the sum totals of both photosynthesis and respiration as the cells are moved up and down. These integrated totals correspond to the *areas under the curves*, as we all learned in math class; in this case they are the filled areas. For example, we can imagine a depth of mixing, a thermocline depth, where a phytoplankton cell is being mixed from the surface to the depth of the thermocline—as the cell and the water it is in move up and down, the cell will experience alternating light levels. First, it will experience high light levels and a brief burst of photosynthesis at a relatively high rate when it is fortunate enough to be near the surface. Then, when it is mixed to its deepest point at the thermocline, it will experience the least light, and would be respiring more than it photosynthesizes. The total amount of photosynthesis by that cell, as it transits vertically between the thermocline and the surface, will correspond to the green-filled area inside the photosynthesis curve, from the surface to the depth of mixing. Total respiration by that same cell, on the other hand, would be represented by the red-filled area inside the respiration curve. For example, if the depth of mixing were deep, if the thermocline were deep, as it is in March, for example (see Figure 7.13) we might have mixing to a depth marked by dashed line 2 in Figure 7.14B. In that case, total respiration would be quite large: e.g., the red-filled area would be large. Therefore, total phytoplankton cell losses would exceed the green-filled area representing total photosynthesis, and the population of cells would not survive.[11] This is the situation that prevails throughout much of the winter in mid- to high latitude regions: because it is dark, the photosynthesis curve is not very large and the depth of vertical mixing is deep. Winter conditions do not allow the total phytoplankton productivity there at that time to equal or exceed its losses. But then along comes spring.

As spring approaches, the depth of mixing shoals as heat is added to the upper ocean and, with increasing light levels, the size of the green-filled curve increases proportionally. So we can imagine a situation where phytoplankton cells would be mixed to shallower and shallower depths as springtime and warming of the surface waters proceed and the thermocline shoals; eventually, the cells are mixed only to some **critical depth** where the area encompassed by the integrated respiration is equal to that for photosynthesis. The cells will just get by under these conditions, but, as spring continues to unfold, that situation changes for the good. For example, we can imagine a late-spring situation where the depth of mixing extends only to dashed line 1 in Figure 7.14B. In that case, the cells' total photosynthetic production would exceed their total respiration (and other losses). And because nutrients are plentiful, the cells can multiply. Therefore, the spring phytoplankton bloom begins some time immediately after the critical depth criterion is met.

The arrival of spring brings with it two related phenomena: first, photosynthesis increases, simply because the light levels are increasing—the light curve and the photosynthesis curve in Figure 7.14B are *stretched* to the right. This means that the cells can photosynthesize more than they respire at greater depths than they could in winter. The *critical mixing depth* has deepened at a rate commensurate with the increase in daylight. Second, and at the same time, the thermocline—the mixing depth—is shoaling. At the point where mixing depth and critical depth become equal, the critical depth criterion is met. At this point, photosynthetic gains exactly match respiratory losses for phytoplankton cells being vertically mixed in that mixed layer. As spring progresses, the mixing depth continues to shoal, following the depth of the thermocline (see Figure 7.13) while at the same time, the total sunlight increases and the critical depth deepens.

The responses of these two phenomena, the mixing depth and the critical depth, throughout the year are given graphically in **FIGURE 7.15**. At the point where the mixing depth and the critical depth are equal, and continuing afterward as the mixing depth becomes shallower than the critical depth, phytoplankton will bloom. It is at this time of year in many locations around the globe—the time of the annual spring phytoplankton bloom—when we see the greatest growth rates of phytoplankton and the highest phytoplankton biomass. In fact, it is not uncommon for this spectacular oceanographic phenomenon to account for nearly half the total annual biological production for these regions (**BOX 7G**).

Once begun, the spring phytoplankton bloom continues in the surface waters, making more and more phytoplankton biomass, usually until the nutrients are all used up. This depletion of nutrients ushers in a summertime

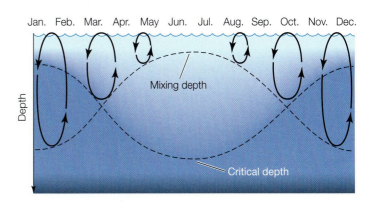

FIGURE 7.15 The seasonal changes in the depth of water column mixing—depth of the thermocline and pycnocline—and critical depth, for a typical Northern Hemisphere temperate-to-high-latitude sea. The mixing depth is generally deepest in winter and shallowest in summer, when the waters thermally stratify, and the critical depth is generally shallowest in summer, when day length is greatest and the Sun is at its highest angle above the horizon.

BOX 7G The Spectacular Spring Bloom

Henry Bryant Bigelow (1879–1967), considered the father of modern oceanography, was one of the early pioneers in the study of plankton blooms. Working from the deck of a schooner nearly a century ago, his observations of the spring phytoplankton bloom in the Gulf of Maine in the northwest Atlantic Ocean prompted him to write:

> Perhaps no phenomenon in the natural economy of the gulf so arrests attention (certainly none is so spectacular) as the sudden appearance of enormous numbers of diatoms [phytoplankton] in early spring, and their equally sud-

den disappearance from most of its area after a brief flowering period.[a]

What would he think if he could see today's satellite images of such blooms?

The satellite image at right shows phytoplankton chlorophyll concentrations, with warmer colors representing higher chlorophyll concentrations and therefore higher phytoplankton cell densities. This image has captured the height of the spring phytoplankton bloom in the Gulf of Maine and on Georges Bank.

---

[a]*Bigelow, H. B., 1926.* Plankton of the Offshore Waters of the Gulf of Maine, *vol 40. Washington, DC: Bulletin of the Bureau of Fisheries. p. 465.*

period of low levels of primary productivity—except where we have nutrients continually being mixed upward into the surface waters. This happens in the upwelling areas we just discussed, in some coastal areas under the influence of tidal mixing, and in estuaries (see Chapter 12). So following the spring bloom, after the slug of nutrients mixed upward by winter convection is used up (which takes about a couple of weeks), phytoplankton have to rely on other forms of nutrient resupply (such as nutrients that slowly diffuse upward, as well as ammonium released by other organisms in the surface waters).

Thus far we have been discussing only the productivity of phytoplankton, or primary production. But of course it is that primary production of biomass that fuels the marine food chain, leading to **secondary production**, and production at still higher levels in the food chain.

Food Chains and Food Webs

Photosynthesis fixes inorganic carbon to form organic carbon, which is used in the biosynthesis of other body materials and in cellular reproduction, both of which result in an increase in biomass. In the case of single-celled phytoplankton, biomass is increased chiefly through cellular reproduction. The rate of that increase in biomass, primary production, is usually given in terms of grams of organic carbon produced per square meter of ocean area per year $(gC/m^2/yr)$.[12] The biomass so produced can be ingested by heterotrophs at the next **trophic level**, thus beginning a **food chain**, and in many cases, leading to the production of fish that sustain a commercial fishery. But this doesn't happen equally everywhere in the oceans. As we have seen, there are noticeable differences in **nutrient fluxes** around the world ocean (that is, differences in the amounts of injections of deep-water nutrients into the surface waters) which means that there are significant regional differences in the total amount of primary production.

We can summarize the productivities in the world ocean by ocean environment, by assigning productivities to very generalized regions in the oceans: those regions are the major ocean upwelling areas, near-shore coastal areas, and the broad expanse of the open oceans (TABLE 7.2). The actual rates of primary production vary greatly, with the upwelling regions being the most productive and the open oceans the least. But the size of the open ocean makes up for the differences in rates. That is, the open oceans exhibit very low rates of primary production, primarily because the upward delivery of deep-water nutrients depends mostly on simple diffusion, which is very slow (the exception is parts of the high-latitude open ocean areas, which exhibit winter convective mixing), but the sheer size of the open oceans over which that low productivity is spread makes up for their low production rates. Conversely,

---

[12] These units vary in the scientific literature, but will always be given as either mass (weight) of organic carbon produced per unit area per unit time, or mass of organic carbon per unit volume of water per unit time.

TABLE 7.2 Patterns of primary production by ocean area			
Area	Productivity $(gC/m^2/yr)$	Percentage of ocean area (%)	Percentage of Total ocean production (%)
Upwelling	300	0.1	0.5
Coastal	100	9.9	18.0
Open ocean	50	90.0	81.5

TABLE 7.3	Comparison of primary production on land and in the oceans		
	Average productivity (gC/m²/yr)	Percentage of Earth's surface	Total production (tons of carbon per year)
Land	160	28	25 billion
Ocean	50	72	20 billion

as highly productive as upwelling regions are, there just isn't a large area over which that high production happens, so their overall importance to the global primary production is very small. As we will see in the coming chapters, however, these upwelling regions are nonetheless extremely important locally, and they often support commercial fisheries.

As productive and as big as the oceans are, we should all be aware that they do not surpass the terrestrial ecosystems with respect to total primary production (**TABLE 7.3**); however, the two are fairly close in total production. The point in making such a comparison is to keep in mind the importance of primary production both land and sea when we return to the role of biological production in the global carbon cycle and global greenhouse warming in Chapter 15.

The fate of primary production in the oceans is, generally, to be either transferred to *higher trophic levels* or to be lost (**FIGURE 7.16**). Notice how the red arrows in Figure 7.16 connecting each trophic level get narrower, which indicates less transfer of food energy. At each step there is significant loss. The terms *energy* and *food* are often used interchangeably in these kinds of discussions, because energy is equated with carbohydrate, or glucose, made in photosynthesis. The Sun's radiant energy is essentially transferred into chemical energy in the form of glucose, which subsequently gets "burned up" in the cells' metabolism, releasing that energy back to the cell for use in driving other chemical reactions. As the food chain in Figure 7.16 illustrates, if there is a gram of organic carbon produced in photosynthesis by a primary producer, it cannot possibly transfer completely without loss to the heterotroph (animal) that eats it; some is lost as heat energy in metabolism. In addition, there are very significant additional losses of *food energy* at each trophic step, the result of a portion of the ingested material passing through the consumer, only to be consumed and remineralized (broken down into its constituents) by bacteria. That is, not all of the food ingested by an herbivore (or a carnivore, or whatever) is completely **assimilated**; some passes through, so to speak. Even we human beings do not use every gram of food we eat, of course; some simply passes through us as well, to be acted upon by bacteria (wherever it goes after we flush). Bacterial decomposition of the unassimilated food also results in nutrients being returned to the sea, and back into the food chain.

There are both food chains and **food webs** in the sea. Let's focus on Figure 7.16 a bit more; we can see there that a food chain is relatively straight: for example, phytoplankton A can be consumed by herbivore B at trophic level 2, which in turn is eaten by carnivore C at trophic level 3. But the path can be less direct, as well. Carnivore C can be at either trophic level 3 or 4, depending on whether the path goes through an intermediate along the way, such as an omnivore, etc. Therefore, the passage of food energy through ecosystems

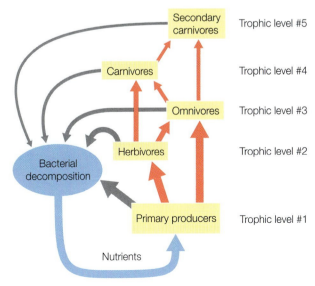

FIGURE 7.16 Direct and indirect flow of energy through a simple food chain (red arrows). Energy losses include metabolic losses as heat, as well as bacterial decomposition of unassimilated organic matter and the recycling of nutrients (gray arrows). Thicker arrows indicate greater transfer of food energy or metabolic losses.

FIGURE 7.17 A marine food web has more complex connections between trophic levels than a simple food chain does.

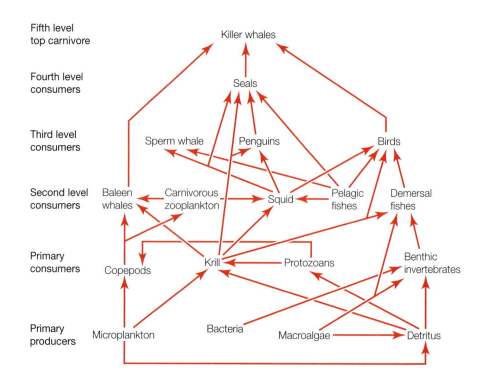

can be very complex, to an extent that the concept of trophic levels becomes almost lost, as in the food web illustrated in **FIGURE 7.17**, which shows multiple possibilities for tracing a gram of organic matter produced by phytoplankton through its consumption by fish and marine mammals.

Adding to the complexity in marine ecosystems is the simple fact that not all of the organic carbon that is made by phytoplankton or other algae, or even that produced by chemosynthetic bacteria, is in *particulate* form. This is because of an interesting complication that has only been studied by scientists in the last few decades. It turns out that a lot of the organic production previously thought to be expressed in cellular tissues—in particulate form—becomes **dissolved organic material** (**DOM**), which is the food of bacteria. This DOM is not shown in the usual illustrations of food webs. The point here is that a lot of the biomass in the oceans is either dissolved or dead—which means that bacteria are very important (**TABLE 7.4**). One might quickly conclude from Table 7.4 that the dissolved organic carbon is effectively lost, even if it goes into bacterial production, but such is not the case. In fact, a significant fraction of that bacterial carbon gets consumed by small heterotrophs and becomes part of the marine food chain (this "microbial loop" will be discussed in greater detail in the following chapters).

TABLE 7.4	How the biomass of organic carbon from primary producers in the oceans is partitioned	
Type of organic matter	**Biomass (gC/m^2)**	**Percentage of total**
Dissolved organic matter	2000	98.3
Particulate (nonliving)	30	1.5
Microscopic plants (phytoplankton)	4	0.2
Microscopic animals (zooplankton)	1	0.05

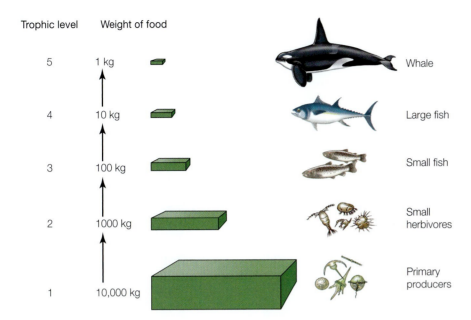

Trophic level	Weight of food		
5	1 kg		Whale
4	10 kg		Large fish
3	100 kg		Small fish
2	1000 kg		Small herbivores
1	10,000 kg		Primary producers

FIGURE 7.18 Losses of energy or carbon through five trophic levels, from the primary producers to the top carnivores.

As we move up the food chain (or food web), the biomass diminishes (see the narrowing red arrows in Figure 7.16). This is because of inefficient trophic transfer, as we just discussed; quite simply, there is a lot of inefficiency and lost energy. Inefficiency is a problem for gasoline engines and electric motors as well, by the way. Engineers are constantly striving to maximize the efficiency of engines in order to minimize lost energy—to get more miles per gallon in our cars, for example. Most engines are about 25% efficient, but in biology the efficiency of transfer between trophic levels is only about 10–20%. This means that at each step in the trophic transfer, 90% of the food energy is lost due to metabolic heat loss and inefficient assimilation. Thus we see **trophic pyramids**. **FIGURE 7.18** is a trophic pyramid that assumes a 10% transfer efficiency.

Often, the length of the food chain or food web, however we wish to characterize it, depends on the nature of the ocean environment, along the lines we already alluded to in the preceding section (coastal, open ocean, etc.). But marine environments aren't categorized only by the level of primary production or whether it is an upwelling region, or a coastal region, for example; within each of these environments, there are specific sub-environments that are important in structuring marine communities. We discussed the general nature of these when we studied the ocean floor in Chapter 3. So let's return to our discussion of the factors that potentially set limits on biological productivity in the oceans—at least with respect to specific zones, or ecosystems, in the oceans.

Factors Controlling the Distribution of Marine Organisms

Not all marine organisms are distributed evenly throughout the oceans. It can be argued that their occurrence, abundance, and general ecology vary according to three overarching factors: depth, latitude, and distance from shore.

Distributions with depth

Depth in the oceans comes in two flavors: depth in the *water column*, and depth on the *bottom*, and within each of these environments, we can distinguish depth zones. The water column is divided two different ways: by depth alone

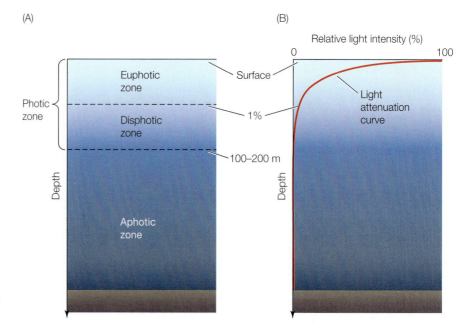

FIGURE 7.19 (A) Vertical zonation in the ocean based on the amount of light that penetrates to a specific depth. The aphotic zone extends from the depth where light is insufficient to enable vision to the bottom. (B) The light attenuation curve in the photic zone, given as a percentage of light at the surface.

and by amount of light penetration. Let's start with light penetration zones. Beginning with the water column, we have just discussed, for example, the dependence of life in the sea on **photoautotrophs**. In turn, they are dependent on light, so we can envision an ecological zonation in the vertical based solely on the amount of light that penetrates to particular depths. The **photic zone** is often defined as the upper 100–200 m or so, or simply the surface waters that are usually "sunlit" (**FIGURE 7.19**). It is meant to indicate the depth range over which there is sufficient light "to see"—i.e., enough for human vision. Within the upper part of the photic zone, there is the **euphotic zone**, where there is sufficient light for photosynthesis; usually, this is defined as the waters down to the depth at which the light level is 1% of the light level at the surface. Notice that photosynthesis requires more light than is required for vision. Below the euphotic zone, but still within the photic zone, is the **disphotic zone**, where there is sufficient light for vision but not enough for photosynthesis. And finally, from the base of the disphotic zone all the way to the bottom of the ocean, where there is insufficient light for vision or photosynthesis, is the **aphotic zone**, by far the largest part of the oceans, and a zone of *almost* complete darkness. But anyone who has ever been down in a submarine and looked out a porthole knows that, even though the Sun's rays seem to have completely disappeared, there are nonetheless flashes of light down there in the form of bioluminescence, which we will discuss later.

This three-dimensional depiction of depth zones in terms of the amount of light present describes what we generally refer to as the **pelagic zone**, which simply means the open water regions of the ocean and includes anything that is in the water and not on the ocean bottom (**FIGURE 7.20**). Remember that the water column can also be defined simply on the basis of *depth* alone—in that vast expanse of water there are a number of zones that have been defined by depth, regardless of the amount of light present. Over the relatively shallow **continental shelves** (extensions of the continents, generally less than 200 m deep), we have the **neritic zone**, which is also referred to simply as **shelf waters**; you'll see both terms used. Offshore, beyond the edge of the continental shelf, we have the **oceanic zone**. The waters that reside between the continental shelves and the open ocean—those that lie over the continental slope—we call continental **slope waters**. Specific depth zones in the pelagic realm have been

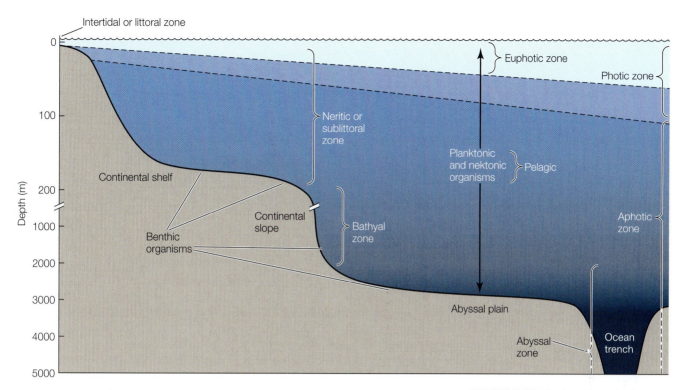

FIGURE 7.20 The benthic and pelagic zones of the ocean environment.

further classified as the **epipelagic zone** (down to about 200 m), **mesopelagic zone** (200–1000 m), **bathypelagic zone** (1000–4000 m), **abyssopelagic zone** (4000–6000 m), and **hadopelagic zone** (>6000 m); we'll revisit these in Chapter 12.

The above terms all deal specifically with zones in the water column, and can be used to describe what lives in those zones, what is dissolved in them, and how they mix and move about. But they do not specifically deal with what may be living in and on the bottom of the ocean—in the **benthic** realm. The benthic zones are classified according to how deep underwater they are (see Figure 7.20). The ocean bottom that extends along the coastline in a narrow strip between high tide and low tide is the **intertidal zone**, also called the **littoral zone**. This zone constitutes an ecologically interesting environment, where organisms face alternating exposure to air and water, as well as extremes in air temperatures (see Chapter 12). Beyond the intertidal zone is the **sublittoral zone**, which includes the ocean bottom all the way out to the edge of the continental shelf (to about 200 m). Beyond that are the **bathyal zone** (200–4000m) the **abyssal zone** (4000–6000m), and the **hadal zone** (bottom depths deeper than 6000 m).

Each of these various zones in the ocean experiences differing types of and extremes in environmental factors that impose biological stresses on living organisms, which are important in determining—setting limits to—their distributions around the world ocean. We have just been discussing how light and nutrients set limits on photosynthesis and the growth of phytoplankton, and we briefly mentioned that temperature can also be important. Both air and ocean temperatures, remember, vary widely with latitude. And because temperature exerts control on the rates of chemical reactions, it also controls the metabolic rates of living organisms; for example, photosynthesis and respiration are chemical reactions, and so their rates are affected by temperature. Increases in temperature speed up reactions, and therefore metabolism, and vice versa. And, of course, it isn't just the primary producers that are affected by temperature; so, too, are the consumers—the animals. In fact, temperature has a greater influence on animals than on plants because respiration is more sensitive to changes in temperature than photosynthesis is. For example, we often see phytoplankton blooms in waters that are at or near the freezing point, but the zooplankton, the animal plankton that graze on phytoplankton, often do not respond to the growing food supply in cold-water blooms until the waters become warmer, with the progression of spring. Most marine animals, with the exception of marine mammals and some large fishes, are "cold-blooded," which means that they do not thermoregulate; rather, the water temperature controls them. When they are subjected to extreme temperatures, they have great difficulty controlling their metabolic rates. Marine species in general have therefore adapted to life in specific temperature ranges; they have preferred environments, within which they are restricted.

DEPTH AND PRESSURE   The deeper one goes in the ocean, the greater the pressure. Recall from our discussion of the atmosphere in Chapter 5 that atmospheric pressure is measured in terms of *atmospheres*: 1 atmosphere pressure is equal to the weight of a column of air from Earth's surface to the top of the atmosphere, which averages about 1 kilogram per square centimeter (km/cm^2) (or 14.7 pounds per square inch, lb/in^2) at sea level. Pressure below sea level is much more impressive. It is equal to the atmospheric pressure at sea level *plus* the additional pressure that results from the weight of water above any given depth. And because water is so much heavier than air, changes in pressure with depth in the oceans are much more dramatic than pressure changes with altitude in the atmosphere. For every 10 meters of depth in the ocean, the pressure increases by 1 atmosphere. For example, at 10 meters, the pressure is twice that on the surface at sea level; it is the sum of atmospheric pressure at sea level, plus the additional 1 atmosphere at 10 meters, giving 2 atmospheres, or $2 kg/cm^2$ ($29.4 lb/in^2$). And at 100 meters, the pressure is 11 atmospheres ($162 lb/in^2$).

Let's go back to units of pounds and feet for a moment in order to use an everyday example that most of us can relate to—automobile tire pressure. Depending on the make and size of the automobile, the optimum tire pressure is about $34 lb/in^2$. This is only 2.3 atmospheres above ambient atmospheric pressure (e.g., $2.3 \times 14.7 lb/in^2 = 34$), which is the same as the pressure at 23 meters depth in the ocean (about 70 feet down). This means that if a scuba diver were to carry an inflated tire down to 70 feet, the tire would be flat. This example illustrates an important principle that explains why scuba divers need to keep depth and pressure in mind with respect to their air supply: air in the diver's lungs is compressed in proportion to depth, so as one dives deeper and pressure increases, more air is required to fill the lungs (**BOX 7H**).

BOX 7H   Pressure Effects on Air Supply for a Scuba Diver

A tank of air at the surface, under 1 atmosphere (atm) pressure, supplies enough air to last about 1.5 hours (hr). But because that air will be compressed in the diver's lungs proportional to ambient pressure, the amount of air effectively available will be reduced as follows:

- At 10 m (33 ft): pressure = 2 atm = 1/2 of 1.5 hr = 45 min
- At 20 m (66 ft): pressure = 3 atm = 1/3 of 1.5 hr = 30 min
- At 30 m (100 ft): pressure = 4 atm = 1/4 of 1.5 hr = 22.5 min

Pressure, like temperature, affects rates of chemical reactions, and, as we can see here, it is also very important when we are talking about things that are compressible, such as air. But because organisms are mostly made up of a watery mass, they are (relatively) incompressible. (Some fish, as we will see in Chapter 11, have an air-filled swim bladder that controls their buoyancy.)

Distributions with latitude

We learned earlier that latitudinal variations in the amount of solar radiation over the course of a year produce an uneven heating of the Earth's surface, and solely on the basis of this annual variation in incident solar energy, geographers have identified the **tropical**, **subtropical**, **temperate** (cold temperate and warm temperate) and **polar** (or high-latitude) climatic zones (**FIGURE 7.21**). The tropics are defined as the single band around the globe between the Tropic of Cancer (23.5°N) and the Tropic of Capricorn (23.5°S), which encompasses the Equator. The subtropics lie to either side of the tropics. The exact delineation is not well defined, but it is generally accepted that the subtropics occur in between but overlapping with the tropics and the temperate latitudes. The temperate zones occur between the Tropic of Cancer and the Arctic Circle (66.5°N) and between the Tropic of Capricorn and the Antarctic Circle (66.5°S) and also overlap with the subtropics; these latitudes are further subdivided into cold and warm temperate regions.

The marine climatic zones do not follow lines of latitude very closely, because, as we have already learned, patterns in ocean water temperatures are also influenced by the major ocean currents, especially on opposite sides of ocean basins. This influence of ocean circulation patterns on latitudinal temperature gradients is revealed in the satellite images of sea surface temperatures in Figure 7.21A,B. In particular, notice how ocean currents distort these regions such that we cannot simply assign lines of latitude to distinguish the regions from one another. This is especially pronounced in the North Atlantic Ocean, where the latitudinal temperature gradients on the eastern side of the Atlantic Ocean are very different from those on the western side as a result of the transport north of warm waters in the Gulf Stream. The North Atlantic Drift, which is the extension of the Gulf Stream into the eastern Atlantic, produces a fanning-out pattern of water temperatures on the eastern side of the Atlantic and a more gradual temperature gradient with latitude than that on the western side, where there is a very sharp latitudinal temperature gradient. A similar pattern can be seen

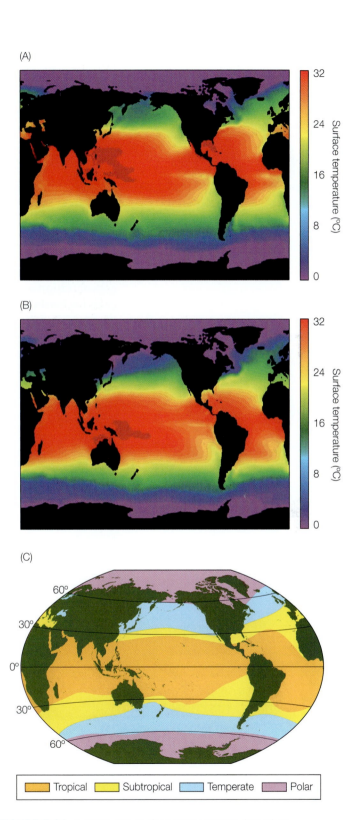

FIGURE 7.21 Average sea surface temperatures, based on satellite data for (A) July and (B) January. (C) Ocean regions based on temperatures.

in the Pacific Ocean, as a result of the Kuroshio Current. These ocean temperature gradients are important in structuring the distributions of marine organisms, and as we'll see later, global warming may already be altering those distributional patterns.

Distributions with salinity: Distance from shore

Generally speaking, the salinity of the oceans increases away from the shoreline. **Salinity** exerts an influence on the distributions of organisms in the ocean, but not in the same way that temperature does. The presence of dissolved solutes in seawater—the salts that give it its salinity—poses a barrier to freshwater species, and the lack of salts poses a barrier to marine species.

The reason for this salinity barrier is that, as we already discussed earlier in this chapter, some dissolved materials (CO_2, O_2, nutrients, metabolic wastes, as well as salts) can and do move across cell membranes, and thus cell membranes are said to be selectively permeable. But if the concentrations of dissolved solutes inside the cell are greater than in the water outside the cell, then water will diffuse across the cell membrane and into the cell along the concentration gradient. This diffusion of water across the cell membrane is called **osmosis** (**FIGURE 7.22**). It is the tendency of the solvent (water) to pass through a semipermeable membrane to a solution of higher concentration. In osmosis, it is the water that moves in and out of the cell—not the dissolved materials. While dissolved materials can diffuse across cell membranes, it is much less pronounced than for water.

FIGURE 7.22 A phytoplankton cell is incapable of osmotic regulation and incapable of transporting certain materials through its membrane, so it responds differently to different levels of solutes in its environment. (A) In a hypotonic solution, in which the concentration of solutes inside the cell is greater than that outside, water molecules from outside the cell move across the cell membrane in order to equilibrate concentrations inside and out, and the cell's volume expands. This case is analogous to a marine phytoplankton cell finding itself in fresh water. (B) When the concentrations inside and outside the cell are equal, there is no net transport of water molecules. (C) When the concentration of solutes is greater outside the cell, water molecules will move out of the cell, causing it to shrink in volume. This case is analogous to a freshwater phytoplankton cell finding itself in seawater.

(A) Lower concentration of solutes outside cell

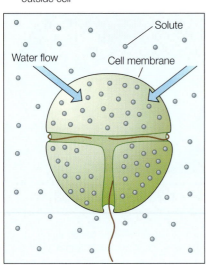

(B) Same concentration of solutes inside and outside cell

(C) Higher concentration of solutes outside cell

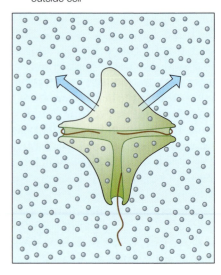

Some organisms are capable of **osmoregulation**; they can alter their internal concentrations of salts as necessary to maintain their internal milieu, maintaining a higher or lower concentration of solutes than the surrounding seawater. Other organisms do not need to maintain a solute gradient; incapable of osmoregulation, they are **osmoconformers**.

Primarily as a result of the abilities of organisms to either osmoregulate or osmoconform, the salinity of the waters is important in determining what organisms can live where. Thus we have: *marine species*, which can tolerate full-salinity seawater as well as somewhat fresher waters, but cannot survive in freshwater; *estuarine species*,[13] which can pass freely between freshwaters and the oceans, but which tend to occupy ecosystems with intermediate salinities (estuaries); and *freshwater species*, which cannot survive (not for very long, anyway) in seawater. Each environment (freshwaters, estuaries, and the ocean) has its own distinctive flora and fauna; relatively few organisms can pass freely between them, as illustrated qualitatively in **FIGURE 7.23**. There are two important features revealed in Figure 7.23: First, there are few estuarine species, compared with the numbers of freshwater and marine species; that is, there are relatively few *euryhaline* organisms—species of plants and animals that can tolerate a wide range of salinities. Second, the point of minimum numbers of species lies much closer to freshwater than to full-strength seawater, strong evidence that freshwater plants and animals are far less tolerant of salt than marine species; relatively few species have evolved to specialize in estuaries. This low species diversity could be simply because estuaries occupy an extremely small fraction of the Earth's surface areas, and as such they do not have as many ecological niches as either freshwater systems or the oceans. However, even though there aren't many species in estuaries, there is usually a high density of individuals on a per area basis. This is because, while estuaries may occupy only a small geographical area, they are, on the whole, highly biologically productive (we will discuss estuaries in greater detail in Chapter 12).

The preceding discussion of marine environments and environmental factors that influence the distributions of marine organisms in the sea has hinted at the broad spectrum of life forms in the sea, and briefly hinted at how various organisms have adapted to their environments. But as we move on to discuss specifics of the various groups of marine life and their diverse environments in the following chapters, we should be aware of just how it is that we organize and categorize them. We don't just give them names—we partition them into categories based on their biological characteristics.

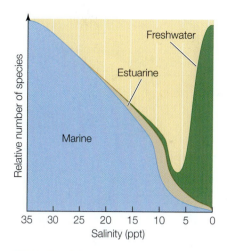

FIGURE 7.23 Plot of the relative numbers of animal species generally found in freshwater, estuarine, and marine waters. The smallest number of species occurs in estuaries, where the salinity may vary from 0 to >25 ppt or so. The greatest number of species is found in freshwater and marine waters. Euryhaline species can exist across a broad range of salinities; it includes fishes that can freely pass from between the ocean and freshwaters. In contrast, stenohaline species are confined to a narrow salinity range, usually at the marine or freshwater extremes of salinity.

Taxonomy of Marine Organisms

Thus far we have treated marine organisms as either primary producers or consumers (animals), almost as if they were simply tiny chemical reactors that process energy and materials in the sea and make food for our tables (as well as make the fascinating life forms we admire in aquaria). But in the chapters that follow, we will be discussing in much greater detail the biology and ecology of the major groups of organisms in the sea, and for that to make sense, we should spend a minute discussing the field of **taxonomy**.

---

[13] The terms brackish-water, estuarine, and euryhaline are often used interchangeably. *Estuarine* refers to estuaries, an environment where fresh waters meet and mix with the sea, *brackish waters* are fresh waters that have had some but usually not much salt added, and *euryhaline* simply means a wide range of salinities.

(A)

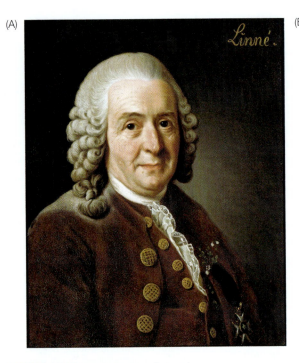

(B)

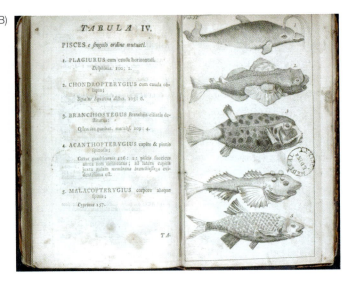

FIGURE 7.24 Swedish naturalist and physician, Carl Linnaeus (1707–1778), and a look inside his *Systema Naturae.*

Taxonomy, in biology, is the organized categorization of life forms. It provides a way not just to give names to organisms, but to place them in an organized structure that allows ecological and evolutionary interpretations. The field of taxonomy got its start with **Carl Linnaeus** (1707–1778) (**FIGURE 7.24**), a Swedish naturalist and physician. While not the first to propose its use, Linnaeus is credited with having established the modern convention of the **binomial system** of taxonomic nomenclature—a hierarchical system based on the observable characteristics of plants and animals. Linnaeus started with three Kingdoms, which were divided into Classes, Orders, and so forth, down to the level of Genus and Species (**TABLE 7.5**). It is the genus and species names that are known as the binomial system. Linneaus produced several editions of his *Systema Naturae*, the first of which was a slim eleven pages, but the tenth, and last, published in 1758, classified some 7700 species of animals and 4400 species of plants; many of those names survive and are still in use today.

TABLE 7.5	Two examples of the binomial nomenclature scheme for classifying living organisms	
Taxonomic level	**Example 1**	**Example 2**
Kingdom	Animalia	Animalia
Phylum	Chordata	Chordata
Class	Mammalia	Osteichythes
Order	Primates	Perciformes
Family	Hominidae	Scombridae
Genus	*Homo*	*Thunnus*
Species	*sapiens* (Human being)	*thynnus* (Bluefin tuna)

The taxonomic system of kingdoms has undergone significant revisions over the years, from an original two kingdoms (plants and animals, proposed by Aristotle), to three (animals, plants, and protists, used by Linneaus) to as many as six kingdoms, a scheme that was proposed in the 1970s. In 2005, a study commissioned by the *International Society of Protistologists* recommended changes to the classification of living organisms, including:

- Protists should have four kingdoms out of six; animals should no longer be a kingdom;

- Both fungi and animals should be in Opisthokonta and plants should be in Archaeplastida;

- The reclassified protists should be grouped as amoebae and slime molds (the Amoebozoa); and

- Various other single-celled organisms should be Rhizaria.[14]

In the last several years, yet another system of classification of organisms in the Eukarya into eight "major groups" was proposed.[15]

More recent approaches to taxonomic hierarchies based on molecular techniques led to the adoption of three **domains** in the 1990s as the highest taxonomic rank (highest, in this sense, means closest to an original ancestor): **Archaea**, **Bacteria**, and **Eukarya** (**BOX 7I**). The most basic tree of life is one that has three Domains that can be traced back to a common ancestor, with braches to numerous other taxonomic groups that exist today, but which are themselves still being revised. In the pages that follow we will refer to Domain as the highest taxonomic category as we continue to study life in the sea. First up will be the primary producers that drive the economy of the oceans.

---

[14] Adl, S. M. et al., 2005. The new higher level classification of eukaryotes with emphasis on the taxonomy of protists. *The Journal of Eukaryotic Microbiology* 52: 399–451.

[15] Baldauf, S. L., 2008. An overview of the phylogeny and diversity of eukaryotes. *Journal of Systematics and Evolution* 46: 263–273.

BOX 7I Modern Taxonomic Structure

I. Domain Archaea (from the Greek *archaios*, ancient) has one kingdom:
  1. The Archaebacteria (ancient bacteria);
II. Domain Bacteria has one kingdom:
  2. The Eubacteria (true bacteria);
III. Domain Eukarya has four kingdoms, which includes all other living things on earth:
  3. The Protista (single-celled organisms);
  4. The Plantae (plants);
  5. The Fungi (molds, mushrooms, yeast—heterotrophs that *absorb* their food);
  6. Animalia (animals—heterotrophs that *ingest* their food)

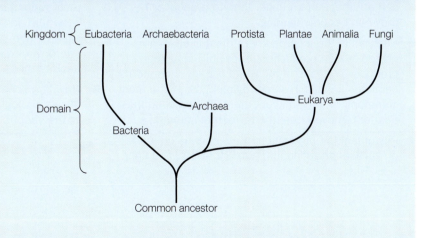

Chapter Summary

- The vast majority of life in the sea depends on **photosynthesis**—the process whereby inorganic carbon (carbon dioxide, CO_2) is converted into organic carbon (simple sugars) by photosynthetic organisms using the Sun's energy. These photosynthetic organisms are the **autotrophs**, or the **primary producers**, which include the larger **macroalgae** and the **phytoplankton**, which are single-celled photosynthetic organisms.

- Photosynthesis has been described using a **chemical equation of life**, a net reaction in which inorganic carbon, CO_2, plus water, H_2O, and the Sun's energy react to produce a simple carbohydrate, CH_2O (or $C_6H_{12}O_6$, glucose), and molecular oxygen, O_2.

- The equation of life also runs in reverse, via **respiration**, in which the carbohydrate is oxidized, producing CO_2 and H_2O. Respiration provides the cell with chemical energy for metabolism, especially **growth**. Growth, which includes **biosynthesis**, utilizes that chemical energy to drive cellular reactions to form new body tissues. Growth can be expressed as an increase in the size of an individual or as an increase in the numbers of individuals. In both cases, growth results in an increase in **biomass**. Biomass is a static measure of how much living organic matter there is, and is distinguished from **production**, the rate at which that biomass is formed.

- The growth of primary producers is **primary production**, the formation of new biomass, and requires more than just the basic elements involved in the photosynthesis net reaction; it requires **macronutrients** and **micronutrients**, each of which can at times set an upper limit to production— they can become **limiting**.

- Nitrogen is the macronutrient most likely to limit phytoplankton production, especially in biologically productive regions of the world ocean; phosphorus becomes limiting in some regions. Iron is the most commonly limiting micronutrient in open ocean areas far removed from the terrestrial source.

- Phytoplankton production in the oceans depends on vertical stratification, nutrient levels, and light.

- Certain requirements by phytoplankton in terms of light and nutrients affect the dynamics that produce seasonal bloom phenomena and geographic patterns of productivity. **Spring phytoplankton blooms** occur in regions that exhibit seasonal alterations in the degree of vertical stratification and mixing; after winter mixing, springtime warming, vertical stratification, and increasing light levels result in the rapid growth of phytoplankton. The phytoplankton bloom until the nutrients run out. These relatively brief bloom phenomena are often followed by summertime periods of nutrient depletion and low phytoplankton production. Where there is upwelling of deep-water nutrients to surface waters, phytoplankton can maintain high levels of production throughout the year.

- The biomass that results from primary production is what drives marine **food chains** (which, when complex, are called **food webs**) and the production of higher-trophic-level animals. But there are losses at each step, as the biomass from primary production is eaten by the primary consumers, and those consumers are in turn eaten by the secondary consumers, etc., through several **trophic levels**.

- Not all ingested biomass is efficiently processed by the consumer; some is lost as metabolic heat and some passes through consumers unassimilated. Therefore, food chains and food webs are inefficient: for example, for every 100 grams of phytoplankton produced in the sea, less than 1 gram of commercially harvestable fish is produced.

- Among the more important factors controlling the distributions of marine organisms in the world ocean, in addition to light, are **temperature** and **salinity**, and to a lesser extent, pressure. The marine environment is sometimes characterized by each of these factors.

- Having a scheme for classifying marine organisms allows ecological and evolutionary interpretations of the various marine taxa.

Discussion Questions

1. Based on the patterns of global winds in winter and summer in Figure 7.9, where would you expect there to be coastal Ekman upwelling and therefore relatively higher concentrations of nutrients at the surface? Why does the water upwell in those particular locations?

2. How do your predictions of locations of upwelling and higher nutrient levels compare with measurements of phytoplankton primary production in Figure 7.10 and the satellite image of phytoplankton biomass (chlorophyll concentrations) in Figure 7.11?

3. How does the cycling of phosphorus and nitrogen in the oceans differ? Where, and why, might you expect iron (one of the micronutrients) to be a limiting factor in phytoplankton production?

4. Under what circumstances might you expect to see no relationship between biomass and primary production?

5. With respect to the Sverdrup model of phytoplankton blooms, compare and contrast compensation depth and critical depth.

6. How do food chains and food webs differ from one another?

7. What do we mean by inefficient trophic transfer?

8. Explain how nitrogen may be limiting to phytoplankton production in the oceans. Under what circumstances might you expect phosphorus to limit phytoplankton production in the sea? Can you think of other factors that might also limit phytoplankton production, besides nutrients?

9. Explain how salinity, depth, temperature, and light can be expected to vary with depth, latitude, and distance from shore. How do these various factors influence the distributions of marine organisms in the sea?

Further Reading

Arrigo, R. A., 2005. Marine microorganisms and global nutrient cycles. *Nature* 437: 349–355.

Buick, R., 2008. When did oxygenic photosynthesis evolve? *Philosophical Transactions of the Royal Society of London, B* 363: 2731–2743.

Carpenter, Edward J., and Capone, Douglas G., eds., 1983. *Nitrogen in the Marine Environment.* New York: Academic Press.

Parsons, Timothy R., Takahashi, Masayuki, and Hargrave, Barry, 1977. *Biological Oceanographic Processes*, 2nd ed. Oxford: Pergamon Press.

Smetacek, V., and Naqvi, S. W. A., 2008. The next generation of iron fertilization experiments in the Southern Ocean. *Philosophical Transactions of the Royal Society of London, A* 366(1882): 3947–3967.

Sverdrup, H. U., 1953. On conditions for the vernal blooming of phytoplankton. *Journal du Conseil International pour l'Exploration de la Mer* 18: 287–295.

CHAPTER **8**

Contents

Life in the oceans depends on the primary producers, which come in all sizes and shapes, from tiny bacteria-sized phytoplankton cells to forests of giant, tree-sized kelp, but they all share at least two common attributes: they provide food that drives the ocean's ecology, and they are beautiful to behold, as this photomicrograph of sessile diatoms (*Licmophora flabellata*) illustrates.

The Primary Producers

All life in the sea depends on the primary producers—the marine autotrophs at the base of the food web. These are the organisms that biologically fix inorganic carbon dioxide into organic carbon, and, along with supplies of the various nutrient elements, create the biomass that feeds consumers, the organisms that make up the higher trophic levels. Without primary production we would have no animals, marine or otherwise, and there most certainly would be no fish in the sea to support commercial fisheries, which would be fine because we wouldn't be here to care, anyway. Quite simply, primary production on land and in the ocean is the very foundation upon which ecosystems are built, and upon which our planet depends for oxygen.

The **primary producers** in the oceans include both the chemoautotrophs and the photoautotrophs, which we introduced in the last chapter. But as we pointed out, the chemoautotrophs, while very interesting and important in specific marine environments, contribute only a tiny fraction of the oceans' total primary production. It is the photoautotrophs that are responsible for the vast majority of the world oceans' primary production. They include the **macroalgae**,[1,2] the seaweeds that we're all familiar with to some extent, and the **seagrasses**, the vascular plants that are common in coastal, near-shore environments. These large plants are fundamental to the ecological functioning of coastal marine ecosystems. Offshore and in the open ocean, it is the much smaller, unicellular **microalgae** and other groups that together make up the **phytoplankton** that assume those ecological roles, and which perform the majority of the world ocean's primary production. The phytoplankton are members of the **plankton**, a general term that refers to all the living forms in the oceans (and in freshwaters) that freely drift about; they are incapable or only very weakly capable of swimming, and are therefore carried passively with water motions. The plankton include both the phytoplankton and the **zooplankton**, the animal plankton, which we will discuss in the next chapter.

The Phytoplankton

Phytoplankton are mostly unicellular aquatic and marine photosynthetic microorganisms. They are a diverse group with thousands of species, all of which are generally small; even the very largest range only from about 0.5 to 1 millimeter (mm) across, while most species are much smaller, and in fact are virtually invisible to the naked eye. However, if you shine a bright light down into the water some night when you happen to be at the shore of the ocean or a lake or a pond, and put your face up close to the water you will see a lot

---

[1] Algae are defined as photosynthetic, oxygen-producing aquatic bacteria or protists. Linda E. Graham, James E. Graham, and Lee W. Wilcox, 2008. *Algae*, 2nd ed. (San Francisco, CA: Benjamin Cummings).

[2] The protists were once defined as the unicellular eukaryotes, but modern classifications schemes refer to them as a broad taxonomic category that includes some 40 diverse phyla, including the macroalgae.

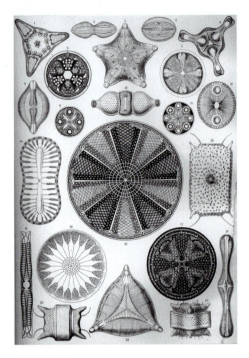

FIGURE 8.1 Early drawings of several diatoms species from Ernst Haeckel, 1904.

of tiny specks, some of which are individual phytoplankton cells (other specks may be zooplankton and some are likely **detritus** particles, bits and pieces of dead organic matter). The particles illuminated by your flashlight look a lot like the suspended dust particles you see in a ray of sunshine that crosses an otherwise darkened room. Those particles are small, and usually you don't notice them. Precisely because phytoplankton cells are small, only the taxonomic groups that have larger cells were studied by early scientists and naturalists who described them in various guides and monographs. Some of the earliest taxonomic descriptions of phytoplankton often included beautiful pen-and-ink drawings that were not only useful for scientists, but they were then and remain today remarkable works of art (**FIGURE 8.1**).[3] The diversity and beauty of the forms represented in the phytoplankton and revealed in these drawings are indeed remarkable, but as striking as these early illustrations are, they fail to capture the vivid colors and multitude of shapes that one can see firsthand by simply concentrating a water sample and placing it under a good-quality compound microscope (**FIGURE 8.2**). The species of phytoplankton in Figures 8.1 and 8.2 are, as we will see later, extremely important in the ecology of the oceans. Phytoplankton encompass all three of the taxonomic domains, the Archaea, the Bacteria, and the Eukarya. The first two, the Archaea and the Bacteria, include some of the smallest cells in the oceans, which are, at the same time, some of the most abundant. Some of these have only been discovered in relatively recent times.

---

[3] Ernst Haeckel, 1904. *Kunstformen der Natur* (Leipzig and Vienna: Verlag der Bibliographischen Instituts).

(A)

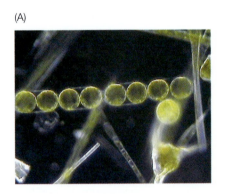

(B)

(C)

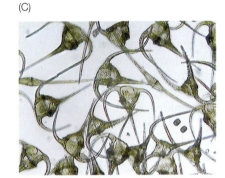

(D)

(E)

FIGURE 8.2 Examples of photomicrographs of live phytoplankton taken immediately after collection at sea. The cells in these photographs are between 30 and 100 μm.

The Archaea

The Archaea, along with the Bacteria, are **prokaryotes**, which are some of the oldest life forms on Earth (**FIGURE 8.3**). They possess a cell wall, but they lack a nucleus and most of the other organelles that are found in the more highly evolved **eukaryotes**. While the primitive but extremely abundant Archaea have only recently been discovered in the oceans, they have been there patiently awaiting our notice for a long, long time, as their name implies.[4,5] In fact, their ancient ancestors played an important role in the evolution of life some 3.8 billion years ago. Once thought to be ordinary bacteria, these single-celled organisms have been found in unusual environments where no living thing would be expected to survive, much less thrive. These include, for example, hot sulfur springs, such as are found in Yellowstone National Park, and in deep sea hydrothermal vents at the bottom of the oceans. The Archaea were once thought to be limited to these extreme environments, but about 20 years ago they were discovered to be common in the relatively shallow depths of the upper ocean;[6,7] some forms have also been shown to be capable of a primitive form of photosynthesis (unlike the photosynthesis of more modern bacteria and algae). Although extremely small, the Archaea are among the most abundant forms of life in the oceans. The role they play—their ecological function and significance—is largely unknown but currently represents an active area of research.

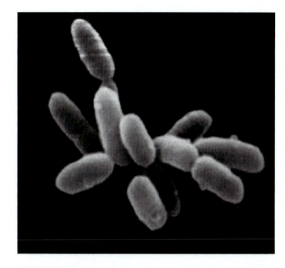

FIGURE 8.3 Scanning electron micrograph of cells of the Archaea genus *Halobacterium*. Individual cells in this image are approximately 5 μm long. The size of the archaea range from as small as several tenths of a μm to as great as 15 μm, with those in the plankton being about the size of the true bacteria, generally less than 1.0 μm.

The Bacteria

Phytoplankton in the domain Bacteria, the *true bacteria*, include the **cyanobacteria**, also known as the **blue-green algae**. In some ways, we owe our very existence to these simple life forms, for it is the photosynthetic oxygen production by ancient members of the cyanobacteria that forever changed Earth's atmosphere, such that by about 2.4 billion years ago, sufficient levels of molecular oxygen (O_2) had been added to the atmosphere and become dissolved in the oceans that aerobic organisms, such as ourselves, could evolve. The chief players in the "Great Oxidation Event" likely included ancient **stromatolites**, rocklike features in shallow seas made of sediment particles cemented together by organic secretions of dense colonies of these single-celled, filamentous, photosynthetic cyanobacteria (**FIGURE 8.4**). Fossil evidence indicates that stromatolites were present more than 3.4 billion years ago, making the cyanobacteria among the earliest life forms on Earth. Living stromatolites can be found in shallow tropical seas today.

FIGURE 8.4 Stromatolites actively growing today in the shallow tropical waters of western Australia.

[4] Woese, C. R., and Fox, G. E., 1977. Phylogenetic structure of the prokaryotic domain: The primary kingdoms. *Proceedings of the National Academy of Sciences* 74: 5088–5090.

[5] DeLong, E. F., 2003. Oceans of Archaea. *ASM News* 69: 503–511.

[6] DeLong, E. F., 1992. Archaea in coastal marine environments. *Proceedings of the National Academy of Sciences* 89: 5685–5689.

[7] Fuhrman, J. A., McCallum, K., and Davis, A. A., 1992. Novel major archaebacterial group from marine plankton. *Nature* 356: 148–149.

(A)

(C)

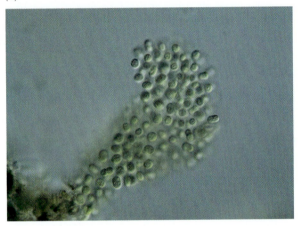

(B)

FIGURE 8.5 (A) A bloom of the freshwater cyanobacterium *Microcystis* in Hamilton Harbor in Lake Ontario. (B) *Microcystis* in South Bass Island, Lake Erie. (C) Photomicrograph of *Microcystis flos-aquae* cells.

Today, cyanobacteria are among the most ubiquitous of all organisms on Earth, inhabiting not only the oceans but also freshwaters, where they sometimes form obnoxious and toxic blooms (**FIGURE 8.5**). They are also commonly found in terrestrial soils, and can even be found growing on bare rocks. They may be unicellular or they may form filaments and colonies of many cells, which are considered to be multicellular organisms.

Cyanobacteria have been known to science for a relatively long time; they have been known to be common throughout the tropical oceans, for example, since the early 1800s and perhaps earlier. This is especially true for the relatively large, filamentous *Trichodesmium* species, which form dense colonies that often collect together in clumps in the near surface waters of the open ocean (**FIGURE 8.6A,B**). *Trichodesmium* species are most commonly found in the nutrient-poor regions of the world ocean, such as around Australia, where they were first noticed by Captain Cook, as noted by Charles Darwin in his journal of the voyage of the *Beagle* in 1839:

> March 18th. – We sailed from Bahia. A few days afterwards, when not far distant from the Abrolhos Islets, my attention was called to a reddish-brown appearance in the sea. The whole surface of the water, as it appeared under a weak lens, seemed as if covered by chopped bits of hay, with their ends jagged. These are minute cylindrical confervae, in bundles or rafts of from twenty to sixty in each. Mr. Berkeley informs me that they are the same species (*Trichodesmium erythraeum*) with that found over large spaces in the Red

FIGURE 8.6 (A) Photomicrograph of tufts of the nitrogen-fixing, filamentous cyano-bacterium, *Trichodesmium erythraeum*. (B) A colony of *Trichodesmium thiebautii*. Some of these colonies are clearly visible to the naked eye; individual colonies measure 3–5 mm across. (C) High concentrations of *Trichodesmium* discolor the surface water in this photograph taken from the deck of a ship off the Great Barrier Reef, Australia, in November 1994.

(A)

(B)
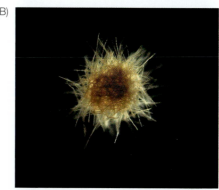

(C)

Sea, and whence its name of Red Sea is derived. Their numbers must be infinite: the ship passed through several bands of them, one of which was about ten yards wide, and, judging from the mud-like colour of the water, at least two and a half miles long. In almost every long voyage some account is given of these confervae. They appear especially common in the sea near Australia; and off Cape Leeuwin I found an allied but smaller and apparently different species. Captain Cook, in his third voyage, remarks that the sailors gave to this appearance the name of sea-sawdust. pp. 12–13.

Trichodesmium are capable of nitrogen fixation—the conversion of atmospheric nitrogen gas (N_2), to the reactive form of inorganic nitrogen, ammonia (NH_3), and upon solution in water, ammonium (NH_4^+). The dense bloom of *Trichodesmium* colonies in **FIGURE 8.6C** is what Darwin and Cook might have seen as they stood at the rails of their respective ships. These blooms are often sufficiently broad in their extent that they are visible even from space; **FIGURE 8.7** is a photograph taken by astronauts aboard the *International Space Station* while in low Earth orbit some 215 miles up. It gives a sense of the potential magnitude of nitrogen fixation that might be occurring in this patch of ocean at the time of this overflight.

While the cyanobacteria have been on Earth as long as, or perhaps longer than, any other group of organisms in evolutionary history, and even though the larger, colonial forms have been recognized for more than 200 years, it was not until the 1970s that technological advances in microscopy allowed the discovery of one of the most abundant phytoplankton

FIGURE 8.7 From space, high densities of *Trichodesmium* off the Queensland coast of Australia appear as streaks on the ocean surface. This photograph was taken by astronauts aboard the *International Space Station*.

FIGURE 8.8 Photomicrograph of a laboratory culture of the cyanobacterium *Synechococcus*.

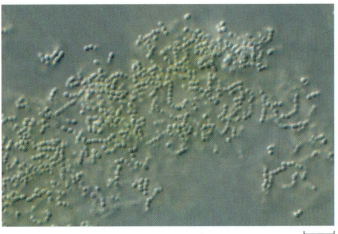

5 μm

taxa in the world ocean, the cyanobacterium genus *Synechococccus* (**FIGURE 8.8**). Microscopic detection of the fluorescence of photosynthetic pigments (discussed below) in tiny cells captured from seawater on specially developed, small pore-size filters, allowed scientists to determine just how abundant some of these cyanobacteria were.[8] These photosynthetic organisms are very small (most are less than 1 μm in diameter) but they can reach densities of some 10,000 cells in a single milliliter of seawater! Soon after their initial discovery, oceanographers and marine biologists began to realize that even though they are small phytoplankton cells, their sheer numbers must mean that they contribute significantly to the ocean's overall biological productivity—as indeed was later shown to be the case.

The *tiniest* and *most abundant* members of the phytoplankton discovered to date are a group of cyanobacteria in the same family as *Synechococccus* that are sometimes referred to informally as the **prochlorophytes**. These minute cyanobacteria use a different light-capturing pigment for photosynthesis and they are significantly smaller than other cyanobacteria; they include some of the smallest and most abundant photosynthetic microbes on Earth. These cells are on the order of 0.6 μm in diameter, and *Prochlorococcus*, a major genus of these prochlorophytes, are as small as 0.3 μm (**FIGURE 8.9**). These extremely abundant photosynthetic organisms were not even discovered until the late 1980s,[9] and their roles in the marine ecosystem are still being sorted out as scientists continue to study them. One thing is sure: their sheer numbers make up for their small size; these cells account for a significant fraction of the oceans' primary production.

As have so many other scientific fields, the field of marine biology has been stimulated by technological advances over the last several decades. New techniques have been developed that have allowed scientists to peer closer and closer at the biological workings of the oceans, one result of which has been the revelation that the smallest of organisms in the oceans are much more widespread, far more abundant, and more important to the overall productivity of the oceans

FIGURE 8.9 Scanning electron micrograph of a laboratory culture of *Prochlorococcus* (Strain MIT9313).

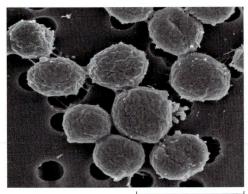

1 μm

[8] Johnson, P. W., and Sieberth, J. M., 1979. Chroococcoid cyanobacteria in the sea: A ubiquitous and diverse phototrophic biomass. *Limnology and Oceanography* 24: 928–935.

[9] *Prochlorococcus* was discovered by S. W. Chisolm and her colleagues at MIT, using a new cell sorting and counting technology commonly used in medical laboratories, a flow cytometer. Chisholm, S. W. et al., 1988. A novel free-living prochlorophyte abundant in the oceanic euphotic zone. *Nature* 334: 340–343.

than many of us had imagined when we were students not that many years ago. Consequently, our ideas about the biology of the oceans have been evolving rapidly. The development of fluorescence microscopes, new micro-pore-sized membrane filters and tissue-specific fluorescent stains led to the discovery of the importance of the cyanobacteria, the equally abundant heterotrophic bacteria, and tiny eukaryotic photosynthetic cells, most of which are less than 1 μm across (**FIGURE 8.10**). In the late 1980s, flow cytometers, developed initially for use in the biomedical sciences, revealed that the cyanobacterium *Prochlorococcus* may be the most abundant organism in the oceans, and more recently, studies using electron microscopes are revealing that marine **viruses** are far more common than once thought. Quite simply, the closer we look, the more we see. But the ecological roles played by these minute marine organisms, while certainly important to overall photosynthetic carbon fixation (and nitrogen fixation) in the oceans, may actually be less important in marine food webs than their less productive but much larger cousins, the eukaryotic phytoplankton.

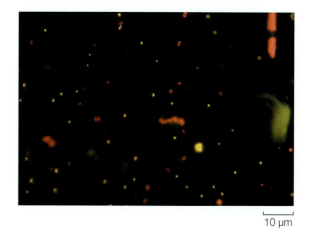

FIGURE 8.10 A fluorescence photomicrograph of phytoplankton cells from the Pacific Ocean. The yellow fluorescing dots are *Synechococcus* cells (cyanobacteria); the red fluorescing cells are tiny eukaryotes.

The eukaryotic phytoplankton

The eukaryotes, in the domain Eukarya, have cellular structures and biochemical characteristics more typical of modern cells, with an organized nucleus and various organelles. It is this group of phytoplankton that includes the more familiar forms that have been studied routinely over the past several hundred years, ever since the development of the microscope.

THE GREEN ALGAE   The *green algae* include not only unicellular forms of phytoplankton, but also some of the macroalgae that we will discuss a bit later. Of the unicellular phytoplankton in this group, most are in the class **Chlorophyceae** (which is in the division Chlorophyta) and are found almost exclusively in freshwater, with only relatively few species occurring in marine waters. But some genera, such as *Chlamydomonas* and *Dunaliella*, can be and often are a major component of the phytoplankton in some coastal areas. Both marine and freshwater *Chlamydomonas* are relatively small, generally on the order of 5–10 μm. They possess **flagella** (singular *flagellum*), which are whiplike appendages that enable them to perform swimming motions (but only across relatively small distances, measured in terms of their body lengths; therefore, they are still considered planktonic) (**FIGURE 8.11**). Also in the Chlorophyta are the class **Prasinophyceae**, the smallest of the eukaryotes. In 1994, the smallest eukaryotic cell ever described, *Ostreococcus tauri*, was discovered in the coastal waters of France (it is only about 0.8 μm across).[10]

The domain Eukarya includes four of the more common, and significantly larger, phytoplankton taxa: the **silicoflagellates**, the **coccolithophores**, the **diatoms**, and the **dinoflagellates**. The first three are sometimes referred to as the chrysophytes. These four taxa of phytoplankton represent diverse taxonomic groups, with some members that are colonial and filamentous, and others that are unicellular, but all have planktonic members. Many, but not all, are photosynthetic and at one time the chrysophytes were grouped with the kingdom Plantae.

FIGURE 8.11 Examples of the various body forms of the unicellular Chlorophyte genus *Chlamydomonas*.

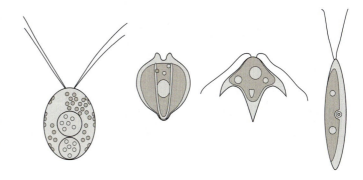

---

[10] Courties, C. et al., 1994. Smallest eukaryotic organism. *Nature* 370: 255.

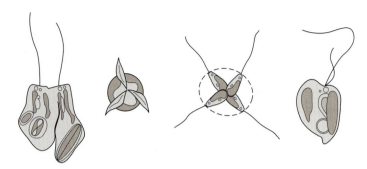

FIGURE 8.12 Examples of motile nanoflagellates in the Chrysophyta.

At some point early in their evolutionary history, ancestral members of these groups ingested a red alga that possessed a chloroplast, the intracellular organelle that contains the photosynthetic pigments. That organelle eventually became a reproductively preserved part of what was originally a protist cell, enabling it to perform photosynthesis.

These eukaryotic phytoplankton also include a broad range of taxa that have traditionally been nicknamed the **nanoflagellates** (FIGURE 8.12; BOX 8A). These are the smaller cells between 2 and 10 μm that are ubiquitous in all seas, and which may or may not possess photosynthetic pigments; they may be autotrophic, heterotrophic, or **mixotrophic** (both). Much attention has been focused on this group, especially the heterotrophic forms that graze on bacteria, with respect to their role in marine food webs. We will return to these nanoflagellates in the next chapter when we discuss zooplankton grazing.

SILICOFLAGELLATES Among the more interesting groups of phytoplankton, not because of their minute size or great abundance in the oceans, but simply because of their beautiful star-shaped skeletons of silica, are the *silicoflagellates*. We have already discussed them in Chapter 3, pointing to their contribution to siliceous marine sediments, and we reintroduce them here as photosynthetic autotrophs and members of the phytoplankton. Unfortunately, there

---

BOX 8A Plankton Nomenclature Based on Size

The nomenclature commonly used to describe planktonic organisms is sometimes based more on the sizes of organisms than their phylogeny. The terms used are **SI Units**, or *Systeme International d'Units*, and are listed in the table here.

Names generally used for the various size ranges of organisms, and which do not necessarily correspond with the strict definitions list in the table, are:

- Femtoplankton (0.02–0.2 μm): includes the viruses
- Picoplankton (0.2–2 μm): also called ultraplankton, includes bacteria and small eukaryotes
- Nanoplankton (2–20 μm): includes the single-celled autotrophs and heterotrophs
- Microplankton (20–200 μm): includes the larger single-celled autotrophs and heterotrophs, as well as multicellular (metazoan) animals
- Mesoplankton (0.2–2 mm): also includes the larger single-celled autotrophs and heterotrophs, as well as the smaller multicellular (metazoan) animals
- Macroplankton (2–20 mm): includes mostly metazoans
- Megaplankton (20–200 mm): includes the larger metazoans and often is used to include organisms larger than 200 mm.

Fraction	Prefix	Symbol	Example
10^{-1}	deci	d	1 decimeter (dm), 10 cm, is about the length of a fish
10^{-2}	centi	c	1 cm is about the length of a ctenophore
10^{-3}	milli	m	1 mm is about the length of a copepod
10^{-6}	micro	μ	1 μm = large bacteria or a very small phytoplankton
10^{-9}	nano	n	1 ng = weight of small phytoplankton cell
10^{-12}	pico	p	1 pg = amount of chlorophyll in small phytoplankton cell
10^{-15}	femto	f	1 fg = amount of ATP in a small cell
10^{-18}	atto	a	You get the idea

are few published studies of this interesting group of phytoplankton with respect to their overall importance in the ecology of the oceans (versus their importance to marine geology), and much remains to be determined. They are fairly common in the plankton, but what one often sees is only their skeletal remains, in preserved phytoplankton samples and also in samples examined live at sea. The photomicrographs in **FIGURE 8.13** show the skeletal structure of one of the more common genera, *Dictyocha*, as well as its internal cellular components, including photosynthetic pigments.

COCCOLITHOPHORES   The coccolithohores are also unicellular members of the phytoplankton, and, like the silicoflagellates, they are important to marine sediments, but instead of silica, they have an ornate outer "skeleton" that is composed of calcium carbonate plates, or coccoliths. While by no means nearly as small as the prokaryotic cells we have just been discussing, one species, *Emiliania huxleyi*, with its armor-like coccoliths of calcium carbonate, is nonetheless one of the smaller phytoplankton forms, measuring only 6–8 μm across. Because of their small size, the coccoliths of *E. huxleyi* and their minute details are virtually indistinguishable under an ordinary light microscope, and only with a scanning electron microscope is their intricate structure revealed (**FIGURE 8.14A**). There are several thousand species of coccolithophores in the world ocean, but *E. huxleyi* is by far the dominant species. It is named for Thomas Henry Huxley, Charles Darwin's famous cheerleader and a behind-the-scenes promoter of the *Challenger Expedition*, who first noticed the presence of coccoliths in his examinations of marine sediments.

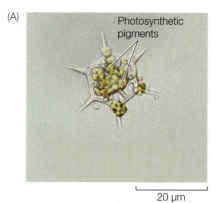

Photosynthetic pigments

20 μm

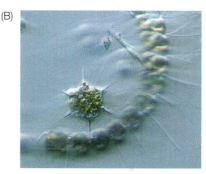

FIGURE 8.13 Photomicrographs of a common silicoflagellate genus, *Dictyocha*, showing the glass-like needles of silica (called spicules) that form its skeleton.

FIGURE 8.14 (A) Photomicrograph of two *E. huxleyi* cells using a scanning electron microscope at 10,000× magnification, showing the ornate plates of calcium carbonate. These cells are only about 6 μm across each. (B) Satellite image of the waters southeast of Newfoundland, Canada, on July 16, 2000, showing the presence of a bloom of highly reflective coccolithophores, most likely *Emiliania huxleyi* (the light blue patch indicated by the arrow). The cell densities, and especially the loose coccoliths shed from the cells make the water a milky blue, almost like an underwater cloud. A deep green bloom of less reflective phytoplankton is evident just to the north of the *E. huxleyi* bloom.

(A)

3 μm

(B)

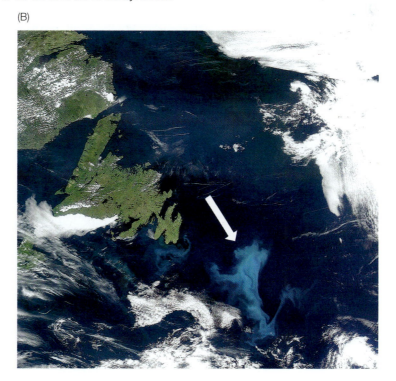

Often, for reasons we still don't understand clearly, these unusual phytoplankton will bloom to great cell densities, jettisoning their coccoliths in the process. With millions of discarded coccoliths in a single liter, the water becomes highly reflective. These blooms of *E. huxleyi* are clearly visible in satellite images as bluish or milky-white clouds in the surface waters (**FIGURE 8.14B**). To the observer aboard a ship at sea, it appears that one is sailing across a shallow, pure white sand bar, even though the bottom may actually be hundreds of meters below the hull and impossible to see from the surface. Eventually the loose coccoliths sink to the bottom and gradually, over millions of years, they accumulate, forming calcium carbonate-rich sediments. One particularly striking and easily visible indicator of the importance of this sedimentation over geological history is the appearance of "chalk cliffs" in a number of locations around the globe—areas that were once ocean bottom, but which have been uplifted by tectonic forces. One of the most famous of these is the White Cliffs of Dover, England (see Figure 3.45). More than 100 m thick in places, the nearly pure white deposits of soft chalk are ancient marine sediments composed almost exclusively of tiny bits of ancient coccoliths.

Phytoplanktonic coccolithophores are, and have been, important to the chemistry of the oceans and the atmosphere. Because they are photosynthetic, they take up carbon dioxide to form organic carbon, removing some of the CO_2 from solution in the seawater surrounding the cells. But, almost paradoxically, they also liberate CO_2 into the surface waters, and from there into the atmosphere, all because of chemical reactions that occur when calcium ions are removed from seawater to form the solid calcium carbonate in the coccoliths. What happens is this: The dissolution of carbon dioxide (CO_2) from the atmosphere into seawater produces carbonic acid (H_2CO_3), which further dissociates in seawater to form bicarbonate (HCO_3^-). This happens independently of the coccolithophores. The chemical reaction that precipitates the solid $CaCO_3$ of the coccoliths is:

$$Ca^{2+} + 2HCO_3^- \rightarrow CaCO_3 + CO_2 + H_2O$$

This reaction adds significantly more dissolved CO_2 into the surrounding ocean waters than is removed as a result of photosynthesis—there is an imbalance. The additional carbon dioxide equilibrates with the atmosphere, the net result of which is to add CO_2 to the atmosphere; therefore coccolithophores are of great interest to scientists studying global greenhouse gasses, of which CO_2 is a major component.[11] Because of their role in water clarity and ocean chemistry, these unusual phytoplankton have been, and continue to be, the subject of much research.

Continuing on our tour of the major phytoplankton taxa, we move on to two groups that we will be discussing in some detail, the *diatoms* and the *dinoflagellates*. These two groups are without any doubt the most-studied and best-known members of the phytoplankton and are important in the seasonal plankton blooms we discussed in the last chapter. Because of their relatively large cell size and their tendency to form dense blooms, they are very important in pelagic food webs. In addition, some species produce what are generally known as "**harmful algal blooms**," or "**red tides**."

---

[11] Over long time scales, $CaCO_3$ (calcium carbonate) is a major form in which carbon is buried in the sediments, and therefore coccoliths are a sink for atmospheric carbon dioxide. But over shorter time scales, during the growth seasons of coccolithophores, precipitation of calcium carbonate in the oceans will export carbon dioxide to the atmosphere. In addition, they and other phytoplankton species produce sulfur compounds that are important as condensation nuclei that allow water droplets to form in clouds.

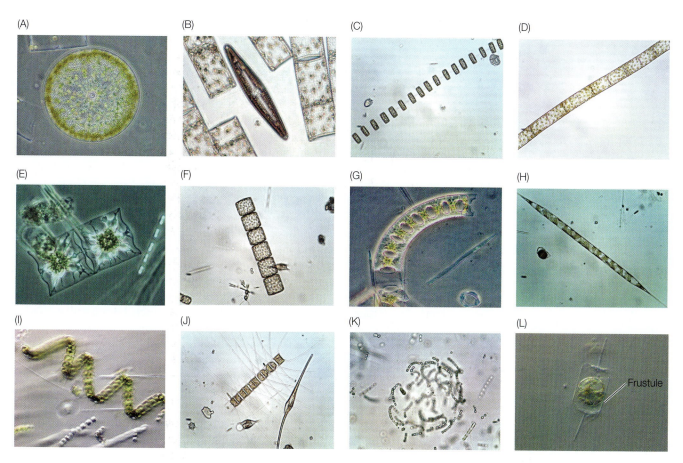

FIGURE 8.15 Photomicrographs of various diatom species, all photographed immediately upon collection at sea. (A) *Coscinodiscus sp.*; (B) *Pleurosigma sp.* (center) and *Guinardia sp.*; (C) *Thalassiosira sp.*; (D) *Leptocylindrus sp.*; (E) *Odontella sp.*; (F) *Lauderia sp.*; (G) *Eucampia sp.*; (H) *Rhizosolenia sp.*; (I,J) *Chaetoceros spp.*; (K) *Chaetoceros socialis*; and (L) *Ditylum sp.*

DIATOMS   The diatoms are members of the same larger taxonomic group as the silicoflagellates and the coccolithophores. Diatoms are among the larger phytoplankton, with cell sizes commonly 20–50 µm, and they are unique among phytoplankton taxa because they are wholly encased in a silica (SiO_2) shell called a **frustule** (see Figures 8.2B and 8.15L). Quite literally, they live in glass houses, and possess, in addition to the normal suite of organelles, both chloroplasts that contain photosynthetic pigments and the additional cellular machinery necessary to perform photosynthesis.

The diversity of diatom sizes and shapes is quite broad (**FIGURE 8.15**). There are two general morphologies represented in the diatoms: the **centric** diatoms, in which the siliceous frustule is radially symmetrical, as in Figure 8.15A, and the **pennate** diatoms, which are bilaterally symmetrical (see Figure 8.15B) . Not all diatoms are planktonic. Many species, especially the pennates, are benthic, with some species residing in interstitial spaces of shallow bottom sediments, or attached to various other substrates, such as other macroalgae and seagrasses, as well as some animals.

Among the planktonic genera of diatoms, some are solitary cells, such as the large *Coscinodiscus* and *Ditylum* cells in Figure 8.15A,L, while others form long chains of individual cells, such as the *Thalassiosira* in Figure 8.15C. Still others not only form chains, but may also have elaborate siliceous spines—**setae**—that extend outward, such as the three species of *Chaetoceros* in Figure 8.15I,J,K. Chains of cells may be tightly connected to one another in some species, as in Figure 8.15D,F,G, in which cells are held together with a polysaccharide mucoid glue, or they may be loosely connected by a delicate chitinous thread, like the chain of *Thalassiosira* cells in Figure 8.15C. Others, such as the various species of *Chaetoceros*, are attached to one another by delicate silica rods. One species of *Chaetoceros* shown (Figure 8.15K) forms colonial masses of short chains, with the setae intertwined in a spherical mass of cells within a gelatinous matrix.

Because diatoms are encased in their silica frustules, which have a density greater than that of the surrounding seawater, their tendency is to sink. But not all do—and those that do, do not always sink very rapidly. Exactly how diatoms manage to stay suspended in the upper water column where there is sufficient light for photosynthesis remains a mystery. A number of ideas or hypotheses have been offered, but none of them is completely satisfactory. One idea is that their spines (setae) and the tendency for some species to form chains of cells may increase their drag sufficiently to slow their sinking rate. An analogy would be a skydiver falling from an airplane: by extending his or her arms outward—like setae on a diatom—the skydiver increases the friction, or drag, between his/her body and the air, and thus significantly reduces the speed of his or her fall. So the morphology of the cells may slow sinking—but it won't prevent it. The diatoms will still sink, just not as fast as they would without setae. Another idea that has been proposed relates to the tendency of diatoms to form internal oil droplets, which, because oil is lighter than water, increases their buoyancy. A third mechanism, which is more of a "habitat selection strategy," is related to their apparent preferred environment. Diatoms are typically most abundant in areas of the oceans where there are injections of nutrients—for example, areas of upwelling and where the waters are vertically mixed, such as in shallow coastal areas under the direct influence of tidal mixing. It has been proposed that these turbulent environments enable the cells to stay suspended for longer than they would if there were no turbulence. But this notion seems to be self-contradictory: turbulent motions in the water go both ways, up and down; unless the cells find a way to avoid being pulled downward, the end result is a zero-sum game. Some species of the pennate diatoms have evolved what is perhaps the best method of staying suspended: they are motile. They are somehow able to propel themselves (this isn't well understood) by secreting mucilage along their longitudinal groove, the *raphe* (pronounced "ray-fee"), giving them the capability to glide across objects, as well as through the water (**FIGURE 8.16**).

If indeed the siliceous frustule imparts such an apparent handicap to diatoms, why did it evolve at all? This is a question without an answer, only speculative conjectures. Some believe that the frustules afford protection from grazers. Silica—glass—is hard, after all. In fact, common glass cutters are made from hardened steel, and some even have diamond cutting edges. It is reasonable then to imagine that zooplankton grazers might be thwarted to some extent in their attempt to bite into these well-armored diatoms. As for long chains of diatoms, one can imagine that ingesting them whole would present a problem.[12] Another potential benefit of siliceous frustules could be related to the efficient capture of light energy for photosynthesis. Some have suggested that diatoms have chloroplasts positioned to receive light focused onto them by the lenslike features of the "glass" frustules. Whatever the reason for their existence, given that diatoms are such successful marine

FIGURE 8.16 Photomicrograph of a chain of pennate diatoms, genus *Pseudo-nitzschia*, "swimming" across the field of view and through the focal plane, from the lower right to the upper left.

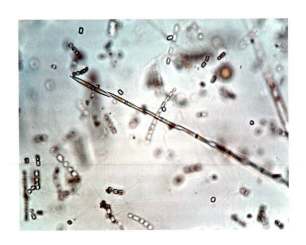

---

[12] As we will see in the next section, some heterotrophic dinoflagellates and various other taxa, such as radiolarians and foraminifera, which we will discuss in Chapter 9, have evolved an interesting way around this apparent limitation imposed by hard diatoms' frustules.

organisms, silica frustules must afford some kind of ecological and evolutionary advantage to this group of phytoplankton.

A major ecological advantage that diatoms have over other similarly sized phytoplankton taxa is they have fast growth rates. By growth, we mean that they can multiply (reproduce) quickly, and in a relatively short period of time they can *grow* to very high cell densities. Diatoms reproduce both sexually and asexually. In **asexual reproduction**, also called **vegetative cell division**, one cell simply divides into two cells, with each daughter cell being a clone—a genetically identical copy—of the parent cell. That is, there is no exchange of genetic material *between* parent cells, as there is in sexual reproduction, only from a parent cell to the daughter cell. The genotype is handed down unaltered generation after generation, with no chance for sharing genetic traits that might have arisen in the population. The solution to this potential problem is sexual reproduction, which diatoms also perform, but not every generation. Diatoms utilize vegetative asexual reproduction more often, which is how they can multiply to great densities: it allows them to *bloom*.

In vegetative, asexual reproduction, the diatom not only needs to divide its internal cytoplasm, but it also has to replicate its external silica frustule as well (**FIGURE 8.17**). The diatom's frustule is constructed of two halves, or valves, which fit together the way a pill box and its cover fit together—one fits down over the other. This means that the diameters of the two halves of a centric diatom are different, with one having a slightly smaller diameter than

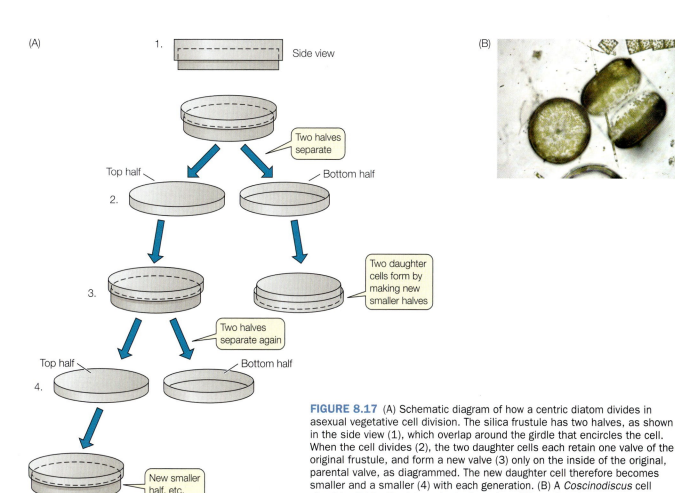

FIGURE 8.17 (A) Schematic diagram of how a centric diatom divides in asexual vegetative cell division. The silica frustule has two halves, as shown in the side view (1), which overlap around the girdle that encircles the cell. When the cell divides (2), the two daughter cells each retain one valve of the original frustule, and form a new valve (3) only on the inside of the original, parental valve, as diagrammed. The new daughter cell therefore becomes smaller and a smaller (4) with each generation. (B) A *Coscinodiscus* cell about to divide; the internal cytoplasm has differentiated.

(A)

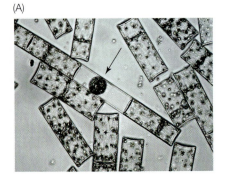

(B)

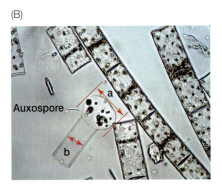

Auxospore

(C)

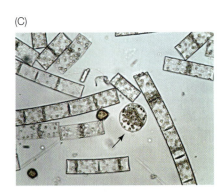

FIGURE 8.18 Auxospore formation in the diatom *Guinardia*. (A) An egg (or developing zygote) forming inside the otherwise empty frustule of a *Guinardia* cell and about to become an auxospore; notice that the cells in that chain are narrower (of smaller diameter) than others in the micrograph. (B) An auxospore attached to one of the two valves of the parent frustule, which have split apart. Notice the larger diameter of the auxospore (a) compared with the parent cell diameter (b). (C) An auxospore that has further developed and is lengthening as it forms a new silica frustule.

the other (see Figure 8.17B). Upon dividing vegetatively, a new half-frustule is formed on the *inside* of each of the original parent halves, producing two new and complete daughter cells. This process continues with each cell division, with new halves of the frustules forming only on the inside of the parent half-frustule.

In addition to not sharing genetic information, there is another problem with vegetative cell division: with each new generation, the diatom cells become smaller and smaller, a process that cannot go on forever. Eventually the cells become so small that they must undergo sexual reproduction. When this happens, pairs of **gametes** from separate individual parent cells come together to form a **zygote**, which then forms a larger **auxospore** (**FIGURE 8.18**). The auxospore develops into a new, larger cell, which then allows the process to repeat through many generations of vegetative cell division. Notice in Figure 8.18A that the parent cell frustule in which the egg or developing zygote is forming is narrower (smaller in diameter) than the other cells in that photomicrograph. Figure 8.18B shows a new auxospore emerging from a parent frustule; notice the difference in diameters between the frustule of the parent cell and the emerging auxospore, which then goes on to develop a new frustule as it elongates into a larger-diameter cell, at which point it can resume vegetative cell division. At times of environmental stress, diatoms also produce, by way of sexual reproduction, **resting spores** with thick cell walls and internal energy reserves that enable them to survive for more than a year in bottom sediments in a state of suspended animation. If resuspended at a time of more favorable growth conditions, resting spores can resume vegetative cell division.

As you can imagine, the process of vegetative cell division, while not ideal in terms of generating genetic diversity, does allow for a geometric increase in the number of cells. Depending on the doubling time—the number of times a cell divides per day—a single cell can "grow," or multiply, to a very large number of cells in a very short period of time. This is where diatoms generally excel over most of the other larger phytoplankton taxa. With their relatively fast cell division rates, they are able to take advantage of short-lived environmental conditions favorable for growth, such as happens in late winter and spring after winter mixing has replenished surface water nutrient levels, and when light conditions are increasing. At such times the diatoms bloom.

An example of how a population of diatoms can quickly exploit ideal growth conditions during a bloom is given in **FIGURE 8.19**. Of course, this geometric progression occurs in environmental conditions that favor rapid growth, meaning that light levels and nutrient concentrations are sufficient; then, cells can divide at rates of one or more doublings per day. But ideal

(A)

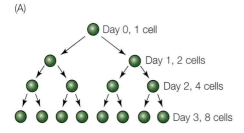

Day 0, 1 cell
Day 1, 2 cells
Day 2, 4 cells
Day 3, 8 cells

(B)

Number of cells = $2^{(\text{Number of Days})}$	
Time	**Number of cells**
3 days	$2^3 = 8$
4 days	$2^4 = 16$
1 week (7 days)	$2^7 = 128$
2 weeks (14 days)	$2^{14} = 16,843$
3 weeks (21 days)	$2^{21} = 2,097,152$

(C)

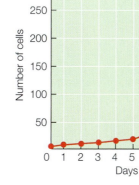

FIGURE 8.19 (A) Diagram of how a single phytoplankton cell dividing once per day for three days will become eight cells. The increase is geometric. If the doubling time is one day—that is, after one day the number of cells will have doubled—then it can be described by the simple exponential equation $N = 2^{(\text{Number of days})}$. (B) The table shows how continuing the trend in the diagram out to three weeks will result in a single cell multiplying to more than 2 million cells! (C) The trend is also shown graphically, but only out to 9 days.

conditions in nature don't last indefinitely. Nutrient concentrations become depleted as the number of cells increase and take up more and more of the available nutrients, and growth rates then slow. The doubling time takes longer, and growth may cease altogether. In addition, cells will simply die from natural causes, usually by being grazed by zooplankton (which we will discuss in the next chapter).

Because of the potential for rapid growth rates, diatoms can be quite abundant; densities of 1000–10,000 diatom cells in a liter of water are common, and even higher densities result during bloom conditions. And because diatoms are among the larger phytoplankton taxa, their high cell densities means that the biomass is also high. For these reasons diatoms are very important in the food web in coastal areas and in higher-latitude regions, wherever there are relatively high concentrations of nutrients resulting from upwelling or winter convective mixing. The spring blooms we discussed in the last chapter consist largely of diatoms. Diatom production in some regions—those that receive injections of relatively high concentrations of nutrients—is often the very foundation of the marine food webs that fuel the production of commercially important fish stocks. Examples include the continental shelves in temperate and high latitudes, and major oceanic upwelling systems of the world, such as off the coasts of Africa, South America, and the Arabian Peninsula.

DINOFLAGELLATES The dinoflagellates are the fourth important member of the larger phytoplankton. Like the diatoms, dinoflagellates are relatively large; cell sizes of 20–50 μm are common. Unlike diatoms, dinoflagellates do not have a silica frustule, but some species have internal plates of cellulose; they are called the **thecate** or "armored" dinoflagellates (**FIGURE 8.20**). Other species may be **athecate** or "unarmored," and are sometimes called the "naked" dinoflagellates. But all dinoflagellates have a pair of flagella: a **transverse flagellum**, which wraps around the cell and when whipped in a wavelike fashion gives the cell a spinning motion, and a **longitudinal flagellum** that trails behind

FIGURE 8.20 General morphology of a thecate dinoflagellate, with both transverse and longitudinal flagella.

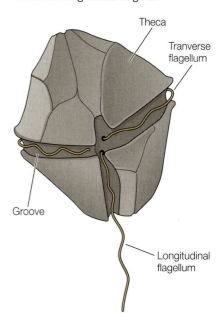

Theca
Tranverse flagellum
Groove
Longitudinal flagellum

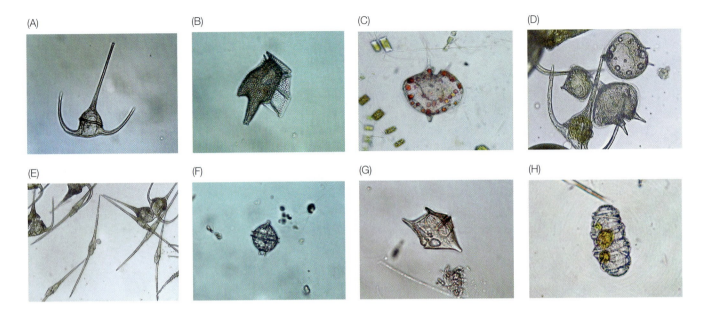

FIGURE 8.21 Photomicrographs of various species of dinoflagellates, all photographed immediately upon collection at sea. (A) *Ceratium longipes*; (B) *Dinophysis tripos*; (C,D) *Protoperidinium spp.*; (E) *Ceratium fusus*; (F) *Gymnodinium sp.*; (G) *Ceratium furca*; and (H) *Polykrikos schwartzii*.

the cell and serves to provide upward (forward) propulsion. Some examples of the diverse shapes of the dinoflagellates are given in **FIGURE 8.21**.

The flagella enable the dinoflagellates to swim, but only weakly. Like the nanoflagellates just discussed, dinoflagellates are capable of swimming distances measured in body lengths. However, because of their relatively large cell size, some species are capable of vertical migrations on the order of 10 meters or so in a 24-hour period. Such species are thought to migrate downward to deeper depths at night to take advantage of the higher concentrations of nutrients and to move upward toward the surface during daylight when photosynthesis is possible.

Although the dinoflagellates are members of the phytoplankton, they aren't always a clean fit. By that we mean that individual species may be autotrophic, possessing chloroplasts and capable of photosynthesis, while other species may lack photosynthetic pigments altogether and function instead as heterotrophs. Still others may be mixotrophic, capable of functioning either way, depending on environmental conditions. Dinoflagellates functioning heterotrophically may feed **phagotrophically**, which means to engulf particulate materials whole, especially other eukaryotic cells. The *Polykrikos* in Figure 8.21H is an example, which appears to have ingested two smaller dinoflagellate cells intact. This particular species is actually a pseudocolony of fused individuals, or zooids. Direct microscopic observations have revealed that dinoflagellates can also feed by capturing cells and ingesting them in a most interesting fashion (**FIGURE 8.22**). Some, maybe most, thecate species capture prey by extending a filament that sticks to the prey organisms, while the cell extrudes a feeding veil, a flexible tissue that encapsulates the prey and allows digestive enzymes to liquefy its cytoplasm for ingestion by the dinoflagellate. As Figure 8.22 shows, this feeding behavior by dinoflagellates is quite versatile; not only are single phytoplankton cells captured and consumed this way, but entire chains, even chains of heavily spined diatom species such as *Chaetoceros*, are susceptible.[13]

Unlike diatoms, which prefer coastal areas, photosynthetic dinoflagellates are more common in ocean environments where mixing (turbulence) is reduced

---

[13] This unusual feeding technique whereby the victim (the phytoplankton) is enveloped by a gelatin-like mass extruded from the dinoflagellate is eerily reminiscent of the 1958 movie thriller *The Blob*. Jacobsen, D. M., and Anderson, D. M., 1986. Thecate heterotrophic dinoflagellates: Feeding behavior and mechanisms. *Journal of Phycology* 22: 249–258.

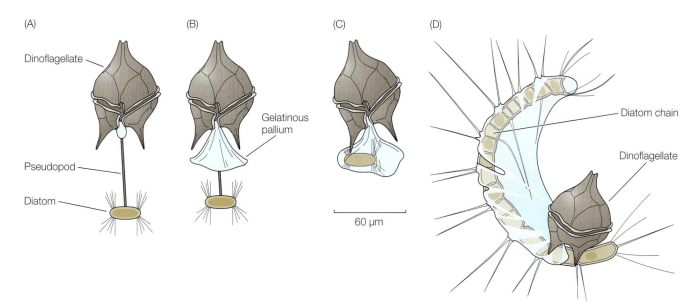

(A)

Dinoflagellate

Pseudopod

Diatom

(B)

Gelatinous
pallium

(C)

60 µm

(D)

Diatom chain

Dinoflagellate

and therefore fluxes of nutrients are relatively low. Dinoflagellates are common in the surface waters of the open ocean in temperate and tropical latitudes, as well as in more productive coastal areas but usually not until after the spring diatom bloom; in some areas of the coastal oceans they may dominate the phytoplankton during the summer months. Dinoflagellates have long attracted the notice of naturalists, as these summertime phenomena are often nuisance, or harmful, blooms. In addition, they have piqued our curiosity because some dinoflagellate species are capable of **bioluminescence**—that is, they give off light. You can see this phenomenon at night during the summer in temperate and higher-latitude coastal waters by just splashing the water with your hands, stimulating the cells enough to flash, giving the water a bluish glow. We will come back to dinoflagellates later (see page 281, Harmful Algal Blooms and "Red Tides") but next we discuss how phytoplankton have adapted to the environmental challenges they face.

FIGURE 8.22 Sketch of the dinoflagellate *Protoperidinium* capturing and feeding on a diatom cell (genus *Corython*) by first attaching a stringlike pseudopod (A) and then encasing the cell in a gelatinous mass, called a pallium (B,C) within which the diatom is liquefied and digested. (D) Another *Protoperdinium* feeding on a chain of *Chaetoceros* cells, showing how the entire chain can be enveloped by the gelatinous pallium.

Ecological challenges faced by phytoplankton: Light and nutrients

In Chapter 7 we discussed in very general terms how phytoplankton require light for photosynthesis and how light levels experienced by phytoplankton cells are affected by the depth of the upper mixed layer in the oceans and by the time of year. But, as we have alluded to in passing, phytoplankton cells may control the amount of light they experience by employing various mechanisms to avoid sinking, and by vertically migrating. But phytoplankton cells don't just need light; they also have to avoid being exposed to too much light. This means that floating right at the surface as a way to avoid sinking won't always work because, among other drawbacks, light levels there may be too high.

Different phytoplankton taxa will have different light affinities, but each will exhibit a response to light that resembles a generic photosynthesis versus irradiance curve (**FIGURE 8.23**). This curve illustrates an important phenomenon: the rate of photosynthesis increases with increasing light intensity, but only up to a point, where it reaches a maximum. This is commonly called P_{max}; at that point the light intensity is ideal and the phytoplankton photosynthesize at their maximum rate. Beyond that, at higher light intensities, photosynthesis becomes inhibited. Ideally, then, phytoplankton will do best at a depth in the water column where the light levels are at or near P_{max}. But when nutrients are also considered, this isn't always the best position to be in.

FIGURE 8.23 A generic photosynthesis versus irradiance curve (P vs. I curve) for a species of phytoplankton. As the light levels increase, the rate of photosynthesis also increases, at first rapidly, and then more slowly as the maximum rate is approached (P_{max}). Beyond that point, increasing light levels produce a decrease in photosynthetic rate; this is photoinhibition.

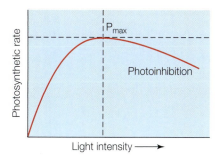

P_{max}

Photoinhibition

Photosynthetic rate

Light intensity ⟶

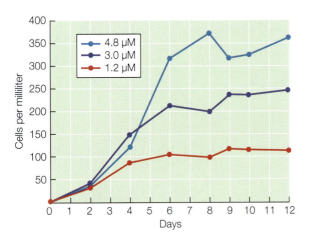

FIGURE 8.24 Example of phytoplankton growth in relation to the concentration of nutrients. A culture of dinoflagellates (*Alexandrium fundyense*) was grown on three different initial concentrations of ammonium (NH_4^+), 1.2, 3.0, and 4.8 micromoles per liter, and the concentrations of cells monitored for 12 days. The total biomass was dictated by the nutrient concentrations available to the cells, being greatest for the highest nutrient concentration, 4.8 micromoles NH_4^+ per liter.

Phytoplankton need nutrients in order to grow—that is, to make new cells. The higher the concentration of nutrients available to the cells, the faster they can grow and the more biomass they can produce. **FIGURE 8.24** shows the growth of a dinoflagellate grown under optimal light levels with varying levels of ammonium in the laboratory. The cells that were exposed to the highest initial concentration of ammonium grew at the fastest rate and reached their highest cell densities over the first six to eight days of the experiment. After the first six to eight days there was no further increase in cell densities because, quite simply, the ammonium ran out; concentrations had dropped to near zero in all three experimental treatments. This experiment illustrates that it is important for phytoplankton populations to be in a position where both light and nutrient concentrations are maximized for their photosynthesis and growth, a situation that is not always possible, which is why not all areas of the world ocean are rich in phytoplankton. Usually one or the other—light or nutrients—is limiting.

Nutrient concentrations vary with depth in the oceans and with location, as we have learned. And the concentration of nutrients exerts controls over which phytoplankton taxa can thrive in a particular area of the oceans; small cells generally do well in low-nutrient environments, and large cells generally do well in high-nutrient environments. This relationship between nutrients and cell size exists because dissolved nutrients are actively transported across the cell membrane and therefore the cell's surface area-to-volume ratio controls the efficiency of that uptake. The greater the surface area a cell has with which to take in nutrients from the surrounding seawater, the more nutrients it can take up. Bigger cells have more surface area than smaller cells, of course, but the internal volume of bigger cells—the cytoplasm that requires the nutrients—is also greater. Therefore, bigger cells do better when seawater nutrient concentrations are higher, while smaller cells can get by with lower concentrations.

This basic principle of surface area-to-volume ratios in relation to cell size is illustrated in **FIGURE 8.25**. For example, a cubic phytoplankton cell that measures 1 cm on each side would have an internal cell volume of 1 cubic centimeter (cm^3), and an outside surface area of 6 square centimeters (cm^2); therefore, its surface area-to-volume ratio is 6:1. Eight such cells would have a total internal volume of 8 cm^3, of course, and a total cell surface area of 48 cm^2; their surface area-to-volume ratio would still be 6:1. However, a single cell composed by lumping together eight of the 1 cm^3 cells would have a surface area of only 24 cm^2 and therefore its surface area-to-volume ratio would be 3:1, which is much less than that of the original eight cells. And because the uptake of dissolved nutrients across a cell membrane is proportional to the cell's surface area relative to its internal volume, the larger cell would be at a disadvantage. Generally speaking, then, small cells are better equipped to take up nutrients, so in

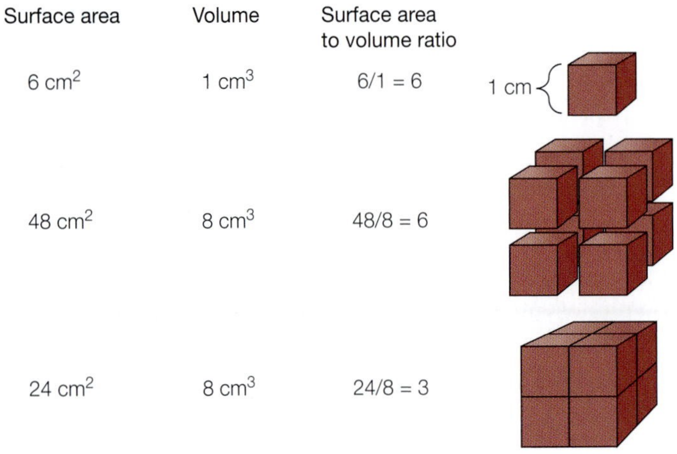

Surface area	Volume	Surface area to volume ratio
6 cm^2	1 cm^3	6/1 = 6
48 cm^2	8 cm^3	48/8 = 6
24 cm^2	8 cm^3	24/8 = 3

FIGURE 8.25 Graphic example of surface area-to-volume ratios. Eight smaller cubes (representing phytoplankton cells) have a larger surface area relative to their combined internal volume than a larger cube with the same volume.

low-nutrient environments like the surface waters of the open ocean, small cells do better than large diatoms, for example.

Methods of studying phytoplankton

Studying phytoplankton involves a number of techniques. The simplest techniques involve collecting cells and preserving them for later microscopic examination. Such collections can be made by concentrating cells in a fine-mesh plankton net or by collecting a whole water sample, preserving it, and then studying the cells when they have settled at the bottom of a chamber. Other methods of study involve more indirect kinds of observations and measurements. Examples include optical techniques related to water color, as influenced by the presence of cells; this "color of the water" approach, which we might call it, is based on the presence of photosynthetic pigments in the phytoplankton, especially chlorophyll. As we learned in the last chapter, Earth-orbiting satellites routinely measure the greenness of the water from space, which can be related to the concentration of phytoplankton chlorophyll, which in turn can be related to the biomass of phytoplankton present in surface waters. A more sensitive method of measuring chlorophyll takes advantage of its fluorescence (see page 280).

One of the oldest techniques used to concentrate and collect plankton—both phytoplankton and the larger zooplankton—is with **plankton nets** (FIGURE 8.26). These are towed through the water, either horizontally while the ship is under-

(A)
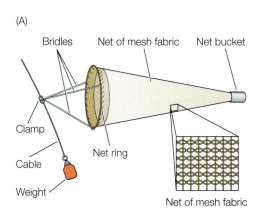

(B)

(C)

(D)

FIGURE 8.26 Examples of plankton nets. (A) Diagram of a standard "ring net," with its weight, bridle for towing, and net bucket, also called the "cod end." (B) A fine mesh phytoplankton net being recovered aboard a research vessel. Upon retrieval of the net, the contents are washed down into the cod end and transferred to a collection jar for preservation and storage. Larger nets with larger mesh sizes include the 60 cm "bongo net" in (C); this net is generally used for larger zooplankton and fish larvae. (D) Shows a 1 m² MOCNESS sampler, which stands for *m*ultiple *open*ing and *c*losing *net* and *e*nvironmental *sampling *system, also used for larger zooplankton and fish larvae. It has nine nets that can open and close, and a CTD system to measure simultaneously the temperature, salinity, and depth where the nets are. Those data are streamed up the wire to the ship's lab, where an operator can decide when and at what depths to open and close the nets.

way, or vertically by using a winch to lower the weighted net to a particular depth and then pulling it back. Plankton nets are used mainly for organisms much larger than phytoplankton, such as zooplankton and small fishes, but they can also collect some of the larger phytoplankton. The net in Figure 8.26B, for example, is made from a material that has a mesh opening of 20 µm, and so it collects the larger diatoms and dinoflagellates. The theory of operation is simple: the net will catch and concentrate anything that doesn't slip past the meshes. And because plankton nets concentrate the sample, they are very useful for collecting taxa that are rare and present in low concentrations.

We now know that even with the finest mesh plankton nets, a significant fraction of the phytoplankton will slip through, especially the picoplankton and nanoplankton, those cells smaller than 20 µm (see Box 8A). Therefore, in order to study the smaller cells, it is necessary to take "whole water" samples. These are usually samples of 50–100 ml seawater, to which a preservative is added. They are brought back to the laboratory and allowed to settle to the bottom of a cylinder. This concentrates the phytoplankton cells, allowing a subsample with lots of cells to be placed under a microscope for identification and enumeration.

Individual cells can also be observed using a fluorescence microscope, which has the capability to view either the natural fluorescence of the phytoplankton pigments, or that of tissue-specific fluorescence stains. This technique also uses special filters to concentrate and collect the cells of interest, which are then examined under the microscope. The image in Figure 8.10 is based on this fluorescence technique.

Scientists are often more interested in how much biomass there is than in how many individuals of a particular species there are. One way to determine biomass quite easily is to measure the amount of chlorophyll in a volume of water. From that measurement one can determine fairly accurately how much phytoplankton biomass is present and therefore how much photosynthesis can be done. Satellites use this technique to measure the "greenness" of the sea surface (see Figure 7.11); their sensitive detectors measure wavelength-specific radiance levels leaving the sea surface. But this technique does not work well with individual water samples. For example, you and I would have a great deal of difficulty discerning differences in glasses of seawater collected from the surface at different locations; our eyes are just not sensitive enough to see small changes. Even laboratory spectrophotometers, which measure the wavelength-specific transparency of a small vial of water, lack the sensitivity needed. A much more sensitive technique is to determine how much the chlorophyll in that water sample **fluoresces**.

Generally speaking, a substance that fluoresces does so by absorbing light energy at relatively short wavelengths and emitting light energy at longer wavelengths. This is how ordinary fluorescent paper works, for example, the bright yellow-green notepaper we all have seen at one time or another. Chemicals in the paper absorb short-wavelength light, usually in the ultraviolet part of the spectrum that we cannot see, and emit light in the yellow-green wavelengths, which we can see. We immediately recognize the effect of this fluorescence in the paper because its color seems to be so different from the normal intensity of colors we are accustomed to seeing. The paper just looks strange in that the yellow-green color seems to be somehow amplified and brighter than normal—which it is.

This basic principle is used to measure chlorophyll, which also fluoresces. Blue light is absorbed by the chlorophyll in phytoplankton cells (the light energy that is absorbed is used in photosynthesis) and red light is re-emitted (**FIGURE 8.27**). This means that slightly more red light is emitted than when chlorophyll is not present, or when there is less chlorophyll present. The more intense the fluorescence, the greater the concentration of chlorophyll. The difference in the

FIGURE 8.27 Schematic representation of the visible spectrum and changes that result from chlorophyll fluorescence. The chlorophyll molecule absorbs some light in the blue part of the visible spectrum, about 425 nm, and emits some light in the red part of the spectrum, about 680 nm, as illustrated here.

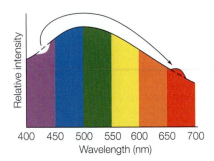

intensity of fluorescence between two samples may be very small and invisible to our eyes, but it is easily detected with a laboratory fluorometer. One way this principle is applied to phytoplankton studies is to filter a specific volume of seawater onto a special filter, which collects the phytoplankton cells, even the picoplanktonic-sized cells, and then place the filter and cells into a solution of acetone which dissolves the chlorophyll. A vial of that solution of acetone and chlorophyll can then be read on a fluorometer, giving a measure of the intensity of chlorophyll fluorescence, as is being done by the student in **FIGURE 8.28**. Higher intensities of chlorophyll fluorescence means higher chlorophyll concentrations and therefore greater phytoplankton biomass. Other fluorometric chlorophyll techniques involve measurements made *in situ*—in the ocean—by exciting the chlorophyll in cells suspended in the seawater surrounding the instrument with a light source and then measuring the fluorescence response. This technique allows the *"in situ* fluorometer" to be lowered through the water column, usually along with a CTD instrument package, to obtain vertical profiles of chlorophyll fluorescence, and therefore phytoplankton biomass.

FIGURE 8.28 A student listening to music while "running" chlorophyll samples at a bench fluorometer aboard a research vessel at sea.

Harmful algal blooms and "red tides"

In certain locations throughout the world ocean, at certain times of the year, usually in the summer, for reasons that are still only poorly understood despite intense research efforts, some species of phytoplankton bloom to unusually high cell densities. Most often these are monospecific, or single-species, blooms which can bring with them various kinds of environmental problems. In some cases, for example, after reaching very high cell densities, the phytoplankton cells simply die and sink to the bottom where they decompose. The process of decomposition consumes oxygen to such an extent that there may be significant oxygen depletion in the bottom waters, which sometimes causes mass mortalities of benthic animals and fish.[14] Another example involves blooms of phytoplankton species that produce **biotoxins**, which affect consumers of those cells, such as shellfish, zooplankton, fish, and in some cases, even marine mammals and humans. These problem phenomena are collectively referred to as "red tides" or harmful algal blooms (HAB). Red tides get their name because sometimes the cells reach such unusually high densities that they can color the water various shades of green, brown, and sometimes bright red, as shown for blooms of nontoxic species in **FIGURE 8.29**. While some bloom events are caused by nontoxic phytoplankton, others can be far more sinister—they can make people very sick.

FIGURE 8.29 (A, left) A bloom of *Noctiluca scintillans*, a nontoxic dinoflagellate, has discolored the near shore waters of Aukland, New Zealand. (A, right) *Noctiluca scintillans* itself, showing ingested *Gymnodinium catenatum* cells. (B, left) A bloom of the photosynthetic ciliate, *Myrionecta rubrum* (shown at right) from coastal Maine. Here, the bloom makes the water look dark purple.

(A)

(B)

---

[14] This is the reverse of our "equation of life"; the decomposition of organic matter involves respiration, which is $CH_2O + O_2 \rightarrow CO_2 + H_2O$, thus consuming O_2.

(A)

(B)

FIGURE 8.30 Blue mussels, *Mytilus edulis* (A), and soft-shell clams, *Mya arenaria* (B), are molluscan shellfish found on both sides of the Atlantic Ocean, and typically are the species involved in HAB-expressed shellfish diseases. While Paralytic Shellfish Poisoning occurs as a result of ingesting these species, as well as scallops, quahogs, oysters, and other species, it has historically been associated most often with mussels.[15]

Common types of HABs and their associated human ailments:

- Amnesic Shellfish Poisoning (ASP), also known as Domoic Acid; caused by a diatom
- Diarrhetic Shellfish Poisoning (DSP); caused by a dinoflagellate
- Neurotoxic Shellfish Poisoning (NSP); caused by a dinoflagellate
- Paralytic Shellfish Poisoning (PSP); caused by dinoflagellates and cyanobacteria
- Ciguatera Fish Poisoning (CFP); caused by a dinoflagellate

There are several types of HABs, each characterized by symptoms that are expressed after people have consumed contaminated finfish and shellfish, especially mussels (**FIGURE 8.30**). The list of human ailments is impressive and underscores the need to understand these events in order to ensure seafood safety and protect public health. This becomes especially important in light of a growing body of evidence suggesting that incidences of HABs are on the increase worldwide, in both the number of incidences and their geographical extent. While experts agree that part of the reason for the increase in reports of HABs has been the result of better detection techniques and a heightened awareness, we must concede that environmental factors may be at work as well.

Shellfish poisonings, most often **Paralytic Shellfish Poisoning** (**PSP**), have been a human health problem throughout recorded history, as have incidences of colored water, or red tides, but the correlation between shellfish poisoning and plankton toxins, especially those toxins in certain species of dinoflagellates, is a relatively recent discovery.[15] The earliest records of red tides and toxic organisms in the plankton date back to the time of Moses, about 1451 B.C., and are described in the Bible.[16] Records of human cases of shellfish poisoning date back to 1689 in Europe, and the medical literature reports numerous cases, mostly in Europe and Great Britain, between the seventeenth and nineteenth centuries, but the cause—the pathogen or toxin that contaminated the shellfish, which were almost exclusively mussels—was unknown.

The first recorded incidence of paralytic shellfish poisoning in the New World was in 1793, as described in detail by Captain George Vancouver (1757–1798) in his memoirs;[17] several of his men became ill and some died after eating mussels collected along the shoreline of the Pacific Northwest. Contaminated shellfish in those waters were known to Native

[15] An excellent early history of shellfish diseases can be found in Bruce W. Halstead, 1965. *Poisonous and Venomous Marine Animals of the World, vol. 1, Invertebrates* (U.S. Government Printing Office).

[16] *Exodus* 7: 20–21.

[17] George Vancouver, 1801. *A Voyage of Discovery to the North Pacific Ocean and Round the World* (London: G. C. and J. Robinson Co.).

Americans long before the arrival of Europeans there, and they apparently understood that there must be some connection with the plankton. It has been written that natives would monitor the waters for bioluminescence, which we now know is caused by nontoxic dinoflagellates of the genus *Noctiluca* (see Figure 8.29A) that frequently accompany the toxic dinoflagellate *Alexandrium catenella*. When the waters became bioluminescent, the shellfish were assumed to be toxic and too dangerous to consume.[18]

The first reported case of paralytic shellfish poisoning in the United States was in 1903 in California, and in 1927 there were 14 cases in the San Francisco area that affected more than 100 people, killing 6. This outbreak marked the beginning of concerted research efforts directed at discovering the underlying cause. Attention initially focused on a suite of possible etiologies and explanations that ranged from pathogenic bacteria to toxic salts in the seawater—but not the possibility that the toxins were in planktonic organisms being filtered out of the seawater and ingested by the mussels. It was not until 1937 that the link between paralytic shellfish poisoning on the Pacific coast of North America and the toxic dinoflagellate *Alexandrium catanella* was first published.[19] In 1949 it was similarly discovered that *Alexandrium fundyense* was the causative organism in Atlantic waters of the U.S. and Canada.[20]

To this day PSP stands out as a recurrent public health issue. It occurs on both the Pacific and Atlantic coasts of North America as well as around the world (**FIGURE 8.31**). In North America, it is the result of toxins in the dinoflagellate species *Alexandrium catanella* on the West Coast and *Alexandrium fundyense* on the East Coast. The PSP toxin is a **saxitoxin** that is one of the most potent biological poisons known to medical science; it acts by disabling sodium channels in nerve cells. Depending on the dosage, PSP intoxication can lead to paralysis and death by respiratory failure. Numerous cases of PSP are reported nearly every year in which humans, after eating contaminated shellfish, again, usually mussels, have become seriously ill, often requiring hospitalization. There have been several reports of fatalities, although not in recent years. Today, state and provincial government agencies carefully monitor shellfish for toxins and do an outstanding job of maintaining seafood safety. When the presence of saxitoxin is detected, officials immediately close affected areas of shoreline to the taking of shellfish.

---

[18] Rachel L. Carson, 1951. *The Sea Around Us* (New York: Oxford University Press).

[19] Sommer, H. et al., 1937. Relation of paralytic shell-fish poison to certain plankton organisms of the genus *Gonyaulax*. *American Medical Association Archives of Pathology* 25: 537–559. These dinoflagellates were later reclassified, moving them from the genus *Gonyaulax* to *Alexandrium*.

[20] Formerly *Gonyaulax tamarensis*. Needler, A. B., 1949. Paralytic shellfish poisoning and Gonyaulax tamarensis. *Journal of the Fisheries Research Board of Canada* 7: 490–504.

FIGURE 8.31 (A) Incidences of Paralytic Shellfish Poisoning around the world through 2006. (B) Several *Alexandrium fundyense* cells and a pennate diatom, genus *Navicula*. The toxic dinoflagellate *A. fundyense* is the culprit in Paralytic Shellfish Poisoning in the Northeastern U.S. and Atlantic Canada.

(A)

(B)

FIGURE 8.32 One of 14 humpback whales on a Cape Cod, Massachusetts, beach in the fall of 1987 that died after eating mackerel that had ingested zooplankton that had in turn ingested toxic *Alexandrium* cells.

Why certain species of phytoplankton, especially certain dinoflagellates, produce toxins as a natural product of their physiology is not well understood. Toxic phytoplankton present a problem because various other marine organisms naturally ingest them; thus, the toxin can be transferred and bioaccumulated up the food chain to higher trophic levels. In the case of PSP, toxic cells are ingested not just by molluscan shellfish, but also by some species of zooplankton. The shellfish and zooplankton are able to ingest these cells and not be susceptible to the toxins—when the toxic cells are not too abundant, that is. Instead of disabling the shellfish and zooplankton that ingest them, the toxins bioaccumulate in their tissues without serious adverse effects. Then, when those consumers of the toxic phytoplankton are in turn eaten by something (or someone) else, we see the toxin's effects—as they adversely affect fish, marine mammals, and humans (**FIGURE 8.32**).

Scientists have launched a number of research programs around the world in attempts to understand better the oceanographic variables that interact to create PSP outbreaks. The goal is to one day be in a position to predict blooms of the causative organisms and issue warnings to consumers and the seafood industry. While we are not there yet, we are nonetheless learning more and more about these plankton organisms and their dynamics. Blooms of *Alexandrium fundyense* in coastal and shelf waters off North America, for example, have been studied intensively in recent years. Blooms occur each year to varying degrees, and in some years there are massive blooms, both in areal extent and in terms of total cell densities. Sometimes the cells reach densities of more than 1 million cells per liter, which colors the waters a reddish brown. But these events are rare.

Like other species of dinoflagellates, *Alexandrium fundyense* has benthic resting cysts, which can survive in the bottom sediments for years, just waiting for the right conditions, or waiting for some amount of time to pass, before they "germinate," sort of like a seed. When the cysts germinate, they grow in number by asexual cell division, which, depending on the local oceanographic conditions, can lead to very high cell densities. Because they are dinoflagellates, *A. fundayense* cells have weak swimming abilities, which enable them to swim downward to get deep-water nutrients, and then to swim back up to the surface where there is sufficient sunlight for photosynthesis, thus helping them to thrive in summer in what are normally nutrient-depleted surface waters. After a certain number of cell divisions, or in response to some unknown environmental cue, the cells will form gametes, which fuse to form a planozygote, which is sort of an enlarged version of itself (this is sexual reproduction). The planozygote then produces a resting cyst once again (**FIGURE 8.33**). The cycle can continue the next season, or years later. But it all begins with a resting cyst that can lie dormant for several years, either in the bottom sediments, or in suspension at deep depths.

Long before they reach the high cell densities necessary to discolor the water, *A. fundayense* cells can cause PSP—that is, not many cells are needed because the shellfish and other organisms filter them out of the plankton, concentrating them and thus accumulating the toxin. But when they do reach high cell densities, which happens over some years, the PSP problem can be

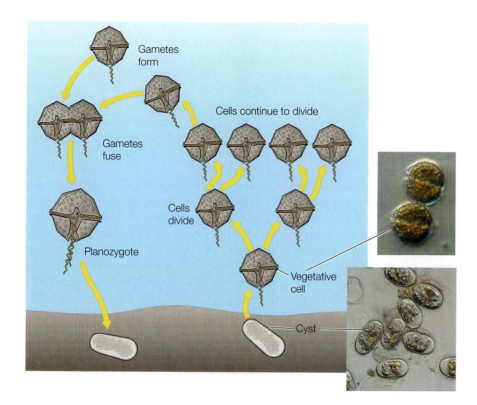

FIGURE 8.33 Diagram of the life cycle of *Alexandrium fundyense*, the dinoflagellate responsible for Paralytic Shellfish Poisoning in coastal waters of the Northeast U.S. and Atlantic Canada. The life cycle begins when a resting cyst germinates to form a vegetative cell; these cells divide vegetatively many times before eventually reproducing sexually by forming gametes, which fuse to form a planozygote. The photomicrograph shows a pair of newly divided vegetative cells. The planozygote resembles the vegetative cell except for possessing two longitudinal flagella; it subsequently transforms into a cyst, completing the cycle. There are several newly formed cysts in the photomicrograph.

greatly exacerbated. Paralytic shellfish poisoning isn't the only HAB problem. While varying significantly in their severity, each HAB has an interesting history and connection with plankton.

Amnesic Shellfish Poisoning (**ASP**) is the most recent HAB to make headlines. It first surfaced in Prince Edward Island, Canada, in the late 1980s when several people died and others suffered what appeared to be permanent amnesia after eating cultured mussels. Only days after the incident, scientists determined that the mussels were contaminated with a toxin known as domoic acid produced by species of the pennate diatom genus *Pseudo-nitzschia* (see Figure 8.16). Subsequent monitoring activities were initiated and since then there have been no further reports of human cases in Atlantic Canada. But on the Pacific coast, there were at least two isolated incidents in the 1990s: one caused significant mortalities of brown pelicans and another incident killed more than 100 sea lions.

Diarrhetic Shellfish Poisoning (**DSP**), as the name clearly indicates, produces gastrointestinal distress. It too is a relatively recent HAB phenomenon, first reported in the early 1960s; it has since been reported in Japan and in European waters. It is considered the least serious of the ailments associated with dinoflagellates in terms of its effects on human health; however, the possible effects of chronic low level exposure have not been evaluated. DSP results from okadaic acid in dinoflagellates of the genus *Dinophysis* (see Figure 8.21B).

Nearly as old a problem as PSP, **Neurotoxic Shellfish Poisoning** (**NSP**), the affliction associated with Florida's red tide, has been occurring off Florida's Gulf Coast since at least 1840. In 1948, the link between fish kills and blooms of the toxic dinoflagellate *Gymnodinium breve* was first established.[21] Renamed *Karenia brevis*, this dinoflagellate produces a **brevetoxin**, a neurotoxin that alters

---

[21] Woodcock, A. H., 1948. Note concerning human respiratory irritation associated with high concentrations of plankton and mass mortality of marine organisms. *Journal of Marine Research* 7: 56–61.

the function of sodium channels and therefore nerve impulses. Shellfish that have been feeding on *K. brevia*, especially during bloom events, can accumulate the toxin, which, if the shellfish are consumed by humans, can produce a range of symptoms, including nausea, numbness of the mouth and tongue, dizziness, and difficulty walking. NSP is also known for causing massive fish kills in Florida coastal waters, which result when the toxin leaks from cells that come in direct contact with the gills. Worse, manatees, which are already under stress in Florida's inland waterways (which we discuss in Chapter 13), have suffered mortalities as a result of NSP. Although there have been no recorded human fatalities, the toxin can leach from the *K. brevis* cells and become dissolved in the surface waters, and aerosols carrying the toxin are known to produce upper respiratory problems in humans during bloom events.

Ciguatera Fish Poisoning (**CFP**) is a seafood-borne disease that afflicts people who have ingested tropical reef fishes that have bioaccumulated toxins produced by dinoflagellates of the genus *Gambierdiscus*. This disease has been known since the time of Captain Cook, whose men contracted symptoms after having eaten, it is believed, red snapper.[15] These dinoflagellates grow attached to algae and corals and are ingested by various reef fishes; when either those fishes, or fishes that have in turn fed upon them, are ingested by humans, ciguatera poisoning results. Symptoms include gastrointestinal and neurological disorders which can recur years later. It is estimated that tens of thousands of cases occur each year in the Caribbean region, and as many as a half million cases worldwide, with few of those cases actually reported.[22]

While these HABs, Neurotoxic, Paralytic, Amnesic, and Diarrheic Shellfish Poisoning, and Ciguatera Fish Poisoning all represent important human health issues, it is NSP, PSP, and CFP that stand out in history, and which may pose the greatest continuing threats to human health. But there is one more kind of harmful phytoplankton species we need to discuss, another dinoflagellate, which may not pose a threat to human health but which may be the strangest yet described in terms of its life history and behavior as a fish predator: that organism is *Pfiesteria*.

PFIESTERIA   This is a most interesting dinoflagellate that was only discovered in the late 1980s. It effectively "lies in wait" in the muddy bottom sediments of estuaries and rivers waiting for schools of fish to pass by. Somehow sensing their presence, the dinoflagellate emerges from its benthic resting stage and attacks the fish, causing open wounds in the fish's flesh, thus causing fish kills. Recent research has shown that rather than kill the fish directly with a toxin, the cells actually attach to the skin of the fish, feeding on the fish's flesh and weakening its defenses such that secondary fungal infections can take over.[23] Originally thought to range from Delaware to Alabama, recent studies based on molecular techniques have shown that the genus *Pfiesteria*, with several species, has a cosmopolitan distribution and is found throughout the world ocean. Many questions remain to be answered about this intriguing dinoflagellate, and research interest continues.

The Macroalgae (the Seaweeds)

Although the pelagic realm of the ocean is dependent on primary production by the single-celled phytoplankton, which form the base of the food web in that

---

[22] Dickey, R. W., and Plakas, S. M., 2010. Ciguatera: A public health perspective. *Toxicon* 56: 123–136.

[23] Litaker, R. W. et al., 2002. Life cycle of the heterotrophic dinoflagellate *Pfiesteria piscicida* (Dinophyceae). *Journal of Phycology* 38: 442–463.

vast environment, some near-shore benthic communities—those on the bottom in shallow water—are more dependent on primary production by the larger multicellular algae, the macroalgae, with phytoplankton making only a relatively minor contribution. We opened this chapter arguing that the majority of the total primary production in the world ocean is accomplished by the phytoplankton, but the larger macroalgae are nonetheless very important in inshore areas. They account for as much as 2–10% of the world's total primary production, and because they do this in an extremely limited area of the world ocean—just a thin line along the shore—those areas of the ocean are very productive indeed. Unlike the planktonic phytoplankton, these primary producers, the macroalgae, are usually attached to the bottom or to some other substrate. That is, they are generally not free-floating, but there are exceptions such as *Sargassum* weed in the open ocean (see page 291). Like phytoplankton, which are single-celled algae, the multicellular macroalgae are also limited by nutrients and light.

There are many species of macroalgae, but the exact number isn't really known and is a matter of some debate among phycologists. To give you some idea of the possible number of species, there are more than 300,000 specimens in collections in the Smithsonian Institution's Museum of Natural History. Certainly, there must be many thousands of species. This section makes no attempt to do full justice to a discussion of the macroalgae, which would need to go far beyond the scope of this introductory text; instead, this is intended only as a very basic overview, and interested students should look into classes on phycology for a more in-depth treatment. We review here the more common and important groups that one is likely to encounter. The major groups of the macroalgae are categorized by the type of photosynthetic **accessory pigments** (light-capturing pigments) they possess in addition to chlorophyll. Using this system we have three major groups: the **green algae**, the **red algae**, and the **brown algae**.

The green algae

The green algae include both single-celled and multicellular species. The multicellular macroalgal examples are, like the single-celled species we discussed earlier, more common in freshwater than in seawater. Of the marine species, most are intertidal or occur in shallow near-shore waters. Their bright green color is due to chlorophyll (**FIGURE 8.34**). The most common example is sea lettuce, genus *Ulva*, which is seen in tide

(A)

(B)

FIGURE 8.34 Examples of a common green alga, genus *Ulva*. (A) A patch of *Ulva* growing on the bottom in the intertidal zone of a beach in Maine. Inset shows a close-up of same patch. These algae are arranged as bundles of elongated tubes, each of which is one cell thick, attached to the bottom by weak holdfasts. The tubes trap air, giving them buoyancy and helping to maintain them upright when covered with water at high tide. (B) Sea lettuce, another species of *Ulva*, is the bright green leafy alga shown here growing in a tide pool in Maine. This species is also attached to the bottom but only weakly, and is often seen floating freely. *Ulva* is commonly associated with areas of nutrient enrichment, either natural, as is the case in the top photos, or as a result of a nearby source of pollution.

FIGURE 8.35 (A) A calcareous green alga, *Halimeda kanaloana*, grows at depths of 20–90 meters in "deep water algal meadows" (B) in the Hawaiian Islands.

(A)

(B)

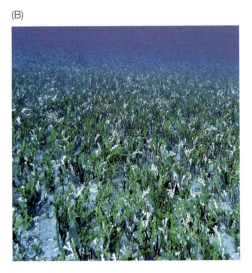

pools around the world's oceans; this is an edible alga that is commonly used in salads and other dishes. The green algae often grow best in response to nutrient enrichment, such as occurs near sewage outfalls, and serious "green tides" are appearing globally. However, the patch of *Ulva* in Figure 8.34A is responding to very localized nutrient enrichment accompanying groundwater that is seeping out of the beach at low tide.[24]

Another group of green algae are the calcareous green algae that are common in relatively shallow tropical waters (**FIGURE 8.35**). These algae incorporate calcium carbonate into their green leaves, which are periodically shed and, upon decomposing, contribute to the sandy carbonate sediments. Much of the carbonate sand on tropical beaches is produced by these algae.

The red algae

These may be the most diverse group of the macroalgae. They occur in all oceans, although they prefer warmer waters, so are most common at lower latitudes. These algae do well in dim light, and they have been collected at more than 200 meters depth in the Caribbean, but they are also common in lagoons and in the intertidal zone. In addition to chlorophyll, they have accessory pigments which effectively mask the green color of chlorophyll and give some species their reddish coloration—at times, that is. Individual species often exhibit a wide range of colors, depending on the ratio of pigments they contain, which in turn may vary with their depth in the ocean and the amount of light penetrating to that depth, and especially in response to nutrient or temperature stress. The same species of red algae may be greener in shallow waters, but increasingly red in deeper, more nutrient-rich waters. Some appear almost black at the deepest depths. The red algae also include *coralline algae*, which have a hard crust made of $CaCO_3$ in the cell walls of the alga, giving it a range of colors from red to pink to almost pure white (**FIGURE 8.36**).

---

[24] A paper published in 2003 announced that two genera of green algae, *Ulva* and *Enteromorpha*, are in fact the same genus. Because the genus name *Ulva* predates *Enteromorpha* (it was assigned by Linneaus), *Ulva* now includes the former *Enteromorpha*. Hayden, H. S. et al., 2003. Linnaeus was right all along: *Ulva* and *Enteromorpha* are not distinct genera. *European Journal of Phycology* 38: 277–294.

FIGURE 8.36 The red algae come in many forms, some of which are encrusted coralline algae that are impregnated with calcium carbonate secreted by the alga, which may have either articulated or nonarticulated forms; still others may be filamentous or leafy in shape. (A) A pink nonarticulated coralline alga in a tide pool, forming a rocklike layer firmly attached to the underlying rock, with a limpet grazing on it. The pink color results from the algal pigments and white calcium carbonate in the cell walls. (B) An example of a branching or articulated coralline alga. (C) A close-up of a filamentous red alga. (D) A patch of a leafy red alga, Irish moss (*Chondrus crispus*), common on both sides of the North Atlantic Ocean, attached to rocks at low tide. It ranges in color from green to a dark red-brown; both green and dark red shades can be seen here.

One species, *Chondrus crispus*, is commercially harvested for its *carrageenan*, a polysaccharide used in food products as a gelling agent.

The brown algae

The brown algae are the largest of these three groups of macroalgae, and include the kelps and common rockweeds we routinely see on both coasts of North America and throughout much of the world ocean (**FIGURE 8.37**). The brown algae get their color from carotenoid pigments and chlorophyll. Abundant in the intertidal zone and at shallow subtidal depths in colder seas at temperate and higher latitudes, kelps are also found at deeper (colder) depths in lower latitudes.[25] The brown algae are extremely important in providing food and habitat for various marine animals, and are harvested for various food products, such as emulsifiers used in making ice cream.

---

[25] Graham, M. H. et al., 2010. Deep-water kelp refugia as potential hotspots of tropical marine diversity and productivity. *Proceedings of the National Academy of Sciences* 104: 16576–16580.

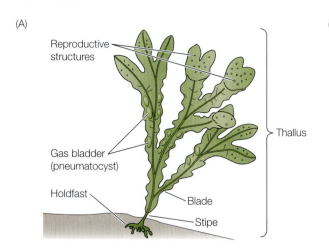

FIGURE 8.37 (A) The giant kelp, *Macrocystis pyrifera*, off the Channel Islands, California. This is the largest species of algae; it can attain lengths of greater than 60 meters. (B) Near-complete coverage of the intertidal bottom by brown algae, *Ascophyllum nodosum* and *Fucus vesiculosus*, on the coast of Maine. (C) A clump of horsetail kelp (*Laminaria digitata*) in a tide pool in Maine.

Anatomically, the brown algae are the most complex of the macroalgae. Like all algae, they are nonvascular plants—they possess no xylem vessels to carry water or phloem cells to move sap. That is, there are no roots, stems, or leaves connected by vessels; they don't need them. They are bathed in fluid that provides them with dissolved inorganic carbon and the nutrients they need for photosynthesis and growth. The main body of any macroalga is the **thallus**, from the Greek word which roughly translates to "sprout-like," and is similar in appearance to a terrestrial plant. The thallus includes the **stipe**, which is similar to the stem of a plant, and **blades**, which are similar to leaves. They are attached to a substrate by a **holdfast**, which is similar to the roots of a terrestrial plant, but there the similarities with terrestrial plants end (**FIGURE 8.38**). The blades

FIGURE 8.38 (A) Diagram of the structure of a brown alga, genus *Fucus*. (B) Two common brown algae on both sides of the North Atlantic Ocean: on the left, *Ascophyllum nodosum*, and on the right, *Fucus vesiculosus*.

(A)

(B)

FIGURE 8.39 (A) Scientists sorting through a recently collected sample of *Sargassum* weed collected in the Sargasso Sea off North Carolina. (B) A floating line of *Sargassum* from the same area of the Sargasso Sea. (C) An underwater photograph of *Sargassum* with a young sea turtle. (D) A close-up of *Sargassum* weed. Note the similar structure to that of *Fucus* and *Ascophyllum* in Figure 8.38, including pneumatocysts on the blades, which are critical for flotation.

on these algae have gas bladders, or **pneumatocysts**, which have no analogue in terrestrial plants; these provide flotation when submerged, allowing for greater exposure to light for photosynthesis.

Often, during periods of high wave action, pieces of these brown algae are torn away from their holdfasts, and begin to drift about. For example, when working from our research ships in the western North Atlantic, we routinely see large rafts of *Fucus* and *Ascophyllum* more than a hundred miles out at sea where they are alive and apparently doing quite well. In fact, it is believed that such "rafted" brown algae were important in recolonizing North America from Europe following the last ice age. Some closely related species of brown algae, species of **Sargassum**, (*Sargassum natans* and *Sargassum fluitans*), reproduce vegetatively in the Sargasso Sea (named for *Sargassum*) in the subtropical North Atlantic Ocean, and host microenvironments that afford refuge and a source of primary production for other organisms, including a variety of juvenile fishes and invertebrates (**FIGURE 8.39**). Still other species of *Sargassum* are attached to substrates in shallow waters and on coral reef systems distributed around the world ocean in temperate and tropical seas. Often they too are freed from their holdfasts and are found drifting freely.

Seagrasses

This last group of marine primary producers is perhaps the least important in terms of their contribution to global primary production. But even with this caveat, anyone who lives near the seashore would bristle at such a callous statement. This is because they are witness to the beauty of the shallow marine ecosystems that are shaped by seagrasses; moreover, they

(A)

(B)

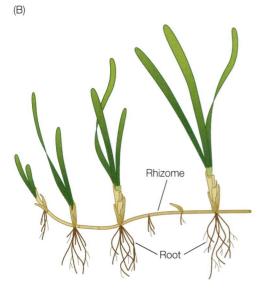

FIGURE 8.40 (A) An eelgrass bed (*Zostera marina*) on Long Island, New York. (B) Diagram of the rhizome system of *Zostera*, which anchors and connects clusters of root nodules that spread beneath the surface of bottom sediments.

have a fundamental sense of their worth, and their fragility. These beautiful marine plants are critical to the stabilization of sediments and provide habitat and food to many organisms. They are indeed much more valuable than any evaluation based solely on primary productivity would indicate.

Seagrasses are submerged vascular plants (also called the submerged aquatic vegetation, or SAVs) that we find in shallow-water environments. They are somewhat unique in that there are few species of vascular marine plants that have evolved to remain permanently submerged as these have (versus salt marsh plants, which we discuss in Chapter 12). The seagrasses include the eelgrass, genus *Zostera* (**FIGURE 8.40**), which is perhaps the best-known example. *Zostera* has twelve species worldwide, with three found in North America. *Zostera marina* is the most widespread seagrass species in the Northern Hemisphere; it occurs in temperate and higher latitudes, in the cooler waters of both the Atlantic and Pacific Oceans, as well as in the Arctic, where it is capable of surviving the winter covered with ice. On the east coast of North America, *Zostera* grows from Maine to North Carolina, and is an important component of shallow water ecosystems throughout that range. It grows best in protected areas, away from heavy wave action, where bottom sediments can accumulate and where it can anchor itself with its root and rhizome system; as such, it also helps to trap sediments and stabilize the bottom sediments. Eelgrass is the favorite food of a number of invertebrates, and eelgrass beds serve as protective refuges for invertebrates and juvenile and adult fishes.

Over the years the distribution and abundance of eelgrass beds have waxed and waned for various reasons, and in recent decades they have suffered severe declines. Today they are at their lowest levels since the 1930s, when they were all but wiped out by what became known as "the eelgrass wasting disease," which destroyed more than 90% of the eelgrass beds in Europe and North America. Gradually over a period of decades, eelgrass beds were recolonized, and by the 1960s they were thought to be in reasonably good health. For years the cause of the 1930s devastation remained a mystery, until researchers discovered that it most probably was the result of a slime mold–like infection, produced by *Labyrinthula zostera* (although the taxonomy of this organism is currently disputed). It wasn't shown to be the probable cause until the late 1980s and early 1990s, which is when it was demonstrated to have caused a similar decimation of eelgrass beds in the early 1980s on the east coast of North America.

The immediate causes of eelgrass population declines today are not limited to diseases; increased pollution and turbidity from human activities are also thought to play an important role, and today monitoring activities are aimed at minimizing such impacts on these important near-shore marine environments. On a positive note, restoration efforts seem to be paying off; reseeded eelgrass beds seem to be doing well, and there are numerous success stories.

Chapter Summary

- The **primary producers** are the marine autotrophic organisms that fix inorganic carbon and produce the organic carbon that constitutes the food at the base of the marine food web. They include chemoautotrophs and photoautotrophs, the latter being far more important in terms of their contribution to total primary production. The photoautotrophs include the **macroalgae and seagrasses**, and the much smaller **phytoplankton**.

- Phytoplankton are single-celled photoautotrophs that comprise a broad range of taxa and thousands of species. The majority are microscopic in size, but some are visible to the naked eye.

- The Archaea and the Bacteria are both **prokaryotes**. The Archaea are found in extreme environments as well as throughout much of the world ocean. The bacteria include some of the most ecologically important members of the phytoplankton, including the **cyanobacteria**, also known as **blue-green algae**, some of which are nitrogen fixers (e.g., *Trichodesmium*). Species of the cyanobacterium *Prochlorococcus* are the smallest and most abundant forms of phytoplankton.

- The **eukaryotic** phytoplankton, single-celled photosynthetic organisms with a modern cell design, are generally larger in cell size. They include a number of major taxonomic subgroups, including: (a) the green algae, which are mostly found in freshwater, with a few species occurring in marine waters; and (b) the **silicoflagellates, coccolithophores, diatoms**, and **dinoflagellates**. The diatoms are the most important group of phytoplankton in terms of their contribution to the marine food web. The dinoflagellates are also important in the marine food web, as well as in harmful algal blooms.

- Phytoplankton reproduce both asexually, by **vegetative cell division**, and sexually, whereby gametes containing genetic information from two individuals combine in producing a daughter cell. High rates of asexual cell division allow some phytoplankton species to achieve high biomass levels at certain times of the year. Thus we see spring blooms of diatoms, and harmful algal blooms of other phytoplankton species.

- **Harmful algal blooms** occur when certain species of phytoplankton grow to unusually high cell densities. When toxic species are ingested by other marine animals and humans, they result in a number of ailments.

- Dinoflagellates and **nanoflagellates** may be heterotrophic, autotrophic, or **mixotrophic**.

- There are three main groups of macroalgae: the **green algae**, the **red algae**, and the **brown algae**, which are distinguished on the basis of their **accessory pigments**. These "seaweeds" are most common in shallow and coastal waters, where they are usually attached to a substrate. They can be very productive and are important to coastal ecosystems.

- Seagrasses are submerged vascular plants that are important in many coastal marine ecosystems, stabilizing sediments and providing habitat and food for many marine organisms. In the 1930s and as recently as the 1980s, populations of seagrasses have been nearly wiped out by disease. They are also vulnerable to pollution and habitat destruction as a result of coastal development activities.

Discussion Questions

1. Why is most of the primary production in the oceans accomplished by the unicellular phytoplankton?

2. Why were the most important primary producers in the world ocean (the cyanobacteria) discovered only in relatively recent times? Are those tiny cells as important to marine food webs as the larger diatoms and dinoflagellates? Why or why not? How can tiny photosynthetic cells such as the cyanobacteria (especially *Prochlorococcus*) possibly be "the most important primary producers in the world ocean"?

3. Why are species of the cyanobacterium *Trichodesmium* so important in the ecology of the oceans?

4. Of what use are flagella to the nanoflagellates and the larger dinoflagellates?

5. Why are coccolithophores of such interest to scientists concerned with atmospheric chemistry? Consider the following: scientists are noticing that as CO_2 continues to be added to the atmosphere, the oceans are becoming more acidic; this is because CO_2 dissolves in seawater to form carbonic acid, H_2CO_3. How might such increases in ocean acidity affect coccolithophores? How might they affect calcareous red and green algae?

6. What are some of the advantages and disadvantages to diatoms of having a siliceous frustule? How do diatoms overcome their tendency to sink?

7. Phytoplankton reproduce both sexually and asexually. What are the advantages of each reproductive mode? Under what circumstances does asexual reproduction predominate? What circumstances promote sexual reproduction?

8. Under what oceanographic conditions are phytoplankton nutrient-limited? Light-limited? Can you think of situations when both nutrients and light are limiting at the same time?

9. Why are the most productive phytoplankton so small? What are some of the ecological advantages and disadvantages of being small? Of being unicellular?

10. Why are macroalgae responsible for only 2–10% of global marine primary production? How are macroalgae and seagrasses important to coastal marine ecosystems?

Further Reading

Anderson, D. M. et al., 2008. Harmful algal blooms and eutrophication: Examining linkages from selected coastal regions of the United States. *Harmful Algae* 8: 39–53.

Baldauf, S. L., 2008. An overview of the phylogeny and diversity of eukaryotes. *Journal of Systematics and Evolution* 46: 263–273.

Chisholm, S. W., Olson, R. J., Zettler, E. R., Goericke, R., Waterbury. J. B., and Welschmeyer, N. A., 1988. A novel free-living prochlorophyte abundant in the oceanic euphotic zone. *Nature* 334: 340–343.

Courties, C., Vaquer, A., Troussellier, M., Lautier, J., Chrétiennot-dinet, M. J., Neveux, J., Machado, C., and Claustre, H., 1994. Smallest eukaryotic organism. *Nature* 370: 255.

DeLong, E. F., 2003. Oceans of Archaea. *ASM News* 69: 503–511.

Fuhrman, J. A., McCallum, K., and Davis, A. A., 1992. Novel major archaebacterial group from marine plankton. *Nature* 356: 148–149.

Graham, Linda E., Graham, James M., and Wilcox, Lee W., 2008. *Algae*, 2nd ed. San Francisco, CA: Benjamin Cummings.

Halstead, Bruce W., 1965. Poisonous and Venomous Marine Animals of the World. Vol. 1., Invertebrates. U.S. Government Printing Office.

Jacobsen, D. M., and Anderson, D. M., 1986. Thecate heterotrophic dinoflagellates: Feeding behavior and mechanisms. *Journal of Phycology* 22: 249–258.

Johnson, P. W., and Sieburth, J. M., 1979. Chroococcoid cyanobacteria in the sea: A ubiquitous and diverse phototrophic biomass. *Limnology and Oceanography* 24: 928–935.

Kirkpatrick, B. et al., 2004. Literature review of Florida red tide: Implications for human health effects. *Harmful Algae* 3: 99–115.

Litaker, R. W., Vandersea, M. W., Kibler, S. R., Maden, V. J., Noga, E. J., and Tester, P. A., 2002. Life cycle of the heterotrophic dinoflagellate *Pfiesteria piscicida* (Dinophyceae). *Journal of Phycology* 38: 442–463.

Needler, A. B., 1949. Paralytic shellfish poisoning and *Gonyaulax tamarensis. Journal of the Fisheries Research Board of Canada* 7: 490–504.

Short, F. T., Muehlstein, L. K., and Porter, D., 1987. Eelgrass wasting disease: Cause and recurrence of a marine epidemic. *Biological Bulletin* 173: 557–562.

Townsend, D. W., Pettigrew, N. R., and Thomas, A. C., 2005. On the nature of *Alexandrium fundyense* blooms in the Gulf of Maine. *Deep Sea Research II* 52: 2603–2630.

Contents

The zooplankton, or animal plankton—the consumers of organic carbon formed by the primary producers—come in an astonishing variety of forms. Some are in the plankton only for a portion of their lives, while others stay in the plankton for their whole life. This larval palp worm hatched from a tiny egg, and will next pass through several developmental stages, eventually acquiring its adult form, when it will leave the water column to assume a completely different life style, confining itself inside a tube it constructs in the bottom sediments.

The Zooplankton

The **zooplankton**, the *animal* plankton, are the principal consumers of the phytoplankton. Sometimes referred to as the **grazers**, most of the zooplankton are **herbivorous** and occupy a position in the second trophic level of the marine planktonic food web. While they are only one of several taxonomic groups in that position, the zooplankton are key links in the food web: as consumers and repackagers of organic carbon produced by phytoplankton, they are themselves products for consumption by organisms at the next higher trophic level. For this reason, zooplankton are usually, but not always, larger than their phytoplankton food and smaller than the carnivores that prey upon them. But as we will see in this chapter, there are exceptions to this generalization: first, not all zooplankton are herbivorous; some are **omnivores**, others **carnivores**, and therefore they can occupy still higher trophic levels. Second, not all are larger than their food organisms; some zooplankton are smaller than the phytoplankon—the heterotrophic dinoflagellates, for example—that they consume. The size range of zooplankton is impressive, from the very large, such as jellyfish—some measuring more than a meter—to the very small, about the same as the phytoplankton.

These interesting and ecologically important marine animals comprise many diverse taxonomic groups, but they all have one thing in common: they are members of the plankton. And like the phytoplankton, zooplankton are incapable of significant swimming. They can only cover distances measured in terms of their body lengths and are therefore mainly carried by water motions—but we need to qualify this statement: Some zooplankton, owing to their larger sizes, can make much better progress than the smaller phytoplankton we discussed in the last chapter, especially vertically, and as we will see in this chapter, this is an important aspect of their behavior.

The zooplankton are highly diverse, with fascinating life histories and ecologies, the major features of which we will review in this chapter. We will begin by separating out the various groups on the basis of how much time—what portion of their life histories—is spent actually in the plankton.

The Meroplankton

Some members of the zooplankton are in the plankton part-time, spending only a portion of their life histories enjoying the freedom of a planktonic lifestyle, usually in their early developmental stages. The remainder of their lives is spent either as **benthic** organisms, living on, in, or somehow associated with the ocean bottom, or as **nektonic** organisms, which are the swimmers. These part-timers are the **meroplankton**, which include the egg and larval stages of various marine invertebrates and vertebrates. Examples include, but are not limited to: benthic worms, corals, shrimp, barnacles, crabs, starfish, sea cucumbers, clams, snails, and most species of fish. In each case, the time spent in the plankton is usually devoted to seeking out and feeding on suitably sized nutritious food particles, such as phytoplankton and other forms of zooplankton, while undergoing morphological changes, often dramatic, as they develop and grow. All the while they are being dispersed, carried by ocean currents away from where they were born. A few examples follow.

The common barnacle, which is found in Northern Europe and on both coasts of North America, begins its life as a planktonic organism before it becomes a benthic, sessile plankton feeder that is glued to a rock or other benthic substrate (such as a dock piling or a boat hull). Barnacles brood their young and release larvae into the water during the spring phytoplankton bloom. Fertilization is internal, with the eggs hatching into **nauplii** (singular *nauplius*), which are held by the adult until just the right time, when phytoplankton food particles are abundant, to ensure the feeding success of the larvae.[1] The meroplanktonic nauplii pass through several developmental stages while feeding in the plankton, eventually molting into the **cyprid** stage, which is the late larval stage, in which the young barnacle resembles a bivalve; the cypris larva will search for just the right substrate on which to attach, and will then metamorphose into the adult form, becoming a benthic organism once again (**FIGURE 9.1**). There it will spend the remainder of its life, stuck in place until either consumed by a predator or squeezed out by other individual barnacles in a sometimes-fierce competition for space. Notice in Figure 9.1 how some of the newly settled barnacles have attached themselves to the older barnacles.

Another common sessile marine invertebrate that has a meroplanktonic phase is the soft-shell clam (*Mya arenaria*), a commercially important bivalve mollusc found on both sides of the North Atlantic Ocean (and as an invasive species on the Pacific coast of North America). The adult clams release eggs and sperm directly into the water column (called broadcast spawning) where external fertilization takes place; the eggs and resulting larval stages then develop in the plankton. Like barnacles, clams also time their spawning to match the seasonal production cycle, the sequence of oceanographic events that culminates in the bloom of phytoplankton—forage for the developing larvae. Evolution of this spawning "strategy" is widespread in the marine environment because of the simple advantage it affords: the larvae stand a better chance of survival when present at a time coincident with high abundances of their planktonic food particles.

FIGURE 9.1 Meroplanktonic larval stages of the common barnacle, *Semibalanus balanoides*. (A) Naupliar stage and cyprid larvae. (B) Both newly settled barnacles (the smaller individuals) and older barnacles from previous years are attached to this rock, exposed at low tide. New barnacles attach to almost any hard substrate, even other barnacles, as can be seen for a few individuals in the photograph.

---

[1] The nauplius stage of larval crustaceans occurs early in their development. The nauplius has three appendages attached to the anterior segment (the head), which are used for swimming and feeding.

(A)

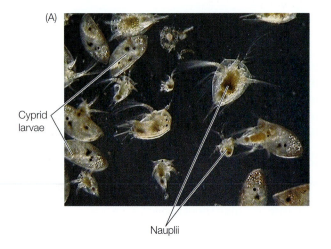

Cyprid larvae

Nauplii

(B)

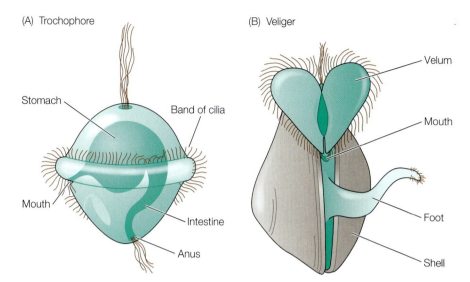

(A) Trochophore

Stomach

Band of cilia

Mouth

Intestine

Anus

(B) Veliger

Velum

Mouth

Foot

Shell

FIGURE 9.2 Sketches of two stages in the early development of a generic bivalve mollusc. (A) The trochophore stage, with a functional digestive system and locomotory bands of cilia; and (B) the veliger stage, in which the mollusc already possesses a shell and a foot.

Larval bivalve molluscs, like the clam (as well as gastropod molluscs, or snails, and annelid worms), pass through two distinct larval stages during the course of their development, both of which are spent drifting freely as plankton in the water column where they feed and continue to develop: these are the **trochophore** and **veliger** stages (**FIGURE 9.2**). Following a short period of time as a trochophore, the young clam transforms into the veliger stage, in which it has a rudimentary bivalve shell. Eventually it develops to the point where it leaves the planktonic realm and descends to the bottom in search of a suitable spot where it can work its way down into the sediment, establishing itself in its new (and only) home, where it will reside for the remainder of its juvenile and adult life (**FIGURE 9.3**).

A motile benthic invertebrate, the common Jonah crab (*Cancer borealis*) broods its eggs following fertilization, carrying them on its underside while they develop; upon hatching, the young larvae find themselves in the water column where they continue to develop as meroplanktonic organisms. During this time they pass through several developmental stages with very different

FIGURE 9.3 (A) Photomicrograph of a veliger-stage larva of a soft-shelled clam (*Mya arenaria*) nearly far enough along in its development to settle to the bottom; note the size of the clam relative to the two smaller dinoflagellates in the frame (*Ceratium* and *Dinophysis*). The velum (a lobed structure with cilia used for swimming and feeding) is retracted into the shell here. (B) A few freshly dug adult soft-shell clams, ready for steaming or a chowder.

(A)

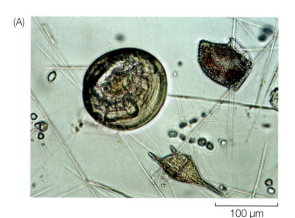

100 μm

(B)

(A)

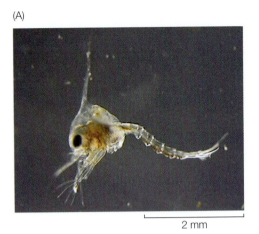

(B)

(C)

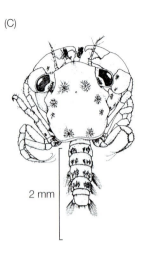

2 mm

2 mm

FIGURE 9.4 The Jonah crab (*Cancer borealis*) passes through several zoea stages (A) during its larval development in which its morphology bears little or no resemblance to that of the adult (B). When it reaches the megalopa stage, it begins to resemble the adult crab. (C) A megalopa stage of a Gulf stone crab (*Menippe adina*).

morphologies before settling to the bottom to begin their adult life (**FIGURE 9.4**). The larvae hatch into the first of five **zoea** stages followed by a **megalopa** stage before metamorphosing into the juvenile which resembles a miniature adult. The juveniles continue to develop, casting off their old shell (molting) several times more; as adults, they will molt on average once each year as they continue to grow throughout the rest of their lives.

It is quite common for the early developmental stages of marine organisms to pass through seemingly unrelated morphologies before eventually assuming a more readily identifiable adult form, as seen for the clam and the crab. Equally striking changes in appearance can be seen in the larval development of the sea urchin (**FIGURE 9.5**), which, like all the echinoderms (starfish, urchins, sea cucumbers, and sand dollars) passes through an interesting larval stage called the **pluteus**.

Meroplankton are most often, but not always, more abundant nearer the shore and in shallow waters, in that this is where the benthic parent stocks are most common and where the progeny originate. But there are exceptions, such as the eggs and larvae of open-ocean fishes. In estuaries, it is not uncommon for the meroplankton to completely dominate the zooplankton community, especially during the productive months of the year.

There are many more examples of meroplankton that we could discuss here (and we will return to this topic when we discuss the Cnidaria in Chapter 10), and for more in-depth information the interested reader should consult any of a number of excellent texts on invertebrate zoology. Instead, we need to discuss the animal plankton that, by far, make up the majority of taxa in

FIGURE 9.5 (A) Photomicrograph of a pluteus larva of a sea urchin, *Strongylocentrotus droebachiensis*, with two dinoflagellate cells and a copepod nauplius also in the frame. (B) The benthic adult sea urchin, photographed *in situ* in a kelp bed; a kelp stipe can be seen lying across these urchins.

(A)

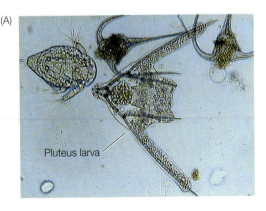

Pluteus larva

(B)

the plankton: the **holoplankton**, zooplankton that spend their entire life cycle in the plankton.

The Holoplankton

The holoplankton include all the rest of the zooplankton and have representative taxa from nearly every animal phylum. They include the single-celled protozoan zooplankton; numerous metazoan taxa of intermediate sizes; and the large jellyfishes, which are the largest zooplankton. We will organize this discussion partly on the sizes of organisms, beginning with the smallest forms, but also preserving to some extent a taxonomic order.

The microzooplankton

The **microzooplankton** are members of the zooplankton that range in size from 20 to 200 μm. They include the protozoans and many of the early developmental stages of the larger zooplankton, as well as many of the meroplanktonic larval forms just discussed.

The **protozoan zooplankton**, or just protozooplankton, are single-celled heterotrophic protists, and in some classification schemes they include the heterotrophic dinoflagellates we discussed earlier. They include the **radiolaria**,[2] a remarkably beautiful group of microscopic protozooplankton with silica skeletons that are in the superclass Actinopoda, and the **acantharians**, a suborder of radiolaria comprising protozoa with a delicate skeletal structure of strontium sulphate. The two groups are also distinguished from one another in having a slightly different organizational structure to that skeleton, with the acantharians possessing less elaborate arrangements of the numerous skeletal rods which appear to emanate from the center of the body. The radiolarians have similar radiating rods and spines (**FIGURE 9.6**) as well as elaborate intersecting needles of silica.

Similar in size to some of the dinoflagellates, with some species significantly larger, the acantharians are quite rare in the plankton as compared with other protozooplankton organisms. They are most abundant in surface waters of the open oceans, with cell densities on the order of tens to hundreds per cubic meter, and they are considered to be virtually absent in coastal waters. The preserved remains of their skeletal structures in marine sediments are used in paleontological studies of past ocean climates.

The radiolarians are far more abundant than the acantharians and can reach densities of more than 10,000 individuals per cubic meter. They are found in all oceans and at both surface and deep depths, but they are most abundant at

---

[2] The Society of Protozologists has suggested that the term *Radiolaria* is no longer justifiable as a taxonomic name, but accepts its informal usage for the class Polycystinea. John J. Lee, Seymour H. Hunter, and Eugene C. Bovee, 1985. *An Illustrated Guide to the Protozoa* (Lawrence, KS: Society of Protozoologists).

(A)

(B)

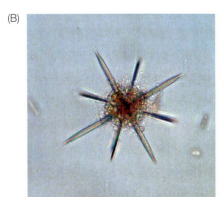

FIGURE 9.6 Examples of radiolarians, single-celled protozoan zooplankton organisms, photographed live at sea immediately after collection. Clearly visible in these images are the siliceous spines and delicate, peripheral layers of cytoplasm surrounding the central capsule of the cell, which contains the various organelles and vacuoles. Notice the differences in the thickness, length, and number of spines; (A) has many thin spines of varying lengths, and (B) has thicker spines of more uniform lengths.

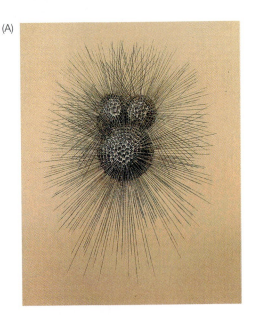

FIGURE 9.7 (A) Painting of a foraminifera cell, genus *Globigerina*, from the HMS *Challenger Reports*. (B) A photomicrograph of a live *Globigerina* cell soon after collection at sea.

FIGURE 9.8 (A,B) Photomicrographs of two species of tintinnid ciliates, each encased in a vaselike structure called a lorica. (C,D) Two species of *Tintinnopsis*, tintinnids with mineral particles coating the lorica. All photographed live at sea immediately after collection.

depths less than 1000 meters. Like the acantharians, these animals are virtually incapable of any swimming motions and can only influence their vertical distributions by changes in buoyancy achieved by altering their biochemical composition. In some parts of the open oceans, they are sufficiently abundant to be important components in the planktonic food web, and we have already learned in Chapter 3 that their siliceous skeletons are what make up the radiolarian oozes of deep-sea sediments.

Another important group of protozooplankton are the **foraminifera** (**FIGURE 9.7**). Originally thought by early naturalists to be gastropod molluscs, these protists are encased in shells of calcium carbonate and are responsible for vast deposits of calcareous marine sediments in the deep sea—the calcareous, or foraminiferan oozes. They feed by way of a net of pseudopodia that entrap phytoplankton cells, in a manner not unlike that of the heterotrophic dinoflagellates we discussed in the last chapter. The planktonic foraminifera (there are also benthic forms, of which there are far more species) are mostly surface-dwelling organisms that are most common in the open oceans. Their abundances vary greatly throughout their distribution; average densities are in the tens to hundreds per cubic meter, but some studies have reported far greater densities, on the order of 100,000 individuals per cubic meter.

The **ciliates** may be the most abundant group of protozoan zooplankton. One subgroup, the **tintinnids**, live inside a vaselike shell, a **lorica**, that they make of a protein material. The lorica in some species is bare and transparent, such as those in **FIGURE 9.8A,B**, while in others it is ornamented with a coating of sediment particles, giving them an opaque, crustose appearance, as in the genus Tintinnopsis (**FIGURE 9.8C,D**). Tintinnids use their ring of cilia around the open end of the lorica both for feeding and for swimming. The cilia create currents that bring food items into the lorica and to the cell for ingestion; notice that the cell in Figure 9.8A clearly has a

(A)

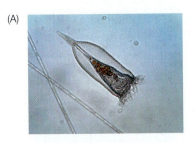

(B)

(C)

(D)

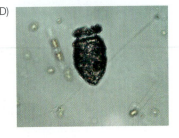

number of ingested phytoplankton cells visible inside it. When the cilia are more active, they propel the tintinnid through the water. Actively beating cilia are evident in the tintinnid in Figure 9.8D (the other has withdrawn into its lorica). In addition to the tintinnids, there are a wide variety of **naked ciliates** (**FIGURE 9.9**). These cells are ubiquitous in both marine and freshwaters.

Ciliates and other protozoan zooplankton feed on both phytoplankton cells and particles as small as bacteria, including the ultraphytoplankton (also called picophytoplankton); along with the heterotrophic nanoplankton, they are important links in what is generally known as the **microbial loop** (**BOX 9A**).

The microzooplankton not only comprise single-celled protists, they also include the smaller (<200 μm) meroplanktonic larval forms that we just discussed, as well as various multicellular, or **metazoan** animals, such as the **rotifers** (phylum Aschelminthes, class Rotifera). Rotifers range in size from about 100 μm to a half millimeter, so not all are considered microzooplankton. More important in freshwater, there are several marine species, with more than 20 in

FIGURE 9.9 The cilia on this naked ciliate can be seen surrounding the cell. They beat in a wavelike pattern, giving the cell a propulsion system for swimming.

BOX 9A The Microbial Loop

Scientists have realized for decades that much of the primary production by phytoplankton in the oceans does not necessarily become packaged into particulate organic matter, such as in the form of phytoplankton cells, which facilitates grazing by, and convenient transfer to, higher-trophic-level organisms. Instead, much of that production is in the form of *dissolved organic matter*, or DOM, which is leaked into the water column. Estimates of the magnitude of this DOM production as a percentage of particulate organic primary production vary, but most scientists agree that the percentage is significant and on the order of half of the total primary production by phytoplankton.

Once thought to be "lost," we now know this DOM as the primary food source of heterotrophic bacteria. Along with the discovery of the smallest forms of photosynthetic organisms based on fluorescence microscope techniques, we also began to realize that the numbers of free-living bacteria in the oceans—those bacteria cells not attached to a substrate—are far greater than had been previously thought, based on older culture plate techniques. Densities of 100,000–1,000,000 heterotrophic bacteria cells per milliliter of seawater are the norm, being highest in the productive coastal waters where DOM production is highest.

But, after consuming this "lost" dissolved organic matter, what becomes of those bacteria? After all, scientists

knew well that those bacteria, on the order of 0.5 μm in size, were too small to be efficiently "grazed" by the metazoan zooplankton. Then a series of studies in the late 1970s and early 1980s revealed that the same microheterotrophic flagellates we discussed in the last chapter, the nanoplanktonic heterotrophic cells in the 2–10 μm size range, graze these bacteria.

This effectively repackages the DOM into bacterial biomass and from there into the nanoflagellates. Because these nanoflagellates are similar in size to the phytoplankton that zooplankton traditionally graze on, they constitute a side loop of sorts in the conventional food chain. The DOM isn't completely lost, after all.

The basics of this microbial loop, as it was called when first described in 1983,[a] are diagrammed here. Later modifications were added,[b] such as inputs from the photosynthetic prokaryotes and consumption by mucus-net feeders, but the basic principle remains: small cells or dissolved materials from primary production are not necessarily lost, not completely, anyway (there are losses at each trophic step, as we discussed in Chapter 7); they are repackaged into larger particles for easier consumption by larger zooplankton.

[a]Azam, F. et al., 1983. The ecological role of water-column microbes in the sea. Marine Ecology—Progress Series 10: 257–263.
[b]Pomeroy, L. R. et al., 2007. The microbial loop. Oceanography 20: 28–33.

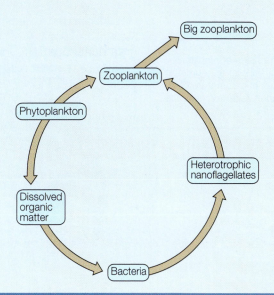

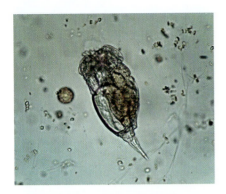

FIGURE 9.10 At first glance, this rotifer, genus *Synchaeta*, resembles a tintinnid, but it is actually a multicellular invertebrate. This individual has closed up its corona, the ciliated structure surrounding its mouth opening, prior to having its portrait taken.

the genus *Synchaeta*, shown in **FIGURE 9.10**. They belong to a group of animals called the **protostomes** ("mouth-first"), and are closely related to flatworms and ribbon worms. These animals have a digestive tract, a primitive brain and nervous system, and they feed by engulfing whole phytoplankton cells and zooplankton of the same size. They are most common in coastal waters.

The metazoan microzooplankton also include the larval stages of the **copepods**, the smaller crustacean zooplankton that we discuss in some detail below. Although most adult copepods are larger than 200 μm, some fall within the 20–200 μm (microzooplankton) size range, and most species are microzooplankton in their early developmental stages. But the majority of copepods have later developmental stages, including the adult stage, that fall within the next larger size class, the mesozooplankton.

The **mesozooplankton** are the next size group, which includes the planktonic animals in the size range 200 μm–2 cm. But at this point, the use of a strict size classification scheme for zooplankton becomes unwieldy, simply because the mesozooplankton organisms often overlap in size with the next larger group, the **macrozoooplankton**, which range from 2–20 cm. Organisms larger than 20 cm, such as some of the jellyfishes, are classed as **megazooplankton**. To preserve some order to the following discussion we are going to lump together the meso-, macro- and mega- size classes, still structuring our discussion as best we can on the basis of organism size, but divided out by taxonomic category. With few exceptions, such as colonial forms of some of the protozooplankton, the majority of zooplankton organisms larger than 200 μm are metazoans. And most of them are crustaceans.

The crustacean zooplankton

Among the various classes of zooplankton in the subphylum Crustacea (in the phylum Arthropoda), several are either relatively rare in the plankton (such as the cumaceans, ostracods, isopods, and most decapods) or are primarily found in freshwaters (such as the cladocerans) and will not be discussed here. Of the remaining groups, some blur the line separating marine organisms that live a benthic life style on the bottom from those that live a planktonic life. Some species, normally considered exclusively benthic, will at times, usually at night, leave the bottom and "swim" about in the plankton. Organisms with this behavior have been given the name **tychoplankton**; they include the cumaceans (along with some annelid worms and various other taxa). In this section we will discuss only the zooplanktonic crustacea that are the most common, which include members of the orders **Copepoda** (the copepods), **Amphipoda** (the amphipods), **Mysidacea** (the mysids) and **Euphausiacea** (the euphausiids).

THE COPEPODS   This group of crustacean zooplankton have a distinctive appearance (**FIGURE 9.11**) and hold the added distinction of being one of the

FIGURE 9.11 Examples of planktonic copepods, photographed live soon after collection. (A) Adult and immature of *Oithona*. (B) An adult *Acartia longiremis*.

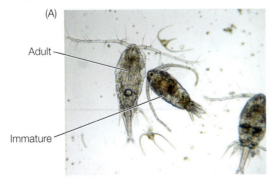

(A)

Adult

Immature

(B)

most ecologically important animals in the sea. They are, by some estimates,[3] the most abundant metazoan on the planet. Entomologists might quibble with this, but they would lose their argument if the most recent estimates are included.[4] One thing is sure: the copepods are certainly the most abundant metazoan *in the oceans* and so they are indeed an extremely important member of the permanent plankton. An ordinary sample collected with a plankton net, from almost anywhere in the world ocean, will include copepods, the exact number and biomass of which will depend on how productive those waters are. **FIGURE 9.12** shows a subsample of a single tow of a plankton net; the catch is overwhelmingly dominated by copepods of a single species, *Calanus finmarchicus*.

There are three general groups, or orders, that comprise the major planktonic copepods (not all are planktonic; some copepods are benthic, and others are parasitic on other animals, especially fish): the **calanoids**, the **harpacticoids**, and the **cyclopoids** (**FIGURE 9.13**). The cyclopoids and harpacticoids have relatively few planktonic species as compared with

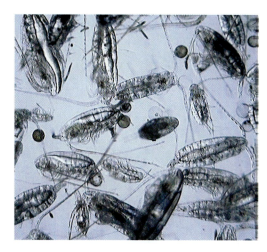

FIGURE 9.12 A low-magnification photomicrograph of an unsorted subsample of the larger plankton collected by a plankton net towed in shelf waters of the Northwest Atlantic in early spring. It contains almost exclusively copepods (stage CV *Calanus finmarchicus*) and large diatoms (the circular objects; genus *Coscinodiscus*).

---

[3] John E. G. Raymont, 1983. *Plankton and Productivity in the Oceans*, vol. 2, *Zooplankton* (Oxford: Pergamon Press).

[4] Horst Schminkee calculates that the copepods outnumber the insects on Earth by a factor of about 1000, and furthermore, that the biomass of copepods is more than 100 times that of all the human beings on Earth. Schminkee, H. K., 2007. Entomology for the copepodologist. *Journal of Plankton Research* 29(Suppl. 1): 149–162.

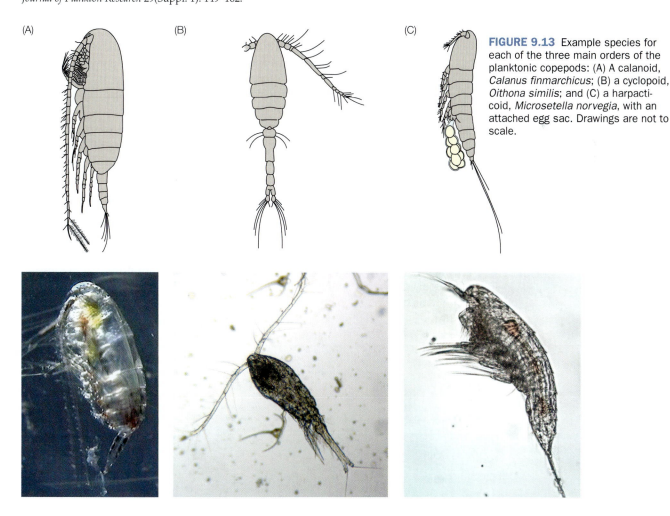

(A)   (B)   (C)

FIGURE 9.13 Example species for each of the three main orders of the planktonic copepods: (A) A calanoid, *Calanus finmarchicus*; (B) a cyclopoid, *Oithona similis*; and (C) a harpacticoid, *Microsetella norvegia*, with an attached egg sac. Drawings are not to scale.

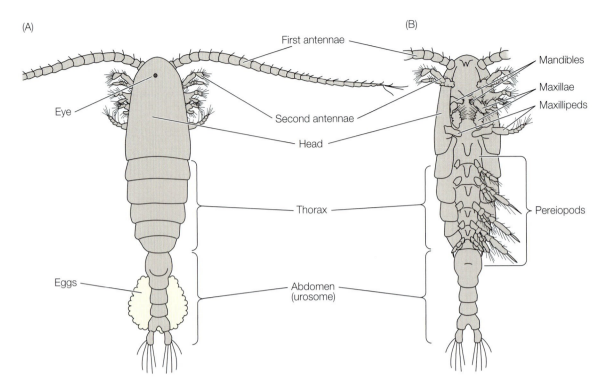

(A)

(B)

First antennae

Eye

Second antennae

Head

Mandibles

Maxillae

Maxillipeds

Thorax

Pereiopods

Eggs

Abdomen
(urosome)

FIGURE 9.14 The general anatomy of an adult calanoid copepod, with a dorsal (top) view (A) and a ventral (bottom) view (B). Each possesses two appendages called antennae, a set of mandibles, and two sets of feeding appendages called maxillipeds all on the anterior segment, or head. The thorax has four or five sets of swimming appendages (pereiopods) and is followed by the abdominal segments, which collectively are called the urosome. Some species will carry eggs in an egg sac, as depicted in (A).

the calanoids, but in total, more than 14,000 species of copepods have been described. Differences among the three orders are based on distinctive anatomical details, such as between which segments the articulation occurs separating the thorax, or the anterior portion of the animal, and the urosome, or abdomen, at the distal end. In the harpacticoids, the articulation is less obvious, and the body has an overall streamlined appearance.

It is somewhat surprising that the copepods, with so many species, all have the same basic architecture, or overall general body shape; there just isn't a lot of variation among species in the three major groups of planktonic copepods. For the most part, they are variations on the general morphological theme given in **FIGURE 9.14**—bullet-shaped organisms with antennae, spinous appendages, and a trailing tail feature. It would appear, given the astonishing success of the copepods, that during their evolution they have happened onto a design that works.

The copepods truly are the "grazers of the sea," as they have been called. Their sheer numbers combined with their relatively large size (0.3–3.0 mm in length, on average), which is greater on average than that of the phytoplankton on which they feed, make them an important link—perhaps the most important link—in the pelagic food web. Their average densities in the plankton range from less than one to tens of individuals per liter of water.[5]

---

[5] Be sure to keep these units clear: When we discussed the phytoplankton and bacteria, we gave cell densities in terms of numbers of cells per milliliter or liter. So, 10 individuals per ml would be 10,000 per liter, since there are 1000 ml in a liter. When we discussed the cell densities of the acantharians, radiolarians, and foraminifera, we used units of cells per cubic meter. A cubic meter is equal to 1000 liters, and 1 million milliliters. So, when we spoke of about 1 million bacteria cells per milliliter, that equates to 1 billion per liter and 1 trillion per cubic meter.

(A)

(B)

(C)

(D)

FIGURE 9.15 Several species of copepod nauplii photographed live immediately after collection at sea. (A) *Calanus finmarchicus*; (B) *Oithona* (a cyclopoid); (C) *Temora longicornis*; (D) *Microsetella norvegica* (a harpacticopid). Each species is distinguished on the basis of size, general body form, and the number and shapes of posterior spines.

The calanoid, cyclopoid, and harpacticoid copepods all reproduce sexually, producing eggs that each hatch into a nauplius (**FIGURE 9.15**), which initially is basically a head with three appendages. Males seek out females by following **pheromone** trails in the water column, chemical cues released by the female that alert males to her presence. The male physically attaches a **spermatophore** onto the female, a sac containing sperm cells, which can be held for later use, fertilizing eggs at the convenience of the female. In some species (such as *Calanus finmarchicus*), the female will release fertilized eggs directly into the water column where they develop while drifting in the plankton, with hatching occurring some time later; in other species, the female holds the eggs in an egg sac (see Figures 9.13C and 9.14A) where the eggs develop and later hatch into free-living nauplii.

The newly hatched nauplius usually passes through six stages, commonly given the abbreviations NI to NVI. The stages are separated by a molt, in which the larva casts off its old shell of chitin-like material for a new one which has been developing inside the old one. In each successive stage, the nauplius adds body segments and becomes somewhat more anatomically complex. The nauplius stage NVI molts into the first of five **copepodid** stages,[6] similarly abbreviated CI–CVI, in which the larvae begin to resemble the adult form; the adult is considerd the sixth copepodid stage. Unlike the nauplii, the copepodites bear some resemblance to the adult form, but they are easily distinguished from the adults in having an unfinished look about them (they still lack certain anatomical features; see Figure 9.11); they develop additional urosome segments and appendage complexity as they molt into later copepodid stages (**FIGURE 9.16**)

Not all species have six naupliar stages or five-plus-one copepodid stages. And in some species, the first three naupliar stages have no functional digestive tract, so they pass through those stages without feeding, relying solely on energy stores in the eggs that were supplied by the female parent. But the later naupliar and copepodid stages actively feed on phytoplankton as they grow and develop, quickly reaching the adult stage when the reproductive cycle can begin again, thus completing a generation. For some species this marks

---

[6] The noun is "copepodite"; the adjective is "copepodid," but some authors make no distinction.

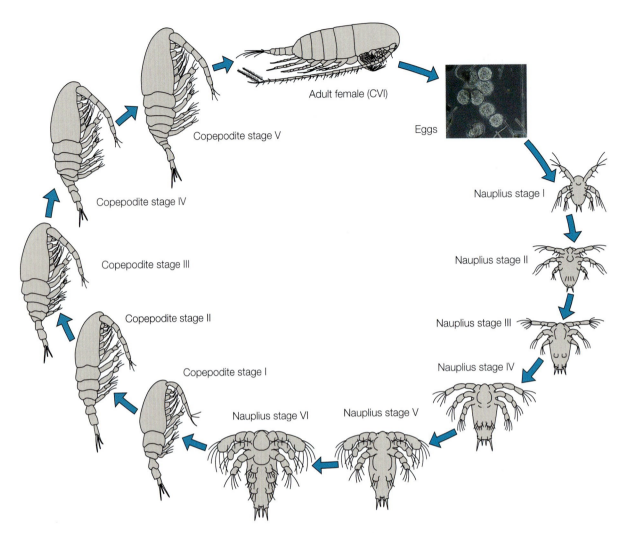

FIGURE 9.16 Development of a generic calanoid copepod, through six naupliar stages and five copepodite stages. The male (not shown) and female copepodite stage V molt into the adult, stage VI. The proportional size of each stage is not conserved.

Labels within figure: Adult female (CVI) • Copepodite stage V • Eggs • Copepodite stage IV • Nauplius stage I • Copepodite stage III • Nauplius stage II • Copepodite stage II • Nauplius stage III • Copepodite stage I • Nauplius stage IV • Nauplius stage VI • Nauplius stage V

the end, and they die, while others may have several generations in a single season, and still others, such as tropical species, may reproduce continuously.

Details of the life histories of copepods vary among species and regions of the world ocean. In some shallow water environments, such as estuaries for example, certain copepod species will undergo a period of **diapause** in the egg stage, when development is arrested. This allows the eggs to survive times of adverse conditions, such as winter periods of cold water temperatures, or exceptionally warm summer periods. Sometime prior to the stressful seasons, the population will produce these resting, or diapause eggs, which are programmed for delayed development; rather than hatching soon after being spawned, they sink to the bottom sediments where they overwinter until conditions become favorable once again. This means that during the dead of winter in temperate and high-latitude environments, there are very few if any copepods present in the water column, a phenomenon that was a bit of a mystery to the scientists studying them until the discovery of this curious life history strategy in the late 1960s and early 1970s.

Still other species of copepods experience delayed development in one of the late copepodid stages, usually the fifth stage. Upon reaching the CV stage, the copepodites feed voraciously, storing copious fat reserves before descending to relatively deep depths of several hundreds of meters where they will reside for the next half year or so. They enter this state of arrested development in spring or summer, after having taken advantage of the spring phytoplankton production cycle to build their lipid reserves. Upon arriving at the right depth, where the waters are very cold (<4°C), they sit almost motionless, not feeding, living off their fat stores. This is what *Calanus finmarchicus* and various other calanoid species do in the North Atlantic and North Pacific Oceans (**FIGURE 9.17**); it is a life history strategy that would seem to make good sense. It allows them to avoid the warmer months of the year, when higher surface water temperatures would needlessly stimulate their metabolism and tax their abilities to find sufficient food at the time of year when phytoplankton production drops to low levels. The following winter-spring period, the CVs will resurface, begin actively feeding once again, and reproduce. Usually, these diapausing copepods will have a single generation of progeny per year, but the same species residing on the continental shelf and in shallower waters may have as many as three generations, entering their overwintering resting stage every two to three generations. On continental shelves they do not overwinter at such deep depths as in the open oceans, of course.

The copepods have acquired the label "grazers of the sea," a title that is well earned. It is they that channel phytoplankton primary production into higher trophic levels with an efficiency unmatched by any other group of marine animals. Their huge numbers and biomass mentioned earlier are the principal reason why this is true. In regions that experience seasonal phytoplankton blooms, such as temperate and high-latitude seas (**FIGURE 9.18**), both on the continental shelf and in the open ocean, the copepods seem to appear almost out of nowhere to feed on the abundant diatoms. In the case of the North Atlantic phytoplankton bloom, *Calanus finmarchicus* leaves its dormant overwintering period spent in the CV stage at great depths and swims to the surface, where it molts into the adult stage, immediately begins to graze the phytoplankton cells, and produces eggs for its annual reproductive cycle there. Their progeny continue to graze the bloom as they grow and develop into CVs themselves, accumulating stores of lipids, and then, after a brief period in the upper ocean, they, too, descend to deep depths prior to the onset of late summer and fall, where they will lie dormant for the next six months or so.

Copepods are typically considered to be **filter feeders**; the expression implies that they feed by a sort of sieving action, sweeping their feeding appendages through the water like miniature lawn rakes, collecting and

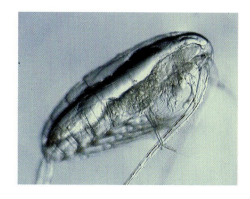

FIGURE 9.17 A stage V copepodite *Calanus finmarchicus* collected in April in the North-west Atlantic Ocean, clearing showing a nearly full oil sac running the length of its body, and beneath that, what appear to be phytoplankton cells in its gut; this animal may be ready to molt into an adult or to enter diapause.

FIGURE 9.18 (A) The seasonal pattern of solar energy (daylight) in megajoules (MJ) received per square meter of Earth's surface per day, ignoring atmospheric effects such as cloud cover for different latitudes, illustrating how the season during which a minimum light level is received varies greatly with latitude. For example, if the minimum light level of interest is 30 MJ/m²/day (dashed line), notice that it occurs for about 6 months at 40° latitude, but for only 4 months at 60° latitude. (B) Monthly composite satellite images for winter and spring of 2011 showing relative concentrations of phytoplankton chlorophyll, and hence, phytoplankton biomass, with warmer colors indicating higher concentrations.

(A)

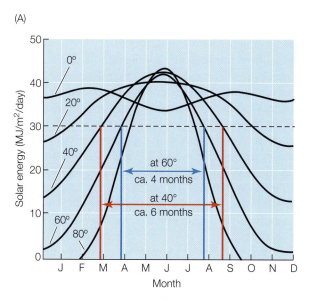

(B)

Winter

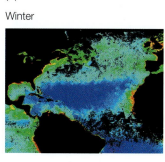

Spring

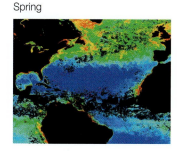

concentrating any phytoplankton cells that become caught between the fine setae, which are then guided to the mandibles and eaten. But this view is now seen as overly simplistic, in that the water is too viscous for the animals to actually sieve cells from it. The situation is a lot like trying to catch an air bubble in thick syrup by swiping at it with a fork; the bubble seems to avoid the fork and is simply pushed out ahead as the fork plows the syrup (and bubble). The syrup is just too thick to slip past the fork's tines. At the much smaller size scales of the phytoplankton and copepod feeding appendages, water is quite viscous, like syrup at our scale, and the copepods are actually unable to sieve things out. So how do they do it? An innovative study of the feeding behavior of a copepod glued to a dog hair and monitored with movie cameras was done by a group of researchers in the late 1970s and early 1980s (this isn't a joke; this was actually done[7]), which demonstrated this fork-in-honey phenomenon. The film showed that instead of filtering the water, copepods guide particles to their mouth parts using their feeding appendages like canoe paddles; it is actually more of a **particle selection** process than a sieving process. The animals are able to locate phytoplankton cells using chemical cues (remember: phytoplankton cells "leak" dissolved organic matter, DOM) and can alter their feeding behavior when a cell is brought in close; when they sense a phytoplankton cell, they switch to a particle manipulation mode and bring the cell to the mandibles. Moreover, movies of glued copepods have shown that not only do copepods select particles by handling them, rather than filtering them, they also can reject individual cells they don't care to eat. One study showed that when copepods were presented phytoplankton cells that were not in an active growing phase (their cell division rates were slowed down) those cells would be rejected in favor of cells with higher growth rates.

Much of this is splitting hairs, in that regardless of the actual mechanism of particle selection, capture, and ingestion by copepods, the end result is the same: the waters are cleared of phytoplankton to some extent by the zooplankton grazers. But there may be even more to the story. A simple set of laboratory observations made in the early 1990s, which turned out to be a major discovery, has caused us all to rethink the role of copepods in the marine environment, or at the very least, it stimulated a renewed interest in the field. Quite simply, that discovery has challenged what we thought we knew regarding the role of spring diatom blooms and copepod feeding (**BOX 9B**).[8] Laboratory studies were showing that diatoms might be toxic to copepods. Almost immediately upon publication of those early experimental results, a raging controversy sprang up among scientists specializing in studies of copepod biology (copepodologists) about just how widespread the phenomenon is, or whether it was a real phenomenon at all; as of this writing the jury is still out.

Not all copepods are strictly heterotrophs; some species are carnivores, feeding on other heterotrophs such as protists and even small metazoans such as developmental stages of other copepods. Many species of copepods, including *Calanus finmarchicus*, are omnivores that often, when selecting a food particle, make little distinction between photosynthetic and heterotrophic protists, for example.

The fact that many calanoid copepod species have resting-stage CV copepodites that spend extended periods in a state of suspended animation at relatively deep depths in the ocean implies that copepods must have a swimming capability—even though they are members of the plankton. Passively sinking to great depths is one thing, but returning to the surface again is another. Indeed, copepods can and do swim, and they can cover significant

---

[7] Koehl, M., and Strickler, R., 1981. Copepod feeding currents: food capture at low Reynolds number. *Limnology and Oceanography* 26: 1062–1073.

[8] Ianora, A., and Miralto, A., 2010. Toxigenic effects of diatoms on grazers, phytoplankton and other microbes: a review. *Ecotoxicology* 19: 493–511.

BOX 9B Diatoms, Copepods, and Teratogenesis

Oceanographers and marine biologists have for a century or more operated under the assumption that copepods eat diatoms. The scientific literature is filled with laboratory studies of copepod feeding behavior that have focused on, for example, how copepod growth rates are affected by food availability—specifically diatom cell densities—and how the production rates of eggs are influenced by diatom cell densities. In virtually all those studies the results were the same: faster growth rates and greater egg production rates at higher diatom cell densities. Then, in the early 1990s, someone took a look at the next generation—that is, they asked a simple question: Just how well do the nauplii hatching from the eggs in those experiments do? The result was a surprise that rocked the small world of copepodologists. Nauplii hatching from eggs from females that fed exclusively on diatoms over expended periods were malformed and experienced mortali-

ties approaching 100%. The discovery of teratogenesis, congenital malformations produced in an embryo or at birth, which occurs in other animals, in copepods sparked a flurry of studies aimed at determining how widespread the phenomenon is and what the ecological significance might be.

Numerous follow-up laboratory experiments with different species of diatoms and copepods produced similar results, but field surveys conducted during diatom blooms in various locales around the world ocean produced mixed results, with most studies not showing any measureable effects of diatom ingestion on naupliar copepod survival.

Later studies have revealed that various organic chemicals produced by the diatoms, including aldehydes and other products of fatty acid degradation, were most likely the agents responsible for teratogenesis in the earlier laboratory experiments. It has been speculated that those same chemicals

have evolved to act both as antibacterial agents and as allelopathic chemicals in competitive interactions among phytoplankton species. One group of researchers has suggested that, since copepods are sophisticated feeders and not the haphazard filter feeders they were once thought to be, it is possible that they can detect which diatoms release the toxic chemicals and which do not, thus avoiding the problem altogether. But in the laboratory setting, the copepod's choices are limited.

This general topic remains a fruitful avenue for research, and indeed there is much work currently ongoing. But whatever the outcome of those researches, we can assume that it *will* make sense in light of ecological interactions among the multitude of taxa involved in the production and cycling of organic matter in the plankton. It *will* make sense because it *has to* make sense, in light of the phenomenal success of the copepods!

vertical distances. Swimming for a period of days over hundreds of meters in the vertical can make a big difference to a copepod; it can bring it from cold and dark waters devoid of food to warmer, well-lit surface waters with a blooming phytoplankton population. But swimming the same distance horizontally is a useless effort; the scales of oceanographic phenomena in the horizontal are measured in kilometers, not meters. These vertical excursions by copepods, usually performed seasonally and in relation to their developmental stage, are called **ontogenetic vertical migrations**.

In addition to the seasonal, life-history-related vertical migrations, some species of copepods also migrate each day. These excursions, called **diel vertical migrations**, are usually downward during the daytime, in order to hide in the dark and avoid visual predators, and upward during the nighttime, for feeding in the surface waters. These migrations are usually on the order of several tens of meters each way, to more than 100 meters for some of the larger copepods, such as *Calanus pacificus*, for which transits of greater than 150 meters have been documented.

Copepods swim by making stroking motions with their appendages. They can maintain their position in the water column against the tendency to sink (like the phytoplankton, they are denser than water) by stroking their pereiopods. Stroking motions can also carry them up or down in their vertical migrations, and propel them through a patch of phytoplankton cells, thus enabling them to search for food. They can also move much more quickly, using their first antennae, but only for short bursts. This swimming technique is primarily employed to escape predators.

We have already stressed the importance of copepods in the marine food web, and the brief discussion that follows is intended only to drive that point

FIGURE 9.19 A common hyperiid amphi-pod, genus *Themisto*, shown here under low magnification. Perhaps the most prominent of its anatomical features are its well-developed eyes. This individual is approximately 3 mm in length.

home. The numerical dominance of copepods over all other metazoan organisms in the sea, and their relatively large size, combine to make them the preferred food item of a number of consumers. Many planktivorous fishes, such as herring, prefer copepods, especially the lipid-rich stages of *Calanus finmarchicus*, as does the northern right whale (we will be discussing fishes in Chapter 11 and marine mammals in Chapter 13). Various marine animals consume copepods, not only the adult stages but their developmental stages as well; many species of fishes have larvae that feed on copepod nauplii almost exclusively, and thus the adult fish time their reproductive cycle to coincide with the season of maximal densities of nauplii in the water column.

Finally, copepods are important in a way that might not seem readily apparent, simply because many species are inefficient feeders, especially when food is abundant. That is, they often ingest far more phytoplankton biomass than they can digest and assimilate, which means that they pass as fecal pellets a significant fraction of the phytoplankton biomass they graze. And because those fecal pellets tend to sink rapidly, they constitute an efficient mechanism for delivering phytoplankton-produced organic carbon to the bottom, where it fuels the benthic food web.

THE AMPHIPODS Although some species of amphipods are terrestrial and some are found in freshwater, most are marine benthic organisms. One group, the **hyperiid amphipods**, are exclusively planktonic. Somewhat larger than the average copepod, they are known to form swarms in surface waters at certain times and locations, and can be locally important in the planktonic food web. While they can be found in the warmer-water regions of the world ocean, they are quite rare in the tropics and subtropics, especially when compared with the copepods, for instance. A few species inhabit higher latitudes—the Arctic and Subarctic waters, as well as some temperate-latitude seas, especially on the continental shelves. One of the most abundant of these hyperiid amphipods is the genus *Themisto* (**FIGURE 9.19**). Recent work in the North Atlantic and Arctic Ocean by researchers from the Alfred Wegener Institute has shown that these amphipods are extremely abundant; their total biomass is estimated at more than 100 million metric tons, on the same order of magnitude as krill (euphausiids; discussed below) in the Antarctic. They are predators on copepods and are in turn believed to be an important food item for fish, sea birds, and whales.

Reproduction in the amphipods differs from that in the copepods in that they have direct development; the young are carried by the female in a marsupium and released into the plankton looking a lot like the adult, except smaller.

THE MYSIDS **Mysids** and euphausiids both have a shrimp-like appearance, and at one time they were classified together; however, they are now recognized to be distinct taxonomic orders. The mysids are more primitive than the eupahisiids. They have a worldwide distribution, with more than 1000 species in 160 genera. Some mysids are benthic and some tychoplanktonic, but others are members of the holoplankton. Most are found in estuaries and in coastal and shelf waters, where they tend to hover just off the bottom in relatively shallow depths (**FIGURE 9.20**). As omnivores, they are important in the estuarine food web, ingesting not only autotrophs and heterotrophs, but also significant quantities of detritus.

THE EUPHAUSIIDS The **euphausiids** are the most important group of macrozooplanktonic crustaceans, and, after the copepods, they are among the most important planktonic animals of any size. They are commonly

FIGURE 9.20 A mysid, the bent opossum shrimp, *Praunus flexuosus*. Originally native to European shelf waters, it is an invasive species in the Northwest Atlantic, where this specimen was collected.

FIGURE 9.21 Two northern humpback whales, *Megaptera noviangliae*, feeding on euphausiids in the Gulf of Maine. The euphausiids, *Nematoscelis megalops*, are just visible here as the red tint in the photo. *N. megalops* (inset) is a common euphausiid found throughout much of the North Atlantic Ocean as well in the Mediterranean and parts of the South Atlantic off Africa and off Antarctica. This specimen was collected on Georges Bank in the northwest Atlantic. The exposed gills are visible, as is the motion of the swimming appendages, the pleopods (blurred area beneath the body).

observed near the surface in swarms of several tens of thousands of individuals per cubic meter; some estimates range as high as 600,000 per cubic meter.[3] Because they are relatively large—from about 8 mm to more than 3 cm in length—the biomass of such swarms is very significant. As mentioned above, the total euphausiid biomass in the Southern Ocean around Antarctica is estimated to be on the order of 100 million metric tons. To place this number in perspective, the total biomass of fishes in world ocean is estimated to be on the order of 250 million metric tons.

The shrimp-like euphausiids are differentiated from the mysids by their exposed gills, which are attached to their thoracic limbs, as well as by other anatomical details. They have five abdominal swimming appendages, the pleopods, the first two of which help to generate a flow of water that ventilates the gills. The beating of the pleopods can been seen as blurs in the specimen in **FIGURE 9.21**.

Most euphausiids are oceanic, but a few species are commonly found in shelf seas and along the continental slope, where they are important food items for whales, sea birds, and some fish, such as herring. They are very abundant in open-ocean areas in high-latitude seas, where they support entire populations of whales, especially in the North Pacific, North Atlantic, and the Southern Ocean around Antarctica. Their numbers and biomass, especially in the Southern Ocean, are so significant that they have attracted commercial interest as a potential source of seafood for human consumption. But taste is important, of course, and apparently these things just don't taste very good; commercial ventures have not yet taken off.

Planktonic molluscs

In addition to the meroplanktonic trochophore and veliger larvae of molluscs, there are three orders of gastropod molluscs that are holoplanktonic, the **Heteropoda**, the **Thecosomata**, and the **Gymnosomata**. The latter two orders are the shelled and naked pteropods, respectively. The first, the Heteropoda, once thought to be members of one of the pteropod orders, is a small group, with only about 8 genera and 100 species. They are most common in tropical waters, but relatively little is known about their biology and distributions in the world ocean and few studies have been published on them. While some species are known to swarm, their contribution to the zooplankton biomass in the world ocean is thought to be minor.

(A)

(B)

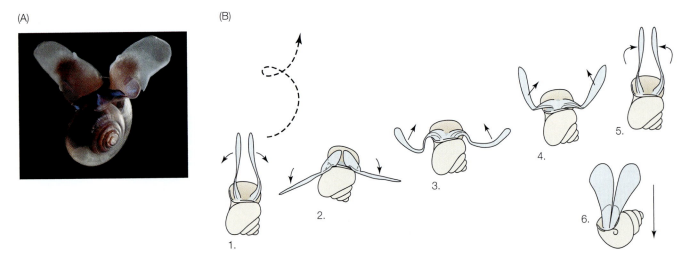

FIGURE 9.22 (A) The shelled pteropod, *Limacina retroversa*. (B) The swimming behavior of *L. retroversa*. (1–5) Wing motions while swimming in an upward spiral; (6) wing positions when passively sinking.

The Thecosomata are the **shelled pteropods**, most of which superficially resemble a common snail with its spiral shell of calcium carbonate. These pteropods are much more abundant than the Heteropoda and are found in the upper water column of all oceans. They are important in forming "pteropod oozes" of calcium carbonate sediments in the deep sea. Some species are known to migrate vertically between the surface and depths of 300 meters. They are able to swim thanks to wing-like modifications to their molluscan foot which they flap—like wings—to propel themselves vertically through the water (**FIGURE 9.22**), but, like the copepods, they are incapable of swimming significant horizontal distances.

One of the smaller species in this group, *Limacina retroversa* (see Figure 9.22), is among the most common; it is found throughout much of the North Atlantic Ocean, as well as in portions of the South Atlantic and Southern Oceans. They are reported to be herbivorous, ingesting primarily diatoms and dinoflagellates. Ranging in size from about 0.3–3 mm, they can at times reach relatively high densities and biomass, and be important components of the planktonic food web: both herring and mackerel are known to feed on them.

The third group, the Gymnosomata, are the **naked pteropods**. Like the Thecosomata, these pteropods also swim by flapping their "wings" (projections of their modified molluscan foot). One of the larger species, *Clione limacina*, is shown swimming in **FIGURE 9.23**. They are distributed in the colder waters of the world ocean, and this particular species, *Clione limacina*, is found in both the North Pacific and North Atlantic Oceans. Not a large taxonomic group (20 or so genera and about 40 species) they are significantly larger than the shelled pteropods, on the order of 1–2 cm in length; they are relatively rare in the plankton, and are not nearly as abundant as their shelled cousins. But their beautiful colors attract the attention of scientists and naturalists who are lucky enough to bring up individuals like the one in Figure 9.23 in their nets. These pteropods are mostly carnivorous, and *Clione limacina*, for example, feeds almost exclusively on its cousin, *Limacina retroversa*.

The next three phyla that we will discuss—the Cnidaria, the Ctenophora, and the Tunicata—are often referred to collectively as the **gelatinous zooplankton** (which may include the naked pteropods as well). They are a diverse, magnificently beautiful, ecologically important, and sometimes dangerous group, as you will see.

FIGURE 9.23 A naked pteropod, *Clione limacina*, caught in the act of swimming gracefully across the field of view here as it was being photographed at low magnification. Below the "wing" of this pteropod, a *Calanus finmarchicus* copepod just 3 mm long is visible.

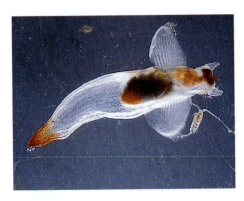

FIGURE 9.24 (A) Sketch of a hydroid with polyps and a new hydrome-dusa in the process of budding off, and a free-living hydromedusa. (B) A colonial hydroid, genus *Ectopleura*, from coastal Maine; the colony is made up of individual stalks that hold a polyp with its food-gathering ten-tacles clearly visible. (C) The same colony. Each polyp is approximately 2 mm across; each stalk is about 3–4 cm in length.

THE CNIDARIA Also known by its former name, the Coelenterata, this phylum includes the hydromedusae and siphonophores, both of which are in the class Hydrozoa, and the more readily recognizable "true jellyfishes," which are in the class Scypho-zoa. The cnidarians also includes the notorious cubomedusae, in the class Cubozoa, certain species of which are among the most venomous animals on Earth.

The **hydromedusae**, in the class Hydrozoa, also called hydrozoan medusae, are actually members of the meroplankton in which the planktonic phase is (usually) a small jellyfish, a hydromedusa (plural *hydromedusae*). Most are quite small, just a few millimeters across, and seldom more than a few centimeters. Their life cycle alternates between a fixed (benthic) **hydroid** stage, in which the hydroid is attached to a substrate and catches food using append-ages from its **polyps**, and a hydromedusa stage, the free-living, planktonic stage, which is more rounded and bell-shaped. The benthic hydroids, which feed on plankton and other suspended particles in the water, usually reside in colonies of interconnected tubes, sometimes in a treelike branching arrangement (**FIGURE 9.24**). The polyp, at the end of the hydroid's hollow tube-like stalks, is a bulbous feature with a terminal mouth opening, usually surrounded by tentacles. Hydroids are commonly seen attached to dock pilings and other submerged objects along the coast, and have a moss-like appearance. By way of asexual budding, the polyps produce free-swimming hydromedusae, such as the one in **FIGURE 9.25**. The hydromedusae reside in the plankton for a period of time before eventually producing gametes that, upon fertilization, produce a **planula larva** (a flattened and elongated, free-swimming larva) that drops to the bottom, attaches once again to a benthic substrate, and develops into another benthic hydroid and polyps, thus completing the life cycle.

Because of this dependence on the benthic hydroid stage, hydromedusae are usually more common in coastal and shelf seas, but not always; periodically, they are collected beyond the

(A)

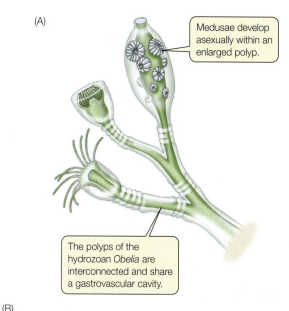

Medusae develop asexually within an enlarged polyp.

The polyps of the hydrozoan *Obelia* are interconnected and share a gastrovascular cavity.

(B)

(C)

FIGURE 9.25 A hydromedusa. The width of the bell is approximately 2 mm.

FIGURE 9.26 A deep-sea siphonophore, *Marrus orthocanna*, photographed *in situ* in the Arctic Ocean. The digestive systems of zooids in siphonophores are sometimes colored bright red and orange as in this species, with other body parts transparent.

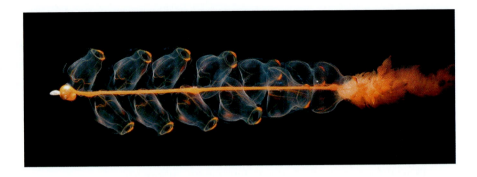

edge of the continental shelf. There are two explanations for their presence in deep waters: the first is that while in the planktonic stage, the hydromedusae of some species are capable of reproducing asexually by way of budding, and the new hydromedusae disperse farther away from shallow waters where hydroids are attached. Another reason is that not all hydroids are benthic. Some (e.g., species of the genus *Clytia*) are apparently themselves planktonic, as has been reported on Georges Bank in the Northwest Atlantic.

Some of the most unusual, and yet better known, examples of the hydrozoa are the **siphonophores**, order Siphonophora. The siphonophores are strange-looking colonies of hydroids, also called *zooids,* arranged in a linear structure. Sometimes zooid colonies form long trails several tens of meters in length, all ordered according to specific functions that the zooids are adapted for, including flotation (the gas bladder in some species), reproduction, feeding, and defense (or predation).

FIGURE 9.27 (A) The Portuguese Man o' War, *Physalia physalis*, is a colony of hydroids. It has a sail, which is a gas-filled bladder that extends above the water line, and a trailing mass of zooids, or tentacles, adapted for feeding and reproduction, with some that have powerful nematocysts that can cause severe injuries. (B) A painting by François Péron from the *Voyage de découvertes aux terres australes* (1807) helps to give a sense of the dimensions of *Physalia* and the depth to which its long trailing tentacles reach; they actually can be many times the length shown here.

Siphonophores are distributed throughout the world ocean, in surface waters and at great depths of several thousand meters. There are believed to be about 150 species, but this number is constantly being revised upward as more direct observational techniques are employed by researchers and, as a consequence, more species are being discovered. The problem in studying them, encountered by earlier scientists and naturalists, is their extremely delicate structure. The long stringlike arrangements of zooids are easily destroyed in plankton nets, leaving only gelatinous gobs of unidentifiable material. It wasn't until they began to be observed and sampled *in situ* by scuba divers and researchers in submersibles that the diversity of siphonophores really began to be revealed. The delicate beauty of these organisms (or colonies of organisms) is exemplified in **FIGURE 9.26**.

One of the most famous (or infamous) as well as one of the largest siphonophores is the Portuguese Man o' War, *Physalia physalis*. This organism can be 30 cm in width (at the surface), while its trail of subsurface tentacles can be more than 30 meters in length (**FIGURE 9.27**). It drifts at the surface under the influence of winds against its sail-like gas bladder that

(A)

(B)

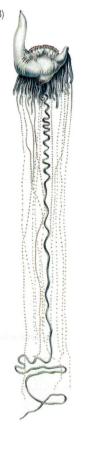

FIGURE 9.28 A *Physalia utriculus* that has been blown onto a beach and become stranded. While dead and dying, the nematocysts can remain viable for days and can still inflict a painful sting if handled. These animals have floats that are 10–15 cm across.

FIGURE 9.29 (A) An *Aurelia aurata*. (B) A small *Cyanea* with a ctenophore entangled in its tentacles.

(A)

(B)

extends above the air water interface. *Physalia* is common in the tropical regions of the Pacific and Indian Oceans, but it is also found in the North Atlantic as far north as the Gulf of Maine, and throughout its range, it is feared because of its toxic stings that can quickly produce a severe reaction in any swimmer so unlucky as to rub up against one.

All of the Cnidaria have **nematocysts**, specialized cells lining much of the integument of the tentacles. The nematocyst functions almost like a miniature spear gun designed to deliver venom with the spear. When innervated by even a slight touch, these cells discharge their spears into the flesh of the offending organism (or person), and in the process they inject their venom. Some species of Cnidaria possess more potent nematocysts and venoms than others, but all have them. In the case of *Physalia*, the venom is particularly powerful. It is not uncommon for 30–50 cells to discharge when a small copepod comes in contact with the tentacles; when a hapless swimmer comes in contact with a colony, hundreds of thousands of these nematocysts will discharge in seconds.[9] They can inflict severe pain and discomfort in the victim, producing welts that last for several days, but fatalities are uncommon. The nematocysts remain active long after the colony seems dead, such as when they are washed up on beaches (**FIGURE 9.28**). Nematocysts serve not only defense purposes for the Cnidaria; they are also useful for disabling prey that get entangled in their tentacles, allowing the meal to be worked up toward the mouth parts without a struggle.

The **Scyphomedusae**, class Scyphozoa, are the true jellyfishes. They, too, undergo an alternating life cycle between a polyp and a medusa, but they are different from the Hydrozoans in that the medusa stage is significantly more developed and much larger than the polyp.

Jellyfish are common in all oceans and can be extremely abundant at certain times, in certain locales. Examples include *Aurelia aurita*, also known as the moon jellyfishes, and *Cyanea capillata*, or the lion's mane jellyfish (**FIGURE 9.29**), both of which are distributed throughout much of the world ocean (*Cyanea* more so in colder waters), and which can seasonally reach very high densities in coastal waters. The Scyphomedusae have no skeletal hard parts, no head or specialized organs systems, and are mostly made of clear or translucent flesh that is more like "Jell-O" than jelly. As cnidarians, they also posses stinging nematocysts, and one should use caution handling them.

---

[9] Halstead (1965), citing an earlier study, points out that loggerhead turtles nonetheless feed on *Physalia*, somehow withstanding the nematocyst assault.

FIGURE 9.30 A mature box jellyfish, *Chironex fleckeri*; this specimen is a particularly large one, measuring 7.5 cm across the bell.

The jellyfish can swim using undulating actions of their bell, an action not unlike a person opening and closing an umbrella. But their progress is slow, and they are always subject to ocean currents, and therefore they are considered part of the plankton. Because they are mostly water, they are not very nutritious as food for other marine organisms; however, sea turtles and some fishes will feed on them.

We cannot leave the Cnidaria without mentioning the **cubomedusae**, class Cubozoa, which include the notorious box jellyfish (*Chironex fleckeri*), also called the marine stinger or sea wasp (**FIGURE 9.30**). These are a small taxonomic group, with only about 36 species having been described. Their name comes from their bell having four flattened sides, giving them a cuboidal shape. The cubomedusae are among the most venomous marine organisms known—stings from these box jellyfish can kill a human being in seconds to minutes. They have well-developed eyes and, it is assumed, fairly acute vision, and they are capable of significant swimming speeds as compared with other medusa.

The distribution of the box jellyfish is confined to the tropical seas of the Indian and Pacific Oceans, and they are infamous in Australia, where most of the serious attacks on humans have occurred. Other species of cubomeducae have been recorded in the North and South Atlantic Oceans, in the Mediterranean Sea, and in California coastal waters.

The ctenophores

The **ctenophores**, phylum Ctenophora, were at one time included in the Cnidaria, but on the basis of their not having stinging cells, in addition to other anatomical differences, they have been separated into their own phylum. Known as the comb jellies, the ctenophores are extremely important members of the zooplankton (**FIGURE 9.31**). Not a large taxonomic group, with about 200

FIGURE 9.31 (A) An *in situ* photograph of an individual ctenophore, genus *Beroe*, swimming with its oral end open. This genus does not possess tentacles. (B) Drawing of the ctenophore genus *Pleurobrachia*, showing its rows of ctenes (cilia bands) and feeding tentacles.

(A)

(B)

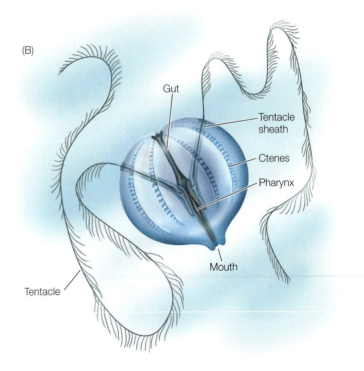

Gut

Tentacle sheath

Ctenes

Pharynx

Mouth

Tentacle

species, they are distributed throughout the world ocean, from the Arctic to the Antarctic and all points in between. These animals are small, as compared with some of the other cnidarians, on the order of 1–3 cm in length. They have a bulbous shape, with eight rows of ciliated bands that are used for propulsion. Some species possess a pair of long branching tentacles (many times their body length) that are retractable. The ctenophores prey on other planktonic animals by trapping them in a sticky mucus on their tentacles (or on their lobes, in species without tentacles), and can impose heavy predation pressure on other zooplankton, especially copepods, when the ctenophores are present in high densities. They have tremendous reproductive potential and can produce about 1000 eggs per individual per day, quickly reaching extremely high densities of between 100 and 400 individuals per cubic meter. Along with their fast growth rates, these animals can comprise a significant fraction of the total zooplankton biomass.

The chaetognaths

The **chaetognaths**, phylum Chaetognatha, are the arrow worms. Their name is well deserved, as can be seen in **FIGURE 9.32**. A small taxonomic group with only six genera and about 120 species, they inhabit all oceans at virtually all depths but are not found in freshwaters. Often they comprise a significant fraction of the zooplankton biomass; at times, their densities rank second only to copepods'. Most chaetognaths are 1–3 cm in length, but one genus is much larger, reaching 6 cm. All genera in the phylum are hermaphroditic, with a single individual possessing both male and female reproductive organs. They are good swimmers and are capable of both slow sustained swimming as well as short high speed bursts, which contributes to their abilities as predators. The chaetognaths are strict carnivores, feeding on almost any form of animal plankton they come across, and they have been implicated as important predators on larval fishes.

FIGURE 9.32 (A) The general anatomy of a chaetognath. (B) Close-up of the head, showing the two eyes and their sickle-shaped feeding hooks.

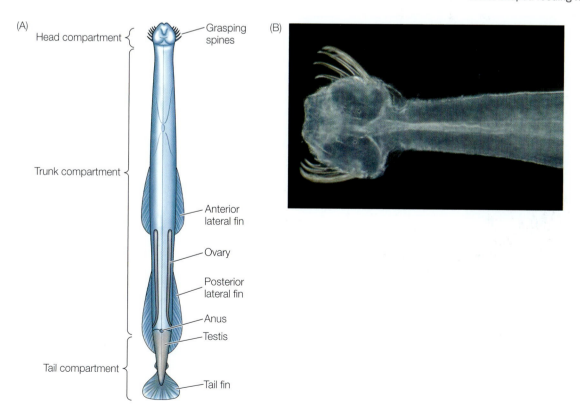

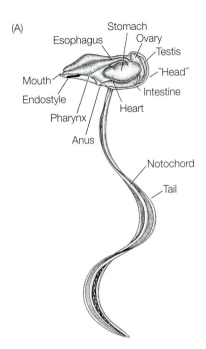

(A)

Stomach
Esophagus
Mouth
Endostyle
Pharynx
Anus

Ovary
Testis
"Head"
Intestine
Heart

Notochord

Tail

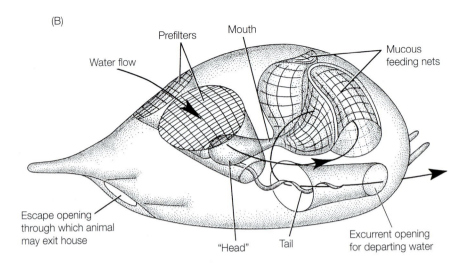

(B)

Prefilters
Water flow

Mouth

Mucous
feeding nets

Escape opening
through which animal
may exit house

"Head"   Tail

Excurrent opening
for departing water

FIGURE 9.33 (A) Drawing of an *Oikopleura* without its house. (B) Diagram of the flow of water through an *Oikopleura* house, created as the animal makes an undulating motion with its tail. Large particles are excluded from incoming water by coarse meshes on the outside surface of the house (prefilters); fine particles as small as one or two micrometers are retained on the mucous feeding nets inside the house.

The appendicularians and salps

The **appendicularians**, also called the **larvaceans**, or simply the pelagic tunicates, are in the phylum Chordata, subphylum Tunicata, class Appendicularia. There are about 70 species, which are found in the surface waters of all oceans. Like the chaetognaths, the appendicularians are hermaphroditic. They are related to the more highly evolved vertebrates in that they possess a notochord, which is a primitive backbone. They are quite small, normally about 2–4 mm in length, but they can be significantly larger in the colder regions at higher latitudes, such as the individual in **FIGURE 9.33A**, which is a specimen of genus *Oikopleura*. These animals are suspension feeders that have evolved the capacity to build a "house" from a mucopolysaccharide material that it excretes; water is pumped through the house, allowing the animal to filter out food particles, as diagrammed in **FIGURE 9.33B**. They frequently abandon their houses for various reasons—when the filter apparatus becomes clogged, for instance, or when they reproduce sexually—and build new ones. Like the delicate siphonophores, which cannot be sampled effectively with nets, the larvaceans are hard to capture entirely; the house is usually destroyed by plankton nets, leaving only the animal itself.

The **salps** are in the order Salpida, which is one of three orders in the class Thaliacea; the other two are the Pyrosomida and the Doliolida. The Pyrosomida are colonial animals made up of zooids, while the Salpida and Doliolida are not. Often these three orders are spoken of as a single taxonomic group. Superficially, they resemble the ctenophores in having a bulbous gelatinous form, except that they are more like a barrel in shape. They too are filter feeders that draw water through their bodies, forced by ciliary actions, filtering out potential food particles. The salps are found in all oceans, usually in surface waters away from the coasts. **FIGURE 9.34** shows an individual *Salpa fusiformis* and a colony of salps. This species occurs in the Atlantic, Pacific, and Indian Oceans and ranges in size from 0.5–4.0 cm in length; it sometimes forms chains of many individuals.

The ichthyoplankton

Ichthyoplankton are the planktonic eggs and larvae of fishes, and are another example of meroplankton. Many species of fishes, both coastal and open ocean species, are pelagic broadcast spawners that release eggs and sperm directly into the water column where the eggs are fertilized. The fertilized eggs are free to drift in the plankton while developing (in some species, however, the

(A)

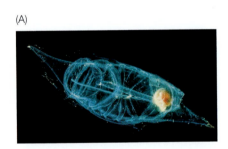

(B)

FIGURE 9.34 (A) An individual *Salpa fusiformis*, photographed *in situ*. (B) A colony of salps.

eggs are negatively buoyant and sink to the bottom where they develop and hatch; the Atlantic herring is one example). Upon hatching, the larvae are often underdeveloped and have retained a significant portion of the egg yolk. Known as yolk-sac larvae, they will continue to develop in the plankton while drawing nourishment from the yolk; often, these yolk-sac larvae have underdeveloped mouth parts and are unable to feed. This stage may last for several days, depending on the water temperature, being longest in colder waters. As the yolk reserves are depleted, the mouth parts develop, as do other anatomical features of the young fish, and once the yolk is depleted, the larvae must begin exogenous feeding on plankton particles.

An example of a post yolk-sac larva is given in **FIGURE 9.35** for the haddock, *Melanogrammus aeglefinus*. This species spawns once each year, usually in late winter and spring, and will produce several million eggs per adult female. But hardly any will survive. The period spent in the plankton by early developmental stages of fishes is a time of heavy mortality, especially if it takes place when planktonic food, typically copepod nauplii, is scarce.

And so ends our brief survey of the planktron, organisms that have held the attention of scientists and naturalists ever since the development of the microscope some 300 years ago. Even today, with technology that allows us to observe them directly in their own realm, we continue to be amazed at the diversity and beauty of some of the largest zooplankton—the gelatinous forms. Discoveries will continue in this fascinating field—but we need to move on to the larger, nonplanktonic marine animals, the benthic and nektonic vertebrates and invertebrates, which, as we will see, rival the plankton in their complexity.

(A)

(B)

(C)

FIGURE 9.35 Eggs (A,B) and a post yolk-sac larva (C) of the haddock, *Melanogrammus aeglefinus*. Image (A) shows a ballpoint pen for scale; the eggs are on the order of 1 mm in diameter. The larva in (C) measures approximately 8 mm.

Chapter Summary

- The **zooplankton** are the *animal* plankton. They are the major **grazers** of the sea, feeding on phytoplankton and other plankton organisms; they may be **herbivorous**, **carnivorous**, or **omnivorous**.

- Zooplankton include the meroplankton and the holoplankton, which are general, not taxonomic, descriptors that refer to whether a portion of an animal's life history is spent in the plankton. **Meroplankton** spend a portion of their life in the plankton, usually during the early developmental stages; examples include barnacles, clams, crabs, and many fishes. **Holoplankton** spend their entire life history in the plankton.

- The zooplankton are also often distinguished on the basis of size. The **microzooplankton** are between 20 and 200 μm; these are mostly the protozoan zooplankton, including the **foraminifera**, **ciliates**, and **radiolarians**, but also some **metazoans**, such as **rotifers**. The **mesozooplanktkon** range from 200 μm to 2 cm; the **copepods** are the most important group in this size fraction. The **macrozooplankton** are between 2 and 20 cm and include the larger crustacean zooplankton such as the shrimps and krills. The **megazooplankton** are greater than 20 cm, and include the jellyfishes.

- The copepods are the most abundant metazoan on the Earth, outnumbering even the terrestrial insects. The copepods are the major grazers. Their life histories include multiple developmental stages, from egg, to **nauplius**, to **copepodite**, to adult. Some species have a **diapause** stage, where they effectively rest and wait out environmentally harsh times. This life history phenomenon is usually performed as resting eggs, but some species enter diapause in a late copepodid stage, spending several months at relatively great depths where they wait out the harsh winter months in the North Atlantic and Pacific Oceans.

- Some copepods are capable of **vertical migrations** of more than 100 meters each way over a 24 hour period, feeding at the surface at night, but descending to avoid visual predators during the daytime. Copepods are not strict **filter feeders**; some are capable of selectively picking out, manipulating, and discriminating among food particles. Given their relatively large size and high abundances, the copepods are not only important grazers, but they are also important food for higher-trophic-level organisms. As such, they are perhaps the most important animals in the sea.

- The larger zooplankton includes several other taxonomic groups: crustaceans, the planktonic molluscs, the gelatinous zooplankton, the ctenophores, the arrow worms, and the appendicularians and salps.

- The non-copepod crustaceans include the planktonic **amphipods**, **mysids**, and **euphausiids**. The amphipods and eupahusiids are both important prey of fish, sea birds, and whales.

- The planktonic molluscs, in the Phylum Mollusca; these include the **naked** and **shelled pteropods**.

- The **gelatinous zooplankton**, which comprise several phyla, include some members of the phylum Cnidaria, which includes the **hydromedusae**, the **siphonophores**, and the **scyphomedusae** (true jellyfishes); the **ctenophores**, or comb jellies, in the phylum Ctenophora; the arrow worms, in the phylum **Chaetognatha**; the **appendicularians** and **salps**, both of which are in the phylum Chordata; and the **ichthyoplankton**, which are the planktonic eggs and larvae of some fishes.

Discussion Questions

1. The zooplankton are the animal plankton, and as such, they are incapable of swimming significant distances. But many forms can and do swim to some degree. Discuss some examples.

2. Discuss the advantages and disadvantages to marine animals that have a meroplanktonic stage early in their development. Consider in your answer aspects of their feeding, dispersion, and risks of predation.

3. What are some of the ecological challenges faced by the protozoan microzooplankton? Discuss some of their food preferences and feeding mechanisms.

4. Why was the development of ideas about a microbial loop so important in marine ecology? Where do the prokaryotic phytoplankton that we discussed in the last chapter fit in that scheme?

5. The are some 14,000 species of copepods, yet superficially they all seem to very much resemble one another in terms of their basic body form. Why do you think they evolved this way?

6. Can you think of any advantages afforded to the copepods by their having so many developmental stages?

7. Explain the ecological advantage to a copepod species that exhibits a diapause stage in the form of resting eggs, and a copepod species with a diapausing copepodite stage.

8. Copepods are known as the grazers of the sea; why? Why is their ability to discriminate among food particles they encounter in the ocean an advantage to them?

9. Explain the life cycle of a hydroid and its dependence on a planktonic existence. Why do you think this life history pattern evolved in this group of animals?

10. What are nematocysts and what is their function? Can you think of any analogous biological adaptations in other marine animals?

11. What are the gelatinous zooplankton? Can you think of any advantages to having a gelatinous body form?

Further Reading

Azam, F. et al., 1983. The ecological role of water-column microbes in the sea. *Marine Ecology Progress Series* 10: 257–263.

Ianora, A., and Miralto, A., 2010. Toxigenic effects of diatoms on grazers, phytoplankton and other microbes: a review. *Ecotoxicology* 19: 493–511.

Koehl, M., and Strickler, R., 1981. Copepod feeding currents: food capture at low Reynolds number. *Limnology and Oceanography* 26: 1062–1073.

Lee, John J., Hunter, Seymour H., and Bovee, Eugene C., 1985. *An Illustrated Guide to the Protozoa.* Lawrence, KS: Society of Protozoologists.

Marshall, S. M., and Orr, A. P., 1955. *The Biology of a Marine Copepod*: Calanus finmarchicus (*Gunnerus*). Edinburgh: Oliver and Boyd.

Pomeroy, L. R., Williams, P. J. leB., Azam, F., and Hobbie, J. E., 2007. The microbial loop. *Oceanography* 20: 28–33.

Raymont, John E. G. 1983. *Plankton and Productivity in the Oceans*, vol. 2, *Zooplankton.* Oxford: Pergamon Press.

Sieburth, John M., 1979. *Sea Microbes.* New York: Oxford University Press.

Contents

The most important, most abundant, and most diverse animals on planet Earth are the invertebrates, the animals without a backbone. They come in all sizes, from microscopic, single-celled heterotrophs to squid that measure several meters in length, and they have evolved various modes of feeding and reproduction that have enabled them to occupy nearly every niche in the oceans, such as this spider crab, which is foraging in a garden-like mat of other invertebrates, soft coral polyps, on a reef in the Fiji Islands.

Marine Invertebrates

We continue our introduction to marine organisms in this chapter with a discussion of the **invertebrates**, animals without a backbone. We covered some of the invertebrates in each of the last two chapters when we discussed the major zooplankton taxa, but we completely skipped over the benthic and neritic forms, which comprise the majority of marine animals. Here we discuss key marine representatives of the most abundant group of animals on the planet.

At the outset of this book, we argued that the oceans are a treasure trove of life, with all types and sizes of plants and animals; and indeed that is the case. One reason for this is the invertebrates, which are far more diverse and abundant in the oceans than they are on land. And most animals are invertebrates; more than 97% of the roughly one million animal species that have been described are invertebrates.[1] Of the 33 or so phyla in the kingdom Animalia, 32 are invertebrates.[2] The remaining phylum, the Chordata, which includes the vertebrates, also includes two important subphyla of invertebrates, the Urochordata and the Cephalochordata. And of the free-living invertebrates (some are parasitic), many have remained exclusively marine throughout their evolutionary history. Quite simply, the invertebrates are very important—both on land and, especially, in the oceans.[3]

A list of the more important marine invertebrate phyla is given in **TABLE 10.1**. Although that list is restricted to phyla with predominantly or exclusively free-living marine taxa, it is still a long list, and it is not possible for us to do justice to all of them in a book that is intended only to survey the field. We discussed certain representatives—the planktonic forms—of several of these phyla in our last chapter; that is, the Rotifera, Cnidaria, Ctenophora, Mollusca, Chaetognatha, Echinodermata, and Arthropoda (class Crustacea). We will revisit and extend our discussions of four of those phyla here, but as for the rest, we will direct most of our attention to the more abundant groups in only some of them. Our focus will home in on the larger metazoans and higher invertebrates.

This chapter will take us on a brief tour—an overview—of the basic biology and ecology of those invertebrate groups we might expect to encounter in our explorations of the oceans. Our approach will be to follow a taxonomic organization while also considering ecological characteristics of each group

---

[1] Estimates of the total number of animal species remaining to be described range between 10 million and 200 million.

[2] Kingdom, as a major taxonomic category, is fading from use, as we discussed in Chapter 7, but it nonetheless persists (see World Registry of Marine Species, at www.marinespecies.org/index.php). The number of phyla in Animalia continues to be revised as well. Edward E. Rupert, Richard S. Fox, and Robert D. Barnes, 2004. *Invertebrate Zoology* (Belmont, CA: Brooks Cole); Funch, P., and Kristensen, R. M., 1995. Cycliophora is a new phylum with affinities to Entoprocta and Ectoprocta. *Nature* 378: 711–714; and Baguñà, J., and Riutort, M., 2004. Molecular phylogeny of the Platyhelminthes. *Canadian Journal of Zoology* 82: 168–193.

[3] The insects have the largest number of invertebrate species; but in terms of numbers of individuals, as we discussed in the last chapter, the copepods are the most numerous of all animals on Earth, vertebrate or invertebrate.

TABLE 10.1	List of the invertebrate phyla with significant representation in the oceans[a]	
Phylum	**Common name**	**Approximate number of species**
Porifera	Sponges	10,000
Rotifera	Rotifers	2000
Cnidaria	Jellyfishes, sea anemones, hydroids, corals	10,000
Ctenophora	Comb jellies	100
Platyhelminthes	Flatworms and tapeworms	18,500
Nemertea	Ribbon worms	1000
Gastrotricha	No common name	400
Kinorhyncha	No common name	150
Gnathostomulida	No common name	80
Priapulida	No common name	15
Tartigrada	Water bears	1000
Nematoda	Roundworms	28,000
Entoprocta	Goblet worms	150
Ectoprocta	Bryozoans	4500
Phoronida	Horseshoe worms	20
Brachiopoda	Lamp shells	350
Mollusca	Clams, snails, octopuses, chitons	94,000
Sipuncula[a]	Peanut worms	250
Echiura[a]	Spoon worms	140
Pogonophora[a]	Beard worms	120
Chaetognatha	Arrow worms	120
Annelida	Segmented worms	17,000
Arthropoda		
Subphylum Crustacea	Crustaceans	>50,000
Subphylum Chelicerata	Horseshoe crabs, sea spider	1300
Echinodermata	Star fish, brittle stars, sea cucumbers, urchins, sand dollars	7000
Hemichordata	Acorn worms	100
Chordata		
Subphylum Urochordata	Classes Thalacia, Ascidiacea and Appendicularia (= Larvacea), Sea Squirts, Salps, Appendicularians	3000
Subphylum Cephalochordata	Lancelets	25

[a]Several disputed phyla are not listed. The Sipuncula, Echiura, and Pogonophora are, as in most textbooks, listed here as separate phyla. However, it has been argued that they belong in the phylum Annelida. Rouse, G., 1998. "The Annelida and their close relatives," in *Invertebrate Zoology*, ed. D. T. Anderson (Oxford University Press); Halanych, K. M., Dahlgren, T. G., and McHugh, D., 2002. Unsegmented annelids? Possible origins of four Lophotrochozoan worm taxa. *Integrative and Comparative Biology* 42: 678–684; Struck, T. H. et al., 2007. Annelid phylogeny and status of Sipuncula and Echiura. *BMC Evolutionary Biology* 7: 57; Sperling, E. A. et al., 2009. MicroRNAs resolve an apparent conflict between annelid systematics and their fossil record. *Proceedings of the Royal Society, B* 276: 4315–4322.

as we distinguish between the benthic and nektonic invertebrates; we further subdivide the benthic animals into infaunal and epifaunal representatives, and within those groups, we distinguish between the motile and the sessile animals.

The Benthos

The marine **benthos** refers collectively to the animals and plants that live on or in the bottom of the sea, which is home to most of the marine invertebrate species. They can be either **infaunal** animals, living wholly or partially beneath the sediment–water interface, or they can be **epifaunal** animals, living on (not in) the bottom. Benthic animals may be **motile**, meaning that they can move about on the bottom the way a crab does, or they may be **sessile**, which means they are stuck in place, the way a barnacle is.

The sponges

The **sponges**, phylum Porifera, are among the simplest and the most primitive of all metazoans and are found in all oceans at all depths, from the Arctic Ocean to the Southern Ocean and the coastal waters of Antarctica (**FIGURE 10.1A**). They are even found in freshwaters, but they are most common in tropical seas, where they often take on vivid colors (see Figure 10.1B). Sponges are sessile (non-moving) animals, which spend their lives stuck to a substrate. Except for the tiny flagella inside them that enable them to filter-feed, they do not move. Prior to 1825, sponges were thought to be plants because they failed to meet the criterion established by Aristotle to distinguish between plants and animals—movement; they do not visibly respond when something brushes up against them, for example.

The sponges differ from other metazoans in that they do not have true tissues; they are held together in a rigid body form by networks of spicules made of either calcium carbonate or silica. Their most obvious feature, apart from lacking tissues and internal organs, is their porous body that, in its most basic form, is shaped into upward-opening tubes (see Figure 10.1C). Primitive though sponges are, they do have a fairly elaborate water pumping system whereby flagellated cells, called *choanocytes*, create internal

FIGURE 10.1 (A) Basic body plan of a sponge; they are sessile animals that attach themselves to a solid substrate at their base. Water flows through the walls of the animal and exits out the opening at the top. (B) A basket sponge, photographed in the Cayman Islands. (C) Skeleton of a glass sponge, *Euplectella aspergillum*, from the Philipines (about 25 cm in length). Inset shows the delicate and intricate arrangement of silica spicules.

(A)

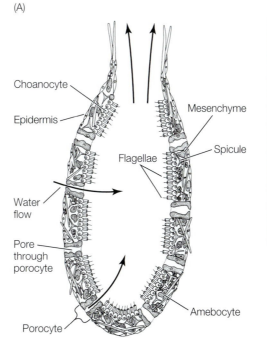

Choanocyte
Epidermis
Mesenchyme
Flagellae
Spicule
Water flow
Pore through porocyte
Amebocyte
Porocyte

(B)

(C)

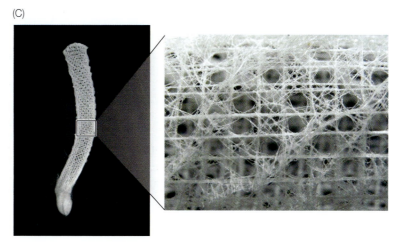

water currents that pull outside water into their body cavity through openings in their pore cells, the *porocytes*, and from there out the top of their tube. In this process, planktonic food particles and oxygen are extracted by the choanocytes and metabolic and digestive waste products are simultaneously expelled.

Sponges come in all shapes and sizes; some species are more than a meter in height, and not all are constructed in the simple tube design just described. In some species, the basic body plan diagrammed in Figure 10.1A is reshaped and multiplied into a number of more complex configurations, all of which retain the same operating principle of pumping water through the animal.

Reproduction in sponges is both sexual and asexual; in sexual reproduction, fertilized eggs develop into flagellated larvae that drift in the plankton for a relatively brief period of time (another form of meroplankton). The larvae eventually settle onto a substrate where they become fastened and develop into the adult form. Asexual reproduction is by regeneration, whereby a portion of the parent separates and grows into a new individual adult; alternatively, if the animal is fractionated (cut up into smaller pieces by whatever means), the individual pieces regenerate into new sponges. Their role in the marine food web is not well studied, but with a few exceptions, most potential predators apparently find them distasteful, perhaps because of their needlelike spicules.

Sea anemones and corals

Unlike the Hydrozoans and Scyphozoans in the phylum Cnidaria (discussed in the last chapter) that have as part of their life cycle a planktonic medusa, or which themselves may be planktonic hydroids, the **anthozoan** cnidarians (class Anthozoa) are either colonial or solitary polyps and do not have a planktonic medusa. These anthozoan cnidarians are the sea anemones, corals, sea fans, and sea pens; and like the sponges, they are members of the sessile benthos. There are three subclasses: the subclass Hexacorallia, also known as the stony corals or the reef builders, as well as all the sea anemones; the subclass Octo-corallia (also known as Alcyonaria, or just octocorals), colonial polyps that include the soft corals, sea pens, and gorgonian corals (the sea fans and sea whips); and the subclass Ceriantipatharia, the black corals and tube-dwelling anemones (also called the cerianthids).

The **sea anemones** are solitary polyps that are generally much larger and of a heavier construction than the hydrozoan polyps we discussed in the last chapter (**FIGURE 10.2**). They are extremely common at the subtidal as well as in the intertidal depths of most rocky shorelines, where they reside with their broad-based pedal disk firmly attached to a rock or other hard substrate. The sea anemones come in a range of colors, from pure white to various shades of green, brown, and red, and when groups of anemones are "open" and their tentacles are fully extended and displayed, they give the appearance of an underwater flower garden. But their outward beauty belies the fact that they are, in a sense, voracious predators. Like all members of the cnidaria, they possess stinging nematocysts, which line their tentacles and are capable of paralyzing small animals that happen to come into contact with them. Anemones are described as "sit-and-wait" predators,[4] patiently waiting for an unlucky prey organism, such as a small fish, to swim into contact with their tentacles and venomous nematocysts. Once the prey is disabled, the anemone's tentacles then contract over and around it and bring it into its oral opening (at the top of the animal) for digestion inside the body cavity. Most anemones, however, are actually opportunists that will also ingest detrital organic matter that happens onto, or into, their feeding mechanism. In processing its captured food items, the anemone's longitudinal muscles and the sphincter and circular muscles may

---

[4] J. Malcolm Shick, 1991. *A Functional Biology of Sea Anemones* (London: Chapman & Hall).

(A)

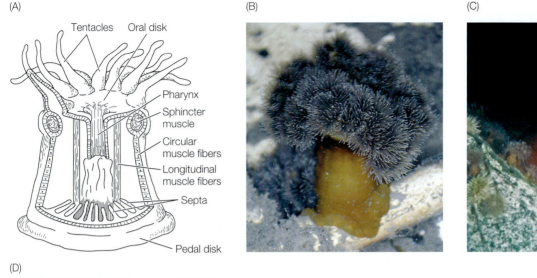

Tentacles   Oral disk

Pharynx

Sphincter muscle

Circular muscle fibers

Longitudinal muscle fibers

Septa

Pedal disk

(B)

(C)

(D)

FIGURE 10.2 (A) Basic anatomical features of a sea anemone. *Metridium* anemones, one of the most common and widespread genera of anemones, (B) growing in a laboratory holding tank at the University of Maine, and (C) photographed *in situ* in the Gulf of Maine. (D) Group of *Metridium senile* anemones and a few larger colonial hydroids growing on a mooring line; the anemones are closing upon being brought to the surface.

all contract, which, along with the anemone's lack of any skeletal hard parts, enables the animal to close up completely and compress itself into what looks like just a lump of flesh on a rock (see Figure 10.2D).

Sea anemones respire by pumping water through their body cavities by way of an internal lining of epithelial cilia; oxygen-rich water is drawn in along the epithelial surface of the pharynx, where it circulates and exchanges with respiratory and digestive waste products in the body cavity, and then leaves the animal through the center of the same oral opening.

Reproduction in the anemones is both sexual and asexual. Asexual reproduction may be by way of what is called basal laceration, where the base pedal disk breaks off pieces of itself by contracting against the substrate on which it is attached, thus tearing away small chunks of the disk which then grow into new individual anemones. This phenomenon can be seen in Figure 10.2C, which shows several tiny anemones growing beside the large one. Asexual reproduction is also accomplished by fission, or budding, whereby the main cylinder or tube divides longitudinally or transversely across the midpoint, for example; it also occurs when a portion of a tentacle buds off a new, small anemone.

Some anemones are hermaphroditic, usually only the smaller species, but hermaphroditism and asexual reproduction are normally mutually exclusive,[4] and in the majority of the larger anemones, the sexes are separate. In sexual reproduction, fertilized eggs and larvae are released into the plankton where they reside for only a relatively brief period before they settle to the bottom and develop.

(A)

(B)

FIGURE 10.3 (A) A coral reef in Guam in the western Pacific Ocean. (B) Coral polyps, which measure just a few millimeters across.

In addition to the sea anemones, the hexacorals (class Hexacorallina) include the hard, reef-building corals—the **stony corals**—that many of us are familiar with to some extent. They build the underlying structures that support the astonishingly beautiful coral reef ecosystems that are common in the tropics and subtropics. Corals are, essentially, very small sea anemones that secrete around themselves a rock-hard skeleton of calcium carbonate (**FIGURE 10.3**). Each of the coral heads seen in Figure 10.3A is built up from thousands or even millions of individual calcium carbonate structures secreted by a collection of polyps that are all genetically identical clones of one another, as in Figure 10.3B. The polyps spread by asexual budding and over time secrete masses of reef that take on the characteristic appearance of that particular coral species.

Corals, like their anemone cousins, are also sit-and-wait predators, which can also sting and capture prey (they, too, are cnidarians and possess nematocysts). But they also have inside their tissues symbiotic algae, which are photosynthetic dinoflagellates. Called **zooxanthellae**, these dinoflagellates provide the coral polyp with at least a portion of its nourishment by way of their photosynthetic products, and they also assist by taking up the polyp's waste products, such as nitrogenous nutrients, which benefits the zooxanthellae. Because of this dependence on their photosynthetic tenants, residing inside of and protected by the polyps, reef-forming corals are therefore limited in their depth distributions and occur only as shallow, near-surface features. We will return to coral reefs when we discuss ecosystems in Chapter 12.

Not all corals are shallow-water colonies of polyps that depend on photosynthetic zooxanthellae; some species of the stony corals reside in much deeper waters, as well as in colder waters outside the tropics and subtropics. For example, the hexacoral *Lophelia pertusa* is a calcium carbonate–secreting reef builder, but it lacks zooxanthellae (**FIGURE 10.4**). *L. pertusa* is found on the bottom of deep

FIGURE 10.4 The deep-water stony coral *Lophelia pertusa* with its polyp tentacles extended, sitting on the ocean bottom in the Mississippi Canyon off Louisiana at a depth of about 450 meters. This photo was taken with a remotely operated vehicle.

continental shelf waters in the Gulf of Mexico and on both sides of the North Atlantic Ocean, as well as beyond the shelf edge to depths as great as 3,000 meters. Deep-water corals (also called cold-water corals) are relatively new to science and have only been studied in more recent times using submersibles and remotely operated vehicles (ROVs). They are now believed to be far more numerous than shallow-water species.

Gorgonian corals, also known as sea whips and sea fans, are a group of **soft corals** in the subclass Octocorallia (**FIGURE 10.5**). Found throughout the world ocean, especially in the tropics and subtropics, gorgonians are similar to the sea pen, another soft coral. They, too, are made up of colonies of tiny individual polyps, but arranged in a flattened and branching shape, like a fan. They are held firmly in place by an infrastructure made of gorgonin, a firm but flexible protein complex secreted by the polyps. These gorgonians can be 2 meters in height but are barely a few inches thick, and are often brightly colored. They are especially popular in aquaria.

Bryozoa, phoronids, and brachiopods

The bryozoans (phylum Ectoprocta), the phoronids (phylum Phoronida), and the brachiopods (phylum Brachiopoda) are a group of sessile organisms that are all characterized by having clusters of ciliated tentacles called **lophophores** used in food capture and in respiration. Thus, they are together considered the **lophophorates**.

Like the corals, the **bryozoans** (phylum Ectoprocta) are also colonial, sessile animals. They are commonly found residing on the surfaces of macroalgae and sea grasses, as well as on rocks and various other substrates, such as dock pilings. With some 4000 species, they are a successful animal phylum that has apparently beat the odds against them in their evolution. They occur in both marine and freshwaters, and in the oceans they can be found at all depths from the surface to as deep as 6000 meters. The lophophore of bryozoans may be circular, spiral, or horseshoe-shaped.

Bryozoans are fairly small metazoans that form colonies often reaching several centimeters in diameter. Superficially, they resemble hydroids and are also known by their common name, the moss animals. They secrete a matrix material, which may be gelatinous or have a firmer consistency, that houses their colony of zooids. The colony sometimes resembles a loosely woven fabric or a network of branching appendages, within which the animals reside and through which they extend their lophophores outward (**FIGURE 10.6**).

The bryozoans have been a target of research for many reasons, one of which is that they are important **fouling organisms** (a term given to unwelcome nuisance organisms that grow on ship hulls and in the water ballast tanks of ships, for instance). As a result of where they grow, they have been carried across oceans and introduced into far-flung regions of the world ocean where they sometimes become established. Once they do become established, these invasive animals can cause problems, with some having been implicated as competitors with native species for substrate space. An example of a nonindigenous bryozoan is the lacy-crust bryozoan, *Membranipora membranacea* (see

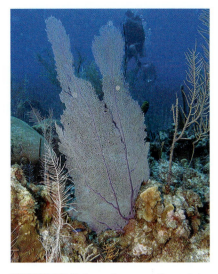

FIGURE 10.5 A sea fan, or soft coral, in the Cayman Islands.

(A)

(B)

5 mm

(C)

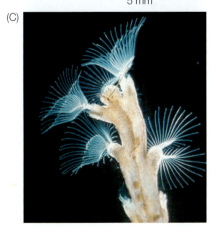

FIGURE 10.6 (A,B) A colony of *Membranipora membranacea*, the lacy-crust bryozoan, which has been introduced to and is now common on the coast of New England, shown here growing on a blade of eel grass. (A) Several individual zooids with their lophophores extended; they quickly withdraw their lophophores when disturbed or when they sense a change in light (making them difficult to photograph). Notice the fan-like pattern created in (B) as the colony expanded in numbers of individuals. (C) Another bryozoan, *Plumatella repens*, with its lophophores clearly visible. The zooids in this photograph are ca. 0.7 mm across.

FIGURE 10.7 A group of phonorid worms, with their U-shaped lophophore, or crown of ciliated tentacles.

Figure 10.6), introduced to the coast of New England, which has been implicated in defoliating the region's kelp beds.

The **phoronids** (phylum Phoronida) comprise an extremely small taxon, with only about 20 species known (**FIGURE 10.7**). Sometimes called horseshoe worms, they are sessile and solitary wormlike animals that attach to or bore into surfaces such as dock pilings and soft sediments, where they build tubes of a chitinous material to give them vertical support. They are generally small animals, usually less than 2 cm in length and about 1.5 mm in width, although some can be much larger, reaching several tens of centimeters in length. They filter-feed using the lophophore that extends out the end of the tube. They are exclusively marine, and are found in intertidal depths and as deep as several hundred meters.

Brachiopods (phylum Brachiopoda), or lamp shells, are solitary, sessile animals that look like a common clam and are often mistaken for molluscs. There are roughly 4000 extinct species known from the fossil record, but there are only about 350 species living today. Most brachiopods reside in crevasses and on the undersides of features that hide them, such as in and amongst coral reef systems where they are very common, but they are also found in a few cold-water areas as well as in deep waters where they are well hidden or protected. They seem to be unpalatable—most potential predators avoid them. The brachiopods resemble small bivalve molluscs with their pair of calcium carbonate shells (valves) within which the animal resides; the shells differ from molluscs' in that one valve has a hole in it through which a pedicle (stalk) runs. In some species the pedicle holds the bivalved animal on one end, suspending it well up off the substrate onto which the other end of the pedicle is attached; in other species, the pedicle may penetrate into bottom sediment where, upon contraction, the animal can withdraw.

The platyhelminthes, nemertians, and nematodes

The **flatworms**, phylum Platyhelminthes, are of interest to biologists for many reasons, one of which is that they mark the evolutionary transition from radial symmetry in the cnidarians to bilateral symmetry. Of the four classes in this phylum, three are wholly parasitic, while only one class, the **turbellarians**, class Turbellaria, is free-living (**FIGURE 10.8**). Turbellarians occur primarily in marine waters but also in freshwaters and in some terrestrial systems; many species are brightly colored in elaborate patterns. The free-living turbellarian flatworms are motile and glide across the bottom on a slimy underside, forced by ciliary action. The digestive tract of nearly all species in this class lacks an anus, and therefore they regurgitate any undigested material back out the mouth opening. But still other species have a complex branching gut with several anuses. The turbellarians are carnivores, and feed by trapping small animals in their slime and then engulfing them in the **proboscis**, which is extended from the underside of their oral end.

FIGURE 10.8 A turbellarian, *Pseudoceros bifurcus*, common in coral reef environments throughout the tropical Pacific Ocean.

To visualize the action of the proboscis, imagine putting your hand inside a sock (the proboscis), grabbing a golf ball (the prey) using your fingers from inside the sock, and then pulling your hand out of the sock while holding onto the ball. This turns the sock inside-out, of course, and leaves the ball engulfed inside it. Some species of turbellarians ingest algae that are maintained alive and, essentially, in culture inside the animal, in a form of symbiosis; as an adult the animal acquires nourishment from its photosynthetic captives, and does not feed. The turbellarians are indeed an interesting group of marine animals.

The **ribbon worms**, phylum Nemertea, are like the turbellarians in possessing a nervous system and a proboscis at their oral end, but they differ in having a digestive tract with an anus at the distal end of the body, and in their possession of a circulatory system (**FIGURE 10.9**). In addition, the nemerteans differ drastically from the Platyhelminthes (and all the other invertebrates) in having a unique proboscis apparatus that lies dorsal to the gut and is surrounded by a fluid-filled chamber called a **rhynchocoel** that is used to fire outward its proboscis to capture prey. The proboscis can be harpoon-like, with barbs, and often delivers a toxin. Once a prey item is captured by the proboscis, the ribbon worm will secrete digestive enzymes into the body of the prey, softening and to some degree liquefying it before pulling it into its digestive tract. Some of the larger ribbon worms will use this action to attack and ingest shelled molluscs, such as clams. The larger ribbon worms may reach several tens of centimeters in length and a few millimeters in width; however, there is at least one report of a specimen that was 54 meters in length.[5] Ribbon worms are commonly found in the intertidal zone under rocks and burrowing in soft sediment (see Figure 10.9); they are also found in shallow subtidal waters. They are motile animals that get around mainly by burrowing and crawling over surfaces, but they have also been observed swimming at the surface with an undulating action similar to that of a snake.

The **roundworms** (phylum Nematoda), usually referred to simply as the **nematodes**, are the most diverse of the three phyla discussed in this section; more than 28,000 species have been described. Their general anatomy is diagrammed in **FIGURE 10.10**. Unlike the flatworms, they possess a digestive tract with orifices located at opposite ends of the animal. The head is distinctively different from the tail end in having what, in many species, appear to be several tooth-bearing lips around its mouth. Nematodes have a well-developed nervous system with sensory bristles that give the animals a sense of touch. The sexes are separate, and reproduction is usually sexual with some species brooding the eggs internally. The young hatch from the eggs looking much like miniature adults.

Nematodes are among the most evolutionarily successful animals on the planet, having colonized marine, freshwater, and terrestrial systems. One author wrote: "Perhaps more truthfully than with any other group, one must say that nematodes occur everywhere, from hot springs, to polar regions, and from deserts to ocean depths."[5] We might also add to this list "the insides of other animals and plants," in that 16,000 of the 28,000 species are parasitic; many are responsible for diseases in plants, animals, and humans.

FIGURE 10.9 A milky ribbon worm, *Cerebratulus lacteus*, uncovered on a tidal flat on the coast of Maine. These delicate worms can reach more than a meter in length; this one was approximately 35 cm total length. Both ends of this worm are still in its burrow.

(A)

(B)

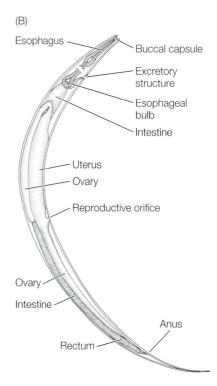

FIGURE 10.10 (A) Photomicrograph of a nematode, *Caenorhabditis elegans*. (B) Anatomy of a female nematode (Genus *Rhabditis*).

---

[5] Paul A. Meglitsch, 1972. *Invertebrate Zoology*, 2nd ed. (New York: Oxford University Press).

In the marine realm, the free-living nematodes inhabit primarily the interstitial spaces among bottom sediment particles where some species prey on other small animals such as protists and other worms (nematodes themselves are small, most are on the order of 1 mm in length, but others get very large; the largest is a parasite found in sperm whale placentas). Different species are adapted to feed on algae and various forms of detritus, or on dead organic matter. They are especially important in benthic microbial food webs and in the recycling of organic matter, and are members of a group called the **meiobenthos**, benthic animals that pass through a 1 mm mesh screen; while most nematodes are longer than 1 mm, they are certainly thin enough to pass through.

The polychaete worms

Commonly called the segmented worms, members of the phylum Annelida include, among other groups, the common earthworms and leeches that most of us are familiar with. As mentioned earlier, this phylum has undergone significant taxonomic revisions in recent years, the details of which are still being sorted out by taxonomists; for example, it has been argued that at least three former phyla (the Sipuncula, Echiura, and Pogonophora), should be included with the annelids on the basis of molecular phylogentic analyses. Strictly speaking, the Annelida should perhaps no longer be referred to as the segmented worms, even though most of them are. Annelida is a large phylum, with about 17,000 species, found in marine environments from the intertidal zone to the deep sea and its hydrothermal vents, as well as in freshwaters and some terrestrial environments. There are four classes of annelids; one of them, the **polychaetes** (class Polychaeta), with about 13,000 species, is the largest group and is predominantly marine.

The polychaetes are one of the most abundant and ecologically important benthic animals; they are found in all oceans and at all depths, although most are shallow-water and intertidal forms that inhabit soft-sediment environments. They generally range in size from about 1 to 15 cm in length, although there are exceptions. With their elongated bodies, they have a typical wormlike appearance; superficially, they resemble centipedes, or millipedes, in that they are divided into segments, with each segment in many species also possessing appendages (but centipedes and millipedes are in a separate phylum, the Arthropoda.) The polychaetes' numerous septa divide the animal's body into segments running along much of its length and are visible as rings in the surface cuticle, or outer covering, which is made of strong but flexible collagen. Most of the segments contain duplicate sets of internal organs, but a single digestive tract, nervous system, and circulatory system run the length of much of their elongated bodies. Most species of polychaetes have **parapodia**, appendages that function as and are analogous to limbs; they extend outward in pairs from either side of each body segment (**FIGURE 10.11**). Formed by extensions of the body wall and under muscular control, they are used in crawling, burrowing, and swimming. A few species, such as in the genus *Tomopteris*, are holoplanktonic and have well-developed and elongated parapodia that they use in swimming.

Most of the polychaetes, like the flatworms and ribbon worms, have a proboscis that can be everted (turned inside-out), and some have evolved sharp, powerful jaws or pincers that are used to capture prey or to tear apart materials for ingestion (see Figure 10.11A). Some species, such as the bloodworm in Figure 10.11B, can inflict a surprisingly painful bite on the hand of an unsuspecting handler; in some polychaete species, including the bloodworm, the bite carries venom.

FIGURE 10.11 (A) Drawing of the head end of a nereid polychaete worm with its proboscis everted, showing the pincing jaws, as well as the parapodia of the first two body segments. (B) A bloodworm, *Glycera dibranchiata*, a common inhabitant of muddy intertidal areas; approximately 7 cm total length.

(A)

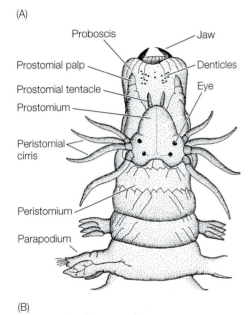

(B)

The polychaetes exhibit both sexual and asexual reproduction, with some doing both in different seasons. In some polychaetes, spawning is a single event after which they die, while in other species, spawning may be nearly continuous. Following successful fertilization, the eggs develop into planktonic trochophore larvae similar to those of molluscs.

The lifestyles of the benthic polychaetes generally fall into one of two categories: the **errant polychaetes**, motile animals that can actively move about, and the **sedentary polychaetes**, sessile species that are usually permanently housed in tubes. (These are not taxonomic categories, but merely lifestyle descriptors.) Active, errant polychaetes have well-developed parapodia, elongated and flipper-like in their appearance, that are used in crawling and burrowing; these species often reside under rocks in the intertidal zone and generally crawl about in and around various objects. They also are able to burrow into the bottom sediments and emerge to feed, and at times they can be found swimming in the water column, but only for brief periods. The sedentary species reside in tubes that they construct, usually with sediment particles glued together with mucus secretions, forming tubes. Some smaller species will also build tubes with calcium carbonate–based secretions that harden into seashell-like tubes that are often seen attached to rocks and other substrates (**FIGURE 10.12**).

The type of feeding differs among the various species of polychaetes. Some are predatory carnivores that capture prey with their proboscis and powerful jaws; the bloodworm (see Figure 10.11) is an example. Other polychaetes are **deposit feeders** that ingest sediment, passing it though their digestive tracts from one end to the other, taking in any digestible organic matter that happens to be in that sediment, the content of which can be substantial. Marine sediments are not just composed of mineral particles, they can be a nutritious mix of microorganisms, including protozoans, bacteria, and other meiofauna, along with detrital organic material in all stages of decomposition. Remember: the fate of much of the planktonic primary production in the overlying water column, if not consumed by zooplankton, is to settle to the bottom. Even phytoplankton cells that are grazed by zooplankton often pass through them relatively undigested and sink as fecal material. The result is that much of the organic matter delivered to the benthos has significant nutritive value, and deposit feeders have evolved to take advantage of that.

Several approaches are employed by polychaetes to feed on those deposits. As just mentioned, some species burrow into the sediments, passing sedimentary material through their digestive tracts as they move through the sediment. Others form tubes and orient themselves head-down, ingesting sediments on

FIGURE 10.12 Two polychaete tube worms of the family Serpulidae (S*pirorbis borealis*) attached to an eel grass blade. These worms reside in a calcium carbonate shell-like tube and feed by extending tentacles that collect suspended particles. Both individuals have withdrawn into their tubes. Each coiled worm tube in these photomicrographs is approximately 1–1.5 mm across. (B) A particularly dense patch of polychaete tubes extending above the sediment–water interface are exposed here at low tide; they are likely tubes of bamboo worms, *Clymenella torquata*.

(A)

(B)

(A)

(B)

(C)

Thin layer of brown, oxygenated surface sediment

Black, anoxic sediment

Polychaete tube

FIGURE 10.13 (A) Pile of fecal material ejected onto the surface sediment by a feeding *Arenicola*. (B) A Nereid worm has begun to burrow down into the sediment. (C) A polychaete tube, most likely one that held a bamboo worm, extending from the thin layer of well-oxygenated, gray-brown surface sediment through the black, anoxic sediment, which begins only about 1 mm beneath the surface. For reference, the periwinkle snail on the surface sediment is approximately 1 cm across.

(A)

(B)

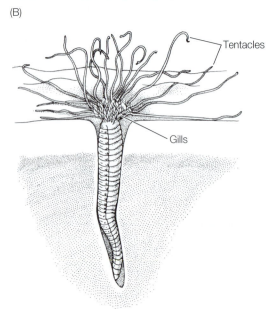

Tentacles

Gills

the bottom of the tube and egesting processed sediment and fecal matter in conspicuous piles which can often be seen on the sediment surface at low tide. Still other species actively pump water through a U-shaped tube, ingesting any suspended particulate material that is drawn in and leaving conspicuous piles of undigested sediment (**FIGURE 10.13A**). Depending on the nature of those sediments, tube-dwelling polychaetes may also need to pump oxygen-rich water through the burrows since the sediments can be anoxic just a few millimeters beneath the sediment–water interface—particularly in fine-grained, muddy sediments (**FIGURE 10.13C**).

The tube-dwelling polychaetes are usually characterized not as deposit feeders but as suspension feeders, but the distinction between the two feeding modes has become blurred, especially with regard to this group of animals. Some species, the sabellid polychaetes (suborder Sabellida), have elaborate plumose appendages around their oral end which they display from the opening in their tubes (**FIGURE 10.14A**). These animals often burrow into the sediment where they sit, perched upright, feeding on particulate materials that collect on their tentacles, which they then guide to their mouths for ingestion. This suspension feeding method is very similar to the way many sea anemones and sea pens feed. Still other species, the terebellid polychaetes (suborder Terebellida), burrow into sediments where they also stay put, extending the long, sticky tentacles that cluster around their oral openings and sweeping up newly settled particles—deposits—off the surface sediment in the immediate vicinity of the tube openings (**FIGURE 10.14B**).

FIGURE 10.14 (A) A group of sabellid polychaetes in a coral reef system in the Caribbean Sea. (B) A terebellid polychaete with its tentacles extended out onto the surrounding surface sediment.

Benthic molluscs

The molluscs, phylum Mollusca, include a number of important infaunal and epifaunal benthic species (animals that reside *in* the bottom and *on* the bottom, respectively). The phylum also includes nektonic species, the cephalopods, which we will discuss later.

The molluscs are one of the largest invertebrate phyla, comprising more than 20% of named marine animals; more than 90,000 species have been described. Molluscs are found in all oceans, as well as in freshwaters and terrestrial environments. There are between 9 and 11 classes of molluscs (two are extinct, and the extant classes are in a state of taxonomic flux), but the four more important classes that we will discuss in this chapter are the gastropods, which comprise about 80% of mollusc species, the bivalves, the chitons, and the cephalopods, all of which have been on Earth since the Cambrian (ca. 600 million years ago).

BIVALVES   Known for a long time as the class Pelecypoda, and now as the class Bivalvea, the **bivalve molluscs** are perhaps the best known to most people; they include the commercially harvested clams, scallops, mussels, and oysters (**FIGURE 10.15**). The bivalves include about 20,000 species that are found at all depths in all oceans, but the detailed taxonomy of the group is still being sorted out. The sizes of bivalves varies significantly, from small infaunal forms on the order of 1 centimeter to the giant clam (*Tridacna* spp.) often featured in old movies as a killer, grabbing hold of pearl divers in their huge shells (they don't actually do that, by the way).

The key anatomical feature of bivalves, of course, is the possession of a pair of hinged valves, or shell halves, made of calcium carbonate that is secreted by the animal. Usually, the valves are identical mirror images of one another (scallops and oysters are exceptions; their top and bottom valves differ in their degree of concavity). The shells are held closed by one or two adductor muscles. Some bivalves, such as the soft-shell clams, which are infaunal, burrowing bivalves, possess a foot that is filled with body fluid (hemolymph); the foot can be extended down into the sediment and then pumped with hemolymph, making it expand, thus forming an anchor that enables the animal to pull itself down deeper into the sediment where it resides (**FIGURE 10.16**). At the opposite end of the foot, facing upward, is

(A)

(B)

(C)

FIGURE 10.15 Some of the more easily recognized, commercially harvested bivalves. (A) A group of blue mussels (*Mytilus edulis*) at low tide. (B) An eastern oyster (*Crassostrea virginica*). (C) A bay scallop (*Argopecten irradians*); notice the numerous blue eyes visible in the mantle tissue.

(A)

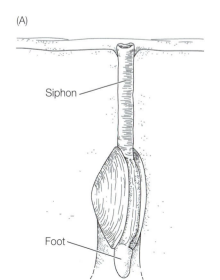

Siphon

Foot

(B)

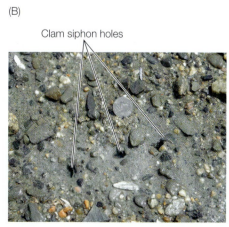

Clam siphon holes

FIGURE 10.16 The soft-shell clam, *Mya arenaria*. (A) Sketch of a clam in sediment with its foot extended downward and its fused inhalent and exhalent siphons extended upward to the sediment–water interface. At low tide, when the sediment surface is exposed, the clam will retract its siphon, leaving only its hole as evidence of its presence. (B) Examples of clam siphon holes in a mud and gravel sediment surface.

(A)

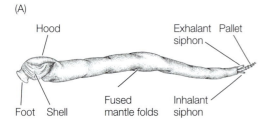

Hood   Exhalant Pallet
         siphon

Foot   Shell   Fused   Inhalant
         mantle folds   siphon

(B)

FIGURE 10.17 (A) External anatomy of the shipworm, genus *Teredo;* their average length is about 2–4 cm. (B) A section of an untreated oak dock piling after only seven months in the water in Belfast Harbor, Maine. The piling had become foam-like, having more air pockets from the shipworm burrows than solid wood.

pair of tubes encased in a single siphon through which the animal filter feeds and exchanges expiratory and digestive wastes.

The bivalves and other molluscs have made the most of having to pass seawater over their gills in order to acquire oxygen—at the same time, they also remove planktonic and detrital food particles that are in suspension in that water. In the infaunal bivalves, the clams, the siphon has both incurrent (inhalant) and excurrent (exhalant) tubes, which may be separate or fused. The animal pumps in seawater and its accompanying load of suspended particles, the potential food items, from above the sediment–water interface. The incoming water and particles pass over the clam's gills, which collect the particulate food, and by way of cilia and mucus action, the animal guides the food to its palps, which in turn direct it to the mouth for digestion. At low tide, when the sediment surface is exposed, infaunal bivalves such as clams will retract the siphon back inside the animal's closed valves, leaving behind a conspicuous hole in the surface sediment, thus revealing the animal's presence—a useful guide for (human) clam diggers (see Figure 10.16B).

One species of boring bivalve, the shipworm, genus *Teredo,* can bore into and feed upon wood, digesting the wood fibers with the enzyme cellulase (**FIGURE 10.17**). (Shipworms also feed on plankton.) This bivalve has evolved to the extent that it hardly resembles its cousins, and as its name implies it looks much more like a worm than a clam. Its vestigial bivalve shell is retained about its head, but it is less than 10% or so of the length of the greatly elongated animal. It bores foot-first into wood—driftwood, wooden ships and dock pilings—with its siphons extended outward. Throughout history humans have reviled these animals for their destructive feeding habits. But try to imagine for a minute what the world would be like without them; what would happen to all the trees that die and get washed down rivers and into the sea?

Not all bivalves are infaunal or burrowing animals. Epifaunal bivalves, such as mussels, oysters, and scallops, for example, reside on the bottom (versus in it), usually on a hard substrate where they filter-feed on suspended particles in the water column. Mussels attach themselves to rocks and other hard substrates using tough byssal threads, and can hold onto their positions even when pummeled by strong surf (**FIGURE 10.18**). They lack the extended siphons that clams have. They don't need them; the animals already reside on top of the

FIGURE 10.18 (A) Close-up of byssal threads of a blue mussel glued to a rock. (B) A group of mussels actively feeding. Notice the slight gap between their valves, exposing their inhalant and exhalent siphons that circulate seawater and particulates across their gills.

(A)

(B)

benthic substrate and are in the water column (at high tide, anyway) along with their necessary supplies of oxygen and suspended particulate food. They simply open their valves and expose their internal siphons. Often these bivalves are found in aggregations, or beds, of many individuals, especially in areas of high plankton production. The filter-feeding efficiencies of extensive mussel beds have been shown to significantly lower the standing stock of plankton, both phytoplankton and zooplankton, in the overlying water column.

Scallops, while considered here to be benthic animals, also fit our definition of nekton They are capable of powerful flapping actions of their two valves, pumping water in a jet-like fashion, which propels the animals up off the bottom and through the water significant distances. Believed to be primarily a defensive action to escape predators, some populations of scallops have been shown by tagging studies to migrate tens of kilometers over a number of years by using this swimming action.

GASTROPODS   The **gastropods**, class Gastropoda, are also easily recognized by their spiral shell (some of them) and, when they are not retracted into their shell, a flattened foot (**FIGURE 10.19**). These are the snails, limpets, sea slugs, and nudibranchs. The gastropoda are the most diverse class of animals in the phylum Mollusca, with more than 60,000 living species (some estimates are as high as 80,000 living species). They are found in all oceans and at all depths, and are especially common in intertidal zones and shallow waters.

The typical shelled gastropod with its spiral shell has a single opening that can be sealed by a stiff flap, called the operculum, isolating the animal inside and protecting it from desiccation at low tide, or from predators at other times. When the motile animal is active, the operculum can sometimes be seen on a portion of its exposed foot. Inside the shell, the animal's body twists back on itself such that the anus has access to the single opening; this anatomical phenomenon is called **torsion**, and occurs only in gastropods.

FIGURE 10.19 (A) General anatomy of a typical gastropod snail. (B) A group of periwinkle snails (*Littorina littorea*) at low tide. (C) The mouth and radula of a *Littorina*. (D) The empty shell of a *Littorina* with a bore hole made by the radula of a carnivorous gastropod.

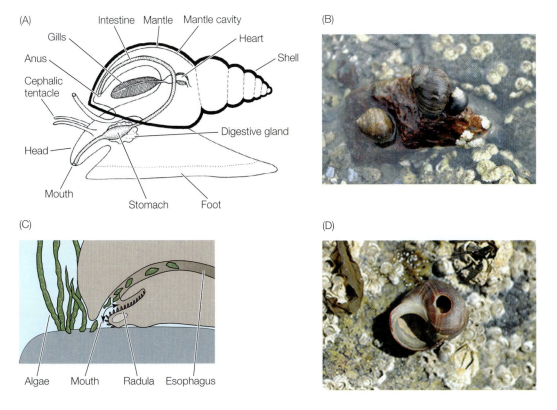

(A) Gills, Anus, Cephalic tentacle, Head, Mouth, Stomach, Foot — Intestine, Mantle, Mantle cavity, Heart, Shell, Digestive gland

(C) Algae, Mouth, Radula, Esophagus

FIGURE 10.20 Cone snails. (A) Major features of a typical cone snail. (B) Shells of *Conus marmoreus* (left) and *C. textile* (right). (C) Scanning electron micrograph of the harpoon, a modified radular tooth of a cone snail.

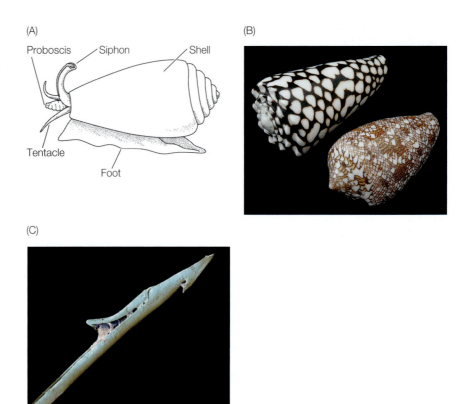

(A)

Proboscis — Siphon

Shell

Tentacle

Foot

(B)

(C)

Gastropods feed on almost everything. Some species specialize on algae, some on detritus, and others are predatory carnivores. But they all possess a **radula**, a tongue-like structure in their mouths embedded with a specialized set of teeth, which is used like a rasp or coarse file to grind and shred their prey. It is often used by gastropods to loosen attached algae from hard substrates, or in the case of predatory carnivores, it is used to grind holes through the shells of other shellfish. One often can find seashells with this characteristic hole neatly drilled all the way through them (see Figure 10.19D). Some predatory gastropods, especially those of the genus *Conus*, or the **cone snails**, are venomous (**FIGURE 10.20**). They hunt other molluscs and marine worms, as well as fishes, using their highly modified, harpoon-like radula, which they can forcefully eject into the flesh of prey organisms, and in the process also inject their venom. In some of the larger cone snails the venom is very potent and can have serious debilitating effects on humans; great care should be exercised when handling live specimens.

Not all the shelled gastropods have the gracefully spiraling shells that we are accustomed to seeing; some, such as the **limpets**, have a greatly flattened shell. Limpets are common in most intertidal and subtidal areas that support coralline algae, one of their preferred foods. The limpets grind the rock-hard algae with their radulas; some species of coralline algae depend on this grazing action in order to stimulate their growth.

Other gastropods, such as the common slipper snail, have a rounded shell and a partial deck-like shell on the inner portion (**FIGURE 10.21**). (Children who find the

FIGURE 10.21 Three individual slipper snails, *Crepidula fornicata*, attached on top of one another. When in this piled arrangement, the bottom individual is usually a female, and the oldest of the three; the individual on top is a male and is the youngest; the one in the middle is a hermaphrodite and will become a female should the original female die.

Snail 3

Snail 2

Snail 1

shells on the shore often call them "boat shells" because the shell resembles the hull of simple sailboat with a deck.) Originally native to the western Atlantic Ocean, the slipper snail is now found on both sides of Atlantic as well as the eastern Pacific Ocean as a result of accidental introductions. These snails attach themselves firmly to a hard substrate, where they filter-feed on plankton and other suspended particles.

Some gastropods, such as the **sea slugs** and **nudibranchs**, have no shell at all (**FIGURE 10.22**). Nudibranchs shed their shells during their larval development and have no shell as adults. About 3000 species of nudibranchs have been described. Most are on the order of 1–2 centimeters in length, although some get quite large, measuring several tens of centimeters. The group gets its name for their exposed gills, which are projections of folded, fleshy tissue that they carry on their dorsal surfaces. Sea slugs, a term also used at times to refer to nudibranchs, are classified separately from the nudibranchs, with each in a separate subclade in the molluscan class Gastropoda (or, according to some authors, in separate clades). The sea slugs include the sea hares and bubble snails; the taxonomy of both of these groups of animals remains a work in progress.

Nudibranchs are most often found crawling along surfaces, but they can swim gracefully by flapping the lateral tissue folds that run the length of their bodies. Most are predatory carnivores, feeding on other, mostly sessile, invertebrates, and some species of nudibranchs are among the few animals known to feed on sponges. In addition, many species feed on various cnidaria, somehow ingesting nematocysts in their undischarged state and passing them to fleshy pockets in their *cerata* (singular *ceras*), extensions of the digestive gland. Some nudibranchs lack gills, but have numerous cerata, which can play a role in respiration; still other species lack cerata and have gills on their posterior ends. If they are disturbed, the nematocysts in the cerata are fully capable of being discharged; thus the nudibranchs have acquired an interesting defense mechanism. Again, it would be wise to use caution should you ever encounter one. Many species, especially those in the tropics, exhibit spectacular patterns of brilliant colors; in most cases, these serve as warnings to potential predators that they are not a tasty treat, and it may be best to leave them alone.

Some species of sea slugs have an interesting lifestyle: said to be "solar powered," the eastern emerald sea slug, *Elysia chlorotica* (**FIGURE 10.23**), in the suborder Sacoglossa, is one of many species that have evolved the capability to acquire chloroplasts from the algae they eat, a phenomenon known as **kleptoplasty**. These slugs have the unusual ability to ingest algae but not digest the alga's chloroplasts. Rather, they are able to sustain the viability of chloroplasts for a period of ten to twelve months (the animal's normal life span), deriving nutrition from their newly acquired photosynthetic capability.

CHITONS **Chitons** are shelled molluscs in the class Polyplacophora. The chitons occur in all oceans and are common on rocky substrates at intertidal and subtidal depths; there are about 1000 species known. They are distinguished by their having a set of eight hard calcium carbonate plates on their dorsal sur-

FIGURE 10.22 An example of the beautiful patterns and colors of the nudibranchs. This is *Berghia coerulescens*, from the Mediterranean.

FIGURE 10.23 The eastern emerald sea slug, *Elysia chlorotica*, feeding on its preferred algal food, *Vaucheria litorea*, a filamentous alga from which it is able to collect and concentrate chloroplasts, thus giving the animal its green leaflike color. The chloroplasts retain their photosynthetic functionality, producing food for their new owner.

FIGURE 10.24 A chiton, *Tonicella lineata*, from the Pacific Ocean.

face, surrounded by a tough mantle material (**FIGURE 10.24**). The animals usually feed on attached algal cells on rocks, and while they appear to be firmly attached to their substrate, they can and do move around, by gliding across rock surfaces on their fleshy foot while grinding free algal cells using their radula. They have drawn interest among researchers because of the curious eyes, hundreds of them, that ring their outer edge; they can apparently distinguish only crude shapes, however.

Echinoderms

It is perhaps the **echinoderms**, and in particular, the starfish and sand dollars, that best symbolize the sea shore. One reason for this association we all have with these easily identifiable marine animals is not only their common occurrence along virtually all shorelines, but because they occur *only* there—in the sea. The echinoderms, phylum Echinodermata, are strictly a marine group of invertebrates; they are intolerant of low salinites and occur in estuaries only if they have very little freshwater input. In fact, high mortalities of starfish and sea urchins have been documented following periods of heavy rains that significantly dilute near shore waters. Nonetheless, echinoderms can be found in all oceans and at all depths, even the very shallow near shore waters; with about 7,000 species, they are the largest taxonomic group found exclusively in marine waters with no terrestrial or freshwater representatives.

As their name indicates, the echinoderms are the "spiny-skinned" invertebrates. They are radially symmetrical and are most often, but not always, divided into five parts or sections; they lack a brain and possess only a decentralized nervous system. They have been around since the beginning of the Cambrian, more than 600 million years ago, and some 20,000 fossil species have been described. Today there are five classes of echinioderms that are still around: class Crinoidea, the sea lilies and feather stars; class Asteroidea, the starfishes; class Echinoidea, the sea urchins and sand dollars; class Ophiuroidea, the brittle stars; and class Holothuroidea, the sea cucumbers (**FIGURE 10.25**). All echinoderms possess a characteristic endoskeleton made up of interlocking platelike structures, called ossicles, which may be heavily impregnated with calcium carbonate, and which is covered by an outer epidermal layer. Their skeletal structures include orifices through which the animals extend outward their numerous **tube feet**—short, tentacle-like protrusions that are under hydraulic control as the animal alters the pressure inside them, and which are used for locomotion and in predatory feeding. The echinoderms control their tube feet with a vascular network for pumping them with coelomic fluid (basically water), changing their shape as pressure is varied in a coordinated fashion such that the animal can walk across the surface of a hard substrate. In addition to tube feet, the sea urchins and some starfishes possess spines, which in many species contain a venom that can inflict a painful sting in humans.

The **crinoids**, or **sea lilies**, class Crinoidea, are among the most beautiful of all sea creatures. Much like several species of the sabellid polychaetes, with which they bear a striking resemblance, groups of some species of crinoids look like a colorful flower garden (a description we also used for sea anemones, but which again seems appropriate here). Some species have stalks that are attached to the bottom substrate, but others are capable of moving about, creeping across the bottom on the base of their stalks with their numerous arms spread open; they can also swim weakly by forcing undulating wavelike arm motions. They feed on plankton by catching particles in the basket-like arrangement of their arms,

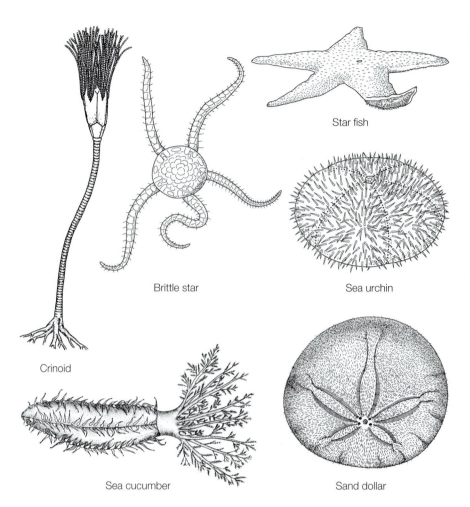

Crinoid

Brittle star

Star fish

Sea urchin

Sea cucumber

Sand dollar

FIGURE 10.25 Sketches of representatives of the four classes of echinoderms: a sea lily (class Crinoidea); a brittle star (class Ophiuroidea); a star fish (class Asteroidea); and a sea urchin, sea cucumber, and sand dollar (class Echinoidea).

using them also to direct any captured particles to their mouth. They reside primarily in deeper waters.

The **starfishes**, class Asteroidea, are common at all depths. Some seem to prefer the shallower depths of rocky coasts with heavy wave action. They range in size from a few centimeters across to more than a meter in the case of the sunflower starfish. Their body wall is covered with delicate papulae, small membranous sacs similar to small tube feet, which are used in respiration. On their dorsal surface they also have jawlike pincers that serve to protect the papillae. If you happen to find a starfish sometime at the seashore, place it upside-down on the back of your wrist; then gently pick it up again. You will feel the pincers, the pedicellariae, that have grabbed and are holding onto your arm hairs (which they will let go of). On the starfish's underside you can watch its tube feet, hundreds of them, arranged in rows along each of the arms, all moving about as the animal varies the hydraulic pressure in each one in an attempt to either walk away from you, or to reach around you as if you were a delicious bivalve mollusc that it is about to devour (**FIGURE 10.26**).

The starfishes are predators and feed on any animals they can find. Some species are able to swallow smaller

FIGURE 10.26 Close-up photo of starfish tube feet. This animal has just earlier been turned over onto its back, and it is searching with its tube feet for a substrate to grasp, in order to right itself. They do this by shaping their bulbous tips, clearly visible here, into suction cups.

prey items whole, and any skeletal remains of their digested prey that won't fit through their anus is ejected out the same way it came in, though the mouth. Larger prey may be digested externally. The starfish can evert its stomach out its mouth and around much of its prey; digestive enzymes soften and partially liquify the prey, which is then ingested through the mouth.

Starfish can often be seen battling with clams, mussels, and oysters. They will creep across their prey, wrapping their arms around it as best they can, and using their tube feet, they apply pressure in an attempt to open the bivalve, which responds by contracting its adductor muscles to stay closed. The starfish's tube feet have bulbous tips that can be formed into suction cups, enabling them to latch onto substrates such as bivalve shells (see Figure 10.26). The starfish eventually wins the test of endurance, as it maintains its internal hydraulic pressure and constant pull on opposite sides of the bivalves' shells. After some time, the bivalve gives in and allows an opening between its valves, at which time the starfish's stomach can be everted through it to begin digesting the prey. Starfish are often **apex predators**—predators at the top of the trophic pyramid. In addition, because the overriding influence that their predatory preferences have in structuring the ecosystem, such as the kelp forests of coastal California waters, they have been labeled **keystone species**. Keystone species reside at the top of the trophic pyramid and influence the species composition and basic ecology at the lower trophic levels.

The **sea urchins** and **sand dollars**, in the class Echinoidea, seem at first glance to be armless versions of starfish. Urchins are rounded in shape, held together by their firm skeleton of interlocking calcium carbonate plates. Like starfish, they possess tube feet, particularly on their ventral surface; on the dorsal surface they have an extensive coverage of spines, which vary in length among species. In some species, the spines carry venom.

Sea urchins feed primarily on algae, although they may at times take advantage of vulnerable animals such as their cousins, the sea cucumbers. Urchins possess a powerful set of jaws—five of them—positioned around the oral opening centrally located on their undersurface; they are essentially five sets of strong teeth made of a hard calcium carbonate. This anatomical feature is known as Aristotle's lantern, following the great naturalist's description of it. The teeth are arranged such that they are self-sharpening and are hard enough that they are capable of cutting scratch marks in rocks. Urchins are important members of coastal ecosystems in that their grazing pressure on the macroalgae, especially kelps, keeps the algae in check. But in situations where the numbers of urchins have grown as a result of reduced predation on them, they are capable of completely denuding broad expanses of ocean bottom. On the other hand, in cases where their numbers have been reduced, as a result of commercial fishing for them (their roe is considered a delicacy), kelp forests have dramatically proliferated.

Sand dollars are flattened versions of urchins, with their spines reduced to a fur-like coating. Their skeletal remains, having lost their furry coat and acquired a sun-bleached white color, are commonly found on sandy beaches. These animals are found subtidally on the surface of sandy and muddy bottom sediments, where they can and do move about with their tube feet. They also burrow into the sediments. They feed on various meiofauna and algae that they encounter.

The **brittle stars**, class Ophiuroidea, are similar in overall appearance to starfishes in that they, too, usually have five arms, and like all the echinoderms they possess a skeleton of calcium carbonate plates and tube feet. There are similar numbers of species of starfishes and brittle stars—about 2000. Brittle stars are found in all oceans from the tropics to the poles. While they occur in shallower waters, they are often hidden from sight, buried in sandy sediments or under rocks. On average, they are found at greater depths than are the starfishes; and they are also very common on the bottom of the sea, where they are among the dominant animal taxa.

(A)

(B)

FIGURE 10.27 (A) A brittle star, *Astrotoma agassizii*, showing its flexible arms. (B) An individual basket star *G. chilensis* collected at the same site.

The arms of brittle stars are much thinner and more elongated than those of the starfishes. Brittle stars also differ from the starfishes in their mobility; not only are they capable of movements under the control of their tube feet, but their long arms are flexible and are under muscular control, giving brittle stars the ability to crawl relatively swiftly across the bottom—much faster than starfishes can with their near total reliance on their tube feed for propulsion (**FIGURE 10.27A**). Given their abundance on the ocean floor, and the fact that they are a significant prey item for some commercially exploited fish populations, there is a growing interest in the brittle stars among fishery scientists. The brittle stars are but one of two clades in the Ophiuroidea; the **basket stars**, characterized by elaborately branched arms which they use to feed on suspended particles, are the other (**FIGURE 10.27B**).

The **sea cucumbers**, class Holothuroidea, stand out among the echinoderms. They look a lot different from starfish and urchins (but so, too, do the crinoids). They are found around the world ocean and are most common at relatively shallow depths, but they can also be abundant in deep waters, with some species found at 6000 meters. Most are of an intermediate size; as one author describes them: "The group as a whole varies from the smallest gherkins to a truly noble watermelon."[5] By watermelon, he was referring to the species *Stichopus variegatus* which reaches a meter in length and 21 cm in diameter.

The sea cucumbers are, essentially, soft bags of water that lie on the bottom. They have an oral end, usually surrounded by retractable feeding tentacles which are modified tube feet, and an anus at the opposite end. They possess skeletal ossicles, like all echinoderms, but in the sea cucumbers, these are very much reduced and are usually microscopic, giving them a soft leather-like body surface held together by flexible collagen fibers that are under the animal's control. They can, at will, loosen their body surface and allow themselves to flow into small cracks and crevasses; likewise, when the need arises, they can stiffen their body surface. They have a vascular system and tube feet, or podia, which in some species run along the body in five rows (**FIGURE 10.28**). Sea cucumbers are opportunists, feeding on any organic material, such as plankton and detritus that happen to drift their way; large populations of some species are often found in areas of swift currents that bring suspended material past their feeding tentacles. Like the brittle stars, they are one of the dominant animal groups in the deep sea.

FIGURE 10.28 A pair of plankton-feeding sea cucumbers, *Cucumaria* sp., from the Adriatic Sea.

Other arthropods

No other phylum comes close to the **arthropods**, phylum Arthropoda, either with respect to the total numbers of animals or with respect to their evolutionary success. Along

FIGURE 10.29 Photograph of a pycnogonid at a depth of 3100 meters in the North Pacific Ocean.

FIGURE 10.30 Horseshoe crabs, *Limulus polyphemus*, that have come in to shallow waters to spawn, where they appear to be pairing off, or trying to. The female will dig into the sandy sediment at the water's edge where she will lay her eggs, which will be fertilized at the same time by the male clasping her back.

with the molluscs, they are one of the largest groups of animals—marine, freshwater, or terrestrial—and constitute some 85% of all described animal species. The arthropods are made up of four extant subphyla (one phylum, the Trilobites, is extinct) and 16 classes, of which two subphyla—the Chelicerata and the Crustacea—include the vast majority of important marine forms.

CHELICERATA   There are three classes in the subphylum Chelicerata: The Arachnida, which includes the air-breathing spiders and scorpions, of which more than 75,000 species have been described; the Pycnogonida, the sea spiders, also called simply the pycnogonids, of which there are about 1300 species; and the Merostomata, which includes an order of extinct sea spiders and the four species of extant horseshoe crabs. Like all the arthropods, the Chelicerata have segmented bodies with jointed limbs, which they shed, or molt, in order to grow, but they are distinguished from other arthropods in possessing two sets of appendages located anterior to their mouth (these are the fangs in spiders).

The **sea spiders**, or **pycnogonids**, are unusual-looking creatures (**FIGURE 10.29**). As one invertebrate zoologist put it: "If one were asked to design a Thing from Outer Space, one might do no worse than to take a pycnogonid as his model. These weird-looking arthropods seem to be all legs."[5] It is difficult to add much to this succinct description of the animals' external anatomy. Interestingly, there appears to be some debate amongst invertebrate taxonomists as to where exactly these animals fit in a classification and evolutionary scheme, and whether they should even be included in the Chelicerata; there are arguments that they may belong to their own phylum, one that evolved in parallel with the Arthropoda.

Most pycnogonids are small, between 1 and 10 mm, but some deep-sea species get quite large, up to a meter across. They are found in all oceans and at all depths but they are especially common in protected areas in shallow waters where they tend to hide. They are carnivorous, feeding on various slow or sessile invertebrates such as sponges and sea anemones; they use their proboscis to puncture and suck out nourishment from some of their prey. Ecologically, they are but a minor component of the marine food web, although admittedly there has been very little research done on their ecology.

The **horseshoe crabs**, of which there are four living species, are also a tad strange-looking (**FIGURE 10.30**). These animals are quite common and abundant in many near-shore sandy and muddy environments, but their distribution in recent years appears to have become more restricted. They are basically a big shell beneath which hides a spider-like animal, with five pairs of walking legs (also used for swimming, which they do upside down, and for guiding food to their mouth). They usually walk about on the bottom feeding on various molluscs, polychaetes, and small crustaceans.

Horseshoe crabs reproduce once a year in a defined breeding season for each specific locale where they are found. The animals move into very shallow waters over a sandy and muddy bottom where the female lays her eggs. Males, which are smaller than females, will grab the female's back while she digs a hole in sediments at the water's edge in which to lay her eggs; the male fertilizes the eggs as they are laid. Shore birds often feast on the eggs, which hatch into planktonic larvae in about two weeks.

The horseshoe crabs have no hemoglobin in their blood to carry oxygen, but instead use hemocyanin, a copper-based compound that gives the blood a blue color. Instead of white blood cells to combat pathogens, they have amebocytes, about which there is some interest in the biomedical research community. Animals are not sacrificed for samples with which to do this research, however, as it is possible to draw blood from them without significant mortalities. Unfortunately, horseshoe crabs appear to be threatened; populations in some parts of the word have been shown to be declining, probably as a result of overfishing (they are used for bait).

CRUSTACEA The **crustaceans**, subphylum Crustacea, include the copepods, barnacles, shrimp, amphipods, isopods, and decapods. With more than 50,000 species, they are by far the most abundant and important benthic arthropods in the oceans (overall, the planktonic copepods, which are also crustacean arthropods, are the most abundant and ecologically important in the oceans). As arthropods, they possess an exoskeleton which is molted as part of their growth process, but unlike other groups, they have biramous—two-part—limbs, and they have a naupliar larval stage. They have a segmented body plan, with a head segment, thoracic segments, and abdominal segments. In some crustacea, the head and throax are fused and may be covered by a carapace, a hard exoskeletal covering that in some ways resembles a saddle. The crustaceans include the copepods and krill that we discussed in the last chapter.

The **amphipods**, class Amphipoda, are extremely abundant crustaceans, occurring in all oceans and at all depths, and are also found in freshwaters. There are about 7000 species, but one group, the suborder Gammaridea, claims about 5500 of them; as seems to be case in most of the invertebrates we've been discussing thus far, their taxonomy is still a work in progress. Amphipods are generally small animals. Most are less than 10 mm in total length, but some deep-sea species are on the order of 30 cm or more. Reproduction is sexual, with the female carrying the eggs in a brood pouch; there are no larval forms, and eggs hatch directly into juveniles that resemble the adults.

The basic body plan of a typical gammarid amphipod is given in **FIGURE 10.31**. Their chief anatomical feature is their lateral compression. They have no carapace covering any of the 13 segments behind their head, as is common in other groups. The head is fused with the adjacent segment, making it one of the seven segments that form the thorax, the main portion of the body that has appendages both for swimming and for crawling as well as for feeding; the gills are also located on those appendages. Three more pairs of appendages, used primarily for swimming, are on the three abdominal segments.

Amphipods are extremely abundant and ecologically important animals. The gammarid amphipods, the most abundant amphipods in the ocean, are mostly epibenthic animals, crawling about on the ocean bottom and scav-

(A)

(B)

Peraeon

Head

Abdomen

Antenna 1

Pleon

Urosome

Antenna 2

Telson Uropods

Gnathopods

Coxae

Peraeopods

FIGURE 10.31 (A) Photograph of a Gammariid amphipod, *Gammarus* sp. on an intertidal mud flat. (B) External anatomical features of a gammarid amphipod (*Gammarus tigrinus*). This genus has seven thoracic segments, the first of which is the head, which bears two antennae. They have seven pairs of walking and swimming legs that are arranged in groups on the thorax, abdomen, and urosome.

enging for food. (We discussed another marine amphipod group in the last chapter, the holoplanktonic suborder Hyperiidae.) Some gammarid species will burrow into the bottom sediment from which it feeds on deposits left on the sediment surface near the burrow; it uses its greatly elongated front legs as a collector to pull particles into a ball and then bring that ball into the burrow where the animal sorts through it for algal food particles. These animals are very important links in the marine food web, serving as food items for various other invertebrates, fishes, and sea birds.

The **isopods**, class Isopoda, are another abundant and important crustacean. There are about 10,000 species of isopods, 5000 of which are terrestrial wood lice, and about 4500 are marine. Some species are wholly parasitic. Of the marine species, most are motile benthic animals. The chief anatomical feature of the isopods is their dorsal–ventral compression, versus that of the amphipods, which they would otherwise closely resemble superficially. Like the amphipods, they lack a carapace covering their segmented body, and they possess seven pairs of walking legs on their thoracic segments. Unlike the amphipods, their abdominal segments are undifferentiated, and the abdominal appendages are flattened and function as gills. They are generally small, like their amphipod cousins, but large species do occur (**FIGURE 10.32**).

The isopods are important in the marine food web, serving as food items for many fishes, especially the late larval and juvenile stages of demersal fishes (fishes that reside and feed primarily on and close to the bottom). Some species are wood borers, which can do significant damage to wood structures such as docks and pilings.

The **barnacles**, class Maxillopoda, infraclass Cirripedia, are also crustaceans, and as anyone who has ever strolled along a rocky seashore already knows, they can be extremely abundant. Barnacles are exclusively marine and are found around the world ocean; the group comprises about 1200 species. A few species are parasitic, and some are found at depths of several hundred meters, but most reside at depths shallower than 100 meters, and are primarily found in the intertidal zone.

FIGURE 10.32 Dr. Kevin J. Eckelbarger of the University of Maine holding a giant isopod, *Bathynomus giganteus*.

There are two orders of free-living barnacles that comprise the majority of species: acorn barnacles and goose barnacles (**FIGURE 10.33**). The order Sessilia, the acorn barnacles, are the sessile forms that are attached head-down directly onto their substrate; the order Pedunculata, the goose barnacles, are suspended on a stalk, known as a penduncle, which is an elongated extension of the animal's head. The goose barnacles are most common in the Eastern Atlantic and Mediterranean Sea, whereas the acorn barnacles occur almost everywhere. Much of what is currently known about barnacles was published by Charles Darwin in two monographs he published in the 1850s.

Both the acorn and goose barnacles are encased in a set of calcareous plates, and because of this, until about 1830, they were considered to be molluscs. In the goose barnacle, two plates are enlarged and enclose the animal and other smaller plates in a fashion that resembles a bivalve. The calcareous plates in the acorn barnacle are glued directly to rocks, pilings, and other substrates (including some fishes and whales). Acorn barnacles often completely cover any exposed, hard surface in the intertidal zone, and are in a constant battle for space with each other as well as with other animals, such as limpets and mussels. They are preyed upon by several molluscs and some fishes.

Barnacles are sessile suspension feeders. They stand on their heads, so to speak, as they extend their modified thoracic appendages, which are elongated and flexible **cirri**, with numerous **setae**. Sweeping through the water, these modified appendages effectively capture, or filter, particles out of the water and pass them to the mouth for ingestion. The animal can extend its cirri during feeding, or contract them inside itself at times of exposure (at low tide) by closing a pair of its calcium carbonate plates (**FIGURE 10.34**). Other appendages assist in the food particle capture by generating water currents.

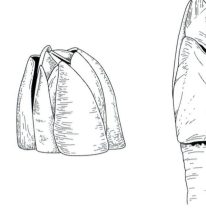

FIGURE 10.33 Two common barnacles: an acorn barnacle (A), genus *Balanus* (*Semibalanuus*), and a goose barnacle (B), genus *Lepas*.

FIGURE 10.34 Barnacles (*Semibalanus balanoides*) on a rock on the coast of Maine; the smaller individuals are young-of-the-year. Inset shows an individual barnacle feeding with its cirri extended.

(A)

(B)

(C)

(D)

FIGURE 10.35 Representative decapods. (A) An American lobster (*Homarus americanus*) photographed *in situ* in a bed of a red alga. (B) The armed nylon shrimp (*Heterocarpus ensifer*), a deepwater species. (C) A green crab, *Carcinus maenus*, an invasive species from Europe that now dominates much of the intertidal zones of both sides of the North Atlantic Ocean. (D) A hermit crab, *Dardanus arrosor*.

The **decapods**, class Decapoda, are the shrimps, crabs, and lobsters that we all recognize, if only from the dinner table (**FIGURE 10.35**). These animals are the all-stars of the crustacea, so to speak; they are the largest and most mobile group of benthic invertebrates, and include a number of species that support important commercial fisheries—both capture fisheries and aquaculture. They are found in all oceans and at all depths, and are mostly marine, although there are some freshwater forms (e.g., crayfish and prawns, for example), and some terrestrial forms (for at least a portion of their lives); approximately 15,000 species have been described.

The decapods get their name because they all have five pairs of thoracic appendages that function as walking legs; the remaining three thoracic appendages are maxillipeds used in feeding. In addition they possess a carapace that covers their thoracic segments. This group begins to blur our organizational structure for this chapter in that the decapods include both the swimming forms—the shrimps and prawns—and the crawling forms, such as the lobsters and crabs. Commercially exploited shrimp do both; they crawl on the bottom as well as swim about well up and off the bottom in dense aggregations. And all decapods have a planktonic larval stage (meroplankton). The scientific literature is filled with ecological and biological studies of the commercially important members of this group, which go far beyond the scope of this overview chapter.

THE HEMICHORDATES The phylum Hemichordata is a small phylum that includes three extant classes, one of which comprises only a single species, extinct. The class Enteropneusta includes the marine, suspension-feeding and wormlike **acorn worms**, of which there are about 100 species; the class

Pterobranchia, also a group of marine wormlike animals, claims only about 30 species.

The acorn worms are the more common of the two groups. These animals possess a well-developed circulatory system with a heart (which has a dual function, serving also as a kidney) and a set of rudimentary gills, very much like those of primitive fishes. Acorn worms have a prominent proboscis on their anterior end just forward of their mouth, a collar behind the mouth where the gills are located, and a trailing abdomen, giving the animal its typical wormlike appearance (**FIGURE 10.36**). But with these evolutionarily advanced anatomical features, the animals are less wormlike than they appear and are argued to be a biologically important transitional group that links the invertebrates to the vertebrates.

Acorn worms are found from the intertidal zone to the deepest ocean depths. They commonly occupy a U-shaped burrow from which they extend their proboscis, which can collect suspended particulates; they can also ingest sediment and function as a typical deposit-feeding worm. Although most are small, on the order of 4–5 cm on average, some species can get quite large, reaching more than a meter in length.

THE CHORDATES   There are two subphyla of marine invertebrates in the phylum Chordata, the lancelets and the sea squirts. The **lancelets**, also called **amphioxus** (a common name from the Greek, intended to mean "both ends pointed"), in the subphylum Cephalochordata, are a group of about 25 species of small, eel-like animals that are commonly found

FIGURE 10.36 (A) Anatomy of an acorn worm, *Saccoglossus* and (B) a specimen of *Saccoglossus kowalewskii* collected in sandy sediments of a tidal flat.

in and on sandy sediments of shallow seas. Like the acorn worms, they are best known among biologists for their for evolutionary links to modern vertebrates and are a topic of discussion in all introductory biology texts. The other group, the subphylum Tunicata (= Urochordata), also known as the sea squirts, is the more important of the two in a marine ecological sense. These are the **tunicates** which include planktonic animals—the appendicularians (class Appendicularia) and the salps (class Thaliacea) that we discussed in Chapter 9—and their benthic counterparts, the **sea squirts** (class Ascidiacea).

The sea squirts are a group of suspension-feeding animals that, like all chordates, have a rudimentary notochord in their early larval development.[6] The basic body form of the sea quirts is that of a sac attached to a substrate. It pumps water through itself in order to extract oxygen and food particles and to expel wastes. There are several variations on this design, giving both solitary and colonial forms of tunicates (**FIGURE 10.37**). Some colonies appear as a gelatin-like mass of soft and spongy slime that grows on almost anything; an example is the genus *Didemnum*, an invasive species of colonial tunicate that has spread throughout much of the world ocean, to the extent that its native waters are not known for sure. Other solitary forms are quite beautiful

---

[6] You will routinely see various terms used interchangeably in reference to these animals: sea squirts, tunicates, and ascidians.

(A) (B) (C)

FIGURE 10.37 Tunicates, or sea squirts. (A) An invasive, colonial tunicate, genus *Didemnum*, growing on a mooring line in Maine along with hydroids and kelp. (B) A solitary tunicate, *Styela clava*. This species shows why the name tunicate is appropriate: its thick, leathery covering resembles a tunic. (C) *Polycarpa aurata* from the Great Barrier Reef. The incurrent and excurrent siphons are clearly visible as is the simple bag-like structure of the main body mass of these animals.

animals; sometimes brightly colored, they all have well-defined incurrent and excurrent siphons used in suspension feeding. Both colonial and solitary forms can be routinely found growing attached to dock pilings and in tide pools.

Nektonic Invertebrates

The cephalopod molluscs

The cephalopod molluscs (phylum Mollusca, class Cephalopoda) comprise four well-known groups of animals: the octopuses, squids, nautiluses and cuttle fish. These animals are excellent swimmers; the nautiluses and squid spend their full time swimming in the water column, but most octopuses spend time both sitting on the bottom and swimming.

The word cephalopod means "head-feet," which is basically how these most interesting animals are built; their body plan is essentially a big head surrounded by a foot with tentacles. They are voracious predators with a highly evolved nervous system and highly developed eyes, with characteristics similar to our own. With about 800 species, the cephalopods are exclusively marine and are distributed throughout the world ocean, in shallow and deep waters; with their unusual anatomy, large brains and sophisticated nervous system, they are the most complex group of invertebrates, both in terms of their general biology as well as their intelligent behavior. So much about their anatomy and behavior is, quite simply, fascinating.

NAUTILUSES The **nautiluses**, also called the *chambered nautiluses*, order Nautilida, are true swimmers. There are only six living species. They are found in

(A)

(B)

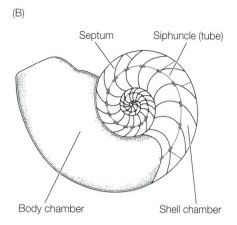

Septum   Siphuncle (tube)

Body chamber   Shell chamber

FIGURE 10.38 (A) Photograph of a nautilus swimming. (B) Diagram of the internal structure of the nautilus shell, with the siphuncle tube connecting the internal shell chambers, which are separated by septa, allowing the animal to vary their air content.

the South Pacific waters of Indonesia and in the Indian Ocean where they are fished heavily; their external shell is unmistakable and is readily recognized in gift shop display cases around the world.[7] The nautiluses are quite long-lived, up to 20 years. The reproduce sexually; the fertilized eggs are attached to a hard substrate where they develop for as long as a year before hatching as large (ca. 3 cm) juveniles that resemble the adults.

Unlike the squids and octopuses, this group has retained its external shell throughout much of its evolution; it is multi-chambered internally, with the chambers functioning as ballast tanks similar to a submarine (**FIGURE 10.38**). By varying the amount of gas and water in those chambers the animals can control their buoyancy and hence determine their vertical distribution. The animal resides in only the outermost chamber. Because of pressure effects on the highly compressible air in their chambers, they are limited in their maximum depth to about 800 meters and are most common in waters much shallower than that. They may possess as many as 90 feeding tentacles encircling their mouths, but unlike the squids and octopuses, those tentacles do not have suckers. Two of the tentacles are longer than the others and used in olfaction. Like the gastropods, the nautiluses have a radula. They are capable swimmers and use a jet-propulsion mechanism whereby they forcefully eject water from their outer chamber in such a way that they can control their direction of movement.

OCTOPUSES   The **octopuses**, in the order Octopoda and with about 300 species, are among the better-known marine animals which everyone has seen either in person at an aquarium, or most certainly, in movies and on television (**FIGURE 10.39**). These molluscs just don't look like molluscs. Like the nautilus, the octopuses possess two highly evolved eyes, but not nearly as many tentacles—only four pair—with a mouth in the center along with a hard, and strong, parrot-like beak. They have no internal or external shell or skeleton (a few species have retained a small vestigial shell in their fleshy head and mantle). They are unique in the animal world in that because of their lack of any significant hard parts, they can

FIGURE 10.39 The common octopus, *Octopus vulgaris*, resting on the bottom, photographed in Bonaire.

---

[7] A word of caution (or compassion) here: there are few if any regulations in place to regulate the taking of these beautiful animals, and their sales in gift shops continue to drive the demand.

distort their bodies to fit into incredibly tight spots. There are numerous clips of octopuses in captivity creeping through narrow-necked bottles in order to gain access to food. Octopuses live more of a benthic life than a nektonic one, but they are capable swimmers. Like their cousins the nautiluses, they propel themselves using a jet-propulsion system, forcing water out their siphon which can be oriented freely in order to steer the animal. Unlike the nautilus, though, they swim head first. Their eight arms, or tentacles, are studded with suction cups, which the animal uses to manipulate prey and various objects it might encounter. They also possess ink sacs, which hold a concentrated solution of melanin which the animal ejects to confuse potential predators, allowing it an opportunity to escape.

The octopuses are not long-lived, and most die fairly soon after reproducing; a few species, though, may live for up to five years. They reproduce sexually. The female guards her eggs in a protected crevasse or burrow until they hatch. The larval octopuses are fairly well developed upon hatching, and they reside in the plankton for some time before dropping to the bottom to begin their adult lifestyle.

Octopuses are known for their high intelligence and are likely be the most intelligent invertebrate. The exact nature of that intelligence is not obvious, however. The octopus has a very complex nervous system, two-thirds of which is located in its arms, and only one-third of which is centralized in its brain. They have good eyesight, and their tentacles have both chemoreceptors and a keen sense of touch. In captivity they have been taught to do simple tricks, and they are notorious for their uncanny ability to figure out how to get at a hidden food item, and to escape from their "secure" holding tanks. Octopuses are excellent predators, and feed mostly on other invertebrates, especially molluscs.

These animals are among the most adept in the animal kingdom at camouflaging themselves. They possess chromatophores, pigment spots in their skin that are under nervous control and can be expanded and contracted at will, altering the appearance of the animal; they also have iridophores, similar to chromatophores except they are highly reflective and, as the name implies, can give the animal an iridescent appearance. They can also alter the texture of their skin, and with their chromatophores, they can take on the appearance of a rock or a chunk of algae. It is believed that these visual displays are used to communicate with each other, as well as for defense: They can even take on the appearance of another potentially dangerous marine animal, such as a lion fish, in order to ward off potential predators. There are remarkable video clips on the web of octopuses displaying this visually stunning ability.

Interestingly, not only can octopuses inflict a painful bit with their beaks, but the bite can be venomous; all octopuses are venomous to varying degrees, but only the blue-ringed octopus is known to be fatal to humans. Nonetheless, it is better not to try to handle these animals should you encounter one while diving.

SQUIDS The **squids**, cephalopods of the order Teuthida, are also true swimmers. There are about the same number of species of squids, 300, as octopuses. Like all cephalopods they have a head, a mantle, and tentacles, or arms, and in some ways they resemble an elongated octopus (**FIGURE 10.40**). Their nervous system and eyes are well developed and they, like the octopuses, are intelligent animals. They too have elaborate choromatophores and iridophores which are used, it is believed, in communication. But unlike the octopus, the squids have ten arms, two of which are longer than the others, and their mantle has two wing-like flaps controlled by muscles that they use in swimming, and which enables them to move easily both forward and backward. Their tentacles also have suction cups.

(A)

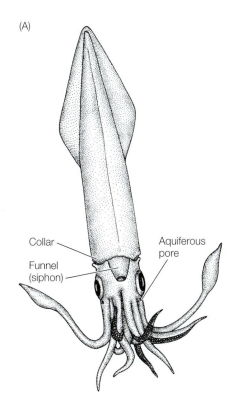

Collar

Funnel
(siphon)

Aquiferous
pore

(B)

Mantle
wings

FIGURE 10.40 (A) The common squid, genus *Loligo*. (B) *L. pealeii* photographed in a laboratory holding tank; its chromatophores and iridophores are evident. This individual, about 10 cm in total length, is swimming backward and forward in the tank using only its mantle wings, which can be seen here as semitransparent in this juvenile.

Squids tend to travel in schools, gracefully swimming in apparent unison as they propel themselves with their mantle wings. Like the octopus and nautilus, the squids also possess a jet propulsion system and are capable of extremely fast and sudden bursts of speed, used primarily as an escape response to avoid predators. They, too, steer with a moveable funnel. Squids also possess a sharp and strong beak and a radula, located in their mouth, which is surrounded by their tentacles. Like their cousins the octopuses, squids have hemocyanin instead in hemoglobin as a blood oxygen carrier, giving their blood a blue color.

These cephalopods are found in all oceans and are especially abundant on continental shelves, where there are commercial fisheries that target them. Most species are on the small side, ranging in lengths from about 15 to 30 cm (mantle length), but some get quite large, such as the Humboldt squid in the eastern Pacific Ocean, which grows to more than 1.5 meters in mantle length. And, of course, there are the giant squids, which for decades had been the subject of maritime lore, and which have begun to draw serious attention from scientists (**BOX 10A**).

CUTTLEFISH   The **cuttlefish**, order Sepiida, are another interesting group of cephalopods. Cuttlefish reside in tropical and subtropical oceans around the world, but they are not found off the shores of North America. They get their name for their only remaining skeletal hard part (evolutionarily speaking), their vestigial cuttlebone, which is commonly found washed up on beaches adjacent to waters where they reside; made of calcium carbonate, the cuttlebone is porous and by varying the amount of gas in it, the animal can control its buoyancy. Their external morphology resembles the squids, to some extent, and cuttlefish have many of the same characteristics as the other cephalopods (**FIGURE 10.41**). They possess ten arms, or tentacles, two of which, like the squids, are longer and more elaborate and are used in capturing their prey items, which are mostly

FIGURE 10.41 The cuttlefish, genus *Sepia*.

BOX 10A Giant and Colossal Squid

At one time the idea of giant squid (*Architeuthis* spp.) was limited to maritime folklore, with various reports by mariners of battles they had witnessed between a giant squid and a sperm whale, but today we know that these animals do indeed exist. There may be as many as 19 species or as few as one; their distributions are becoming better documented with several hundred specimens having been examined, most of which are stranded carcasses; they are found in all oceans, but more sightings and strandings occur along both coasts of the temperate North Atlantic.

The history of how and when these animals were first spotted and subsequently verified as credible is a particularly rich one. The first reliable sighting of a giant squid was by a French warship in 1861, the *Alecton*, which hauled in part of the animal (**FIGURE A**). Before this, they were considered a myth.

Roper and Boss (1982) reported a giant squid carcass that washed ashore in Massachusetts in 1980, which measured 10 meters in total length. But they may get much bigger, as much as 20 meters in total length and up to 500 kilograms, as based on a specimen washed ashore in New Zealand in 1880, although such accounts have been challenged, as no fresh specimens have been that large (**FIGURE B**). We now know that sperm whales prey on giant squid, and instances of scars on landed sperm whales have been routinely reported; the scars are inflicted by the large suckers of giant squid and the surrounding rim of sharp teeth, as shown in **FIGURE C**.

FIGURE A The *Alecton* struggling with a giant squid.

FIGURE B A giant squid weighing 250 pounds and measuring 25 feet caught off the coast of New Zealand in 1997.

Now there appears to be an even larger squid, the colossal squid (*Mesonychoteuthis hamiltoni*), reported and described in 1925 and then again in 1970 (juvenile of the species). These animals have a larger mantle than the giant squid, but their tentacles are not as long. They reside primarily in the Antarctic.

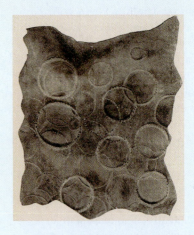

FIGURE C Photograph of scars left on a sperm whale by the sharp teeth that ring the suckers on tentacles of giant squid.

other invertebrates. These animals are perhaps even more impressive than the octopuses with respect to their abilities to camouflage themselves and mimic other animals. Like the octopuses, they are intelligent and have excellent eyesight, but with few exceptions, they reside exclusively in the water column, more like the squids and nautiluses, swimming gracefully with their long mantle wings, but accelerating rapidly, using their jet-propulsion system, when the need arises.

Chapter Summary

- The **invertebrates** are the most numerous animals in the oceans and on land. Of the 33 or so phyla of animals, 32 are invertebrates; they make up about 97% of the approximately 1 million animal species that have been described.

- Invertebrates have either a pelagic or benthic life style. The benthic invertebrates may exist as **infaunal**, residing in or beneath the sediments at the ocean's bottom, or **epifaunal** animals, residing on top of the bottom, and they may be **motile**, capable of moving about, or they may be **sessile**, attached to a bottom substrate.

- The **sponges** are the simplest and most primitive marine invertebrates; they are sessile filter-feeding animals. They lack tissues, and their skeletal structure is made of calcium carbonate or siliceous needlelike spicules.

- The **sea anemones** and **corals** are also sessile invertebrates, but, in addition to feeding on suspended particles in the water column, they possess, as do all cnidarians, nematocysts for use in prey capture. Some corals and anemones have symbiotic, photosynthetic dinoflagellates, called **zooxanthellae**. The **stony corals** secrete calcium carbonate structures that build up over time to form vast reef systems. Scientists are discovering deep-water corals that may be more numerous than those in shallow waters in the tropics and subtropics.

- The **lophophorates** are a group of three phyla, the **bryozoans**, **phoronids**, and **brachiopods**. These sessile suspension-feeding invertebrates possess a **lophophore**, a cluster of ciliated tentacles used for feeding.

- One class of platyhelminthes, the **turbellarians** or **flatworms**, are motile predators, possessing a **proboscis** that they use to engulf prey. The nemerteans, or **ribbon worms**, also possess a proboscis that can be used like a harpoon with venom. The **nematodes**, or **roundworms**, are found in virtually all environments, many as parasites, and are among the most abundant animals in the world.

- The **polychaete** worms, in the phylum Annelida, are among the most ecologically important animals in the sea; they are found almost everywhere, but are most common in soft-sediment environments of shallow and intertidal depths. They

are capable of burrowing, crawling, and swimming, and may function as predators, suspension feeders, or **deposit feeders**. Most species possess a proboscis and some have jawlike pincers that may also carry venom. Some species are motile predators, while others are sessile tube dwellers. They are very important food organisms for various fishes and other invertebrates.

- The molluscs are one of the two largest taxonomic groups of marine invertebrates (with the **arthropods**). The **bivalve molluscs** comprise the well-known commercially harvested species such as clams, mussels, scallops, and oysters, as well as many other non-commercial species. They possess paired valves (shells) of calcium carbonate. Some species are epifaunal, such as the mussels and oysters, and others are infaunal burrowers, such as some of the clams. Most filter food particles out of the water as it passes though them. **Gastropod** molluscs are easily recognized by the coiled shells in many taxa, but they also include forms without shells: the **sea slugs** and **nudibranchs**. All gastropods possess a **radula** that is used to scrape algae off rocks or to bore into other shelled invertebrates.

- The **echinoderms** include the **starfishes**, **sand dollars**, and **sea cucumbers**. These are radially symmetrical, spiny-skinned animals that possess tentacle-like **tube feet**. They are found in all oceans and at all depths. Some are suspension feeders and others are predatory carnivores. All are exclusively marine.

- The larger, nonplanktonic arthropods include the **sea spiders**, **horseshoe crabs**, and several important groups of **crustaceans**: the **amphipods**, **isopods**, **barnacles**, and **decapods**.

- The hemichordates include the **acorn worms** and two groups of chordates, the **lancelets** and the **sea squirts**.

- The cephalopod molluscs include the **octopuses**, **squids**, **nautiluses**, and **cuttlefishes**. These larger, nektonic invertebrates are highly developed and are relatively intelligent. They are interesting in terms of their behavior and camouflage capabilities. They are all capable swimmers, and only the octopus spends appreciable time on the bottom.

Discussion Questions

1. Discuss the differences in the lifestyles of the various marine invertebrates with respect to their being benthic or neritic; motile or sessile; infaunal or epifaunal. Include in your discussion the various feeding modes common to each lifestyle.

2. How are anemones and corals alike in terms of their anatomical features and their modes of reproduction and feeding? How are the two groups different from one another?

3. What are deep-water corals, and how are they similar to shallow-water reef-forming corals? Why has their abundance only just recently been discovered?

4. The lophophores in the bryozoans and phoronids, for example, resemble the feeding tentacles of terrebellid and sabellid polychates, even though these groups are only distantly related to one another. Discuss how they might coexist with one another in the same area of ocean bottom.

5. Discuss the possible survival advantages for the bivalve molluscs of possessing a pair of shells (valves).

6. Suspension feeding and filter feeding are two modes often ascribed to the various marine invertebrates, but in some cases the distinction is blurred. What do we mean by this?

7. What are tube feet? Which animals possess them? And what functions do they perform?

8. Why are nautiluses restricted to depths shallower than about 800 meters? What does that group of cephalopods have in common with submarines?

9. What are the different modes of locomotion among the cephalopods?

Further Reading

Bousfield, E. L., 1973. *Shallow-Water Gammaridean Amphopoda of New England*. Ithaca: Comstock Publishing.

Brusca, Richard C., and Brusca, Gary J., 2003. *Invertebrates*, 2nd ed. Sunderland, MA: Sinauer Associates.

Collins, M. A., and Rodhouse, P. G. K., 2006. Southern Ocean Cephalopods. *Advances in Marine Biology* 50: 191–265.

Ellis, Richard, 1998. *The Search for the Giant Squid: The Biology and Mythology of the World's Most Elusive Sea Creature*. New York: The Lyons Press.

Halstead, Bruce. W., 1965. *Poisonous and Venomous Marine Animals of the World. Vol. 1., Invertebrates*. Washington, D.C.: U.S. Government Printing Office.

Hayward, P. J., and Ryland, J. S., eds. 1990. *The Marine Fauna of the British Isles and North-West Europe*, vol. 1. Oxford: Clarendon Press.

McSweeny, E. S., 1970. Description of the juvenile form of the Antarctic squid *Mesonychoteuthis hamiltoni* Robson. *Malacologia* 10: 323–332.

Meglitsch, P. A., 1972. *Invertebrate Zoology*, 2nd ed. New York: Oxford University Press.

Robson, G. C., 1925. On *Mesonychoteuthis*, a new genus of oegopsid Cephalopoda. *Annals and Magazine of Natural History* 9: 272–277.

Roper, C. F. E., and Boss, K. J., 1982. The giant squid. *Scientific American* 246: 96–105.

Shick, J. Malcolm, 1991. *A Functional Biology of Sea Anemones*. London: Chapman & Hall.

Contents

The fishes are among the most successful vertebrates on Earth and have been around for hundreds of millions of years. They are successful for many reasons. One reason for their success is the ability of some species to remain inconspicuous to potential predators, as exemplified in this leafy sea dragon from Australia. A relative of the seahorse, the sea dragon has evolved spectacular features that give it the appearance of just another piece of sea weed.

The Fishes

We continue our introduction to marine organisms in this chapter with a discussion of the fishes. Fish are **vertebrates**, a subphylum of the Chordata, and as such are related to the invertebrates that we discussed in Chapter 10: the lancelet, in the subphylum Cephalochordata, and the appendicularians, salps, and sea squirts in the subphylum Urochodata. The appearance of the vertebrates and their proliferation between 400 and 500 million years ago represented a marked anatomical improvement, or at least an evolutionary sophistication, that clearly set them apart from the invertebrates; for the first time, a group of animals possessed a fully developed backbone, a vertebral column, rather than a much simpler and more primitive notochord. That design change allowed for the continued evolution of more complex modes of movement and lifestyles which eventually enabled vertebrates to encroach onto land. Today the vertebrates are the most recognized and visible animal group on the planet. Even though they do not dominate in terms of their total number of species nor their sheer numbers of individuals, it is the vertebrates we see every day—birds, mammals, reptiles, amphibians, and, of course, if we peer beneath the surface of the water, fishes.

Everyone knows what a fish is, sort of anyway. But because there are so many different kinds of fishes—because of their great species diversity—attempting to describe a fish to someone who has never seen one could prove to be a frustrating experience. Nonetheless, one definition that has been offered is: a fish is a vertebrate that possesses a cranium, is aquatic throughout its life, and therefore relies mainly on gills for gas exchange.[1] As we will see in this chapter, there is much more to the fishes than this simple definition entails.

The fishes were the first vertebrates. They appeared during the Cambrian explosion more than 500 million years ago, a time when species richness—the total number of species on Earth—took off, as evidenced by the rich fossil record from that time. By the Devonian Period, also known as the Age of Fishes, more than 400 million years ago, the fishes were well established and comprised three taxonomic groups of vertebrates that we recognize today (**FIGURE 11.1**): the Agnatha, which are the jawless fishes; the Chondrichthyes, which are the cartilaginous fishes—the chimaeras, sharks, skates, and rays; and the Osteichthyes,[2] which are the bony fishes, the most diverse group of modern fishes, which include tuna, salmon, and most of the fish species we normally encounter (and consume). In fact, the bony fishes are not only the most diverse group of fishes, they are by far the most diverse group of vertebrates. The agnathans, however, are not nearly so diverse; the number of agnathan species declined drastically after reaching their height in the Devonian, and today there are but two groups, the lampreys and the hagfishes. The diversity of the Chondrichthyes falls in between. But all three groups have been

---

[1] Michael Barton, 2007. *Bond's Biology of Fishes* (Belmont, CA: Thomson Brooks).

[2] These taxonomic groups are often referred to as classes, within the phylum Chordata, but in recent times they more often referred to as a clade or group. The Osteichthyes were renamed the Teleostomi in 1994, but that name has not been universally accepted; to avoid any confusion we are retaining use of Osteichthyes.

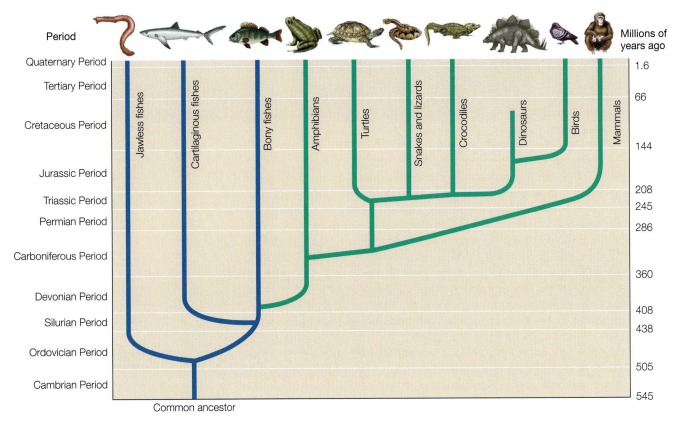

Period ... Millions of years ago

Period	
Quaternary Period	1.6
Tertiary Period	66
Cretaceous Period	144
Jurassic Period	208
Triassic Period	245
Permian Period	286
Carboniferous Period	360
Devonian Period	408
Silurian Period	438
Ordovician Period	505
Cambrian Period	545

Jawless fishes · Cartilaginous fishes · Bony fishes · Amphibians · Turtles · Snakes and lizards · Crocodiles · Dinosaurs · Birds · Mammals

Common ancestor

FIGURE 11.1 Geological time scale with major periods and the time before present for each, in millions of years. Superimposed is an evolutionary tree of life for the vertebrates, showing that the fishes were the first vertebrates. Between 400 and 500 million years ago, they differentiated from a common ancestor into the three main groups we have today, the jawless fishes, the cartilaginous fishes, and the bony fishes.

established for a long time; notice in Figure 11.1 that the fishes had already been around for some 200 million years by the time the dinosaurs and other reptiles first appeared in the fossil record.

The vertebrates, you will recall, have far fewer species than the invertebrates, as they comprise only 3–5% of the total number of species of animals (in the kingdom Animalia), which might lead one to presume that they are unimportant. But even that small percentage means that there are some 55,000 species, more than half of which—some 30,000—are fishes. And nearly 60% of the fishes are exclusively marine. The fishes are indeed an extremely important group of animals that are instrumental in structuring marine ecosystems; some occupy lower trophic levels, feeding directly on phytoplankton, and others are top carnivores with virtually no natural enemies other than human beings. They are found everywhere in the oceans, from the shallowest of tide pools to the oceans' deepest depths, and they are objects for our enjoyment in large public aquaria and in small home aquaria. And, of course, they have been throughout history and continue today to be an important source of protein for human consumption.

In this chapter we will discuss the major biological features and ecological importance of the three main groups of fishes, as well as the more important and interesting aspects of the biology and ecology of marine fishes in general.

The Jawless Fishes

The **jawless fishes**, the **Agnathans**, are the most primitive fishes alive today. They had their heyday back in the Devonian Period (see Figure 11.1), when they reached their greatest diversity, but now comprise only about 100 species.

Hagfishes

The **hagfish**, order Myxiniformes, are exclusively marine and are the more primitive of the two groups of Agnatha (**FIGURE 11.2**). These eel-like fishes are relatively small, growing to a maximum length of about 50 cm, with the largest species reaching 1.5 meters. They lack actual vertebrae in their backbone, which for them is just an unadorned notochord, and the rest of their skeleton is mostly cartilage with some chitinous hard parts. Neither the hagfish nor the lampreys (see below) possess paired fins, and have only a single continuous tail fin that wraps dorso-ventrally around the tip of their tails. They have individual gill openings on each side of their bodies. Rather than articulating jaws, they have a pair of cartilaginous plates with toothlike structures at the ends that they use to pull away bits of their food items.

The hagfishes are scavengers; they are best known for their tendency to feed on dead or dying fish and marine mammals, including whales, by boring into them or entering through an orifice and feeding on them from the inside out. On the Internet, you can find dramatic footage of hagfishes completely devouring a dead whale on the bottom of the deep sea. They have a series of slime glands running along the sides of their bodies, giving them their other common name, the slime eels; these glands produce copious volumes of thick mucus which may assist in their burrowing into and feeding on carcasses. They have poor vision; their eyes are very small and are covered by a layer of skin, presumably to protect them as they burrow into their food, or into muddy bottom sediments where they are also known to feed on benthic polychaetes and small decapods. These eel-like fishes have been observed tying themselves in a simple knot which they force along the length of their bodies, wiping excess slime off themselves; the maneuver is believed to be a predator escape response.

Hagfishes are found in both the deep sea and on continental shelves, where, when not actively feeding, they reside in temporary burrows. There are commercial fisheries that target hagfish which are mostly exported to Korea and other Asian markets, where they are used both for food and for their skins (commonly known as "eel skins") which are processed into a fine leather used in making wallets. The demand for hagfish has resulted in the depletion of local stocks in the Pacific Ocean, and a fishery has recently developed in the eastern North Atlantic Ocean.

Lampreys

The other group of Agnatha, the **lampreys**, in the order Petromyzontiformes, reside in coastal marine waters as well as in some freshwaters (there is a landlocked population in the North American Great Lakes), but because they prefer cool water, they are not found in the tropics and subtropics. Like the hagfishes, they lack jaws and paired fins and have a cartilaginous skeleton. The lamprey is considered to be a parasitic fish that feeds by attaching itself onto another fish with its circular mouth, forming a tight seal and cutting away the host fish's tissue with its circles of sharp teeth in order to suck its body fluids (**FIGURE 11.3**). Because the host fishes of lampreys often do die from their injuries, lampreys are not considered true parasites (which do *not* kill the host).

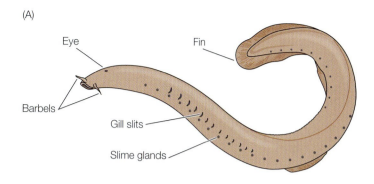

(A)

Eye

Fin

Barbels

Gill slits

Slime glands

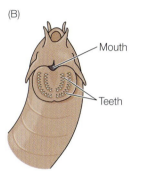

(B)

Mouth

Teeth

(C)

FIGURE 11.2 (A) External anatomy of the hagfish, *Eptatretus polytrema*, from off Chile and (B) ventral view of mouth region exposing the two rows of teeth. (C) A hagfish relaxing on the ocean bottom.

FIGURE 11.3 (A) A sea lamprey attached to a salmon. (B) A sea lamprey with one side of its seven pairs of gill slits, a well-developed eye, and circles of teeth clearly visible.

These fish became a major headache for fisheries managers in the Great Lakes of North America back in the 1930s, when the deepening of the Welland Canal connecting Lake Erie with Lake Ontario was finally completed. Built in stages beginning a hundred years earlier in the 1820s, the canal was intended for commercial ship traffic, but, with its connection to the St. Lawrence Seaway that leads to the ocean, it also allowed free passage for sea lampreys to invade the Great Lakes. Eventually sea lampreys were observed in Lake Ontario. Following further improvements to the canal, sea lampreys began to show up in greater numbers, and by the 1940s and 1950s their population levels in the Great Lakes had taken off and they had already begun to deplete stocks of lake trout and white fish. Management programs were initiated in the 1950s, including the construction of barriers to the passage of sea lampreys into and out of the natal streams where they spawn, an effort that capitalized on a bottleneck in their life history.

The sea lamprey is an anadromous fish (see page 384); it leaves the sea (or the Great Lakes) to spawn in freshwater streams where its wormlike larvae reside in burrows in the sediments; they remain in this larval stage for as few as 2 years to as many as 18 years before they move downstream to the ocean, or to the lakes, to begin their adult predatory lifestyle. Today the Great Lakes problem is being managed, but it has not gone away.

The Cartilagenous Fishes

The **cartilaginous fishes**, the **Chondrichthyes**, are also among the most primitive, having evolved from the same ancestor as the jawless fishes more than 400 million years ago. They are one of two groups of gnathostomes, or jawed fishes (as distinct from the agnathans, or jawless fishes); the Osteichthyes is the other group. The chondrichthyans include three major subgroups: the sharks, the rays (and skates), and the chimaeras. Despite their primitive roots, they have been very successful, and today there are more than 1000 species.

With the chondrichthyans, for the first time in evolutionary history, we see articulated jaws and paired fins. However, their articulated jaws have the upper jaw *not* attached to the skull (except in the chimaeras), as it will be later in the evolution of higher vertebrates. In addition, they have dorsal and ventral fins that no longer run continuously around their tails but are configured as isolated, triangular rudder-like structures. Their skeleton is still cartilaginous, like that of the more primitive jawless fishes, although it may be heavily calcified in some taxa. Their integument is a tough coating of **denticles**, or placoid scales, oriented backward (see Figure 11.7). Placoid scales are not true fish scales but are similar in embryonic origin to modern vertebrate teeth. This gives the sharks, for example, their characteristic rough, sandpaper-like feel (the term "shark's skin" is used to describe a number of manufactured items in order to convey the potential for a tight, sure grip).

The three groups within the Chondrichthyes belong to two subclasses, the elasmobranchs and the chimaeras. The elasmobranchs, which includes the sharks, skates, and rays, are the more diverse of the two subclasses, with about 20 orders and more than 1000 species, whereas the chimaeras have but a single order with about 40 species.

Chimaeras

The **chimaeras** (**FIGURE 11.4**), subclass Holocephali, split off from the elasmobranchs 400 million years ago but remain similar to them in their having, among other features, a cartilaginous skeleton; they are dissimilar in having an upper jaw that is fused to the skull. The skin of chimaeras is smooth and without any scales; they posses only a single gill slit; and they lack the teeth of sharks, having only what can be best described as upper and lower grinding plates. Their caudal fin is **diphycercal** (meaning it tapers to a point), with the notochord running to the tip, but in some species it is more **heterocercal** (meaning that the dorsal lobe of the caudal fin is larger than the ventral lobe), almost an exaggerated version of that of sharks (see Figure 11.7). Most chimaeras have a venomous spine at the leading edge of their first dorsal fin. Their mouth is on their ventral surface, removed from the immediate front end of the fish, and so respiratory currents of water enter through their spiracle (an external opening that allows water to pass over the gills). Chimaeras are usually deep-sea fishes, but some species are commonly encountered in colder but relatively shallow, continental shelf waters.

Elasmobranchs: Sharks, skates, and rays

The **elasmobranchs**, subclass Elasmobranchii, are the readily recognizable sharks, rays, and skates that we have all seen either in person or at the very least on television (**FIGURE 11.5**).

SHARKS   The modern sharks, in the superorder Selachimorpha, with about 450 species, are certainly the best-known group within the elasmobranchs. Distributed throughout the world ocean, the pelagic sharks have evolved a streamlined body form designed for swift swimming speeds that have enabled

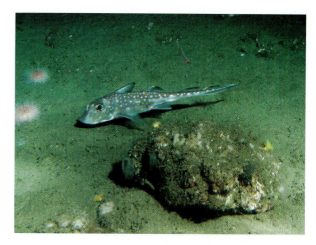

FIGURE 11.4 A chimaera, the spotted ratfish, *Hydrolagus colliei*, photographed at a depth of 175 m in the Cordell Bank National Marine Sanctuary.

FIGURE 11.5 Elasmobranchs. (A) A great white shark (*Carcharodon carcharias*). (B) A roundel skate (*Raja texana*).

FIGURE 11.6 Examples of shark teeth. (A) A sand tiger shark (*Carcharias taurus*) showing its impressive teeth. (B) A shark's tooth showing the pointed, triangular shape and serrated edges.

these fishes to become the master predators of the oceans. Their numerous sharp teeth are almost legendary (**FIGURE 11.6**); arranged in rows on both the upper and lower jaws, they are continuously being replaced as the front teeth are lost. Maybe it is their ominous-looking front ends with all those teeth that promote anxiety in so many people. It certainly isn't that so many of us are killed by sharks; only 30 people worldwide are killed by shark attacks each year. In fact, the opposite is far closer to the truth—more than 800,000 tons of sharks are harvested each year in commercial fisheries (some just for their fins). They should fear us.

The sharks lack fin rays, the firm, rod-like support structures in the fins that most bony fishes have, but they have well-developed pectoral fins and a caudal fin, which not only provide propulsion and maneuvering while swimming, but in pelagic species also provide lift to the negatively buoyant fish (**FIGURE 11.7**). The sharks are significantly denser than their seawater surroundings, and unlike the modern bony fishes, they do not have an air-filled swim bladder with which to control buoyancy. Having a cartilaginous skeleton helps, though; the density of cartilage is significantly lower than the density of bone. Sharks also have a lot of oil in their large livers; the density of that oil is only about 0.86. Even with these adaptations, most sharks are still negatively buoyant. In order not to sink, most pelagic sharks must keep moving through the water, generating hydrodynamic lift with their broad pectoral fins and their heterocercal caudal fin. The pectoral fins function almost like the wings of an aircraft, and the top half of the heterocercal caudal fin generates a downward thrust that helps keep the tail end up. But this isn't the only reason that the pelagic sharks have to keep swimming; they also need to generate water flow to ventilate their gills. Unlike the modern bony fishes, the sharks lack a gill cover, which the bony fishes use in pumping water across the gills to exchange respiratory gases carried in the blood. Instead, each of the shark's five to seven gills has its own opening in the form of external gill slits. As the shark swims, water is forced into its mouth and also into its spiracle; both openings facing forward into the direction they are swimming. This forces water into the shark's buccal cavity and across the gills, which are just behind it, where oxygen and carbon dioxide are exchanged; the water leaves through the external gill slits. This is why the pelagic sharks always seem to have their

FIGURE 11.7 External anatomical features of a typical shark.

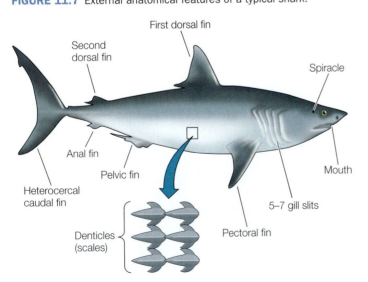

First dorsal fin

Second dorsal fin

Spiracle

Anal fin

Pelvic fin

Mouth

Heterocercal caudal fin

5–7 gill slits

Denticles (scales)

Pectoral fin

FIGURE 11.8 A whale shark, *Rhincodon typus*, photographed at the Georgia Aquarium.

mouths open in photographs: it isn't to show off their sinister-looking teeth; it is so they can breathe.

Not all sharks swim constantly, however. Some species are built for a different lifestyle, one that is more of a benthic existence. One example is the nurse shark. These sharks manage to pump water across their gills without the need to be constantly moving through the water; they spend significant periods of time sitting quietly on the bottom, waiting for their next meal to pass within striking distance, and their respiratory demands are far less than those of the pelagic sharks. Their mouths are generally smaller than those of pelagic species, but their buccal cavities are relatively large, enabling these fish to capture prey by quickly opening their mouths, producing a sudden and powerful suction that pulls prey organisms into their mouths. And with their smaller mouths and large buccal cavities, they are able to pass the gulped water across their gills at a rate that is sufficient to meet the relatively meager respiratory demands of their sessile lifestyle.

Most sharks are predatory carnivores and sit atop the trophic pyramid; their preferred food is other fishes. But not all sharks are carnivores; some are **planktivores** (plankon-eaters). The world's two largest sharks, the basking shark and the whale shark, for example, are planktivores, feeding on zooplankton as well as on small fishes and other nektonic invertebrates they sieve out of the water (**FIGURES 11.8** and **11.9**).

The sizes of sharks vary widely, from the smallest species, the dwarf shark (genus *Squaliolus*), which measures only about 20 cm in length when reproductively mature, to the whale shark, the largest fish there is, which can reach 15 m (**FIGURE 11.10**). Approximately half of all sharks are less than a meter in length; but on average, the sharks are all quite large, with lengths of two meters being the norm.

"Man-eaters" do exist, unfortunately, with the great white shark being the most notorious. Great whites, the largest of the pelagic predatory sharks, are often featured on cable TV programs recorded off South Africa, where they congregate at certain times of the year. But these shows can be misleading

FIGURE 11.9 A basking shark, *Cetorhinus maximus*, swimming slowly through the water with its massive mouth fully opened such that the plankton-containing water passes over its gills. Gill rakers collect the particles, which are then swallowed.

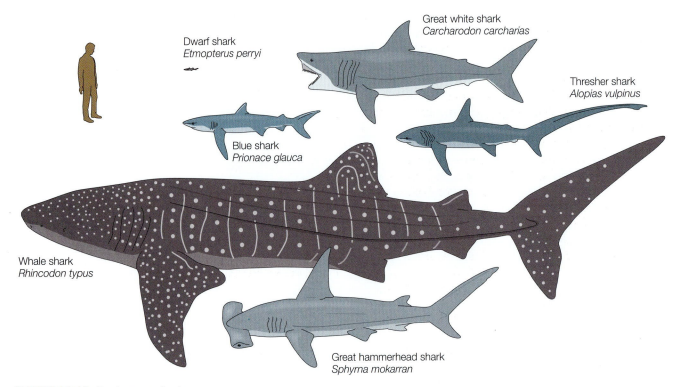

Dwarf shark
Etmopterus perryi

Great white shark
Carcharodon carcharias

Thresher shark
Alopias vulpinus

Blue shark
Prionace glauca

Whale shark
Rhincodon typus

Great hammerhead shark
Sphyrna mokarran

FIGURE 11.10 Sharks range in size from the dwarf shark to the whale shark, with a human shown for reference.

in that the great white is a cosmopolitan species and is not restricted to South Africa or even the subtropics—they are distributed around the world and can be found in all oceans.

The great white shark gets to be quite large, and there are numerous (mostly unsubstantiated) claims for the largest great white shark ever caught, but actual scientific verification is usually lacking. Claims for the record include one that was landed in Taiwan in 1997, which was reportedly almost 7 m in length and weighed 2500 kilograms. Another contender was one caught in Cuba in 1945 that measured 6.4 m and weighed 3175 kilograms (about 7000 lbs!). These reports have photographic evidence to back them up, in the form of poor-quality black-and-white snapshots, as well as various newspaper accounts and of course, a number of blogs. Ironically, one of the biggest fish stories was published in a scientific journal in 1935, which reported that a great white shark caught in the Gulf of Maine, inside the Bay of Fundy, measured 37 feet (!).[3]

The **skates** and **rays** are also elasmobranchs, but they are in the superorder Batoidea and do not superficially resemble the sleek pelagic sharks at all. Rather, these fishes are flattened, with pectoral fins that are greatly expanded into wing-like structures, and long thin tails at their ends (the name Batoidea was coined based on their bat-like wings). There are about 500 species of the Batoidea in four orders. They include such well-known examples as the electric ray, the manta ray, and the stingray. Electric rays possess organs that are capable of discharging a potent shock of more than 200 volts, which they use to stun and paralyze their prey (and which can easily kill an adult human).

[3] As reported in Valydykov, V. D., and McKenzie, R. A., 1935. The marine fishes of Nova Scotia. *Proceedings of the Nova Scotian Institute of Science* 19(1): 17–133. The length of 37 feet (>11 m) has been questioned by many.

Manta rays are the largest rays, with some reaching a wingspan of some 7 meters. These magnificent animals are limited to warmer ocean waters where, like the basking and whale sharks, they feed on plankton. The stingrays get their name for the sharp barbs on their tails, which the animals use for defense and which may carry venom. The stingray doesn't use its barbs offensively; but whoever accidently steps on one will get a very painful sting. Stingrays are bottom feeders, searching out and finding their food by detecting the electrical fields produced by potential prey animals—a sensory adaptation common to most or all of the other elasmobranchs as well.

The elasmobranchs exhibit a number of adaptations that have contributed to their evolutionary success over the past few hundreds of millions of years; in addition to their possession of jaws and paired appendages, they possess several very advanced sensory systems. Two of the more important sensory systems that evolved are the lateral line system and the ampullae of Lozenzini (**FIGURE 11.11**).

The **lateral line system** detects physical disturbances and vibrations in the water. That capability is also present in a rudimentary form in the hagfishes and lampreys in the form of individual **neuromasts**, receptors just beneath the skin that detect minute water motions through tiny pores. Neuromasts can detect disturbances created by low-frequency noises and animal movements. But in the elasmobranchs and the bony fishes, the neuromasts are organized into more sophisticated, organ-like canals that run along the sides of the fish, hence the name lateral line, and extend onto the head as shown in Figure 11.11. Another sensory capability of the elasmobranchs is their detection of electrical impulses given off by muscular actions in potential prey animals. Called **ampullae of Lorenzini**, these subcutaneous sensors are also found in other primitive fishes, such as sturgeons and lungfishes. These two sensory systems for the detection of movements of potential prey have enabled sharks to become efficient predators. They can even locate prey that have buried themselves in sediment and are invisible; electrical impulses generated when prey organisms move a muscle reveal their location.

Pelagic sharks are well known for their olfaction—their sense of smell—which allows them to detect phenomenally low concentrations of substances in the water, such as blood. Some sharks are known to be capable of homing in on such chemical cues in the water by differentiating which side of their head receives the stronger signal. In experiments with sharks in which their nostrils are sealed, the animals tend to swim in circles, never locating the olfactory target source. It is believed that the hammerhead sharks (**FIGURE 11.12**), whose nostrils are widely separated on either end of their hammer-shaped heads, can detect even smaller chemical gradients than other sharks can. Not only are the nostrils separated in the hammerheads, so too are their eyes, which facilitates their three-dimensional vision. The sensory pores of their ampullae of Lorenzini and lateral line which are distributed across their hammer-shaped head, add to their arsenal of prey-seeking capabilities.

FIGURE 11.11 The numerous pores visible on this shark are a mix of ampullae of Lorenzini and lateral line pores.

FIGURE 11.12 A hammerhead shark.

(A) Coelacanth (*Latimeria chalumnae*)

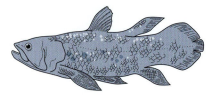

(B) Australian lungfish (*Neoceratodus fosteri*)

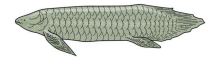

FIGURE 11.13 Examples of the two groups of living sarcopterygians, lobe-finned fishes.

The Bony Fishes

The **bony fishes**, the **Osteicthyes**, are the modern fishes. Most of the fishes alive today belong to this group, and in fact, they are the largest group of living vertebrates. Of the 30,000 or so fish species, more than 96% of them are bony fishes; about 41% of the bony fishes are found in freshwaters and 59% in marine and brackish waters. Unlike the cartilaginous fishes, which come in only two basic body styles—the relatively large (average 2 meters) torpedo-shaped, fast-swimming pelagic sharks, and the flattened and otherwise less-streamlined benthic sharks, skates, and rays—the bony fishes come in many forms. With the development of skeletons made of bone, which, although denser than cartilage, is much stronger and therefore requires less mass, the bony fishes were free to evolve a multitude of more complex shapes and a wider range of sizes. No longer were the fishes limited in how small they can be (many modern species are only a centimeter or two in length) or how large they can be (tunas and catfishes, for example, can exceed 3 or 4 meters in length). The fishes now have a far wider variety of basic body forms than do the cartilagious fishes, as well as a multitude of modes of locomotion—burrowing, crawling, walking, hovering, and even flying.

There are two basic groups, or clades, of bony fishes: the **lobe-finned fishes**, the Sarcopterygii, and the **ray-finned fishes**, the Actinopterygii. It is from the lobe-finned fishes that the tetrapods (the higher vertebrates) evolved—the four-legged (four-appendaged) animals that include amphibians, reptiles, birds, and mammals. Only eight species of lobe-finned fishes still exist today: they are the lungfishes and the coelacanths (pronounced "seel-a-kanth"; **FIGURE 11.13**; **BOX 11A**).

BOX 11A The Coelacanth: A Fish Believed To Be Extinct Turned Out Not To Be, after All

One of the greatest discoveries, and surprises, in the field of modern zoology was the accidental finding of a living coelacanth. This marine member of the sarcopterygian fishes, the lobe-finned fishes, had long been thought to be in the same straits as most of its cousins in that group—extinct for the last 70 million years or so. In the back rooms of museums where their fossils were studied, scholars tried to sort out the details of how these unusual fishes eventually invaded land and evolved into the earliest amphibians—how fishes evolved into tetrapods, the four-legged animals that walk the earth today.

Then, a few days before Christmas in 1938, a fishing boat pulled into port in East London, a town in South Africa and off-loaded its catch. As they usually did when they had caught an odd-looking fish, the crew notified Marjorie Courtenay-Latimer (1907–2004), the curator of the local museum. Recogniz-ing the significance of their find, Latimer sent a sketch of the fish to her colleague at Rhodes University, who published the finding in 1939, naming the fish for Latimer and the river that empties into the Indian Ocean where the fish was caught: *Latimeria chalumi-nae*.

Soon the hunt was on for additional specimens, since all that Latimer was able to preserve of the large fish (>1 m) was its head and skin. Fourteen years later, in 1952, another was caught in the Indian Ocean off the Comoros Islands, between Madagascar and East Africa. Another 50 or so specimens would be collected from this part of the Indian Ocean over the next 15 years.

The coelacanth occupies deep waters, from about 100 to 600 m, and is difficult to observe in nature. Nonethe-

less, numerous specimens became available for study in the 1960s and 1970s, and the field exploded with publications. In 1997, another coelacanth, a different species, was discovered in a fish market in Indonesia, some 6000 km east of South Africa. These remarkable fish are apparently more widely distributed than was once thought, which is not saying much, since they were so recently thought to be extinct!

The taxonomy of the fishes is not much cleaner than that of the invertebrates. The groups of fishes we have been discussing, the Agnatha, the Chondrichthyes and the Osteichthyes, have been placed in the taxonomic categories of class, superclass, or subclass, with little apparent agreement among ichthyologists. Generally, the two groups of bony fishes in the Osteichthyes, which itself has been renamed Teleostomi, have been grouped as clades: the Sarcopterygii and the Actinopterygii.

The actinopterygians, the ray-finned fishes, comprise the remaining bony fishes. They have been by far the most successful throughout their evolutionary history and today they can be found virtually wherever there is water—freshwater, saltwater, or anywhere in between. They have flourished for two reasons: first, they successfully invaded and became established in freshwaters, where some 41% of all species are found; second, they have an astonishing capacity to evolve and adapt to almost any environment, with a bewildering array of feeding mechanisms, means of locomotion, and body forms (**FIGURE 11.14**).

There are three subclasses in the clade actinopterygian fishes: the **Chondrostei**, a group of fishes that are mostly cartilaginous but exhibit significant ossification; this group includes the sturgeons and bichirs and, the **Neopterygii**, which includes the gar fishes. Both these subclasses are primarily freshwater, with the exception of the sturgeons, which move effortlessly between marine and freshwater environments. The third subclass is by far the most important: the **Teleostei**, the "perfect bone" fishes, which comprise some 96% of all named fishes. Quite simply, they include virtually all the commonly known fishes—trouts, salmons, tunas, mackerels and swordfishes; cod and haddock; flounders, sunfish, guppies, goldfish; if you can name it, it's probably a teleost.

The characteristics of the bony fishes, and the teleosts in particular, include: an internal skeleton of strong bone which, unlike cartilage, is a network of vascularized cells; a vertebral column with vertebrae made of bone; relatively thin and flexible epidermal scales; slender and flexible **fin rays**; an **operculum**, or bony gill cover; a gas-filled

FIGURE 11.14 A very small sample of the variety of forms of the bony fishes. (A) The common seahorse (*Hippocampus erectus*), found in shallow waters of the western North Atlantic. (B) A lionfish (genus *Pterois*), native to tropical waters of the Indian and Pacific Ocean. These fish have highly venomous dorsal fin spines, clearly visible here, which can cause severe pain and death should a person be pricked by one. At least two species of this fish have been introduced into the subtropical western Atlantic, and they can now be found in coastal waters of the southeast United States. (C) The lumpfish (*Cyclopterus lumpus*), found on both sides of the North Atlantic. (D) The more classical-looking Atlantic salmon (*Salmo salar*). Notice the "adipose" fin behind the dorsal fin, which is characteristic of the trout and salmon as well as certain other species. (E) A starry flounder, *Platichthys stellatus*.

(A)

(B)

(C)

(D)

(E)

FIGURE 11.15 Basic external anatomical features of a teleost fish. Not shown are the adipose fins of trout, salmon, and certain other species; see Figure 11.14D.

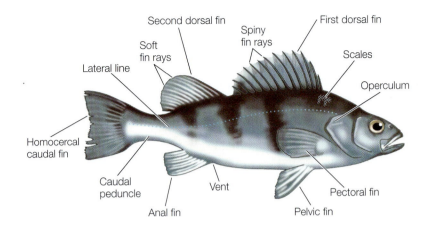

swim bladder for buoyancy control; and maintenance of hypoosmotic body fluids in seawater (**FIGURE 11.15**).

In the following section we will elaborate on some of the biological characteristics and adaptations in the marine fishes, as well as the more important aspects of the behaviors and life histories of a number of key species.

General Biology of Marine Fishes

Respiration

Fishes are heterotrophs; they consume organic matter—food—and burn it, consuming oxygen and liberating carbon dioxide (and water) in the respiration process. This means that they must acquire oxygen, somehow. For air-breathing animals, the source is atmospheric oxygen, which we learned several chapters back is 21% of the volume of air. But the percentage of oxygen dissolved in water is so low that it is measured not as a percentage, but as parts per thousand, or milliliters of dissolved oxygen gas, O_2, per liter of water. The concentration is almost always on the order of 5–10 ml/L. In order to extract enough oxygen for their metabolic needs from the water, fishes have evolved efficient gills. And of course, to carry that oxygen, most fishes have blood cells that contain hemoglobin, a compound with a high affinity for oxygen. For example, a liter of blood plasma without red blood cells and hemoglobin can carry only a few milliliters of dissolved oxygen, but with red blood cells and hemoglobin, a liter of blood can carry a hundred times that much oxygen.

As we alluded to earlier, the operculum, or gill cover, has given most of the bony fishes freedom from the need to swim constantly, as most sharks must, in order to acquire sufficient oxygen to meet their high metabolic demands. Sharks ventilate their gills by allowing water to enter their buccal cavity through their open mouths and spiracle while they swim; this water then flows over the gill filaments, exchanging oxygen and carbon dioxide with the blood, and then exits out through their five to seven gill slits. The bony fishes, on the other hand, possess an operculum which (for most of them) precludes their having to actively swim in order to pass water across their gills (the large pelagic fishes such as the tunas are exceptions). A bony fish simply opens its mouth, expanding its buccal cavity while the operculum is sealed, which creates suction, thus filling the buccal cavity with water; the fish then closes its mouth and forces that water out the now-open operculum and

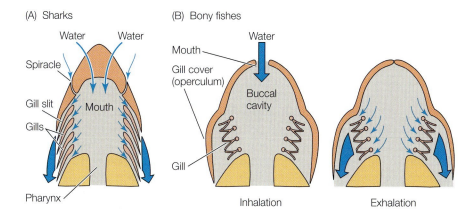

(A) Sharks

Water   Water

Spiracle

Gill slit

Gills   Mouth

Pharynx

(B) Bony fishes

Water

Mouth

Gill cover (operculum)

Buccal cavity

Gill

Inhalation   Exhalation

FIGURE 11.16 Mechanics of gill ventilation in sharks and bony fishes. (A) Sharks ventilate the gills passively as they swim, with water entering the mouth and spiracle and exiting through the 5–7 gill slits. (B) The bony fishes take in water through the mouth (inhalation), filling the buccal cavity. The mouth closes, and the water is forced out across the gill filaments (exhalation).

over the gills (**FIGURE 11.16**). This inhalation and exhalation in the bony fishes gives the "gulping" action we are have all seen in aquarium fish. This greater control over its gill ventilation in the bony fishes is another reason for their greater species diversity; it has allowed them to adapt to various lifestyles, and therefore assorted environments and niches.

The gills themselves are remarkable organ systems that utilize a counter-current exchange mechanism that maximizes the efficiency of respiratory gas exchanges between seawater and the fish's blood (**FIGURE 11.17**). A gill is a

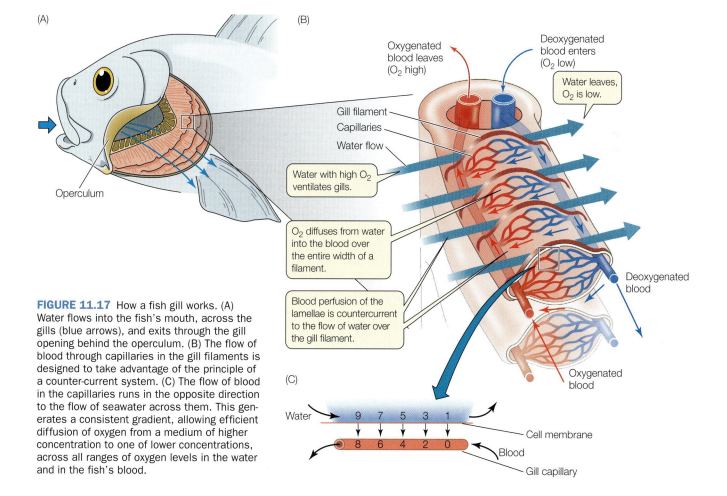

(A)

Operculum

(B)

Oxygenated blood leaves (O_2 high)

Deoxygenated blood enters (O_2 low)

Water leaves, O_2 is low.

Gill filament

Capillaries

Water flow

Water with high O_2 ventilates gills.

O_2 diffuses from water into the blood over the entire width of a filament.

Blood perfusion of the lamellae is countercurrent to the flow of water over the gill filament.

Deoxygenated blood

Oxygenated blood

(C)

Water

9   7   5   3   1

8   6   4   2   0

Cell membrane

Blood

Gill capillary

FIGURE 11.17 How a fish gill works. (A) Water flows into the fish's mouth, across the gills (blue arrows), and exits through the gill opening behind the operculum. (B) The flow of blood through capillaries in the gill filaments is designed to take advantage of the principle of a counter-current system. (C) The flow of blood in the capillaries runs in the opposite direction to the flow of seawater across them. This generates a consistent gradient, allowing efficient diffusion of oxygen from a medium of higher concentration to one of lower concentrations, across all ranges of oxygen levels in the water and in the fish's blood.

plumbing system that exposes blood carried in the fish's lamellae (thin capillaries) within their hundreds of gill filaments, to external seawater that flows across them in the opposite direction; oxygen is exchanged by diffusion across the thin membrane, only 1–2 μm thick, separating the fish's blood and the seawater. The direction of dissolved oxygen diffusion is from the medium of higher concentration (seawater) to the medium of lower concentration (the fish's blood), with carbon dioxide diffusing in the opposite direction. The key feature of the counter current system is a design that enables the flow of low-oxygen blood to be carried adjacent to a flow of seawater moving in the exact opposite direction; in the case of oxygen, this means that the fish's blood will always be adjacent to seawater of a higher oxygen concentration, which diffuses into the blood. The counter-current mechanism maintains the concentration gradient.

Interestingly, we mammals also use oxygen only after it is dissolved in water, in the thin watery layers of our lungs, after which it is exchanged with our blood. So, then (in theory), if our lungs and breathing mechanics were strong enough to handle the much denser water (versus air), we too could "breathe water."

FIGURE 11.18 Osmoregulation in marine fishes. (A) Cartilaginous fishes are slightly hyperosmotic, with tissue solutes slightly greater than that the surrounding seawater; this condition is maintained by elevated concentrations of urea in the blood. (B) Bony fishes are significantly hypoosmotic, with significantly lower concentrations of solutes in their body tissues, which requires the retention of water and excretion of excess salts.

(A) Cartilaginous fishes

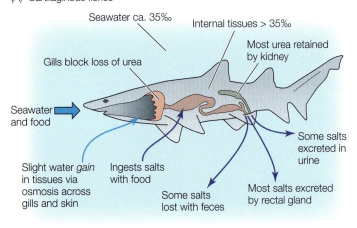

(B) Bony fishes

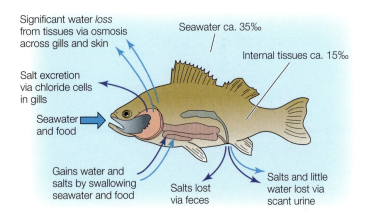

Osmoregulation

As we briefly discussed in Chapter 7, marine fishes are faced with problems of osmosis, the passage of water across their tissues. The extent and direction of that passage depends on the difference between the concentration of salts and other solutes in their tissues and that of the external environment (**FIGURE 11.18A**). The cartilaginous fishes maintain an internal tissue concentration of solutes that is about the same as seawater, thus reducing the osmotic pressure and the tendency to either lose or gain water across their tissues. This is accomplished by their maintaining an elevated concentration of **urea** in their blood. Urea is a compound used by animals to rid the body of nitrogen wastes, such as those resulting from the breakdown of proteins. Water is absorbed mostly from their food and across their gills, and excess salts that are taken in are excreted by way of their kidneys, their digestive tract, and their **rectal gland**, a specialized salt-secreting gland unique to the cartilaginous fishes.

The bony fishes maintain a much lower concentration of salts in their body fluids than the cartilaginous fishes do, less than half that of the surrounding seawater, and so they are constantly faced with an osmotic pressure gradient that forces a significant loss of water out of their tissues (**FIGURE 11.18B**). They overcome this water loss by ingesting seawater, passing much of the ingested salt through their digestive tract without absorbing it; salt that is absorbed is excreted by the kidneys and by specialized cells in the gills, called **chloride cells**. The kidneys also reduce water loss by minimizing the production and excretion of urine.

Fish propulsion

Fish, of course, swim. But they don't all swim the same way. Some fishes swim in a manner that is far more efficient than others, in terms of the energy expended, while others don't swim much at all, but rather spend much of their time simply lying on the bottom or in a burrow. Still others, like the pelagic sharks we have been discussing, swim continuously. There are even flying fishes. In this brief section we will consider the mechanics of fish propulsion, or swimming.

In each case, swimming involves the fish moving in an undulating fashion—moving its entire body, its tail (and caudal fin), or just a few specific fins (**FIGURE 11.19**). Each mode has its specific locomotory purpose, and each has an associated energetic cost. Primitive fishes, like the lampreys, hagfishes, even the modern eels and a few of the more slender species of bony fishes such as the sand lances, swim by an undulating motion of their entire body, a swimming mode that is energetically inefficient. Called **anguilliform** swimming (named after the eel genus *Anguilla*), the entire body moves, producing undulations that form one or more waves along the length of the body and generate forward thrust. At the opposite extreme from anguilliform swimming, the more advanced fishes such as the tunas, jacks, pompanos, and mackerel, for example (plus some of the large pelagic sharks), are more energetically efficient, at least in terms of the energy expended to swim a specific distance; they use rapid, powerful yet almost imperceptible motions of their caudal fins. This swimming mode is called **thunniform** (named for the tuna genus *Thunnus*), and is characteristic of the fastest-swimming fishes (**BOX 11B**).[4]

Anguilliform
(least efficient)

Thunniform
(most efficient)

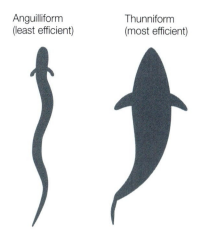

FIGURE 11.19 Swimming modes, from the most energetically efficient tunas, to the least energetically efficient eels.

[4] Wardle, C. S. et al., 1989. The muscle twitch and the maximum swimming speed of giant bluefin tuna. *Journal of Fish Biology* 35: 129–137.

BOX 11B A Word on "Top" Swimming Speeds in Fishes

There are various accounts of just how fast the fastest fish in the sea is.

Some textbooks list the wahoo, with a top speed of about 44 miles per hour, as the fastest fish, and give honorable mention to the barracuda. Other sources that are based on estimates of line-stripping speeds (how fast they draw line off the fishing reel) of game fishes caught by rod and reel list 75 miles per hour for the swordfish, 68–70 mph for the sailfish, and 80 mph for the black marlin. But a reliable scholarly journal credits the bluefin tuna with a maximum burst speed of 71.5 miles per hour.[4]

To put these speeds into perspective, the men's world record in the 100 meter freestyle is just over 47 seconds; the wahoo can do it in 5 seconds!

Sailfish

Tuna

Wahoo

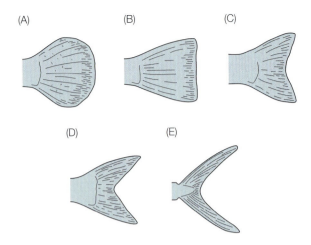

FIGURE 11.20 Caudal fin aspect ratios. (A) A flounder, with an aspect ratio of about 1; (B) a cod; aspect ratio about 2; (C) a salmon, about 3; (D) a bluefish; about 4; and (E) a bluefin tuna, about 7.

The fastest-swimming fishes tend to have caudal fins that are highly lunate (in a crescent shape, with the upper and lower parts symmetrical) with a high aspect ratio (the ratio of the height of the fin to its width) along with sleek, muscular body forms (**FIGURE 11.20**). An extreme example is the bluefin tuna (*Thunnus thynnus*), which has a sudden burst speed of more than 70 miles per hour (32 meters per second). In general, the caudal fin aspect ratio is a good indicator of fish swimming speed.

Not all fish swim with the classical undulating motion of their bodies; some use only certain fins, and they may or may not use their caudal fin. Examples of these swimming modes are given in **FIGURE 11.21**. Some fishes, such as the triggerfishes (family Balistidae) use only their second dorsal and anal fins, which they use in a wavelike, undulatory pattern to move both forward and backward, without the use of their caudal fin which, with their pectoral fins, is used in steering and stopping. It is also the mode used by the electric eel, which possesses only a long ventral fin running nearly the entire length of it body, which it undulates, and small pectoral fins that it uses to orient itself. Some fishes use only their pectoral fins to hover and maneuver (but not to swim very fast). This mode is common in the family Labridae, the wrasses, such as the tautog (see Figure 11.21B). The tautog is commonly found around steep rock ledges and dock pilings where it tends to hold its position, almost like treading water. Still other swimming modes include those employed by the skates and rays, which use their large pectoral fins in wavelike motions for the skates and wing-like flapping motions for the rays (e.g., see the roundel skate in Figure 11.5).

Given the great diversity of the bony fishes, it should not be surprising that there are numerous examples of unusual modes of swimming, but perhaps the most unusual use of fins in propulsion is that employed by the flying fishes (**FIGURE 11.22**). These are a family of fishes with about 60 species that

(A) Triggerfish First dorsal fin

Second dorsal fin

Anal fin

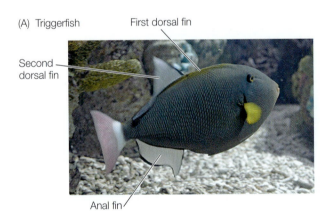

(B) Klunzinger's wrasse

Pectoral fin

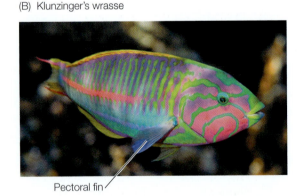

(C) Electric eel

Ventral fin

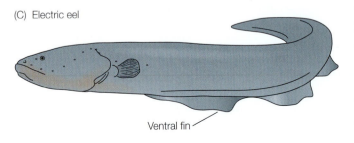

FIGURE 11.21 Examples of various modes of swimming. (A) A pinktail triggerfish (genus *Melichthys*), native to the tropical Pacific Ocean. This individual swims using wavelike undulations of its dorsal and anal fins, motions that are visible here, and steers using its caudal and pectoral fins. Its first dorsal fin (laying flat on back in photo), with three moveable spines, is characteristic of this family of fishes; it is in the lowered position in this photograph. (B) Some fish, like this Klunzinger's wrasse (*Thalassoma rueppellii*), use primarily their large pectoral fins in an oscillatory action. (C) The electric eel, *Electrophorus electricus*, has only a ventral fin, which it uses in an undulatory fashion.

(A)

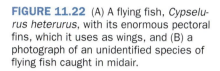

(B)

FIGURE 11.22 (A) A flying fish, *Cypselurus heterurus*, with its enormous pectoral fins, which it uses as wings, and (B) a photograph of an unidentified species of flying fish caught in midair.

can leave the water at a speed sufficient to glide significant distances in the air, several hundred meters, while staying aloft for as long as 45 seconds. These relatively small fishes, usually less than 30 cm, are sometimes found on the decks of research ships at sea depending on where the ship is (this is possibly because of the low deck height characteristic of research ships versus that of commercial ships; the fish come down on the decks after clearing the low rails). They are thought to glide rather than actually fly, but anyone who has watched them at sea might disagree; they certainly seem to be flapping their pectoral and caudal fins, apparently gaining lift and maintaining air speed as a result.

A close second for the most unusual fins and mode of swimming might be the ocean sunfish (**FIGURE 11.23**). These planktivorous fish get to be very large and are argued to be the largest bony fish, averaging about 2 m in length and weighing about 1000 kg; there are only two or three species, *Mola mola* being the most common and found in all oceans. Lacking a caudal fin, for all intents and purposes, and possessing only small pectoral fins, they swim by flapping their large dorsal and anal fins from side to side. Often encountered at sea, they are easy to spot, as they tend to reside near the surface such that their flopping dorsal fin breaks the surface, producing conspicuous splashes.

FIGURE 11.23 The ocean sunfish, *Mola mola*, off the coast of California.

Shoaling and schooling behavior

The terms shoal and school are often used interchangeably but they are not synonymous. A **shoal** is a group of fish. It is made up of a number of fish that have come together or are residing in the same area for social reasons; usually, but not always, they are all of the same species. A shoal is analogous to a flock of birds. A **school**, on the other hand, is a term that refers to shoals of fish, almost always of the same species, that have a uniform orientation and swimming speed, and which move as one, so to speak. Schooling is a special case of shoaling. Loose aggregations or shoals of fish may reform into schools for a variety of reasons, usually when there is a predator about. When under actual attack by predators, these schools may further contract into **pods**, sometimes called balls, or bait balls. These social behaviors are diagrammed in **FIGURE 11.24**.

FIGURE 11.24 Diagram of a fish shoal, school, and pod, as compared with solitary fishes. A shoal is a group of fish not orientied in any one direction or forming any organized structure in terms of distances between one another. A pod of fish may be oriented in the same direction and swimming together, but they are not uniformly spaced, the way a school is.

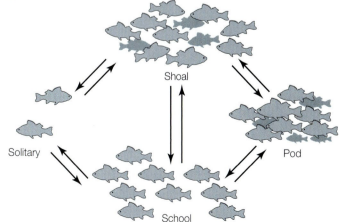

(A)

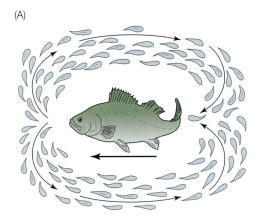

(B)

(C)

FIGURE 11.25 (A) Drawing of the response of a school of fish to a predator; the fish in front of the predator retreat to the back of the school. (B) A school of pilchards, *Sardina pilchardus*. (C) A school of bait-fish exhibiting schooling behavior as a shark approaches.

Shoaling is not common to all fishes, but a significant number, about 25% of all species, do exhibit the behavior, usually during migrations and during spawning. They apparently learn the behavior shortly after metamorphosing from the larval stage to the juvenile stage.

The chief advantage of shoaling (and schooling) has traditionally been thought to be to confuse predators. For instance, a shoal of feeding fish may be only loosely organized, but when a predator approaches they will begin to school, swimming as a unit, tightening and organizing themselves into a pod, decreasing the spaces between individuals; when under actual attack some species will form a "bait ball" with a mass of fish circling one another (**FIGURE 11.25**). The initial response of the fish as the predator encroaches into the school is for the fish at the front of the school, ahead of the predator, to circle to the back and reestablish themselves behind the others in the school. As the photograph shows, the school of bait fish avoids the shark, but not through a mad scattering of the fish—there is some symmetry, or some coordinated order apparent in their swimming behavior.

Schooling fishes depend on vision to maintain their formation, and at night most schools break up into shoals. They also use the their lateral line system to sense each other's movements; abrupt motion by one or more fish in the school produces a nearly instantaneous transmission of compression waves through water, to which, upon being sensed by their lateral line systems, the other fish seem to respond simultaneously.

The overall advantage afforded to fishes by their shoaling and schooling behaviors, traditionally thought to confuse predators, is not entirely clear, and fish biologists have argued that it is most likely a combination of several specific advantages. One advantage that has been suggested is hydrodynamic efficiency, especially when swimming significant distances such as during migrations; this would be analogous to the way race cars, runners, and cyclists bunch up behind the leader in order to reduce drag. Elite competitive (human) swimmers also do this, and even though the trailing swimmer may be in a adjacent lane, he or she gains some advantage by "drafting" behind the leader at some distance. Interestingly, drafting also aids the lead fish in that trailing fish will fill the eddy created behind the leader, which reduces some of the drag. Efficiency in locating food may be another advantage, in that many eyes and other senses can be put to work, but one can also imagine drawbacks. Tight associations of individuals, especially plantivorous fishes that are known to shoal and school, would seem to result in a competition for food; for example, fish in the front of a school or shoal would be the first to swim through a patch of plankton, feeding on the plankton as they swim, which would deplete the food density for those fish at the back. Reproductive success is an undeniable advantage, in that mates will be easy to locate in shoals. But of all the potential advantages, most scientists agree that the confusing of potential predators that schooling affords may be the most significant advantage, which probably outweighs any disadvantages.

Fish feeding

The great diversity of the modern bony fishes is reflected in the wide variety of their feeding mechanisms and food preferences.

Some fish eat other animals and are therefore predatory carnivores; those carnivorous fishes that prey specifically on other fishes are **piscivores**. Some fish eat algae and are therefore herbivores. Others feed on plankton, both phytoplankton and zooplankton, and are omnivorous planktivores, and still others feed on detritus, and are detritivores. The only trait they all share is that they are heterotrophic; they all have to eat something. And the total amount of food that fish require is determined by their metabolic rates, which in turn are correlated with activity, temperature, and body size, with food consumption going up as each factor increases.

We have already alluded to the large predatory sharks and their impressive teeth, swimming abilities, and sensory systems used to locate prey, and we have briefly mentioned the feeding modes used by the hagfishes and lampreys. The modern bony fishes also include predatory species such as the tunas and barracudas, but in addition, there is a multitude of fish species with specialized feeding adaptations, both anatomical and behavioral, that have enabled the group to be successful. Many species have obvious mouth adaptations that suit their preferred foods, such as those shown in **FIGURE 11.26**, as well as variable body forms, fin arrangements and swimming capabilities, all matched to their environment and to enable them to feed most efficiently. Bottom feeding fishes, for example, will have mouths that are more ventrally positioned than fishes that feed on near-surface organisms, which have their mouths more dorsally positioned.

The planktivorous fishes, including both the bony fishes and the elasmobranchs, have specialized **gill rakers** that are longer, more numerous, and more finely spaced than those in other fishes, which enable them to sieve particles out of the water (**FIGURE 11.27**). On the other hand, gill rakers in most other fishes are short and more widely spaced than in the planktivores, and function primarily to keep their prey items from escaping through the operculum opening. The planktivorous fishes, such as Atlantic and Gulf of Mexico menhaden (*Brevoortia tyrannus* and *Brevoortia patronus*), are capable of collecting small phytoplankton cells on their gill rakers, while the whale shark collects much larger items such as small fishes and squids. For example, the basking shark in Figure 11.9 is swimming with its large mouth wide open; its gill rakers, which sieve the larger zooplankton, small fishes, and squid as seawater passes over them, are not visible in the photograph, but they are there.

Like the basking shark, most planktivorous fishes swim with their mouths open, passively allowing water to enter their mouths and exit the gill opening,

| Barracuda | Butterflyfish | Parrotfish |

FIGURE 11.26 Examples of different mouth morphologies. (A) A barracuda (genus *Sphyraena*) mouth is full of sharp teeth designed to hold onto prey fishes. (B) A butterflyfish (family Chaetodontidae) has a tiny mouth designed to feed on coral polyps. (C) A parrotfish, (family Scaridae) has a small but powerful mouth designed to chip away at coral skeletons in order to ingest the polyps.

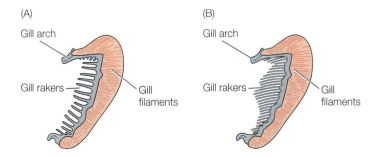

(A)
Gill arch
Gill rakers
Gill filaments

(B)
Gill arch
Gill rakers
Gill filaments

FIGURE 11.27 Gill rakers. (A) A cutaway view through the operculum of a standard bony fish, revealing the gill arch and gill filaments, where respiratory gases are exchanged between the fish's blood and seawater, and the gill rakers, which are bony protrusions that extend from the gill arch inward toward the fish's pharynx. (B) The gill arch of a planktivorous fish, which has much longer and more finely spaced gill rakes, for use in straining small food particles such as plankton out of the seawater as it flows over the gills.

FIGURE 11.28 Northern anchovies (*Engraulis mordax*) swimming with their mouths wide open, allowing water to pass over their gills in order to feed on the plankton in the water.

sieving plankton out of the water as it passes over their fine gill rakers. Note the gaping mouths on the anchovies and Pacific sardines in **FIGURE 11.28** as they swim through the water straining plankton. This is similar to the behavior of the basking shark in Figure 11.9.

Fish reproduction

All fishes—all animals, for that matter—are faced with two overriding challenges to their survival as a species: living long enough to reach reproductive age, and then actually reproducing once they get to that point. Reaching reproductive age, of course, involves effectively finding food and avoiding predation. To deal with the second challenge, actually reproducing, the fishes have evolved a number of what are called **reproductive strategies**.

Nearly all fishes reproduce sexually (although there are a handful of reports of parthenogenesis, development of unfertilized eggs). Some fishes have evolved strategies whereby they invest most of their reproductive effort into the production of large numbers of eggs, which are fertilized as they are released into the ocean, where they are left to survive the multitude of challenges the oceans place on them. Some species produce fewer eggs that, egg for egg, are larger and exhibit a higher proportion of survivors over time than the smaller, more numerous eggs of other species. Still other groups have evolved greater parental care of the developing embryos, such as seen in many of the sharks whereby fewer eggs are produced and are brooded internally for relatively long periods of time, producing well-developed pups at hatching; this condition is known as ovoviviparity. Other reproductive strategies involve the construction of nests by some fishes (such as the sticklebacks) and/or parental care of the young after they hatch.

Apart from various anatomical features, the cartilaginous fishes also differ significantly from most of the modern bony fishes in their reproduction. Unlike the bony fishes, the sharks and rays have extremely low **fecundity**—they produce very few young at a time. This means that their reproductive potential is far more limited than that of fishes such as the cod, for example, a single adult female of which can produce millions of eggs per spawning. Instead, the sharks and rays have on the order of four to ten young per spawning, which are born well-developed. With such a low number of offspring, the chondrichthyes need to ensure the survival of enough young such that the spawning adults can, at least on average, replace themselves.

The chimaeras and nearly half of the sharks and rays are **oviparous**, meaning that they lay eggs. They usually lay only a few eggs at a time, and they do so only once a year; the eggs are simply dropped from the female to the ocean floor where they develop on their own, usually in a tough and leathery egg shell (**FIGURE 11.29**). Unlike the egg shells of higher vertebrates, such as reptiles and birds, the egg cases of the chondrichthyans have openings that allow seawater to pass freely into and out of the egg in response to movements of the developing embryo or fetus. Most of these eggs contain only a single developing embryo (often called a pup), but in some species, there may be more. In each instance, however, the young pup hatches at an advanced stage of development and is fully capable of making it on its own after hatching.

Other species of Chondrichthyes, the sharks especially, retain the eggs in the mother's reproductive tract where the

FIGURE 11.29 Egg case of skates are often found washed up on the seashore after the pup hatches.

embryos develop to an advanced stage prior to being born alive, often carrying an attached yolk reserve with them (**FIGURE 11.30**); this is known as **ovoviviparity** (as distinguished from live births in mammals, for example, and even in some fishes, which is known as **viviparity**). The young pup is capable of swimming, and with its yolk reserve still attached, it has some time to get used to the new world around it, especially the need to find a meal once the yolk reserve is exhausted.

The low fecundity and long development, or gestation, period of the Chondrichthyes means that the group as a whole has low reproductive potential (**TABLE 11.1**). In the case of the sharks, in particular, this presents problems when there are fisheries focused on them; even recreational shark hunting is being shown to have an impact on their populations. There is a growing recognition, and concern, that when (not if) certain species of exploited sharks become depleted to low stock levels, their phenomenally low reproductive potential may prove inadequate to their survival as a species.

Because of these various reproductive strategies the marine fishes vary greatly in their fecundity and reproductive potential, which we alluded to in our discussion of reproduction in the Chondrichthyes. The sharks, as we have mentioned, usually reproduce only once a year, have a long gestation period and have very few offspring at a time. In such cases the young are born well developed and have a relatively high survival rate. Most other marine fishes, with some exceptions, are far more prolific spawners, producing hundreds of thousands, even millions, of eggs per spawning as dramatically illustrated for several fish species in Figure 11.32. For example, a large female cod (*Gadus morhua*), historically one of the most important commercially harvested food fish in the North Atlantic Ocean, will produce as many as six million eggs. But of these millions of eggs spawned, most will die long before reaching adulthood; more specifically, about 99.9999% of the six million eggs will die before reproductive age, leaving only six survivors—which is okay, and in fact it is better than okay, it is an excellent outcome, because on average only two offspring, not six, need to survive to reproductive maturity in order to sustain the population.

Fish growth and mortality

Most of the modern bony fishes exhibit a life history that begins with an egg, of which the adult females of many species produce and broadcast huge numbers into the ocean, while being fertilized externally by the male as they are released. The egg usually hatches into a **yolk-sac larva**, which upon resorbing its yolk reserves metamorphoses into a **post-larva** (also called simply a larva) which eventually metamorphoses into a juvenile that closely resembles the adult. The juveniles of many species, but not all, do not become reproductively mature until they are several years of age. An example of this development of larval and juvenile stages of a modern bony fish is given in **FIGURE 11.31**.

FIGURE 11.30 A newly born pup of a spiny dogfish shark (*Squalus acanthias*) from off the coast of Alaska. Notice the large yolk sac that is still attached, and that the young shark already resembles its parents.

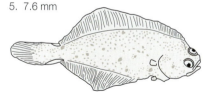
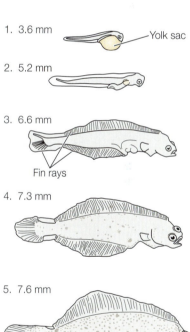
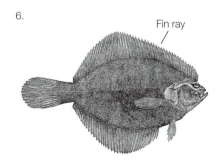

FIGURE 11.31 Larval development of the smooth flounder, *Liopsetta putnami*, a coastal flounder that is common in the North Atlantic Ocean. The flounder spawns hundreds of thousands of negatively buoyant eggs that are only 1.2 mm in diameter; after floating around in the plankton for 3–4 weeks, they hatch into planktonic yolk sac larvae that are about 3.6 mm in length (1). After a few weeks, the larvae will have absorbed their yolk sac, during which time their jaws will have developed; they begin feeding on plankton at this point (2). Over time the planktonic post-larvae develop pigmentation and fin rays (3), and when they are just over 6 mm in length and still only a few weeks of age, their left eye begins to migrate to the right side of their head (4). At this point the larva no longer swims upright the away a trout does; instead, the larva is transformed into a juvenile flounder that resembles the adult (5). It will reside on its side, on and near the bottom, for the rest of its life (6).

TABLE 11.1 Shark versus bony fish fecundity

Sharks	Average number of offspring
Sand tiger	2
Porbeagle	5
Basking	6
Thresher	4
Frilled	15
Hammerhead	40
Bony fishes	
Trout	200
Herring	30,000
Cod	6,000,000
Ling	30,000,000–60,000,000

Source: After Parker, S., and Parker, J., 1999. *The Encyclopedia of Sharks*. Buffalo, NY: Firefly Books.

A graph of the number of survivors over time that might be expected for a typical **year class** of fish, that is, a group of fish that are all spawned at the same time, is called a survival curve (**FIGURE 11.32**). The mortality that operates on the population of fish in that year class is constant; that is, the loss of fish each day is a constant percentage of the number of fish present at the time. This results in an initial drop in the number of survivors that is quite dramatic, as Figure 11.32 shows, necessitating that the data be plotted on a log scale to be readable. The initial mortalities are the result of predation on the eggs and defenseless larvae, as well as starvation, or at the very least, debilitating weakness and greater predation on larvae as a result of low food concentrations. The reproductive strategy employed by the cod is to flood the environment with eggs and newly hatched larvae with the expectation that some, at least a few, will ultimately survive (of course, no conscious decision-making process is implied here).

A classical theory of the biology of fishes is that species that adopt this mode of reproductive strategy—species with lots of offspring that experience high mortalities versus that of the sharks and some bony fishes that produce far fewer, but more well-developed offspring that experience much lower mortalities—is that the variability in subsequent mortality of the new larvae comes down to timing. That is, the moment when the yolk-sac larvae exhaust their yolk reserves they are faced with having to capture plankton particles—food particles—in order to eat. If they cannot, for whatever reason, then they will not survive. For example, if the larvae deplete their yolk reserves at a time of year when the densities of their planktonic food organisms are low, then there may be even greater mortality than that plotted in Figure 11.32. But if the densities of planktonic food particles are great, then survival would be expected to be better than normal. The key point here is that in any given year, stocks of fishes may experience extremes in survival, which translates into good and bad year classes. In the case of cod, one can immediately see that if mortality is

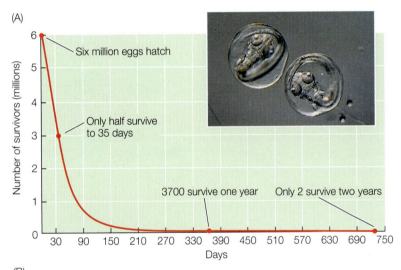

(A)

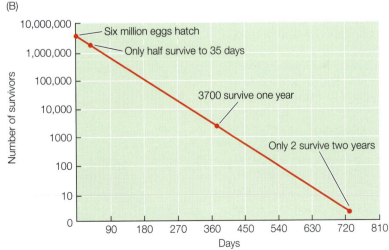

(B)

FIGURE 11.32 A hypothetical survival curve for Atlantic cod, *Gadus morhua*, from the egg to two years of age, assuming that six million eggs are spawned by a single adult female, and that the mortality of the offspring is 2% per day. The number of survivors at any time after hatching, in days, is plotted both on a linear scale (A), which curves downward quickly and is difficult to read, and on a log scale (B). Of the six million eggs spawned, only half survive for 35 days, and after one year there are only about 3700 survivors. After two years, only two can be expected to survive. Inset shows cod eggs about ready to hatch.

not 99.9999% over the two-year period we just discussed above, but rather it turns out to be closer to 99.9998%, then there will be twice as many survivors. Conversely, there can be years of poor survival, with subsequent low stocks of adult fishes.

Recruitment is a term that refers to the survivors and is defined by fishery scientists as the number of fish that survive to a particular age (**BOX 11C**), usually 1–3 years old, when the fish first become large enough that they are available to the fishery (i.e., when the fish are big enough to be susceptible to the fishing gear). As you might imagine, fisheries managers are very concerned with the various processes that control recruitment, both the natural mortality we just discussed and the mortality due to commercial fisheries, and how to tell the difference between the two. Only by knowing each of these two mortality processes can fisheries scientists predict the level of a stock of commercially-exploited fish in a particular year class.

An exceptionally good year class—a year in which recruitment is exceptionally large—can sustain a fishery for many years even though the commercial catch may be focused primarily on that one age group. A classic example of this phenomenon is given in **FIGURE 11.33** based on data from more than 100 years ago.

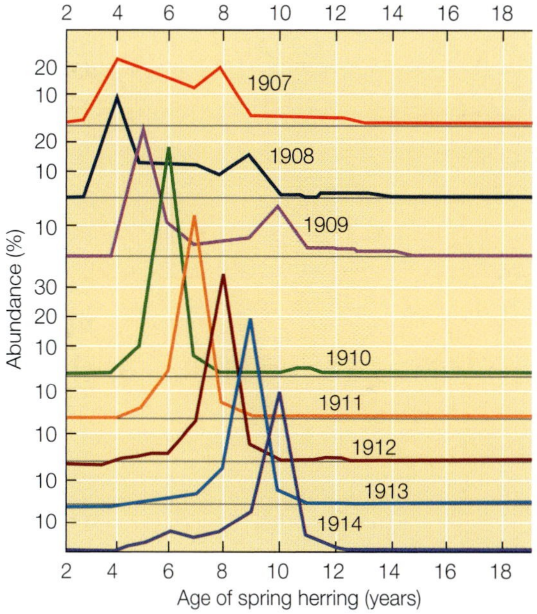

This numbers game is at the root of fisheries science. A particularly important question is: If a fish stock is fished to near depletion, will it recover? The answer is often a disappointing, Who knows? The Canadian cod stocks in the western North Atlantic collapsed in the early 1990s and have yet to recover, even though commercial fishing was halted. But no one can deny that the potential is there; cod are prolific spawners, and there are still a few cod in those Canadian waters.

You may be wondering why are we using examples from a century ago or even a few decades ago. The fact is that very few if any stocks of commercially exploited fishes have been unaffected by fishing pressures. Many fisheries are either closed or so tightly managed that catch data such as those in Figure 11.33 are now rare. We will come back to the plight of commercial fisheries around the world in Chapter 14.

FIGURE 11.33 The percent abundances of herring (*Clupea harengus*) caught in the North Sea fishery between the years 1907 and 1914, plotted by the fish's age (see Box 11C). The commercial catch in 1907, for example, comprised mostly fish between the ages of four and eight years. The catch the very next year, however, revealed the presence in the North Sea of a very strong year-class of fish that were born in 1904 (these herring from the 1904 year-class were four years old when they were caught in 1908). That year-class peak in percentage of total catch can be followed as a progression of peaks each year from 1908 to 1914.

Fish migrations

In our discussion of shoaling and schooling behavior of fishes we mentioned that migratory fishes often move in such formations. The feats of migration by certain species of fishes are truly spectacular, and the most spectacular ones are still only poorly understood by scientists.

Fish have been known for a long time to move about, to alter their distributions *en masse* from one habitat to another for a variety of reasons. Some of these migrations cover distances on the order of thousands of kilometers and cross entire ocean basins, while others are far shorter. Many of the herrings, for example, migrate seasonally over a few hundred kilometers up and down coastlines in search of food, as well as to spawn. Several mackerel species also perform similar excursions. Most fish migrations are round-trip excursions taken on a seasonal or annual basis for the purpose of seeking out a new feeding ground, for reproduction, or as a response to seasonally changing ocean temperatures. There are many instances where coastal fishes will move less than a kilometer seasonally to avoid cold winter temperatures.

Some fishes move between the ocean and freshwaters for purposes of spawning; these are generally termed **diadromous** fishes. They include the **anadromous** fishes, which spend most of their lives in the oceans and migrate into freshwaters to reproduce (the salmons of both the Atlantic and Pacific Oceans are classic examples), and the **catadromous** fishes, which do just the opposite; they reside most of their lives in freshwaters and migrate to the sea to breed (the American eel is an example; we discuss both these examples below.) There are also fishes that move freely between the ocean and freshwaters for reasons unrelated to spawning; these are the **amphidromous** fishes.

THE PACIFIC AND ATLANTIC SALMON MIGRATIONS   Seven species of salmon in the Pacific Ocean spend their adult lives in the open North Pacific Ocean, migrating thousands of kilometers along the west coast of North America, along the Aleutian Islands of Alaska, with some moving north into the Bering Sea and the Arctic Ocean (**FIGURE 11.34**). Others will routinely venture as far west as Kamchatka and Japan on the other side of the ocean. Only after a number of years, sometimes making several circuits around this counter-clockwise migration route in the North Pacific Ocean, do they begin to home in on their birthplace, their natal freshwater stream from which they hatched, where they, too, will spawn—their last act, having completed their life cycle. That is because most of the salmons die after spawning, a phenomenon called **semelparity**. But some do return to spawn more than once; this is, called **iteroparity**. An example is the steelhead trout (actually a salmon, *Oncorhynchus mykiss*), which may return to spawn for three successive years.

The Atlantic salmon (*Salmo salar*) don't have quite such spectacular migration routes as their Pacific Ocean cousins, but they, too, return to their natal streams in either North America or Europe after spending one or more adult years in the open North Atlantic Ocean. Like the Pacific salmons, the Atlantic salmon may spend as long as three years in the freshwater streams where they are hatched before leaving for the ocean.

It isn't known how these fishes make their remarkable trips; much research on Pacific salmon has shown convincingly that they use olfaction to identify their natal stream—but only once they are in the vicinity. How they and the Atlantic salmon navigate while at sea remains a mystery, but it is thought that they can use the angle of the sun, and even the Earth's magnetic field.

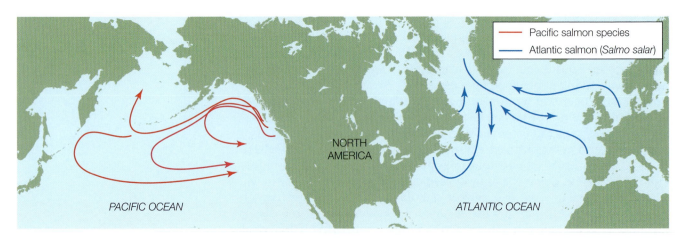

FIGURE 11.34 The extent of migrations by several species of Pacific salmon and by the Atlantic salmon *Salmo salar*. These salmons spend several years in the open ocean before returning to spawn in rivers of North America and Europe (Atlantic salmon only).

BOX 11C Determining the Age of a Fish Using Otoliths

Most fishes have three pairs of otoliths, sometimes called ear stones, that are found in the labyrinth of the inner ear, as illustrated for an Atlantic cod (*Gadus morhua*) in **FIGURE A**. Used for balance and orientation, these otoliths grow over time by laying down concentric layers of calcium carbonate within a protein matrix, each layer precipitated out of the labyrinth fluid. Because the layers alternate between carbonate-rich and carbonate-poor over the course of a year, with summer growth being faster than winter growth, rings are produced, comparable to the growth rings of trees. Those otolith growth rings allow scientists to determine the ages of the fish. When the largest otolith, the sagitta, of an adult fish is dissected out and either sectioned or simply illuminated as shown here, it will reveal concentric rings of opaque and translucent otolith material, giving the age of the fish in years. The example of annual growth rings shown in **FIGURE B** for a North Atlantic cod reveals that the fish was about seven years old when it was captured.

The otolith aging technique was further developed in the late 1970s and 1980s, when it was discovered that not only are there *annual* growth rings in otoliths; there are *daily* growth rings as well, because the growth of the otolith also varies over the day–night cycle. Therefore, otoliths can also be used to determine the age of very

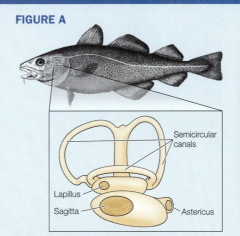

FIGURE A

Semicircular canals

Lapillus

Sagitta

Astericus

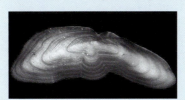

FIGURE B Annual growth rings (cod)

young larval fish—in days. **FIGURE C** shows an otolith from a larval haddock (*Melanogrammus aeglefinus*), very closely related to the North Atlantic cod, viewed with transmitted light under a compound microscope showing light and dark rings. This larva was about 24 days old when captured. The photomicrograph in **FIGURE D** is that of a North Atlantic cod egg with developing larva (see Figure 11.32). The arrow

FIGURE C Daily growth rings (haddock)

FIGURE D

points to the labyrinth capsule on one side of the head with its pair of developing otoliths (white dots). Upon hatching, the otolith is left with a "hatch mark" as shown in the larval haddock otolith as a prominent ring, outside of which are the post-hatch daily rings (see arrow in Figure C).

THE EUROPEAN AND AMERICAN EEL MIGRATIONS Without doubt, the most impressive spawning migration of any fish has to be that of the American and European eels (*Anguilla rostrata* and *A. anguilla*). The eel is a catadromous fish, which lives most of its life in freshwater but returns to the ocean to spawn (**FIGURE 11.35A**). Adult eels can be found in estuaries and freshwaters throughout the Mediterranean Sea and Back Sea areas, and along the Eastern Atlantic coast from Africa to Northern Europe. On the other side of the Atlantic Ocean, they occur from South America to Labrador. Almost anyone who has ever put a baited hook in a freshwater pond east of the Mississippi River knows what an eel is; these curious fish are commonly caught by anglers in freshwaters throughout much of the eastern half of North America and southeastern Canada. But despite their common occurrence, their life history was a mystery until the 1920s, and even today significant questions remain unanswered. The eel undergoes two morphological changes in body form during its life cycle

FIGURE 11.35 (A) The migratory, or oceanic drift, routes of the American and European eel, *Anguilla rostrata* and *A. anguilla*. (B) The leptocephalus larval stages spent at sea (top); the glass eel stage (center); and the adult stage (bottom).

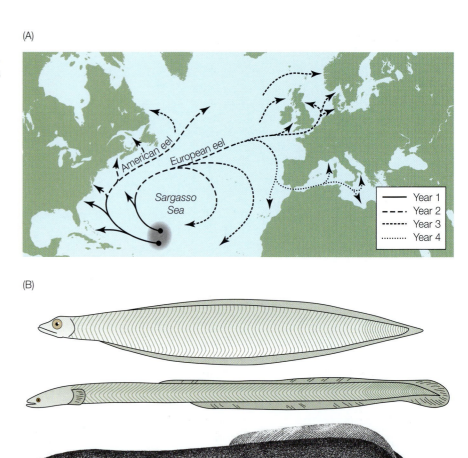

(A)

(B)

(**FIGURE 11.35B**). At one time, their leptocephalus larvae were thought to be a separate species, until 1896 when the larvae were observed to metamorphose into glass eels in the laboratory. Leptocephalus larvae are not unique to the American and European eels, however; all the "true eels" in the order Anguilliformes have leptocephalus larvae, as do a number of other species. The name means "thin head." These leaf-like larvae are transparent. They are carnivores, feeding on other zooplankton as they drift about, eventually metamorphosing into elvers, immature eels. But where the leptocephalus larvae of the American and European eels collected out in the open Atlantic Ocean came from was unknown until the early twentieth century, when a more thorough evaluation of their distributions at sea revealed that they must all be coming from the open ocean waters of the eastern subtropical North Atlantic—in particular, from the Sargasso Sea (see Figure 11.35). It is apparently in those waters, at a depth of about 400 meters, that the eels spawn—although to this day, no one has ever caught a spawning eel in the Sagasso Sea, or anywhere else. Upon hatching from their presumed eggs (again, never seen) in their presumed spawning area, the leptocephalus larvae drift north with the North Atlantic gyre and the Gulf Stream. The American eels split off and move toward shore in North America, while the European eels stay at sea for a third or fourth year before moving onto shore and seeking our freshwaters. Once they reach coastal waters and begin to enter rivers and streams, the leptocephalus larvae metamorphose into elvers, or transparent "glass eels." Eventually, when they arrive at their freshwater destination the elvers develop into the adult eel.

Perhaps even more spectacular than the migration of leptocephali as they drift with the currents of the open ocean, or the subsequent upstream movements of the elvers, is the return trip by the adults to the Sargasso Sea to spawn. Almost nothing seems to deter them; they even crawl like a snake through moist forests in order to escape apparently landlocked ponds and lakes on their way to the sea. At one time the status of the American and European eels as separate species was questioned: they appear to spawn together, and apart from differences in the numbers of **myomeres** (the segments of body musculature, which corresponds to the number of vertebrae; the American eels have fewer), there are virtually no morphological differences between them. But more recent genetic testing has confirmed that they are, indeed, separate species.

TUNA MIGRATIONS   Migrations of the tunas in both the Pacific and Atlantic Oceans rival those of the American and European eels. These migrations are also related to spawning.

The Atlantic bluefin tuna (*Thunnus thynnus*) is one of the biggest and certainly one of fastest, if not the fastest fish in the sea. Its close relative, the Pacific bluefin tuna (*Thunnus orientalis*) has the distinction of having fetched the highest price ever paid for a single fish: about $396,000 was paid in Tokyo in 2011 for a 754 lb Pacific bluefin tuna. Prices paid for Atlantic bluefin tuna, while fetching somewhat lower prices, have approached $200,000 for a single fish.

Tuna live, reproduce, and die in the sea, but they, too, undergo dramatic migrations for spawning and feeding. Tuna species in the Pacific have extensive migration patterns that cover several thousands of kilometers. The giant bluefin tuna in the Atlantic has a dramatic migration as well; after reproducing in the Gulf of Mexico and in the Straits of Florida and the Bahamas, the spent adults migrate north to the rich summer feeding grounds in the Gulf of Maine and the Gulf of St. Lawrence (**FIGURE 11.36**). They are known to move across the Atlantic Ocean and back again in a matter of just a few weeks.

(A) Bluefin tuna

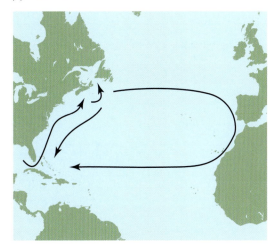

(B)

FIGURE 11.36 (A) Migration routes of Atlantic bluefin tuna (*Thunnus thynnus*) from spawning grounds in the Gulf of Mexico and the Straits of Florida to summer feeding grounds in the Gulf of Maine and shelf waters of maritime Canada; some tagged fish have crossed the Atlantic ocean and returned. (B) Atlantic bluefin tuna weighing 758 lb landed by writer Zane Grey (1872–1939).

The Field of Ichthyology

To keep within the bounds of an introductory textbook, we have left a lot out of this chapter—so much that it seems that in closing, we should at least mention those aspects of ichthyology (the study of fishes) and fisheries science that we have had to skip over, but which the interested student might want to explore further. For instance, we barely mentioned the lungfishes and their capacity to breathe air, which was of course an important step in the evolution of tetrapods. In our discussion of fish gills and respiration, we emphasized the importance of hemoglobin in blood and that blood plasma alone can carry very little oxygen, but we neglected to mention that many Antarctic fishes, in their sub-zero environments, do not have such pigments, and they appear to do quite well without them. Sensory systems in the fishes were stressed, but again, we left out that some cave fishes have no functional eye at all, and rely solely on their other senses. While fishes are indeed ectothermic—cold-blooded—with their body temperatures controlled by the water temperature, the large tunas are effectively warm-blooded. They have a counter-current circulatory system, similar to that in gills, which keeps their core temperature warmer than that of the seawater surrounding them. The tunas and other migratory fishes in particular have their trunk muscles divided into two groups: the red muscle, which is concentrated nearer their backbone, which fishes use for slow, steady swimming, and the white muscle, which makes up more of the body musculature, which fishes use for their powerful, short-term, high-speed bursts of swimming. We alluded to, but did not get into, the subject of toxic and venomous fishes, of which there are many interesting examples. Our discussion of reproduction left out the phenomenon of sex reversals, which is common in fishes, and we left out the elaborate courtship behavior exhibited by many species. We also neglected to discuss protective and cryptic coloration. Finally, our discussion of fish population dynamics was also very brief, but we will revisit those population-level issues when we discuss fisheries and aquaculture in Chapter 14.

Chapter Summary

- Fishes have been around for more than 400 million years. They were the first **vertebrates**, and today they comprise more than half the total number of vertebrate species. Of the 30,000 or so species of fishes, more than 60% are exclusively marine.

- There are three major groups of fishes: the **Agnatha**, the **jawless fishes**; the **Chondrichthyes**, the **cartilaginous fishes**; and the **Osteichthyes**, the **bony fishes**.

- The agnathans are the **hagfishes** and **lampreys**, the most primitive group. These small, eel-like fishes have a notochord but no actual vertebrae, no paired fins, and no jaws. The hagfishes are primarily scavengers. The lampreys may be parasitic, attaching to and feeding on other fishes, or they may feed on benthic invertebrates.

- The chondrichthyans have cartilaginous skeletons and do not possess true bone tissues. They include the **chimaeras**, the most primitive of the cartilaginous fishes, and the **elasmobranchs**, the **sharks**, **rays**, and **skates**.

- The osteichthyans are a diverse and abundant group that comprises about 96% of the extant fish species. They include the primitive **lobe-finned fishes** (which include the lungfishes and coelacanths) and the modern **ray-finned fishes**.

- Fishes have evolved a wide range of biological characteristics and capabilities that have enable them to thrive in virtually all marine and freshwater environments.

- Fishes possess gills for exchanging oxygen and carbon dioxide between the water and their blood. The bony fishes possess an **operculum**, or gill cover, that facilitates the pumping of water past the gills;

but the cartilaginous fishes, including many species of sharks, lack an operculum, and therefore must swim constantly to keep water passing over their gills.

- The tissue fluids of cartilaginous fishes have nearly the same osmotic pressure as seawater, due to elevated **urea** concentrations in their blood. The bony fishes osmoregulate primarily by excreting excess salts via **chloride cells** in their gills.

- Fishes have evolved various modes of swimming, from the inefficient eel-like undulations of some species to the highly efficient mode enabled by lunate caudal fins and well-developed body musculature exemplified in the **thunniform** fishes. Some fishes use primarily their pectoral fins for propulsion, and some use wavelike undulations of their dorsal and ventral fins.

- Some species of fishes have evolved social behaviors: **shoaling**, or the formation of aggregations, and **schooling**, the highly organized and synchronized swimming of many individuals as a group.

- Feeding mechanisms of the fishes range from the parasitic (e.g., lampreys) to highly predatory (e.g., sharks). In between are the herbivorous fishes that feed on algae, the specialized coral feeders, and **planktivorous** fishes.

- Fishes reproduce sexually, and individual species vary widely in their **fecundity**. Some of the sharks, rays, and skates have only a handful of offspring per year. Other species of bony fishes may spawn millions of eggs annually, which are fertilized externally. Still other fishes may produce relatively few young at a time, but may reproduce throughout the year.

◼ Some fishes pass through several developmental stages, from egg, to **yolk-sac larva**, to **post-larva**, to juvenile, and to adult; others hatch directly into a well developed stage that resembles the adult.

◼ Many species of fish undergo extensive migrations, which may or may not be related to spawning, and which may be between fresh waters and the ocean. Examples include Pacific and Atlantic salmon, American and European eels, and the bluefin tuna, each of which exhibits migrations that can span entire ocean basins.

◼ There are numerous general areas of fish biology and ecology, that because of the limited focus of this book, were not discussed.

Discussion Questions

1. What are the main anatomical differences among the agnathans, the chrondrichthyans, and the osteichthyans? What are some of the characteristic lifestyles of these groups?

2. Discuss the anatomical features of the sharks versus those of the skates and rays.

3. Why did the sea lamprey become a fisheries management problem in the Great Lakes? What measures are being taken to control the problem? How useful to managers was the fundamental understanding of the ecology of this species in this case?

4. How do the bony fishes and the cartilaginous marine fishes osmoregulate? How are the two osmoregulatory mechanisms similar and dissimilar to one another?

5. What are some of the sensory capabilities that enable the elasmobranchs to be effective predators?

6. Why are cartilaginous fishes generally larger than bony fishes? How did the evolution of bone lead to a greater diversity of forms?

7. How can you reconcile the fact that the largest fish in the sea, the basking shark and the whale shark, feed on some of the smallest particles in the sea (plankton)?

8. What anatomical features enable fishes to swim fast? Why do you think those body features are not common to all fishes?

9. What are the differences between the mechanisms of respiration in the sharks and the bony fishes?

10. What is the difference between shoaling and schooling behavior? What are the survival advantages of each behavior?

11. Discuss the major modes of reproduction in the fishes, especially differences between the sharks and rays versus the bony fishes.

12. The hagfishes and lampreys have been around for more than 400 million years, and are the most primitive fishes alive today. Why do you think they have survived for so long?

13. Sharks are not only denser than bony fishes, but they are denser than seawater. Why? How do they overcome the resulting tendency to sink?

Further Reading

Barton, Michael, 2007. *Bond's Biology of Fishes*. Belmont, CA: Thomson Brooks.

Moyle, Peter B., and Cech, Joseph J., 2004. *Fishes: An Introduction to Ichthyology*. Upper Saddle River, NJ: Prentice Hall.

Parker, Steve, and Parker, Jane, 1999. *The Encyclopedia of Sharks*. Buffalo, NY: Firefly Books.

Pitcher, Tony J., and Hart, Paul J. B., 1982. *Fisheries Ecology*. Westport, CT: Avi Publishing Co.

Contents

Not all marine organisms are found everywhere in the oceans; each species is adapted to a set of physical and biological conditions that limits its distributions to a particular type of habitat. Often, several species share some of the same environmental limitations and are found together, comprising assemblages or communities of organisms that are typical of specific marine environments. While there are no agreed-upon or standard descriptions for all the various marine environments, a few stand out and have attracted the attention of scientists and naturalists for centuries. One of the most obvious and interesting marine environments of all is also the one that is accessed most easily: the thin strip of shoreline between high and low tides, the intertidal zone, such as this tide pool along the Pacific coast. Marine life forms in the intertidal zone have fascinated naturalists for centuries, for reasons obvious to anyone who has ever explored them.

Marine Environments

In this chapter our discussion moves from organisms themselves to the environments in which they live. And so we begin to talk about *ecology*, the study of the interrelationships among living organisms and the multitude of environmental factors that influence their distributions and abundances. We have already introduced in our earlier chapters a number of very basic ecological principles, such as how temperature and salinity influence the distributions of organisms in the oceans, and how light and nutrients serve as limiting factors for the primary producers. These and other factors act together to create a highly diverse range of marine environments and ecosystems throughout the world ocean.[1]

Because the physical characteristics of marine environments tend to vary in a fairly predictable manner, there are a number of ways they can be subdivided and characterized. The simplest approaches have incorporated *latitude*, which factors in the role of average annual water temperatures, and *water depth*. Taken together, these factors allow us to make distinctions between shallow- and deep-water tropical, subtropical, temperate, and polar environments. Within that framework we have, in turn, distinguished between the benthic and pelagic realms, discussed in Chapter 7 (see Figure 7.20). But missing from this framework are a number of specific environments that include some of the most interesting and important marine systems in the world ocean. The study of both general frameworks and specific ecosystems helps us to gain an appreciation of how the organisms we have been discussing in the previous few chapters interact with the physical environment covered earlier in this book. And so in this chapter we explore some of the features and important processes operating in a few of these more interesting marine environments.

The Intertidal Zone

Much of what we have been discussing thus far has concerned globally significant, large-scale phenomena, many of which are best studied using satellites and ships at sea. But one of the simplest and yet most enjoyable ways to explore ocean life is to poke around in the **intertidal zone**, also referred to as the littoral zone (**FIGURE 12.1**).

Generally treated as a benthic ecosystem, the intertidal zone comprises the bottom substrate and the organisms living in and on that substrate between the highest and lowest extent of the tide. But the organisms in this zone are not only benthic; **tide pools** of various sizes, which are common in intertidal areas, harbor nektonic and planktonic organisms, as well. At low tide, various birds and terrestrial mammals forage for food there; and of course, when the intertidal zone is submerged, it hosts various other marine organisms that

---

[1] The terms *ecosystem* and *environment* are sometimes used interchangeably, but there is a subtle difference between them. The word *environment* generally refers to the physical surroundings; when we include the living organisms in reference to that set of physical surroundings, we are talking about an *ecosystem*.

FIGURE 12.1 An intertidal zone on the coast of Maine, where a boulder field of various-sized rocks, deposited following the retreat of the last glaciation, is exposed. Some of the rocks are covered with brown algae and barnacles. Between the rocks are sandy and muddy sediments.

may move in from the **subtidal zone**, beyond the extreme low-tide mark.

It is not an exaggeration to say that the intertidal zone is teeming with life; benthic biomass per square meter there is some ten times that on the continental shelves, and a thousand times that of the deep sea. High in species diversity, the intertidal zone is an environment that, almost paradoxically, is illustrative of many of the oceanographic processes and phenomena we have discussed so far—but on a small scale. Because of its biological richness and easy access, much of what we know today about marine ecology had its beginnings in early studies conducted in the intertidal zone; certainly this was at the heart of the "marine biology craze" in Victorian England, which, as we discussed in Chapter 1, helped set the stage for the *Challenger* expedition. It is in the intertidal zone where terrestrial and oceanographic influences are squeezed together in a thin strip that marks the edge of the sea, and operating within that thin strip are important physical and biological processes that change with the tides.

This narrow marine environment runs along the shorelines of all land masses that border the ocean; the tides may be larger in some areas than others, but there are tides everywhere in the oceans—except at the amphidromic points—and so there are also intertidal zones everywhere. The width of that strip is primarily a function of how steep or how gradual the bottom slope is: correcting for variations in tidal ranges, the most steeply sloped bottoms have the narrowest intertidal zones; and conversely, the most gradually sloped bottoms have the widest. Within the intertidal zone itself, however narrow it might be, there are a variety of habitats available to organisms. In fact, the intertidal zone spans a continuum of sub-environments—from fully terrestrial to fully marine.

The limits that define the intertidal zone—the high- and low-water marks—are not always exact. They vary fortnightly in relation to the spring–neap tidal cycle we discussed in Chapter 6, and they may be blurred somewhat by wave action. For example, an intertidal area exposed to high-energy ocean waves will have an upper limit that extends beyond the high-tide mark; the splash from waves in the **supratidal zone** helps to keep this area sufficiently wet that it can support some of the organisms that are common in the upper intertidal zone, such as a few marine algae and lichens. And, of course, the lower limit of such intertidal areas may also be somewhat blurred by wave splash (ocean waves will crash the shore at low tide as well as at high tide). On the whole, the upper and lower limits of the intertidal zone are more sharply delimited in low-energy areas, those areas protected from ocean waves, and are less sharply defined on wave-swept, high-energy shorelines.

Life in the intertidal zone can be stressful. Over the course of a tidal cycle, benthic organisms that are sessile or choose to remain in place experience alternating exposure to air and submergence underwater. When exposed, they face stresses associated with desiccation (simply drying out), as well as extremes in air temperature, which vary far more widely than ocean water temperatures. Seasonally, depending on geographic location, organisms exposed at low tide may experience periods of especially hot daytime temperatures or especially cold nighttime temperatures. And regardless of geographic location, rainfall at low tide can drastically lower the salinity of shallow tide pools and dilute the surface moisture coating exposed organisms, causing osmotic stress.

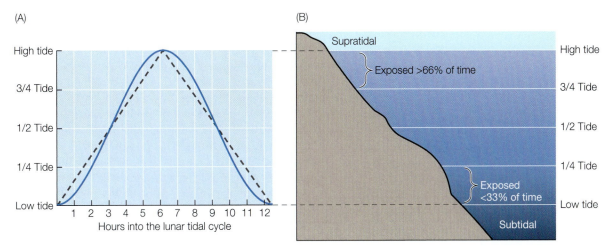

(A)

High tide

3/4 Tide

1/2 Tide

1/4 Tide

Low tide

1 2 3 4 5 6 7 8 9 10 11 12
Hours into the lunar tidal cycle

(B)

Supratidal

Exposed >66% of time

High tide

3/4 Tide

1/2 Tide

1/4 Tide

Exposed
<33% of time

Low tide

Subtidal

FIGURE 12.2 The relative durations of submergence and exposure in the intertidal zone over a lunar tidal cycle. (A) A sine curve (solid line) of the lunar tidal cycle, beginning at low tide and continuing for one complete cycle (12.42 hours). The dashed line is drawn to represent the tidal cycle as if it were linear and not sinusoidal. (B) The percentage of time that organisms occupying the upper fourth and bottom fourth of the intertidal zone are exposed to air.

The duration of exposure and submergence varies with an organism's position in the intertidal zone. Generally speaking, those in the upper part of the zone will be exposed to air for significantly longer periods than those in the lower portions. Organisms in regions of the world with solar tides that come and go once a day will have periods of exposure and submergence that are about twice as long as those in regions dominated by the twice-daily lunar tides. The tidal cycle of exposure and submergence for an intertidal shoreline under the influence of a lunar tide is diagrammed in **FIGURE 12.2**. The lunar tide comes and goes about twice a day (about every 24 hours and 50 minutes) with a pattern of tidal height versus time that can be described with a sine curve repeating every 12.42 hours: the tide will flood for 6.21 hours and ebb for another 6.21 hours. But because the tides conform to a sine wave,[2] the change in water depth is not constant; that is, beginning at low tide, the tide rises slowly at first, flooding only a small percentage of the total tidal range during the first hour. About three hours later, at the halfway point, the water level is rising at its fastest rate; then, as it approaches high tide, it begins to slow down again. As the tide turns and begins to ebb, the water level drops slowly at first, then reaches its maximum rate at the half-tide point, and slows down as it approaches low tide, completing the tidal cycle. Thus, the tide levels (water levels) stay at and near high tide, as well as at and near low tide, for a longer period of time than they would if the tide were to come and go in a simple linear fashion, illustrated by the dashed line in Figure 12.2A. While intertidal organisms residing at the half-tide mark are exposed to air 50% of the time and submerged 50% of the time, organisms at the three-quarters tide mark are exposed not 75% of the time, as they would be if the tidal flow were a linear function, but only 66% of the time. At the quarter-tide mark, they are exposed 33%, not 25%, of the time. Nevertheless, it is a fair generalization to say that intertidal organisms nearest the high-tide mark are exposed to air the longest, and organisms nearest the low-tide mark are exposed the least.

Because the period of a lunar tide is 12.42 hours, the times of high or low tide are out of phase with the 24-hour day–night cycle; each day the tides are about 50 minutes later. For example, if the high tide on one day is at 3:00 A.M. and again at 3:25 P.M., the very next day they will occur later, at 3:50 A.M. and 4:15 P.M. Therefore an organism that is exposed to air during the middle

---

[2]The sine curve describes the ideal pattern; in reality, each intertidal environment will exhibit variations resulting from the coastal geomorphology (its general shape and depths) and the influence of the solar component of the tide.

of the night on the first day, for example, will after a week be exposed to air during mid-day. This means that sessile organisms experience fortnightly cycles in the time of day when they are exposed and submerged, which can be important, especially during hot summer days and cold winter nights. On the other hand, intertidal zones in areas of the world ocean that exhibit a solar tide (which, like the lunar tide, is also described by a sine wave; see Figure 6.27) will experience one high tide and one low tide every 24 hours, with the time of high and low tides not necessarily changing each day as they do for a lunar tide, although they do exhibit significant local variability in timing.

Biological productivity in the intertidal zone is generally quite high. One reason for this is that algae in this zone are either submerged in shallow water or exposed to air, so algal photosynthesis there is almost never light-limited, except on a seasonal basis. With ample light and what is often an abundant supply of dissolved nutrients delivered from the coastal waters beyond the intertidal zone as well as from land, attached algae typically attain very high rates of primary production. While phytoplankton production in the near-shore waters may also be relatively high, plankton are available as food for intertidal animals only during times of submergence; therefore, it is various forms of attached algae, especially drifting fragments of those primary producers (organic detritus), that constitute the most important food source in most intertidal environments.

There are several types of intertidal zones. These environments differ from one another in a number of ways, but principally in terms of their substrate: whether it is a rocky shoreline, or a soft-bottom benthic environment, such as sandy beach, or a mud flat.

The rocky intertidal zone

Rocky coastlines offer some of the most spectacular scenes in all of nature (**FIGURE 12.3**), and it is easy to overlook the fact that for all their splendor, therein lies a hidden but no less fierce struggle for existence for those organisms that make it their home. Not only is the rocky intertidal zone an environment of alternating exposure and submergence, but because it is usually exposed to the open ocean, it is subject to pounding surf from large ocean waves, which keeps sediments from accumulating and allows only the hardiest of marine life forms to thrive. Because the rocky intertidal zone is virtually free of bottom sediments, intertidal organisms that reside there must either attach to the rock surfaces to avoid being swept away in the surf or be sufficiently motile that they can seek out suitable protection from wave action. Organisms that attach themselves to rocks include, among others, various species of macroalgae, sessile barnacles, and mussels. Among the most common motile organisms are the snails and other gastropod molluscs. And in addition to the physical demands associated with life in the intertidal zone, there are ongoing biological interactions in the forms of predation and competition for basic living space.

Of the physical stressors in the rocky intertidal zone, desiccation has to be the most fundamental. To avoid desiccation, intertidal organisms must be able to either tolerate loss of water from their tissues or be able to ward against water loss in the first place—strategies

FIGURE 12.3 An example of the rocky intertidal zone typical of northern Europe, the northeast United States, and maritime Canada. These exposed coastlines are not only subject to the everyday problems of exposure and submergence as the tides come and go, but they are also subject to the brunt force of ocean waves arriving on shore, where they break as surf. This photograph was taken on a relatively calm summer day; waves get to be much larger. Although these exposed, high-energy intertidal shorelines often seem devoid of ocean life, they are not; notice in this photograph the persistence of the attached green algae growing on the rocks (arrow). Other microhabitats exist in the various nooks and crannies among the rocks, as well as in tide pools.

some authors refer to as the "run-and-hide" and "clam-up" methods. The run-and-hide strategy is employed by many motile organisms, such as crabs and snails, which seek out tide pools or hide beneath tufts of rockweed (see Figure 12.1) where they are able to retain moisture while the tide is out. The clam-up method is used by organisms such as barnacles, which attach themselves firmly to a rock surface, and by littorinid snails (periwinkles), which can close their opercula to guard against water loss. Littorinid snails are air-breathers, similar to their cousins, the terrestrial snails, and need only to maintain a moist surface on their highly vascularized mantles in order to respire in air; as a result, periwinkles can be found at the upper levels of the intertidal zone, where exposure time is greatest. In fact, they can remain exposed for days before drying out; but if submerged for extended periods, they may drown. However, they are apparently less capable of forming a tight seal with their opercula than some other gastropods, and as a result they are often seen in cracks and crevasses and in shady spots in order to better retain moisture at low tide (**FIGURE 12.4**). Barnacles close up tight but still breathe air at low tide and hope for the best, as they are easy prey for those intertidal animals that can grind through their calcareous plates, such as the carnivorous snails in Figure 12.4.

FIGURE 12.4 A group of littorinid snails (*Littorina* spp.) are crowded into a narrow rock crevasse during low tide one August day on the Maine coast, while a group of dog whelks (*Nucella lapillus*), which are less susceptible to water loss, feed on the young-of-the-year barnacles.

Some species of algae can withstand significant water loss at low tide on warm and dry summer days and may appear to be near death, changing color as they lose moisture; but upon being resubmerged on the next tide, they recover with no apparent or long-lasting ill effects. Some studies have shown that intertidal species of macroalgae can tolerate losses of as much as 90% of their tissue water content and still recover upon being resubmerged.[3]

Avoiding desiccation at low tide is one thing, but the macroalgae and other sessile intertidal organisms in the rocky intertidal zone also have to withstand the full power of the sea in the form of crashing surf. Anyone who has ever explored these environments has to be impressed with how organisms somehow manage to hold on in the face of powerful wave actions. Rockweeds, such as those shown in Figures 12.1 and 12.4, are permanently attached to rocks by their holdfasts. They develop from fertilized eggs that settle out of the plankton and, upon encountering the right substrate—in this case, rocks—they become firmly anchored as zygotes, and grow in place. Animals, too, anchor themselves on the rocks; late larval barnacles, for example, settle out of the plankton and glue themselves to hard substrates, usually rocks, but also other organisms, where they develop into adults and remain for the rest of their lives. Other animals, such as mussels, similarly arrive as larvae and become attached with their strong byssal threads (see Figure 10.18). But in time, even the strongest attachments may fail and, as a result, there is constant turnover, with new algae and animals colonizing newly opened spaces.

In addition to coping with the physical stresses of temperature and salinity extremes, desiccation, and wave disturbance, intertidal organisms are also in competition with each other for space. Life on rock surfaces in the rocky intertidal zone can become very crowded, indeed, as can be seen in **FIGURE 12.5A**, where, earlier in the year, the underlying rock was completely covered with patches of barnacles and mussels. But, after a season of wave action, sharing food resources with so many neighbors, and suffering losses from predation,

[3] Yen-Chun Liu, 2009. *Mechanism for Differential Desiccation Tolerance in Porphyra Species*, Ph.D. Dissertation, Northeastern University.

(A)

(B)

(C)

FIGURE 12.5 (A) Densely crowded patches of blue mussels and barnacles in late November in Maine. Notice the patches of bare, uninhabited rock; these were probably also covered by mussels and barnacles earlier in the year, but through a combination of wave actions, predation, and competitive interactions, they have been freed up for new occupants. Because these animals reproduce only once each year (in spring), the space remains vacant; but the following year, young-of-the-year mussels and barnacles will again settle on any open spots. (B) This rock was photographed in the spring, following settlement of new barnacles among the older barnacles from the previous year(s). Also notice here the young barnacles on top of other barnacles. (C) Space cleared of barnacles by whiplash from the rockweed (*Ascophyllum nodosum*).

patches of open space have appeared, only to be recolonized by newly-settled sessile replacements the following season. Barnacles create more space for themselves by undercutting their neighbors (often other barnacles) as their attached base grows. Small, newly settled barnacles cannot possibly grow to the same size as the adult barnacles already there without somehow eliminating some of their siblings (**FIGURE 12.5B**); there just isn't enough space for all of them to grow to adult size. So, using their edges almost like shovels, they cut beneath and dislodge adjacent barnacles as they grow and expand their footprint. Even fully mature barnacles are not immune to this attack by others, as they too can be dislodged by the new cohorts. Older barnacles are also susceptible to a more direct attack by newly settled barnacles that attach to them and eventually overgrow and smother them (see Figure 12.5B). Mussels, too, can smother barnacles. Attached mussels are not as sessile as one might think; some are capable of releasing their byssal threads and then attaching new ones, allowing them to move over new ground, so to speak, and compete with barnacles by growing over them, thereby denying them access to their suspended food source.

These kinds of inter- and intraspecific competitive interactions are important to patchy distributions of mussels and barnacles (see Figure 12.5). It is these same phenomena—competition, predation, and the vertical gradient in physical stress factors—that often produce discrete zones in the rocky intertidal zone, where assemblages of species seem to dominate in horizontal bands. Called **intertidal zonation**, these patterns are most common in the rocky intertidal zone (**FIGURE 12.6**). Generally speaking, the upper limit of a species' distribution in the intertidal zone is determined by that organism's ability to tolerate physical stress, whereas the lower limit in the intertidal is determined more by an organism's ability to compete for food and habitat space.

The rocky intertidal zones on both sides of the Atlantic Ocean and along the Pacific coast of North America exhibit similar bandings of organisms at different heights. While differences exist, and in fact, the intertidal zone on either side of even a small peninsula will exhibit marked differences in the vertical zonations of dominant algae and animals, we can give some general examples.

In the rocky intertidal of the North Atlantic Ocean, the upper intertidal and supratidal zones often exhibit a black, soot-like

(A)

Black lichen zone

Brown and red macroalgae

Barnacles

(B)

Barnacles

Brown and red macroalgae

(C)

Mud flat   Brown algae   Black lichen zone

coating on the rocks, which is actually a black encrusting lichen (often *Verricaria maura*). Called black tar lichens, and sometimes mistaken for weathered remains of oil spills, they are clearly visible in Figure 12.6A,C. Still higher in the supratidal zone is an area of less wave splash but more salt spray carried in the wind; often found in this zone are salt-tolerant white, gray, and sometimes yellow or orange lichens (see Figure 12.6C). Just below the black tar lichens, in the part of the upper intertidal that is briefly submerged at high tide, there is often a band of filamentous cyanobacteria, most of which are capable of nitrogen fixation. Their coating of gelatinous film (which is what makes the rocks in the intertidal zone so slippery, as anyone who has ever ventured onto them can attest) makes it possible for them to withstand fairly long exposures to air. These blue-green algae, as well as benthic diatoms, are grazed upon by periwinkles (littorinid snails) and some species of limpets.

Still farther down in the middle intertidal zone is a discrete band above which barnacles cannot live (see Figure 12.6B). Only below that line are the barnacles submerged long enough to feed on suspended food particles (plankton and detritus); the lower limit of barnacles

FIGURE 12.6 Intertidal zonation in the rocky intertidal zone on the Maine coast results in a horizontal banding pattern of dominant organisms. Notice how these three examples differ from one another. (A) A gently sloping rocky intertidal zone, where the bands are stretched horizontally, with brown and red macroalgae in the lower intertidal, a sporadic coverage of barnacles above, and a band of black lichens above that. (B) A vertical rock face exposed to high-energy ocean waves, with little evidence of black lichens, but a sharp line marking the upper limit of dense barnacles. Some brown algae (*Fucus vesiculosus*) are also evident. The lower band is dominated by red algae (Irish moss, *Chondrus crispus*). (C) A low-energy rocky intertidal zone adjacent to a mud flat, with a distinct black lichen band above a band of brown algae growing on the rocks next to the soft bottom; above the black lichens is bare rock.

is often set by the grazing pressure of carnivorous whelks (see Figure 12.4), which feed on them by drilling through the barnacles' plates using their radula. The whelks, in turn, are limited to this deeper portion of the middle intertidal zone by their susceptibility to desiccation; both whelks and periwinkles are also limited by wave action; they are able to hold onto the rock surface using suction created by their mantles only in areas of relatively low wave energy. This middle intertidal is also where mussels sometimes abound, usually below the level of highest barnacle abundances, with patches established by competition for space. Abundant brown macroalgae (rockweeds, or seaweeds) also occur here, often coating the rocks in thick mats (see Figure 12.6A).

The lower intertidal zone is where we see the red algae begin to dominate (e.g, Irish moss; see Figure 8.36), along with kelps (see Figure 8.37), predatory sea stars, and whelks. Whelks have a low tolerance for desiccation and do not extend very far into the middle intertidal. Sea stars tolerate exposure even less, and seldom venture above the lower extent of the intertidal zone. Because of their nearly constant submergence, sea stars can feed almost all the time—so barnacles and mussels (their prey) are relatively rare in the lower intertidal zone. Coralline algae are common in the lower intertidal, as are the limpets that graze them. And it is in the lower intertidal zone that we begin to see various crabs that reside subtidally, but which venture into the intertidal zone, finding shelter from desiccation beneath patches of algal turf or in tide pools.

The soft-bottom intertidal zone

Unlike the rocky intertidal zone, regions of the coastal ocean where wave action is low can accumulate sediments, producing **soft-bottom** benthic environments, not just in the intertidal zone, but subtidally as well (**FIGURE 12.7**). Attached algae and animals are missing here because, apart from the odd rock or boulder, there is no hard substrate to attach to; the sediments here are unstable and are easily moved about by tidal currents and by foraging animals. As a result, few macroalgae become established; instead, it is the vascular seagrasses, with root and rhizome systems that can hold onto loose sediments, that sometimes form dense beds in this zone. Other than these seagrass beds, and thin layers of benthic diatoms, the soft-bottom intertidal zone lacks significant abundances of primary producers. While near-shore plankton are available to consumers

FIGURE 12.7 (A) An intertidal mud flat submerged at high tide and (B) exposed at low tide. Notice the ice at low tide, where the wet surface became frozen when exposed. (C) Low tide over a subtidal soft-bottom environment, with fragments of mussel and clam shells visible.

(A)

(B)

(C)

during times of submergence, the principle source of food is detritus that drifts in with the tides, or is delivered from adjacent terrestrial sources.

The ecological structure of these soft-bottom benthic communities is governed principally by sediment grain size. Fine-grained sediments made up of silts and clays (mineral particles smaller than about 62 μm; see Chapter 3), which have the smooth feel of mud if you rub some between your fingers, are generally richer in detrital organic matter content. This is because the tendency of organic detritus to settle out, or to be resuspended, is similar to that of the silt and clay sediment particles. Soft-bottom environments with coarse-grained sediments are typically devoid of significant quantities of particulate organic matter for the same reason that the smaller grain sizes are absent: the turbulent energy of that environment does not allow fine particles to settle, and any that do are soon washed away. So it is the muddy intertidal zone—the low-energy environments protected from wave action—that accumulates the fine-grained sediments, and which is richest in detrital organic material that serves as the main food source for consumers there.

The fine-grained sediments—the muddy intertidal sediments, also called mud flats—retain **interstitial water** (water between the sediment grains) better than coarse-grained sediments do. Organisms residing in these muddy sediments (infaunal animals) are less susceptible to desiccation than they would be in coarse-grained sediments, which drain more completely at low tide. But this retention of water, a result of reduced water flow in the more closely packed sediments, also means that oxygen levels in that interstitial water can be depleted easily. Without photosynthetic organisms below the sediment–water (and sediment–air) interface, it is only by way of direct exchange of interstitial water with oxygen-rich water above the sediment–water interface that oxygen can be delivered to subsurface sediment layers. Better retention of interstitial water means that such motions are suppressed, which is why the mud flats stay wet in the first place. And coupled with the relatively high sedimentation rates of organic detritus, those sediments can become hypoxic (low in oxygen) or anoxic (oxygen-free). Just a small distance below the surface, consumption and respiration of that organic matter by heterotrophs, especially bacteria, use up the available oxygen. An example of this oxygen depletion just below the sediment surface can be seen in Figure 10.13, where only a millimeter or so beneath the surface, the sediment turns from a light gray-brown, indicating oxygenated conditions, to black. This color change reflects the absence of oxygen and the presence of hydrogen sulfide (H_2S), the product of sulfate reduction by bacteria. It is this bacteria-produced hydrogen sulfide that turns the mud black and gives mud flats their characteristic odor of rotten eggs (or so it is said; I have never actually put my nose up to a rotten egg). The depth of this black anoxic layer, where H_2S is being produced, depends on the sediment grain size, which controls the movement of water and dissolved oxygen. The anoxic layer begins deeper in coarse-grained sediments, and it may be absent altogether, as in well-mixed beach sand. Higher-energy environments, such as exposed beaches, will tend to promote water flow through the sediments, oxygenating the sediments to a greater depth; here, the lack of significant sedimentation of organic matter means that there is little heterotrophic respiration occurring in the sediments, anyway.

Because there is no firm substrate on which to attach themselves, the animals of the soft-bottom intertidal zone are a mix of sessile infaunal and motile epifaunal species. The infaunal animals include those we discussed in Chapter 10, such as the bivalve molluscs, polychaete worms, and amphipods. They cope with hypoxic and anoxic sediment conditions by ventilating either themselves or their burrows with oxygen-rich water from above the sediment surface. Some burrowing animals disturb the sediments sufficiently that they and their surrounding sediments become flushed with oxygen-rich water; in effect, they vertically mix the subsurface layers of sediments, a phenomenon called **bioturbation**. Most of these infaunal benthic animals are deposit feeders that either collect suspended particulates that fall on the sediment surface, such as is done by some amphipods, or pass sediments through their digestive tracts, effectively eating sediment at their oral ends as they burrow. This latter type of feeding is practiced by some polychaete worms; digestable organic material is absorbed in the gut, with the undigested sediment passing through and out the back end of the animal, often showing up as fecal deposits on the surface of the sediments (see Figure 10.13). Other polychaetes pump overlying water through their burrows in the sediment, collecting suspended food particulates from the flow. Bivalve molluscs such as clams are suspension feeders that reside below the sediment surface and extend their siphons out where they pump water through their bodies and back out again, capturing and processing suspended food particles. But there are predators in the sediments as well, such as some of the nereid polychaete worms (see Figure 10.13). There are also much smaller organisms residing in the intertidal soft-bottom sediments, including various protozoans that reside in the interstitial spaces between sediment grains, and other benthic meiofauna (animals smaller than 1 mm), such as nematode worms, and other taxa that occur nowhere else.

Coarse-grained soft-bottom intertidal zones include, of course, our beloved beaches (**FIGURE 12.8**). The grain sizes of beaches vary with wave energy, being largest in areas of high wave energy. Extreme examples are cobble beaches with "sediments" consisting of stones the size of baseballs or larger. Because of their relatively high energy wave environments, which is what forms beaches in the first place, there is very little detrital organic matter that accumulates, which in turn makes these environments fairly unproductive. Although there is photosynthetic primary production by benthic diatoms in the upper few millimeters of interstitial spaces of the sand, called **psammon**, the main source of food to beach-dwelling organisms is the organic debris brought ashore in the waves (via Stokes drift; see Figure 6.6). It is common for the beach breakers to bring ashore ahead of them significant loads of kelp and other macroalgal fragments, which tend to be further fragmented into smaller pieces as the waves grind that material into the beach sand. And as the tide floods, subtidal fish and invertebrates will follow the lead edge of the flooding water to feed on any infaunal animals that might be revealed. These macroalgal deposits are often left behind in strand lines at the high-tide

FIGURE 12.8 (A) A typical New England white sand beach at low tide. The slope of this particular beach is slightly greater nearer the high-tide mark (to the left and out of the frame) than at the lower level. At low tide, water drains from the sand on the upper beach first, while the lower beach retains water longer. (B) A conspicuous track left in the sand at low tide by a hermit crab. After exploring the sediment surface for while, it buried itself again in the sand, to safely await the return of high tide.

(A)

(B)

FIGURE 12.9 American beachgrass, *Ammophila breviligulata*, is native to the Atlantic coast, and is an invasive species on the U.S. west coast, where it was introduced in the 1930s. The expanse of beach grass extends about 100 meters to the left in this photograph, and out of the frame. It is capable of spreading 2–3 meters a year by sending out subsurface rhizomes, or runners, a process that appears to be underway here, as evidenced by lines of plants extending into the fore-dune area. These salt-tolerant plants are important in stabilizing the sand against wind-driven erosion.

Vertical shoots

mark. Terrestrial amphipods called sand hoppers, or beach fleas, as well as some isopods, abound in this beach wrack. As the tide ebbs, sea birds congregate in the swash zone—where the waves wash up the beach face—ready to pluck any benthic animals, such as sand crabs, that might be exposed.

Beyond the high-tide mark on many beaches are the dry sand dunes, with their coverings of beachgrass. These hardy grasses, with their extensive root and rhizomes systems, are capable of spreading rapidly, colonizing the **foredune**, or ocean-facing portions of the upper beach and sand dune system, and protecting against wind erosion of sand (**FIGURE 12.9**). They are firmly anchored in place in the shifting sand by root systems that extend a half-meter deep, and can withstand complete burial under windblown sand, sending up new vertical shoots that become new plant stems.

Estuaries

Estuaries are coastal environments where freshwater rivers and streams encounter and mix with the sea. They have been defined a number of ways, but one definition that is generally accepted is: "An estuary is a semi-enclosed coastal body of water which has a free connection with the open sea and within which seawater is measurably diluted with fresh water derived from land drainage."[4] Estuaries may be large or small, but all are basically some form of a semi-enclosed embayment within which the transition from freshwater to full ocean salinity occurs. Exceptions to our definition would include the largest rivers, such as the Amazon, that discharge such enormous volumes of freshwater that the mixing with seawater occurs out in the ocean, well beyond the mouth of the river (see Figure 4.22, for example). Regardless of the exact definition, it is this physical act of mixing of fresh and salt waters that produces interesting vertical and horizontal circulation patterns within estuaries, which are important in driving what are commonly very high rates of biological production.

Estuaries are categorized (1) by their geomorphology, or general shape, and (2) by their salinity distributions. The **drowned river valley estuary**, also known as a **coastal plain estuary**, a category based on geomorphology, is an elongated embayment where river waters and ocean waters come together in a valley that may have been carved more than 10,000 years ago by rivers and,

[4] D. W. Pritchard, 1967. "What is an estuary: Physical viewpoint," in *Estuaries,* ed. G. H. Lauff (Washington, DC: American Association for the Advancement of Science). 83: 3–5.

depending on location, glacial scour, during and just after the last ice age when sea level was much lower than it is today. As sea level began to rise, the valley became flooded, and today the sea mixes with the river water in a relatively narrow channel. One of the largest of these estuaries is the Columbia River estuary between the states of Oregon and Washington, where the Columbia River empties into the Pacific Ocean. Coastal ocean conditions at the mouth of this powerful estuary are extremely treacherous, with turbulent wave actions produced as immense volumes of surface outflow pour into the coastal waters. It was the United States Exploring Expedition (the U.S. Ex. Ex. discussed in Chapter 1) that first mapped the mouth of the Columbia and reported on the extreme ocean conditions there (which sank one of the expedition's ships). There are many drowned river valley estuaries on both coasts of North America, including the Chesapeake Bay, Delaware Bay, and Hudson River estuaries.

Another common type of estuary based on its geomorphology is the **bar-built estuary**; these estuaries are more common in the Gulf of Mexico and along the southeast coast of the United States. They were formed as **barrier islands** built up from accumulations of terrestrial sediments in the shallow continental shelf waters just offshore of low-lying coastal areas. The barrier islands are just that—they are barriers to high-energy, open ocean wave actions, and they create and protect an enclosed embayment on their landward side where relatively small rivers and streams dilute the seawater to well below open-ocean salinities. Examples of these estuaries are the Pamlico and Albemarle Sounds in North Carolina, which together comprise one of the two largest estuaries in North America (Chesapeake Bay being the largest), and the numerous bar-built estuaries along the south Texas coast, such as Galveston Bay (**FIGURE 12.10**).

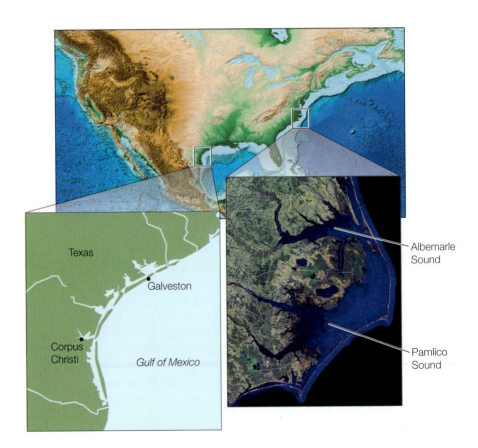

FIGURE 12.10 The bar-built estuaries on the south Texas coast (left), and Pamlico and Albemarle Sounds (right), bar-built estuaries in North Carolina.

Categories of estuaries based on salinity distributions span a range that is determined by the mixing patterns of fresh water and seawater—especially how the volume of freshwater being discharged by the river compares with the degree of tidal mixing, which in turn, is a function of the tidal range (**FIGURE 12.11**). In **well-mixed estuaries**, the incoming freshwaters are of sufficiently low volume that vertical mixing by the tides overcomes the buoyancy resulting from the fresh water flow. The result is only a relatively slight vertical gradient in salinity between the surface and bottom, and instead there is a gradual longitudinal salinity gradient from the head of the estuary to the mouth at the seaward end. In a **partially mixed estuary** there is some vertical mixing by tides, but it is insufficient to completely overcome the buoyancy of the significantly greater volume of incoming freshwater at the head; as a result, there is a halocline separating relatively fresh water at the surface from a layer of higher-salinity water beneath it. There is also a gradual longitudinal salinity gradient along the length of the estuary. In a **salt wedge estuary**, the intensity of tidal mixing is very weak compared with the large volume of buoyant freshwater entering the estuary, such there is little vertical or horizontal mixing of freshwater and seawater, except at the seaward end. The fresh water outflow extends from top to bottom in the upper estuary, and pushes seaward some distance down the channel, where it abruptly meets the high-salinity seawater. At that point there is a strong salinity gradient, across the inclined (wedge-shaped) halocline, where the fresh waters overrun the deeper seawater layer. Mixing of freshwater and seawater occurs mostly at the shallower depths.

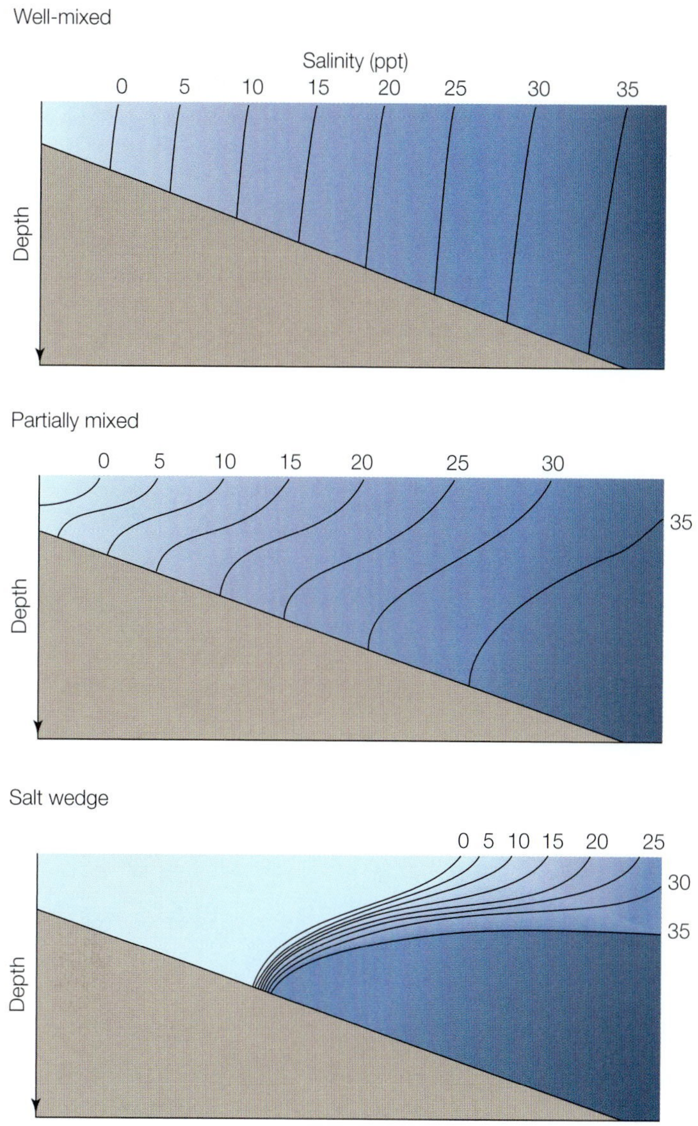

Biological productivity in estuaries is very high for several reasons. For one, estuaries serve almost as collection basins for detrital organic material delivered from both the freshwater and ocean ends, providing a food source for various suspension and deposit feeders. Another reason is that they are rich in dissolved nutrients that support macroalgal and

FIGURE 12.11 The three types of estuaries based on the pattern of mixing: a well-mixed estuary, a partially mixed estuary, and a salt wedge estuary.

phytoplankton production. Unfortunately, many of the rivers emptying into estuaries are artificially enriched with nutrients from pollution as a result of human activities (agricultural runoff and human sewage), and thus there are significant deliveries of dissolved nitrogen and phosphorus into the estuarine ecosystem. But many rivers come by their high nutrient loads naturally. Nutrients are typically high in the terrestrial runoff that feeds the rivers and tributaries leading to the estuaries, and because rivers are usually light-limited as a result of turbulent vertical mixing, phytoplankton production is low and nutrient concentrations remain relatively high. Upon encountering saline waters in the estuaries, the buoyant, nutrient-rich fresh waters are suspended in the surface layers where light levels are more favorable for phytoplankton production. These terrestrially derived nutrients are important, but the most important source of dissolved nutrients to many estuaries is that carried in deep saline waters entering from the ocean, which is a result of the nature of estuarine mixing and circulation.

Most of the drowned river valley estuaries, especially the partially mixed estuaries, have a pronounced two-layered flow, with a surface flow of freshwater seaward and a return flow of deep waters up the estuary in the opposite direction. These flows are magnified in partially mixed estuaries because of the degree of mixing across the halocline (**FIGURE 12.12**). The outgoing, fresher surface waters continuously mix across the halocline with the deeper, higher-salinity waters as they flow seaward, such that by the time those waters reach the mouth of the estuary, the volume flow in the upper layer may be several times that of the original freshwater flow at the head. The higher-salinity water

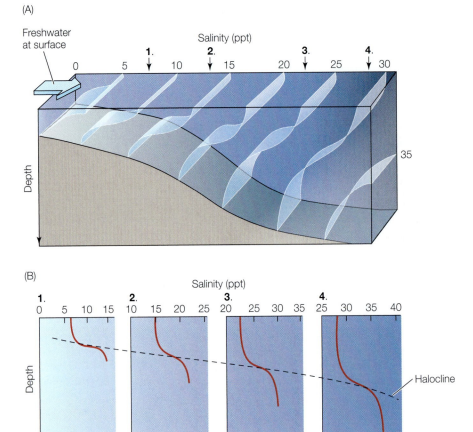

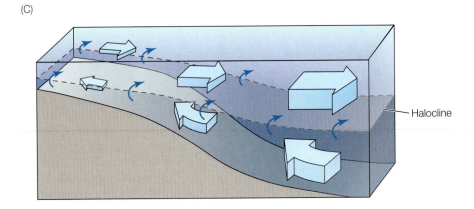

FIGURE 12.12 Estuarine circulation. (A) Salinity distribution from the head to the mouth in a partially mixed estuary. Freshwater enters at the surface (left) and mixes with the higher-salinity water beneath. (B) Vertical salinity profiles at the four locations in (A) reveal sharp haloclines, where salinity increases markedly across a narrow depth interval; the depth of the halocline increases from the head of the estuary to the mouth. (C) The relative volume of inflowing water below the halocline from the mouth of the estuary to the head, along with the outflowing water volumes at the surface. The halocline mixing between the two layers is indicated by the curled arrows.

mixing upward into the surface flow has to come from somewhere; in particular, the *salt* that is flowing out at the surface, after mixing upward across the halocline, has to be replaced by an up-estuary deep water flow. The result is a similar, greatly magnified deepwater flow directed up the estuary, a flow that is greatest at the mouth and least at the head. And because those deep waters moving up the estuary have their origins in deeper coastal waters, the nutrient loads are relatively high. Partially mixed estuaries, then, are *nutrient pumps* that bring deep-water nutrients from the coastal ocean up into the estuary, mixing them into the surface where they support phytoplankton production.

The planktonic productivity of estuaries, coupled with the significant fluxes of detrital organic material, make these environments very productive indeed, supporting large populations of deposit-feeding and suspension-feeding invertebrates, which in turn support large populations of fishes. As a result, some estuaries serve as important **nursery grounds** for many species of commercially important fish species, such as menhaden in the mid-Atlantic estuaries of the U.S., which spend a portion of the early life histories there.

Salt Marshes

Salt marshes, also called tidal marshes, develop along gently sloping shores of temperate and higher-latitude areas that are protected from high-energy wave action, and where fine sediments can collect (**FIGURE 12.13**). Salt marshes are usually associated with the banks of estuaries, but they also occur adjacent to protected beaches and mud flats (see Figure 12.7). Residing at the high-tide mark in low-energy coastal areas, there are rich expanses of salt-tolerant plants, especially grasses and rushes. Salt marshes are biologically productive ecosystems that support a number of terrestrial animals, birds, and local estuarine and marine organisms residing in and among the marsh grasses and sediments as well as in the tidal streams and near-shore waters that nourish them.

These environments are the product of accumulations of sediments, primarily from nearby terrestrial sources, which become trapped and stabilized among the marsh grasses and their extensive root and rhizome systems. The sediments build up over time, extending their coverage to surrounding low-lying areas, sometimes creating broad expanses of marsh grasslands. Salt marshes are restricted to those low-lying areas where sea level is rising at a rate that keeps pace with the accumulations of sediments and which also allows the

FIGURE 12.13 (A) New England salt marsh with its mud flats at low tide, and (B) evidence of the tidal stream eroding the marsh and exposing the underlying peat.

(A)

(B)

(A)

(B)

Roots and rhisomes   Peat   Beach sediment

FIGURE 12.14 (A) Smooth cordgrass, *Spartina alterniflora*, growing at the edge of a small salt marsh, with plants expanding onto a protected gravel beach. (B) An eroding edge of a salt march, exposing a tangle of roots and rhizomes of *S. alterniflora*. Just below that is exposed peat, which may extend several meters deeper, reflecting the history of vertical growth of the marsh.

marsh to be periodically flooded at high tide, or at extreme spring high tides. This regular flooding of the marsh prevents invasions of terrestrial plants that are not salt-tolerant. For these reasons, salt marshes are most common on passive continental margins, such as the east coast of North America and the Gulf of Mexico, and they are less common on the Pacific coast, where the shoreline is generally steeper.

The typical salt-marsh grasses are species of *Spartina* (and, indeed, salt marshes are sometimes called Spartina salt marshes). Smooth cordgrass, *Spartina alterniflora*, is a pioneer species, found from Newfoundland to Florida and Texas and as an invasive species on the Pacific coast; it tends to grow along the edges of the marsh closest to the shoreline or tidal creek, where it sends out rhizomes where new plants will sprout, thus expanding its coverage (**FIGURE 12.14**). It usually stands taller along these edges than the other grasses in the marsh, dominating the other species and inhibiting their growth. Behind this leading edge, the *S. alterniflora* take on a smaller form, which eventually become replaced by the more finely textured salt-meadow cordgrass, also called salt-meadow hay, *Spartina patens*. Still farther back in the salt marsh are other species of grasses and salt-tolerant plants, such as rushes (*Juncus* spp.). On the Pacific coast, California cordgrass, *Spartina foliosa*, dominates the leading edge of the marsh, usually adjacent to mud flats, which it gradually invades. The landward limit of a salt marsh is characterized by several species of plants that are tolerant to salt spray, but not to immersion.

The development and growth of salt marshes proceed as muddy layers of sediments and the detrital remains of marsh grasses build up over time. Where the contribution of marsh grass detritus is substantial, the organic content of the sediment accumulations is very high. Like intertidal mud flats, salt marshes exhibit oxygen depletion just below the sediment surface. This lack of oxygen effectively preserves the sedimentary organic matter as **peat**, which is a compacted mass of partially decayed marsh grass remains and mineral sediments (see Figure 12.14). Peat is the major contributor to the buildup of the marsh itself, which is less reliant on terrestrial (mineral) sediments. The underlying layer of peat in some marshes, depending on their ages, can be extensive; peat layers as deep as 10–20 m are not uncommon.[5]

One of the reasons for the high rates of production in salt marshes is nitrogen fixation both by photosynthetic blue-green algae on the surface of the marsh sediments, as well as by nitrogen-fixing bacteria around the roots and rhizomes of the marsh grasses. But the base of the food web is not algae or phytoplankton, it is detritus from the marsh grasses themselves, broken down by grazers, fungi, and bacteria.

---

[5] Redfield, A. C., and Rubin, M., 1962. The age of salt marsh peat and its relation to recent changes in sea level at Barnstable, Massachusetts. *Proceeding of the National Academy of Sciences* 48: 1728–1735.

Mangrove Forests

Mangrove forests are the low-latitude counterpart of salt marshes in the tropics and subtropics, growing in similarly protected intertidal shorelines (**FIGURE 12.15**). There are more than 100 species of mangrove trees, which often occur in forests that are also known as **mangals** (**FIGURE 12.16**). These woody flowering plants range in size from shrubs to medium-sized trees that may grow to more than 10 m height.

Similar in ecological function to salt-marsh grasses, mangroves trap sediments in their extensive root systems, which are adapted to the anoxic muddy sediments in which they are most common. They have snorkel-like aerial roots, called **pneumatophores**, that emerge from the anoxic mud and, essentially, breathe air, thus oxygenating the buried roots. The roots themselves are impressive physical structures that function like the flying buttresses used to support large stone buildings; they flare out from well above the water level, providing support to the tree in the generally unstable muddy sediments. Their numerous branching tangles of roots trap sediments and allow the forest to expand; the dense coverage of roots interferes with the flow of water as the tides come and go, which allows suspended sediments to settle out and accumulate.

Mangrove trees produce seeds that germinate while still attached to the tree, growing into long, pea-pod-like seedlings that grow heavier with time until they eventually drop from the tree to the water. There they float upright like a spar buoy and are dispersed with the currents. Eventually they ground out in the shallow waters, growing roots that anchor them in place where they grow and develop. Examples of newly established seedlings of mangroves in the Indian Ocean are shown in Figure 12.16B.

Mangrove forests constitute important tropical marine environments, providing habitat protection for birds and other terrestrial animals in the

FIGURE 12.15 World map showing the distributions of salt marshes along temperate and higher-latitude coasts and mangrove forests in the tropics and subtropics. Their distributions overlap in some regions, such as the Gulf of Mexico.

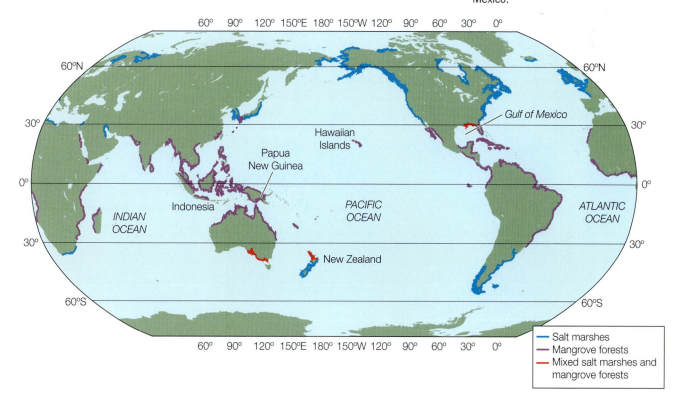

(A)

(B)

FIGURE 12.16 (A) A mangrove forest with thickly tangled root systems in the Florida Keys National Marine Sanctuary. (B) Mangroves at low tide on Havelock Island, one of the Andaman and Nicobar Islands in the Bay of Bengal, India. Mature trees with their branching root systems can be seen in the background; newly settled seedlings are in the foreground.

forest canopy, and they harbor numerous marine and estuarine fishes and invertebrates. They are particularly important for the protection of low-lying areas from tsunamis and storm surges.

Coral Reefs

As we learned earlier, the tropical oceans of the world are vertically stratified throughout most or all of the year, and as a result, the warm surface waters are isolated from the dissolved nutrients that reside at deeper depths. Without upwelling and mixing processes to bring those nutrient-rich waters to the surface, significant levels of primary production by phytoplankton are not possible, and as a result the tropical oceans remain biologically unproductive. This is why they have such spectacularly beautiful, clear blue waters—because they are virtually devoid of planktonic life. In fact, the tropical oceans are often thought of as biological deserts. But there are two exceptions: one is the areas where upwelling does occur, which we discussed in Chapter 7, and the other is the **coral reefs** (**FIGURE 12.17**).

FIGURE 12.17 Some examples of different forms of corals that make up pristine coral reef ecosystems. (A) A reef in the Republic of Palau in the eastern tropical Pacific Ocean some 500 miles east of the Philippines. (B) Palmyra, an unoccupied coral reef atoll that is one of the Northern Line Islands just north of Hawaii.

(A)

(B)

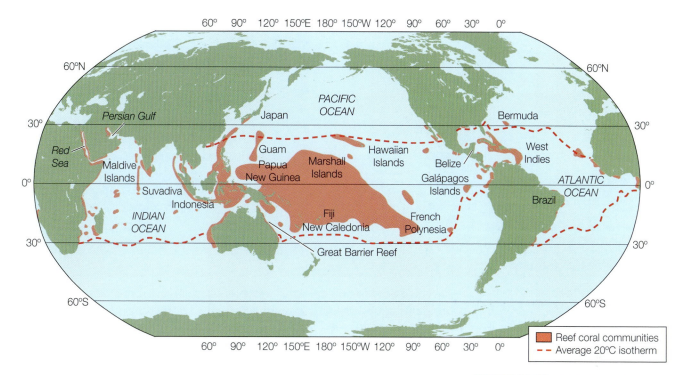

Anyone who has ever snorkeled in these colorful underwater gardens would agree that coral reefs are perhaps the most beautiful underwater environments on the globe. The reef-forming coral species are categorized as **massive corals** (spherically-shaped, but generally not exceeding a diameter of 50 cm), **encrusting corals**, **branching corals**, **columnar corals**, and **foliaceous corals**. All feature a great variety of nooks and crannies—habitats for other reef-dwelling organisms.

Coral reefs constitute an interesting intersection of geology and biology. There are basically two kinds of coral reef systems: **shelf reefs** that grow in shallow coastal waters on continental margins, and **oceanic reefs** that grow around the edges of islands. But all are geographically restricted by surface-water temperatures: these must remain higher than about 18–20°C, and they cannot rise much above 28°C without negative effects (**FIGURE 12.18**).

The oceanic reefs may be further classified as **fringing reefs**, **barrier reefs**, and atolls. Fringing reefs are the most common, developing along coastlines in tropical seas; there are extensive examples along the shores of the Indian Ocean and the Red Sea. Barrier reefs also form along coastlines, but they are generally much farther offshore, nearer the edges of the continental shelf, and are separated from the mainland by a shallow lagoon. The most famous barrier reefs, and by far the largest reef system in the world, is Australia's Great Barrier Reef, which runs the length of Australia's northeast coast (**FIGURE 12.19**). Much smaller versions of these barrier reefs are also found in the Caribbean Sea. We have already discussed the formation of the ring-shaped atolls and how Charles Darwin explained the mechanism of their formation, as volcanic islands sink in the asthenosphere (see Chapter 3). These reef atolls are found throughout much of the tropical Pacific Ocean in association with volcanic islands.

Coral reefs can be massive geological structures. They are made up of remains of the skeletons of the coral polyps, the anemone-like colonial animals that precipitate around themselves a calcium carbonate ($CaCO_3$) structure that is their base, and which is the actual reef. The photographs in Figure 12.17 show

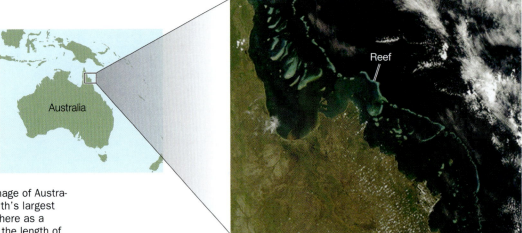

FIGURE 12.19 Satellite image of Australia's Great Barrier Reef, Earth's largest biological structure, visible here as a faint line of islands running the length of Australia's northeast coast. A close-up of the area shows the line of the reefs more clearly. This spectacular image, taken in late February 2007 by NASA's Terra satellite.

several species of corals and their elaborate skeletal base forms. They are capable of building upward and leaving behind thick structures of carbonate rock, primarily ancient coral skeletons, but also reef debris made up of carbonate deposits from numerous other invertebrates and protists that help to cement the reef's structure (**FIGURE 12.20**). The bases of some reefs extend hundreds of meters below sea level, reflecting the upward growth of the deposits as the host land mass subsided over time.

The coral polyps are more than just the builders of these massive carbonate structures: they also are the reason that these features are so biologically productive in an ocean that would otherwise be a biological desert. The coral polyps are heterotrophic animals, consuming organic matter that is produced somewhere else, by something else. They are suspension feeders that collect suspended particulates that happen to fall onto their tentacles (as is done by their close cousins, sea anemones); and they are also predators. Like all cnidaria, they possess stinging nematocysts that line their tentacles and which can disable small animals that happen to come into contact with them. But, they are unusual in that they (and various other cnidaria) also have within their bodies their own primary-producing endosymbionts, photosynthetic dinoflagellates,

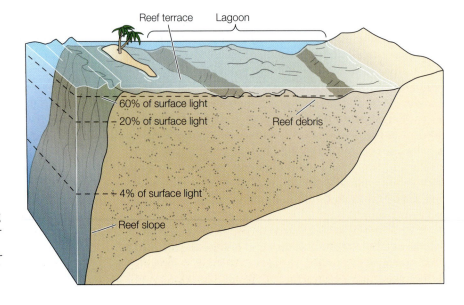

FIGURE 12.20 Cross section through a fringing coral reef, showing the built-up remains of ancient coral skeletons of calcium carbonate against the coastline of a tropical island or mainland. The photosynthetic zooxanthellae require light, and so the living and growing portion of the reef is relatively shallow waters where more than 20% of the surface light is available. Other (nonphotosynthetic) corals can grow along the outer edges to depths approaching 150 m.

(A)

FIGURE 12.21 (A) Photomicrograph of the dinoflagellate *Symbiodinium microadriaticum*, a species of coral zooxanthellae. (B) Anatomy of a coral polyp, showing the location of zooanthellae beneath the integument (center inset), and the exchanges between the host polyp and the symbiotic zooxanthellae of essential nutrients and products of photosynthesis (right inset).

(B)

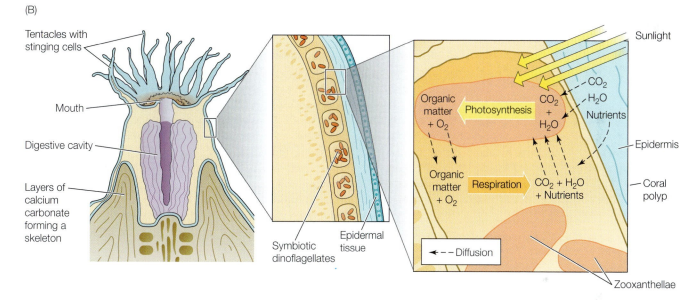

commonly called **zooxanthellae** (**FIGURE 12.21**). These dinoflagellates have lost their flagella and thecal plates, and reside in the inner of the polyp's two layers of cells, where, depending on the depth of the polyp itself, they are exposed to sufficient sunlight to perform photosynthesis. These algae are extremely abundant in the polyps, with as many as one million cells per square centimeter of polyp colony surface area; such phenomenally high cell densities just below the surface give the polyps, and the coral reefs, their characteristic coloration. Their photosynthetic potential is extraordinary: these symbiotic dinoflagellates can photosynthesize between 10 and 100 times more organic carbon than is needed to meet their own metabolic and growth requirements. Thus, they constitute a primary production system that supplies carbohydrate to the coral polyps, in addition to that supplied by their external food sources.

The zooxanthellae are well adapted to be symbiotic with the polyp. The polyp supplies the zooxanthellae with needed CO_2 and nutrients, such as ammonia, via its own metabolic processes (i.e., respiratory and metabolic waste products), and in return, the zooxanthellae provide the polyp with a valuable service: they remove these toxic waste products for the polyps, and at the same time, they manufacture food for the polyp via photosynthesis. The metabolic activity of the zooxanthellae also increases the pH, which makes the thin layer between the polyp's tissues and the carbonate reef less acidic, thus facilitating the polyp's secretion of more calcium carbonate, which continues the reef-building process. The zooaxanthellae benefit in that they acquire greater supplies of inorganic nutrients and CO_2; they are protected from grazers; and they are held near the surface, with plenty of sunlight. But, they can also be held too near the surface, where they are exposed to harmful UV radiation; for this reason, corals have the unusual capability to produce UV sunscreens with which to protect themselves. The corals also are susceptible to the harmful effects of abnormally warm ocean water temperatures. When water temperatures do rise above the upper limit of about 28°C, such as during El Niño events (see Appendix B) a phenomenon known as **coral bleaching** may occur (**BOX 12A**).

BOX 12A Coral Bleaching

As seems to be the case for so many seemingly pristine environments these days, there is concern for the viability of coral reefs.

In 1983, after the 1982–1983 El Niño, a strange phenomenon began to appear in corals of the eastern Pacific Ocean and in the Caribbean Sea: the coral polyps apparently lost their zooxanthellae, causing the corals to lose their greenish-brown color and expose their white calcium carbonate skeletons—hence the name coral bleaching (**FIGURE A**). Subsequently, about 70% of the corals in these areas died. Most of the coral reefs eventually recovered, but only after several years.

The reason for the geographically extensive bleaching is believed to have been the unusually warm waters associated with El Niño. And there is growing concern that bleaching, both localized and widespread, may be worsening, as a result of long-term global warming. Following the last major El Niño (the 1997–1998 event) was another major coral bleaching event that impacted the Great Barrier Reef in Australia in 1998. The impact was extensive. A map of the sea-surface temperature anomaly—the difference in temperatures, in °C, from the long term average—associated with the severe 1998 bleaching event is shown here for February 1998 (**FIGURE B**); the green points indicate reported bleaching hotspots. Notice the anomalously warm water against the east coast of Australia (the Great Barrier Reef) and in the Indian Ocean.

Worldwide, reef corals are already living within only a few degrees of their upper thermal limits. Recent studies have produced evidence that tropical

FIGURE A Recent bleaching of corals in Guam.

FIGURE B

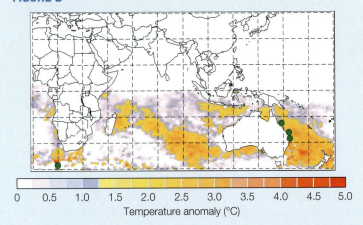

Temperature anomaly (°C)

ocean surface-water temperatures have increased by about 0.7 to 0.8°C over the past 100 years; numerical model simulations predict those temperatures to increase at a rate of about 1–2°C per century. Superimposed on this increase from global warming is the additional abnormal warming

seen in El Niño years. The prognosis: future trouble for coral reefs. But this isn't the only crisis facing our coral reefs: they must also deal with an ocean that is becoming more acidic because of increases in atmospheric CO_2, which we'll discuss in greater detail in Chapter 15.

Coral reef systems are highly efficient biological engines, forming the base of a rich food web—albeit a fairly small one in comparison with the whole tropical ocean. Estimates of their rates of primary production approach 5000 gC/m^2/yr, a rate that is some ten times that of phytoplankton production in upwelling systems. But as we have emphasized earlier in this book, primary production is not growth; for there to be growth in numbers of zooxanthellae and for growth of new polyps, nutrients are required. Because the polyps are also predators and suspension feeders, they acquire nutrients from their food organisms, some of which (e.g., ammonia) are recycled to the zooxanthellae. Still, this cannot go on for long, this recycling back and forth of nutrients within the coral polyps, and an external supply of nutrients is needed. But, these coral reefs, as we explained above, are located in nutrient-poor surface waters of the tropical oceans, and without a new source of nutrients, their biomass will reach a maximum, beyond which there can be no further growth; quite simply, the system will run down. So, how do they do it? How can they be such biologically productive systems in such a nutrient-poor environment? The answer lies in the nature of the coral reef as a habitat, and in the high ecological efficiencies of the coral polyps themselves. On the one hand, once acquired, nutrients are recycled rapidly and efficiently between producer (zooxanthellae) and consumer (polyp) in the coral colonies. And on the other, nutrients are brought to the reef from outside, in the form of other organisms. Coral reef ecosystems are biological islands in an otherwise unproductive tropical ocean, supporting a high biomass and richly diverse community of fishes and invertebrates; because the coral reefs attract other animals, they serve as "habitat." It is because the reef system harbors a wide variety and large biomass of organisms that it can maintain itself; but not all of those organisms can derive all of their needs from the reef ecosystem. A net positive flux of nutrients to the reef, in the form of organism tissues, must be brought to the reef from elsewhere. Whether brought by pelagic animals that are attracted to the reef for its abundance of potential prey organisms, or by tropical planktonic organisms that drift by, the nutrients must come from elsewhere. The coral symbiosis with their zooxanthellae is largely self-sufficient in terms of energy (sunlight) and carbon, but the nutrients must come from external sources, such as the polyp's prey. But the details of this balance—the finer, intricate details of coral reef trophic dynamics—are not well known.

The Deep-Sea Environment

The **deep sea** is a term that has taken on many meanings, but in oceanography it generally refers to those ocean depths beneath the photic zone, where light from the surface has all but disappeared (see Chapter 7). From there to a depth of 1000 m, the **mesopelagic zone**, what little light remains disappears completely, even in the clearest ocean waters. Water temperatures vary only slightly over that depth range, dropping to about 2°C. Beneath that layer, in the **bathypelagic zone**, from 1000–4000 m, the average bottom depth for the world ocean, it is not only dark, but it is cold (see Appendix C). Temperatures are uniformly about 2°C from 1000 m all the way to the bottom, even to the oceans' greatest depths, and fluctuate very little from ocean to ocean, except at the poles, where the temperatures drop to near zero. Beneath the bathypelagic zone, from 4000–6000 m, is the **abyssopelagic zone**, and below that, down to the ocean's deepest depths (>11,000 m) is the **hadopelagic zone**.

With no light, there can be no photosynthesis at depths below the photic zone, and so animal life in the deep sea, whether on the bottom (the benthos, where the terms **abyssal zone** and **hadal zone** apply) or in the pelagic realm,

(A)

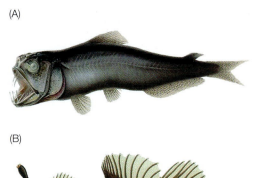

(B)

(C)

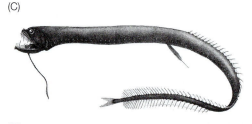

(D)

FIGURE 12.22 Examples of deep-sea fishes. (A) A sabretooth fish, *Coccorella atrata*. (B) A humpback anglerfish, *Melanocetus johnsonii*. (C) A black dragonfish, *Idiacanthus atlanticus*. (D) Photograph of an anglerfish.

is dependent on food brought in from elsewhere—from the surface waters, by way of sinking. Because consumption of food particles produced in the surface waters is fairly complete there, very little escapes and sinks. And so with the supply of food resources being very meager, so too are the abundances of animals at greater depths. But at the extremely low temperatures down deep, metabolic rates are slowed, which reduces the food requirements.

Deep-sea fish adaptations

Fishes of the mesopelagic zone, sometimes called midwater fishes, which reside to depths of about 1000 m, begin to take on an odd-looking appearance, as they have evolved to make the most of their dark, cold and food-poor environment. Many will have extremely large eyes in proportion to their relatively small bodies (most are less than 10 cm in length), enabling vision by capturing what little light may penetrate to these great depths. Many species also have bioluminescent organs along their ventral portions that help to offset their silhouette as viewed by predators from below, such as the rows of white dots that can be seen along the ventral surface of the dragonfish in Figure 12.22C. The lantern fishes of the family Myctophidae are the best examples.

Deep-sea fishes, especially those that inhabit depths below the mesopelagic, are good examples of how animals have evolved to deal with the scarcity of food in deep ocean waters; almost caricatures of themselves, many species possess enormous mouths and long teeth that can hold onto and devour any prey they are lucky enough to encounter (**FIGURE 12.22**). The viperfish, for example, has specialized jaw hinges that allow it to open its mouth much wider than would otherwise be possible, enabling it to latch onto another fish with its dagger-like teeth (which are the longest teeth in the world relative to an animal's head size). Some of the anglerfishes have evolved an unusual way to enhance their chances of feeding: a special dorsal fin ray with a bulbous, bioluminescent end dangles in front of its large mouth and acts as a bioluminescent fishing lure, attracting other fish close enough to be engulfed (see Figure 12.22B).

Not only is finding food in the deep sea a challenge; so, too, is reproducing. Because of the extremely low densities of animals, especially animals of the same species, finding a mate is perhaps even more difficult than finding a meal. Some species of anglerfishes have evolved a way around this predicament, as well. Upon finding a female, the much smaller male, whose sole purpose in life is to fertilize the females eggs, bites onto the female and simply never lets go (or, till death do they part). Their tissues eventually fuse, and she is never again without a mate to fertilize her eggs.

We could go on to give numerous accounts of fascinating aspects of the ecology of deep sea fishes, but that is beyond the scope of this introductory text. Much more complete treatments are given in the fish biology texts cited in Chapter 11.

The deep-sea benthos

The bottom of the deep ocean is an even tougher place to make a living than is the deep-water pelagic realm above it. Also cold and dark, and under even more pressure, this environment offers even less food—only that delivered from above after the pelagic consumers remove their portions. But animals of the benthos have more time to locate and eat that food once it arrives on the bottom.

Most of the deep-sea bottom is covered in a fine muddy sediment that characterizes the abyssal plains, which constitute most of the deep-sea floor. These sediments are very low in organic content, but relative to terrestrial silts and clays, they are rich in biogenic oozes, the organic and inorganic remains of planktonic organisms that settle to the bottom. While much of that organic material (such as chiton) is not digestible by most marine animals, much of it is digestible by bacteria, which serve as food for the benthic meiofauna, which serve as food for still larger invertebrates. Thus the dominant feeding mode among benthic animals in the deep sea is deposit feeding, and most of the animals are infaunal and epifaunal invertebrates. The polychaete worms are the most abundant macro-invertebrates in the deep sea benthos; they are prey for the few predators there, which are mainly echinoderms and small crustaceans.

The abundances of animals on the deep sea floor are, not surprisingly, very low. But the species diversity is much higher than one might expect it to be, and this diversity has been the subject of much study and discussion by marine ecologists over the years. Prior to the 1960s, it was thought that the harsh living conditions there should be able to support just a handful of highly adapted species. Instead, researchers found that benthic species diversity increased the deeper they sampled (to as deep as 5000 m).[6] The Stability–Time Hypothesis, proposed by Howard Sanders in 1968, argued that the constancy of the deep sea benthic environment would, over long periods of time, result in very finely partitioned resources that would increase the numbers of species that occupy those similar, but nonetheless different, niches.

Hydrothermal vent communities

The fields of marine biology and oceanography were caught by surprise in 1977 by the discovery of a totally new deep-sea ecosystem at the site of hydrothermal vents off the Galápagos Islands in the eastern Pacific Ocean. Located along the spreading ridges where oceanic crust is being formed, these vents are sites where seawater that has percolated into hot magma becomes superheated to as high as 350°C (without boiling, which is suppressed under such extreme pressures), and spews out in high volumes of hot water rich in dissolved metals and hydrogen sulfide. The sulfate ions in seawater (SO_4^{2-}, one of seawater's most abundant ions, remember) react with hot seawater to form hydrogen sulfide (H_2S), which in turn reacts with metals such as iron to form metal sulfides. Upon contact with the cold ocean-bottom waters, those metal sulfides precipitate out as fine black particles, forming what appear to be thick black clouds of smoke (**FIGURE 12.23**). Bacteria oxidize the H_2S back to sulfate ions (SO_4^{2-}) using the electrons from H_2S to reduce CO_2, and in the process form carbohydrates (CH_2O):

$$CO_2 + H_2S + O_2 + H_2O \rightarrow CH_2O + H_2SO_4$$

This is *chemosynthesis*, a form of primary production at the bottom of the ocean that uses, instead of sunlight, the chemical potential energy in reduced inorganic molecules such as H_2S to produce organic carbon compounds.

FIGURE 12.23 Active hydrothermal chimneys, also known as a black smokers. This photo was taken by scientists during a study of hydrothermal vents in the Pacific Ocean off New Zealand as part of the New Zealand American Submarine Ring of Fire 2007 expedition.

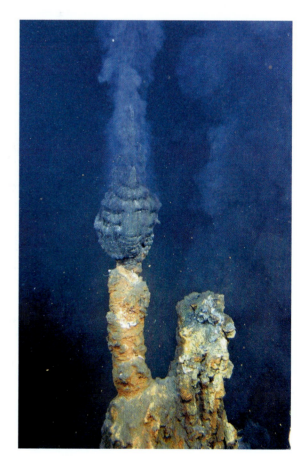

[6] Hessler, R. R., and Sanders, H. L., 1967. Faunal diversity in the deep sea. *Deep-Sea Research* 14: 65–78; and Sanders, H. L., 1968. Marine benthic diversity: A comparative study. *American Naturalist* 102: 243–282.

FIGURE 12.24 Giant tubeworms of the genus *Riftia* growing in a vent community off the Galápagos Islands. The heated water is causing convection currents which distort the photograph.

The primary production by these chemosynthetic bacteria form the base of a deep-sea food web that is unlike anything discovered before. The researchers, upon approaching these vent communities were stunned by what they saw: clustered about the vents were tube worms more than 2 m in length, giant clams and mussels more than 30 cm in length, and new species of crabs. New research expeditions were launched over the years to other parts of the ocean rift valley system in both the Atlantic and Pacific Oceans, with each expedition discovering something new.

The red-plumed *Riftia* tubeworms were among the biggest surprises (**FIGURE 12.24**). They completely lack a digestive track and are unable to ingest food. They rely on an internal system of endosymbiotic bacteria that oxidize hydrogen sulfide, and produce carbohydrate.[7]

It was later learned that rift valleys of the Pacific and Atlantic Oceans, which have very different spreading rates, also support very different biological communities. Chemosynthetic tube worms were for a long time thought to be all but absent in the Atlantic Ocean, for example, where white, eyeless shrimp predominate; however, in 2011, some chemosynthetic tube worm species were discovered in the Caribbean Sea of the western Atlantic Ocean. And the research on these fascinating deep-sea environments continues.

[7] Their bright red plume is a form of gill for uptake of O_2 and H_2S; the red color is from hemoglobin that transports not only O_2 for the worm's respiration, but also H_2S and O_2 for delivery to the endosymbiotic sulfide-oxidizing bacteria.

Chapter Summary

■ The **intertidal zone** is principally a benthic marine environment between the high-tide and low-tide marks. It is rich in biomass and has high species diversity. Living conditions there are harsh, with organisms exposed to alternating periods of submergence and exposure to air, and with that, periods of desiccation and extremes in temperatures and salinity. Exposure is greatest nearer the high-tide mark, and submergence greatest at the low-tide mark. Types of intertidal zones are based on substrate type.

■ The rocky intertidal zone occurs in areas of high wave energy and no sediment accumulation. Desiccation and exposure to waves are the biggest physical stresses; predation and competition for space are the greatest biological stresses. At the base of the food web are attached algae and plankton.

■ Competition, predation, and the gradient in physical stress factors create **intertidal zonation**, where a banding pattern of dominant species is expressed. The upper limit of an organism in the intertidal zone is determined by its ability to tolerate physical stresses, while the lower limit is determined more by biological processes.

■ **Soft-bottom** intertidal environments are areas where low wave energy allows sediments to accumulate. Attached algae are usually lacking here; the main food source for intertidal animals is detritus, with plankton also important at times of submergence.

■ The ecological structure in soft-bottom intertidal environments is determined by sediment grain size. The finest sediments, found in the lowest energy environments, are made up of silts and clays; these sediments are also the richest in detrital organic matter. Water is retained better in fine-grained sediments at low tide; as a result, these sediments become anoxic just below the sediment surface. Infaunal and epifaunal deposit-feeding invertebrates dominate here. Higher-energy soft bottom environments, such as beaches, have less organic detritus and do not retain water as well as fine-grained sediments, and do not become anoxic as readily. Biological activity is significantly lower in coarse-grained sediments.

- **Estuaries** are where fresh water meet and mix with seawater. They are categorized according to their geomorphology (with **drowned river valleys** and **bar-built estuaries** being the two most common), and according to their salinity distribution, which is determined by the volume of freshwater additions and the degree of tidal mixing, giving **well-mixed**, **partially mixed**, and **salt wedge estuaries**.

- Biological productivity is high in estuaries because they trap sediments and receive nutrients from both the freshwater and seawater ends. The seawater source is far more significant in some estuaries as a result of a greatly magnified two-layer flow, especially in partially mixed estuaries.

- **Salt marshes** are salt-tolerant grasslands that develop along low-lying, gently sloping shorelines, usually along the edges of estuaries, in temperate and higher latitudes. They form by trapping primarily terrestrial sediments in areas where sea level is rising at a rate that matches the buildup of sediments. These areas are also periodically flooded by high tides, which allow only salt-tolerant species, such as *Spartina* spp., to grow. Detrital remains of these grasses, along with terrestrial sediments, build up over time and become anoxic below the surface. This preserves the organic matter as **peat**, which may form thick layers beneath marshes, depending on their ages. Salt marshes are important, biologically productive systems in which the base of the food web is marsh grass detritus.

- The tropical and subtropical counterparts of salt marshes are **mangrove forests**, which also trap sediments in their root systems. They provide important habitat for numerous species, marine and terrestrial.

- **Coral reefs** are confined to tropical surface waters warmer than about 18–20°C. They are massive calcium carbonate structures built up over time from the remains of skeletal structures secreted by the coral animals, the polyps. Many have photosynthetic symbiotic **zooxanthellae**, which are dinoflagellates, and which provide the coral polyps a source of organic carbon while removing wastes. The polyps are suspension feeders, deriving nutrition from particles that collect on their tentacles, and they are also predators, disabling and ingesting small animals that encounter the stinging nematocysts lining their tentacles.

- Coral reefs may shed their zooxanthellae when exposed to temperatures exceeding about 28°C, a phenomenon called **coral bleaching**, which has become a problem in recent decades, as warm El Niño events elevate temperatures above already high temperatures as a result of global warming.

- The **deep-sea** environment extends from about 1000 m, or the lower limit of the photic zone, to the ocean's greatest depths. It is a cold (ca. 2°C), dark environment under extreme pressures. Food is scarce, delivered primarily from overlying planktonic production; predatory deep-sea fishes tend to have large mouths full of long, sharp teeth. Midwater fishes, at the lower limits of penetration of daylight, may have bioluminescent organs along the ventral surfaces to counter their silhouettes as viewed by potential predators from below.

- In 1977, hydrothermal vent communities were discovered by accident in the vicinity of deep-sea spreading ridges. Here, the base of the food web are chemosynthetic bacteria, which support numerous, previously unknown animals.

Discussion Questions

1. Describe how a sessile benthic organism is exposed and submerged in an intertidal zone that experiences: (a) a lunar tide; (b) a solar tide; (c) a mixed tide.

2. Why are intertidal zones considered stressful environments? What are the physical stresses? What are some of the biological stresses?

3. Can the availability of open space on a rock surface in the intertidal zone be considered a limiting factor? How are competitions for space waged?

4. What are some of the ways that sessile animals and attached

macroalgae deal with desiccation when exposed to air at low tide? What approaches do motile animals employ?

5. Explain why the distribution of organisms at the upper intertidal zone is controlled primarily by physical factors, while at the lower intertidal it is controlled by biological factors.

6. Why is desiccation less of a problem for intertidal animals at low tide in muddy sediments than it is for animals in coarser-grained sediments? Why is anoxia a problem in fine-grained, muddy sediments? Why are coarser-grained sediment environments not as rich in animal life as mud flats are?

7. Explain the relationship between volume of freshwater discharge into an estuary and the degree of tidal mixing.

8. What is tidal pumping in an estuary? What is meant by an estuary having a two-layered flow? What factors are involved in making an estuary biologically productive?

9. Why are salt marshes almost always found only in low-lying areas where sea level is rising? Does the same constraint hold for mangrove forests? Why or why not?

10. Describe conditions in the deep sea and how they affect animal populations. Why is the biomass of animals in the deep sea so much lower than in surface waters? Why is the dominant feeding mode in the deep sea deposit feeding?

Further Reading

Ballard, Robert D., 1995. *Explorations: A Life of Underwater Adventure*. New York: Hyperion.

Denny, Mark W., and Gaines, Steve (eds.), 2007. *Encyclopedia of Tidepools and Rocky Shores*. Berkeley, CA: University of California Press.

Dyer, Keith R., 1973. *Estuaries: A Physical Introduction*. London: John Wiley & Sons.

Lewis, J. R., 1964. *The Ecology of Rocky Shores*. London: The English Universities Press, LTD.

Little, C., and Kitching, J. A., 1996. *The Biology of Rocky Shores*. Oxford: Oxford University Press.

Matsen, Brad, 2005. *Descent: The Heroic Discovery of the Abyss*. New York: Vintage Press.

Ricketts, E. F., Calvin, J., and Hedgpeth, J. W., 1992. *Between Pacific Tides*, 5e. Rev. by D. W. Phillips. Stanford, CA: Stanford University Press.

Contents

Terrestrial vertebrates crawled out of the ocean and onto solid land about 350 million years ago. Having lungs and four legs with which to move about, these ancient air-breathing tetrapods were the first amphibians. Today's higher vertebrates evolved from these primitive and now-extinct animals. But along the way, a few of them did an about-face and reinvaded the sea: the marine reptiles, sea birds, and marine mammals. While marine invertebrates are the most numerous and diverse animals in the oceans, and fishes are the most abundant and diverse vertebrates, it is this third group of marine tetrapods, the marine mammals, that seem to attract peoples' interest the most. We just seem to find them interesting, not for scientific reasons, but because we feel a connection, such as with this Hawaiian monk seal, which desperately needs our attention—it is one of the most critically endangered marine mammals in the world.

Marine Reptiles, Birds, and Mammals

While the fishes are by far the dominant marine vertebrates, there are other less diverse and less abundant but no less interesting groups that make the oceans their home. In this chapter we round out our survey of the marine flora and fauna with discussions of some of these other marine vertebrates: the marine reptiles, sea birds, and marine mammals.

These three groups of marine vertebrates are tetrapods (which means "four-footed") that evolved from early lobe-finned fishes with rudimentary lungs, not unlike today's lungfishes and coelacanths. Terrestrial life was harsh for these early tetrapods; maintaining optimum amounts of water in their cells was especially difficult. This remains a problem for today's amphibians, which must keep their integuments moist at all times; and they are still dependent on water for reproduction, returning to lay eggs that develop as aquatic animals before crawling out as young adults. There are no truly marine amphibians today, expect perhaps the crab-eating frog (*Fejervarya cancrivora*), also called the mangrove frog, that lives among the mangroves in estuaries of Southeast Asia, the Philippines, and parts of India. It copes with extended periods in brackish waters by elevating urea concentrations in its blood such that it is isosmotic with its surroundings, similar to the approach taken by elasmobranchs.

While marine reptiles are relatively uncommon—we do not routinely encounter them when we visit the seashore—the same cannot be said of marine mammals and sea birds. The sounds of sea gulls, for example, are instant reminders that we are near the shore (or a shopping mall);[1] and while we don't routinely see them, many of us have been on whale-watching and seal-watching boat tours, or have watched and been captivated by these same animals on television.

Some of the marine tetrapods that we discuss in this chapter are wholly marine; the sea snakes, the whales, and the sirenia (the dugongs and manatees) have made a clean break from land, and spend their entire lives at sea. But most of the marine tetrapods spend a portion of their lives on land; for instance, seals, sea birds, and sea turtles, which spend the majority of their lifespans in the sea, must return to land in order to reproduce and rear their young.

Included among the marine vertebrates are a number of groups that are in trouble, as we'll discuss. While the same cannot be said for most sea birds, thankfully, it is an unfortunate fact that all species of sea turtles and many marine mammals are either threatened, or are in danger of extinction—and the more we all learn about them, the better chance we have of taking actions to ensure their survival.

Marine Reptiles

Of the more than 8000 species of reptiles alive today, only a handful are considered truly marine; these include the sea snakes, the marine iguanas, the sea turtles, and the marine crocodiles. Like the fishes, reptiles are cold-blooded animals, or

---

[1] Or a fast-food restaurant or a landfill. Gulls are remarkably well adapted to life in today's world; we'll return to this discussion below.

FIGURE 13.1 A banded sea snake (*Laticauda colubrina*) photographed in Fiji.

poikilotherms—they do not control their own body temperatures; their environment does. They are dependent on their environment for body heat—they are therefore **ectotherms**, acquiring heat from sources external to their own bodies, either directly from the Sun's rays by sunning themselves, or from their warm-water surroundings. For this reason, they are mostly restricted to tropical and subtropical waters.

Sea snakes

Having descended from terrestrial ancestors that once had four limbs, snakes are considered to be tetrapods, even though they have lost their legs. Since becoming terrestrial, about 60 species of snakes have returned to the sea. All species are found in tropical and subtropical waters of the Indian and Pacific Oceans; none in the Atlantic (**FIGURE 13.1**). Sea snakes are close relatives of cobras, coral snakes, and mambas, making them among the most venomous snakes in the world. They are not aggressive, however, and even if they do get provoked, they have such small mouths that the likelihood of humans receiving a fatal bite is remote (but it has happened).

Sea snakes have evolved a number of physiological and morphological adaptations to a life in the ocean. The cross-sectional shape of their body tapers from circular to more laterally flattened toward the tail end, and culminates with a paddle-like tail that facilitates swimming. But they are still air-breathers, and as such must surface to breathe; however, some are thought to be capable of acquiring at least some dissolved oxygen across the skin. Like other marine vertebrates, they must be able to rid themselves of excess salt in their blood and body tissues; for this purpose, sea snakes have glands near the base of their tongues that secrete excess salt, which is then expelled by tongue actions (i.e., they spit). They have much shorter olfactory tongues than their terrestrial cousins. They have evolved nostrils with a fleshy closure valve that keeps water out when they are not breathing at the surface. Their single lung is much larger than that of a terrestrial snake, running along much of the animal's length; its size is thought to assist in buoyancy regulation, or perhaps to allow them to hold their breath longer. Sea snakes are most often found near shorelines and in shallow waters, where they feed on small fishes and crustaceans. With the exception of one genus, which lays its eggs on land, most sea snakes reproduce at sea and are ovoviviparous, giving birth to well-developed live young at sea.

Marine iguanas

One species of marine lizard, the marine iguana (*Amblyrhynchus cristatus*), is found only on the Galápagos Islands in the eastern Pacific Ocean. Charles Darwin, upon first encountering marine iguanas during his voyage on HMS *Beagle* in 1839, was unimpressed. He wrote: "It is a hideous-looking creature, of a dirty black colour, stupid, and sluggish in its movements."[2]

FIGURE 13.2 A group of marine iguanas (*Amblyrhynchus cristatus*), photographed on the Galápagos Islands.

---

[2] Charles Darwin described the marine iguana in 1839, in his *Journal of Researches in the Geology and Natural History of the Various Countries Visited by H.M.S. Beagle*, which has been reprinted several times since Darwin's death in 1882. Charles Darwin, 1909. *The Voyage of the Beagle*, vol. 29 (New York: P. F. Collier & Son). p. 377.

Marine iguanas live in groups that congregate near the shoreline (**FIGURE 13.2**). Darwin commented that he never saw one more than ten yards from the water's edge. They feed almost exclusively on subtidal algae attached to the island's volcanic rocks; hence, they are proficient swimmers and can dive to as deep as 10 m. But being lizards, and ectothermic, they need to spend much of their time sunning themselves to warm up after swimming in the cold, newly upwelled waters of the eastern tropical Pacific.

There are thought to be as many as several hundred thousand of them in the Galápagos population, varying in color and size among the many islands. Their main enemies are rats and dogs, which feed on their eggs; they are currently protected by law.

Saltwater crocodiles

Only two of the 23 species of crocodiles routinely enter ocean waters: the American crocodile (*Crocodylus acutus*), which ranges from the south Florida coast throughout the Caribbean and the northern coast of South America (**FIGURE 13.3**); and the saltwater crocodile (*Crocodylus porous*), which is found in the estuaries and mangroves of tropical Indo-Pacific waters. Saltwater crocodiles are not only the largest crocodiles, but they are perhaps the most aggressive and dangerous; two people are attacked and killed in Australia each year, on average, by these animals.

Sea turtles

Sea turtles are among the most beloved sea creatures. Their gentle demeanor and, let's admit it, their sweet expressions, tug at our heartstrings (**FIGURE 13.4**). And it is a sad fact that all species of sea turtles are at risk of becoming extinct, and are officially listed as either threatened or endangered.[3] All six species that occur in U.S. waters are protected under the U.S. Endangered Species Act.

There are seven species of sea turtles: the green, hawksbill, loggerhead, leatherback, Kemp's ridley, olive ridley, and the flatback; an eighth, the black sea turtle, is a subspecies of green turtle. All are in the family Chelonidae except the leatherback, which is the sole member of the family Dermochelyidae.

Sea turtles, like freshwater turtles and terrestrial tortoises but unlike the leatherback, have a bony shell, or carapace, that is fused with their vertebrae. They are unable to retract their head into these shells and their limbs have evolved to resemble and function as flippers. As reptiles, all sea turtles (except the leatherback) are ectotherms and are therefore found almost exclusively in tropical and subtropical waters.

Leatherback sea turtles (*Dermochelys coriacea*) have a cosmopolitan distribution and are found in all oceans, ranging as far north as Alaska and Norway, and as far south as South Africa and New Zealand. They can tolerate colder water temperatures for a couple of reasons: First, they benefit from the dark, almost black coloration of their backs, which helps them to heat up better as they bask in the sun at the surface. Second, simply because they are large animals (in fact, they are the largest of the

FIGURE 13.3 A saltwater crocodile (*Crocodylus acutus*).

FIGURE 13.4 A loggerhead turtle (*Caretta caretta*).

[3] CITES, Convention on International Trade in Endangered Species of Wild Flora and Fauna; www.cites.org/.

FIGURE 13.5 A green sea turtle sleeping on a beach in Hawaii.

sea turtles, with adults reaching lengths greater than 2 m and weighing more than 850 kg) they have a reduced surface area-to-volume ratio. Thus, they radiate less heat than smaller animals. For the same reason, they are better able to retain heat from their own metabolism. They are also different from other sea turtles in the nature of their carapace: it is not bony, like that of most turtles, but a firm, leather-like skin with seven longitudinal ridges or keels that run fore and aft, which gives them their name and their distinctive, unmistakable appearance (see Figure 13.6).

The green sea turtle (*Chelonia mydas*) is one the best known, and is found throughout the tropical and subtropical Atlantic, Pacific, and Indian Oceans (**FIGURE 13.5**). It and its close relative, the black sea turtle (*Chelonia mydas agassizii*), which is found in the Pacific Ocean, are unique among the sea turtles in that they are herbivores, feeding on seagrasses and algae. The largest of the hard-shelled species, they grow to a length of more than a meter, and can weigh more than 200 kg. These turtles were once very abundant, with a population estimated as high as 500 million in the 1800s in the Caribbean Sea alone. Estimates today are scattered and incomplete, which is not surprising, given their worldwide distribution. There are only about 1100 nesting females in Florida; but in other parts of the world, such as in the Philippines, there may be as many as 1.4 million.[4] One thing is certain: their numbers have been drastically reduced, in large part as a result of commercial harvesting of their meat and eggs.[5]

Sea turtles can survive for months without food or water, which made them ideal natural food storage systems for long sea voyages back in the days before refrigeration. Turned onto their backs, they are unable to right themselves, so it was easy to store them live aboard ships. Sea turtles continue to be harvested legally in some parts of the world for their meat as well as for their shells; the hawksbill shell, for example, is prized as a source of tortoiseshell used to make various articles of jewelry and the like.

The leatherback and hawksbill sea turtles are both listed as critically endangered,[6] but the Kemp's ridley (*Lepidochelys kempii*) may be the most endangered. Named for Richard M. Kemp, a fisherman from Key West, Florida, who in 1906 suspected it might be an undescribed species, it is the smallest of the sea turtles, with adults only reaching 60 cm or so and about 45 kg. Kemp's ridley turtles range throughout the Gulf of Mexico and up the Atlantic coast, where they have strayed as far north as Nova Scotia.

Kemp's ridleys have a synchronized nesting behavior, in which the females spawn en masse at prime nesting sites. One of these sites is a beach near Rancho Nuevo, Mexico, where they come ashore in waves during daylight hours to lay their eggs, a behavior known as *arribada*, which means "arrival" in Spanish. The timing of these events is believed to be related to the phase of the moon. Their numbers are dwindling; in the year 2000, an estimated 2000 females came ashore to nest in Mexico, far fewer than the estimated 42,000 in 1947. And in Texas, there were on the order of 200 nests each year during the early and mid-2000s, but historical population levels are not known with any certainty.

---

[4] National Marine Fisheries Service and The U.S. Fish and Wildlife Service, August 2007. *Green Sea Turtle 5-yr Review*.

[5] According to the U.S. Fish and Wildlife Service and NOAA's National Marine Fisheries Service.

[6] IUCN, International Union for Conservation of Nature, Red List of Threatened Species™.

All sea turtles are long-lived, with ages of 80 years not uncommon, and they become reproductively mature late in life. Although the hawksbills become sexually mature as early as 3 years of age and the leatherbacks between ages 6 and 10, the loggerheads do not reach sexual maturity until they are 12 to 30 years old, and the green sea turtle, between 20 and 50 years.[7] Mating occurs at sea. The female comes ashore to lay her eggs; she digs a hole in soft beach sand above the high-tide mark and covers the eggs with sand (**FIGURE 13.6**). Females may have several clutches, returning to the beach over a period of weeks to lay between 50 and 200 eggs at a time; however, most sea turtles reproduce only every 2 to 5 years. After an incubation period that lasts between 45 and 75 days, depending on the species and the temperature, the young turtles hatch from the soft, leathery-shelled eggs and make their way to the ocean. They suffer significant losses on the way, however, as they are easy prey for crabs and birds that feed on them. Hatching at night—under the cover of darkness—helps, but their survival is mostly a matter of luck. The temperature at which the eggs incubate determines the sex of the hatchling; warmer temperatures favor the development of females (interestingly, the opposite is true for crocodiles).

FIGURE 13.6 A female leatherback sea turtle (*Dermochelys coriacea*) digging her nest on a sandy beach.

Sea turtles are prone to problems of excess salt intake; the jellyfish and other marine animals on which many of them feed, and on which they depend for water intake, are isotonic with seawater. Sea turtles possess excretory glands in the corners of their eyes that concentrate salts from their blood and excrete them almost like salty tears; when out of water, they often appear to be crying, as their eyes well up with the salty excretions.

Known for their prolific migrations, sea turtles are excellent swimmers and undertake impressive open-ocean excursions about which very little is known. For example, several species of sea turtles are thought to spend a "lost year" at sea, carried by the Gulf Stream around the North Atlantic basin, as they are unaccounted for soon after they hatch and enter the sea along the coasts of Florida and the Gulf of Mexico.[8] Some populations of green sea turtles are known to migrate between nesting sites on Ascension Island in the middle of the South Atlantic Ocean and feeding grounds on the coast of Brazil, more than 2000 km away.

Sea Birds

General characteristics

Birds are also tetrapod vertebrates, which at first might seem odd since they walk on two feet; they possess four limbs, but their front limbs, of course, have evolved for purposes other than walking or crawling. Among the more than 10,000 species of birds in the world, fewer than 500 species are considered sea birds. With 15 families in four orders, they are those birds with life histories adapted to the marine environment that descended from a number

---

[7] According to the U.S. Fish and Wildlife Service and the NOAA's National Marine Fisheries Service. Casale, P., Antonios, A. D., and Freggi, D., 2011. Estimation of age at maturity of loggerhead sea turtles, *Caretta caretta* in the Mediterranean using length-frequency data. *Endangered Species Research* 13: 123–129.

[8] Archie F. Carr, 1967. *So Excellent a Fishe: A Natural History of Sea Turtles* (Gainesville, FL: University Press of Florida).

of diverse taxonomic groups of land birds. They spend a significant portion of their lives at sea, where they feed on marine organisms. But all sea birds nest on land, most of them doing so in colonies ranging from dozens to millions of individuals. For some species, that is the only time spent on land; the rest of their lives is spent at sea in a constant search for food. With this common life history, sea birds are examples of **convergent evolution** in action, having evolved a number of morphological and physiological adaptations to life at sea that are remarkably similiar to those acquired by other, unrelated species.

Unlike reptiles, all birds are warm-blooded animals, or **homeotherms**, which means that they control their own body temperature within a fairly narrow range, usually higher than the temperature of their surroundings. They do this by generating their own internal metabolic heat. Their bodies are covered with feathers, with an especially thick layer of **down**, the fine-textured feathers beneath the tougher outer layers, providing additional insulation against the cold. Their feathers also have a waterproof coating, which is achieved by oils secreted by their **preen glands**, located near their tails. Birds are often seen preening themselves, rubbing their heads against the oil and spreading it to feathers on other parts of their bodies. Present in most birds, not just sea birds, preen oil not only keeps the feathers dry by preventing water from being absorbed, but it also keeps the feathers, and the bird, more buoyant, allowing them to float more easily. Also, all true sea birds have webbed feet (unlike **shorebirds**, such as herons and sand pipers), and they all lay eggs with hard shells. All sea birds also possess salt-excreting organs, or **salt glands**, positioned on each side of their heads, above the eye sockets, with which they concentrate and excrete excess salt that accumulates as they feed and drink seawater. Also present in some terrestrial birds, these glands concentrate salt from the blood (mostly sodium chloride) and excrete it via tubular connections with the nostrils.

Sea birds possess a wide range of wing types, depending on their life style. The flightless penguins and auks, for example, have wings that are modified to function primarily as flippers, while the albatross (*Diomedea exulans*) has the widest wing span of any extant (not extinct) bird, averaging more than 3 m from tip to tip. Longer wing spans are typical of pelagic sea birds that spend most of their time in the air, and which migrate great distances, while shorter wing spans are associated with sea birds that dive for their food and spend less time soaring.

Many of the pelagic birds are capable of **dynamic soaring**, whereby they take advantage of the different vertical components of winds on opposite sides of ocean waves, similar to the way manned gliders maneuver to positions over hills and valleys, taking advantage of updrafts to stay aloft for extended periods. Observed at sea, these pelagic sea birds appear to glide almost continuously, flapping their wings only infrequently. Some species, such as the shearwater, follow ships and search for food at and just beneath the water, stopping only to rest or to feed (**FIGURE 13.7**). These graceful birds are masters

FIGURE 13.7 (A) Greater shearwaters (*Puffinis gravis*) soar just above the sea surface, almost touching the water, in search for food. (B) To feed, they land on the water and look around them, often poking their heads beneath the surface to get a better look.

(A)

(B)

of dynamic soaring, and put on captivating shows for anyone fortunate enough to watch them from a ship at sea. Gliding inches above the water, they appear to just miss colliding with the next wave, as they ride the vertical air flows generated by the undulating sea surface.

Rather than discuss each of the 15 families in a taxonomic order, it is perhaps more instructive to treat sea birds collectively according to their feeding mode, which follows taxonomic structure somewhat. There are four fundamental feeding modes practiced by sea birds, which can be described as: **surface feeding**; **pursuit feeding**; **plunge diving**; and, an all-encompassing group that includes **scavenging**, **predation**, and even **stealing** food from others.

Surface feeders

Surface-feeding sea birds prey on small fishes and invertebrates, such as krill, a feeding mode practiced by many of the pelagic sea birds such as the shearwaters and gulls. This group includes some of the most spectacular flyers (e.g., Figure 13.7) which, once they suspect food is around, land on the water and slowly swim around, periodically poking their heads beneath the surface for a better look—called dipping—or to snatch a morsel. The black skimmer (*Rynchops niger*), common on the Florida coast, will fly close to the water surface with its elongated lower beak cutting through the water, searching for food by making contact with it (**FIGURE 13.8**).

Pursuit feeders

Pursuit diving is practiced by birds diving below the surface and pursuing their prey by actively swimming after it. Some, such as the penguins and the auks, for example, use their wings for propulsion underwater, and others, such as the cormorant, use their feet (**FIGURE 13.9**). The cormorants, nicknamed "shags," are a bit unusual in that their feathers, unlike those of other sea birds, are partially wettable. They isolate a thin layer of air next to their bodies, which serves to insulate the bird, but water penetrates and soaks into the outer portions of their feathers, making them wet and therefore less able to trap air, which reduces their buoyancy.[9] They sit lower in the water than

FIGURE 13.8 A black skimmer (*Rynchops niger*) performing its aerobatic feeding move.

FIGURE 13.9 (A) A double-crested cormorant (*Phalacrocorax auritus*) swims at the surface. (B) It dries its wings on a rock while a ring-billed gull (lower left in photo) preens itself.

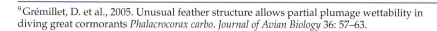

[9]Grémillet, D. et al., 2005. Unusual feather structure allows partial plumage wettability in diving great cormorants *Phalacrocorax carbo*. *Journal of Avian Biology* 36: 57–63.

(A)

(B)

other sea birds or ducks (e.g., compare the water line of the cormorant in Figure 13.9 with that of the shearwater in Figure 13.7), but at the same time, this makes it easier for them to stay submerged. They just duck the heads beneath the surface, as you and I might do while swimming, and their bodies follow in a graceful motion as they swim about underwater chasing small fishes. But before they can take flight again, they need to perch for a while and hold their wings out to air dry (see Figure 13.9B).

Penguins—every bit as beloved as sea turtles—use their forelimbs, their flippers, for swimming and pursuing prey. Except for the Galápagos penguin, they are found only in cold climates of the Southern Hemisphere. They comprise between 17 and 22 species (their taxonomy is still being debated) in a single subfamily, the Spheniscinae. They range in size from the little blue penguin (*Eudyptula minor*), which stands only about 40 cm fully grown, to the emperor penguin (*Aptenodytes forsteri*), which grows taller than 1 m.

All penguins are excellent swimmers, fast and agile, and feed on fish, squid, and krill. They do not normally dive to great depths, feeding instead near the surface and staying down for just a few minutes at a time. The emperor penguin, however, is an exception; it can dive to more than 500 m and stay down for more than 20 minutes.[10] On land, penguins are far less accomplished; walking upright like other birds, they rely on their tails for assistance, and swing their wings to keep balance. On ice and snow, they sometimes "toboggan," sliding on their bellies and pushing themselves with their hind feet. They are caricatures of themselves, which has only helped to endear them to us.

In addition to being well adapted to a life of pursuing food from the sea, penguins are also equipped with both a thick layer of feathers and a layer of subcutaneous fat that provide insulation from the cold. They have excellent vision underwater, and like other sea birds, they possess salt glands, which allow them to drink seawater.

With few exceptions, penguins breed in large colonies that can number several hundred thousand strong. The adelie penguins (*Pygoscelis adeliae*) are particularly social animals that form large colonies that forage and nest together. They breed during the austral summer, between October and February, on ice-cleared rocky surfaces around the Antarctic continent, in nests they build from small stones (**FIGURE 13.10**). To find these windswept clearings, they may walk or toboggan as far as 100 km. After their eggs are laid, usually two in a season, both parents assume responsibility for incubating them and keeping them warm. They take turns; while one guards the nest, the other returns to the water to feed and may not return for a week or two. The absent parents gorge themselves on krill, their preferred food, putting on as much weight as possible before returning to the nest. Once the chicks hatch, after about five weeks, the newly refreshed parent will regurgitate undigested stomach contents that the chicks eagerly devour. By the age of two months, the chicks are ready to go to sea themselves.

By comparison, emperor penguins (*Aptenodytes forsteri*) would seem to prefer doing things the hard way; it is the only species of penguin to breed in the austral winter—in Antarctica, which has the coldest winters on the planet (**FIGURE 13.11**). The breeding season begins

FIGURE 13.10 Individual adelie penguins *(Pygoscelis adeliae)* attending their nests and eggs.

---

[10] Kooyman, G. L. and Ponganis, P. J., 1997. The challenges of diving to depth. *American Scientist* 85: 530–539.

FIGURE 13.11 A colony of emperor penguin (*Aptenodytes forsteri*) adults and chicks.

during the austral autumn (March) when the adults begin the migration from the ice edge, where they have been feeding, to nesting areas which may be from 50 to 100 km inland. There they form colonies that can number in the thousands, and begin their courtship rituals to find mates. (While emperor penguins remain faithful to their mates throughout the breeding season, most will find a new mate the following year.) The female lays a single egg, which she immediately transfers to the male, an act that has to happen quickly so that the egg isn't fatally frozen. Having done this, the female heads back to the ice edge to feed for the next two months, while the father stands with the egg carefully cradled in its brood pouch, down by its feet. About 64 days later, the egg hatches, usually about the end of July and early August. Throughout this incubation period, the males all stand huddled together in order to preserve as much of their body heat as possible, carefully switching places with one another to share the hardship of standing exposed at the perimeter. With luck, the mothers return to the colony about this time and take over caring for the chicks, feeding them regurgitated food she has stored in her stomach. Now it's the males' turn to head for the ocean to feed, after more than 100 days of fasting. They will be gone for about three weeks before returning and resuming care of the chicks. They will continue this ritual of switching back and forth, one caring for the chick while the other feeds, for another 50 days or so. By that time, the chicks can be left alone, to huddle with their peers, while their parents head back to sea to feed. By November, the chicks have molted into their juvenile plumage (see Figure 13.11), and they all head for the sea.

Plunge divers

Plunge divers, such as gannets, terns, and brown pelicans, do just that: they plunge head first into the sea (beak first, actually), often from a significant height above the water, and almost always at a very high speed (**FIGURE 13.12**). At least that is what appears to be the case to anyone who has ever had the privilege to watch these magnificent birds. Gannets, for example, hit the water with such force and at such high speed that it looks almost suicidal. This technique is more energy efficient than only swimming underwater, in that the momentum of their dive brings them to a significant depth without having to fight against the opposing force of their natural buoyancy. This feeding mode also allows the birds to survey a relatively broad area of the sea surface from some elevation above the water, searching for prey, which are almost always small fishes. This feeding behavior is more common among sea birds

(A)

(B)

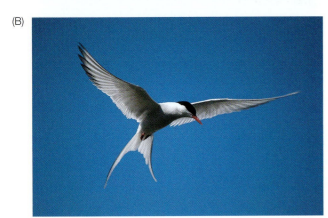

(C)

FIGURE 13.12 (A) A northern gannet (*Morus bassanus*) about to dive into the sea. (B) An Arctic tern (*Sterna paradisaea*). (C) A brown pelican (*Pelecanus occidentalis*).

in the clearer, more transparent waters offshore and in the tropics. But terns can often be seen feeding this way near shore and in harbors.

Scavengers, predators, and others

The fourth feeding mode is a category that includes almost every other feeding mode, but especially scavenging, predation, and stealing food from others, which is known by a more formal term, **kleptoparasitism**. Kleptoparasitism is practiced by frigatebirds (*Fregata minor*), for example (**FIGURE 13.13**). Frigatebirds prefer flying fish as food, which they pluck out of the air; in fact, they seldom light onto the water, and when they do, they have difficulty taking off. They also feed on other small fishes and squid,[11] but they are best known for stealing other sea bird chicks from their nests.

Gulls will also steal food, and not just at sea; they are notorious for stealing unattended picnic lunches at beaches. Gulls, which are opportunists and practice almost all types of feeding behaviors, are the most common sea bird. They are scavengers that will feed on almost anything, from molluscs and fish to terrestrial rodents and reptiles, to carrion, and other birds, and they can unhinge their jaws and swallow surprisingly large prey items whole. They are found on the coastlines of all continents, from the Antarctic to the Arctic, and some, such as the California gull (*Larus californicus*), are even found at times far from the coast.

FIGURE 13.13 A great frigatebird (*Fregata minor*) in a courtship display.

[11] Weimerskirch, H. et al., 1994. Foraging strategy of a top predator in tropical waters: great frigatebirds in the Mozambique Channel. *Marine Ecology Progress Series* 275: 297–308.

The California gull both nests and feeds far inland on lakes, but returns to the coast in winter.

Gulls are magnificent fliers, but they are equally at home on land, displaying walking skills that aren't seen in other sea birds. They are intelligent, can apparently communicate amongst themselves, and have a highly ordered social structure.[12] In teams, they will defend each other's nests from predators, and they are among the few birds that display tool use behavior; they have been seen using bread as bait to attract goldfish and are well known to use gravity by dropping shellfish from some height onto rocks to open the shells.[13]

There are about 57 species of gulls in 11 genera, but they have been informally divided into two groups, the "large white-headed gulls," which include the herring gulls, ring-beaked gulls, and the great black-backed gulls (**FIGURE 13.14**), and the "white-winged gulls," which are the smaller gulls that normally breed in the Arctic, such as the black-headed gull (Figure 13.14C). But these do not include all the species of gulls. Just a few years ago (2005), some 53 gull species were placed in a single genus, *Larus*, but, like many other groups of marine organisms we have discussed in this book, the genus is currently in a state of revision by taxonomists.[14] Gulls have also been grouped according to the number of years, or annual molts, it takes for them to attain their adult plumage. The great black-backed gull and the herring gulls, for example, are four-year gulls; the ring-billed gull is a three-year gull; and the black-headed

---

[12] John Alcock, 1998. *Animal Behavior: An Evolutionary Approach*, 7th ed. (Sunderland, MA: Sinauer Associates, Inc.).

[13] Henry, P.-Y., and Aznar, J.-C., 2006. Tool-use in Charadrii: Active bait-fishing by a herring gull. *Waterbirds* 29: 233–234; and Lefebvre, L., Nicolakakis, N., and Boire, D., 2002. Tools and brains in birds. *Behaviour* 139: 939–973.

[14] Pons, J.-M., Hassanin, A., and Crochet, P.-A., 2005. Phylogenetic relationships within the *Laridae* (Charadriiformes: *Aves*) inferred from mitochondrial markers. *Molecular Phylogenetics and Evolution* 37: 686–699.

(A)

(B)

(C)

FIGURE 13.14 (A) An adult great black-backed gull (*Larus marinus*). (B) An immature great black-backed gull, scavenging. (C) A black-headed gull (*Larus ridibundus*).

gull, a two-year gull. Immature gulls have a mottled or spotted coloration (see Figure 13.14).

Most gulls are long-lived; the herring gull (*Larus smithsonianus*) has been reported to live to 49 years in the wild, the great black-backed gull 27 years, and the black-headed gull as long as 62 years.[15] Several species fell to low population levels in the 1800s when they were hunted for their eggs and their feathers, but in the last century their numbers have recovered. All species of gulls in the United States are currently protected by law.

Shorebirds, raptors, and sea ducks

We close out this section on sea birds with a brief overview of several groups of birds that aren't considered true sea birds, but which nonetheless feed on marine organisms and are therefore at least partially dependent on the sea; and, of course, they exert some predatory impacts on sea life. These include the shorebirds, sea ducks, and some raptors.

Shorebirds include a number of groups that feed at the seashore, but which do not necessarily spend much time actually at sea, such as the herons and sandpipers, to name but a few. These are wading birds that are equally at home in freshwater and marine environments. The least sandpiper (**FIGURE 13.15A**)

---

[15] The Animal Ageing and Longevity Database, genomics.senescence.info/species/.

(A)

(B)

(C)

FIGURE 13.15 (A) Least sandpipers (*Calidris minutila*) foraging along the edge of a gravel beach. (B) A great blue heron (*Ardea herodius*) standing patiently in shallow waters on the Maine coast. (C) An osprey (*Pandion haliaetus*) in its nest built of sticks atop a channel marker.

is just one of many species of sandpipers in the family Scolopacidae. This family of birds has a cosmopolitan distribution, with representatives found in wetlands and along seashores of all continents except Antarctica. They feed mostly on small invertebrates. The great blue heron (*Ardea herodius*) frequents the seashore as well as wetlands throughout much of North America, Central America, the Caribbean, even the Galápagos Islands and, at times, Europe (**FIGURE 13.15B**). They use their long beaks to feed mostly on small fishes in shallow waters, where they patiently stand, almost motionless, waiting for potential prey to swim by. The osprey (*Pandion haliaetus*), also called a fish hawk, is a raptor, a large predatory bird, also found on all continents except Antarctica. It feeds almost exclusively on freshwater and marine fish and is well known for its large nests made of sticks that it builds on tall objects, such as channel markers (**FIGURE 13.15C**). American bald eagles also nest along the seashore and feed on marine fishes.

FIGURE 13.16 A flock of male (black and white) and female (brown) common eiders (*Somateria mollissima*).

Finally, there are the sea ducks—a subfamily, the Merginae, in the family Anatidae, which comprises ducks, geese, and swans. There are about 20 species of sea ducks, most of which assume the lifestyle of a marine bird. Many of them have salt glands which develop as they become adults. Most of them, about 18 of the 20 species in this group, are higher-latitude species; an example is the common eider (*Somateria mollissima*), one of the largest ducks (**FIGURE 13.16**). Eiders breed in the Arctic but range throughout the Northern Hemisphere, preferring rocky shorelines and islands, as well as the Arctic tundra. They feed on crustaceans and molluscs; they are able to swallow mussels whole, crushing the shells in their gizzards. They are highly social and are almost always seen in flocks of several dozen. They are known for their down, which is an exceptionally soft and efficient insulator; the eiders use it to line their nests. Called eiderdown, it was once widely used in clothing and bedding; it could be collected from the nests after the chicks had hatched and left without harming the eiders, but it has been replaced with down from farm-grown geese.

Marine Mammals

The mammals are one of the most successful groups of animals on the planet. They dominate all environments; they can fly, swim, climb, burrow, survive in all climates, and, perhaps most importantly, they are intelligent. And, of course, they are also found in the oceans. Important characteristics of all mammals, terrestrial as well as marine, include: they have lungs and they breathe air; they have hair; they give birth to live young and nurse them with mammary glands; they are relatively long-lived, with 20–40 years typical; and, they are homeotherms. But, of course, it is their intelligence that most clearly sets them apart and makes them so interesting.

There are four major groups of marine mammals: the sea otters; the sirenia, which are the manatees and dugongs; the seals, sea lions, and walruses; and the cetacea, which are the whales, dolphins, and porpoises.

The sea otter

The **sea otter** is the smallest marine mammal, but the largest species in the family Mustelidae, which includes about 70 species of river otters, skunks, badgers, weasels, and the like. There is only a single species of sea otter, with

(A)

(B)

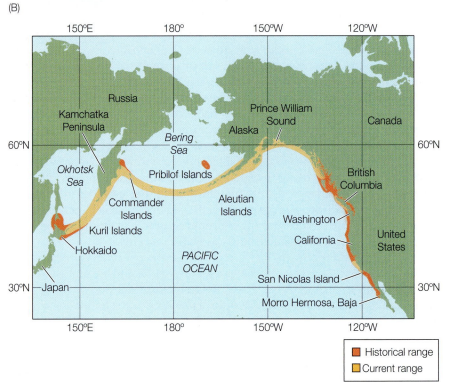

three subspecies: the common, or Asian, sea otter, *Enhydra lutris lustris*, which ranges from the coast of northern Japan to Kamchatka in the western North Pacific; the Northern, or Alaskan sea otter, *Enhydra lutris kenyon*, which ranges from the western end of the Aleutian Islands in the North Pacific to northern Washington State; and, the California sea otter, *Enhydra lutris nereis*, found off the coast of California. Sea otters are restricted to the waters of the northern Pacific Ocean (**FIGURE 13.17**).

Adult males may weigh as much as 45 kg. Unlike most of the other marine vertebrates we have discussed, the sea otter is a relatively recent expatriate from the terrestrial and freshwater realms, having returned to the sea as recently as two million years ago, and it retains many of its ancestors' characteristics. Nonetheless, these animals have made a near-complete transition to the sea, feeding exclusively there and usually resting and giving birth at sea, as well. This sets them apart from five other species of otters that spend at least a portion of their time feeding on marine organisms.

Throughout their range, sea otters reside in coastal, near-shore waters, usually shallower than about 30 m, and within a kilometer of shore. They feed primarily on sea urchins, but also on a number of bivalve molluscs and other marine invertebrates, making brief dives to the bottom that usually last less than two minutes. There they collect food items and store them in pouches of loose skin located on the sides of their chest and beneath each forearm, where they also store a rock that they use to break open shellfish. They feed while floating on their backs at the surface, breaking shells on their chests with their rocks. The sea otter is able to drink seawater because their large kidneys produce a concentrated urine, which prevents osmotic problems. They can often be seen clutching their front paws in the fur on their chest in order to keep them warm as they float on their backs (see Figure 13.17A). They manage to stay warm in the cold waters of the North Pacific because of their unique fur, which is the densest fur of any animal, marine or otherwise. It is so thick that water can't penetrate between the densely packed hairs, which maintains a dry layer of air that both isolates them from the water and insulates them from heat loss.

Sea otters are social animals, and will often congregate in *rafts* of a dozen to more than 100 individuals, segregated by sex. They can be seen resting on their backs at the surface or grooming their fur coats. To avoid drifting out to sea, they intentionally entangle themselves in kelp, which serves as an anchor.

Although sea otters have been hunted by aboriginal peoples for a long time, as revealed by archeologists, it was not until 1750–1900, when they were hunted by Europeans for the fur trade, that they came close to extinction. In 1911, there were only an estimated 1000 animals surviving.[16] They have been protected under U.S. law since 1911, and today they have mostly, but not completely, recovered throughout portions of their former range (see Figure 13.17B). Reintroduction programs in the 1960s and 1970s have had some successes, although their numbers off California remain low. The biggest threats to sea otters today come from killer whales and oil spills.

The sirenians

The **sirenia**, order Sirenia, also known as the **sea cows**, have two families, the Trichechidae, which are the manatees, with three (living) species, and the Dugongidae, which has a single species, the dugong. This order gets its name from the sirens in Greek mythology. The dugong and manatees are gentle giants; they are herbivores that lazily graze in shallow coastal, estuarine, and riverine waters, bothering no one, and apparently without natural enemies—though they are occasionally attacked by sharks, killer whales, and crocodiles. They are characterized by their large, rotund, but nonetheless fusiform (streamlined) body forms, with down-turned snouts, short pectoral (front limb) flippers, and a large tail fin, or fluke. They have no pelvic limbs. These animals are closely related to elephants and are the only herbivorous marine mammals.

The dugong is described as looking like a cross between a rotund dolphin and a walrus.[17] It grows to about 3 m in length and is strictly a marine species, unlike the manatees, which spend time in fresh and brackish waters. It also differs from the manatees in having a split fluke, or tail fin, similar to the cetacea (the whales, dolphins, and porpoises, which we will discuss below), and in having tusks. It ranges from the tropical and subtropical waters of the West African coast of the Indian Ocean to the Coral Sea west of Australia in the South Pacific. The dugongs' preferred food is the rhizomes of seagrasses, which they uproot with their snouts as they plow through seagrass beds, leaving long furrows in their wake. They are long-lived animals, with one

[16] Estes, J. A., and Bodkin, J. L., 2002. "Otters," in *Encyclopedia of Marine Mammals*, eds. W. F. Perrin, B. Wursig, and J. G. M. Thewissen (San Diego: Academic Press). pp. 842–858.
[17] Marsh, H., 2002. "Dugong," in *Encyclopedia of Marine Mammals*, eds. W. F. Perrin, B. Wursig, and J. G. M. Thewissen (San Diego: Academic Press). pp. 344–347.

FIGURE 13.18 A Florida manatee, *Trichechus manatus latirostris*.

individual having lived an estimated 73 years. They do not reproduce until they are between 10 and 17 years of age, giving birth to a single calf every 3–7 years. Thus, their reproductive potential is low.

There are three species of manatees. The West Indian manatee, *Trichechus manatus*, resides in coastal, estuarine and riverine (fresh) waters throughout the Caribbean and western tropical and subtropical Atlantic as far south as southern Brazil. The Florida manatee, *Trichechus manatus latirostris*, is a subspecies of the West Indian manatee (**FIGURE 13.18**). It ranges from Texas to as far north as North Carolina, although there has been at least one incident, in the summer of 2010, when one individual made its way into Cape Cod Bay, Massachusetts, in the Gulf of Maine. This is the largest of the four species of sirenians, attaining lengths of 4 m and weights of 1500 kg, with the females larger than the males. They tolerate a wide range of salinities, but they must enter fresh waters to drink. The Amazonian manatee, *T. inunguis*, is found primarily in freshwaters of the Amazon River and its tributaries. It is the smallest sirenian, reaching a maximum length of about 3 m, but usually weighing less than 500 kg. The third species, the West African manatee, *T. enegalensis*, is found in coastal, estuarine, and riverine waters of tropical and subtropical West Africa.

There was a fourth species, the now-extinct Steller's sea cow (*Hydrodamalis gigas*). It apparently had a very limited range, around the Commander Islands at the far western end of the Aleutian Islands in the Bering Sea off the Kamchatka Peninsula of Russia. It was described in 1741 by **Georg Wilhelm Steller** (1709–1746), a German naturalist serving on a Russian Navy expedition to the area that became shipwrecked on Bering Island, one of the Commander Islands. Half the crew died while overwintering on the island, but Steller survived and reported his detailed observations of the flora and fauna there; he was probably the only scientist or naturalist to have seen the Steller's sea cow alive. But by then its numbers were already in decline, and a mere 27 years later, in 1768, the last one was apparently killed. This was a particularly large sea cow, estimated to weigh some 3000 kg (6600 lb, about 6 times the average weight of a horse).

The dugongs and manatees are highly vulnerable to human impacts. Dugongs depend on seagrasses that commonly undergo natural fluctuations in abundance, and they are vulnerable to human activities such as encroachment and accompanying habitat loss, as well as to collisions with boats. As is the case with the manatees, these sea cows come by their nickname naturally, in that they lumber about, feeding at and just below the surface, without a care in the world, and without any natural enemies, other than humans, by way of inevitable boating accidents. Dugongs are legally protected in most of the countries throughout their range, but some aboriginal peoples are allowed to hunt them. Their population today is estimated at about 85,000 animals.

Manatees are also protected by laws, to varying degrees, in countries throughout their range. The Florida manatees are perhaps the best studied; there are an estimated 3300 animals remaining, a population size that scientists believe may be stabilizing, or even increasing slightly. Like the dugongs, they, too are vulnerable to boat collisions, which account for about a quarter of all documented fatalities, or about 70 per year (there are more than 750,000 registered boats in the State of Florida), as well as habitat destruction, and harmful algal blooms, as we discussed in Chapter 8.

The pinnipeds

The **pinnipeds** ("fin-footed") are a group of marine mammals in the order Carnivora (with cats, bears, and the like), comprising three families: the family Odobenidae, the walruses; the family Otariidae, the fur seals and sea lions; and the family Phocidae, the "true" seals. All the pinnipeds possess flippers and are excellent swimmers, but they have not completely cut their ties to land, where they rest and breed. Most are found in cold waters, but there are exceptions, such as Hawaii's monk seal (*Monachus schauinslandi*). They all have a bristly coat of fur and a layer of blubber, both for insulation from cold water and for buoyancy. And as mammals go, they are large—not as large as the sirenians, but certainly larger than most terrestrial mammals.

Fur seals and sea lions

The fur seals and sea lions, in the family Otariidae, are not considered true seals. They have external ears; true seals do not. They also have more flexible flippers, front and back, which, along with a pelvis that can rotate, give them much greater mobility on land than the true seals, allowing them to walk and even run, and they tend to have longer necks than the true seals. These anatomical differences are reflected in their swimming styles; the fur seals and sea lions use their front flippers to propel themselves, while both the walruses and the true seals use their rear flippers.

Sea lions are found in both hemispheres; there are seven species in five genera, with one genus found exclusively in the Northern Hemisphere, three exclusively in the Southern Hemisphere, and one in both hemispheres. There are no sea lions in the North Atlantic Ocean.

The California sea lion (*Zalophus californianus*) is one of the best known. These animals are extremely fast and graceful swimmers, capable of maneuvering tight turns with their front flippers as they play and as they pursue their preferred food—fish and squid. They routinely haul out on land, assembling together in large numbers, and call out with loud barks that sound almost like those of a dog. They are intelligent, adapt well to life in captivity, and are common in zoos and marine parks. Well known for their circus performances, balancing balls and the like, they have also been trained by the U.S. Navy to perform various tasks. California sea lions range from Vancouver Island to Mexico, preferring open waters and kelp beds. Like all pinnipeds, they are protected by law, but their numbers are considered to be stable.

The Steller sea lion (*Eumetopias jubatus*), also known as the northern sea lion, is the largest member of the family Otariidae (**FIGURE 13.19**). They range throughout the coastal waters of the North Pacific Ocean from Japan to California, and throughout the Bering Sea and the Okhotsk Sea between Kamchatka and Russia (see Figure 13.17B). Like the Steller's sea cow, these animals were also first described by Georg Wilhelm Steller in 1741. The Steller sea lions are endangered, and are listed under the U.S. Endangered Species Act; in recent decades their numbers have been declining, especially throughout portions of their range in Alaska. The cause of the decline remains unknown, but scientists suspect it may be related to reductions in food availability coincident with declining stocks of

FIGURE 13.19 A group of Steller sea lions (*Eumetopias jubatus*) hauled out on a rock near Kodiak Island off the coast of Alaska. Notice how they use their pectoral and pelvic fins for support; compare this with the harbor seals in Figure 13.22, which are more bullet shaped and have much shorter fins.

(A)

(B)

Alaskan pollock, one of the sea lion's preferred food fishes. Male Steller sea lions are significantly larger than females, reaching body lengths of nearly 3.5 m and weighing more than 1100 kg; females are almost all less than 3 m and 350 kg—smaller than the males, but still large by any measure. They breed in rookeries on remote beaches and on islands; the males stake out and guard specific territories where they mate with harems of females. Although endangered, native Americans in Alaska are still allowed a subsistence harvest of about 300 animals per year.

Close relatives of the sea lions, the fur seals are found in high latitudes of both hemispheres. Of the nine species, only two, the northern fur seal, *Callorhinus ursinus* (**FIGURE 13.20**), and the Guadalupe fur seal, *Arctocephalus townsendi*, are found in the Northern Hemisphere. Thought in the 1800s and early 1900s to be extinct, the Guadalupe fur seal population around Guadalupe Island off the coast of Baja California is today estimated at 7000. Northern fur seals, however, are far more abundant; there are estimated to be about 1.1 million northern fur seals throughout their range in the North Pacific Ocean; their range is nearly identical to that of Steller sea lions, extending from Japan to California. Although they coexist, Steller sea lions are known to prey on small fur seals, as do sharks and killer whales. The northern fur seals spend most of their time at sea, going ashore for 35–45 days a year to breed. Like the Steller sea lion, they breed in colonies on islands and remote beaches. Unlike the Steller sea lion, the males aggressively defend a harem of as many as 50 females.

As their name implies, the northern fur seals were hunted for their fur. They were first described by Georg Wilhelm Steller in 1742, and soon after that they were hunted for their pelts for making garments. Their numbers declined quickly, and in 1911 a treaty was signed by the U.S., Russia, Japan, and Great Britain (on behalf of Canada), which put limits on the harvest, but hunting didn't end until 1985. The population today is about 1.5 million.

FIGURE 13.21 A female walrus (*Odobenus rosmarus*) and her pup beside another adult, on an ice floe in the Chukchi Sea off the northwest coast of Alaska.

Walruses

Walruses (*Odobenus rosmarus*) reside in the Arctic seas, with three populations (considered subspecies) in the Arctic waters of the Atlantic Ocean, Pacific Ocean, and the Laptev Sea, in the Arctic Ocean north of Siberia. Whether there are two or three subspecies is disputed. These animals are easily recognized by their tusks, which are present in both sexes, their short-haired fur, which may be absent in spots, their prominent and heavily whiskered upper lips, and their sheer size (**FIGURE 13.21**). Walruses are big animals, but Pacific walruses tend to be larger and have bigger tusks than Atlantic walruses; adult males can weigh more than 1500 kg (3300 lb). Walruses are slow swimmers but can dive to more than 100 m and hold their breath for about 25 minutes. They reside primarily over shallow continental shelves, shallower than about 80 m, where they feed on benthic invertebrates. Bivalve molluscs are their preferred food, which they find by plowing through bottom sediments with their snouts, using their front flippers and tusks as sled runners, leaving behind pronounced furrows in the bottom sediments. The bivalves are detected by their sensitive whiskers; the walrus seizes the exposed siphon between its powerful lips and sucks the fleshy morsel out of its shell, leaving behind a wake of empty shells. In their search for food, they can break through ice as thick as 20 cm, but when it gets thicker than that, they move on to areas of open water and ice, known as **polynyas**. When not feeding, they haul out—the males onto rocks and the females onto ice floes—and rest, often in such a deep sleep that they are difficult to wake up. They are relatively long-lived, reaching ages of 30–40 years, and are very social animals: with their well-developed facial muscles they communicate with one another by way of facial expressions.

Walruses were hunted for their ivory tusks, oil, and hide in the eighteenth and nineteenth centuries, which drastically reduced their numbers. The Pacific population appears to have recovered, with an estimated 200,000 animals today. But the Atlantic population, with just over 10,000 animals, and the Laptev Sea population, with fewer than 5000, have not; the reasons are unknown. Walruses are protected today, but aboriginal peoples are still allowed a take.[18]

The true seals

The true seals, in the family Phocidae, comprise about 19 species; they are distinguished primarily by their lack of visible outer ears (the pinnae), which the walruses, sea lions, and fur seals all have. They cannot rotate their pelvis and hind limbs to a forward position or flex their small forelimbs the way the other pinnipeds can in order to facilitate moving about on land, and so they can only crawl clumsily about when out of water. The seals are confined to the colder temperate and Arctic regions; they are found throughout the Atlantic and Pacific Oceans in both hemispheres, but not in the Indian Ocean. The phocids range in size from the Baikal seal (*Pusa sibirica*), which only grows to about 1 m in length and about 45 kg, to the elephant seal, which is the largest pinniped, reaching 5 m in length and more than 3200 kg. They are relatively long-lived, with 30–45 years not uncommon.

The harbor seal (*Phoca vitulina*), also known as the common seal, is perhaps the most common phocid (**FIGURE 13.22**). It is limited to the Northern Hemisphere, but it nonetheless has the widest distribution of the pinnipeds. There are five subspecies, with a combined distribution in a coastal belt that extends across the North Pacific from the southern tip of Japan in the west to Baha California in the east, and from Florida in the North Atlantic to the Arctic Ocean, across to Greenland, Iceland, and Europe to as far south as northern Portugal. Like their cousins, these pinnipeds also are superb swimmers; they can dive deeper than 500 m and are generalist feeders, adapting to what is available.[19] They migrate seasonally, but generalizations are difficult to make given their extremely broad distributions. They live about 35 years on average, and begin breeding between 3 and 5 years of age. They do not form breeding colonies.

FIGURE 13.22 A group of harbor seals (*Phoca vitulina*) hauled out on a rock island in Casco Bay, Maine, accompanied by cormorants (top) and eiders (bottom).

[18] Kastelein, R. A., 2002. "Walrus" in *Encyclopedia of Marine Mammals*, eds. W. F. Perrin, B. Wursig, and J. G. M. Thewissen (San Diego: Academic Press). pp. 1294–1300.

[19] Burns, J. J., 2002. "Harbor seal and spotted seal" in *Encyclopedia of Marine Mammals*, eds. W. F. Perrin, B. Wursig, and J. G. M. Thewissen (San Diego: Academic Press). pp. 552–560.

(A) Baleen whales

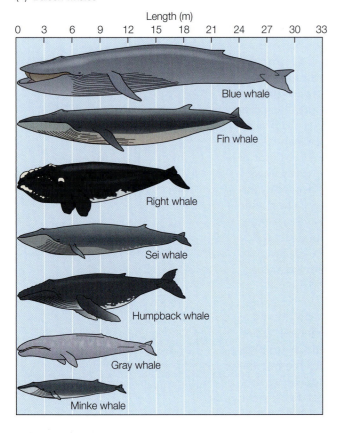

(B) Toothed whales

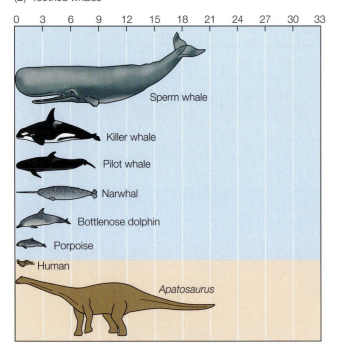

FIGURE 13.23 The size range of representative cetacea compared with the largest dinosaur, *Apatosaurus*.

Females give birth in late spring or early summer to a single pup, usually in the intertidal zone.

Harbor seals have fallen victim to viral infections in recent decades that have inflicted extensive mortalities on various populations. Historical hunting pressures are now greatly reduced, as they are legally protected throughout much, but not all, of their range. Population estimates vary by region, and are incomplete. But to give some examples, they are considered to be abundant in the Pacific waters off North America, where there may be as many 200,000 between Alaska and California; in coastal Atlantic waters of North America, the numbers are far less certain. There were thought to be somewhere between 40,000 and 100,000 throughout the western North Atlantic in 1993, with 30,000–40,000 of them in Canadian waters, where they are still harvested for their hides, but only about 4700 in U.S. waters. Human exploitation, pollution, climate change, disease, and interactions with fisheries all combine to influence their numbers.

Cetaceans: The whales, dolphins, and porpoises

There are some 90 species in the order **Cetacea**, in two major groups: the suborder Odontoceti, which are the **toothed whales**, and includes the dolphins and porpoises; and the suborder Mysticeti, also known as the **baleen whales**. All but five of those 90 species are marine, the exceptions being the freshwater dolphins. The baleen whales are the largest cetaceans. In fact they are the largest animals that have ever lived (**FIGURE 13.23**). The largest whale, the blue whale (*Balaenoptera musculus*), can reach more than 30 m in length, which is about the same size as a Boeing 737 passenger jet, except that the whale weighs much more. The smallest whales are the dolphins and porpoises.

This group of marine tetrapods has made the cleanest break from their terrestrial ancestors of any, with the possible exception of the sirenia. The cetacea are fully adapted to a life in the ocean and spend no time on land at all—nor could they; their sheer weight cannot be supported out of water and even if that were not the case, they retain only vestigial pelvic appendages internally, making mobility on land an impossibility. They still breathe air, but their nostrils, or **blowholes**, are located on the tops of their heads, which allows them to get a breath of air with only minimal emergence above the surface, thus conserving energy. And to swim most efficiently, they, like the pinnipeds and sirenia, have evolved a streamlined body, with front limbs modified as flippers, and a layer of blubber for insulation and buoyancy. But unlike their cousins, some of the cetacea have a dorsal fin, and all of them have a dorso-ventrally flattened posterior end that is their tail fin, or **flukes**. The two groups of whales have very different lifestyles, and with the exception of the sperm whale (*Physeter macrocephalus*), the baleen whales are much bigger than the toothed whales. More importantly, the two groups feed in entirely different ways.

Cetacean feeding

The mysticetes, or baleen whales, can be generally characterized as filter feeders that concentrate and ingest plankton and other smaller members of the nekton, including small fishes. The gray whales (*Eschrichtius robustus*) are a bit unusual among the mysticetes in that they are actually soft-bottom, benthic-feeding, filter feeders. This is a long way of saying that they feed on small amphipods and other benthic invertebrates that they are able to sieve out of mouthfuls of bottom sediment. The baleen whales can filter-feed because, instead of teeth, they possess **baleen plates**, specialized anatomical mouth parts, also called **whalebone**, that are downward-projecting plate-like structures around their top palate, made of keratinous material similar to our fingernails. Lining the edges of those plates are strands of hairlike fibers, which are formed as the plates wear away from constant rubbing against one another, as well as against the whale's huge tongue; in effect, the plates become frayed like the torn edges of an old pair of blue jeans. These stiff fibrous strands act like the tines of a comb, allowing water to pass, but collecting any objects too large to fit between them, such as planktonic and nektonic organisms, the baleen whales' staple food.

Different members of the mysticetes have different baleen characteristics, giving three basic feeding types: **sediment straining**, such as that performed by the gray whale; **skimming**, or skim feeding, such as that performed by the right whales and the bowheads; and **gulping**, or gulp feeding, which is performed by the **rorquals**. The rorqual whales comprise 9 of the 15 mysticete species and have fairly short baleen plates lining their extraordinarily large mouths. These whales include, for example, the blue, fin, sei, humpback, and minke whales (see Figure 13.23). Derived from Old Norse, the word *rorqual* means "furrow whale," and it is the furrows on their throats that distinguish these gulp-feeding whales. The furrows are like pleats, and they stretch apart, allowing their buccal cavity to greatly expand in volume. An example of these groves can be plainly seen in Figure 9.21, where a humpback whale has broken the surface, and rotated onto its back and expanded its accordion-like furrows, exposing its mouthful of water and euphausiids.

The mechanics of gulp feeding are shown in **FIGURE 13.24**. The whale swims into a high-density patch of plankton or nekton with its mouth wide open at a 90° angle; its throat grooves stretch, increasing the volume of water it can gulp into its mouth as much as four times the amount that would be possible without the grooves. Then, the tongue and the elastic throat grooves force that water back out of the mouth through the baleen plates, which retain the food organisms. The larger whales can gulp phenomenal volumes of water, as much as 60 m³, or 70 metric tons. Some whale species perform a variation of this feeding behavior called **lunge feeding**. In lunge feeding, instead of swimming along at a constant depth, as illustrated in Figure 13.24,

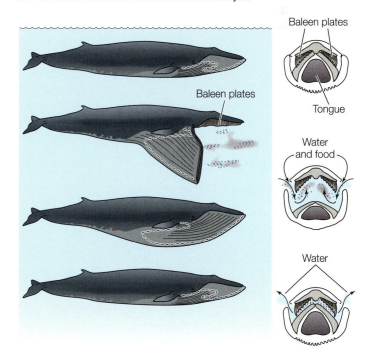

FIGURE 13.24 Lateral and cross-sectional views of gulp feeding in a rorqual baleen whale. The dashed line in drawings on the left outline the buccal cavity. The baleen plates and tongue are diagrammed in cross section on the right, showing the flow of water and food into the mouth and the sieving of food from that water as it is forced out. Notice how much the whale can extend its lower jaw.

Baleen plates

Baleen plates

Tongue

Water and food

Water

the whale swims up into the patch of plankton or school of small fish from beneath and breaks the surface with its upward momentum (this is the action that resulted in the photograph in Figure 9.21).

But you may be wondering, how can whales possibly ingest enough of their relatively tiny prey items to meet their energetic needs? Some whales diversify their diets, like the humpback whale (*Megaptera novaeangliae*), which is a lunge feeder, but also at times will bottom-feed, like the gray whale. This problem was brought into sharper focus when a paper, published more than 30 years, ago reported on the stomach contents of a fin whale harvested off Nova Scotia years earlier, in 1965.[20] The report showed that the amount of krill in the whale's stomach was much more than could theoretically be obtained by the whale's simply gulping large volumes of water to get krill. That is, there was not enough krill in the area (number of krill per m^3) to account for the quantity of krill in the whale's stomach. This report raised the question: What do the whales know about krill that we don't?

The answer, we now know, is that these whales find dense patches of krill, which are known to concentrate in hydrographic fronts—areas where different water masses come together and mix with one another—and in upwelling areas. And, it was later shown that some whales use a "bubble-feeding" technique; in other words, in the absence of natural oceanographic processes that concentrate prey, the whales do it themselves. Whales have been observed feeding cooperatively in groups, practicing a technique that concentrates their nektonic prey. In this group feeding, an individual positions itself at depth and blows bubbles while swimming in a circle toward the surface; this forms, essentially, a bubble net that corrals shoals of small fishes and euphausiids, causing them to congregate closer together, increasing their density. The other humpback whales in the group lunge upward, engulfing the concentrated prey. The discovery of this technique, whereby whales actively concentrate their prey items, helped to settle the question of how whales can ingest enough tiny prey to satisfy their metabolic needs.

Skim feeding is practiced by the bowheads and right whales (**FIGURE 13.25**), which do not possess throat grooves; they don't need them. These baleen whales have much longer baleen plates than the rorquals, with much finer filtering capabilities that allow them to sieve plankton-rich waters while they swim along at, or just beneath, the surface with their mouths open. Water and plankton enter the whale's buccal cavity through a gap in its baleen plates and water exits out the sides, first passing through the baleen plates, which filter out the plankton. Because this feeding method targets the smaller zooplankton, particularly the calanoid copepods, these whales do not need to swim as fast as the rorquals, which target the more elusive nekton.

Unlike the mysticetes, the odontocetes (the toothed whales), actively pursue individual prey items, usually fish and squid. This, of course, requires fast swimming speeds, as well as good vision. While they certainly locate and capture prey as individuals, dolphins are also known to feed cooperatively, effectively herding shoals of fish into tighter aggregations to make their capture easier. They use their acoustical abilities both to locate prey and to communicate among themselves. It has also been speculated that

FIGURE 13.25 Drawing of a right whale skimming.

[20] Brodie, P. F., Sameoto, D. D., and Sheldon, R. W., 1978. Population densities of euphausiids off Nova Scotia as indicated by net samples, whale stomach contents, and sonar. *Limnology and Oceanography* 23: 1264–1267.

they use their sonar to stun and disable prey. Killer whales (*Orcinus orca*) hunt both cooperatively—particularly when they are hunting dolphins and other whales—and individually, when smaller prey are involved. Sperm whales feed individually, mostly on large squid at great depths, in either total or near-total darkness, and more than likely, they use **echolocation** to find their prey.

One of the most unusual-looking whales is the narwhal (*Monodon monoceros*) (see Figure 13.23). It is a relatively small-toothed whale that reaches only about 3 m in length (not including its tusk). They are highly social and often aggregate in groups of 30 or more animals in the waters of the Arctic Ocean, where they reside year round. The narwhal is distinguished by its elongated, spiral tusk in males, long thought to be the basis for the unicorn legend. Once thought to be used to break open ice and spear prey, it turns out that the tusk, which can exceed 2.5 m in length, is highly innervated and may have important sensory functions.[21]

Whale breathing

Almost everyone has seen whales spouting—in movies, if not in person. A "blow," which results when a whale rapidly empties it lungs before immediately refilling them, is the most easily spotted evidence of a whale at the surface. It is often visible as a mist-like spray that shoots up out of an otherwise calm sea. All whales breathe through their blowhole, which is actually their nostril (toothed whales have a single opening, but baleen whales have a pair), located on the top of their heads, usually at a point just forward of and above the eyes; however, it is located near the front of the head in sperm whales. The spray we see when a whale spouts, or blows, is actually a mixture of seawater in and around the blowhole that is entrained in the rushing stream of air, along with mucus and condensed water vapor in the animal's breath (**FIGURE 13.26**). It is claimed by some commercial fishers that there is a distinctive smell to air in the vicinity of whale spouts. If your ship's engines are silent, you can often hear whales blowing off in the distance.

Whales are adapted to make the most of their time at the surface, exhaling and inhaling rapidly. Large mysticetes, such as the fin whale (*Balaenoptera physalus*), which is the second-largest whale, can empty and refill their huge lungs in as little as 1–2 seconds, which is much faster than you or I can do it.

---

[21] Malkin, C., 2005. Nothing but the tooth. *New Scientist* 185: 46–49.

FIGURE 13.26 A whale spouting.

Blowhole

Blowhole

FIGURE 13.27 Sequence of four photographs of Atlantic white-sided dolphins swimming in front of the bow wave of a research ship; total elapsed time is about 0.4 sec. The dolphin begins exhaling before breaking the surface (first frame). The blowhole is visibly open for inhalation only in the third frame for about 0.1 seconds, before it is again submerged (bottom frame).

This exchange is especially fast in the dolphins. **FIGURE 13.27** shows how quickly an Atlantic white-sided dolphin (*Lagenorhynchus acutus*) can refill its lungs. Notice that it begins to exhale before it breaks the surface with its blowhole, and once above the surface, it inhales in a fraction of a second.

Breathing patterns vary among species of whales. The baleen whales will seldom dive for longer than 15–20 minutes, and will spend most of their time near the surface and breathing every few minutes. The right whales can stay down for a maximum of about 50 minutes, and the bowheads, about 80 minutes. Dolphins exhibit a similar pattern of breathing every couple of minutes; the smaller dolphins stay down for a maximum of 7–12 minutes, and the larger dolphins (such as the killer whale) for about 20 minutes. The sperm whale on the other hand, is a deep diver; it may stay near the surface for extended periods, taking a breath every few minutes, but it can dive to depths greater than 2200 m and stay down for well over an hour, and perhaps as long as 2 hours, 18 minutes;[22,23] either way, this feat brings with it some interesting complications.

Deep diving

Whales and pinnipeds are generally known for their impressive deep-diving and breath-holding abilities. As we just mentioned, the sperm whale can dive to well over a mile down, and the elephant seal is not far behind (**FIGURE 13.28**), but these are just estimates. A recent study of two species of beaked whale (*Ziphius cavirostris* and *Mesoplodon densirostris*) reported actual measurements taken with time-depth recorders that were attached to the whales with suction cups; they documented a dive to 1888 m for close to one hour.[24]

So, how do they do it? How is it that cetaceans can perform such impressive dives? First, they have very efficient respiratory systems. Not only can they empty and fill their lungs rapidly, they do so with remarkable efficiency: they can exchange about 90% of the oxygen in each breath, versus about 20% for humans. Second, they also have a

[22] Kooyman, G. L., and Ponganis, P. J., 1997. The challenges of diving to depth: The deepest sea divers have unique ways of budgeting their oxygen supply and responding to pressure. *American Scientist* 85: 530–539.

[23] Annalisa Berta, James L. Sumich, and Kit M. Kovacs, 2006. *Marine Mammals: Evolutionary Biology*, 2nd ed. (Amsterdam: Academic Press).

[24] Tyack, P. L. et al., 2006. Extreme diving of beaked whales. *The Journal of Experimental Biology* 209: 4238–4253.

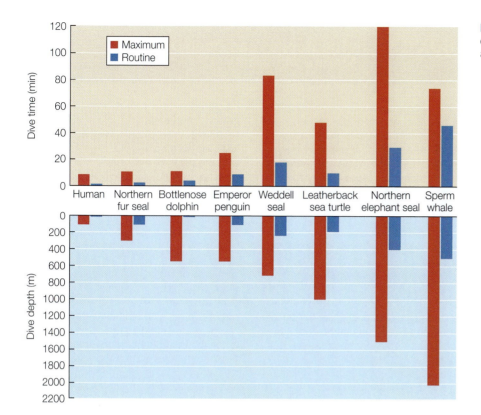

FIGURE 13.28 Maximum and routine diving depths and submergence times for a variety of swimming organisms.

higher concentration of red blood cells, each of which carries more hemoglobin, which helps account for that 90%. Like all other marine animals, they can slow their heart rate during a dive, and can restrict blood flow to their extremities during dives, making oxygen available to parts that need it most (the heart and brain, etc.). Finally, they apparently do not get "the bends" (**BOX 13A**).

The phenomena of reduced heart rate and restricted blood flow to their extremities is also exhibited by small (human) children. Known as the **mammalian diving response**, young children retain a vestige of this capability from our ancestors; there have been reports of near-drownings of children in icy-cold waters, sometimes when they break through thin ice, for example, from which they recover—even after as much as 30 minutes underwater.

Swimming speed

Cetaceans are fast swimmers because their bodies are made for efficient movement through water. Their streamlined shape—with minimal surface area—produces little drag, or friction, against the flow of water. One way that cetacea have minimized their surface area is by having lost their hair. They also have minimized their surface area by having very smooth skin, which is especially pronounced in the dolphins; smooth skin produces less drag than rough skin. The smooth, streamlined bodies of cetaceans produce minimal wakes as they move through the water, and this also aids their speed; fewer wakes mean less drag.

There are two ways of considering the swimming speed of animals: (1) according to their sustained swimming speeds, and (2) according to their burst, or sprint, swimming speeds. The baleen whales are capable of sustained swimming speeds of between about 2–13 km/h (1.2–8 mph), with the bigger whales, such as the blue whale and the fin whale, sustaining the highest speeds.

BOX 13A "The Bends" and Deep-Diving Marine Mammals

Gases are soluble in liquids; and they are more soluble in a liquid under pressure, such as the blood of a person, or a marine mammal, that is underwater and therefore under higher pressure than at the surface. In the case of a person diving to depths of several tens of meters while breathing from a scuba tank, more gases will dissolve in one's blood (especially nitrogen gas, N_2, which is 78% of the volume of air). When the diver surfaces quickly, releasing that pressure, the dissolved nitrogen will come out of solution, making bubbles in the blood. (An analogy is the bubbles that form when you open a carbonated beverage, reducing the pressure; in this case the gas is carbon dioxide.) When this happens in the blood of a scuba diver, it is called "the bends," and can cause severe pain (causing one to bend over in pain) and even death. Marine mammals are susceptible to the same phenomenon; so, how do they avoid getting "the bends"?

During a dive, marine mammals keep air isolated and out of contact with the parts of their lungs that could transfer it to the blood. Whales, for example, do not have a sternum connecting their ribs, and so under great pressure the rib cage collapses causing the lungs to collapse. This forces the air out of the alveoli (the small air sacs in the lungs where air is exchanged with blood). The air instead goes into the central part of the lungs, the bronchial tubes and the trachea, which are not in contact with the blood-exchange mechanism. This then results in only a minimum amount of nitrogen getting into the blood when the whale is at depth and under great pressure. Some seals, which do the same thing, actually exhale before a dive, to further reduce this risk of "the bends."

FIGURE 13.29 A group of Atlantic white-sided dolphins (*Lagenorhynchus acutus*) alongside the bow of a research ship moving at about 11 knots (about 20 km/h).

The fastest speeds reported for the baleen whales range from about 7 km/h to as much as 48 km/h (4–29 mph); maximum reported swimming speeds of trained dolphins are on the order of 30–40 km/h (19–25 mph).

Dolphins are often seen riding the bow waves of ships at sea (**FIGURE 13.29**), which they appear to do just for the fun of it. Often appearing almost out of nowhere, they will rush to the bow where they line up, and they don't seem to care which direction the ship is headed; they're just there for a ride. They take advantage of the slight elevation of water immediately beside and slightly ahead of the front tip of the bow (which is invisible to the observer craning over the rail to watch), catching the rising water with their flukes, only infrequently needing a tail beat to maintain their positions, usually as a result of their having to break the surface for a breath of air.[25]

Echolocation

It is believed that most species of the toothed whales, dolphins, and porpoises can echolocate, using a kind of sonar to locate their prey. They produce low-frequency clicks and buzzes that travel quickly through the water with little attenuation (**FIGURE 13.30**).

[25] Scholander, P. F., 1959. Wave-riding dolphins: How do they do it? *Science* 129: 1085–1087.

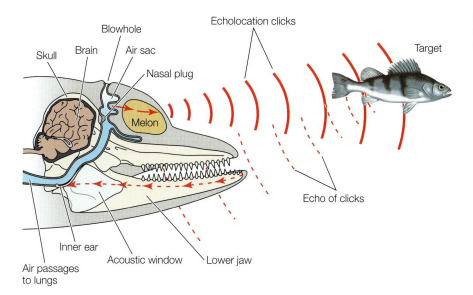

FIGURE 13.30 Diagram of the mechanism that enables a dolphin to echolocate.

(Cetaceans do not have vocal cords; they make their sounds by manipulating their air passages while the blowhole is closed, sort of like a person making a nasal snort.) The outgoing sound waves are focused by the **melon**, a fatty blob on their forehead; these sound waves bounce off of objects in the water and return as echoes to the cetacean. The incoming sound is received by the lower jaw area; the jaw bones are filled with fat or oil that focuses the sound to the two sensitive inner ears.

Whales are also thought to use this sound-generating capability to stun their food: live squid have reportedly emerged from the guts of stranded whales without any teeth marks on them, suggesting that they had been swallowed whole after being "knocked out."

Whale migrations

Some of the great whales are known for their lengthy migrations. The mysticetes, for example, migrate between their biologically productive feeding grounds, usually at higher latitudes, to their breeding grounds, usually in the tropics and subtropics, where they calve. The migrations of the humpback and Pacific gray whales are among the better-known examples (**FIGURE 13.31**). Pacific gray whales (there are no longer any Atlantic gray whales; they were hunted to extinction in the 1700s) feed from May to September in the Bering Sea, putting on weight in the biologically productive waters. In October they leave, swimming south at about 100 miles per day, feeding very little, and using up about a fourth of their body weight as they migrate to the Gulf of California off Mexico, where they calve. The humpbacks perform similar migrations. There are several populations, with feeding grounds in the North Atlantic, North Pacific, and Southern Ocean, that mate and calve in the tropics and subtropics. The blue whale has also been known to migrate from pole to pole between seasonal feeding grounds.

Cetacean intelligence

We have all heard that the whales and dolphins are intelligent, and they are. In fact, it is probably their high level of intelligence that most of us find so fascinating. We have seen them perform at marine parks and we have heard recordings of their soulful songs, and we wonder what they mean. We could

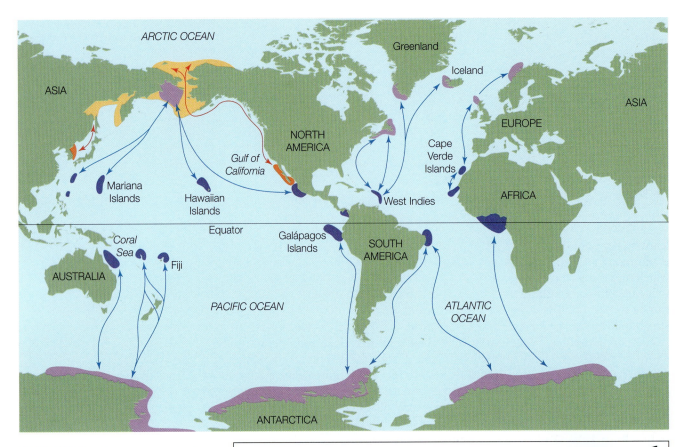

FIGURE 13.31 Migration routes for the gray whale and humpback whales between their high-latitude feeding grounds during the warmer months of the year and their mating and calving grounds in winter.

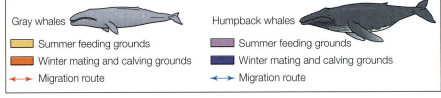

Gray whales		Humpback whales	
🟨 Summer feeding grounds		🟪 Summer feeding grounds	
🟧 Winter mating and calving grounds		🟦 Winter mating and calving grounds	
← → Migration route		← → Migration route	

go on to give many examples to make this point, but anyone who has ever had a close encounter with one of these magnificent animals understands.

The cetaceans do have large brains (**FIGURE 13.32**), but just how smart are they? Well, we know that whales and dolphins in captivity can learn to do simple tricks, with a reward of a snack from their trainer. And, it is also known that they (like humans) have an enlarged cortex, which is the part of the brain where memory and sensory perception are located; this is the part of the brain that thinks. The enlarged cortex is the section of the brain used by cetaceans for hearing and processing of sound signals, while for us it is devoted to vision and hand-eye coordination. The brains of dolphins (and other cetaceans) are larger than ours (**TABLE 13.1**), but not as a ratio of brain weight to body weight. So does their larger brain give them more intelligence? Well, their brain has more neurons (brain cells) than ours. But more to the point, dolphins in captivity have exhibited extraordinary signs of intelligence, including self-recognition in reflections of themselves,

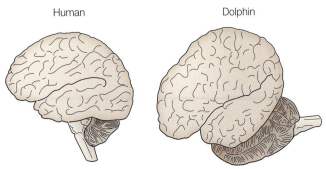

Human   Dolphin

FIGURE 13.32 Comparative sizes of a human brain and a bottlenose dolphin brain.

TABLE 13.1	Comparison of brain weights of some cetaceans and primates
Speces	**Brain weight (g)**
River dolphin	200
Sperm whale	>9000
White-sided dolphin	1200–2000
Human being	1000–1800
Chimpanzee	ca. 350
Gorilla	ca. 600

learning sign language, and picking up instructions from their trainers, which is impressive for animals that aren't accustomed (in their native habitat) to watching a human being make hand gestures. But more importantly, they seem to have an understanding of word order and how the meaning of a sentence can be so altered. It is precisely this intriguing, high-level intelligence that stirs passions and makes the killing of whales so disturbing to many people.

Whaling

People have been harvesting whales throughout recorded history. Bowhead whales were taken by native peoples in the Arctic as long ago as 500 A.D., and right whales were taken by the Basques in Europe as early as the beginning of the second millennium. Whaling slowly spread throughout the world's oceans and its peoples over the centuries, and by the late 1600s, Americans entered the industry and eventually dominated it (**FIGURE 13.33**). Originally, whales were hunted not for food, but for other products: their blubber, which was used to make "**train oil**" for soap and for burning in lamps; baleen, which had properties similar to plastics today, and was used for corsets, umbrellas, and various other goods; and **ambergris**, a waxy substance that accumulates in the intestines of sperm whales for unknown reasons that was used as a preservative in the manufacture of perfumes.

Although whales have been hunted for centuries, the industry did not become efficient until 1868, with the development of the harpoon gun (**FIGURE 13.34**). With its explosive harpoon tip, it could kill or at least disable a whale, making it much easier to haul back to the ship. Technological innovation marked the end of the famous "Nantucket sleigh ride" days, in reference to the whaling port of Nantucket Island in Massachusetts. The expression describes what happened when whales were harpooned from small boats, called whaleboats, to which the line from the harpoon was attached. The whale would try to escape and in the process it would tow the whaleboat until the whale became exhausted— this was the Nantucket sleigh ride. The whale would then be killed with the *killing lance*, and hauled aboard the mother whaling ship. The industry also got a boost in 1925, when whaling factory ships entered the scene, after which the harvest of whales climbed to its highest levels. Even the fastest whales, the blue whale and the

FIGURE 13.33 Historical painting depicting American whalers battling a sperm whale by William Heysman Overend (1851–1898).

(A)

(B)

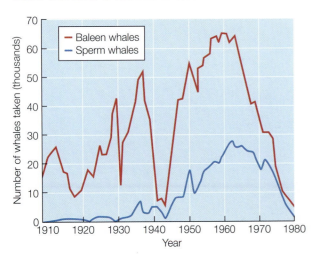

FIGURE 13.34 (A) A harpoon gun in the 1890s. (B) A modern Japanese whaling ship.

FIGURE 13.35 Annual harvest of baleen whales and sperm whales from 1910 to 1980 for all oceans.

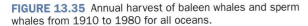

fin whale, could now be hunted down. Populations began to plummet, because although whales are relatively long lived, with a lifespan of 20 to 40 years not uncommon, they have a very low reproductive rate. A female whale has a long gestation period—ten to twelve months for the mysticetes and sixteen months for the sperm whale—and will give birth to only one calf at a time. The calf is then cared for by the mother for more than a year.

The impact of commercial whaling took its toll on the world's whale populations. **FIGURE 13.35** shows how many whales were taken between the years 1910 and 1980. Commercial whaling indeed took off in the 1920s and 1930s; it dropped off during World War II, and reached its peak in the 1950s and 1960s before abruptly declining. The reason for the drop was simply that there were fewer whales in the ocean.

Concerned with the possibility of over-harvesting the world's whales, a number of countries formed the **International Whaling Commission** (**IWC**) in 1946. Initially, membership in the IWC was only the 15 "whaling nations," countries actually involved in whaling, but there was then, and remains today, no such requirement to be a member. By 2011 there were 89 member nations. At the First Annual Meeting of the IWC, in Washington, DC in December 1946, the member nations signed an International Convention for the Regulation of Whaling. Prior to formation of the IWC, there were only informal agreements among some countries. The purpose of the 1946 Convention was: "to provide for the proper conservation of whale stocks and thus make possible the orderly development of the whaling industry."[26] The specific charge to the IWC was, and remains: "to provide for the conservation, development, and optimum utilization of the whale resources." Note well that this was a pragmatic act, and not one of compassion.

From 1946 to the late 1960s, the IWC tried to limit whaling by setting quotas, with the goal of allowing individual whale stocks to replenish their numbers. But it didn't work. Despite their efforts, the harvest of whales continued, even after the 1946 agreement, as can be plainly seen in Figure 13.35, and so the great whales continued to decline in numbers. In 1979, faced with clear evidence of dramatic declines (the drop in whale harvest by 1979 was not because of reduced effort, but because of dwindling stocks), the IWC placed a moratorium on all whaling in the Indian Ocean, and banned factory ships. Whaling continued nonetheless. In 1982, the Commission adopted what was to have been a pause in whaling on all whales except dolphins and porpoises, to begin in 1986, in order to study their predicament. That ban remains in effect today, and is sometimes referred to as the 1986 Moratorium. A few countries objected, notably, Japan, Norway, Iceland, Korea, Peru, and the former U.S.S.R. (now Russia).

---

[26] The International Whaling Commission, iwcoffice.org/commission/iwcmain.htm.

The 1986 Moratorium granted two exemptions: scientific, or research whaling, as it was called, was allowed in order to study whale population numbers, and subsistence whaling was allowed for aboriginal peoples that had traditionally hunted whales for food. The IWC and its Moratorium have been largely ineffective because member nations, under the Articles of the Convention, can be exempted by simply lodging an objection to any IWC resolution it doesn't like. The IWC had no power to prevent any nation from doing what it wants, and so whaling (commercial whaling, not just subsistence or research whaling) continues to this day, at a much reduced level from what it was in the middle of the last century.

All but Japan and Norway formally ended commercial whaling in 1989. Iceland resumed commercial whaling in 2006, and so there are currently only these three recognized whaling nations, if we exclude research whaling and takes by aboriginal peoples. Those three countries continue to harvest several hundred minke whales and other species each year.

Delegates from the IWC member nations meet once each year to consider the status of the world's whale populations. One of the contentious issues is that the current ban, the 1986 moratorium, which is still in effect today, was based on political and emotional factors, not science. The recent rapid growth in membership in the IWC is seen by pro-whaling nations as a reaction to the anti-whaling nations' opposition to reopening commercial whaling; that is, a three-quarters majority vote was needed to invoke the moratorium, and the same majority vote is needed to lift it. And so a number of pro-whaling nations (nations that do no whaling themselves, but are not opposed to whaling) have also joined, in order to—as anti-whaling nations argue—vote for lifting the ban in return for foreign aid from pro-whaling nations. And, of course, identical counter-claims have been lodged by the pro-whaling nations, that the U.S., for example, is buying votes. In fact, they have a good point: it is clear that a number of anti-whaling nations were recruited by the U.S. and other anti-whaling nations as new members for the sole purpose of voting for the moratorium back in 1982.

Rankling continues every year at the IWC meetings, and little is accomplished. One accomplishment, though, is the establishment in 1994 of the **Southern Ocean Sanctuary**, an area where commercial whaling is banned. The ban is subject to a vote every ten years. But whales continue to be taken by a number of nations, including the United States. At their 2010 meeting, the IWC reported that a total of 1864 whales were taken by Denmark, Iceland, Norway, St. Vincent and the Grenadines, Japan, Korea, Russia, and the U.S. That total included 1421 minkes, 136 sperm whales, 116 gray whales, and others. Those nations were allowed to harvest whales under the various provisions of the IWC. Norway continues whaling "under objection" to the provisions of the IWC.

Some case histories

THE VAQUITA PORPOISE AND THE CHINESE RIVER DOLPHIN   The vaquita porpoise (Spanish for "little cow"), also known as the Gulf of California porpoise (*Phocoena sinus*),[27] is a small animal, less than 1.5 m in length, and it occurs only in the northern part of the Gulf of California (**FIGURE 13.36**). It is one of the most endangered cetacean species in the world today. Estimates of the number of

---

[27] The words porpoise and dolphin are often used interchangeably, but they are different animals. The dolphins are in the family Delphinidae, and the porpoises in the family Phocoenidae. There are four distinct anatomical differences between them: (1) dolphins have cone-shaped teeth while porpoises have flat or spade-shaped teeth; (2) dolphins usually have a pronounced beak, but porpoises do not have a beak; (3) dolphins have a more curved dorsal fin, while porpoises have a triangular dorsal fin; and (4) dolphins are generally larger than porpoises.

FIGURE 13.36 Painting of a vaquita porpoise (*Phocoena sinus*).

remaining vaquita are between 100 and 300. Never hunted, its depletion is attributed to accidental entanglement in fishing nets.

The Chinese river dolphin (*Lipotes vexillifer*), a freshwater dolphin, may already be extinct. In the 1950s there were approximately 6000 animals, but in 1997 there were estimated to be only 13 remaining; one individual was spotted in 2004 and another possible sighting was reported in 2007 (**FIGURE 13.37**). Its numbers were depleted by hunting and by accidental entanglements in fishing gear.

THE RIGHT WHALE   There are three species of right whale, the North Atlantic right whale (*Eubalaena glacialis*) (**FIGURE 13.38**); the North Pacific right whale (*Eubalaena japonicas*) and the southern right whale (*Eubalaena australis*). All three were listed as endangered in the U.S. Endangered Species Act of 1969. Right whales have the dubious distinction of being the first whales to be taken by a regular, organized whaling industry—by the Basques in the twelfth century—and the first to be threatened with extinction as a result of commercial whaling. They are also the first to be protected by international agreement.

Despite being protected since the 1930s, the right whales have been slow to recover, if they have recovered at all. The North Atlantic and North Pacific right whales are the most endangered species of large whales in the world today. In contrast to their Northern Hemisphere counterparts, Southern Hemisphere right whales have shown signs of recovery, although only a few hundred remain. There are fewer than 500 North Atlantic right whales left in the world, and only about 100 North Pacific right whales, of which only 30 are believed to be mature females.

The maximum length of right whales is about 18.6 m (60 feet), and the maximum weight is about 100 metric tons. These whales are skim feeders, feeding primarily in coastal ocean regions where there are high densities of copepods. They were hunted for their baleen and train oil. The right whale comes by its name for unfortunate reasons: it is slow swimmer; it floats when dead; and it has coastal distribution. These attributes made it easy to hunt—they made it the "right whale" to kill.

Right whales have very low reproductive potential; females have only one calf every 3–5 years, so recovery will be difficult. The major causes of death these days are collisions with ships and entanglements with fishing gear. For example, more

FIGURE 13.37 Chinese river dolphin (*Lipotes vexillifer*).

than half of all the remaining North Atlantic right whales have scars from one or the other of these encounters. At the 2011 meeting of the IWC, it was reported that there were four fatal entanglements of right whales off the US coast between November 2009 and October 2010.

In the Northwest Atlantic, there are an estimated 350 right whales remaining. They migrate from their calving grounds off the coasts off northern Florida and Georgia to their feeding grounds in the Gulf of Maine during the spring to fall period, feeding on rich populations of the copepod *Calanus finmarchicus*.

THE FIN WHALE   The fin whale (*Balaenoptera physalus*), is the second-largest whale. It was declared endangered in 1970. These whales are cosmopolitan in their distribution, ranging throughout the world ocean, except for the Mediterranean, and they are exceptionally long-lived, reaching ages of 80 to 90 years. They are social animals and are often seen in groups of 2 to 7, and are prime examples of gulp-feeding rorquals, feeding primarily on krill, small schooling fishes and squid. More than 700,000 fin whales were taken in the last century, and their current population status is not well known, and is based on few estimates; the best available population estimates suggest that there are about 53,000 remaining. Although it is depleted, the fin whale is considered abundant compared with some of the others.

FIGURE 13.38 A North Atlantic right whale (*Eubalaena glacialis*), skim feeding at the surface in the Gulf of Maine, with its long baleen plates clearly exposed.

THE HUMPBACK WHALE   The humpback whale (*Megaptera novaeangliae*) was listed as endangered throughout its range in 1970. Named for their long pectoral fins, which can be more than four and a half meters long, *Megaptera novaeangliae*, means "big-winged New Englander" (**FIGURE 13.39**). These animals are famous for slapping the water with their pectorals, and for their spectacular leaps out of the water, called breaches. They too are rorquals, and feed in the productive waters at higher latitudes, migrating between there and their calving grounds each year. They are very social animals, and are also famous for their songs, with individuals often singing for twenty minutes or more. Prior to whaling the world population was estimated at about 125,000, and today its numbers are increasing in many parts of the world. There are no agreed estimates of current population, although there are some signs of recovery in parts of the southern hemisphere, where there are thought to be more than 30,000. But, back in the 1960s there were only about 1000 remaining worldwide. The current abundance estimate for the North Pacific is about 20,000 whales, and for the North Atlantic about 11,570 whales. Like the right whales, it is also vulnerable to coastal development, especially shipping, recreational boating, and various forms of coastal pollution.

FIGURE 13.39 A breaching humpback whale (*Megaptera novaeangliae*).

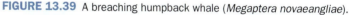

THE BLUE WHALE   The blue whale (*Balaenoptera musculus*) is the largest whale. It has been listed as endangered since 1969, but its population status in the Northern hemisphere is "unknown." A best guess offered at the 2004 June meeting of the IWC was between 400 and 1500 world-wide. But there may be as many as 500 in the Antarctic Stock. Sightings have increased in the Pacific. The U.S. NOAA Office of Protected Resources estimates that there may be as many as 5200 worldwide. It was suggested at the 2010 IWC Meeting that they may be increasing at a rate of 8% per year. These huge rorquals feed almost exclusively on krill in the Southern Ocean (off Antarctica) and on small-sized schooling fish in the Northern Hemisphere. These were also a prized whale to hunt: one large individual could yield 9000 gallons of oil. Commercial harvest was banned in 1966. It is hoped that they can recover, but the likelihood of that happening is not known.

THE SPERM WHALE The sperm whale (*Physeter macrocephalus*) is the largest of the Odontocetes (see Figure 13.23). They are deep divers and feed mainly on squid, including the giant squid. The gestation period for the sperm whale is 15–16 months. They are found off the continental shelf—very seldom in waters less then 300 m deep. Their teeth, which line both sides of the narrow lower jaw only, were symbols of the whaling heyday. Works of art called scrimshaw were carved into their teeth by sailors aboard whaling ships (**FIGURE 13.40**).

The sperm whale was listed as endangered in 1969. Over the past two centuries, whalers have taken more than one million sperm whales. Despite this pressure, the sperm whale is thought to be one of the most abundant today; while estimates of the current population size vary, there are probably somewhere between 200,000 and 1.5 million sperm whales today, according to the NOAA Office of Protected Resources. If the number is closer to the larger estimate, then there are far more sperm whales today than the combined abundances of the other six endangered large whale species. The main threat to sperm whales today is the 150 or so whales accidentally taken per year in the California pelagic gill net fishery.

(A)

(B)

FIGURE 13.40 (A) A sperm whale skeleton. (B) An example of scrimshaw showing the whaleship *William Tell* and a sperm whale. This piece was carved by second mate Edward Burdett, ca. 1830.

Chapter Summary

- Marine reptiles, sea birds, and marine mammals are tetrapod vertebrates that evolved from lobe-finned fishes to become terrestrial animals, continued to evolve, and returned to the sea.

- The marine reptiles, like all reptiles, are **poikilotherms**, or cold-blooded animals. They reside mostly in the tropics and subtropics. They include sea snakes, the marine iguana, saltwater crocodiles, and sea turtles. All seven species of sea turtles are either threatened or endangered. The Kemp's ridley sea turtle is the most endangered. Sea turtles are long-lived, and reach reproductive maturity at relatively late ages; they all come ashore to lay their eggs on beaches. Most species undergo long migrations at sea.

- Sea birds, like all birds, are **homeotherms**, or warm-blooded animals. Sea birds spend a significant portion of their lives at sea, where they feed on marine organisms, but they nest on land. They include the flightless penguins and the pelagic birds that spend most of their time flying. Penguins feed at sea and breed ashore, often in colonies, and reside only in the Southern Hemisphere. Pelagic birds come onshore only to breed. All sea birds possess **salt glands** that enable them to drink seawater. Seabirds can be characterized as **surface feeders**, **pursuit feeders**, **plunge divers**, and **scavengers/predators**. **Shorebirds**, such as herons and sandpipers, feed at the seashore, as do some raptors, such as eagles and ospreys. About 20 species of sea ducks, such as eiders, spend significant time on the ocean and feed on marine organisms.

- There are four groups of marine mammals: the **sea otters**; the **sirenia**—manatees and dugongs; the **pinnipeds**—seals, sea lions, and walruses; and the **cetacea**—whales, dolphins and porpoises.

- There is a single species of sea otter, which ranges across the coastline of the North Pacific Ocean. Sea otters feed, rest, and reproduce on the water close to the shoreline. They feed on benthic invertebrates, especially sea urchins. They have the thickest fur of any animal.

- The sirenia, or **sea cows**, include three species of manatees and one species of dugong. All are herbivores. They reside in shallow riverine, estuarine, and coastal waters.

- Most pinnipeds are found in colder climates. The seals, also called true seals, differ from the sea lions and fur seals in their lack of external ears and in the restricted mobility of their pectoral and pelvic fins. All feed on fishes, except the walrus, and some breed in large colonies. Walruses feed on benthic invertebrates.

- There are two groups of cetaceans: the odontocetes, or the **toothed whales**, and the mysticetes, the **baleen whales**. All whales breathe through a **blowhole** at the top of the head.

- The mysticetes are filter feeders, concentrating plankton and nekton with their baleen plates. They have been divided into two groups based on the lengths of the baleen plates and their methods of feeding: the **rorquals**, with shorter baleen plates, are **gulp feeders**, and the other group, with longer baleen plates that filter smaller zooplankton, are skim **skim feeders**.

- The odontocetes are predators, feeding mostly on fishes and squids, either as individuals, or in groups. Most are capable of **echolocation**. These whales have several adaptations that facilitate deep diving and allow them to hold their breath for as long as two hours.

- Commercial whaling has reduced the population sizes of many species of the great whales, although they are still hunted under provisions of the **International Whaling Commission**, which permits scientific whaling and limited takes by aboriginal peoples. Three nations continue to hunt whales commercially: Norway, Japan, and Iceland.

Discussion Questions

1. Compare and contrast poikilotherms and homeotherms; ectotherms and endotherms. Give marine examples of each.

2. Why are the marine reptiles mostly restricted to the tropics and subtropics? Are there any exceptions? Why are there no marine amphibians?

3. What are the advantages and disadvantages of sea turtles' laying their eggs on sandy beaches? Of what advantage is it for most sea turtles to be very large? Why do you think they are larger than their freshwater cousins?

4. Sea birds exhibit several morphological adaptations that are examples of convergent evolution: What are they? Why do most species of sea birds breed in colonies of large numbers of birds?

5. Discuss the major types of feeding modes among the sea birds. What are some of the advantages and disadvantages of each?

6. Why do you think the emperor penguins breed in the cold of winter, in the coldest environment on Earth?

7. What are salt glands, and what marine organisms possess them? Are there any marine tetrapods that do not have them?

8. What are some of the current threats to the sirenia? What are some management actions you might consider if you were to find yourself suddenly in charge of their protection?

9. Explain the anatomical differences among the three groups of pinnipeds discussed in the chapter. Of the three groups, which is the more highly evolved in terms of its adaptations for a marine life, and why do you think so?

10. Explain the differences, anatomical and behavioral, between the mysticetes and the odontocetes. Why is it advantageous for the mysticetes to be so much larger, on average, than the odontocetes?

11. Should all whaling be ended, for all nations and all peoples? Why or why not? What are some of the arguments would you bring to the governments of the world in defense of your position?

Further Reading

Berta, Annalisa, Sumich, James L., and Kovacs, Kit M., 2006. *Marine Mammals: Evolutionary Biology*, 2nd ed. Amsterdam: Academic Press.

Carr, Archie F., 1967. *So Excellent a Fishe: A Natural History of Sea Turtles*. Gainesville, FL: University Press of Florida.

Dunson, William A. (ed.), 1975. *The Biology of the Sea Snakes*. Baltimore, MD: University Park Press.

Evans, Peter G. H., 1987. *The Natural History of Whales and Dolphins*. New York: Facts on File, Inc.

Kooyman, G. L., and Ponganis, P. J., 1997. The Challenges of Diving to Depth: The deepest sea divers have unique ways of budgeting their oxygen supply and responding to pressure. *American Scientist* 85: 530–539.

Spotila, James R., 2004. *Sea Turtles: A Complete Guide to Their Biology, Behavior, and Conservation*. Baltimore, MD: Johns Hopkins University Press.

Welty, Joel C., and Baptista, Luis, 1988. *The Life of Birds*, 4th ed. Belmont, CA: Thomson Brooks/Cole.

Contents

The oceans produce vast sources of protein that we have been harvesting for a long time. At first, this resource must have seemed inexhaustible; after all, how could we possibly catch all the fish in the sea? Well, over the past hundred years we have found out how; not only have we learned that is it possible to effectively catch all the fish in the sea, but we have been doing precisely that for some stocks. These trawlers, tied up at the dock in New Bedford, Massachusetts and prevented by law from fishing—on a beautiful August day when they *should* be fishing—are reminders of this last point.

Marine Fisheries and Aquaculture

The oceans have provided us with food since before recorded history, and as civilizations developed, so too did organized fisheries, which were coordinated efforts to maximize the catch of food fishes. The word **fishery** (plural fisheries) itself refers to the business of catching fish; it can also refer to the specific fish being caught (such as the "tuna fishery"). It can be a commercial fishery, in which the fishers are in it for money (but also for a way of life), or it can be a recreational fishery, such as the recreational rod-and-reel fishery for striped bass. Usually, a fishery targets a specific species of invertebrate (such as the Maine lobster fishery), finfish (such as the California sardine fishery), or particular ecological or taxonomic grouping of fishes, such as **groundfish**, which are the cod, haddock, and founders that live on and near the bottom (also called **demersal fishes**). All fisheries have the goal of catching something that is, almost always, a higher-trophic-level organism that is the beneficiary of the ocean's primary productivity (in the case of marine fisheries), and which serves as food for us.

At the heart of all marine fisheries, regardless of the species or group of fishes targeted, is the extraction of living marine resources (mostly fish), and the replacement of these resources by natural processes. The fish we harvest are the end result of primary production that is passed up the food chain to higher trophic levels, with inefficiencies that cut into that passage each step along the way (see Chapter 7). Despite the losses, a large biomass of higher-trophic-level organisms is created, some of which constitutes a food source for each other in marine food webs and some of which serves as food for us. We have been extracting that food from the oceans throughout human history. Ever since our Stone Age ancestors learned to fashion fishhooks out of bone, we have been making incremental advances in the efficiencies with which we harvest living marine resources—to the point where we may now be taking too much.

Today, fisheries around the world are chasing fewer and fewer fish; stocks of commercially exploited species have been fished down to levels unimaginable only a few generations ago.[1] Quite simply, we have been taking fish and other living marine resources out of the oceans at rates that are unsustainable—that surpass the oceans' capacity to replace them. As stocks shrink, the cost per pound of some species of fish and shellfish is becoming so high that what was once a reliable and inexpensive source of protein for the masses is fast becoming a delicacy for the well-to-do. Not only has over-fishing depleted stocks around the world, but the ecology of the seas is changing as a result, in ways no one ever anticipated. The commercial fishing industry, resource managers, and

---

[1] You will see the terms fish *stock* and fish *population* used interchangeably in reference to commercial fisheries. A stock is generally considered a population or subpopulation of fish that is discrete for management purposes, with little or no mixing with other stocks by way of immigration or emigration. An example is the North Sea cod stock, and the New England (U.S.) cod stock. The same species is distributed across the North Atlantic, but they don't all intermix with one another. There are distinct spawning populations that can be managed as distinct units.

marine scientists are thus faced with two (at least two) fundamental, overriding questions: Is it possible for depleted fish stocks to recover? And, if so, can steps be taken and policies implemented that will facilitate such a recovery? To answer these questions, we need to understand how fish populations fluctuate, both naturally and as a result of fishing, and to do that we need to step back and explore in greater detail some of the marine ecological principles we have been covering, specifically as they relate to **fisheries production**, the reproduction and growth of fish targeted by that fishery.

In this chapter we are going to review some examples of fisheries that have run into trouble and how they got to that point. We will discuss the fundamentals of how fishes are harvested by commercial fisheries, how fish and populations of fish grow, and how commercial fisheries are managed. We will conclude with a discussion of where things stand, including a look at some of the questions that still challenge fisheries scientists today, and we will examine the role that innovative approaches to the artificial propagation of living marine resources—aquaculture—can play. We begin with a discussion of the trophodynamics involved in the production of commercially-harvested fishes.

FIGURE 14.1 A simplified representation of how photosynthetic primary production fuels commercial fisheries production. At each step in the food chain, only 10–20% of the energy is transferred, such that by the time the initial primary production gets to the commercially harvested fish in this diagram, which is after three trophic steps, only 0.1 to 0.8% of the initial energy from primary production remains (e.g., 0.1 × 0.1 × 0.1 = 0.001, which is 0.1%, and 0.2 × 0.2 × 0.2 = 0.008, or 0.8%).

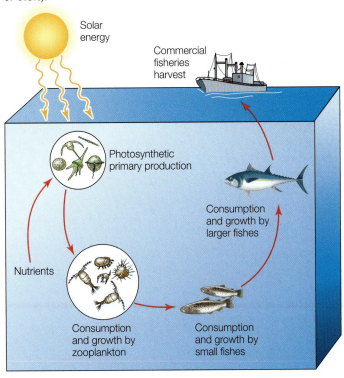

Fisheries Trophodynamics

The basics of how a fishery works can be summarized as the transfer of carbon (or the equivalent energy)[2] through progressively higher trophic levels to commercially harvested fishes or invertebrates (**FIGURE 14.1**). But there is much more that happens in the marine food chain leading to fish production, in terms of losses along the way, and in terms of the dynamics of the fish populations that are the beneficiaries of that food chain prior to their capture in the fishery.

More than 40 years ago, **John H. Ryther** (1922–2006) wrote an influential paper that expanded on the basic principle illustrated in Figure 14.1.[3] He used the rates of primary production in the world's oceans to calculate the theoretical maximum production of commercially harvested fishes. His analysis was based on assumptions we reviewed in Chapter 7, that the rates of primary production vary throughout the world ocean, but are greatest in upwelling systems, which occupy only a small portion of the global ocean's area, and least in the open oceans, which constitute the majority of the global ocean's area. But his analysis also factored in the losses between trophic levels from the primary producers to the fish, as well as how the number of trophic levels varies among the three general ocean regions—the vast expanses of

---

[2] Carbon and energy are interchangeable in the sense that the organic carbon from primary production has caloric value to the consumer. For example, 1 mg of protein has 5.65 calories of energy; 1 mg carbohydrate as 4.1 calories, and 1 mg fat has 9.45 calories.

[3] Ryther, J. H., 1969. Photosynthesis and Fish Production in the Sea. *Science* 166: 72–76.

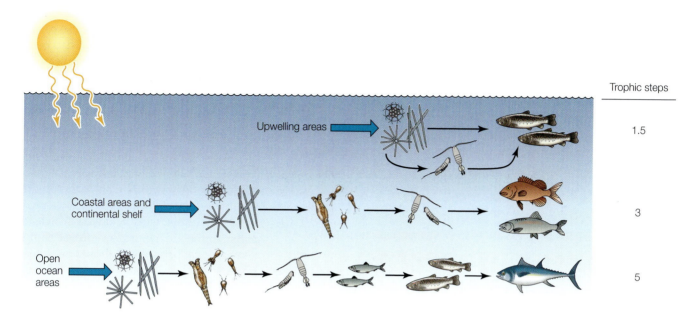

Trophic steps

Upwelling areas   1.5

Coastal areas and continental shelf   3

Open ocean areas   5

FIGURE 14.2 Example of the different number of trophic steps in different ocean environments, from highly productive upwelling systems to less productive open ocean areas.

the open ocean, coastal and continental shelf waters, and the world's major upwelling regions (**FIGURE 14.2**).

The open oceans are **oligotrophic**, which means that they have very low rates of primary production and therefore offer little to sustain higher trophic level organisms. Open-ocean ecosystems have low concentrations of nutrients in surface waters because of a general lack of physical mechanisms to bring significant quantities of nutrients up from deeper waters. The vertical **nutrient flux** is low, and primary production is thus limited. Not only are rates of primary production low, but the phytoplankton that are produced are generally small-celled taxa. Those smaller cells, the nanoflagellates and picophytoplankton (see Box 8A), are not grazed efficiently by larger zooplankton, and so they have to be channeled through an intermediary—an extra trophic level (or two)—before their biomass can be passed up the food chain. So, the oligotrophic open oceans not only have low rates of primary production, but they also have longer, less efficient planktonic food chains. This situation contrasts sharply with the more productive coastal and upwelling areas that benefit from higher nutrient levels. Nutrient-rich environments not only have higher rates of primary production, but the phytoplankton tend to be dinoflagellates and diatoms, larger cells that can be grazed more efficiently by zooplankton, making for a shorter planktonic food chain. Moreover, planktivorous fishes in upwelling areas, such as anchovies in the Peruvian upwelling waters of the eastern tropical Pacific Ocean, can and do feed directly on these larger phytoplankton cells, in addition to feeding on the zooplankton. Indeed, the upwelling areas of the world ocean have the highest rates of fish production (of planktivorous fishes, especially).

Based on these considerations, Ryther calculated that the total production of fish that is theoretically possible in the world ocean is about 240 million metric tons (MMT) of fresh weight annually. But, as he was quick to point out, we can't fish it at that rate. For one thing, of that total fish production, much is consumed by other top-level carnivores in the sea; and for another, the commercial fisheries must leave enough fish to maintain the populations through reproduction. And so he predicted that the **maximum sustainable yield** of marine fish to commercial fisheries is about 100 MMT per year for the world ocean.

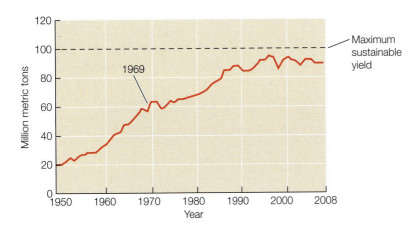

FIGURE 14.3 World fish landings from 1950 to 2008 in metric tons. The predicted maximum sustainable yield is 100 million metric tons. Notice where we were in 1969, when Ryther's paper was published.

Back in 1969 when Ryther published his paper, the total world fish catch (also called *landings*) were about 60 MMT and had been increasing by several percent each year for a number of years; he warned that such increases in fisheries landings could not be sustained. Well, he was right. As **FIGURE 14.3** shows, world fisheries landings continued to increase until about 1995, when they approached Ryther's predicted maximum of 100 MMT, and since then, landings have been decreasing.[4] So how did we get to this point? How have fisheries advanced to where it would appear that we can, indeed, "catch all the fish in the sea"? A review of the history of commercial fishing will show how.

A History of Commercial Fisheries

Throughout much of human history, the world's population remained nearly stable. Then about 11,000 years ago, when there were some four million of us, agriculture first appeared in the Fertile Crescent, a region of favorable climate for crop growth that stretched approximately from northern Egypt to Mesopotamia (modern-day Syria and Iraq). With more efficient production of food, the world population began to increase slowly but steadily. In more recent times the world's population growth has accelerated, growing from about 275 million people a thousand years ago to about twice that number by the year 1500, and to some 1.6 billion by 1900 (**FIGURE 14.4**). By the twentieth century, high birth rates were no longer kept in check by high death rates, as our quality of life improved with better technologies, medicine, and, of course, food production—but such rapid population growth places corresponding demands on our food supply.

FIGURE 14.4 World population growth from 2.5 million years ago, the beginning of the Stone Age, to today, with projected growth through 2025. Notice the scale break at 8000 years before present, after which the scale is linear.

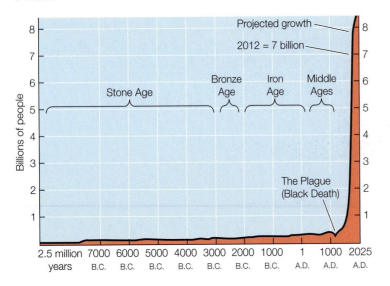

Fish from the sea has always been an important source of high-protein food, and so as the world's population grew, so too did the world fish catch. As early as the 1400s, cod, a favorite food fish among Europeans, was already becoming scarce in the eastern North Atlantic, and the discovery by British explorer John Cabot of huge abundances of cod in the cold waters on the western side of the North Atlantic was an important stimulus to the settlement of North America.[5] By the early 1600s, about 50 British boats were fishing cod in those waters off New England, salting their catch to preserve it for the return trip to England. Cod remained relatively abundant in the eastern North Atlantic into the 1800s and early 1900s, though estimates of population sizes up to that time are,

---

[4] Food and Agriculture Organization, 2012. *The State of World Fisheries and Aquaculture, 2010.* United Nations.

[5] The Basques may have already been making trips to North America for cod by that time. Mark Kurlansky, 1997. *Cod: A Biography of the Fish that Changed the World* (New York: Penguin).

of course, just guesses. But the fishing pressure on cod, and other species, had begun.

As technology advances over time, so too does the efficiency with which fisheries catch fish. Two advances in particular had a great impact on the total quantity of fish harvested from the ocean: (1) the advent of steam-powered fishing vessels in the early 1900s and diesel engines soon thereafter; and (2) on-board freezing and refrigeration systems. All of a sudden, it was possible to bring a fresh—not salted—catch to market. As fresh fish became more popular, the market demand was met by increased commercial landings. By the middle of the last century it was becoming evident that the populations of cod, the main target of the fishery, simply could not withstand the growing fishing pressure. Today we are witnessing the near-total collapse of the North Atlantic cod fishery in Canadian waters, and the number of fish remaining in U.S. waters is also dangerously low.

North Atlantic cod were not the only fish stocks to be depleted under heavy fishing; the California sardine fishery (the Pacific sardine) is probably the most dramatic example of a fisheries collapse (**FIGURE 14.5**).[6] Landings in California reached their peak in the late 1930s and early 1940s, totaling nearly 700,000 metric tons a year, but soon thereafter they all but disappeared. By the 1950s, what was once a thriving industry was in ruin; landings had dropped to almost nothing, producing severe economic hardship in the region (times made famous by John Steinbeck's novel *Cannery Row*). The fishery was closed in 1960. In the 1980s, the fishery experienced a comeback, with landings only half of what they had been in the 1930s—only to disappear once again in the 1990s. Most recently, there have been large fluctuations in landings, and researchers are now trying to sort out the role of fishing from that of oceanographic oscillations in the Pacific, discovered only in recent years, in controlling these population fluctuations; it now appears that both are important. In addition, there are likely competitive interactions between Pacific sardines and anchovies, which alternate in abundance. Overfishing was clearly a driving force behind the collapse of the sardine fishery, but other factors are important; this is but one example of how difficult it is to understand all the variables that contribute to fluctuating abundances of commercially exploited fish stocks.

Another dramatic example of overfishing and the collapse of a fishery is the Northwest Atlantic redfish fishery (**FIGURE 14.6**). Redfish, also called ocean perch, were first fished in the 1930s, following the development of freezing technologies that allowed this relatively lean, flaky white-meat fish, which freezes well, to be distributed around the U.S. It became very popular, and landings soon reached more than 50,000 metric tons in the 1940s. But then, the landings dropped abruptly, reflecting the drop in abundances as the stocks began to feel the intense fishing pressure. The catch fell to about

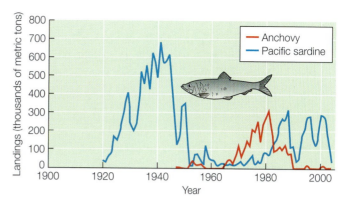

FIGURE 14.5 California landings of Pacific sardine (*Sardinops sagax*) from 1920 to 2004 showing the apparent alternating abundance of northern anchovy (*Engraulis mordax*).

FIGURE 14.6 Landings of redfish (*Sebastes marinus*) from the Gulf of Maine and Georges Bank region of the northwest Atlantic, from 1934 to 2000.

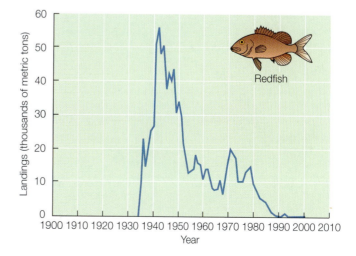

[6] Takasuka, A. et al., 2008. Contrasting spawning temperature optima: Why are anchovy and sardine regime shifts synchronous across the North Pacific? *Progress in Oceanography* 77: 225–232.

10,000 metric tons between the 1950s and about 1980, and by 1988, landings were less than 1000 metric tons. Nearly identical boom-and bust-landings were seen in Newfoundland, Canada, for the Grand Banks redfish stocks as well as the Gulf of St. Lawrence and Nova Scoatian Shelf stocks. Unlike the California sardine fishery, the redfish fishery has not yet made a comeback; but there are signs that stock levels are rebuilding.[7]

There are numerous other examples of fisheries collapses over the past century that we could discuss, and while some of the collapses can be explained by oceanographic phenomena, nearly all of these dramatic declines in fish stocks around the world are the direct result of overfishing. Increased fishing effort and better fishing technologies advanced faster than did our abilities, or our will, to manage and conserve the stocks.

Types of fishing gear

Commercial fisheries today have gone far beyond the early days when fish were caught almost exclusively on baited hooks, either single hooks on hand lines or *long lines* with many hooks, also called *tub trawls*. Until the mid-nineteenth century, most fishing was done this way by one or two fishermen in open boats, set off from shore or from a larger ship out on the offshore fishing banks (**FIGURE 14.7**). Long lines continue to be used today to catch tuna, swordfish, halibut, and other species, but the majority of fish landed by commercial fisheries today are caught with nets. There are many kinds of nets, from small *cast nets* thrown by one person over a few fish in shallow water to various kinds of *fish traps* and *stop seines*, nets that close off a cove or stream, trapping fish where they're easily caught. The most commonly used commercial fishing nets are gill nets, purse seines, and trawls.

Gill nets are constructed of nylon monofilament line, the same type used in spin-casting reels on ordinary fishing rods, which makes the net nearly invisible in water. It has a mesh opening designed to allow the target fish to just begin to swim though the net. As it does so, the mesh tightens around the fish's body just behind its head, but the fish can't escape, because when it tries to back out of the net, the mesh catches against its gill openings. Gill nets are set either at the surface or along the bottom (**FIGURE 14.8**). Called "drift nets" when deployed at the surface, gill nets can be extremely long—several tens of kilometers. They are set adrift on one day, allowed to fish overnight, and then tended the next day; the catch is removed from the meshes and the **bycatch**,

FIGURE 14.7 (A) "Halibut dory and crew hauling the trawl, gaffing and clubbing the halibut," 1887. (B) Preparing long lines, or tub trawls, on the deck of the fishing schooner *Mayflower*, 1922. The men are cutting and baiting hooks with herring. Each dory will take two tubs, with each containing 10 trawl lines that are 2000 feet long with 500 baited hooks; they will be fished on the bottom for one hour before being hauled up.

[7] Mayo, R. K. et al., 2002. Biological characteristics, population dynamics, and current status of redfish, *Sebastes fasciatus* Storer, in the Gulf of Maine–Georges Bank region. *Northeast Fisheries Science Center* Reference Document 02–05.

(A)

THE FRESH HALIBUT FISHERY.
Halibut dory and crew hauling the trawl, gaffing and clubbing the halibut. (Sect. v, vol. i, p. 16.)
Drawing by H. W. Elliott and Capt. J. W. Collins.

(B)

the unintended capture of nontargeted species, is discarded. It is this unintended catch by drift nets that has drawn criticisms from marine ecologists and environmentalists; it is seen as not only wasteful, but tragic. Gill nets are notorious for drowning dolphins and sea turtles that become entangled in them.

Purse seines are used to catch fishes that shoal, such as the herrings, sardines, and tunas. Once a shoal is spotted, the captain of the seiner will steer his boat around the shoal, releasing the skiff, a smaller boat carried on the stern of the seiner, to which one end of a long net, the seine, is attached. The skiff maintains position (it has a motor), while the larger, main boat encircles the shoal, all the while deploying the seine behind it, eventually completely encircling the shoal of fish (see Figure 14.8B). Before the fish can swim down and out of the encircling net, a draw line running along the bottom of the net is taken up, thus pursing the catch.

Towed nets, or **trawls**, were used aboard sail-powered fishing boats in the 1800s, but they became much more efficient after steam-powered fishing boats came on the scene in the early 1900s (steam was eventually replaced by diesel). There are both mid-water trawls, cone-shaped nets towed behind a boat to catch fish in the middle and upper depths, and bottom trawls, called **otter trawls**, that are designed to catch **demersal fishes**, species that feed on and just off the bottom. Demersal fishes include cod, haddock, and flounders, for example. This form of fishing is also called dragging. The otter trawl is a conical net with a wide opening that is kept open while being towed by two otter boards, positioned such that they pull the nets apart to either side (see Figure 14.8C). The boards usually drag along the bottom and resuspend bottom sediments, creating a "mud cloud" that helps to direct fish into the center of the net opening. The net itself has a chain called a footrope on the lower half of the net opening and a row of floats along the top, called a headline. The footrope weights the bottom of the net, and the headline buoys the top, so that the net stays open as it is towed. The footrope is usually equipped with "rock hoppers," wheels made of heavy rubber, that allow the net to ride up and hop over large rocks on the bottom, which might otherwise tear or completely destroy the net. The end of the net, where the fish collect, has traditionally been called the *cod end*.

The invention of otter trawls in the 1900s was a significant technological advance in commercial fishing, but their use has drawn criticisms from marine ecologists and environmentalists, not just because they make overfishing so easy, but because of their impact on the benthos. That impact became even more pronounced following the development of rock hoppers, which allow draggers to fish in areas not possible before. Not only are more fish being caught, but the physical actions of the nets on the bottom disturbs the marine benthic communities that constitute habitat and food for the very same fishes being fished, which, ecologists argue, ultimately reduces production of commercially exploited species. Several studies have documented the destructive

(A) Gill nets
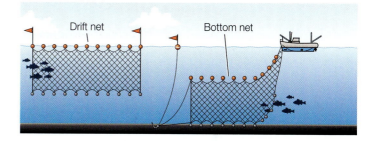

(B) Purse seine
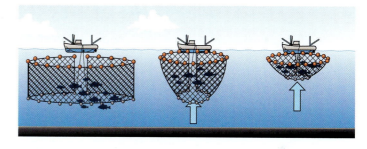

(C) Trawl
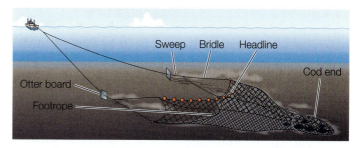

FIGURE 14.8 The basic commercial fishing methods. (A) Gill nets. A surface drift net (left), and an anchored bottom net (right) being deployed. (B) A purse seine surrounding a shoal of fish, then being pursed with a bottom drawn line. (C) An otter trawl.

impact of trawls on delicate benthic organisms such as sponges, anemones, tube worms, and soft corals. Benthic habitat destruction became a hot-button issue a few years ago as more evidence was collected by researchers monitoring bottom habitat using ROVs, which showed convincingly the before-and-after effects of bottom trawling. Some concerned environmentalists were calling for a complete ban on otter trawls, or, at the very least, closing specific areas to fishing, making them marine sanctuaries.

The rapid advancement in fishing technologies, the development and widespread use of the otter trawl, and the development of at-sea refrigeration and freezing technologies all contributed to greater commercial harvests of fish. This was good, because the population was growing rapidly, and the demand for fish was growing as well. But with a near-complete lack of effective management actions, we quickly began to see case after case of stock depletions around the world. Perhaps the most egregious example of overfishing coupled with mismanagement of the resource was the Georges Bank and New England groundfish fishery over the past century. That fishery was not only fished hard by the U.S. domestic fleet, but it was also fished by a foreign fleet of even more efficient fishing vessels; the results were nearly catastrophic for fish stocks on Georges Bank, which have only recently begun to show signs of recovery.

THE GEORGES BANK GROUNDFISH FISHERY The Georges Bank groundfish fishery—one of the most important and biologically productive in the world—was showing signs of impending collapse during the 1960s. The landings for the region (fishery landings records are combined for New England coastal and offshore waters, which include Georges Bank) were dropping precipitously, apparently as the result of overfishing. For decades, the fishing effort—mostly on Georges Bank—had been growing steadily. The post-World War II years, especially the late 1950s and 1960s, witnessed a veritable explosion in fishing activity on Georges Bank, mainly by the distant water fleet, from Eastern Europe and the Soviet Union (also known as the "foreign fleet"). Their huge factory ships and trawlers were catching and processing at sea enormous numbers of fish (**FIGURE 14.9**). Georges Bank, especially, was fished hard. Although it was recognized by the U.S. fishing industry that these foreign vessels were removing vast quantities of fish, there was very little, if anything, that could be done to curtail that effort; the management policies at that time, which were little more than international agreements, were largely ineffective. Those management policies were developed in 1950, when the various nations fishing on the continental shelves off Canada and the United States recognized the potential to deplete the fish stocks. In response, they formed the International Commission for the Northwest Atlantic Fisheries, or ICNAF, which was to protect and conserve the fish resources of the Northwest Atlantic using principles of modern fishery science. Well, it didn't work; one only has to look at what happened to the U.S. landings between the years 1950 and 1970 to draw that conclusion (**FIGURE 14.10**). It was obvious that the fishery was not really being managed at all. The foreign trawlers

(A)

(B)

FIGURE 14.9 (A) A Soviet factory ship and smaller trawler off the U.S. East Coast in the 1960s. (B) A cod end of a trawl net full of Alaskan pollock being emptied.

(A)

(B)

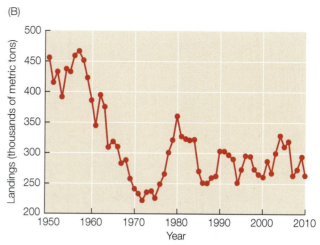

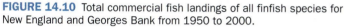

FIGURE 14.10 Total commercial fish landings of all finfish species for New England and Georges Bank from 1950 to 2000.

went after almost everything that was out there, although their focus was on haddock, cod, flounders, and herring.

The point here is simple: a lot of fish were being caught, and they were being taken faster than the populations were able to replace themselves through growth and reproduction. The commercial catch was falling off dramatically, and the U.S. domestic fishing industry was calling for action. Unfortunately, there wasn't much that we could do, because we could not legally regulate fishing on Georges Bank—or anywhere else beyond the 12-mile limit of the legal jurisdiction of the United States. Twelve miles is not very far: ships 12 miles out are still visible from shore. And so without effective regulations in place, we saw what generally became known as a **Tragedy of the Commons**.[8] Simply stated, this means that although *all* of us, the general public and everyone in the commercial fishing industry, shared the *losses* associated with fishing down this important natural resource, we did not all share in the *profits* from that catch. Fishing vessel owners needed only to compare their unshared profits with the overall losses to the fishery shared among everyone else to be motivated to continue fishing. With joint ownership, joint control, joint responsibility for the resource, no one felt compelled to self-regulate and reduce fishing in order to preserve the stocks for future generations; voluntarily doing so would have simply meant that someone else would catch the remaining fish. And so things continued to get worse into the early 1970s. Unless something were done, the U.S. fishing industry was convinced that the stocks would continue to free fall under the heavy foreign fishing pressure. They wanted to prevent—to outlaw—the Soviets and Eastern Europeans from fishing off our shores.

In 1976, the United States expanded its Exclusive Economic Zone (EEZ) from 12 miles to 200 miles offshore as part of the Magnuson–Stevens Fisheries Conservation and Management Act (MFCMA). Foreign fishing vessels could no longer fish anywhere within 200 miles of the U.S. coastline without a permit. The law also laid out a new way to manage fisheries, by creating eight Regional Management Councils which were charged with developing fishery management plans and management measures for the fisheries within their EEZ. The NOAA Fisheries Service approves the plans and is responsible for implementing them. One of those councils is the New England Fisheries

---

[8] Hardin, G., 1968. The tragedy of the commons. *Science* 162: 1243–1248.

(A) Landings

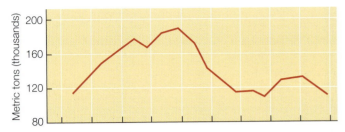

(B) Fishing effort

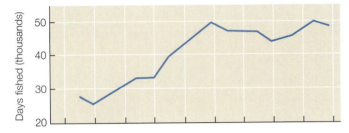

(C) Catch per unit effort

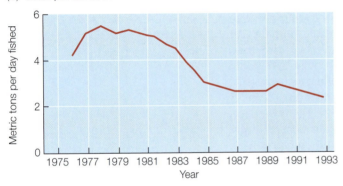

FIGURE 14.11 (A) Total commercial landings of all fish species. (B) Fishing effort in days fished per boat. (C) The ratio of the landings (catch) in metric tons per unit of fishing effort for the New England and Georges Bank fishery from 1976 to 1993.

Management Council, which regulates fishing in New England and on Georges Bank.

After the MFCMA was passed by Congress in 1976, the fishery should have started to recover; the foreign fleet was gone, and we were in control. We could finally act on our fears that the Georges Bank fish stocks were in trouble. And at first glance, that would seem to have been the case. Beginning that year, 1976, and continuing until about 1980, the landings increased dramatically (see Figure 14.10B). But what actually accounted for those increased landings after the U.S. expelled the foreign fleet was not a reduction in fishing pressure and simultaneous recovery of the stocks—it was an intensification of our own fishing efforts. In anticipation of new, un-exploited resources, we expanded and modernized our domestic fishing fleet using existing low-interest government loan programs to build newer and more efficient stern trawlers. It wasn't until analyses were done years later that it became clear that we still had big problems on Georges Bank. Although the New England and Georges Bank landings in the late 1970s had indeed begun to increase, the **catch per unit effort**, the tons of fish landed per fishing vessel per day, was actually declining (**FIGURE 14.11**).[9] U.S. landings went up initially because of the greater domestic fishing effort following expulsion of the foreign fleet, but after reaching a peak in the early 1980s, the landings leveled off and then declined—even though the fishing effort had increased and remained high through the early 1990s. This was a clear and worrisome trend that the newly created Management Council had apparently not addressed. The stocks were not recovering, as it first appeared in the late 1970s. They were perhaps even more depleted than they had been before.

Despite such clear warning signs—declining catch per unit effort in the 1970s and 1980s—the New England Fisheries Management Council remained unable to manage the fishery; fishing continued, virtually unregulated. It was not until the Council was sued in 1997 in federal court by the Conservation Law Foundation that any real management actions were taken to reduce the fishing pressure. Prior to this, it was just too politically painful to regulate fishing to the degree that was actually needed.

One of the problems in formulating effective fisheries management plans then and today is that the Regional Management Councils include political appointees and fishing industry representatives, and there are inevitable conflicts of interest. Fisheries management is just a technical term for reducing or stopping fishing effort and catching fewer fish. What else could it mean? The Council, for whatever reason, did *not* manage the fishery in this sense, as evidenced by the continued dwindling of the stocks.

[9] NOAA, 1995. Status of the Fishery Resources of the Northeastern United States. *NOAA Technical Memorandum* NMFS-NE-108.

Then, to everyone's surprise, the reduced stock sizes exhibited a curious ecological wrinkle with respect to the species composition: the commercial catch changed from "desirable" species like cod, haddock, and flounders to "undesirable" species like skates and dogfish (**FIGURE 14.12**). The species composition had flip-flopped, apparently because of indirect ecological effects of reduced stock sizes. The cod, haddock, and flounders were fished to low numbers, and other species became abundant and took their place in the Georges Bank ecosystem.

We now know, in hindsight, that this flip-flop should not have come as a surprise. If we deplete the total biomass of some species of fish, then what happens to all the primary production that had once been channeled into them? After all, the sun is still shining, providing energy for photosynthesis in the sea, and if anything, the total concentrations of nutrients in the oceans are staying constant, if not increasing (as we will discuss in Chapter 15). This should mean that the total annual primary production in the oceans has not changed, even though the numbers of some species of fish have—so other species respond to that newly available food resource. Of course the question then becomes: once the desirable species have lost their niches to other species, will management efforts—reduction or cessation of fishing—allow them to recover? We don't know.

The cod fishery in Newfoundland, Canada, crashed in the late 1980s and early 1990s, which prompted—perhaps too late—a 1992 ban on fishing in the coastal waters of Newfoundland and on the Grand Banks (see Figure 14.10A), creating heavy unemployment in the region. After some 20 years, the cod fishery remains closed, but the fish have not yet rebounded; still the hope is that the stocks will recover. Will cessation of fishing allow the cod stocks in Canada to rebound? A dramatic example of what could happen to stocks when fishing pressure is minimized is given in **FIGURE 14.13**. Depleted cod and haddock stocks in the North Sea, in the northeastern Atlantic between Great Britain, Scandinavia, and northern Europe, appeared to recover soon after two protracted periods of minimal fishing pressure—when fishing was all but halted by World Wars I and II.[10]

[10]J. A. Gulland, 1974. *The Management of Marine Fisheries* (Seattle: University of Washington Press).

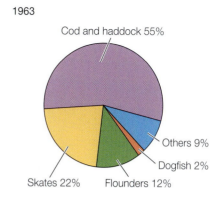

1963

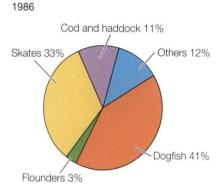

1986

FIGURE 14.12 The change in species composition on Georges Bank from 1963 to 1986.

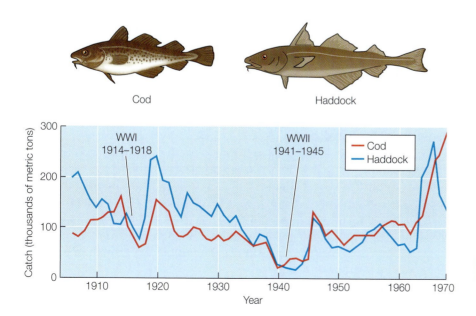

Cod

Haddock

FIGURE 14.13 Landings of cod and haddock in the North Sea from 1906 to 1969. The years of World Wars I and II are shown.

Principles of Fishery Science

Fluctuations of fish stocks in relation to natural variability and human exploitation of the resources are the subject of the field of **fishery science** (also called fishery biology). It is a complex field that includes the basic biology of fishes, population dynamics, and oceanography, but we can break it down into a few fundamental premises:

- We know that young fish grow, but many die before they reach either catchable size or reproductive maturity.

- We also know that natural causes of death include starvation, predation, and combinations of the two.

- If the fish do not die at a young age, they may eventually grow large enough to be captured by fishing gear, and thus they are "recruited" to the fishery.

- Recruitment—that is, the number of fish of the youngest age class to reach harvestable size—is often highly variable between and among years, and it is very difficult to predict future levels of recruitment. We aren't sure what causes good and bad year classes.

- Fisheries catch fish. The problem is they sometimes catch too many fish, so the question facing fishery scientists and fisheries managers, is: How many fish can be safely caught without depleting the stocks?

Important aspects of these premises can be reduced to an equation generally known as the *basic fish population model*, which describes the growth (positive or negative) of a population of fish in terms of its total biomass (weight). The total biomass of a population of fish changes from year to year as a result of (1) the growth of individual fish, **somatic growth**; (2) the addition of new fish by **recruitment**, which are younger fish that have become large enough to be captured in the fishery; and (3) fish mortality. The equation is:

$$\Delta P = (G + R) - (M + C)$$

Here, ΔP is the change in population biomass (weight) over a time interval of (usually) one year. The term G is the growth in biomass of the population over the one-year period as a result of somatic growth of individuals, and R is the biomass increase that results from recruitment of new individuals to the population. That is, each year, assuming that the fish spawn once a year (most species do), there will be a year class, or cohort (these words are often used interchangeably) of individuals that have grown big enough to be captured by, or recruited to, the fishery. Those two growth terms $(G + R)$ will be reduced by fish losses over the course of the year, which results from natural mortality (M) and the fish that are caught by the fishery (C).

Growth and mortality

Individual fish grow, both in terms of their length and their weight; this is somatic growth. Populations (or stocks) of fish grow in numbers. If we multiply the number of fish in a population by the average weight of the fish that make up that population, we get a value for what fisheries scientists call the *stock biomass*, which is an important number used in fisheries management decisions. Comparing that stock biomass with the same value computed for the previous year gives us a measure of the population growth in biomass, which can be positive or negative.

Growth of an individual fish begins inside the egg, immediately after fertilization, and continues until the fish dies. That is, unlike most mammals, the majority of cold-blooded animals, including fish, exhibit **indeterminant growth**; their growth slows down, but it doesn't stop. An example is given in **FIGURE 14.14A** for North Atlantic cod; the growth is fastest when the fish are youngest, and slows down more and more as they grow older. Growth curves such as that in Figure 14.14A are often given in terms of fish length, because it is much easier to measure the length of a fish than it is to weigh it on a moving ship at sea. But the relationship between length and weight is usually well known, and fish lengths can easily be converted to fish weights.[11] An example of the growth in weight for the European plaice (a flounder) is given in **FIGURE 14.14B**; notice that the growth rate in weight is greater than the rate of growth in length in Figure 14.14A. Because the increase in weight is a power function of the length (usually the length cubed) the weight of the plaice increases from about 1 milligram as a newly hatched larva, to nearly 10 grams in the first year—a factor of more than 1000! The rate of weight increase (steepness of the curve), like growth in length, drops off noticeably with age, but it doesn't stop.

Populations of fish grow in numbers by reproducing, usually once a year, which, depending on their fecundity (see Figure 11.29), can result in an extremely large number of new offspring that initially, at least, greatly increase the *number* of individuals in the population. But, they die fast. As we learned in Chapter 11, very few newly hatched larvae survive their first year; some do, of course, and they eventually grow to a size that makes them susceptible to fishing gear—at that point, they become recruited to the fishery.

As just mentioned, the growth in biomass of a population of fish is the product of somatic growth of individual fish and the changes in total numbers of those individual fish over time. For example, in **FIGURE 14.15**, we have plotted both a hypothetical growth curve, similar to that in Figure 14.14B, to describe the average growth in weight of individual fish over a six-year period, and a plot of the change in numbers of individuals in a hypothetical year class over the same six-year period, beginning with about 1 million fish some time after hatching as larvae. The change in numbers of individuals follows the principle we discussed in Chapter 11 (see Figure 11.33), in which

---

[11] The relationship is usually given as: $W = aL^b$, where L is length, W is weight, a is a coefficient and b is an exponent, which usually has a value close to 3.

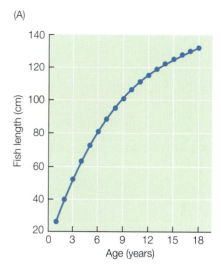

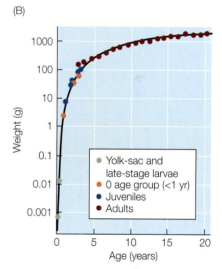

FIGURE 14.14 (A) The average growth curve for North Atlantic cod (*Gadus morhua*), from age 1–18 years. (B) Growth in weight of a different species, the North Sea plaice (*Pleuronectes platessa*), from yolk-sac larvae and late-stage larvae through age 21 years.

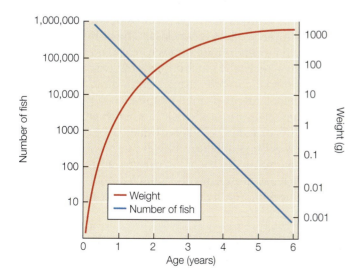

FIGURE 14.15 Hypothetical example of changes in numbers of individual fish in a year class over a six-year period, beginning as about 1 million larvae, plotted on a log scale (blue), and the somatic growth in weight over that period (red), also plotted on a log scale.

there is a constant mortality rate that reduces the population numbers over time. That mortality rate varies by species, and is generally inversely related to the fish's fecundity. The more eggs (and larvae) produced by an adult female, the higher the mortality rate. In the example in Figure 14.15 we are using a mortality rate of about 1% per day. This means that 1% of the fish in our hypothetical population dies on day one; then, of the survivors, 1% die on day two, and so on. The mortality rate applied in the plot here is assumed to be constant over the six-year period, and does not change with the age of the fish; therefore it produces a straight-line decrease when plotted on a log scale. It shows that every year about 90% of the fish die. For example, to begin, at year 0, there were just under one million young fish larvae, but a year later, that number is only about one-tenth the starting value; there are only about 100,000 fish surviving to the next year, which means that 900,000 died. The number of fish surviving to age 1, multiplied by the average weight of an individual fish at age 1, will give a stock biomass for that age 1 year class. After another year there will be about 10,000 survivors, for example; that number multiplied by the weight of age 2 fish gives the stock biomass for that year class, and so forth.

Notice in Figure 14.15 that, although the numbers of fish dropped by about 90% each year, the initial growth in weight easily compensates. In the first year, the numbers decreased to one tenth their starting value, but the weight of an individual fish increased from about 1 mg to more than 1 g, more than a 1000-fold increase. Therefore, biomass of that year class increased greatly. The next year, the growth in weight is not nearly as great, but it is still greater than tenfold, which again compensates for only 10% of the population surviving; the biomass of two-year-olds is therefore greater than that of the one-year-olds a year earlier. Unless the mortality rate is especially severe, or the somatic growth rates of individual fish are especially low, we would expect to see an initial increase in the total biomass of a new cohort over the first year or two. That is, while natural mortalities reduce the numbers of fish in a year class, the somatic growth of survivors more than compensates for loss in numbers, so the biomass of the cohort increases over time, but only for a while. Eventually, the rate of somatic growth slows to a point where the mortality rate begins to win; biomass will begin to decrease with time for that year class, and eventually that year class will disappear altogether (of course, or the sea would fill up with old fish). In our hypothetical example, the increase in biomass for that year class slows down by year two, and then begins to decrease.

Fisheries scientists follow such changes in the biomass of year classes over the life span of the fish to estimate the total biomass of a fish stock, and to determine the mortality in each cohort (year class). The numbers used in the example just given would be determined for each year class in the population by conducting surveys—actually catching fish the way commercial fishing boats do, but in a systematic fashion using standardized fishing gear and procedures, and then calculating their total abundance in the particular fishery area.[12] They determine the ages of the fish they catch either directly, by counting annuli in otoliths or scales, or they estimate ages based on the fish lengths, using data such as those in Figure 14.14A.

Mortality rates are also determined for each year class based on the changes in numbers over time for each age group, or cohort, such as that given by the slope of the blue line in Figure 14.15 for the numbers of fish in our hypotheti-

---

[12] In addition to systematic surveys, fisheries-dependent sampling also is done, whereby fish are sampled at ports as the fishing boats come in with their catch. Sometimes sampling at sea is also done, whereby scientists go to sea with the fishing boats and count and measure fish in the catch. This at-sea sampling also allows the estimating of bycatch.

cal fish population. When applied to all the age groups in the population, plotting the abundances of fish by their ages, scientists can get an estimate of the average mortality rate of the entire stock, and not just one particular year class. An example of the total mortality rate for the North Sea haddock stock is shown in **FIGURE 14.16**, based on the numbers of each year class caught in standardized survey trawls. The older the fish, the fewer that are caught; every year some die. The slope of the line in Figure 14.16 gives the mortality rate for that population.

Maximum sustainable yield

One of the goals of fishery science is to be able to determine how many fish, or more precisely, how many tons of fish, can be taken by the fishery each year, while leaving enough fish to conserve the population. In order to conserve the stocks, the fishery should remove only some portion of the new biomass created each year as a new year class recruits to the fishery and as fish in other year classes grow. That is, the fishery scientists need to determine the level of fishing mortality that, along with natural mortality, constitutes the overall mortality factor $(M + C)$ in the above population dynamic equation, such that ΔP does not become negative. In other words, scientists need to determine how much biomass a stock of fish can yield and still remain healthy and viable. One of the classical ways to assess this yield (there are several other ways) is by using what is called a **stock–recruitment model**. This model estimates a maximum sustainable yield based on the relationship between the biomass of reproductive adult fish (the "spawning stock size") in the population and the numbers of progeny that eventually recruit to the fishery ("recruitment"). The basic principle is fundamentally sound—that the more reproductively-mature adults there are in a population, the more progeny they can potentially produce, and therefore, the more recruits that can be expected to enter the fishery some time later.

An example of this principle is given in **FIGURE 14.17**. It is a plot of the spawning stock biomass of a hypothetical fish population each year over a period of twenty years. Each year, for twenty years, we conduct a survey to determine the biomass of the spawning stock; and then we follow up each estimate a year or two (or three) later, depending on the age at which that species recruits to the fishery, with a survey to estimate how many progeny from that spawning stock have been recruited. The 20 data pairs are then plotted.

Recruitment values in our hypothetical fish population over the twenty-year period range from a low of about 0.5 biomass units to a maximum of about 1.8, and spawning stock biomass values of the reproductive adults range from less than 0.1 to more than 1. A curve is fitted to the data that indicates, initially, increasing recruitment with increasing spawning stock biomass. This makes sense intuitively: the more spawning adults there are, the more progeny they will produce, and therefore the greater the likelihood that a greater number of them will recruit to the fishery. But the stock-recruitment curve in Figure 14.17 does not sustain its initial upward trajectory; the curve reaches a peak and then declines, indicating that there is an intermediate spawning stock size that produces the greatest recruitment. Larger stocks than that optimum produce fewer recruits, which

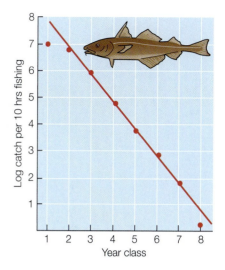

FIGURE 14.16 Abundances of haddock (*Melanogrammus aeglefinus*) in the North Sea for each year class, averaged over the years 1923 to 1938. Abundance is given as the natural log of the number of fish caught in ten hours of trawling. The relatively low abundances of age 1 fish is likely due to their not having grown large enough to be caught efficiently. The line of decreasing abundances of fish in older year classes is a measure of the annual mortality rate.

FIGURE 14.17 Stock–recruitment plot for a hypothetical fish stock, where the biomass of recruits is plotted against the biomass of the corresponding adult biomass that produced those recruits.

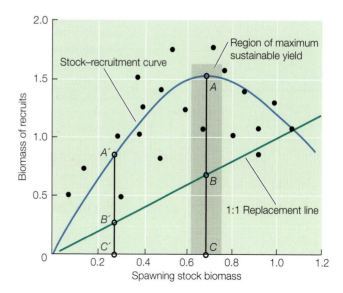

may seem counterintuitive. This reduction in recruitment at larger stock sizes is thought to result from cannibalism—adults feeding on their progeny at times of high population densities—as well as from other population density–dependent mortalities that accompany high population levels.

Figure 14.17 also shows the 1:1 replacement line, which identifies the recruitment biomass required to replace the parent stock over the course of the life of a fish in order for the population to be maintained. Recruitment levels above that line represent surplus fish that could theoretically be safely removed by the fishery while not impacting the overall stock size. An analogous situation would be like a flock of sheep eating only the growing tips of blades of grass as they exceed some predetermined length. The total biomass of grass present at any one time will therefore remain unchanged, while the field continuously grows and produces a supply of food for the sheep. In this grass analogy, the sheep keep the plant's total biomass constant, eating only the growth that exceeds that. In the case of our hypothetical fish stock, the theoretical maximum yield that will allow the population to be sustained is only the recruited biomass above the 1:1 replacement line, which is greatest at a spawning stock biomass of about 7 units (point C in Figure 14.17). We can safely remove a biomass equal to the difference between point A and point B. On either side of that optimum on the stock-recruitment curve, the theoretical yield is less, such as shown for points A′ and B′ at the lower stock biomass level at C′, for example. Maintaining the adult spawning stock size at a biomass level equal to 7 units in our example would be the goal of fishery managers. When the estimated stock size varies from that target value, then the managers would react by making an additional compensation in the allotted catch the following year.

In theory, the stock–recruitment model should work fine. But in practice, there are problems. For one thing, notice the noise in the data in our example in Figure 14.17, which is very similar to actual data for most fish stocks. The stock–recruitment curves that we plot for most actual fish stocks have large errors associated with them. This means that an allocated catch for a particular year, a value set by fisheries managers that places an upper limit to the fisheries landings for that species for that year, could be in error—by a lot. Establishing allocations based on noisy survey data could mean that allocations are set too high, which could make ΔP in our equation negative—we would be fishing down the stock. Worse, we would not even know about this error until some years later, when the newly hatched progeny of that year eventually recruits to the fishery.

For many species, the stock–recruitment relationship is not just noisy, it is almost nonexistent. All we can be sure of is that the relationship, the plotted curve, must pass through the origin, of course—a spawning stock size of zero will produce zero recruits—and we can be sure that at some larger stock size, there will be some recruits. This leaves the question: Why is the logical relationship between parents and offspring so unreliable? In other words, why is there so much variability in recruitment?

Recruitment variability

If we look back at the Pacific sardine landings off California (see Figure 14.5) and the total fisheries landings for New England and Georges Bank (see Figure 14.10), and focus on the more recent years plotted, we cannot help but notice the large variability in landings, which differs among years by anywhere from 30 to 100%. Those landings are directly proportional to ΔP in our fish population dynamics equation—the differences in total biomass of the stocks between years—which means that the three factors in our fish population dynamics equation, G (growth), R (recruitment), and $M + C$ (mortality) must themselves vary from year to year. And they do, but they don't all vary equally, and the variability of each is dependent on many environmental and ecological factors that are difficult, if not impossible, to predict.

Growth (G) would seem to be less likely to vary much between years. Fish growth is primarily a function of water temperature (growth rates are generally faster in warmer waters) and food supply. Average annual water temperatures, while variable, of course, do not normally vary significantly over short time scales within the areas where the stocks are fished. As for food supply on fish growth, that can be important, and it can be highly variable for planktivorous fishes, especially, such as the herrings, sardines, and anchovies. These species are likely to experience year-to-year variability in their growth because the production of their planktonic food is closely tied to local oceanographic processes, which in turn are often tightly coupled to variable weather phenomena. On the other hand, other fishes that feed on higher-trophic-level organisms, as most demersal fishes do, such as the cod, haddock, and flounders on Georges Bank or in the North Sea, may eat organisms that are more than one or two years old, and the variability in abundances of those prey organisms will be damped and less likely to track closely the year-to-year fluctuations in plankton production. Nonetheless we have seen that growth does vary among year classes and years as a function of fishing pressure. Curiously, when fishing pressure is intense and population levels are reduced, the fish reach reproductive maturity at younger ages and at smaller sizes and the overall average growth rate is reduced.

The mortality factor has two components: natural mortality (M) and fishing mortality, or catch (C). Fishing mortality is, in theory, managed by fishery managers, and should be fairly well understood for each fishery. Natural mortality is most likely to be the result of predation, which is unlikely to vary significantly between years, because it is directly proportional to the abundances of predators, which depend on the numbers of their prey species, the commercially exploited fish in question. And so there is likely a predator–prey oscillation that is damped in populations with several year classes. Although there are cases of unusual natural mortalities of fishes, such as during El Niño and other climate-related oscillations (see Appendix B), these are at least recognizable to fishery scientists when they do occur. But unusual peaks in natural mortalities of adult fish are just that, unusual.

Recruitment (R), on the other hand, is highly variable, as the scatter of data points in Figure 14.17 attests. In other words, the numbers of the youngest age classes of fish entering the fishery are variable from year to year, which makes sense. After all, for most fish species, this age group will have just managed to get past a series of developmental hurdles: hatching from eggs to begin immediately a life in the plankton relying initially on remaining yolk reserves for nutrition; and then, once that reserve is depleted, having to capture planktonic food particles while undergoing various anatomical and physiological transformations. After developing as larvae in the plankton, these young fish metamorphose to the juvenile stage, at which time they resemble miniature adults; that transformation also brings with it, for many species, another lifestyle change, as some may drop out of the water column to begin life as bottom, or demersal, fish, while others begin their adult pelagic lives. Regardless, they all switch food sources, from smaller to larger forms of plankton and nekton. Along the way from egg to young adult fish, it shouldn't surprise us that there will be episodes—critical periods of transition—that bring high mortalities. In particular, the numbers of larvae that survive the transition from the yolk-sac larval stage to the first-feeding larval stage may set the recruitment level for that year class, and as such, the dynamics of the larval stage have been the focus of much of the research on recruitment over the past century.

Recruitment variability in some fish stocks runs the gamut from all-out recruitment failure in some years, with very few recruits, to exceptionally strong recruitment in other years. For some stocks, low recruitment almost seems to be the norm, but every few years there is exceptionally strong recruitment. In some cases, those strong year classes sustain the fishery for several years afterward (as Figure 11.33 shows for the North Sea herring fishery). But the underlying causes of strong year classes and year class failures in various fisheries are not well understood.

The idea of a **critical period** in the early life history of fishes was first offered in 1914 by **Johan Hjort** (1869–1948), who suggested that during the transition from a yolk-sac larva to a first-feeding larva there could be, and usually are, very high mortality rates.[13] As we discussed in Chapter 11, most fish hatch with underdeveloped mouth parts, unable to feed on exogenous food particles; most won't even have a mouth for several days after hatching. For a while, they rely on yolk reserves that remain from their egg stage. Eventually that yolk reserve is exhausted, the larva's mouth is developed, and it must now find food particles (plankton) to capture and ingest, or it will starve. Those planktonic food particles for larvae of many species of fishes are the developmental stages of copepods, especially nauplii, and it is this connection—between first-feeding fish larvae and their zooplankton prey—that is the make-or-break point for survival.

Building on Hjort's ideas, **David H. Cushing** (1920–2008) of the Lowestoft Fisheries Laboratory in England, offered a particularly attractive idea in his 1975 book.[14] He pointed out that many species of fishes time their reproductive cycle to coincide with the plankton production cycle, such that their progeny hatch and begin to feed at a time of abundant food items. This basic principle was not new; it has long been known to aquarists that once larval fish hatch from the eggs and exhaust their yolk supplies, the presence or absence of adequate densities of food particles will dictate their survival rates. The same ideas seem to apply to marine fishes in that they time their reproductive cycle to match the timing of the spring phytoplankton bloom, such that larvae have available to them at the time of first-feeding adequate densities of zooplankton, especially copepod nauplii. Should the timing of the spring phytoplankton bloom and the zooplankton response to that bloom not coincide with the timing of first feeding, there will be catastrophic mortalities of fish larvae. And, it was already known that the timing of the spring phytoplankton bloom did vary from year to year, as influenced by variable light levels (cloudy years versus sunny years), or years in which the waters are late to stratify because of stormy conditions (e.g., greater vertical wind mixing). Known as the **match-mismatch hypothesis**, it accounted nicely, in theory, for good and bad years of recruitment. But it didn't always work. Variable plankton dynamics didn't always explain recruitment numbers. We later learned that variable survival rates of young fish extend beyond the larval period to the juvenile stage. Although the full story of what controls variable recruitment has yet to be told, and research into the "recruitment problem" in fisheries continues, it is abundantly clear that recruitment is determined by a number of environmental and biological factors, not just by the size of the adult spawning stock.

---

[13] Hjort, J., 1914. Fluctuations in the great fisheries of northern Europe viewed in the light of biological research. *Rapports et Proces-Verbaux des Conseil International Pour L'Exploration de la Mer* 20: 1–228.

[14] D. H. Cushing, 1975. *Marine Ecology and Fisheries* (Cambridge, UK: Cambridge University Press).

Current Status and Management of Fisheries

Fisheries today are a worldwide enterprise, and with the exception of open ocean fisheries in international waters, each coastal nation manages the fisheries within its EEZ, which extends 200 nautical miles from shore. Some 90% of the world's fish catch is from coastal and shelf waters that fall inside that limit. Among more than 150 coastal nations, each with a fishery of some size, there are obviously a wide range of management practices; some have none at all, while others enact strict measures, including full-out closures of at-risk fisheries. The need for more effective management practices throughout the world ocean will only grow in importance as we continue to fish the seas. As we can see in Figure 14.3, the world's fisheries landings doubled between the years 1965 and 1995, but since then, the landings have been declining. A significant fraction of that catch is increasingly being made by developing countries. Fisheries statistics compiled by the Food and Agriculture Organization (FAO) of the United Nations show that the majority of the world's landings come from the ten regions, identified in **FIGURE 14.18**, based on landings data for 2008.[4] Of those ten regions, the majority of fish landings (including squid and shrimp) are from the Northeast Pacific Ocean, and are landed primarily by China and Japan. The fisheries landings off the shores of North America accounted only for about 4.6 MMT in 2008, as compared with the China's 14.8 MMT; the world total in 2008 was just over 80 MMT. International management agreements will be needed, but they probably won't come easily, if our experiences are any indicator; it has been hard enough for us to manage our own domestic fisheries.

Management of marine fisheries in the EEZ of the United States falls under the purview of the NOAA Fisheries Service, also known as the National

FIGURE 14.18 Map of the principal marine fishing areas of the world ocean, as defined by the FAO of the United Nations, with marine fish landings in 2008 (in MMT) listed for the top ten areas, along with the predominant species.

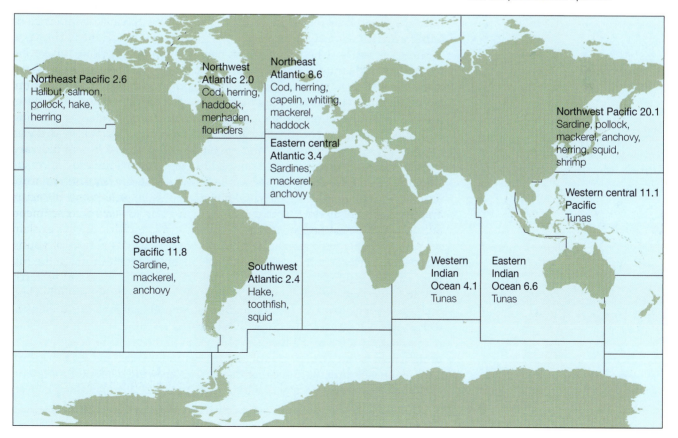

Northeast Pacific 2.6
Halibut, salmon, pollock, hake, herring

Northwest Atlantic 2.0
Cod, herring, haddock, menhaden, flounders

Northeast Atlantic 8.6
Cod, herring, capelin, whiting, mackerel, haddock

Eastern central Atlantic 3.4
Sardines, mackerel, anchovy

Northwest Pacific 20.1
Sardine, pollock, mackerel, anchovy, herring, squid, shrimp

Western central 11.1 Pacific
Tunas

Southeast Pacific 11.8
Sardine, mackerel, anchovy

Southwest Atlantic 2.4
Hake, toothfish, squid

Western Indian Ocean 4.1
Tunas

Eastern Indian Ocean 6.6
Tunas

Marine Fisheries Service; this is the federal government agency responsible for managing the nation's marine fisheries in federal waters (from 3 to 200 miles offshore). The states are responsible for managing fisheries in nonfederal coastal waters inside 3 miles (Texas state waters extend out to 9 miles, as do Florida state waters in the Gulf of Mexico). NOAA management policies must comply with directives spelled out in the Magnuson-Stevens Fisheries Conservation and Management Act (MFCMA) of 1976 and amendments made in its reauthorizations in 1996 and 2006. But it is the regional councils, created under that act, that produce the Fisheries Management Plans, with scientific advice from NOAA and input from the general public. The plans must not only account for biological aspects of the fishery, but they must also consider social and economic concerns. And therein lies the problem.

Management of fisheries is really the management of the fishing industry, of actions by people, and the goals of management—to conserve the fish stocks for future generations, while at the same time maximizing the yields—have proven to be very difficult to achieve. Management plans in the 1970s were mostly limited to the protection of spawning grounds (closing the areas to fishing during spawning seasons) and some size limits (e.g., by regulating the mesh sizes of trawl nets, to allow smaller fish to escape), but they didn't work. They didn't work because they didn't go far enough. Effective fisheries management means making hard social and economic choices; it means reducing the fishing pressure (reducing the catch) of depleted stocks. And that is best done by setting limits, or allowable catch quotas, above which no more fish from that stock can be landed in a given season. The setting of allocations, or catch quotas, requires balancing the economic interests of those involved in the fishing industry with conservation needs. Hard management decisions are becoming more accepted today than they would have been a few decades ago, and there are bright spots on the horizon for a number of once-decimated fish stocks.[15] We have to operate under the assumption that the stocks can indeed be conserved for future generations, even though it is clear that those fish stocks will not meet future demands for seafood.

In recent years, fisheries managers have been directed by law (under the reauthorized Magnuson Act) to begin what is being called **ecosystem-based management**. This approach requires managers to include all aspects of the environment of a targeted species that might affect the population, such as the role of predation on that species, competitive interactions with other species, and the dynamics of prey organisms of the target species. It also includes possible oceanographic effects, and of course, the effects of fishing pressure itself. Although this approach is only a little more than a decade old, it has still to be incorporated into fisheries management plans in a meaningful way; part of the problem, as you may already have guessed, is that most scientists and managers really don't understand how to go about such a complicated and all-inclusive management scheme.

Seafood today makes up only about 1% of the world's food supply, but it supplies about 16% of the animal protein consumed by the world's population. Of that seafood, the majority, about 84%, is fish (termed *finfish* in the seafood industry) and the remainder shellfish, which includes molluscs and crustaceans. The demand for this high-protein food is expected to grow, but the prospects for any future growth in landings are dim. While it is difficult to predict how the world's fisheries will adapt to the growing need for high-protein food, as the world's population grows and the demand for fish grows with it, it is not difficult to see that the world's growing demand for seafood cannot come from the oceans' wild stocks. Not only have landings leveled

---

[15] Worm, B. et al., 2009. Rebuilding Global Fisheries. *Science* 325: 578–585.

off, but the capacity for expansion is limited; some scientists estimate that the majority of the world's marine fish stocks are already being exploited to their maximum extent.

Aquaculture

With more effective fisheries management, there is hope for the world's fish stocks, that we have not yet fished them to the edge of extinction. Conservation measures must continue, not just in the United States and Canada, but around the world. But one only has to look at the rate at which our population is growing to realize that, no matter how we manage our fisheries, we still must face the sobering conclusion drawn by John Ryther in 1969: total fish production in the oceans is finite. Our total landings are limited. And as we have shown at the start of this chapter, we appear to have been brushing up against that limit since 1995. If we want to increase the world's production of seafood, we must look to **aquaculture**, the artificial production of marine organisms for human consumption.

The world's aquaculture industries comprise the fastest-growing sector in the food production industry (**FIGURE 14.19**). Aquaculture production has grown from almost nothing in 1950 (less than 1 MMT) to more than 55 MMT in 2008, which is now well over half of the world's marine fisheries landings. Actually a very old practice—the Chinese have been raising freshwater fish in ponds for more than 1000 years—the expansion of this industry has been remarkable, and it is growing at a rate that is three times the growth rate of the world's meat production. In addition to the production of marine and aquatic animals, the production of aquatic plants, more than 99% of which is seaweeds, was nearly 16 MMT in 2008, primarily in China and Southeast Asia.

More than half of the marine and aquatic animal production in aquaculture is freshwater fishes, mostly carp, but production of marine molluscs, such as oysters, clams, mussels, and scallops, is increasing steadily. Marine fish species (mostly flatfishes such as turbot, halibut, and sole, but also cod), and diadromous fishes (dominated by Atlantic salmon) remain only minor components of the overall aquaculture industry.

Maintaining large volumes of seawater away from the coast is impractical; recall our example of how much salt it takes just to make two cubic meters of sea water—about 70 kg. Thus, it is logical that the majority of aquaculture species are freshwater species, which makes it more feasible to have inland operations, away from the coast. In the U.S., catfish, a freshwater fish, dominates overall aquaculture production in terms of both weight and dollar value, with crawfish and trout trailing far behind.

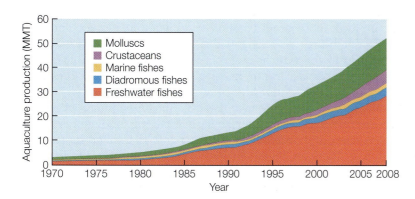

FIGURE 14.19 World aquaculture production from 1970 to 2008, given in metric tons.

FIGURE 14.20 (A) A line of oyster cages of growing oysters suspended at the surface. (B) A rope choked with blue mussels ready for harvest. (C) An oyster grower cleaning and sorting his product for market.

Marine aquaculture operations in the U.S. today grow mostly shellfish; oysters lead the industry in weight and dollar value, but clams and mussels also contribute. The technology is fundamentally simple: animals can be grown either suspended in cages (**FIGURE 14.20**) or on the bottom in shallow waters, a practice that is regulated with special bottom lease site permits. Either way, the filter-feeding shellfish rely on the on the abundant natural production of plankton in near-shore waters.

Marine finfish aquaculture lags behind other forms, in the U.S. and around the world. The technical difficulties are far greater than those for shellfish or for freshwater fishes. For one thing, fish are motile and can swim away, whereas shellfish tend to stay put. Most marine and diadtromous fish aquaculture operations use net pens to hold fish in protected, near-shore areas that allow easy access (**FIGURE 14.21**). The pens are usually covered with nets to keep sea birds out, and to prevent the fish from jumping out. The fish are fed processed food until they reach a marketable size.

Finfish aquaculture, both freshwater and marine, is also fundamentally different from shellfish aquaculture in that the fish have to be fed; food for the cultured animals is added, rather than produced in situ, as is the case for shellfish. Shellfish aquaculture can be viewed, then, as extractive, in that the animals feed on the natural production of phytoplankton and zooplankton, while finfish aquaculture can be viewed as additive, requiring food to be added to the fish enclosure, which adds another layer to technical difficulties. Not only must food be added, but there will always be an imbalance between the amount of nutrient elements added to the system in the fish feed and the amount actually incorporated into new fish flesh that is eventually removed when they are harvested. Not all the food given to them (e.g., scattered as pellets into the pens) is eaten, and even of the food eaten, not all is incorporated into new fish flesh (e.g., there are metabolic losses, such as urine and feces); that is, nutrients will be excreted by the animals. In the case of freshwater ponds and other enclosures, the potential buildup of nutrients has to be monitored and those nutrients removed. In the case of net pens in marine waters, the nutrient flux is released directly into the environment which, depending on the mixing rates in those environments, may or may not lead to local eutrophication, or nutrient enrichment and nuisance phytoplankton blooms. This is one of the reasons that net-pen fish farms are best located in coastal areas that have sufficient natural circulation and mixing so that the lost nutrients are harmlessly dispersed. For example, Atlantic salmon farms in Maine and New Brunswick, Canada, are located in waters that are tidally well mixed, where there is sufficient turbulent dispersion of the nutrient fluxes from the pens.

Potential nutrient enrichment is only one of the reasons that marine finfish aquaculture has been scrutinized; the industry faces a number of technical challenges.[16] One of them is the need to grow fish in pens that sit adjacent to the shoreline, which, in the U.S., has sometimes been met with opposition from coastal property owners who object to obstructed views. There are also objections that such pens interfere with other fishing activities, especially near-shore lobster and crab fishing activities. And there are questions of chemical additives, such as the use of antibiotics in fish feed, and the rapid spread of various diseases among the penned fish, whether this poses a threat to the wild populations. But the sharpest criticisms leveled against marine fish aquaculture is that we are catching wild fish to feed the aquaculture finfish. And that problem stems from the fact that the marine species targeted by the finfish aquaculture industry are carnivores, such as salmon, flounders, sea bass, sea bream, cod, and tuna, which require animal protein in the form of fish meal as well as fish oil. Most freshwater finfish species raised in current aquaculture operations, on the other hand, are herbivores or omnivores such as carp, talapia, and catfish, that can be raised on terrestrial plant-based feeds, and are therefore not dependent on the very same marine finfish resources we are trying to supplement. For the marine finfish aquaculture industry to grow beyond its current production levels, which are far lower than those of freshwater species (see Figure 14.19), we must develop alternate feeds that do not depend on wild-caught fish to feed aquaculture fish.

So what is the future of aquaculture, particularly marine finfish aquaculture? Well, to make the most of aquaculture, for it to meet our needs and expectations, we must find a way to reduce our dependence on fishing itself in order to feed the fish we propose to raise via aquaculture, and to do that, more research will be needed. But despite what may at first appear to be a long list of problems facing the growth of aquaculture, the future of the industry is bright and it is expected to continue to grow at a phenomenal rate. It has to. The future of aquaculture, especially marine finfish aquaculture, will be driven by a growing demand for fresh seafood, a demand that is currently outrunning supply; world population growth is not matched by increases in aquaculture production, which is already growing at a phenomenal rate, and the growing demand for fish most certainly cannot be matched by fisheries landings.

(A)

(B)

(C)

FIGURE 14.21 (A) Net pens on a salmon farm in eastern Maine. (B) A fully grown Atlantic salmon ready for market. (C) Salmon food pellets.

---

[16] Naylor, R. L. et al., 2000. Effects of aquaculture on world fish supplies. *Nature* 405: 1017–1024; and Powell, K., 2003. Eat your veg. *Nature* 426: 378–379.

Chapter Summary

- Marine fisheries have shown signs of overfishing and stock depletions for several hundred years in Europe, and the search for fisheries resources was part of the stimulus for European settlement of North America. Today, fisheries around the world are showing signs of overfishing.

- Fisheries production can be explained as the passage of organic carbon created by primary producers through marine food chains to commercially exploited fish and invertebrates.

- Based on a trophodynamic estimate of the upper limit to total theoretical fish production, the total **maximum sustainable yield** that fisheries can harvest in the world ocean is 100 million metric tons (MMT).

- World fisheries landings reached a maximum of about 95 million metric tons in 1996 and have been dropping since then. The human population of the world is at 7 billion and growing, and the demand for seafood is fast outpacing fisheries landings.

- Examples, among others, of collapsed fisheries are the California sardine and the northwest Atlantic redfish, which both succumbed to heavy fishing pressure, but each is showing signs of recovery.

- Types of fishing gear include **long lines**, or **tub trawls**, of multiple baited hooks, an old technique still used today; **gill nets** (also called drift nets), **purse seines**, and **otter trawls**.

- Commercial fisheries became much more efficient after the development of steam, and then diesel power, on board refrigeration, and otter trawls with **rock hoppers**.

- The Georges Bank and New England demersal fish fishery was dramatically reduced by the foreign fleet in the 1960s and 1970s, before passage of the Magnuson-Stevens Fisheries Conservation and Management Act in 1976, which extended the U.S. Exclusive Economic Zone to 200 miles. United States fisheries are now managed by eight Regional Fisheries Management Councils.

- As a result of diminished stock levels, catch per unit effort on Georges Bank dropped under increased domestic fishing efforts in the 1970s and the catch shifted from desirable species to undesirable species.

- Principles of fishery science can be expressed in a fish population dynamics equation that related stock growth to **recruitment**, **somatic growth**, and mortality; Fish that grow large enough to be captured by fishing gear have recruited to the fishery.

- Maximum sustainable yield can be described with a **stock-recruitment model**, relating recruitment levels to size of the adult spawning stock that produced it; the stock-recruitment curve shows a maximum recruitment at intermediate adult stock levels; fisheries yields are the difference between that curve and a 1:1 replacement line; the stock-recruitment relationship is highly variable.

- Recruitment is highly variable, because of highly variable mortalities or survival of young-of-the-year fishes, especially in the larval and juvenile stages; a **critical period** of survival occurs at the time of yolk-sac absorption and transition to exogenous food sources for fish larvae.

- Fisheries management is difficult because of the need to balance economic interests with conservation of the stocks.

- Aquaculture produced more than 55 million metric tons in 2008, compared with marine fish landings of less than 90 MMT; Most finfish aquaculture is freshwater fish; Most marine aquaculture is shellfish; Aquaculture growth has been rapid since 1950, but it must continue to advance, and to develop new feeds for carnivorous fish to meet the growing demand for fish.

Discussion Questions

1. Why are the coastal and upwelling areas of the world ocean sites of higher fish production than the open oceans? What does the length of a food chain have to do with fish production?

2. What is the maximum theoretical production of marine fishes per year, and how does that compare with current and past commercial fisheries landings? How do the world commercial fish landings compare with the world population growth?

3. What do we mean by the collapse of a fishery? What are some examples?

4. What are the major types of commercial fishing gear? How have they evolved and improved over time? What were some of the most important advances in fishing technology?

5. Why do some ecologists and environmentalists object to the use of otter trawls and drift nets?

6. Why were fisheries not managed properly before the drastic declines in fish stock sizes, on Georges Bank and on the Grand Banks of Newfoundland? How are fisheries managed today? What is the role of the federal government, the states, and the general public?

7. What is meant by "catch per unit of effort"? What are some of the early warning signs that a fish stock in under heavy fishing pressure?

8. How do individual fish grow? How do populations of fish grow, in terms of total numbers and in terms of weight, or biomass? How do fishery scientists determine the biomass and age structure of a population of commercially exploited fish?

9. What is a stock–recruitment curve? What does it tell us about a population of fish?

10. What is recruitment? Why is it so variable from year to year? What are some of the ideas about fish survival that help to explain recruitment variability?

11. What does the future hold for aquaculture? What are some of the social, economic, and technical problems associated with marine and diadromous finfish aquaculture?

Further Reading

Beverton, Raymond J. H., and Holt, Sidney, J., 1957. On the Dynamics of Exploited Fish Populations. *Fisheries Investigations*, Series II, Volume XIX. London: Ministry of Agriculture, Fisheries and Food.

Cushing, D. H., 1968. *Fisheries Biology: A Study in Population Dynamics*. Madison, WI: The University of Wisconsin Press.

Cushing, D. H., 1975. *Marine Ecology and Fisheries*. Cambridge, UK: Cambridge University Press.

Gulland, J. A., 1974. *The Management of Marine Fisheries*. Seattle: University of Washington Press.

Kurlansky, Mark, 1997. *Cod: A Biography of the Fish that Changed the World*. New York: Penguin.

Hannesson, Rognvaldur, 1996. *Fisheries Mismanagement: The Case of the North Atlantic Cod*. Oxford, England: Fishing News Books, Blackwell Science.

Royce, William F., 1996. *Introduction to the Practice of Fishery Science*. San Diego: Academic Press.

CHAPTER **15**

Contents

These beachgoers playing in the waves on a warm summer day probably aren't thinking much about what is in the water. Most of us assume that the oceans are clean, unspoiled, and too big to be adversely impacted by human activities. Why else would so many people be in the water?

Human Impacts

As we have seen throughout this book, we human beings have been using the oceans to meet our various needs throughout recorded history. Their sheer size, covering some 71% of the Earth's surface and extending to depths greater than the Himalayas are high, has made the oceans seem an inexhaustible source of food until relatively recently. We have come to realize that such is not the case—that we can indeed catch too many fish, driving stocks to levels that can no longer be fished commercially, and in so doing, alter the basic ecology of the seas. For centuries, we have also been using the oceans, and the rivers that run into them, as dumps for our refuse—solid trash, human sewage, and industrial wastes of untold kinds. What is ironic is that in so many of these cases, the oceans seem to have absorbed it all, and still they are beautiful—if we don't look too closely. You won't see many people swimming in the world's major harbors, where signs of discharge are clearly visible, but nearby beaches and other coastal areas are a different matter; distance from the source somehow seems to cleanse the oceans, and as long as they look clean, well, then, they must be. The truth is that much of what we have been dumping into the oceans is invisible, detectable only with sophisticated instruments; but the more we look, the more we find, and the more concerned many of us become.

Another, perhaps less obvious, way that we have been changing the ecology of the oceans is by our unintentional transport and introduction of species from their native waters to far-flung areas of the world ocean where they may become established and thrive. The introduction and proliferation of these nonindigenous species sometimes leads to competitive interactions with the native flora and fauna, which in some cases leads to all-out displacement of the native species.

The oceans are not just dumping grounds for our solid and dissolved wastes, they are also sponges that absorb energy; much of the heat from global warming, which has accelerated over the past century, is being stored in the oceans—but not harmlessly. Barely measureable, and completely invisible, the rising heat content of the oceans can alter gyre currents, raise sea level,[1] and affect our weather patterns in ways that most of us are completely unaware of. But the oceans don't just passively absorb the impacts of global climate change, to which they respond according to the laws of nature; they are important players in a number of feedback controls that have damped the effects of global warming—at least so far. And one of the driving forces behind global warming, the increased release of carbon dioxide (CO_2) into the atmosphere, is having another direct and immediate adverse effect on the oceans: those increasing loads of CO_2 in the atmosphere also dissolve in seawater, making the oceans more acidic, to the point that many of the fascinating marine life forms we

---

[1] Eustatic sea level is global sea level; it is a function of the volume of water in the oceans. It is monitored by a global array of tide gauges and by satellite altimetry. Isostatic sea level is a local phenomenon; it too responds to the volume of water in the oceans, but it also responds to isostacy of the continents and other land masses, which, for example, can sink into the asthenosphere, thereby raising local sea level along the coast of that land mass.

have extolled in the earlier chapters of this book are becoming threatened. In this chapter we will review some of the more important examples of marine pollution and global climate change, and what the future holds for the oceans and our intimate connections with them.

Marine Pollution

Marine pollution takes many forms; the most commonly recognized are industrial chemicals, agricultural fertilizers, solid wastes, petroleum products, and human sewage. When the point of origin—for example, a discharge pipe—can be identified, the contaminant is called **point-source pollution** (FIGURE 15.1). When the source is widespread—carried by rain, rivers, snowmelt, or groundwater seepage—the contaminant is called **nonpoint-source pollution**; this includes contaminants such as agricultural runoff and seepage from household septic tanks, which eventually enter streams and rivers, which carry them to the sea.

A number of definitions of pollution have been offered. One of the simplest is: "the introduction by humans, directly or indirectly, of substances or energy to the marine environment, resulting in deleterious effects." The terms *substances* and *energy* would seem to encompass all the possibilities, but it is the *deleterious effects* that hits at the heart of the problem. It prompts the obvious questions: Is their introduction to the marine environment actually harming anyone or anything? If so, how? And how does that harm compare with the cost of discontinuing the practice, or remediating the harmful effects? Here is where politics and science clash with a force that makes fisheries management issues look like quibbling over answers to a homework problem. And things will only grow worse: all of the above-mentioned forms of pollution in the marine environment are increasing in step with our growing world population; how could they not? In the United States alone, some 53% of the population lives in coastal counties—and that number is expected to rise in the future. This means that more than half of our population lives in coastal areas that account for only about 17% of the total U.S. land area (not counting Alaska). So, it is only natural that, throughout history, and continuing today, we would look to the oceans as a convenient, nearby repository for our refuse.

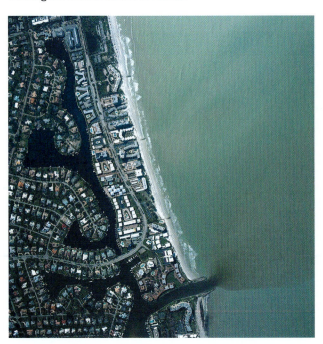

FIGURE 15.1 A visible example of nonpoint-source pollution entering the coastal waters in Florida.

Nutrient enrichment and coastal eutrophication

One of the oldest problems in the world is what to do with our wastes—our own human wastes. The earliest solutions were simple, when there weren't too many of us on the planet: we just threw it outdoors, somewhere. But as we advanced from rural agrarian societies to urban industrial societies, and as cities grew, we began to experience plagues of one kind or another, and we learned that just throwing everything outdoors—into the street gutters—was not such a good idea. At some point we decided to send our wastes down rivers or directly into the ocean. Most cities, after all, were sited on major rivers or coastal harbors. And when the Industrial Revolution hit, Europeans used this natural conduit—the rivers, especially—to rid their local areas of human refuse and unwanted industrial wastes of all kinds. In the 1800s, the Thames River in London is said to have smelled like an outhouse. So, what's the problem with this approach to sewage treatment? After all, don't all animals in the sea do the same? They do. But not all in one area near the coastline or in an estuary where rivers meet the sea. One of the problems with dumping human sewage into rivers is nutrient enrichment.

Sewage, even some forms of treated sewage, is rich in the nutrients nitrogen and phosphorus. Emptied into estuaries, those nutrients are held near the surface in the low-salinity buoyant layer which, with the high light levels, stimulates phytoplankton production. If the waters are shallow, as in the Chesapeake Bay and the Delaware Bay, there can be a stimulation of benthic algae production. Kaneohe Bay, a coral reef embayment on the coast of Oahu, Hawaii, one of the most beautiful marine environments in the world, is a case in point. It became heavily developed after the arrival of the U.S. military in the late 1930s and early 1940s; by the 1960s the population had grown considerably, and sewage was being dumped directly into the bay. But in the early 1970s, the corals started to die. They were being smothered by a green macroalga growing on top of them, the growth of which had been stimulated by the excess influx of nutrients, and the overgrowth of algae was preventing sunlight from reaching the coral's zooxanthellae. Phytoplankton were blooming, too, further reducing the water clarity. Following public outcry, those sewage outfalls were redirected from the bay to offshore in the late 1970s, and by 1983, scientists surveying the area reported that the corals were making a remarkable recovery.[2] At least that was the early diagnosis. In the 1990s, the recovery seemed to have leveled off (before it was complete), and today, algae stimulated by nutrient fluxes from fertilizers used on lawns and golf courses, as well as runoff from roads, have again blanketed the bay's corals, but this time the culprit algae are non-native. While the best method of removing the algae is hotly debated today, nobody disagrees that it is unthinkable to allow this stunningly beautiful marine ecosystem to degrade.[3]

Nutrient enrichment and the resulting stimulation of algal production—whether macroalgae or phytoplankton—is called **eutrophication**, and it is a growing problem around the world. To review all the issues here would be to write another entire book on the subject, but we can briefly discuss a few examples that stand out.

Coastal eutrophication continues to present problems in Europe, especially along the edges of the North Sea, where some of Europe's major rivers empty—rivers that have been polluted for centuries with enriched concentrations of nutrients from human sewage. A particularly striking example is given in **FIGURE 15.2** which shows how the nutrient loads in the Rhine River increased over a 30-year period, and the response to those nutrients by algae in the receiving coastal waters.[4] The nutrient loads in the river doubled between the 1950s and 1980s, as did primary production and biomass of algae in the coastal waters between 1968 and 1982. The good news is that those trends are being addressed by the European nations responsible for polluting the Rhine, and recent reports show that progress is being made and that the nutrient loads in all of northern Europe's major rivers are in fact decreasing.[5]

FIGURE 15.2 (A) Total dissolved phosphorus and (B) total dissolved nitrogen discharged by the Rhine River from 1954 to 1985. (C) Primary production and (D) biomass (as chlorophyll concentration) by benthic algae from 1968 to 1981 in the coastal receiving waters.

(A)
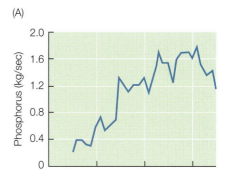

(B)
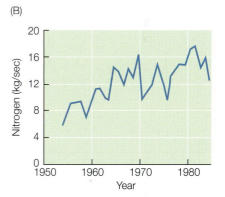

(C)
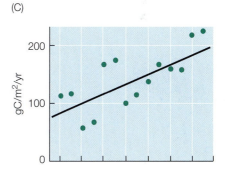

(D)
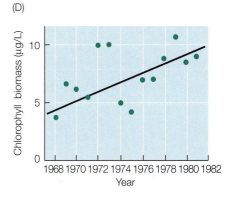

[2] Maragos, J. E., Evans, C., and Holtus, P., 1985. Reef corals in Kaneohe Bay six years before and after termination of sewage discharges (Oahu, Hawaiian Archipelago). *Proceedings of the Fifth International Coral Reef Congress* 4: 189–194.

[3] For more on this, see Smith, J. E. et al., 2008. Response to "Fighting Algae in Kaneohe Bay," *Science* 319: 157.

[4] Nienhuis, P. H., 1993. Nutrient cycling and foodwebs in Dutch estuaries. *Hydrobiologia* 265: 15–44.

[5] Nienhuis, P. H., and Gulati, R. D., 2002. Ecological restoration of aquatic and semi-aquatic ecosystems in the Netherlands: an introduction. *Hydrobiologia* 478: 1–6. Also as reported by the European Environment Agency.

Not only does the overall algal production and biomass respond to increases in nutrients, but the species of phytoplankton that respond to those nutrient fluxes in the receiving coastal waters may be altered by the concentration ratios of nutrients. For example, in the North Sea, the proportions of nitrogen and phosphorus have been altered by pollution, which has stimulated blooms of a gelatinous, colony-forming phytoplankton (*Phaeocystis* sp.). These phytoplankton sometimes reach such high cell densities that their gelatinous matrix, within which they form their colonies, creates obnoxious volumes of foam when they get whipped up by surf on beaches surrounding the North Sea (**FIGURE 15.3**).[6] Nontoxic to humans, the foam is just an annoying side effect of anthropogenic nutrient discharges to the sea, as far as we know. But there are likely to be indirect effects on the basic ecology whenever and wherever the species composition of the primary producers is upset.

Coastal eutrophication is not a problem unique to Europe. The Mississippi River, the largest river in North America, which empties into the Gulf of Mexico, is also polluted. The "muddy Mississippi" has always carried huge loads of suspended sediments to the Gulf of Mexico, but in the 1800s and 1900s, those loads increased as a result of soil erosion that accompanied the expansion of agriculture throughout America's heartland. Then, extensive dam construction in the 1950s, particularly on the Missouri River, which feeds the Mississippi, trapped sediments behind the dams, allowing the particles to settle out, thereby reducing the loads of suspended sediment that reached the Gulf of Mexico. Along with the reduced suspended sediment load, concentrations of dissolved silicate, the nutrient required by diatoms to make their siliceous frustules, declined as well, as the eroding sediments are themselves a source of silicate (as siliceous minerals in the terrigenous sediments dissolve). It was also about this time that concentrations of nitrogenous nutrient loads, especially nitrate, increased in the Mississippi. Concentrations of nitrate (and silicate) in the lower Mississippi had been stable and at presumed normal levels from about 1906 to the 1960s, and probably before that as well (although data are lacking). The increase in nitrate was a result of widespread use of nitrogen- and phosphorus-rich agricultural fertilizers that began in the 1930s and peaked in the early 1980s. The result has been that since 1955, loads of

---

[6] Riegman, R., Noordeloos, A. A. M., and Cadée, G. C., 1992. *Phaeocystis* blooms and eutrophication of the continental coastal zones of the North Sea. *Marine Biology* 112: 497–484.

(A)

(B)

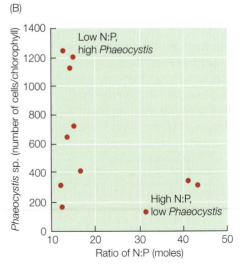

FIGURE 15.3 (A) Foam on a North Sea beach. (Inset) A photomicrograph of a gelatinous colony of the phytoplankton *Phaeocystis globosa*. (B) The relationship between *Phaeocystis* cell abundances, expressed as cell numbers per unit of chlorophyll, and the ratio of dissolved nitrogen to phosphorus.

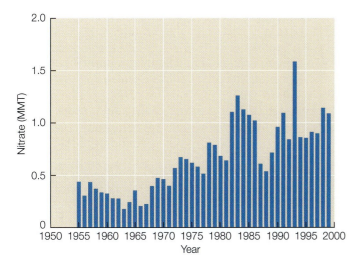

FIGURE 15.4 (A) Inputs of nitrogen, as nitrate, in millions of metric tons per year, to the Mississippi River from 1955 to 1999. (B) The drainage basin of the Mississippi River.

silicate have declined by half, and the nitrogen loads reaching the Gulf of Mexico have doubled (**FIGURE 15.4**).[7] Both nutrient elements (N and Si) are required for diatom growth, remember, but non-diatom phytoplankton do not need silicate. And so the response in the receiving waters in the coastal Gulf of Mexico has been both an increase in phytoplankton production and an altered phytoplankton species composition, with a relatively lower proportion of diatoms. And this ecological change in phytoplankton dynamics in the shelf waters of the Gulf has caused an unusual and serious problem since the 1980s.

DEAD ZONES First documented in 1985, and recurring every summer since then, the northern Gulf of Mexico has experienced extreme depletions of dissolved O_2 in bottom waters, which is a direct result of the dramatic increase in phytoplankton biomass produced in the shelf waters. The phytoplankton populations grow to unusually high cell densities in the nutrient-enriched waters delivered by the Mississippi outflow. Some of that biomass is grazed by zooplankton, but much of it simply dies and sinks slowly to the deep and bottom waters on the shelf. And as all that organic matter sinks, it is decomposed by bacteria—it is respired as described by our "chemical equation of life":

$$CH_2O + O_2 \rightarrow CO_2 + H_2O$$

As you will recall from Chapter 7, this chemical equation is the reverse of photosynthesis, and it consumes oxygen (O_2). Large quantities of decaying organic matter consume large quantities of dissolved O_2 in those waters; this consumption of O_2 is called **biological oxygen demand**, or **BOD**. During the summer, less O_2 can dissolve in the warmer waters anyway, and without the turbulent vertical mixing that is more common in winter, the O_2-deficient deep and bottom waters remain isolated from the O_2-rich surface waters. As a result of the large biological oxygen demand, shelf waters of the northern Gulf of Mexico become hypoxic (low in O_2) or anoxic (O_2-depleted), especially the bottom waters. The phenomenon of hypoxia, when the O_2 concentrations reach levels that suffocate fishes and benthic invertebrates, has been given the

[7] Turner, R. E., and Rabalais, N. N., 1991. Changes in Mississippi River water quality this century: Implications for coastal food webs. *BioScience* 41: 140–147; and Goolsby, D. A., and Battaglin, W. W., 2000. Nitrogen in the Mississippi Basin—Estimating Sources and Predicting Flux to the Gulf of Mexico. *U.S. Geological Survey, USGS Fact Sheet* 135–00, December.

FIGURE 15.5 The extent of the 2011 "dead zone," an area of bottom-water hypoxia in the Gulf of Mexico in July 2011. The black line indicates the limit of the area within which the dissolved O_2 level is less than 2 mg/L, i.e., where hypoxia occurs.

(A)

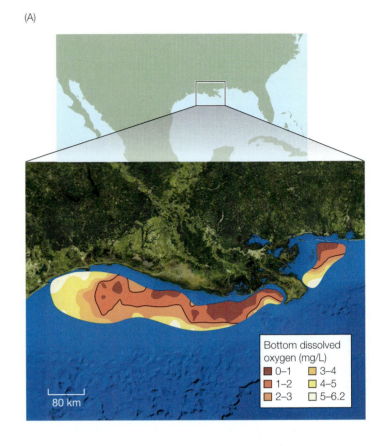

Bottom dissolved oxygen (mg/L)
- 0–1
- 1–2
- 2–3
- 3–4
- 4–5
- 5–6.2

80 km

name **dead zone**. No oxygen, no animals—it's as simple as that (**FIGURE 15.5**). Later in the year, as O_2 levels are refreshed, the waters do recover, only to face the same insult the very next summer.

Since the 1980s, the nitrogen loadings to the Gulf of Mexico have leveled off, although they remain highly variable from year to year (see Figure 15.4). But as agriculture begins to shift to the production of corn for use in ethanol production, there are concerns that, given the greater nitrogen yield from corn fields, nitrogen runoff into the Mississippi will increase in future years.[8]

The Gulf of Mexico is not alone in dealing with anthropogenic nutrient enrichment problems; so too has Boston Harbor, for example. Polluted by sewage for more than a century, it may have been one of the deciding factors in the 1988 Presidential Election, when George H. W. Bush's campaign made it an issue against then Massachusetts Governor Michael Dukakis. Construction of a massive new sewage treatment plant on Deer Island, at the mouth of the harbor, soon followed and was completed in 1995. This plant currently discharges treated sewage from 46 surrounding communities through an underground tunnel that extends nine miles offshore. At the time, this was the most expensive public works project in U.S. history. Its effects on the local ecology of that part of the Gulf of Maine appear to be minimal, and it has made a big difference in the quality of the water inside Boston Harbor, but the notion of pumping our sewage into the ocean at all is a troubling one and it remains a hotly debated topic.

An interesting example of a dead zone occurred in the 1970s in Maine's largest river, the Penobscot. Maine and its rivers are normally thought of as

[8] Turner, R. E., Rabalais, N. N., and Justic, D., 2008. Gulf of Mexico hypoxia: Alternate states and a legacy. *Environmental Science & Technology* 42: 2323–2327.

pristine, unaffected by the pollution associated with more urban and industrialized areas. But the cause of the dead zone wasn't nutrient enrichment and eutrophication; it was the result of Maine's log drives. For more than 100 years, during the heyday of Maine's lumbering industry, its rivers were used as free, efficient transport systems to get logs to coastal ports, where they were loaded onto ships. But not all the logs made it downstream. Some became waterlogged and sank; but more importantly, as rafts of thousands of logs made their way downstream, they produced tons of small wood chips as they constantly ground against one another. A large proportion of those chips sank to the bottom where they were decomposed by bacteria. As a result, a portion of the Penobscot River just above the salt wedge (where the fresh and salt waters meet), where the wood chips accumulated, became anoxic throughout much of the 1960s. Oxygen concentrations in that part of the estuary went to zero just below the surface. One consequence was that few if any Atlantic Salmon could get past this point on their annual spawning migrations up the river, and so the salmon eventually disappeared from the Penobscot. The last log drive on a Maine river was in the early 1970s. Since then, the Penobscot River has recovered, but we are still waiting for the river's once-famous salmon runs to return.

Shipping, oil pollution, and nonindigenous species

It is no news that the world runs on petroleum—oil—that is captured by oil wells that are drilled down into underground deposits. But, then what happens? Most of us don't worry about that, as long as we get the refined products, gasoline for our cars and heating oil for our homes. Much of the world's oil production, about 62%, is in nations in Africa and the Middle East, which produce far more than they need. And so much of that surplus production, over 2 billion barrels of oil per year, is transported around the world in oil tankers via the routes shown in **FIGURE 15.6**. The rest of the world's crude oil moves by trucks and pipelines.

FIGURE 15.6 Major routes for crude oil shipping around the world. Line thickness indicates the relative volume flow.

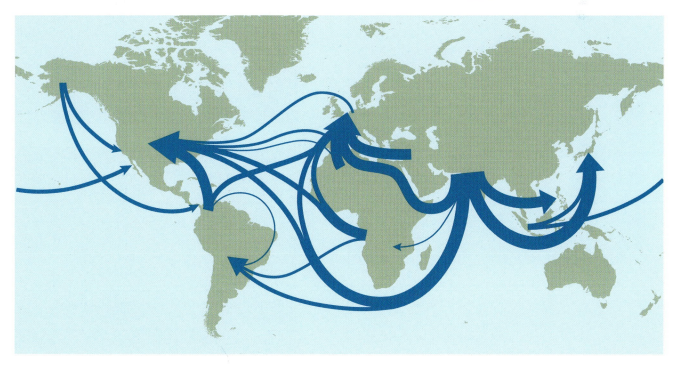

The world's tanker fleet numbered more than 7500 ships in 2009, sailing the world's oceans on the routes in Figure 15.6; unfortunately, those voyages have not been without mishaps. One of the most famous accidents in U.S. waters was the *Exxon Valdez* oil spill. In March of 1989, the 300-meter-long super tanker was loaded with crude oil that had been pumped from the Prudhoe Bay oil field on Alaska's north shore through the Alaska Pipeline to the port of Valdez on Alaska's south shore, and from there onto the waiting ship. On March 24, while just underway from Valdez and headed for Long Beach, California, the tanker went aground on Bligh Reef, a rocky ledge off Alaska's Prince William Sound. The collision opened a gash in the hull, and 50 million gallons of thick, black crude oil spilled into the Sound. The thick crude oil spread to the shoreline of the Sound, coating the rocky intertidal zone and every creature in that intertidal zone (**FIGURE 15.7**). The remote Prince William Sound was accessible only by water or air, which made the response extremely difficult. Until that time, the region had been one of the most beautiful and pristine of all marine environments—home to sea birds, sea otters, seals, and various other marine animals.

(B)

(A)

(C)

FIGURE 15.7 (A) Off-loading oil from the *Exxon Valdez* to another (smaller) ship; a floating containment boom can be seen running across the photograph. (B) The crude oil that leaked from the *Exxon Valdez* coated the intertidal zone of Prince William Sound, Alaska. (C) High-pressure hoses being used to loosen and wash away oil from the intertidal zone after the spill.

The clean-up effort was phenomenal, involving some 1600 vessels and more than 10,000 people working more than 20.5 million hours. With the best of intentions, approximately a third of the shoreline in Prince William Sound was scrubbed with high-pressure, hot-water hoses. But subsequent studies suggested that this aggressive clean-up may have caused more harm than good: while half of the organisms in the intertidal shoreline survived being oiled, in the aggressively treated areas, nearly all were killed.[9]

For more than two decades, the *Exxon Valdez* spill stood as largest oil spill in U.S. history—but in the spring and summer of 2010, we would learn that massive oil spills can happen before crude oil is ever loaded onto a tanker. On April 20, 2010, in the northern Gulf of Mexico, the *Deepwater Horizon*, a floating offshore drill rig capable of drilling oil wells at depths beyond the edge of the continental shelf, exploded, killing eleven men working on the rig (**FIGURE 15.8**). That accident itself was tragic enough, but what followed in the next several months would capture the public's attention like no other world event except war. Beginning that day, and continuing until July 15th, the undersea wellhead, at a depth of about 1500 m (nearly a mile down), spewed some 4.9 million barrels, or 210 million gallons, of crude oil directly into the Gulf of Mexico; this was more than four times the volume of oil spilled by the *Exxon Valdez*. Throughout the summer, engineers struggled with the problem of how to cap the flow—how to maneuver a fixture over the damaged pipe, which was sticking up just above the bottom, and then tighten that fixture, as if turning off a faucet, to stop the flow of oil.

FIGURE 15.8 Fireboats battle the blazes of the *Deepwater Horizon* on April 21, 2010.

While the engineers struggled, massive clean-up efforts were launched to retrieve as much as possible of the gushing oil that had floated to the surface, creating massive slicks that extended for miles. Some of the slicks were set afire, to burn them away; floating containment booms were deployed to concentrate slicks and facilitate the use of skimmers to physically scoop up as much oil as possible; various barricades were set up along the shorelines of salt marshes in Texas, Louisiana, Mississippi, Alabama, and Florida, in an effort to protect them; sand bags lined beaches; and, chemical dispersants (over one million gallons of it) were sprayed on slicks at sea—almost any method that anyone could think of was considered. But none of them worked. Oil still made it to the salt marshes, the beaches, and, perhaps worse, there were extensive undersea plumes of dispersed and dissolved oil detected by oceanographers from a number of universities later that summer.

After the well was finally capped in July 2010, reports were issued about what had happened, who was at fault, and what the potential environmental consequences might be. The National Oceanic and Atmospheric Administration (more commonly known simply as NOAA), reported in August 2010 that half of the oil that had leaked still remained in the Gulf of Mexico, but below the surface. Estimates by university scientists differed; they concluded that only 25% of the spilled oil could be accounted for by the clean-up and

[9] Mearns, A. J., 1996. "Exxon Valdez shoreline treatment and operations: Implications for response, assessment, monitoring, and research," in *Proceedings of the Exxon Valdez Oil Spill Symposium* (American Fisheries Society Symposium, vol. 18). pp. 309–328.

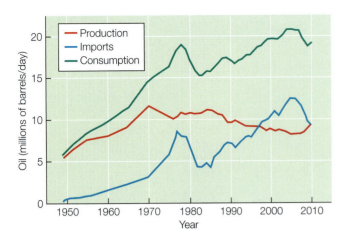

FIGURE 15.9 United States oil production, consumption, and imports, in millions of barrels per year.

burn-off efforts, and that 75% of the spilled oil remained unaccounted for, probably in dissolved or dispersed form. British Petroleum (BP), the oil company that owned the well, argued the following winter that NOAA had overestimated the size of the spill by 20–50%.

This catastrophic oil spill didn't have to happen: it could have been prevented, according to findings of the U.S. House of Representatives' Committee on Energy and Commerce. More than 130 lawsuits were filed within one month of the wellhead blowout; BP even filed its own lawsuit against the owner/operator of the *Deepwater Horizon*, which was operating under contract to BP. No doubt, the courts will stay busy for years, and reports of sightings of oil slicks, presumed to be from the spill, continue to come in. Though both accidents described here were preventable, they happened, and future accidents are inevitable. One only has to look at the ever-increasing rate of consumption of oil by the United States to see that (**FIGURE 15.9**).

Oils spills aren't the only way that petroleum finds its way into the ocean. Oil naturally makes its way to the surface through fissures in the Earth's crust, whether on land or under the ocean. Such **oil seeps** are analogous to natural springs, which bring water to the surface. Oil seeps account for nearly half (46%) of all the oil that reaches the marine environment each year. In other words, more oil seeps naturally into the oceans each year than from the *Exxon Valdez* and *Deepwater Horizon* spills combined. But they aren't all concentrated in one area the way these disastrous spills were. Other sources of oil in the marine environment include discharges from ships, boats, and land-based sources (about 37% of the total each year) and accidental spills from ships (about 12%). Only 3% is associated with extraction, or oil wells.[10] This last datum could be misleading: these are yearly averages, and massive accidental oil spills don't occur every year; but when they do, they make up far more than 3% for that year.

Solid waste

How many of us have found a food wrapper, a soft drink container, or beer can while on a walk in a park or in "the country"? Evidence of our presence on this Earth is all around us, and future archeologists won't be short of material with which to study our habits. But these are just examples of refuse that we can see on the ground; imagine what must be on the bottom of the oceans, where we can't see it. And of course, not all refuse sinks. Much of it stays afloat, finding its way across oceans—this is the principle behind leaving messages in a bottle, after all. Those floating objects are called **flotsam** and **jetsam**; flotsam is accidental junk floating about (e.g., from a shipwreck), whereas jetsam is intentionally jettisoned material. In centuries past, flotsam and jetsam was mostly biodegradable. But now, much of it is made of plastic, and floating pieces of nearly indestructible, lightweight plastic trash can be carried long distances by ocean currents.

The "Great Pacific Garbage Patch" has received some attention in the popular media in recent years. The ocean basin gyres, you'll recall from Chapter 5, are created by the Ekman currents converging at the centers of the

---

[10] National Research Council, 2002. *Oil in the Sea III: Inputs, Fates, and Effects* (Washington, DC: National Academy Press).

ocean basins (actually just west of center), where the surface waters then sink and spread, giving us the ocean gyres. So it is logical that floating debris will collect in the vicinity of those "hills in the ocean" where surface waters converge. In addition, anything floating on the surface will be blown downwind while riding atop the Ekman currents. The end result of all this is the garbage patch. And close by that patch are the Hawaiian Islands, Midway, and other islands in the Pacific Ocean that tend to collect this stuff as Stokes drift brings it ashore (**FIGURE 15.10**). Some of the materials that collect are comical—curiously lost articles of clothing, for example—but many constitute hazards to marine life, ensnaring sea turtles, for example, and concentrating oil slicks.

INVASIVE SPECIES Although we don't usually think of living organisms as pollutants, the introduction of **nonindigenous species** to an area where they can become **invasive** fits our definition of pollution. The mode of delivery of these invasive species is often by ship traffic. Oil tanker traffic is extensive, as we just discussed, with a fleet of more than 7500 tankers, but the total number of commercial ships of more than 1000 gross tons is more than 30,000. Nearly all of those ships have ballast tanks that can be filled with water to stabilize the ships for transits across long spans of open ocean. Ships are safer when they have a deeper draft (when they sit deeper in the water) because of the nature of ocean waves: remember, the wave orbits decrease in size exponentially with depth. The deeper a ship sits in the water, the less jostling it will experience from ocean waves, and the safer it is. Most ships are safest, then, when fully loaded. For this reason, should a ship offload its cargo and not take on cargo for a return or follow-on trip, it will pump water into its ballast tanks to mimic the effect of a full load of cargo. Then, upon arriving at its next destination, the ship must offload that ballast water by pumping it out before it can take on new cargo (**FIGURE 15.11**). And this creates a problem, because along for the ride in that ballast water are whatever marine organisms were in the water when it was pumped aboard. They get a free ride to wherever in the world they are pumped out, and are released into a new environment where they may or may not flourish.

More than 120 exotic species introduced into the Great Lakes, for example, were most likely transported there in ballast water. The most notorious was probably the zebra mussel, native to the Black Sea and the Caspian Sea. The zebra mussels appeared in the Great Lakes in 1988 and became firmly established, out-competing other filter-feeding invertebrates, and covering almost any surface they land on. They continue to be a nuisance, and have begun to spread to other U.S. lakes.

Another example is the periwinkle snail, *Littorina littorea* (see Figure 12.4), one of the most common intertidal snails on both coasts of the United States—so common, in fact, that it is hard to imagine what the shoreline would look like without them. But they haven't always been here. They were actually introduced from Europe during colonial times, possibly with rock ballast used in ships back then, but nobody knows for sure. And nobody knows what ecological effect

FIGURE 15.10 Materials washed up on a beach in Hawaii.

FIGURE 15.11 A commercial ship in port pumping ballast water from its tanks prior to, or while, being loaded with cargo.

their invasion has had, either. The Asian shore crab, *Hemigrapsus sanguineus*, a small crab only about 20 cm carapace width, is another invasive invertebrate species that came to the United States more recently, from Japan and Korea. It first appeared in New Jersey in 1988, and since then has been expanding up the east coast as far as Maine, where there are fears that it may compete with Maine's beloved lobsters. The Asian shore crab is thought to have come in ballast water as eggs or larvae. These are just a few examples of the phenomenon of invasive species—the introduction of nonindigenous species—which today has become a subject of intense study, with ramifications that touch upon marine biodiversity, as well as international shipping regulations.

A less-well-known aspect of ballast water is its potential to transport disease, such as cholera. Cholera is caused by a bacterium that is known to thrive in standing water, and is responsible for many pandemics throughout history. It causes severe diarrhea, which, if not treated, can cause death in a matter of hours. This bacterium lives in and on plankton organisms; copepods are the natural "host" organism, and in Bangladesh, where cholera is a big problem, outbreaks appear to be correlated with the timing of the plankton production cycle.[11] The first widespread outbreak of cholera in the New World, which occurred in Peru in 1991, was traced back to ballast water discharged from a ship that had arrived from the Indian Ocean.

Global Climate Change

Global climate change has been making headlines for decades. It is an area of intense research by university and government scientists that is routinely given extensive coverage by the popular press; and, unfortunately, it is a hot topic that has become highly politicized. The subject of climate change is actually much broader than most people realize, in that changes in Earth's past climate are well known to have occurred; about this there seems to be little debate, either in scientific circles or among the general public. That is, we know that in the distant past there have been **ice ages**, or **glacial periods**, that have alternated with **interglacial periods**; this is why there are no longer any woolly mammoths still roaming North America, for example; they were all killed off by the end of the last ice age, about 11,000 years ago. And we also seem to accept the fact that there are climate oscillations, with periods of colder-than-normal winters that may last for several years, and periods of hot summers, that also may persist for several years; these are well documented in history. Our parents and grandparents have lived through periods of extreme weather, and can share their firsthand experiences as evidence. And as we have discussed in Appendix B, there are shorter-term fluctuations that last only a year or two that clearly affect our weather patterns worldwide. But the subject of **global warming** seems to be much different, one that provokes heated exchanges in some circles, because the ramifications of a globe that is getting progressively warmer touch upon important economic issues that affect us all. Decisions will be made that affect prices of virtually all goods and services and our current way of life, whether we take actions in the near future to mitigate the warming or continue waiting until we are forced to take actions to deal with its effects. And so we need to discuss in some detail here just what this issue of global climate change is all about, and to examine why it is, despite the political rancor, that atmospheric scientists, oceanographers, and the majority of scientists in other fields are convinced by the evidence

---

[11] Colwell, R. R., 1996. Global climate and infectious disease: The cholera paradigm. *Science* 274: 2025–2031.

that not only is global warming real, but that human activities are currently having a significant influence on that warming.

The greenhouse effect

To begin our discussion, we need to explain what the **greenhouse effect** is, because it is this *greenhouse effect* and the role played by **greenhouse gas** emissions stemming from human activities that are behind much of the current political debate. So what is it all about? Well, it is important to realize that the greenhouse effect is real, and it is natural. If there were no greenhouse effect here on Earth, life as we know it would not exist either. Without it, Earth would be more like Mars—a planet without much of an atmosphere, which makes it a cold, dry, and lifeless rock in space. But if the greenhouse effect were to run rampant, and become very much amplified beyond what it is today, Earth would become more like Venus, a planet with a thick atmosphere and a runaway greenhouse effect, making it an uninhabitable hothouse. And so, here we are on Earth, a habitable planet where the greenhouse effect appears to be operating "just right," which is why Earth is sometimes called the "Goldilocks planet."

The basic physics behind the greenhouse effect are well understood and follow the fundamental scientific principles we have touched upon earlier in this book. It gets its name, of course, from actual greenhouses—houses made mostly of glass, where plants are cultured—which are famous for their tendency to become warmer on the inside than it is outside, because of the way light rays behave when they pass though glass. As visible wavelengths of sunlight penetrate the glass and strike whatever is inside, those objects absorb the Sun's energy and become warmer. As they get warmer, they reradiate longwave radiation, light in the infrared range. While infrared radiation is invisible, we can nonetheless detect it. For example, as we discussed in Chapter 4, you can sense this longwave radiation yourself by holding your hand up close to your face; you can feel the heat coming off your hand—your hand is radiating infrared radiation that strikes your face, making your skin warmer. This phenomenon is called radiant heat. In a greenhouse, some of the Sun's incoming light has been converted to infrared radiation, which is being reradiated off everything inside the greenhouse that had become warmed by the Sun's rays—the same way your warm hand is radiating infrared radiation. But the kicker here is this: that reradiating long-wavelength radiation, infrared light, does not penetrate glass nearly as easily as incoming, short wavelength (visible) radiation does, and so only a fraction of it can escape the greenhouse. The heat is therefore trapped on the inside of the glass. Of course, some of that heat is lost by *conduction*, as the glass itself gains heat from the warm inside air, which is then free to radiate away from the outside edge of the glass (for this reason, actual greenhouses need another heater inside to assist in cold weather, but in summer, they usually need to be ventilated with big fans and open windows to prevent the buildup of too much heat). This greenhouse effect causes your car to become hot inside when you leave the windows up on a sunny day; your car windshield and windows act as a greenhouse. And this is why objects stored in your car trunk (without windows), for example, don't get nearly as warm as those stored inside the passenger compartment.

The Earth's atmosphere functions like the glass in a greenhouse or your car windows; it traps heat from the Sun, which warms the Earth (**FIGURE 15.12**). The atmosphere's greenhouse effect is the result of certain **trace gases** in the atmosphere that are responsible for blocking the passage of outgoing infrared radiation from the Earth's surface—these are the **greenhouse gases**, which make up less than 1% of the atmosphere. Some of them are naturally occurring, including (listed in order of their importance): water vapor, carbon dioxide, methane, nitrous oxide, and ozone; still others are of our own making, such

FIGURE 15.12 The atmospheric greenhouse effect. Solar energy penetrates the atmosphere and heats objects on the Earth's surface, which then reradiate long-wavelength radiation incapable of penetrating back through the atmosphere because of greenhouse gases, such as CO_2. Carbon dioxide is being added to the atmosphere by the burning of forests and fossil fuels and by the decomposition of land fills. Carbon dioxide is absorbed by actively growing biomass on land and in the sea. Some of this CO_2 is actively respired back into the atmosphere, and some is stored away, by burial on land and by sinking in the ocean.

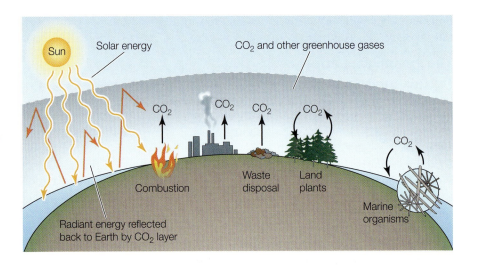

as CFCs (chlorofluorocarbons) that are used in refrigerators and as aerosol propellants in spray cans, for example. As long as these greenhouse gases do not disappear or accumulate too much, we will have an optimal-temperature "Goldilocks planet." And there's the rub: some of those greenhouse gases, especially CO_2, are indeed increasing in concentration.

Not only do we have evidence that the total amount of CO_2 in the atmosphere is on the rise, the evidence shows that the rate of increase has been accelerating ever since the Industrial Revolution, which began about 150 years ago. Those CO_2 concentrations, while still only present in trace amounts, have been dramatically increasing for the simple reason that we have been burning **fossil fuels**—coal, oil, and natural gas—at an unprecedented rate, and cutting down the rainforests, which either rot or are burned, releasing CO_2. And as the rainforests disappear, so too does their capacity to absorb CO_2 from the atmosphere into new plant biomass. This fundamental phenomenon, whereby this organic carbon is released to the atmosphere as CO_2 from burning fossil fuels and **deforestation**, is accounted for by our chemical equation of life that we have been referring to throughout this book (and which also explains biological oxygen demand, BOD, as we just discussed). Carbon dioxide and water, along with energy, combine to form organic carbon and molecular oxygen (O_2); this is the basics of photosynthesis. But when the reaction runs in reverse, we have respiration, or the *burning* of organic matter to release energy (which is needed for basic metabolism), which rereleases the basic starting ingredients, CO_2 and water. For example,

$$CH_2O + O_2 \rightarrow CO_2 + H_2O + energy$$

This burning of CH_2O (carbohydrate, or organic carbon), which releases CO_2, is the very heart of the problem, because fossil fuels *are* organic carbon. We burn oil, coal, and natural gas in order to *extract energy*, to use that energy for our various needs—to drive our cars, to heat our homes, to generate electricity, to run our factories. And in the process we release CO_2, which is added to the atmosphere. This is important because fossil fuels are just that—they are organic carbon that has been locked away, in long-term storage, you might say, chemically holding onto carbon that was once CO_2 that became "fixed" by photosynthesis. It is stored as oil, or coal, or natural gas, far underground and under the sea bed where it cannot be respired or burned to release its carbon as CO_2. But, as our world population has grown, requiring more sources of energy—more materials to burn—we have been extracting those fossil fuels and burning them, adding CO_2 to the atmosphere in the process. And, we have been clearing tropical forests to create farmland. Look once again at Figure 15.9; the United States is burning four times as much fossil fuel now as we did in 1950. What is key here is that this burn rate is presently occurring at a rate that is not matched by photosynthesis, which would chemically bind that carbon dioxide back into organic carbon, removing it from the atmosphere. That is, we are *out of steady state*—we are producing more CO_2 than the planet can lock away as organic carbon via photosynthesis. And so, more and more CO_2 is being added to the atmosphere, and that means that the Earth's greenhouse effect is gaining strength (see Figure 15.12).

The evidence

So, where is the evidence that such tiny amounts of CO_2 in the atmosphere can influence Earth's greenhouse effect, making Earth warmer? Carbon dioxide is a trace gas, after all, representing less that 1% of the gases in the atmosphere; can changes in such tiny amounts really matter? And where is the evidence that the concentrations of CO_2 are actually changing, or have changed in the recent past, thus affecting Earth's temperatures? In the sections that follow, we will answer these questions as we review the scientific evidence—evidence based on paleontological data that allow us to reconstruct Earth's temperatures and atmospheric CO_2 concentrations in the distant past, and evidence based on modern-day direct measurements of both variables.

The Earth's major ice sheets in Greenland and in Antarctica were built up over a period of hundreds of thousands of years, as snow fell to earth but did not melt; rather, it accumulated and became compacted into solid ice. Below the surface are layers of ice that originally fell as snow long ago, and which ever since have been locked away, isolated from the atmosphere. Air bubbles trapped in the ice have also been locked away and out of contact with the atmosphere. And so the air inside those bubbles includes all the gases that were in the atmosphere long ago—in the same concentrations that they were when the snow originally fell. Therefore, by drilling down into the ice sheet and recovering undisturbed samples of ice that was formed many thousands of years ago, we can bring up those bubbles and analyze their gas composition. This chemical analysis, which reveals the concentrations of CO_2 long ago, can be matched with other analyses that reveal the Earth's surface temperature at the time the air bubbles were trapped (**FIGURE 15.13**). And this is exactly what

FIGURE 15.13 (A) Scientists from the University of Maine drilling an ice core in Antarctica. (B) One end of the core with the ice sample exposed.

(A)

(B)

FIGURE 15.14 Record of global mean temperatures given as temperature anomalies, departures in °C above or below today's average temperature, and the concentrations of CO_2 in the atmosphere, in parts per million by volume as determined from analyses of the Vostok, Antarctica ice core.

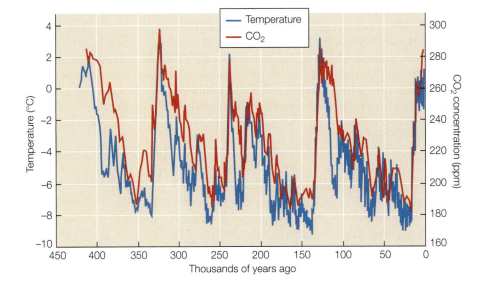

has been done to examine how Earth's past concentrations of CO_2 in the atmosphere have varied, and how those variations influenced global temperatures.

One of the most important examples of these ice core analyses was based on the 3300-meter-long Vostok, Antarctica, ice core—a core obtained by drilling down through more than two miles of ice. Samples were taken at regular intervals along the core; the deeper the core, the older the ice, of course, and this particularly long core sampled ice that at its deepest point was formed about 420,000 years ago. Chemical analyses were performed on the trapped gas bubbles for CO_2 concentrations; and, in order to get the corresponding ambient temperatures, deuterium concentrations in the ice were also measured.

The results of those analyses reveal a fascinating history of Earth's global mean temperatures and concentrations of atmospheric CO_2 (**FIGURE 15.14**).[12] The Vostok ice core showed that the Earth has undergone four glacial–interglacial cycles over the past 420,000 years, during which the average global mean surface temperatures on Earth varied by 10°C, from about 2°C above today's average, which occurred during the four earlier interglacial periods (we are in one today), to temperatures about 8°C colder than today, during the four glacial periods, or ice ages. Notice that it took a long time, as long as 100,000 years, for the Earth to cool down and enter a glacial period, but that we warmed up and came out of the glacial periods much faster, in just a few thousand years. But notice especially how the rise in CO_2 and the rise in temperature are almost perfectly aligned, and overlap one another. This is an exceptionally strong correlation that, coupled with the fact that CO_2 is a known greenhouse gas, provides strong evidence that rising concentrations of CO_2 in the atmosphere was responsible for the rise in temperatures.

A higher-resolution analysis of the changes in atmospheric CO_2 that correspond to just the last two centuries was performed on another ice core taken in Antarctica, which showed that CO_2 concentrations have been rising at an

[12] Petit, J. R. et al., 1999. Climate and atmospheric history of the past 420,000 years from the Vostok ice core, Antarctica. *Nature* 399: 429–435.

increasing rate (**FIGURE 15.15**).[13] Notice also that although the concentration of CO_2 in the atmosphere never exceeded 300 ppm (parts per million by volume) over the past 420,000 years (see Figure 15.14), we exceeded that threshold for the first time early in the last century, about 1910. This recent and rapid increase in atmospheric CO_2—to levels that have not been seen on Earth for the past 420,000 years—is the direct result of the Industrial Revolution that began in the 1800s. That was when we began to burn fossil fuels (especially coal) at unprecedented levels. We are continuing this practice today, burning oil (and automobile gasoline), coal, and natural gas, and we are cutting down the forests, especially the tropical rainforests, releasing all the CO_2 locked up in that biomass as those trees are burned to make room for farmland.

You may be wondering: If these two examples represent chemical analyses of CO_2 in air bubbles of ice cores and the corresponding temperatures, which are analyses of what appears to have been conditions in ancient times, what about measurements of atmospheric CO_2 and temperatures made in modern times? Well, actual measurements tell the same story (**FIGURE 15.16**). Since 1958, atmospheric CO_2 concentrations have been measured directly at a sampling site atop Hawaii's Mauna Loa volcano, out in the middle of the Pacific Ocean, where the air should be relatively clean and a reflection of global averages. The records from this sampling site are known as the *Keeling plot*, for David Keeling, who had the foresight to begin this long-term data set.[14] The Keeling plot shows that over the last half-century, CO_2 concentrations have increased from about 315 ppm to more than 390 ppm as of late 2011, an increase of 24%. Those data also show an annual oscillation of relative highs and lows (the wiggle in the plot) that corresponds to the annual growth cycle in the Northern Hemisphere: each season, marine algae and terrestrial plants grow

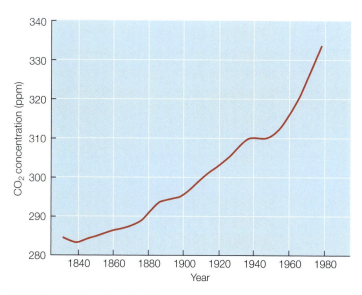

FIGURE 15.15 Carbon dioxide concentrations in parts per million by volume, corresponding to the period 1832–1978 as determined by Antarctic ice core analyses.

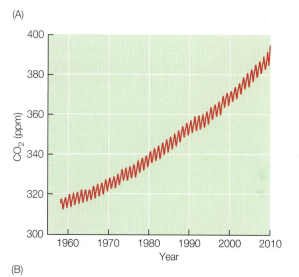

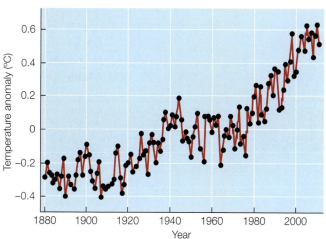

[13] Etheridge, D. M. et al., 1996. Natural and anthropogenic changes in atmospheric CO_2 over the last 1000 years from air in Antarctic ice and firn. *Journal of Geophysical Research* 101: 4115–4128. (Note that "firn" is partially compacted snow.)

[14] For more information, see www.esrl.noaa.gov/gmd/ccgg/trends/. Dr. Pieter Tans, NOAA/ESRL (www.esrl.noaa.gov/gmd/ccgg/trends/) and Dr. Ralph Keeling, Scripps Institution of Oceanography (scrippsco2.ucsd.edu/).

FIGURE 15.16 (A) The Keeling plot is a record of CO_2 measurements made on the top of Moana Loa in Hawaii, begun in 1958 and continuing today. The concentration on December 31, 2011, was 391.8 ppm. (B) Global temperature anomalies, given as departures from °C above or below the average temperature over that time period, from 1880 to 2010, based on actual measurements.

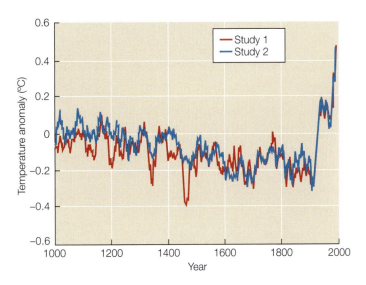

FIGURE 15.17 Reconstructed Northern Hemisphere temperature anomaly over the last 1000 years from two separate studies, given as departures in °C above or below the average temperature over that period.

anew, incorporating CO_2 into new biomass; this alternates with a period of winter die-off and dormancy, when much of that biomass decomposes, releasing the CO_2 absorbed by photosynthesis the previous year. But that wiggle, that oscillation of growth and decay, is riding on a baseline that is not only rising, but curving upward, reflecting the accelerating rate of CO_2 accumulation in the atmosphere—today, we are at unprecedented levels.

To go along with those actual measurements of atmospheric CO_2, we also have direct temperature measurements. Recorded at meteorological stations around the world since 1880 (see Figure 15.16), those temperature records show that the global average temperature on Earth has warmed by about 0.7 to 0.8°C over the 130-year period.[15] Global mean temperatures are indeed rising today. And that rise is a relatively recent phenomenon that began in the early twentieth century, lagging just behind, and responding to, the rise in atmospheric CO_2 that began several decades earlier, as the Industrial Revolution got underway.

The abrupt rise in global mean temperatures that began in the early twentieth century is shown in a reconstruction of global air temperature going back 1000 years. Published in 2000, and based on analyses of tree rings, corals, and ice cores by earlier workers, were two sets of data (**FIGURE 15.17**).[16] This plot is a more recent version of what has become the infamous "hockey stick" plot (based on the shape of the plot, with a long shaft on the left and the upward-turning blade on the right) that has been the subject of political attacks ever since it was cited in an international report.[17] Details of the statistical analyses in that report were called into question by several scientists following the initial publications in the late 1990s, and those criticisms were subsequently seized upon by politicians who claimed that "man-made global warming is the greatest hoax ever perpetrated on the American people."[18] All of a sudden, the science of global climate change was a political issue, and the credibility of thousands of scientists was being called into question. Most scientists ignored the rancor and continued their work. But the rhetoric just grew hotter. In an attempt to unravel the nature of the controversy, a study group was formed by the nonpartisan, nonpolitical National Research Council of the Unites States National Academy of Science, the most prestigious group of scientists in the world. They reported that there had, indeed, been flaws in the statistical methods employed in some of the earlier studies, but—and this is critical—those errors did not affect the overall conclusions. They affirmed that the shape of the curve is valid, and that the twentieth century has indeed been the warmest on record, as shown graphically in the hockey stick plot.

[15] Hansen, J. E. et al., 2001. A closer look at United States and global surface temperature change. *Journal of Geophysical Research* 106: 23947–23963.

[16] Crowley, T. J., 2000. Causes of climate change over the past 1000 years. *Science* 289: 270–277.

[17] Houghton, J. T. et al., 2001. *Climate Change 2001: The Scientific Basis. Contribution of Working Group I to the Third Assessment Report of the Intergovernmental Panel on Climate Change* (Cambridge, UK: Location: Cambridge University Press).

[18] James M. Inhofe, U.S. Senator, Chairman of the Committee on Environment and Public Works, "The Science of Climate Change" (speech, United States Senate, Washington, DC, July 28, 2003).

Based on these various studies of Earth's past climate, past and present concentrations of CO_2, and relationships between those concentrations and Earth's past and present temperatures, scientists have built sophisticated computer models that use projections of fossil fuel consumption, deforestation, and various other human practices that produce greenhouse gases to make predictions of what we can expect to see in the future. An example of these projections is given in **FIGURE 15.18**, which is based on various scenarios of future human practices, from stopping CO_2 emissions altogether, to large increases in CO_2 emissions. The models show that we face continued future warming of the Earth to a degree that depends on our rate of CO_2 emissions, and that even if we were to cease all emissions, the globe will continue to warm, because of the time it takes for the ocean–earth–atmosphere system to equilibrate.

Sea level

As the Earth's temperatures continue to warm, sea level will rise as the temperature of the oceans equilibrate with the rising temperatures in the atmosphere, not just the surface waters, but throughout the water column. As the oceans warm, they will expand—a phenomenon called **thermal expansion**. The basic principle at work here is identical to that in our thought experiment in Chapter 4, where we changed the temperature of water in a laboratory flask and noted the change in volume. And just as the water level in that flask rose with rising temperatures, sea level too will rise. But unlike our experiment with a laboratory flask, the volume of water in the oceans will be increased even more as Earth's major ice sheets melt—but only the ice sheets on land, principally on Greenland and on Antarctica. The melting of floating ice sheets will have no effect; you can convince yourself of this by watching the level of ice water in a glass, as that ice melts. The level of water in your glass will remain unchanged.

Direct observations have revealed that the rising temperatures on Earth have already warmed the oceans enough to raise sea level. Changes in sea level, as measured by tide gauges around the world, along with measurements made using satellite altimetry, are shown in **FIGURE 15.19** for the period 1860–2009.[19] Over the last 150 years, sea level has risen about 25 cm. This rise is primarily the

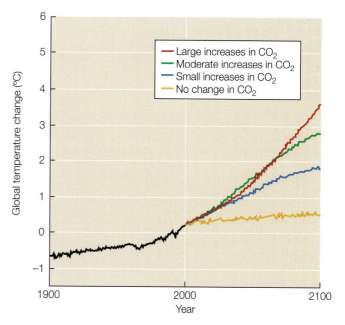

FIGURE 15.18 Global temperature change from 1900 to 2000 (black plotline), with model predictions of future global temperature changes based on four scenarios of future CO_2 increases (colored plotlines), from no increase to large increases. Temperatures are given as anomalies (departures in °C) above or below the global mean temperature in the year 2000.

FIGURE 15.19 Change in sea level since 1860 as measured by tide gauges around the world and, for more recent years, satellite altimetry. Sea level is given as departures above and below global sea level in 1990.

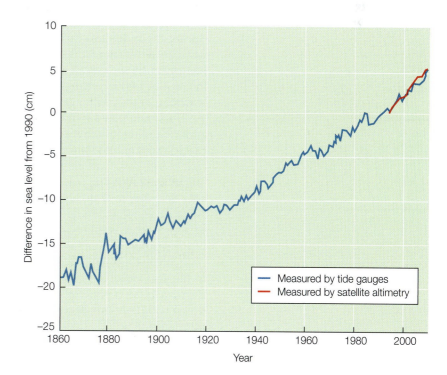

---

[19] Church, J. A., and White, N. J., 2011. Sea-level rise from the late 19th to the early 21st century. *Surveys in Geophysics* 32: 585–602.

FIGURE 15.20 Bar graphs showing the warming of the oceans between 1948 and 1998, shown as annual values of heat content (in units of 10^{22} joules) in the upper 3000 m for the Atlantic, Indian, Pacific, and combined world ocean. Red lines show the statistical trends.

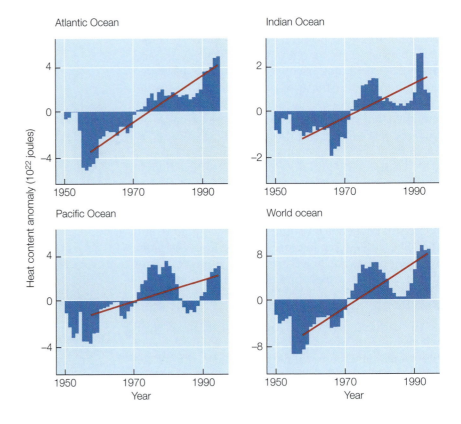

result of thermal expansion of the oceans as the upper ocean has warmed over that time period. The warming of the oceans over the 50-year period between 1948 and 1998, based on actual measurements, shows that the heat content of the oceans (which is the energy equivalent of water temperature) has indeed warmed, as can be seen in **FIGURE 15.20**.[20]

The main point here is that global warming is real. The surface of the Earth, its atmosphere and oceans, is warming; and, that warming is being driven primarily by increases in atmospheric greenhouse gas emissions that began during the Industrial Revolution in the mid-nineteenth century. The rise in CO_2 concentrations in the atmosphere and the relationship to temperatures have been shown based on ice cores, and actual measurements from the mid to late nineteenth century, thus confirming not only the rise in of CO_2, but the warming of the Earth's surface, and as a result, the rise in global sea level.

The oceans play a significant role in all this, and not just by becoming warmer and expanding to increase sea level. A large proportion of the atmospheric CO_2 that has been emitted since the Industrial Revolution is now stored in the oceans.

THE OCEANS AND CO_2 The world's oceans take up atmospheric CO_2 the same way that they take up O_2, by simple diffusion across the air–water interface. As cold surface waters sink as part of the thermohaline circulation (remember: gases are more soluble in cold water, such as at high-latitudes), that load of CO_2 becomes "stored" in the deep ocean until those deep waters resurface, probably in the Pacific Ocean several centuries from now. Also, dissolved CO_2 in the surface waters is taken up by phytoplankton, which may sink to the deep sea (which is the source of the biogenic oozes, remember) taking carbon down with them for deep, long-term storage. The contributions of these two mechanisms of CO_2 storage in the deep sea are very important in computer

[20] Levitus, S. et al., 2000. Warming of the World Ocean. *Science* 287: 2225–2229.

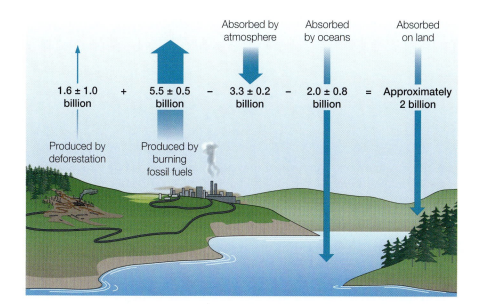

FIGURE 15.21 The annual global carbon budget, presented as a diagram of inputs to the atmosphere, and removals into long term storages. Notice that there is uncertainty associated with each of the estimates presented.

models of what we can expect as CO_2 levels in the atmosphere continue to rise. We have a pretty good idea how much CO_2 we are adding to the atmosphere as we burn fossil fuels, for example, based on records kept by world energy markets; but of that, how much stays in the atmosphere, influencing the greenhouse effect, and how much gets "pumped" into the oceans by dissolution and deepwater formation and from photosynthetic uptake and subsequent sinking? The answer is: a lot, but we aren't too sure just how much; in other words, the estimates have large uncertainties.

A balance sheet of the global carbon budget is given in diagrammatic form in **FIGURE 15.21**. The components of that annual budget are as follows:

- About 1.6 billion tons of carbon, in the form of CO_2, is added to the atmosphere each year by deforestation. Trees are cut and burned, or allowed to rot; either way, their organic carbon is respired which releases the carbon to the atmosphere as CO_2.

- Another 5.5 billion tons of carbon, again in the form of CO_2, is released by the burning of fossil fuels—oil, coal, natural gas.

The total input each year to the atmosphere is thus 7.1 billion tons of CO_2. Of this,

- About 3.3 billion tons stays in the atmosphere.

- About 2 billion tons is stored in the oceans.

- About 2 billion tons is taken up by vegetation on land.

The oceans are important players in that budget, but notice that there is a lot of uncertainty associated with each of the estimates presented. Not only will it be important to sort out just how much of the CO_2 is going into the oceans as atmospheric concentrations of CO_2 continue to climb, so that we can generate accurate computer models of climate change, but it is also important to know how much dissolved CO_2 is added to the oceans because there is another problem, quite apart from global warming and the greenhouse effect.

THE OTHER CO_2 PROBLEM   The other CO_2 problem is **ocean acidification**. This uptake of CO_2 by the oceans is changing the pH (acidity) of the oceans.

When CO_2 moves from the atmosphere into the oceans, where it becomes dissolved, it undergoes some interesting chemistry; it forms carbonic aid:

$$CO_2 + H_2O \rightarrow H_2CO_3$$

and that carbonic acid further dissociates in seawater to give bicarbonate (HCO_3^-) and carbonate (CO_3^{2-}):

$$H_2CO_3 \rightarrow H^+ + HCO_3^- \rightarrow H^+ + CO_3^{2-}$$

This increases the H^+ concentration in the oceans, which, of course, makes the oceans more acidic—it lowers the pH.[21] The increasing acidity makes oceans caustic to calcium carbonate–forming organisms, which our review of just some of the planktonic, nektonic and benthic marine organisms showed is a lot of organisms. And then there are the coral reefs of the world, which are themselves massive structures made of calcium carbonate precipitated by the coral animals. Remember our earlier discussion in Chapter 4 about what happens when you place a drop of a weak acid on a seashell (which is made of calcium carbonate): it dissolves.

Direct measurements of dissolved CO_2 concentrations in the surface waters of the oceans show that they are, indeed, increasing—and the increase matches the increases in CO_2 in the overlying atmosphere, as does the decrease in pH (**FIGURE 15.22**). Notice also in Figure 15.22 the CO_2 beginning to enter

[21] The pH is a shorthand notation for acidity; it is equal to the negative logarithm (base 10) of the hydrogen ion concentration. A pH of 7.0 is neutral; pH values less than that are acidic, and pH values above that are basic.

(A)

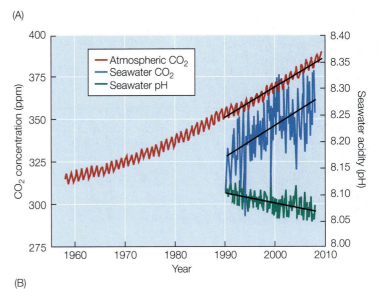

(B)

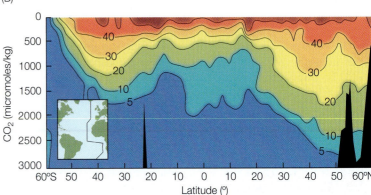

FIGURE 15.22 (A) Recent changes in Pacific Ocean surface water CO_2 concentrations and seawater acidity (pH) alongside the increase in atmospheric CO_2 (in parts per million) from Figure 15.16. (B) CO_2 concentrations in the North Atlantic Ocean, given in micromoles per kilogram, which show sinking with deep water formation at the higher latitudes.

the thermohaline circulation of the world ocean, as sinking surface waters in the North Atlantic bring with them the load of CO_2. The world's oceans are becoming more acidic, and that increase in acidity will continue in step with the increases in atmospheric CO_2. Of this, there is no dispute; there are no political refutations. How can there be? It's basic chemistry.

MELTING GLACIERS AND ICE SHEETS Evidence of global warming is expressed not only in rising sea level, but also in the melting of alpine glaciers, as well as ice in Greenland, the Arctic Ocean, and the Antarctic. Alpine glaciers have been disappearing over the past several decades, as can be seen in **FIGURE 15.23** for the Muir Glacier in Alaska and the Main Rongbuck Glacier on the side of Mount Everest. These magnificent features took millennia to create, as snowfall accumulated and became compacted into flowing rivers of ice, but it is taking only decades for them to melt away. Even the Greenland ice sheet is melting. This massive feature covers an area of more than 1.7 million square kilometers (more than 650,000 square miles)—most of Greenland. It is the second-largest ice sheet in the world (behind Antarctica), and if it were to melt completely, which some scientists fear will happen over a period of a few hundred years, it would raise global sea level by more than 7 meters. The rate of melting has increased in recent decades as the Arctic has warmed, and the current rate of melting of Greenland's ice is on the order of more than

(A)

(B)

(C)

FIGURE 15.23 Photographs showing the melting of Muir Glacier in Alaska in (A) 1941, (B) 1950, and (C) 2004.

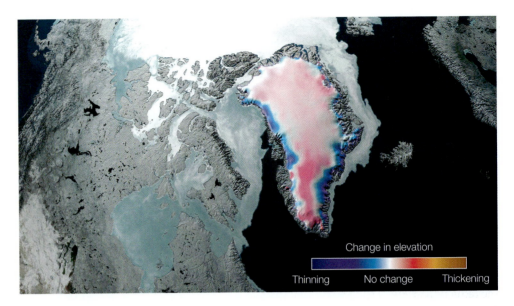

FIGURE 15.24 Relative changes in the thickness of the Greenland ice sheet between 2003 and 2006.

60 cubic miles of ice loss per year, which is raising global sea level by 0.7 mm per year (**FIGURE 15.24**).[22]

The Arctic Ocean ice cap is also melting at an alarming rate. It was first reported by scientists in the 1990s that the ice cover had thinned by as much as 25% over a 20–30 year period. The melting is most noticeable in the spatial coverage of the ice cap as it waxes and wanes with the annual cycle of melting in summer and refreezing in winter. Satellites have been monitoring the extent of ice cover since 1979, and over that short period of time, the decrease in the minimum extent of the ice in late summer has been dramatic (**FIGURE 15.25**). The long-sought Northwest Passage (see Chapter 1) is now open for ship traffic in late summer.

Melting of the Arctic Ocean ice cap will not change sea level, as we discussed above, because it is floating ice, analogous to the ice melting in a glass of water; its melting does not change the water level. But the continued melting of the Arctic ice will have other important effects on the oceans. As the ice

[22] Mernild, S. H. et al., 2009. Greenland Ice Sheet surface mass-balance modelling and freshwater flux for 2007, and in a 1995–2007 perspective. *Hydrological Processes* 23: 2470–2484.

(A)

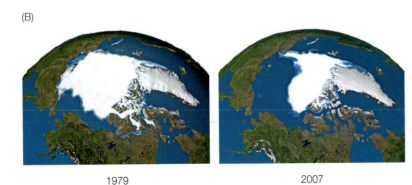

(B)

1979   2007

FIGURE 15.25 The extent of melting of the Arctic Ocean ice cap is shown here as the extent of ice cover in September, in late summer when ice extent is least, each year from 1979 to 2011, along with composite satellite images of the minimum ice extent in 2007 as it compared with 1979.

sheet melts each summer and refreezes the following winter, the Arctic functions as a still, making freshwater at the surface out of seawater. Sea ice, remember, is pure freshwater. Only the water molecules themselves bond together to form the hexagonal crystalline lattice we know as ice; the salt is excluded, sinking as a brine down through the layer of sea ice and increasing the salinity of subsurface waters in the process. Then, when that sea ice melts, it produces a freshwater layer that floats on the surface, gradually mixing with deeper layers. The volume of freshwater created each year, along with meltwater from Greenland, could interfere with winter convection in the Arctic; the reduced salinity of the surface waters gives them additional buoyancy, and thus requires colder temperatures to give them the density to sink and enter the global thermohaline circulation. Some scientists have predicted that should there continue to be widespread melting in the Arctic, there will likely be a complete breakdown of this important phenomenon of winter convection, which will have drastic effects on ocean circulation and climate in the Northern Hemisphere.

The Antarctic ice sheet is melting as well, and as it does, giant icebergs the sizes of New England states are breaking free. The Antarctic ice sheet is **fast ice**, or grounded ice, about three miles thick, built up over hundreds of thousands of years. It is a massive glacier that flows from the center of the continent outward toward the sea, and at the edge of the continent, the sheet extends out over the ocean as **ice shelves**—sheets of floating ice—that are between 100 and 1000 meters

FIGURE 15.26 (A) The Ross ice shelf in Antarctica from a distance. (B) Close-up of the Ross ice shelf. (C) Satellite photo of icebergs breaking off the ice shelf in 2003; the largest one in the center (arrow) of the image is approximately the size of Connecticut.

thick (**FIGURE 15.26**). The rate at which the fast ice on the continent moves toward the ocean is speeding up, as meltwater seeps downward through the ice sheet and effectively lubricates the interface between it and the continent, facilitating its movement. As the ice sheet extends outward over the continental shelf waters, it tends to break apart into giant, flat icebergs, so large that their sizes are compared with states; the long thin one floating freely in Figure 15.26 is about the same size as the state of Connecticut. Their fate is to drift with the prevailing currents as they break apart further and eventually melt, a process that will take many years. As Antarctica's ice slowly extends out over the sea, sea level will rise; this ice was once over land, and now it is displacing water.

THE FUTURE   The subjects we have begun to address in this brief concluding chapter should give you pause. Our precious oceans, and everything that lives in them, are changing before our very eyes—because of our own actions.

If anything, this book is intended to show how important the oceans are and to encourage anyone who reads it to realize how fascinating it is to learn more about how the oceans work, and especially to think broadly about how to manage our growing marine environmental problems. The oceans' future is in our hands. It will be up to all of us to find ways to conserve its fish stocks for future generations, to find ways to rid our growing population centers of waste products without polluting rivers and the seas, to engineer agricultural systems and flood protection mechanisms that protect the health of our coastal seas, to develop aquaculture techniques that don't require catching fish to feed fish in culture, and to find ways to deal with the prospects of a warming planet. We will have to find safe ways to extract petroleum and natural gas from beneath the sea and to transport it safely to refineries. And, yes, I did just say that. We environmentalists cannot simply protest against the continued extraction and burning of fossil fuels without redoubling our efforts to find and promote the use of alternative sources of clean energy, such as harnessing ocean tides. We cannot yet simply turn off one source of energy, one that we and the rest of the world, especially, have become so dependent on, and just turn on the other; there will have to be a transition that will take time, and patience. Protests awaken interest across all sectors of our society, but they must be constructive if we are to make meaningful changes. And we will all have to reach compromises that will seem to many of us like giving in, or giving up. This kind of attitude, the willingness to listen to all sides of arguments about the future health of the natural environment will be required if we are going to face up to global warming; but in doing so we cannot compromise science. And so we must find ways to educate others in the basic workings of science and to recognize the significance of the science that supports what is virtually an undeniable conclusion—that the Earth is indeed warming, and it is doing so in large part because of what we are doing to it. Denying the existence of this global phenomenon won't make it go away, or mitigate its effects, which, as we have seen in this brief review, are presently well underway.

In 2004, a new U.S. Ocean Policy was sent to Congress, written over a period of several years by a thoughtful team of scientists, managers, educators, and representatives for numerous marine industries, in which a number of recommendations were made that address the issues we have just raised, and many more.[23] It's time to take that plan seriously and implement its recommendations.

---

[23] For more on this, see www.oceancommission.gov/documents/full_color_rpt/welcome.html. U.S. Commission on Ocean Policy, 2004. *An Ocean Blueprint for the 21st Century: Final Report.* Washington, DC.

Chapter Summary

- **Marine pollution** is defined as "the introduction by humans, directly or indirectly, of substances or energy to the marine environment, resulting in deleterious effects."

- Forms of pollution include industrial chemicals, agricultural fertilizers, solid wastes, petroleum products, and human sewage. They may be **point-source pollution**, with a specific point of origin, or **nonpoint-source pollution**, with widespread sources, such from the atmosphere, agricultural runoff, or household septic tanks, with the pollutant eventually carried to the sea.

- Pollution from sewage contains elevated concentrations of nutrients, which stimulate algal production in the marine environment; this phenomenon is called **eutrophication**. Most large rivers, including the Mississippi, are polluted. The elevated nutrient loads carried by the Mississippi from its tributaries are dumped in the shelf waters of the Gulf of Mexico, creating **dead zones** when large quantities of algal biomass decompose, depleting dissolved O_2. Altered nutrient ratios as a result of pollution can alter the species composition of the responding algae.

■ More than 60% of the world's oil moves by tankers from producing nations to consumer nations, and mishaps have created disastrous oil spills; the supertanker *Exxon Valdez* incident is an example. Oil also can be spilled in the oceans at offshore production sites, as we learned in 2010 when the *Deepwater Horizon* oil rig exploded in the Gulf of Mexico, killing eleven men, and leaking four times as much oil as the *Exxon Valdez*.

■ The transport of marine organisms in the ballast water of commercial ships transports **nonindigenous species** to distant marine environments where they can become established, and are considered **invasive species**. The cholera bacterium, whose natural host is zooplankton organisms, is thought to have been brought to South America from the Indian Ocean in ballast water in 1991, causing the first outbreak of cholera in modern times.

■ Floating solid waste accumulates at the convergence zones of the ocean gyres; this creates the Great Pacific Ocean Garbage Patch, which sometimes washes trash ashore on Pacific islands, such as Hawaii.

■ **Global warming** is driven primarily by Earth's **greenhouse effect**, which is growing stronger as we add more **greenhouse gases** to the environment, especially CO_2. We are presently out of steady state; more CO_2 is being added to the atmosphere than is being locked away via photosynthesis because of our burning of **fossil fuels**, which increased during and after the Industrial Revolution, and from **deforestation**.

■ Atmospheric CO_2 concentrations have increased about 25% since the 1800s, and are at levels higher than any time in the past 420,000 years, based on ice-core analyses. The increase in CO_2 since the 1800s has led to a global warming of 0.7 to 0.8°C.

■ Much of the Earth's heating has been in the oceans, which because of **thermal expansion**, is raising sea level; global sea level has risen about 20 cm since 1860.

■ The other CO_2 problem is **ocean acidification**. Increased CO_2 in the atmosphere has increased its concentrations in the oceans, producing carbonic acid, lowering pH, and threatening all calcium carbonate–secreting marine organisms, including the world's tropical coral reefs.

■ In modern times, global warming is melting the world's alpine glaciers, as well as the major ice sheets in Greenland and Antarctica. The Arctic ice cap has become about 25% thinner since the late 1970s, and each summer there is now an open, navigable Northwest Passage connecting the Atlantic and Pacific Oceans.

Discussion Questions

1. What is coastal eutrophication? Why are rivers and coastal marine environments experiencing nutrient enrichment? Where do those nutrients come from? What are some of the consequences of nutrient enrichment of marine environments?

2. What are some examples of point-source pollution? Nonpoint-source pollution?

3. What are dead zones? How do they get their name? What are some examples? What can be done to alleviate this environmental problem?

4. What can be done to reduce the incidences of massive oil spills by oil tankers and offshore oil drilling rigs, short of banning them altogether? What are the effects of oil dispersants that are sprayed on oil spills? What happened to all the oil that spilled into the Gulf of Mexico in 2010?

5. What steps can be taken to reduce the incidence of solid waste pollution in the oceans by individuals? By corporations?

6. Why is ballast water important to ship safety? What can be done to reduce the incidence of transport of nonindigenous species across oceans in ballast water? What should be done, if anything, once an invasive species becomes established in a marine ecosystem?

7. What is the difference between global climate change and global warming? What is the greenhouse effect? How does it work? Why does the production and emission of CO_2 enhance the greenhouse effect?

8. Discuss the lines of evidence, both paleontological and modern-day measurements, that have led scientists to conclude that Earth's climate is warming.

9. What are the sources and sinks of CO_2 on Earth? Discuss the Earth's carbon cycle: how inorganic CO_2 becomes organic carbon and how it is returned back into inorganic form (CO_2 in the atmosphere).

10. How do we know that the Earth has undergone numerous glacial and interglacial cycles in the past? Why do you think that some people today recognize this and accept it as a scientific fact, but do not accept the notion of global warming?

Further Reading

Clark, R. B., 2001. *Marine Pollution*, 5e. Oxford: Oxford University Press.

Doney, S. C., Fabry, V. J., Feely, R. A., and Kleypas, J. A., 2009. Ocean acidification: The other CO_2 problem. *Annual Review of Marine Science* 1: 169–192.

Lynas, Mark, 2008. *Six Degrees: Our Future on a Hotter Planet*. Washington, DC: National Geographic.

National Research Council, 1999. *Global Environmental Change: Research Pathways for the Next Decade*. Washington, DC: National Academy Press.

Pachauri, R. K., and Reisinger A., eds., 2007. *Climate Change 2007: Fourth Assessment Report (AR4). Contribution of Working Groups I, II and III to the Fourth Assessment Report of the Intergovernmental Panel on Climate Change*. Geneva, Switzerland: IPCC.

U.S. Commission on Ocean Policy, 2004. *An Ocean Blueprint for the 21st Century: Final Report*. Washington, DC.

APPENDIX A
Satellite Remote Sensing

In Chapter 1, we briefly reviewed some of the difficulties early ocean scientists faced with respect to the limited technologies available to them (e.g., how to determine water depth, how to measure water temperature at different depths, and how to obtain uncontaminated water samples from beneath the surface, among many others). Those days are long gone; the advanced technologies available to scientists today are almost mind-boggling. This is especially true for the numerous satellites that orbit the Earth. Their sophisticated ocean-observing capabilities yield detailed images which we use throughout this book. And so we need to examine just what satellites are, how they orbit the Earth, and what they can do.

Satellites fly in orbit around the Earth after being launched by rockets to just the right altitude and speed such that they do not (for some time, at least) fall back to Earth. The principle of Earth-orbit is fundamentally the same as swinging a bucket of water over your head (**FIGURE A**, left). If you swing the bucket with the right speed, you can complete a full circle without the bucket and its contents falling onto your head. The bucket (and the water in it) is pulled outward and remains away from your body. It doesn't fall because of a balance of forces: the force of the bucket pulling outward (*centrifugal force*) and the force of your arm pulling inward (*centripetal force*). The outward-pulling centrifugal force is actually an *apparent force* because if you were to let go of the bucket, centrifugal force will not send the bucket straight outward, in the opposite direction that your arm is pulling. Instead, the bucket will fly off on a tangent. (This is also the principle behind throwing a baseball.)

The balance between centrifugal and centripetal forces explains how satellites can orbit the Earth. If we launch a satellite on a rocket that flies parallel to the Earth's surface (after climbing to some altitude above the surface, of course), its straight-line tendency over the curved surface of the Earth (the tangent line **FIGURE A**, right) produces a centrifugal force that pulls it up and away from Earth. All the while, gravity is pulling the object inward and back

FIGURE A (Left) An everyday example of the balance between centrifugal and centripetal forces. (Right) The same balance of forces as at left, but instead of a person's arm representing centripetal force, it is gravity countering the centrifugal force of a satellite in orbit. In both examples, centrifugal force is an apparent force, because if either the person lets go of the bucket or if gravity were somehow turned off, the bucket and satellite would not fly off in the direction they are being pulled by centrifugal force, but rather they would fly along a tangent line to the orbital curve, as shown.

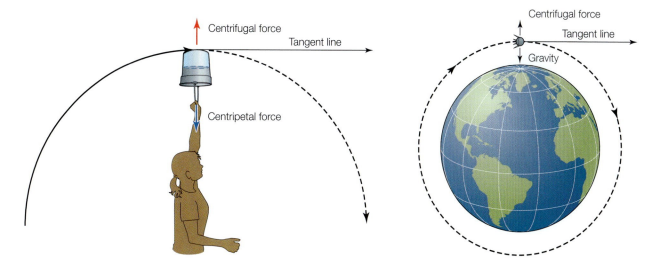

FIGURE B A Soviet Union engineer working on the world's first artificial satellite, Sputnik, which was launched on October 4, 1957, and orbited the Earth for three months, until it fell to Earth on January 4, 1958. It completed an orbit in just over 1 hour and 36 minutes. Sputnik transmitted a radio signal that was picked up by amateur radio operators around the world as it flew overhead, but the transmitter's batteries died after about three weeks.

FIGURE C The principle behind communications satellites. Radio transmissions are generally limited to a line-of-sight path, and, with the exception of radio waves that bounce off layers in the atmosphere, one cannot send a radio signal beyond the horizon (blue arrows). The distance to the horizon can be extended by using a radio transmission tower (height greatly exaggerated here), but using a satellite to relay signals can do much better (red arrows), depending on the altitude of the satellite.

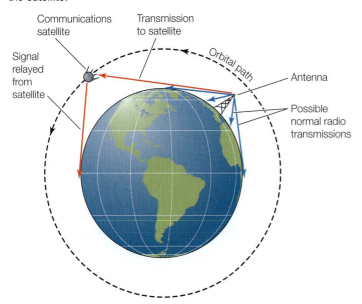

toward Earth. By varying the speed of the object (the rocket and its cargo, the satellite), we can vary the centrifugal force. So if the rocket and satellite accelerate to just the right speed, they can achieve a perfect match between centrifugal force and the pull of gravity; they will then be in orbit. The rocket can release the satellite, and, because space is almost a perfect vacuum, the satellite will effectively coast, losing very little speed over time. Eventually (years later, usually), the speed of the coasting satellite does slow down, however. With that reduction in speed, centrifugal force is reduced, and gravity wins the struggle; thus, the satellite eventually falls back to Earth. But if the satellite were to speed up, with rocket thrusters, for example, then it would move away from Earth and into a higher orbit. And if it were to go fast enough, it would escape Earth's gravity altogether and fly off into space.

With advances in rocket technology after World War II, it became possible to launch small cargoes into space and into Earth orbit. The first artificial satellite, Sputnik, was launched by the Soviet Union in 1957 (we already have a natural satellite, of course, the Moon). Sputnik didn't do much, except transmit a weak radio signal, and in so doing instill a sense of anxiety in most Americans, who were already distrustful of the Soviet Union (**FIGURE B**). Sputnik was a proof-of-concept experiment; the intent was only to demonstrate that a cargo could in fact be launched into space, achieve the balance of forces we just discussed, and, therefore, orbit Earth. This would enable a whole host of possibilities, from extending the range of radio communications to observations of the Earth from space. The successful launch of Sputnik into orbit by the Soviets spawned the *space race* with the United States during the already tense Cold War, and was the impetus behind the creation of NASA, the U.S. National Aeronautics and Space Administration, in 1958 and the suite of satellites we have today.

Satellites have two fundamental applications: communications and observations of Earth and space. Communications satellites were the first to be developed and launched by the United States in order to extend the range over which radio waves can be transmitted. While some radio waves can bounce off layers in the atmosphere and return to the Earth's surface at a point beyond the horizon, the range of radio communications—the distance over which transmitted radio waves can travel to a receiver—is generally limited to the line-of-sight across the surface of our spherical Earth. In other words, radio waves generally cannot be transmitted efficiently beyond the horizon. While the distance between a radio transmitter and a receiver can be extended by constructing tall antennas, relaying signals via a satellite (which is analogous to a *very* tall relay antenna) back to a receiver on Earth allows radio communication across far greater distances than is otherwise possible (**FIGURE C**, not to scale).

There are a number of types of orbits around the Earth that a satellite can follow that vary in altitude or direction of flight relative to the Earth's surface, for example, but they all are based on the same principle of a balance between centrifugal force and gravity. The satellites most commonly used in Earth- and ocean-observing applications are either *geostationary orbiting satellites*, which are in a high-altitude orbit (36,000 km or 20,000 miles) or *polar orbiting satellites*, which are in a low-altitude orbit (around 600 km or 330 miles; **FIGURE D**). Geostationary satellites orbit directly over the Equator at a speed that exactly matches the rotation of the Earth; therefore, to an observer on the ground, these satellites don't seem to move. They stay fixed over a point on the rotating Earth and look down on one half of the globe at all times. The NOAA GOES satellites are examples (see Figure 4.16), as are the satellites used for satellite television. Because these satellites are orbiting directly over the Equator, you will notice that satellite dishes mounted on the sides of houses in the Northern Hemisphere are always pointed toward the south, and vice versa. Polar orbiting satellites, on the other hand, are in a much lower altitude, North–South orbit that takes them over both poles. The plane of their orbit is fixed in space and the Earth rotates beneath them; this enables them to see the entire globe after a number of orbits (**FIGURE E**, top). Some of these Earth- and ocean-observing satellites use a scanning imaging system, whereby their imaging cameras sweep side to side as the satellite flies overhead, like a paint brush action, thus observing a swath, or a wide area path on the Earth's surface beneath them. Plotted on a flat map of Earth, the flight paths of polar orbiting satellites trace out an S-shaped flight path (**FIGURE E**, bottom).

Both types of satellites (geostationary and polar orbiting) used to observe the Earth and the oceans carry a wide array of both active and passive sensors. Passive sensors measure Earth-leaving radiation, much the way a camera records visible radiation (visible light) leaving the subject being photographed. These sensors include radiometers that measure infrared radiation (yielding surface temperatures of the land and oceans) and color (yielding a measure of phytoplankton chlorophyll on land and in the oceans), as well as ordinary cameras that take photographs, like terrestrial dust being blown out over the oceans (see Figure 3.47). Active sensors include radar transmitters and receivers, such as those used in radar altimetry (see Box 3D), as well as others used to discern wind directions over the oceans

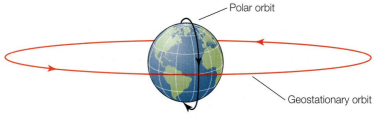

FIGURE D The paths of polar orbiting satellites and geostationary orbiting satellites.

FIGURE E The swath (area being imaged) on the Earth's surface beneath a polar orbiting satellite as the Earth rotates. (Top) This vertical swath is the field of view of the satellite's sensors, which is the same as the field of view in an ordinary camera. (Bottom) The orbital path of the polar-orbiting satellite shown above results in a sigmoidal (S-shaped) path on the Earth's surface directly beneath the satellite.

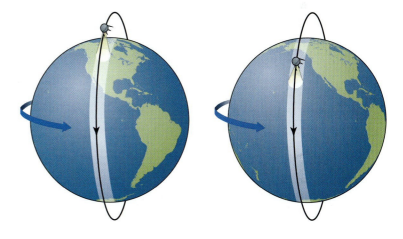

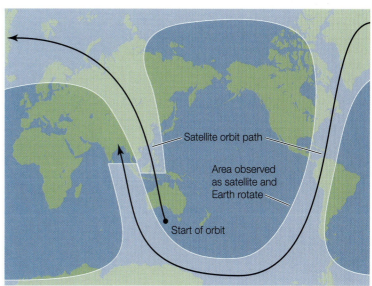

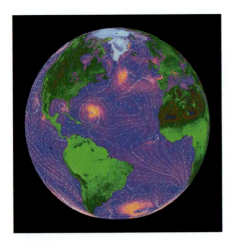

FIGURE F An active radar image showing the ocean wind field, measured on September 20, 1999. This shows two ocean storms in the North Atlantic and one in the South Atlantic.

based on their surface wave patterns (**FIGURE F**). Unlike passive radiometers, active radar sensors can function day and night and can penetrate clouds.

Today, oceanographers, meteorologists, and Earth scientists depend on these satellites for what has become almost routine applications: informing us in real time what the surface water temperatures are in the Equatorial Pacific Ocean; whether the Northwest Passage is open or closed because of ice cover; the exact locations of Gulf Stream eddies and the core of the Gulf Stream current itself; where hurricanes are about to develop from tropical disturbances; where, when, and how intense the spring phytoplankton bloom is in any particular area; what the wave heights are anywhere in the world ocean; and so much more. They have revolutionized our fields of research to the point that we cannot imagine what we would do if they were to be discontinued.

APPENDIX B
El Niño and La Niña

The term "El Niño" was coined more than a century ago by fishermen along the South American coasts of Ecuador and Peru to refer to an anomalously warm ocean current that appears every few years, usually around Christmas (hence the Spanish word *el niño*, the Christ Child). The fishermen took notice because the unusually warm waters brought with them a virtual collapse of their fishery—the fish just seemed to disappear—a situation that would sometimes last all winter and into spring. Accompanying that warm-water current was usually very heavy rainfall in the Western Equatorial areas of South America, areas that normally are among the driest on Earth. Eventually, scientists learned that this oceanographic phenomenon in the Pacific Ocean off South America affects not only the local fisheries and regional weather, but is responsible for drastic weather changes around the world.

The periodic warming of the surface waters off the west coast of South America is the result of changes in the Trade Winds and Equatorial Currents in the Pacific Ocean. As we have discussed, both the North and the South Pacific Equatorial Currents flow from east to west, with each current comprising a portion of the North and South Pacific gyre systems, respectively (see Chapter 5). Between these two currents, the Pacific Equatorial Countercurrent flows in the opposite direction, from west to east. Each of these three major ocean currents owes its existence to the Northeast and Southeast Trade Winds. Normally, the Southeast Trade Winds drive an upwelling of cold waters that reach the surface along the coast of South America (Ekman upwelling), but upwelling of colder waters also occurs along the Equator far out into the Pacific Ocean as a result of divergent upwelling along the Equator-straddling South Equatorial Current. As long as the Trade Winds blow normally, we see cold surface waters off South America in addition to cold surface waters extending to the west along the Equator, as seen in **FIGURE A** (top), a composite satellite image

FIGURE A (Top) Normal conditions, December 1993. (Bottom) El Niño conditions, December 1997.

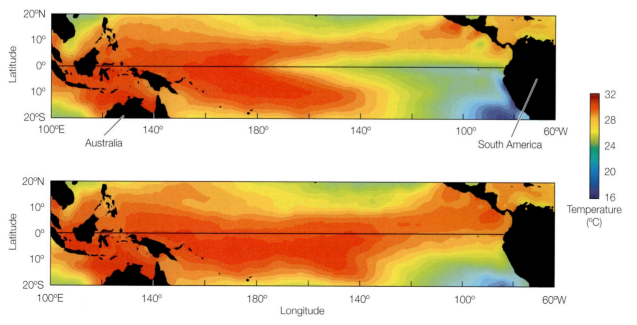

of sea surface temperature for the month of December during 1993, which was "a normal year." Much farther to the west, in the western tropical Pacific Ocean, we see some of the warmest ocean waters in the world.

But every few years the Trade Winds relax. When they relax, the wind-driven currents flowing to the west—the North and South Equatorial Currents—also relax, and transport far less surface water to the west. This relaxation allows a huge volume of warm water in the western Pacific to flow back to the east and toward South America as a series of large, wave-like phenomena. The result is significant warming of surface waters in the eastern Equatorial Pacific (which is El Niño); this can be seen in **FIGURE A** (bottom), which shows sea surface temperatures in December during the 1997 El Niño. With the Trade Winds relaxed and the South Equatorial Current much reduced, what little upwelling continues along the Equator and along the western coast of South America now brings only relatively shallow water to the surface—water that is both warmer than normal and low in the dissolved nutrients necessary for plankton growth. The result is that the fish populations collapse or move away, and weather systems change. This El Niño condition usually lasts the better part of a year, well into the following spring.

It was Sir Gilbert Walker, a British scientist working in India in the 1920s, who first noticed that atmospheric pressure in the eastern and western tropical Pacific (specifically, in Tahiti and in Darwin, Australia) reversed every few years. Normally, the eastern Equatorial Pacific is an area of relatively high atmospheric pressure due to the colder surface waters there, compared to much warmer waters farther to the west. (Remember, however, that on average, the Equator is a band of low pressure, but along it there can be areas of relatively higher or lower pressure.) In some years, atmospheric pressure in the east would drop unexpectedly, accompanied by a rise in atmospheric pressure in the west. And when this flip-flop in pressure occurred—specifically, when pressure in the east dropped—Walker noticed that weather patterns around the world changed in a predictable manner: there would be floods in South America; droughts in Australia, India, Indonesia, and parts of Africa; and western Canada would have mild winters. He coined the term *Southern Oscillation* to describe this phenomenon of oscillating atmospheric pressure in the South Pacific and he came up with indices to characterize the trends. His High Index described the normal condition, which corresponds with relatively higher pressure in the east and lower pressure in the west; his Low Index described the lower-than-normal atmospheric pressure in the east with relatively higher pressure in the west. His Low Index is what we now know is the El Niño condition. (You will also see the term *ENSO* in reference to the coupled phenomena, *El Niño–Southern Oscillation*.)

When the atmospheric pressure changes along the Equatorial Pacific Ocean, so too does the atmospheric convection cell above the Equator. Because there is no Coriolis effect directly on the Equator, relative differences in atmospheric pressure along the Equator result in atmospheric convection cells that are oriented east–west, rather than north–south (the way the Hadley Cells are, for example). This atmospheric convection cell directly on the Equator is known as a **Walker Cell**, named for Sir Gilbert. The Walker Cell, under normal conditions, produces heavy rains in the western Pacific, where atmospheric pressure is low, and warm, humid air is rising and condensing into rain (**FIGURE B**, top). In the east, however, only cool, dry air falls back to Earth, helping (along with the coastal mountains there) to create a very dry climate in western South America. But during El Niño events, when the warm water in the west moves eastward, the Walker cell is displaced as the low atmospheric pressure follows the warm water. Now it rains much farther east—out over the Pacific Ocean and even over South America—while in the west, there may be droughts; this results in the Walker Cell splitting (**FIGURE B**, bottom).

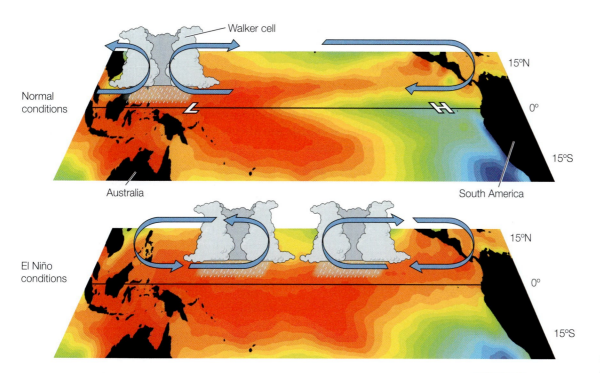

Often, this period of El Niño conditions lasts throughout the winter and into the spring. In some cases, it is followed not only by a return to normal conditions, but also by what might be described as exaggerated normal conditions, in which the Equatorial Pacific not only becomes colder in the east once again, but that cold water extends much farther to the west, as shown in **FIGURE D** for sea surface temperatures for the month of December 1998, following the 1997 El Niño year. This condition is known as a La Niña. Now the Walker Cell is compressed toward the west, as shown in the diagram.

FIGURE B Under normal conditions, the Walker Cell is over the warmest waters in the western Equatorial Pacific, but during El Niño conditions, when the warmest waters extend out over the central Equatorial Pacific, the Walker Cell becomes split.

FIGURE C La Niña conditions, December 1998.

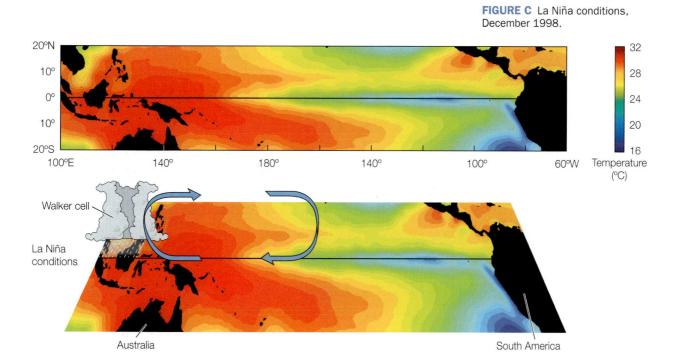

In each of these extreme cases of both water temperatures and atmospheric pressures shifting in the Equatorial Pacific Ocean, the paths of the Jet Streams are altered too, in the process altering weather around the world (also noted by Sir Gilbert).

You can follow oceanographic observations being made in the Pacific Ocean today as scientists monitor water temperatures with an array of moored instruments, satellites, and shipboard measurements. You can also watch how long-term weather forecasts are affected (visit www.pmel.noaa.gov/tao/elnino/nino-home.html).

Exploring the Deep, Dark Ocean

For centuries the ocean depths have fascinated naturalists, and indeed, as we have already discussed, questions about what, if anything, lives in the deep sea were among the prime motivations behind the *Challanger* expedition. But, our curiosity is not be piqued by simply bringing up bottom samples with dredges; instead, we feel compelled to actually go down to the ocean depths ourselves.

In the 1930s, engineer Otis Barton and explorer, naturalist, and writer William Beebe teamed up to build a deep diving submersible, named the *Bathysphere* (**FIGURE A**), with which they intended to do what no human being had done before: observe marine life forms that reside in the ocean's deepest and darkest depths firsthand. In 1932, after many months spent designing and constructing an underwater vessel that could withstand the immense pressure at great depths, the *Bathysphere* was bolted closed with Barton and Beebe inside, and they were lowered by crane from a surface ship into the clear blue Atlantic Ocean waters off Bermuda. They narrated their descent, which was broadcasted live on national radio; this was an event not unlike the first Moon landing in 1969 in the interest it generated. From their 5-inch porthole, they observed the disappearance of light at about 450 m and the presence of various bioluminescent organisms that drifted by (which became the subjects of famous illustrations from the time). They reached a depth of 932 m (3028 feet)—exceeding their goal of a half mile—and in so doing, they set a deep diving record that would stand for another 21 years.

In 1953, the Swiss scientist Auguste Piccard and his son Jacques Piccard descended to 3050 m (10,000 feet) in the Mediterranean Sea in the Italian-built, un-tethered *Trieste* (**FIGURE B**). The submersible had two main parts: a large cylindrical chamber filled with 22,000 gallons of gasoline, used to regulate buoyancy, and a much smaller pressure sphere for the occupants. They later sold it to the U.S. Navy and, on January 23, 1960, Jacques Piccard and U.S. Navy officer Don Walsh, took the *Trieste* to the bottom of the Challenger Deep, the deepest part of the Mariana Trench in the western Pacific Ocean. They reached 10,911 m, setting a deep diving record that has not been duplicated, and which cannot be broken, because the Challenger Deep is deepest spot in the world ocean. The *Trieste* was only on the bottom for about 30 minutes. Through its porthole, the two occupants

FIGURE A William Beebe and the *Bathysphere*.

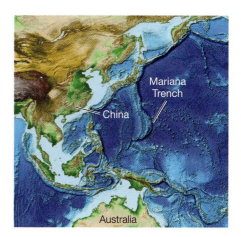

FIGURE B (Left) The *Trieste*, hoisted out of the water. The top part is the gasoline chamber and the much smaller bottom sphere is where observers sit. (Middle) Don Walsh and Jacques Piccard in the *Trieste*. (Right) The Mariana Trench.

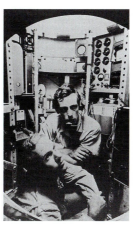

FIGURE C (Left) The DSV *Alvin* at the surface, ready to be retrieved aboard the mother ship, and (right) *Alvin* being deployed.

were able to observe fishes (with eyes) and jellyfish swimming about—finally dispelling the notion that nothing is alive at the oceans' deepest depths. When they hit the bottom, they created a cloud of fine, white, dust-like particles, the biogenic ooze that is the remains of siliceous frustules of diatoms.

Deep sea exploration like these early heroic feats has all but ended. Such explorations today are limited to relatively shallow depths compared to the depths to which the *Trieste* voyaged. But the deep submergence vessel DSV *Alvin*, perhaps the most famous of the several manned deep sea submergence research vessels currently in operation around the world, still carries scientists to the bottom of the oceans (**FIGURE C**). Built in 1964 and overhauled numerous times over the years, the *Alvin* is the oldest deep-sea submersible in use today, but even after nearly 50 years, it remains an active research submersible. It is owned by the U.S. Navy and operated by the Woods Hole Oceanographic Institution. The *Alvin* can reach depths of about 4500 m (**FIGURE D**), and has logged more than 4000 dives. Its claims to fame include locating a lost hydrogen bomb in the Mediterranean sea in the 1960s, discovering fascinating life forms in and

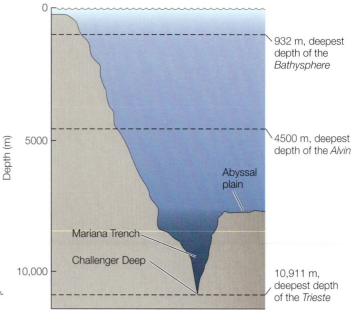

FIGURE D Diving depths of the *Bathysphere*, *Alvin*, and *Trieste*.

FIGURE E (Left) The RMS *Titanic* in 1911. (Right) An ROV exploring a piece of the RMS *Titanic* at 3800 m.

around deep-sea hydrothermal vents in the late 1970s, and exploring the wreck of the RMS *Titanic* in 1986.

Carrying human beings to the great depths of the oceans is dangerous and is not always necessary. For many applications, remotely operated vehicles (ROVs) meet the needs of scientists and industry (**FIGURE E**). While still tethered to the surface ship, these vehicles can perform many functions at depth using their robotic arms, such as manipulating objects, collecting bottom samples, and taking photographs. It was with this technology in 1985 that Robert Ballard discovered the wreck of the *Titanic*, which sank in 1911 in about 3800 m of water in the North Atlantic Ocean.

Glossary

A

abiogenesis The supposed creation of living organisms from nonliving matter; this has apparently happened only once.

absorption The taking up of electromagnetic energy by molecules, such as when an object absorbs sunlight and becomes warmer.

abyssal zone The marine benthic environment between 4000 and 6000 m depth.

abyssopelagic zone The pelagic marine environment between 4000 and 6000 m depth.

acantharians Single-celled organisms, in a suborder of Radiolaria, comprising protozoa with a skeletal structure of strontium sulfate.

accessory pigment Light-capturing photosynthetic pigments in algae that are in addition to chlorophyll *a*.

acid A chemical that donates protons (hydrogen ions) and/or accepts electrons.

acorn worms Hemichordate worm-like organisms that possess a proboscis.

active continental margins Tectonically active margins of continents that face a plate boundary.

adenosine diphosphate (ADP) An organic molecule involved in cellular metabolism that stores chemical energy upon the acquisition of a phosphorus atom, converting it to adenosine triphosphate (ATP).

adenosine triphosphate (ATP) An organic molecule involved in cellular metabolism that releases chemical energy upon the loss of a phosphorus atom, converting it to adenosine diphosphate (ADP).

ADP *See* adenosine diphosphate.

aerobic To require air, which generally means to require molecular oxygen.

Agassiz, Alexander (1835–1910); son of famous Harvard professor, Louis Agassiz, and proponent of John Murray's theory of coral atoll formation.

Agassiz, Louis (1807–1873); Swiss paleontologist, glaciologist, and geologist; a famous Harvard professor who opposed Darwin's theory of evolution.

Agnathans A primitive group that includes the hagfishes and lampreys. Also called jawless fishes.

algae A large and diverse group of photosynthetic, autotrophic organisms, including unicellular and multicellular forms.

ambergris A waxy substance secreted in the intestines of sperm whales; used as a preservative in the manufacture of perfumes.

amino acids Molecules that are the chemical building blocks of proteins.

ammonium NH_4^+; one of the forms of dissolved inorganic nitrogen; a nutrient.

Amnesic Shellfish Poisoning (ASP) Memory loss that results from ingesting shellfish that have concentrated the toxin domoic acid, produced by diatoms of the genus *Pseudo-nitschia*.

amphidromic point The central point in a rotary tidal wave where the tidal range is zero.

amphidromous Refers to a form of fish migration between fresh and salt water at some life cycle stage other than the breeding.

amphioxus Another common name for a lancelet, marine invertebrate in the subphylum Cephalochordata.

Amphipoda An order of crustacean invertebrates that lack a carapace and have laterally compressed bodies.

amplitude The height of a wave above a position at rest; equal to one half the wave height.

ampullae of Lorenzini Network of pores that are sensory organs in elasmobranchs that detect electric fields.

Amundsen, Roald (1872–1928); Norwegian explorer and the first to navigate between the Atlantic Ocean and Pacific Ocean by way of the Northwest Passage.

anadromous Refers to fish that swim into rivers from the sea to breed.

anaerobic In the absence of molecular oxygen.

anammox Stands for *anaerobic ammonium oxidation*, the conversion of ammonium to nitrogen gas.

anguilliform Eel-like.

anthozoan A class in the phylum Cnidaria that includes the sea anemones and corals.

apex predator The top of the trophic pyramid; top carnivores.

aphotic zone The depth in the ocean below which there is insufficient light for vision or photosynthesis.

appendicularians Suspension feeders that build a "house" from a mucopolysaccharide material that it excretes; water is pumped through the house allowing the animal to filter out food particles. Also called the larvaceans, or simply the pelagic tunicates, they are in the class Appendicularia.

aquaculture The artificial rearing of marine or freshwater organisms for commercial purposes.

Archaea A group of prokaryotes that constitute one of the three taxonomic domains.

Archimedes' Principle An object placed in water or any other fluid will be buoyed up by a force equal to the weight of the displaced fluid.

Aristotle (384–322 B.C.); Greek philosopher and scientist.

arthropod An invertebrate animal that possesses an exoskeleton with a segmented body and jointed appendages.

asexual reproduction In reference to phytoplankton, whereby a cell divides into two daughter cells. Also called vegetative cell division.

assimilated Ingested matter that has been digested and incorporated into new body tissue.

asthenosphere The upper layer of the mantle that has fluid-like properties.

athecate Absent a theca, or external cuticle-like shell.

atmospheric pressure The pressure on the surface of the Earth, or at other elevations or altitudes, that results from the weight of the air above that level.

atoll Encircling coral reefs with an interior lagoon that form around a volcanic island as it subsides into the asthenosphere.

atom The smallest unit of a chemical element that retains the chemical properties of the element.

ATP *See* adenosine triphosphate

attenuated The loss of light as it penetrates down into the ocean.

autotrophs Living organisms that produce their own food by photosynthesis or chemosynthesis.

autumn equinox When the sun is directly over the Equator as summer transitions to winter.

auxospore A life history stage in diatoms that results from sexual reproduction in diatoms; the auxospore will develop into a new diatoms that is larger than its parent cells.

Avogadro, Amadeo (1776–1856); Italian chemist who determined the number of elementary particles (atoms or molecules) in a mole; known as Avogadro's Number, 6.02×10^{23}.

azoic theory The idea proposed by Edward Forbes (1815–1854) that the ocean below 300 fathoms (ca. 550 m) is devoid of life.

B

Bacon, Francis (1561–1626); English philosopher, statesman, scientist, lawyer, jurist, and author, who speculated on the curious fit between Africa and South America.

bacteria Prokaryotic organisms without an organized nucleus that constitute one of the three taxonomic domains.

Baker, Howard B. A geologist who published an article in *The Detroit Free Press* (1911) proposing that an original supercontinent broke apart to give us today's continents.

baleen plates Specialized anatomical structures in the upper jaws of baleen whales, made of keratin, which allow them to filter feed. Also called whalebone.

baleen whales The group of whales characterized by possessing baleen plates.

bar-built estuary Embayments behind barrier islands, where freshwater meets seawater.

barnacle Crustacean invertebrate in the subgroup Cirripedia; as adults are sessile filter feeders built of overlapping plates, and glued to rocks and other hard substrates.

barrier islands Relatively narrow islands that run parallel to the mainland coast, built up from accumulations of terrestrial sediments in the shallow continental shelf waters just offshore of low-lying coastal areas.

barrier reef Coral reefs that grow along the coastline some distance offshore.

basalt Volcanic rock, which makes up oceanic crust.

base A chemical that is a proton acceptor.

basket stars A group of brittle stars characterized by their elaborately branching arms.

bathyal zone The benthic marine environment beyond the edge of the continental shelf, generally between the depths of 200 and 4000 m.

bathypelagic zone The open water, or pelagic, marine environment generally between the depths of 1000 and 4000 m.

Beaufort Wind Force Scale An empirical measure based on observations at sea that relates wind speed to ocean wave heights.

Beer's Law A mathematical expression that relates the attenuation of light with depth in the ocean: $I_z = I_0\, e^{-kz}$ where I is the light intensity; z is the depth in meters and k is the attenuation coefficient.

Benioff, Hugo (1899–1968); American seismologist and Cal Tech professor, who in 1940 plotted the locations of earthquakes around the edge of the entire Pacific Ocean, which revealed a pattern that became known as the Pacific Ring of Fire.

benthic Referring to the bottom of the ocean.

benthos Referring to living organisms on the bottom of the ocean.

Big Bang The explosion that created the Universe; the term was coined by astronomer Fred Hoyle.

Bigelow, Henry Bryant (1879–1967); Considered the father of modern oceanography, he was a Harvard professor and the first director of the Woods Hole Oceanographic Institution; he is best known for his pioneering studies of the physical oceanography, plankton and fishes of the Gulf of Maine and adjacent waters.

binomial system The genus and species names of organisms.

biogenesis The tenet that states that life arises only from pre-existing life, not from nonliving materials.

biogenous sediment Marine bottom sediments that are accumulations of the hard parts of once living organisms.

biological oxygen demand The consumption of oxygen by decaying organic matter or by respiratory demands of living organisms.

bioluminescence The production and emission of light by chemical processes in a living organism.

biomass The mass of tissues or living organisms.

biosynthesis The chemical synthesis by living organisms of organic molecules and tissues.

biotoxin Chemical substance produced by organisms that are toxic to other organisms that ingest them.

bioturbation The physical action by animals in the benthos that overturn bottom sediments.

bivalve mollusc Mollusc that possess a pair of calcium carbonate shells.

blade The flattened, leaf-like structure of macroalgae.

blowhole The breathing hole on the top of the head in the cetacea.

blue-green algae Common name for cyanobacteria.

BOD *See* biological oxygen demand.

bony fishes The group of fishes with ossified skeletons in the Class Osteicthyes; these are the modern fishes.

brachiopods An invertebrate phylum of suspension feeding lophophorates, characterized by a pair of calcium carbonate valves that resemble bivalve molluscs.

branching corals A group of stony corals characterized by their elaborately branching calcium carbonate base structures.

brevetoxin The biotoxin produced by the dinoflagellate, *Karenia brevis*; responsible for Florida's red tides.

brittle stars Sea stars, or star fish, in the class Ophiuroidea characterized by their long, thin and mobile arms that can be used for locomotion.

brown algae Common macroalgae, better known as kelp and seaweed.

bryozoans A phylum of lophophorates; a group of marine invertebrates also known as the moss animals.

Bullard, Sir Edward (1907–1980); Cambridge University scientist who, with colleagues J. E. Everett and A. Gilbert Smith, used a computer to find the best statistical fit of the continents, lending support to earlier suspicions that the continents were once connected to one another.

buoyancy The upward-directed force exerted by a fluid that opposes an objects weight, and is equal to the weight of displaced fluid.

bycatch The unintentional catch of marine organisms by a fishing operation that are usually discarded.

C

calanoids The most common and abundant family of copepods.

calcareous Pertaining to being made of calcium carbonate.

calcareous ooze Marine bottom sediment composed of calcium carbonate skeletal materials of once living organisms.

calcium carbonate compensation Refers to the calcium carbonate compensation depth, where the waters become too acidic for calcium carbonate sediments to accumulate.

calorie The energy required to raise the temperature of 1.0 gram of water by 1°C, which is equal to 4.1868 joules.

capillary wave A surface water wave less than 1.73 cm wavelength, for which the restoring force is the surface tension of water.

carbonate Calcium carbonate ($CaCO_3$).

carnivore An animal that consumes other animals. Also known as a meat eater.

Carpenter, William Benjamin (1813–1885); an officer in the Royal Society who, with Charles Wyville Thomson, promoted the *Challenger* expedition; it was they who first coined the term "oceanography" in its present context.

cartilaginous fishes The class Chondrichthyes; the group of fishes with a cartilaginous skelton.

catadromous Refers to a fish that lives in freshwater and enters saltwater to breed.

catch per unit effort The commercial fish catch per fishing boat per unit time; used as an indicator of the health of commercially exploited fish stocks.

centric Referes to the radial symmetry of centric diatoms.

centrifugal force The apparent, outward-pulling force on an object traveling in a curved path.

centripetal force The inward-pulling force that counteracts centrifugal force.

Cetacea A class of vertebrates that includes the whales and dolphins.

CFC *See* chlorofluorocarbons.

chaetognaths A group of marine planktonic invertebrates in the Phylum Chaetognatha. Also known as the arrow worms.

chemical equation of life $CO_2 + H_2O \leftrightarrow CH_2O + O_2$.

chemosynthesis A form of primary production based on the oxidation of inorganic molecules (e.g., hydrogen gas, hydrogen sulfide) or methane as a source of energy, rather than sunlight, as in photosynthesis.

chimaeras A group of fishes in the Subclass Holocephalii that split off from the elasmobranchs 400 million years ago, but remain similar to them in their having, among other features, a cartilaginous skeleton.

chitons A group of marine molluscs with a dorsal shell made up of eight separate shell plates or valves.

chloride cells Specialized cells in the gills of bony fishes used in osmoregulation by excreting excess salts.

chlorofluorocarbons (CFC) A group of manufactured greenhouse gases.

Chlorophyceae The green algae.

chlorophyll The photosynthetic pigment found in most plants and algae that absorbs sunlight and uses its energy to synthesize carbohydrates from CO_2 and water.

Chondrostei A group of mostly cartilaginous-like fishes that exhibit significant ossification; this group includes the sturgeons and birchirs; and, the Neopterygii, which include the gar fishes.

chronometer An accurate clock used in navigation aboard ships at sea.

Ciguatera Fish Poisoning (CFP) A disease that afflicts people who have ingested tropical reef fishes that have bioaccumulated toxins produced by dinoflagellates of the genus *Gambierdiscus*.

ciliates A group of protozoans characterized by their possession of hair-like organelles called cilia used in feeding and locomotion.

cirrus (plural cirri) Long, conical structures made of fused cilia, used in locomotion and feeding in many invertebrates.

cluster effect The phenomenon of expanding volume when clusters of hydrogen bonds form among water molecules.

coastal plain estuaries A geomorphological classification of estuaries, where a river valley carved by glacial scour is flooded as the sea level rose after the last ice age, and where today, the sea mixes with the river water in relatively narrow channels. Also called drowned river valley estuaries.

Coats, Robert R. (1910–1995); U.S. Geological Survey scientist who described the origin of the Aleutian Islands off Alaska in a 1962 paper; also described mechanism of plate subduction.

coccolithophores A group of single celled phytoplankton that precipitate around themselves plates of calcium carbonate.

cohesion The attraction between water molecules that creates hydrogen bonds.

Columbus, Christopher (1451–1506); Italian explorer credited with the modern discovery of the New World in 1492.

columnar corals A group of stony corals that form skeletal structures of calcium carbonate that form column-like features.

compensation light intensity The light intensity sufficient to enable photosynthesis at a rate that matches respiratory loses.

compensation point The point where photosynthesis and respiration are equal; may be a point in time or depth in the water.

Condrichthyes The class of cartilaginous fishes, includes sharks, skates, rays, and chimaeras.

conduction The passage of heat through an object.

cone snails A group of predatory marine gastropod snails of the genus *Conus* with cone shaped shells, often brilliantly colored. All are venomous.

constructive interference The phenomenon when waves interact such that the crest of one wave encounters the crest of another, producing a wave height equal to the sum of their individual wave heights.

continental crust Crustal materials that make up the world's land masses, comprising mostly granite.

continental drift The movements of lithospheric plates on the surface of the Earth.

continental shelves Extensions of the continents that have become flooded by the ocean, usually extending to the depth of about 200 m.

convection The movement of a heated or cooled object or material that results from a change in density.

convection cells Coupled high and low pressure systems whereby upward and downward motions of a fluid are joined together into a continuous flow circuit.

convergence The point where fluid motions come together and either rise or fall.

convergent evolution The independent evolution of similar features among organisms of different ancestral lines.

convergent plate boundary The boundary between lithospheric plates that are coming together with either one riding over the other, or both plates pushing up mountain ranges between them.

Cook, James (1728–1779); British explorer who undertook three major voyages of discovery in the late 1700s, and is credited with discovering and charting New Zealand and Austrailia's Great Barrier Reef.

Copepoda The subclass of crustacean zooplankton, the copepods, which are the most abundant in the oceans.

copepodid One of six developmental stages of copepods that follows the naupliar stages.

coral bleaching The phenomenon whereby corals expel their zooxanthellae, causing the coral to lose its coloration.

coral reef A massive calcium carbonate structure secreted by coral in shallow tropical waters.

de Coriolis, Gaspard-Gustave (1792–1843); French scientist and mathematician who worked out the nature of motions in a rotating frame of reference; today called the Coriolis effect.

Coriolis effect The tendency of moving objects on a rotating sphere, such as the Earth, to be deflected.

Coriolis force Synonymous with Coriolis effect, but in the context of a physical force that deflects moving objects on rotating spheres.

cosmogenous sediment Sediment particles that fall from space as tiny fragments of meteorites.

covalent bond The strong chemical bond that results when two atoms share an electron.

crest In reference to waves, the highest point, or top, of a wave.

crinoids Echinoderms in the class Crinoidea are characterized by a stalk

attached to the bottom with a mouth at the other end surrounded by tentacles. Also known as sea lilies.

critical depth The depth at which water and phytoplankton can be mixed and still receive an average illumination such that photosynthesis equals respiratory loses.

critical period With reference to fish survival, the period when a larva has exhausted its yolk reserves and must begin exogenous feeding on plankton; a time that is usually associated with high mortalities if planktonic food abundance is low.

critical steepness The wave angle at which the surface gravity wave cannot support itself, and the wave breaks. Defined as when the ratio of wave height to wavelength exceeds 7:1.

crust In reference to the Earth's internal structure, the thin upper layer; may be continental or oceanic crust. Often used synonymously with lithosphere.

crustaceans A large group of arthropods, distinguished by possession of biramous limbs (two parts), and by a nauplius stage as larvae. Includes crabs, barnacles, etc.

ctenophores Members of a phylum of marine "gelatinous" zooplankton, which possess characteristic "combs" or rows of cilia used in swimming.

cubomedusae Jellyfish in the class Cubozoa; known for their potent venom, which is among the deadliest of any creature on Earth. Also called box jellyfish.

Cushing, David H. (1920–2008); British fisheries scientist best known for the match-mismatch hypothesis of larval fish survival.

cuttlefish A group off cephalopod molluscs in the order Sepiida.

cyanobacteria A group of photosynthetic aquatic bacteria. Also known as bluegreen algae.

cyclone An atmospheric system characterized by a low pressure center and circular wind motions around it; winds circle counterclockwise in the Northern Hemisphere and clockwise in the Southern hemisphere.

cyclonically Refers to the direction of rotating winds around a low pressure cell.

cyclopoids A family of copepods.

cyprid Larval stage of barnacles, following the naupliar stage; characterized by a pair of hinged valves.

D

dark energy The hypothesized energy that is responsible for the accelerating expansion of the universe.

Darwin, Charles (1809–1882); English naturalist and author of the *Origin of Species*; originator of the modern theory of evolution.

dead zone An area in the ocean or an estuary where bottom waters have been depleted of dissolved oxygen.

decapods "Ten-footed"; an order of crustaceans that includes among other lobsters, prawns and shrimp.

deep sea General term that refers to waters beyond the continental shelf and deeper than about 2000 m.

deep sound channel The depth in the ocean where sound waves are trapped as they refract toward the base of the permanent thermocline.

deep water gravity waves Surface gravity waves in waters where the depth is greater than 1/2 the wavelength.

deforestation The cutting of forests, usually for the creation of land for agriculture.

demersal fishes Fishes, usually on continental shelves, that feed and reside on and just off the bottom. Also called groundfish.

density Mass per unit volume; symbol is the Greek letter Rho, ρ.

denticles Scales on cartilaginous fishes; structurally similar to vertebrate teeth.

deposit feeder Feeding on sediment deposits, either surface sediments or subsurface sediments; examples are amphipods and polychaete worms.

destructive interference The phenomenon when waves interact such that the trough of one wave enounters the crest of another, producing a sea surface elevation that, if the two original waves are the same wave heights, the two waves cancel out one another, leaving a flat sea surface.

detritus The particulate remains of once living organism(s).

dew point The temperature at which the air can hold no more water vapor, and it condenses into droplets (dew).

diadromous Fishes that migrate freely between seawater and freshwater.

diapause A period of suspended development or growth, especially in copepods (also in insects).

Diarrhetic Shellfish Poisoning (DSP) A human disease resulting from the consumption of shellfish that have accumulated the toxin okadaic acid in dinoflagellates of the genus *Dinophysis*.

diatom ooze Deep-sea bottom sediment of which 30% or more is made up of the siliceous remains of diatom frustules.

diatoms Unicellular, eukaryotic autotrophs characterized by a siliceous frustule; usually planktonic (members of the phytoplankton).

diazotroph Organism that metabolically fixes, or converts, atmospheric nitrogen gas (N_2) into a biologically usable form, such as ammonia (NH_4^+), usually performed by bacteria and archaea.

diel vertical migrations This expression refers to organisms such as fishes and zooplankton, but also some flagellated phytoplankton, that exhibit vertical excursions over the course of a 24-hour day, often swimming to shallow depths at night, and to deeper depths during the nighttime.

diffraction The phenomenon whereby a wave bends as it passes around the end of an obstruction, such as an ocean wave passing the end of a jetty, which is a rock wall that extends out into the ocean.

diffuse attenuation coefficient The value k in Beers Law, $I_z = I_0 e^{-kz}$; it can vary between 0 and 1.0.

dinoflagellates A group of unicellular, eukaryotic autotrophs characterized by possession of a pair of flagella; usually planktonic (members of the phytoplankton).

diphycercal Refers to the form of caudal fin where the vertebral column extends all the way to the end of the tail, such as in the coelacanths and chimaeras.

dispersed The phenomenon whereby waves of different wavelengths travel at different speeds, and upon propagating away from the area where they are formed, will sort themselves out into groups of similar wavelengths.

disphotic zone The depth range in the water column below the euphotic zone, but still within the photic zone, where there is sufficient light for vision but not for photosynthesis; usually extends no deeper than 200 m.

dissolved inorganic nitrogen (DIN) Refers to most common, inorganic, reactive forms of dissolved nitrogen, used as nitrogenous nutrients by phytoplankton; includes nitrate (NO_3^-), nitrite (NO_2^-) and ammonium (NH_4^+).

dissolved organic material (DOM) A broad range of organic molecules dissolved in seawater that come from either once living organisms, or excretions from living organisms, and include materials that color the water, similar to the coloring one sees when making a cup of tea.

dissolved organic nitrogen (DON) A range of organic molecules that includes urea, uric acid, amino acids and others, all of which are dissolved in seawater and come from once living organisms, or excretions from living organisms.

disturbing force A force that causes the sea surface to be displaced, making a wave; can include earthquakes, wind, and objects thrown into the water.

diurnal Opposite of nocturnal, refers to an activity during daylight hours; also used more loosely to indicate an activity that occurs daily.

divergence To move apart. In reference to atmospheric circulation, refers to air masses that descend over a surface high pressure cell and then spread laterally; also refers to air masses that rise above a surface low pressure cell, and diverge at high altitudes by spreading laterally. In the oceans, refers to the spreading of waters that, for example, rise to the surface as in the Pacific Equatorial Upwelling system.

divergent plate boundary Tectonic plates that are moving apart from one another, results from the formation of new crustal material from rising magma, such as occurs in the mid ocean ridge systems.

Doldrums The oceanic waters in the vicinity of the Intertropical Convergence Zone, where air masses rise, creating an area of little horizontal winds for use in sailing.

domain The system of three taxonomic groups, on molecular techniques, that are the highest taxonomic ranks: the three domains are Bacteria, Archaea, and Eukarya.

Doppler Effect The shift in frequency and wavelength of waves that results when a wave source is moving with respect to the medium within which those waves are propagating.

drowned river valley estuary A geomorphological class of estuary; produced by rising sea level flooding ancient valleys carved by glaciers and ancient rivers during times of lower sea level. Also called coastal plain estuaries.

dynamic soaring The phenomenon whereby sea birds take advantage of vertical air motions created by surface waves on the ocean.

E

echinoderms A group of radially-symmetrical invertebrates possessing a coelom, or body cavity; also characterized by spiny skin. Includes the starfishes, sea urchins, sand dollars and sea cucumbers.

echolocation Used by the toothed whales, and other animals, to locate prey and other obstacles by producing sound waves that reflect off objects and return to the animal, which interprets the signal patterns.

ecosystem-based management A relatively new approach to the management of commercial fisheries, whereby the all components of the ecosystem, not just the targeted species in the fishery, are included in management plans.

ectotherms Animals that control their body temperature through external means; their body temperatures take on the temperature of their environment.

Ekman current The top few tens of meters of the ocean under the influence of the wind; the integrated flow of which is exactly 90° to the right of the wind direction in the Northern Hemisphere and to the left in the Southern Hemisphere.

Ekman depth The depth to which the wind influences the surface Ekman current.

Ekman layer That top layer of ocean above the Ekman depth.

Ekman spiral The downward twisting of the direction of current layers in the Ekman layer.

elasmobranchs A group of cartilaginous fishes that includes sharks, rays, and skates.

electric charge A property of matter that departs from neutrality at a point or in an object upon the accumulation or acquisition of electrical particles (electrons and protons).

electromagnetic spectrum The complete range of electromagnetic radiation frequencies from the lowest, radio waves, to the highest, gamma rays; includes visible light.

electrons The negatively-charged particle in an atom.

element A pure chemical substance made up of a single type of atom, defined by its atomic number, which is the number of protons in its nucleus.

encrusting corals Refers to corals that exhibit growth across the surface of rocks and other substrate upon which they grow, with little or no upward growth.

epifaunal Refers to animals that grow on top of something; for example, epifaunal benthic invertebrates reside on the bottom of the ocean, versus beneath the sediment surface.

epipelagic zone The portion of the water column down to a depth of about 200 m.

equinox The phenomenon of the Sun being directly over the Equator; occurs twice a year: on the first day of spring, which is the vernal equinox, and the first day of fall, which is the autumnal equinox (with reference to the Northern Hemisphere).

Eratosthenes of Cyrene (276–194 B.C.); a Greek astronomer, geographer, mathematician and third librarian of the Great Library at Alexandria. He was the first person to calculate accurately the circumference of the Earth.

errant polychaetes A general group of poylchaetes based on life style; includes those that are motile animals and can actively move about.

estuary A semi-enclosed coastal body of water that has a free connection with the open sea and within which seawater is measurably diluted with fresh water derived from land drainage.

Eukarya One of the three domains of life; includes organisms whose cells contain complex structures enclosed within membranes, such as a nucleus.

Euphausiacea An order of shrimp-like crustaceans that are among the macrozooplankton, and include krill.

euphotic zone The depth in the water column above which there is sufficient light for photosynthesis; sometimes defined as the depth to which only 1% of the surface light penetrates.

eutrophication The condition of nutrient enrichment and resulting increased biological productivity and biomass.

evaporites Refers to the materials left behind, especially salts, upon the evaporation of seawater; these are important constituents of marine sedimentary deposits.

exoplanets Planets outside our own solar system; planets that orbit other stars.

exponential decay The exponential loss of something; the decrease is proportional to the amount remaining at any one time, such that it behaves according to the equation: $dN/dt = -N$; the solution is: $N_t = N_0 e^{-t}$ where N is the amount of something, and t is time.

extinction coefficient The attenuation coefficient, k, in Beers Law; *see* Beers Law.

extratropical cyclone A storm that develops outside the tropics when waves form along thermal fronts, usually along the boundary between the Polar Easterlies and the Prevailing Westerlies, and grow into low pressure cells.

F

fall overturn Refers to the convective mixing of freshwaters when heat losses at the surface in fall and winter increases the density of surface water there, which sinks, resulting in a concomitant upward mixing of deeper water.

fast ice Sea ice that is grounded; versus ice sheets that float.

fecundity The relative reproductive potential as measured by the numbers of eggs produced per breeding cycle.

Ferrel Cell Named for William Ferrel (1817–1891), it is the atmospheric circulation cell that is predicted in theory between approximately 30° and 60° latitude in each hemisphere.

fetch The distance over which a wind acts to build ocean waves.

filter feeder Refers to marine and aquatic animals that extract suspended particulate food materials (e.g., they filter) out of the water.

fin ray Bony rodlike structures that lend support to fins of the bony fishes.

fisheries production The amount, or biomass, of commercially exploited fishes or invertebrates, usually given in tons fresh weight, that can be produced per unit area of ocean; depends on the primary production and number of trophic linkages.

fishery Refers to the business of catching fish, and usually also refers to the targeted fish, fish group, or invertebrate.

fishery science The scientific study of fluctuations of fish stocks in relation to natural variability and human exploitation of the resources.

flagella A tube-like structure in bacteria and protists, usually equal to or greater than the cell length, that can be produce wave-like motions used in locomotion.

flatworms Members of the phylum Platyhelminthes, only one class of which is free-living, the turbellarians (class Turbellaria); they get their name from their flat, ribbon-like or leaf-like shapes; most are predators, and all live in water or in moist terrestrial environments.

flotsam Anything that is floating at sea, especially the wreckage of a ship and its cargo.

fluke The tail fin of whales and dolphins; oriented laterally, versus the tail fin of most fishes, which are oriented dorsoventrally.

fluoresce The action of a fluorescent substance, which absorbs light energy at one wavelength, or range of wavelengths, and emits light of longer wavelength (and lower energy).

foam line Line of foam that develops along the surface of the ocean in areas where water masses meet, with the less dense one riding up and over the other denser mass.

foliaceous coral Coral that develops thin, delicate leaf- and flower petal-like structures.

food chain The chain, or straight-line character, that describes the passage of food from primary producers to primary consumers, to secondary consumers, moving up one trophic level each step in the chain, etc.

food web Analogous to food webs except that the passage of food up to higher trophic levels may be direct or indirect, passing first through a number of intermediaries along the way.

foraminifera A phylum of single-celled protists that possess a snail-like shell of calcium carbonate; may be benthic or planktonic, the skeletal remains of which are important components of marine biogenic sediments.

Forbes, Edward (1815–1854); an Edinburgh University professor who argued that there is no life in the sea below a depth of 300 fathoms (ca. 550m); known as the azoic theory.

foredune A ridge of sediment, usually beach sand, that runs parallel to the shoreline and which is stabilized by vegetation, such as beach grass.

fossil fuels Ancient organic carbon compounds, formed originally by living biomass, including oil, coal, and natural gas, which are burned for energy.

fouling organism Organism that becomes established on surfaces of various structures, such as dock pilings and ship hulls, often considered nuisances.

Francis Bacon (1561–1626); English philosopher, statesman, scientist, lawyer, jurist, and author, who speculated on the curious fit between Africa and South America.

fringing reef One of the three main types of coral reefs (along with barrier reefs and atolls) that grows parallel to the shoreline facing the open sea and may or may not protect a shallow lagoon behind it.

front The intersection of two air masses, or water masses, whose physical properties differ from one another (for air masses: temperature, and humidity; for water masses, temperature, and salinity).

frustule Name given to the siliceous structures that diatoms precipitate around themselves.

fully developed sea A sea state that has reached the maximum wave height for a particular wind speed.

G

galaxy A collection of about 200 to 400 billion stars and remnants of stars, all kept together by their gravitational attractions to one another.

gametes The reproductive cells that fuse together in sexual reproduction to form a zygote.

gastropods A large group of invertebrates in the phylum Mollusca commonly known as snails and slugs.

gelatinous zooplankton The larger form of zooplankton that includes salps, ctenophores, chaetognaths, and various medusae; characterized by their "Jello"-like consistency.

geostrophic currents Density driven currents where there is a balance of forces between the density gradient and the Coriolis force.

gill net Fish net based on the principle of a fish being caught up in meshes as it tries to swim through it, only to be unable to pull back out as the mesh becomes caught on the gill openings.

gill raker Bony protrusion on fish gills that prevent unwanted objects from passing down the fish's throat; in some species they are greatly enlarged and are used to filter planktonic food particles.

glacial marine sediment Sedimentary particles that become deposited as glaciers melt, and release terrestrial materials that are embedded in the ice.

glacial periods Ice ages; the period between warm, or interglacial periods; may last for more than 100,000 years.

global climate change The change in conditions on Earth throughout Earth's history, as well as projected into the future; includes changes such as sea level and global temperatures.

Global Positioning System (GPS) The system of Earth-orbiting satellites used in navigation; are capable of triangulating one's geographic locations to a few meters.

global warming The phenomenon of increasing Earth temperatures that began in the late 1800s; also refers to warming episodes following glacial periods in Earth's distant past.

Gondwanaland The term coined by Eduard Suess for the ancient supercontinent that comprised the once-connected but now widely separated Southern Hemisphere land masses. Also spelled "Gondwana-land."

gorgonian corals Also known as sea whips and sea fans, these are group of soft corals in the subclass Octocorallia that are found throughout the world ocean, especially in the tropics and subtropics; they are named for the firm but flexible protein complex, gorgonin, secreted by the polyps that gives them their structure.

granite The principle rock that comprises continental crust; rich in the aluminum, silicon and oxygen-containing minerals; density is about 2.7, making it a lighter rock than basalt.

gravitational force The attractive force due to gravity, which is a property of anything that has mass.

gravity wave A wave on the surface of water, or at a density interface such as a pycnocline, where gravity serves as the restoring force.

grazers A general term given to heterotrophs that feed on primary producers.

great ocean conveyor belt Name given by Wallace Broecker (1931–) to the phenomenon where the deep ocean's thermohaline circulation transports heat and materials throughout the world ocean.

green algae A large group of both unicellular forms of phytoplankton and multicellular macroalgae in the class Chlorophyceae; most are freshwater with relatively few marine forms.

greenhouse effect The phenomenon where visible wavelengths of light freely pass through glass, or other substances such as greenhouse gases in the atmosphere, but upon being converted to longer wavelength infrared radiation, are less capable of passing in the other direction; this is how a greenhouse is heated by the sunlight.

greenhouse gases Trace gases in the atmosphere that allow passage of visible wavelengths of light but are less transparent to long wavelengths, thus trapping heat in the atmosphere and the Earth's surface; gases include water vapor, carbon dioxide, methane, nitrous oxide, and ozone, as well as CFCs (chlorofluorocarbons).

groundfish Nickname given to species of demersal fishes that feed and reside on or near the ocean bottom, usually on continental shelves.

group speed The speed of surface gravity waves that travel as a group, with individual waves in that group traveling at twice the group speed; this is possible only because as the wave outruns the group, it disappears and another wave appears at the back of the group, thus conserving energy.

growth The process of biosynthesis of new living biomass; may be expressed as somatic growth, which is the production of new body tissues in individual organisms, and reproductive growth, which is the increase in numbers of individuals by way of reproduction.

gulping A feeding behavior of the rorqual baleen whales, where the whale opens its mouth while swimming into a patch of planktonic or nektonic prey, closing its mouth around a large volume of prey and water, made possible by the pleat-like throat furrows that expand, forcing water out of its buccal cavity and past its baleen plates, which retain the food organisms.

guyot A flat-topped undersea mountain formed by volcanoes that emerged above the sea surface and were eroded flat before subsiding.

Gymnosomata An order of naked pteropods, shell-less planktonic molluscs. Also called sea angels.

gyres The large-scale general current systems operating in the ocean basins, to depths of 800 to 1000 m.

H

hadal zone The marine benthic environment at depths greater than 6000 m.

Hadley Cell Named for George Hadley (1685–1768), it is the atmospheric circulation cell that is predicted in theory between the Equator and approximately 30° latitude in each hemisphere.

hadopelagic zone The pelagic marine environment at depths greater than 6000 m.

Haeckel, Ernst (1834–1919); German biologist, naturalist, philosopher, physician, professor and artist, best known today for his beautifully-detailed illustrations of various marine organisms.

hagfish One of the agnathans, or jawless fishes, in the order Myxiniformes; they are exclusively marine, lack actual vertebrae in their backbone, and lack paired fins.

Haldane, J. B. S. (1892–1964); British biologist who suggested in a 1929 paper that Earth's primitive atmosphere was exposed to enough lightning and ultraviolet radiation to produce some of the complex organic (carbon-based) molecules that make up living things.

halocline The depth in the water column where salinity changes abruptly, usually from shallow waters that are fresher to deep waters that are saltier.

harmful algal blooms They are oceanographic phenomena where certain species of phytoplankton, which may be toxic or nontoxic, bloom to high cell densities and produce various harmful environmental effects, such as oxygen depletions and fish kills; toxins may also affect humans who eat contaminated seafood, especially shellfish. Also known as red tides.

harpacticoids A family of copepods, most common in the benthos, but with members in the plankton as well.

Harrison, John (1693–1776); British clock maker who developed the first chronometer, or clock, that could keep accurate time aboard a ship at sea.

He, Zheng (1371–1435); Chinese explorer from the province of Yunnan who conducted numerous ocean voyages between 1405 and 1433, during which may have been the first use of a magnetic compass.

heat A form of energy proportional to motions of atoms or molecules, and which can be transmitted through solid and fluid media by way of conduction or convection, and through a vacuum by radiation.

heat capacity A physical property of material that is characterized by the amount of heat required to change a substance's temperature by a given amount.

Henry the Navigator (1394–1460); a Portuguese prince responsible for stimulating maritime exploration and trade with other continents; advanced modern navigational methods.

herbivorous To feed on plants, or algae.

Hess, Harry (1906–1969); Geologist and Princeton professor who developed one of the first theories of sea floor spreading; also discovered guyots during his service in the U.S. Navy during World War II.

heterocercal Refers to the asymmetrical caudal fin of sharks, where the dorsal lobe is larger than the ventral lobe.

Heteropoda A small order of planktonic gastropod molluscs.

heterotrophs Animals; organisms that acquire their nutrition from exogenous sources, from feeding on other organisms.

Hjort, Johan (1869–1948); Norwegian fisheries scientist who focused attention on the early life history of fishes as possible critical periods in their survival.

holdfast Refers to the sinuous, tube-like structures that attach macroalgae such as the kelps to a substrate.

Holmes, Arthur (1890–1965); British geologist who described the basic mechanisms of sea floor spreading and plate subduction.

holoplankton Plantonic organisms that spend their entire life cycle in the plankton.

homeotherms Organisms that maintain their body temperature by metabolic activity, usually at a constant level that is warmer than ambient temperatures.

Horse Latitudes Oceanic regions at latitudes of about 30° north or south, where high atmospheric pressures result in little wind for use in sailing; origin of term is uncertain.

horseshoe crabs Strange-looking marine crustaceans related to spiders, which are common in near shore sandy and muddy environments; they are basically a big shell beneath which hides a spider-like animal, with five pairs of walking legs.

hot spots Geographically stationary area in the upper mantle where magma escapes to the surface, which can form undersea volcanic mountains and island chains, such as the Hawiian Islands and Emperor Seamounts in the Pacific Ocean; also responsible for terrestrial features such as the hot springs and geysers in Yellowstone National Park.

Hubble, Edwin (1889–1953); American astronomer who discovered that distant galaxies in all directions are flying away from us at speeds proportional to their distance, thus providing evidence for the Big Bang.

Hubble Space Telescope Named for Edwin Hubble, a sophisticated telescope in Earth orbit, carried into space by the Space Shuttle in April of 1990, where it has been providing stunning images of space ever since.

hurricane A tropical cyclone with sustained winds that exceed 64 knots (74 mph).

Huxley, Thomas Henry (1825–1895); British biologist and staunch supporter of Darwin's theory of evolution; helped to promote the *Challenger* expedition.

hydrogen bond A type of weak chemical bond that is the result of attractive forces between two particles with partial electric charges of opposite polarity, such as between water molecules.

hydrogenous sediment Marine sediment that is the result of chemical precipitation of solutes, such as salts.

hydroid A group of carnivorous, colonial cnidarian polyps that have a reproductive medusa stage, called a hydromedusa.

hydromedusa The reproductive, free-living medusa stage of a hydroid; a small jellyfish.

hyperiid amphipods A group of planktonic crustacea, with no carapace and generally with laterally compressed bodies.

I

ice ages Periods in Earth's history lasting tens of thousands of years when global mean temperatures were reduced and the extent of continental ice coverage, polar ice sheets and alpine glaciers were expanded.

ice shelves Floating extensions of a glacier or ice sheet that flows to a coastline and onto the ocean surface; most common in Antarctica.

ice-rafted Terrestrial materials that have become trapped in the frozen mass of glaciers, which upon melting in the ocean, drop their loads into the sea.

Ichthyoplankton The egg and larval stages of fishes that are planktonic.

in phase Meaning for two or more wave phenomena to be lined up such that their crests occur together, and their troughs occur together.

indeterminant growth The phenomenon whereby an organism grows throughout its life.

inertial oscillation A phenomenon where a moving object under the influence of the Coriolis effect will continue to deflect as it moves, eventually tracing out an arc and thereby returning very near to its starting position; these complete loops are referred to as oscillations.

infaunal Refers to organisms that reside below the sediment-water, or sediment-air interface, versus living on the top of the sediment surface.

intensification of Western Boundary Currents A phenomenon that results because of changes in the Coriolis force with latitude and causes ocean currents on the western edges of ocean basins to be intensified, to become narrower and swifter, regardless of the hemisphere or the direction the currents are moving.

interglacial period The warm climatic condition on Earth between glacial periods, lasting tens of thousands of years.

intermediate gravity wave Surface gravity wave that occurs in water depths that are between one-half and one-twentieth of their wavelengths.

internal wave Ocean wave that propagates beneath the air-water interface, such as those that move along a pycnocline.

International Whaling Commission (IWC) Eighty-nine nations that meet each year to review the status of whale populations and make recommendations for regulating the taking of whales; formed in 1946 among 15 whaling nations.

interstitial water The water that resides in the spaces between sediment particles.

intertidal zonation The vertical patterns of distributions of algae and animals in the intertidal zone controlled by physical stresses from wave energy and amount of time submerged or exposed, and biological stresses from predation and competition for space.

intertidal zone The portion of ocean bottom between high tide and low tide. Also called the littoral zone.

Intertropical Convergence Zone (ITCZ) The meteorological Equator; the low pressure convergence zone between the two Hadley cells; migrates north in the Northern Hemisphere summer and south in winter.

invasive species A non-native species that has been introduced to a new area, or which has extended its geographic range.

invertebrates Animals without a backbone.

ionic bond The chemical bond that results when one atom acquires one or more electrons from another atom; an example is the bond between sodium and chlorine atoms in ordinary table salt, NaCl.

isopods Crustacean invertebrates in the class Isopoda; their chief anatomical feature is their dorsal-ventral compression.

isostasy The buoyant suspension of the Earth's lithosphere in the asthenosphere.

isostatic equilibrium The attainment of a balance between gravity and buoyancy of the lithosphere, such as when the mass changes under the influence of ice between glacial and interglacial periods, for example.

isothermal No change in temperature; for example, the lack of measureable change in temperature in the top several tens of meters of the upper water column during winter convective mixing events.

iteroparity The reproductive strategy of an organism that reproduces more than once over the course of its lifetime.

J

jawless fishes Members of the class Agnatha, a primitive group of fishes that lack articulated jaws; includes the hagfishes and lampreys.

jetsam Pieces of a boat in distress that have been intentionally thrown overboard to avoid sinking by lightening their load; also refers to anything intentionally discarded at sea, such as solid waste.

Johansen, Frederick (1867–1913); Norwegian arctic explorer who accompanied Fridtjof Nansen in 1893 on a four year unsuccessful expedition to reach the North Pole.

K

keystone species A species whose predatory and competitive activities exert significant control, disproportionate to the species' abundance, on the structure of an ecosystem.

kleptoparasitism A feeding behavior whereby one animal steals another's prey or stored food supply.

kleptoplasty A form of symbiosis in which intracellular organelles from algae are sequestered by a host organism that has ingested that alga, such as happens when chloroplasts are retained and continue to function, giving the new host organism photosynthetic capability.

Kuhn, Thomas (1922–1996); American philosopher of science who proposed that science advances not by incremental steps, but by great leaps in understanding, which he called paradigm shifts.

L

lampreys One of the two groups of extant jawless fishes; the other group is the hagfishes.

LAN *See* local apparent noon.

lancelets A group of primitive, fish-like marine cephalochordates. Also known as amphioxi.

larvaceans Solitary, free-swimming members of the class Appendicularia. Also known as tunicates.

latent heat Hidden heat; there is no change in temperature as latent heat is either acquired or released.

latent heat of melting Hidden heat associated with melting ice; refers to the heat required to break water's hydrogen bonds such that it changes phase from a solid to a liquid; no change in temperature occurs as this heat is added to the ice.

latent heat of vaporization Hidden heat associated with evaporating water; refers to the heat required to break water's hydrogen bonds such that it changes phase from a liquid to a gas; no change in temperature occurs as this heat is added to the water.

lateral line system A sensory system of fishes and some amphibians that detects pressure changes from physical disturbances in the water; made up of individual receptors just beneath the skin on the head; often arranged in a line along the sides of animals.

latitude The position north or south of the Equator on the surface of the Earth; measured in degrees.

latitudinal heat pump The phenomenon whereby atmospheric circulation patterns and ocean currents transport heat from the tropics and subtropics to higher latitudes.

Laurasia One of two ancient supercontinents, the other being Gondwanaland, that included the Northern Hemisphere land masses.

Law of Constant Proportions Maintains that the proportions of the major ions in seawater do not vary with location in the world ocean or with total salinity; salinity may vary greatly, but the proportions of salts will not.

Lemaître, Georges (1894–1966); a Belgian priest and scientist who proposed that the Universe began with "the explosion of a primeval atom" based on observations of a "red shift" in distant galaxies, which he related that to a model of the Universe based on Einstein's theory of relativity.

light intensity The amount of sunlight striking the Earth's surface or the amount of light penetrating to a particular depth in the ocean.

light limited Situations in which there is insufficient light for photosynthesis, such as winter in high latitudes and in deep waters everywhere.

limiting factor A factor, such as light, nutrients, or temperature, which may set an upper boundary on the amount of biological activity that can occur, such as in phytoplankton photosynthesis or growth.

limiting nutrients The macronutrient elements nitrogen, phosphorus, or silicon; also refers to micronutrients, such as iron; these can be in short supply in the oceans, and therefore set a limit to primary production and growth of phytoplankton.

limpets A group of gastropod molluscs that have a greatly flattened shell; common in most intertidal and subtidal areas that support coralline algae, which is one of their preferred foods.

Linnaeus, Carl (1707–1778); Swedish naturalist and physician credited with having established the modern convention of the binomial system of taxonomic nomenclature based on observable characteristics of plants and animals.

lithosphere The rigid outer layer of the Earth; includes the outermost portion of the mantle with similarly rigid physical properties.

lithospheric plates The distinct pieces of the Earth's lithosphere. Also called tectonic plates.

littoral zone *See* intertidal zone.

lobe-finned fishes The Sarcopterygii, a clade of bony fishes; characterized by possession of fleshy lobes at the proximal ends of the pelvic and pectoral fins; the tetrapods evolved from this group of fishes.

Local Apparent Noon (LAN) The time of day when the Sun is at its highest point in the sky, midway between sunrise and sunset.

longitude The position east or west of the Prime Meridian on the surface of the Earth; measured as degrees.

longitudinal flagellum One of the two flagella in dinoflagellates; extends posteriorly and provides forward propulsion for swimming short distances.

longshore current The current produced along a shoreline by surf arriving at an angle to that shoreline.

longwave Wavelengths of light in the electromagnetic spectrum that have longer wavelengths than visible light; includes infrared wavelengths.

lophophorates A group of marine invertebrates that includes the bryozoans, the phoronid worms, and the brachiopods, all of which possess a lophophore.

lophophore A cluster of ciliated tentacles used in food capture and in respiration in some marine invertebrates, including the bryozoans, the phoronid worms, and the brachiopods; may be circular, spiral, or horseshoe shaped.

lorica The external skeletal structures of some protozoans such as tintinnids; from the Latin word that means "body armor."

lunar Refers to the Moon.

lunge feeding A feeding mode exhibited by some of the rorqual baleen whales, such as the humpback, in which the whale swims upward with its mouth open into a patch of planktonic or nektonic food, often breaking the surface.

M

macroalgae Microscopic algae; phytoplankton.

macronutrients Nutrients that comprise 1–5% of tissue weight of phytoplankton, including nitrogen, phosphorus, and silicon.

macrozoooplankton Zooplankton measuring 2–20 cm.

Magellan, Ferdinand (1480–1521); Portuguese explorer, who with five ships under his command led the first expedition to circumnavigate the globe.

magnetic compass Used in navigation; an instrument in which a magnetic needle is delicately balanced such that it points to the magnetic North Pole.

magnetometers An instrument that can measure the intensity and orientation of magnetic fields; can be towed behind ships to record magnetic fields in the ocean floor.

mammalian diving response The phenomenon whereby marine mammals reduce their heart rate and constrict blood vessels, limiting flow of blood to their extremities during deep ocean dives; a vestige of this is retained in human children.

mangals Forests of mangrove trees.

manganese nodule Rock that is rich in manganese that precipitates on the deep ocean floor; one of the forms of hydrogenous sediments.

mangrove forest Common along gently sloping, protected intertidal shorelines in the tropics and subtropics; comparable to salt marshes.

mantle The inner layer of the Earth between the crust and the outer core; it is about 2900 km (1800 mi) thick and is similar in chemical composition to oceanic crust in that it is made up of compounds of oxygen and silicon but has more of the metals magnesium and iron.

marine pollution The introduction by humans, directly or indirectly, of substances or energy to the marine environment resulting in deleterious effects.

mass A quantity of matter that has weight in a gravitational field.

massive corals Stony corals that form large masses, usually in simple nearspherical shapes.

match-mismatch hypothesis The early life history stage in fishes in which food resources in the yolk have been exhausted and the larva must begin to feed on plankton, which may be present in high densities (match) or not (mismatch); a mismatch is thought to result in greater mortalities and reduced recruitment; proposed by David Cushing.

Maury, Matthew Fontaine (1806–1873); U.S. Naval officer who wrote *The Physical Geography of the Sea* in 1855; considered the father of physical oceanography.

maximum sustainable yield The theoretical maximum catch of fishes that is possible for a stock that will still allow the population to be maintained through reproduction.

megalopa The last larval stage in decapod crustaceans; resembles the adult.

meiobenthos Benthic organisms that range in size from 0.1–1.0 mm.

melon The enlarged front portion of the head in toothed whales; used in focusing sound waves for echolocation.

meridian A line of longitude; usually refers to each 15° of longitude, corresponding to one hour's rotation of the Earth.

meroplankton The planktonic stage of an organism that spend only a portion of its life history in a planktonic stage, usually an early life history stage.

mesopelagic zone The open ocean between the depths of 200 and 1000 m.

mesosphere The layer below the asthenosphere in the Earth's interior; it is the less fluid-like, rigid part of the middle and lower mantle, and extends to a depth of about 2900 km (1800 mi).

mesozooplankton The zooplankton in the size range 200 µm to 2 cm.

metazoan A multicellular heterotrophic organisms, as opposed to a unicellular heterotrophic organism such as a protozoan.

meteorological equator *See* intertropical convergence zone.

microalgae The unicellular algae, which may be planktonic, or benthic.

microbial loop The cycling of dissolved organic carbon produced in photosynthesis through heterotrophic bacteria to nanoflagellates which, like phytoplankton, are consumed by zooplankton, forming a loop for that dissolved organic carbon to cycle back into the main food chain.

micronutrients Nutrient elements that make up less than 0.05% of the biomass of living organisms, especially phytoplankton; includes S, Na, Cl, Mg, B, Mg, Zn, Si, Co, I, F, Fe, Cu, and others.

microzooplankton Zooplankton in the size range 20–200 µm.

mid-ocean ridge The spreading ridges where new ocean floor is formed; generally found in the centers of the oceans, especially the Atlantic Ocean.

Miller, Stanley (1930–2007); University of Chicago graduate student who, in 1953, conducted a laboratory experiment with his advisor, Nobel Prize-winning chemist Harold Urey, where amino acids (building blocks of proteins) were formed in a mixture of water, methane, ammonia, and hydrogen gas subject to electric sparks.

mix vertically Waters that are vertically homogenized by way of physical mixing processes, which can include mixing by wind, waves, tides, and winter convection.

mixed tides Tides that show the influences of both solar and lunar components.

mixotrophic Organisms capable of functioning both as autotrophs and heterotrophs.

molecule A chemical compound made of two or more atoms of the same or different elements joined by chemical bonds.

moons Planet-like celestial bodies that orbit a planet, not a star.

motile Capable of movement or locomotion.

Murray, John (1841–1914); Second in command on the HMS *Challenger* expedition; proposed an alternative theory, later proved wrong, of coral atoll formation to that of Charles Darwin.

myomere A segment of body musculature of fishes that corresponds to the vertebrae.

Mysidacea The family of shrimp-like crustaceans to which the mysids belong.

Mysids Shrimp-like crustaceans, resembling eupahussids; some are benthic, some tychoplanktonic, and most are holoplanktonic.

N

naked ciliates Ciliated protozoans that lack an outer cuticle or lorica.

naked pteropods Planktonic molluscs in the order Gymnosomata that lack a calcium carbonate spiral shell; capable of weak swimming actions by flapping their wings which are projections of their modified molluscan foot.

nanoflagellates Unicellular autotrophic, heterotrophic, or mixotrophic protists in the size range 2–20 µm.

Nansen, Fridtjof (1861–1930); Norwegian explorer, scientist, diplomat and Nobel Peace Prize laureate, who discovered that there is an ocean beneath the Arctic ice cap.

nauplius (plural nauplii) Larval stage of crustaceans such as copepods characterized by a head and three attached limbs used for swimming and feeding.

Nautilus Common name of marine cephalopods in the family Nautilidae; characterized by a coiled, chambered shell and numerous tentacles.

neap tide The fortnightly period of tides when the Moon and Sun are approximately at 90° angles from the Earth; a period of reduced tidal ranges.

nebula Interstellar cloud of dust and gases that is the remnant of a super nova, the explosion of a star.

nektonic Organisms that swim; as opposed to planktonic organisms, those that drift.

nematocyst A specialized stinging cell in all Cnidaria that lines much of the integument of tentacles and functions almost like a miniature spear gun with venomous spears; may be innervated by even a slight touch, which causes the cell to discharge its spears into the flesh of the offending organism (or person) and in the process inject its venom.

nematodes The round worms in the phylum Nematoda; possess a digestive tract with orifices traditionally located at opposite ends; the head end, with teeth lining its mouth, is distinctively different from the tail end; have a well developed nervous system with sensory bristles that give the animals a sense of touch; sexes are separate.

Neopterygii A subclass of actinopterygian fishes; includes the freshwater gar fishes.

neritic sediment Marine bottom sediment that occurs on the continental shelf in water depths less than 200 m.

neritic zone The waters overlying the continental shelves. Also knows as shelf waters.

neuromast Sensory receptor just beneath the skin in fishes that detects minute water motions through tiny pores, capable of detecting such disturbances as are created by low-frequency noises and animal movements; usually arranged in the lateral line system.

Neurotoxic Shellfish Poisoning (NSP) The affliction associated with Florida's Red Tide, caused by a dinoflagellate, *Karenia brevis*, which produces the neurotoxin brevetoxin.

neutron Uncharged subatomic particle in the nucleus of atoms.

NH_4^+ Ammonium; one of the forms of dissolved inorganic nitrogen.

nitrate NO_3^-; the oxidized form of dissolved inorganic nitrogen; the final form in which dissolved inorganic nitrogen is found in greatest concentrations in the deep ocean.

nitrification The oxidation by bacteria of ammonium to nitrite and nitrite to nitrate.

nitrite NO_2^-; one of the oxidized forms of dissolved inorganic nitrogen; an intermediary between ammonium and nitrate.

nitrogen fixation The biological conversion of molecular nitrogen gas (N_2) to particulate organic nitrogen and ammonium.

nitrogen fixers The microorganisms that fix nitrogen. Also called azeotrophs.

NO_2^- Chemical formula for nitrite.

NO_3^- Chemical formula for nitrate.

nonindigenous species A species that is not native to the area in which it is found.

nonpoint-source pollution Pollution that cannot be traced back to an originating spot; includes agricultural runoff and atmospheric deposition, for example.

normal temperature effect The effect of temperature in controlling the molecular speed of vibrations of molecules of water, which controls the volume occupied and therefore its density.

Northeast Trade Winds The prevailing surface winds between 30° north latitude and the Equator, which blow toward the Equatorial band of low pressure, but become deflected by Coriolis, causing the winds to blow from the northeast to the southwest.

Northwest Passage The long sought-after northern sea route from the Atlantic Ocean to the Pacific Ocean; normally blocked by Arctic ice cover, except in late summer in recent years.

nuclear fusion The fusion of two hydrogen atoms to become a helium atom, a process that releases huge amounts of energy; the energy source in stars.

nucleus The proton(s), and neutron(s) when present of an atom.

nudibranchs Marine gastropods characterized by a lack of a shell during larval development and possessing exposed gills.

nursery grounds Areas on the oceans and estuaries where fishes tend to spawn such that their young have a favorable feeding environment.

nutrient elements The macronutrient elements N, P, and Si.

nutrient flux The delivery by upwelling, mixing, or diffusion of dissolved inorganic nutrients into surface waters where they can be taken up by phytoplankton.

nutrients The dissolved organic and inorganic nutrients in all their various chemical forms.

O

ocean acidification The increasing concentrations of carbonic acid in seawater that is accompanying increasing concentrations of carbon dioxide in the atmosphere.

oceanic crust The top layer of the Earth's surface that forms the ocean floor; composed primarily of basalt.

oceanic reef Coral reef system that occurs away from the shoreline.

oceanic zone The marine environment that includes the water column beyond the edge of the continental shelf.

octopus A cephalopod mollusc characterized by possession of eight tentacles.

oil seep Natural seepage of crude oil from the Earth's interior through cracks and fissures in the Earth's crust to the surface.

oligotrophic Marine and freshwater environments of low nutrient concentrations and therefore low rates of primary production and biomass.

omnivore An animal that ingests both plants and other animals.

ontogenetic vertical migration Seasonal vertical migrations as part of some marine organism's normal breeding cycle.

ooze Biogenic marine bottom sediment that contains 30% or more of a specific type of sediment material, such as siliceous oozes.

Oparin, Alexander (1894–1980); Russian biochemist who speculated on the origins of life more than 3.5 billion years ago.

operculum The bony gill covering in the bony fishes.

Orion Nebula A nebula in the Orion constellation about 1500 light years away from Earth that is visible to the naked eye.

Ortelius, Abraham (1527–1598); A Flemish cartographer, or map maker who was the first person to speculate that the continents were once connected.

osmoconformer An organism in which solute concentrations in its body fluids match those of its surroundings because it is unable to metabolically regulate solute concentrations in its internal body fluids.

osmoregulation The metabolic alteration of internal concentrations of solutes as necessary to maintain a constant internal milieu, different from that of their surroundings.

osmosis The tendency of water to pass though cellular membranes from media of low solute concentrations to media of higher concentrations.

Osteicthyes The taxonomic class of the bony fishes; the modern fishes.

otter trawl A commercial fishing net towed behind a boat that consists of a conical net with a wide opening kept open while being towed by two otter boards positioned such that they pull the nets apart to either side. Also called a trawler or a dragger.

out of phase Surface water gravity waves that come together such that the crest of one wave coincides with the trough of another.

outer core The outer, liquid portion of the Earth's central core.

oviparous Animals that produce eggs that hatch externally.

ovoviviparity Animals that produce eggs that hatch internally, thus giving live birth.

oxidation A chemical reaction involving the combination of a substance with oxygen; also a chemical reaction in which atoms lose electrons to another element.

P

P waves Primary waves; seismic compression waves.

Pacific Ring of Fire The pattern of earthquakes and volcanoes around the Pacific Ocean rim.

paleomagnetism A phenomenon discovered in the early 1900s whereby volcanic rocks, when heated and then cooled, acquire a stable magnetization that is oriented with the Earth's magnetic field.

Pangaea Name coined by Alfred Wegener for the ancient supercontinent that broke apart more than 225 million years ago to form Laurasia and Gondwanaland.

paradigm shift A phenomenon discussed by Thomas Kuhn whereby an important discovery or realization forever changes the way we view the world.

Paralytic Shellfish Poisoning (PSP) The affliction that results when humans ingest shellfish that have bioaccumulated saxitoxin from toxic dinoflagellates of the genus *Alexandrium*.

parapodia The paired appendages connecting to body segments in various invertebrates that function as limbs.

partially mixed estuary A classification of estuary based on the degree to which the freshwaters and seawater are mixed; there is some vertical mixing by tides, but not sufficient to overcome completely the buoyancy of a significantly greater volume of incoming freshwater at the head; there is a halocline separating relatively fresh water at the surface from higher salinity water beneath it; there is also a gradual longitudinal salinity gradient along the length of the estuary.

particle selection Copepod feeding behavior where apparent filter feeding also involves particle selection (and rejection) before ingestion.

particulate organic nitrogen (PON) The nitrogen contained in tissues of living organisms or in organic detritus.

passive continental margins Edges of continents that do not abut an oceanic plate boundary and therefore are not tectonically active.

peat The deposits of organic detrital material that, along with mineral sedimentary materials, build up into thick layers over time in marshes.

pelagic sediment Marine bottom sediment beyond the edge of the continental shelf.

pelagic zone The marine environment in the water column beyond the edge of the continental shelf.

pennate A type of diatom that is bilaterally symmetrical.

phagotrophically To engulf or ingest whole, as in one organisms feeding on another.

pheromone A chemical substance released by one organism that influences the behavior of another individual of the same species.

phoronids One of the lophophorate invertebrates, characterized by possession of a lophophore.

phosphate PO_4^{3-}; a form of dissolved inorganic phosphorus that is one of the nutrients.

phosphorite nodule A class of hydrogenous sediments that is a rock that has grown by chemical precipitation processes on the ocean floor.

photic zone The depth in the water column, or on the bottom, to which enough sunlight penetrates to allow photosynthesis; often defined as the depth to which 1% of the surface light penetrates.

photoautotrophs The photosynthetic primary producers that biologically synthesize new organic matter via photosynthesis.

photosynthesis The biologically mediated production of carbohydrates from water and carbon dioxide in the presence of photosynthetic pigments such as chlorophyll and light as an external energy source.

photosynthetic pigment Chemical compound that is responsible for absorbing and trapping light energy in photosynthesis.

phytoplankton The (usually) microscopic, single-celled photosynthetic algae and protists in the plankton.

pinniped "Fin-footed"; the group of marine mammals that includes the walruses, the fur seals and sea lions, and the seals.

piscivore Fish-eater; many fish are themselves piscivores.

planet A large celestial body that orbits a star and that is large enough that its own gravity has made it spherical and attracted to itself other smaller objects in the vicinity of its orbital path.

planetesimal A small planet formed by the collision in space of asteroids and other solid objects.

planktivore An organism that feeds on plankton.

plankton net A conical-shaped net of fine mesh used to collect plankton.

plankton Living organisms that are incapable of significant swimming and drift with the prevailing currents.

planula larva A short-lived developmental stage in the early life history of cnidarians.

plate One of many of Earth's lithospheric plates that makes up the surface layer of the Earth.

Pliny the Elder (A.D. 23–79); published *Natural History*, a work that stood as the authoritative science treatise for more than 1000 years and which included observations of ocean tides and the ocean's saltiness.

plunge diving A form of at-sea feeding performed by some sea birds whereby they dive into the water from the air at speeds that allow them to penetrate to depths sufficient to capture prey.

pluteus A developmental stage in the early life history of echinoderms in which the larva has prominent arm-like features.

pneumatocyst A gas chamber in macroalgae used for flotation of their blades for maximum exposure to sunlight for photosynthesis.

pneumatophore The root system of mangrove trees that have adapted to the anoxic muddy sediments by extending above the sediment surface, thus helping to oxygenate the buried roots.

pod A tightly packed school of fishes, such as a bait ball.

poikilotherm An organism that does not control its body temperature metabolically, but which acquires a body temperature the same as its surroundings. Also called cold-blooded.

point-source pollution Pollution from a source that can be identified or traced back to a specific source, such as an effluent pipe from a factory or a ship.

polar Geographic North or South Pole; also the portion of the globe above the Arctic Circle and below the Antarctic Circle.

Polar Cell The atmospheric circulation cells between 60° north or south and the poles.

Polar Easterlies The surface winds that blow out of the east between 60° north or south and the poles as part of the polar cells.

Polar Jet Stream The high-altitude, ribbon-like wind that flows west-to-east between the Hadley and Ferrel cells in each hemisphere.

polar molecule A molecule with opposite electric charges on it ends.

polychaetes One of the four classes of annelids; strictly marine, most have a wormlike appearance with elongated, segmented bodies, but some excrete shells of calcium carbonate.

polynya High-latitude areas of open water surrounded by sea ice.

polyp One of two life history stages of members of the phylum Cnidaria, the other being the medusa; in this stage, the organism attaches to a substrate and may form colonies.

PON *See* particulate organic nitrogen.

Popper, Sir Karl (1902–1994); German philosopher of science who is credited with establishing the modern scientific method and hypothesis testing.

post-larva A stage in fish development following absorption of the yolk sac.

Prasinophyceae A class of generally small (picoplanktonic size range) unicellular algae.

predation The act of animals capturing and ingesting other animals.

preen gland An oil producing gland common to birds used to keep feathers hydrophobic and dry.

Prevailing Westerlies The surface winds associated with the Ferrel Cells in each hemisphere.

primary producer An autotrophic organism at the base of the food chain.

primary production The act of fixing inorganic carbon into organic carbon by either chemosynthesis or photosynthesis.

Prime Meridian The 0° longitude line, or meridian, that runs through Greenwich, England.

proboscis An elongated extension, or appendage, from the head of an animal, either an invertebrate or a vertebrate, used in engulfing prey organisms.

prochlorophytes An informal name for photosynthetic prokaryotic phytoplankton.

progressive wave A wave that propagates across the surface of the ocean, as opposed to a standing wave.

prokaryote Bacteria and Archaea; cells that lack an organized, membrane-bound nucleus.

proton The positively charged subatomic particle.

proton acceptor A base with a negative charge; it chemically bonds with hydrogen ions (protons).

proton donor An acid with a positive charge; it provides hydrogen ions (protons) that chemically bond with bases.

protoplanet A large celestial object in the process of accreting mass which will eventually become a planet.

protostome A clade of animals in which the mouth develops first in embryonic stages; literally, "mouth first."

psammon Microorganisms that reside in the interstitial spaces between aquatic and marine bottom sediments.

pteropod A suborder of gastropods that are either shelled or naked planktonic animals capable of weak swimming abilities with their wings, which are modified extensions of their foot.

Ptolemy (ca. A.D. 90–168); Greek mathematician and astronomer.

purse seine A form of fish net with lines along the bottom that is deployed and then snatched tightly closed around a shoal of fish.

pursuit feeding A type of feeding in sea birds in which the birds actively chase prey underwater.

pycnocline The depth interval over which there is a change in density.

pycnogonid A class of marine arthropods. Also known as sea spiders.

Pythagoras (ca. 570–495 B.C.); Greek philosopher and mathematician.

R

radioactive decay The process whereby the nucleus of an atom loses a nuclear particle and emits energy.

radiolaria Microscopic amoeboid protozoa with elaborate siliceous spines.

radiolarian ooze Marine bottom sediments made up of 30% or more of siliceous radiolarian remains.

radula In gastropods, a tongue-like structure that is embedded with specialized teeth used to shred algae and other prey and to grind holes through the shells of other shellfish.

ray-finned fishes The actinopterygians, a subclass, or clade, of bony fishes characterized by the bony supporting structures running though their fins; this group includes the majority of modern bony fishes.

rays Cartilaginous fishes in the superorder Batoidea that are characterized by their flattened body, greatly expanded, wing-like pectoral fins, and long thin tails.

recruitment The process whereby young fishes have grown sufficiently large as to be susceptible to fishing gear, and are not part of the fishery for that species.

rectal gland A gland in sharks and rays located near the rectum that removes excess salt from the blood.

red algae A diverse group of macroalgae characterized by the accessory pigments phycoerythrin and phycocyanin, which effectively masks the green color of chlorophyll and gives some species their reddish coloration.

red tide A phenomenon that occurs when some species of phytoplankton or zooplankton reach high enough densities such that they discolor the waters; may or may not be toxic.

redox Short for *reduction-oxidation* reactions; coupled chemical reactions in which two atoms interact, such that one gives up electron(s) (it is oxidized) to another atom, which is reduced.

reduced The chemical condition attained upon the acquisition of one or more electrons.

refraction The bending of waves that result when a wave form encounters a medium that has a different of speed of wave propagation, such as light arriving at an angle to glass, or surf arriving at an angle to a beach.

reproductive strategy The particular reproductive biology and behavior of a species, implying an associated survival advantage afforded by such.

resonant frequency The frequency of a wave that matches the size of some structure within which that wave is oscillating, such that the wave is reinforced; an example is the resonance that results from a musical instrument.

respiration The oxidation, or burning, of organic matter by living organisms.

resting spore A cell formed by way of sexual reproduction by diatoms and other organisms during times of environmental stress that is capable of surviving prolonged periods of time in a state of suspended animation.

restoring force The force that acts to return a surface water wave back to its position of rest; may be surface tension, in the case of capillary waves, or gravity, in the case of surface gravity waves.

rhynchocoel A body cavity in nemertean worms that contains the proboscis.

ribbon worm Common name for worms in the phylum Nemertea (the nemerteans).

ridge-push, slab-pull model The two forces believed to produce motions of lithospheric plates away from the spreading ridges.

rift valley The valley that runs down the middle of mid-ocean spreading ridges.

rogue waves Unusually large ocean waves that have been responsible for sinking large ships.

rorquals A group of gulp-feeding baleen whales characterized by expandable furrows on their throats, making their buccal cavities elastic and greatly expandable; from French and Norwegian meaning "furrow whale."

Ross, Sir James Clark (1800–1862); British Naval officer and Arctic explorer famous for having discovered the Ross Sea and Victoria Land in Antarctica.

Ross, Sir John (1777–1856); Scottish explorer and Royal Navy Commander who searched unsuccessfully for a Northwest Passage, a northern sea lane connecting the Atlantic and Pacific Oceans, in 1818; he sounded to a depth of 1919 m near Greenland, and recovered a bottom sample.

rotifers A group of aquatic and marine microscopic multicellular or metazoan zooplankton in the class Rotifera.

roundworms Common name for members of the phylum Nematoda, usually referred to simply as the Nematodes, a group of highly successful invertebrates.

Ryther, John H. (1922–2006); American oceanographer who wrote an influential paper in 1969 that illustrated the limited potential of the oceans to produce fish with respect to the commercial harvest.

S

S waves Secondary waves; transverse seismic waves which cannot pass through liquids.

salinity The total quantity, by weight, of dissolved inorganic solids in seawater.

salps Planktonic, filter-feeding gelatinous zooplankton in the order Salpida in the class Thaliacea that are either solitary of colonial animals; they superficially resemble the ctenophores in having a bulbous gelatinous form, except that they are more barrel-shaped.

salt A chemical compound that is a metal bound with a non-metal.

salt gland A specialized organ for excreting excess salts in sea birds, marine reptiles, and elasmobranchs.

salt marsh Rich expanses of salt-tolerant plants, especially grasses, that develop at the high tide mark along gently-sloping shores of temperate and higher latitude areas that are protected from high energy wave action, and where fine sediments can collect.

salt wedge estuary A classification of estuary that is dominated by the volume of freshwater flow, such that there is little mixing between seawater and freshwater, but rather, a sharp horizontal and vertical salinity gradient.

sand dollar A group of flattened, burrowing echinoderms.

Sargassum A genus of pelagic, floating marcroalgae often found in extensive mats in the Sargasso Sea of the North Atlantic Ocean.

saxitoxin A neurotoxin produced by some species of dinoflagellates that causes Paralytic Shellfish Poisoning (PSP).

scattering The reflection of light off of particles and air bubbles in water.

scavenging A feeding mode whereby animals search for and ingest almost any kind of food, including carrion.

school A shoal of fish, almost always the same species, that have a uniform orientation and swimming speed, and which move as one.

scientific method The organized conduct of science that generally follows a progression from observations, to hypotheses, to experiment.

scurvy A human disease, common before the late 1700s on long sea voyages, that results from vitamin C deficiency.

Scyphomedusae The true jellyfish, in the class Scyphozoa, which undergo an alternating life cycle between a polyp and a medusa, in which the medusa stage is significantly more developed and much larger than the polyp.

sea Locally generated wind waves characterized by their asymmetry and steeply sloping wave fronts.

sea anemone Solitary cnidarian polyps that have a relatively large and heavy body form, and are attached to a substrate by a broad base pedal disk.

sea cows Common name for the sirenia; the dugongs and manatees.

sea cucumber An echinoderm that resembles a fleshy bag of water with retractable feeding tentacles.

sea grasses Submerged vascular plants, also called the submerged aquatic vegetation, common is shallow waters; includes the common eelgrass, genus *Zostera*.

sea lilies Echinoderms that, along with the feather stars, are members of the class Crinoidea.

sea otter The only marine otter and the smallest marine mammal; related to river otters, skunks, badgers, weasels, and the like.

sea slugs Common name for gastropod molluscs that have lost their shell.

sea spiders Common name for the pycnogonids, a group of marine arthropods.

sea squirts Common name for a group of suspension-feeding chordates in the subphylum Tunicata that, like all chordates, have a rudimentary notochord in their early larval development.

sea urchin An echinoderm characterized by its rounded shape and covering of moveable spines and tube feet.

sea wall Man-made walls, usually of stone, erected along shorelines in front of homes intended to protect property from storm waves.

seamount An undersea mountain of volcanic origin that never emerged above the sea surface.

secondary production The production, usually in grams of carbon per meter squared per year ($gC/m^2/yr$), of consumers of primary producers.

sedentary polychaetes Polycheate worms that reside in burrows or tubes.

sediment bed transport The tumbling and rolling of bottom sediment particles in a current.

sediment denitrification The conversion by bacteria of nitrate to molecular nitrogen (nitrogen gas).

sediment deposition The phenomenon of forming bottom sediments by the settling of particulates from the overlying water column.

sediment erosion An erosion process whereby a current resuspends and carries bottom sediments.

sediment straining The action by some whales that engulf bottom sediments and strain food organisms out of the sediments using their baleen plates; analogous to filter-feeding.

sedimentation rate The rate at which bottom sediments accumulate, usually given in millimeters per year in near shore environments, and in millimeters per 1000 years in the open ocean.

seismic waves Huge, low-frequency sound waves usually generated by earthquakes.

selective attenuation The wavelength-specific attenuation of light with depth in the ocean.

semelparity A reproductive strategy of some organisms in which all reproductive effort goes into a single breeding episode, reproducing only once in their lifetime, and usually dying soon afterward.

semidiurnal Twice a day, as in the case of semidiurnal lunar tides.

sensible heat Heat that can be measured as a change in temperature.

sessile Not motile; refers to organisms that burrow or are attached to a substrate.

setae The stiff bristle-like hairs of some organisms.

settling velocity The speed at which a sediment particle sinks through the water column.

shadow zone The location and depth beneath the surface where sonar signals will miss an object, such as a submarine, because of the refraction of sound waves as a function of depth and temperature.

shallow water gravity wave A surface gravity wave when it enters water with a depth that is one twentieth or less than the wavelength.

shelf reef A coral reef that forms shelf-like platforms.

shelf water Water overlying the continental shelves.

shelled pteropods Planktonic gastropod molluscs that have retained their coiled shells.

shoal An aggregation of fishes; distinct from a school of fishes.

shorebirds Numerous taxa of birds that spend significant time feeding on marine organisms in shallow waters but do not possess webbed feet.

shortwave The wavelengths of light in the visible range, about 400 to 700 nm.

SI Units The International System of Units, which is the modern metric system of measutrements.

siliceous Pertaining to silicon dioxide, or silica (SiO_2).

siliceous ooze Marine bottom sediments containing more than 30% siliceous remains of organisms.

silicoflagellate Unicellular phytoplankton that have a star-like skeletal structure of silica.

siphonophores Hydrozoans in the order Siphonophora that are colonies of hydroids arranged in a linear structure that sometimes form long trails of zooids several tens of meters in length, all ordered according to specific functions that the zooids are adapted for, including flotation (the gas bladder in some species), reproduction, feeding, and defense (or predation).

sirenia A group of marine mammals that includes the manatees and dugongs (the sea cows).

skates A group of elasmobranchs that closely resemble, and are related to, the rays.

skimming A feeding mode used by bowhead and right whales, whereby they swim through swarms of copepods allowing water, with the copepods in it, to enter their buccal cavities, and then push the water out, past the baleen plates, retaining the copepods.

slab-pull A hypothesis of tectonic plate movements and subduction, whereby the downward motion of plates in subduction zones pull the plate downward, while the downhill-flowing plates away from spreading ridges provide the push.

slope water The water column above the continental slope, beyond the edge of the continental shelves.

Snider-Pellegrini, Antonio (1802–1885); French scientist who proposed that Africa and Europe were once connected to the Americas, but then separated to form the Atlantic Ocean.

soft corals A group of cnidarian corals which do not produce calcium carbonate skeletons.

soft-bottom Muddy or sandy bottom sediments; to be contrasted with rocky bottoms.

solar Refers to the Sun.

somatic growth Growth of tissues, as opposed to growth in numbers of individuals through reproduction.

Southeast Trade Winds The surface winds associated with the Southern Hemisphere Hadley cell, between 30° south and the Equator; these winds blow from southeast to northwest.

Southern Ocean Sanctuary An area of the Southern Ocean around Antarctica that is a sanctuary for whales, where they may not be commercially harvested, by agreement of the International Whaling Commission in 1994.

specific heat The amount of hear required to raise one gram of a substance one degree centigrade.

spermatophore A sac containing sperm cells.

sponges Members of the phylum Porifera; simple and primitive suspension-feeding metazoans that do not have true tissues, but are held together in a rigid body form by networks of spicules made of either calcium carbonate or silica.

spontaneous generation The notion that life can be or was at one time created out of nonliving materials.

spring The season between winter and summer, beginning when the Sun crosses the Equator during the vernal, or spring, equinox.

spring phytoplankton bloom The rapid growth to high cell densities of phytoplankton in the spring at higher latitudes, when conditions of light and nutrients become optimal.

spring tide The fortnightly occurrence of above average tidal ranges in response to the additive gravitational effects of the Sun and Moon at times full and new phases of the Moon.

squid One of the cephlapod molluscs, characterized by possessing ten tentacles; relatively large size (especially the giant and colossal squids) with excellent swimming abilities.

standing wave A non-progressive wave; a seiche is an example, as is the vibrating guitar string.

star A massive body held together by gravity in which sustained hydrogen fusion reactions generate immense quantities of heat and light.

starfishes Echinoderms that have a star-like shape of five (sometimes more) arms.

steady state A balance between inputs and outputs, as in Earth's heat budget where the amount of incoming solar energy equals the amount of outgoing radiant energy.

Steller, Georg Wilhelm (1709–1746); German naturalist who served aboard a Russian navy ship that was shipwrecked on Bering Island, in the Bering Sea off Kamchatka; he survived and described numerous species named for him, including the Stellar sea lion and the now extinct Stellar sea cow.

stipe The stem-like portion of a brown macroalga.

stock–recruitment model A theoretical model that predicts the maximum catch of fishes that can be safely harvested without depleting the stock based on the level of recruitment and the size of the adult spawning stock.

Stokes Drift The phenomenon of net transport of water and flotsam in water where there are surface waves; results from the vertical distortion of wave orbits between the crest and the trough.

stony corals Corals that form calcium carbonate skeletal structures.

storm surge The elevated sea level associated with storms whose winds push a surface flow of water toward shore, which, in addition to the rise in sea level resulting from the low atmospheric pressure, can cause coastal flooding.

stromatolites Rocklike features in shallow seas made of sediment particles cemented together by organic secretions of dense colonies of these single-celled, filamentous, photosynthetic cyanobacteria.

subduction The phenomenon where one lithospheric plate slide beneath another.

subduction zone A location where one lithospheric plate sides beneath another, creating a deep sea trench.

sublittoral zone That marine environment beyond the intertidal zone but on the continental shelf. Also called a subtidal zone.

subtidal zone *See* sublittoral zone.

subtropical Those latitudes between the Tropic of Cancer and the Tropic of Capricorn, but some scientists extend that definition to include slightly higher latitudes, based on the tropical nature of the local climate.

Subtropical Jet Stream The thin, ribbon-like high velocity wind at high altitudes where the Hadley Cells and Ferrel Cells meet in each hemisphere.

Suess, Eduard (1831–1914); a 19th century British geologist who discovered the now-lost Tethys Ocean and the ancient supercontinent, Gondwanaland.

supernova The explosion that occurs when a star has exhausted its hydrogen and helium fuel.

supratidal zone The area immediately adjacent to the high tide mark kept moist by wave splash.

surf The breaking of waves upon encountering waters shallower than 3/4 their wavelength.

surface feeding Feeding at or near the surface of the ocean.

surface tension The skin-like effect at the air–water interface created by the cohesive attractions among water molecules.

Sverdrup, Harald Ulrik (1888–1957); Norwegian oceanographer and meteorologist who made several significant scientific contributions, including his studies of ocean currents and the dynamics of phytoplankton blooms.

swash zone The area of a beach where the breaking surf runs up and back down the beach face.

swell A symmetrically-shaped surface gravity wave that was created some distance away.

swim bladder An air-sac in some fishes that helps in buoyancy regulation.

Sykes, Lynn (1937–); Scientist who provided evidence of J. Tuzo Wilson's prediction that seismic activity in the vicinity of mid-ocean ridges would be found only on parts of transform faults that lie between the spreading ridges.

T

taxonomy The scientific identification and classification of living organisms and assigning names.

Taylor, Frank Bursley (1860–1938); American geologist who was one of the first to propose actual movements of the continents, in a talk he first gave in 1908 and later published in 1910.

tektite Cosmogenic sediment formed by the impact of meteorites, comets, or asteroids hitting Earth and ejecting fragments of melted terrestrial rock into the atmosphere, where they solidify before falling back to Earth.

Teleostei The "perfect bone" fishes; the most abundant group of modern bony fishes.

temperate Latitudes between the subtropics and polar latitudes higher than the Arctic and Antarctic Circles.

temperature A measure of the kinetic energy of a substance, which is proportional to the speed of atomic and molecular vibrations; usually measured in degrees centigrade.

temperature-salinity diagram An X-Y plot of temperature and salinity data pairs, useful in describing water mass properties.

terrane A complicated continental land form made up of fragments of continental crust (granite), island arcs, and ancient marine sediments.

terrigenous sediment Marine bottom sediments that are of terrestrial origin, brought to the sea by rivers, streams, and winds.

Tethys Ocean Ancient ocean that once separated Laurasia and Gondwanaland; the Mediterranean Sea is a remnant.

tetrahedron A molecular structure formed by one atom bonded to four others, such as in ice.

Thales of Miletos Greek philosopher who first suggested that there were laws controlling nature and that those laws could be discovered; marked the birth of science.

thallus The main body of a macroalga; from the Greek word which roughly translates to "sprout-like."

thecate To possess a theca, which is the outer shell-like structures of chitinous material in certain dinoflagellates.

Thecosomata A suborder of planktoic gastropods.

Theory of Plate Tectonics The general theory of continental drift that incorporates the Theory of Sea Floor Spreading and ideas about lithospheric plates and their movements.

Theory of Seafloor Spreading The phenomenon where new sea floor is being created in the mid-ocean ridge systems; proposed by Harry Hess in the early 1960s.

thermal expansion The phenomenon whereby the volume occupied by a mass increases with increasing temperature, a result of increasing atomic and molecular vibrations.

thermocline The depth over which there is a change in temperature.

thermohaline circulation The density-driven deep and bottom water circulation of the world ocean.

Thomson, Charles Wyville (1830–1882); Scottish naturalist and chief scientist on the Challenger expedition.

thunniform Tuna-like; refers to the general muscular body form.

tidal range The difference in sea level between high tide and low tide.

tide The alternation in sea level at a particular location or region in response to the Earth's rotation and the gravi-

tational forces of the Earth itself and celestial bodies, especially the Moon and the Sun.

tide pools Pools of seawater remaining at low tide in the intertidal zone.

tintinnids A group of marine microzooplanktonic ciliates.

du Toit, Alexander (1878–1948); South African geologist and supporter of Alfred Wegener's ideas about continental drift.

toothed whales The Odontocetes, one of the two suborders of cetaceans, which includes the sperm whale, beaked whales, killer whales, dolphins, and porpoises.

torsion The twisting of the body early in the development of gastropods.

trace gases Gases that usually comprise 1% or less of the atmosphere.

Tragedy of the Commons Cultural phenomenon suggested by ecologist Garrett Hardin to account for wasteful treatment of common property, such as fishery stocks, in that resource losses are shared among the greater public, but the benefits accrue solely to the few who exploit the resource.

train oil The fine oil obtained from the blubber of whales once used to lubricate watches and for burning in lamps.

transform fault A geological fault that runs perpendicular to mid ocean spreading ridges, described by J. Tuzo Wilson.

transverse flagellum One of two flagella in dinoflagellates that encircles the cell and produces a spiral action when swimming.

trawl A commercial fish net, towed either in the water column or along the bottom.

trench The deep ocean features where one lithospheric plate subducts beneath another, creating some of the world oceans' greatest depths.

trochophore A free-swimming larval developmental stage of several invertebrate taxa.

trophic level The position in a food chain or food web, assigned a number to indicate level, or position away, from the primary producers.

trophic pyramid The phenomenon of decreasing biomass with increasing trophic level, from the primary producers at the base to the carnivores at the top.

tropical The region near the Equator, sometimes defined as extending to the Tropic of Cancer and the Tropic of Capricorn.

tropical depression An area of low atmospheric pressure that develops over warm tropical waters, which may develop into a tropical storm; wind speeds are greater than 38 kt.

tropical disturbance The predecessor to a tropical depression; an area of low atmospheric pressure that develops over warm tropical waters and persists for 24 hours.

tropical storm A tropical cyclone that develops from a tropical depression where wind speeds range from 34–63 kt.

troposphere The lowest levels of the atmosphere, to about 20 km altitude, within which most weather phenomena occurs.

trough The point at which a wave is at its lowest depression below the average sea surface.

T-S diagram *See* temperature-salinity diagram.

tsunami A large scale ocean wave with a wavelength on the order of one hundred kilometers but a wave height on the order of 1 m; caused by undersea earthquakes or landslides.

tube feet The hydrostatically-controlled tubular appendages of echinoderms used in locomotion and in grasping prey.

tunicates Marine filter-feeding invertebrates in the subphylum Tunicata in the phylum Chordata characterized by their sac-like body form.

turbellarians A class of flatworms in the phylum Platyhelminthes.

turbidite Marine sedimentary material that results from undersea landslides.

turbidity current The bottom current of water and sediments that flows down a slope as a result of an undersea earthquake.

tychoplankton Aquatic and marine benthic organisms that leave the bottom at night and become part of the plankton.

typhoon Name for tropical cyclones in the western North Pacific Ocean.

U

upper mixed layer The upper water column between the surface and the pycnocline; generally of uniform density, maintained by wind and wave mixing.

urea A metabolic waste product from the breakdown of protein.

Urey, Harold (1893–1981); Nobel Prize winning chemist who was Stanley Miller's academic advisor in graduate school.

V

vegetative cell division The asexual reproductive mode where one cell splits into two cells with the same genetic makeup of the parent cell.

veliger A larval stage in the development of some molluscs.

Venerable Bede (A.D. 673–735); British monk who published in A.D. 730 a work describing the ocean tides.

vernal springtime; refers to the vernal equinox.

vertebrates Animals with a well-developed backbone.

Vikings The first group to reach the New World who sailed to North America as many as two dozen times around A.D. 950.

virus A small infectious life form capable of self replication only inside another living cell.

viviparity Having the characteristic of giving live birth.

W

Wadati, Kiyoo (1902–1995); Japanese seismologist who presented convincing evidence of deep earthquakes and undersea volcanoes around Japan and argued that they were likely the direct result of continental drift.

Walker Cell The atmospheric convection cell on the Equator that is oriented east–west, rather than north–south, disruptions of which occur during El Niño and La Niña events.

water molecule A molecule with two hydrogen atoms and one oxygen atom held together by covalent bonds.

wave height The vertical distance from trough to crest of a surface gravity wave.

wave orbit The elliptical path of water particles in a surface gravity wave.

wave speed The rate at which a wave passes by a point, given in meters per second.

wavelength The horizontal distance between two adjacent wave crests or two adjacent wave troughs; the length of an individual wave form.

Wegener, Alfred (1880–1930); German scientist and early proponent of continental drift; his 1915 book, *The Origin of Oceans and Continents*, started the scientific debate.

well-mixed estuary A class of estuary defined by the relative importance of freshwater discharge to tidal mixing where there is both a vertical and a horizontal salinity gradient.

whalebone Common name for a baleen plate.

whitecaps Breaking waves in deep water; they result in a fully developed sea state in which the energy imparted to the sea surface by the wind is matched by the energy dissipation of breaking waves.

Wilkes, Charles (1798–1877); U.S. Navy Lieutenant who led the U.S. Exploring Expedition (1838–1842).

Wilson, J. Tuzo (1908–1993); University of Toronto professor who suggested that the surface of the Earth is made up of a system of several large, rigid tectonic plates that are in motion; also introduced a new kind of geological fault, the transform fault.

wind wave A wave on the surface of a body of water where the wind is the disturbing force.

wind-driven circulation The upper ocean currents that are the result of energy from the winds transferring momentum to the surface ocean as Ekman currents.

winter convective mixing The sinking of dense waters at high latitudes in winter as a result of density increases accompanying colder water temperatures.

Y

year class A group of fishes or invertebrates that were spawned in the same year.

yolk-sac larva Refers to a larval fish that has hatched from an egg but which retains a sac with remaining egg yolk.

Z

zoea A larval stage in the development of marine crabs.

zooplankton The animal plankton; planktonic heterotrophs.

zooxanthellae Symbiotic dinoflagellates in coral polyps and other cnidarians.

zygote A diploid cell that results from the fusion of a male and a female gamete.

Illustration Credits

The following figures use elements originally rendered for Sadava et al., *Life: The Science of Biology* (Sinauer Associates and W.H. Freeman, 2011): 4.1, 4.3, 4.12, 9.24, 9.32A, 11.1, 11.15, and 11.17.

Cover

© Flip Nicklin/Minden Pictures, Inc.

Preface Photo

Photograph taken on December 22, 1968, by the astronauts aboard Apollo 8 on its way to the Moon.

Chapter 1

Opener: From Seba, Albertus, 1758. *Locupletissimi Rerum Naturalium Thesauri.* 1.8A: Courtesy of Danny Ewing Jr./U.S. Navy. 1.11B: Courtesy of Racklever/Wikipedia 1.11C: Courtesy of BabelStone/Wikipedia. 1.15B: Courtesy of NOAA Photo Library. 1.23A: Courtesy of the Library of Congress. 1.27A: Courtesy of NOAA Photo Library. 1.28B: Courtesy of NOAA. Box 1D: Globe provided by Trent Schindler, NASA/Goddard Scientific Visualization Studio.

Chapter 2

Opener: Courtesy of NASA,ESA, M. Robberto (Space Telescope Science Institute/ ESA) and the Hubble Space Telescope Orion Treasury Project Team. 2.3: © UK History/Alamy. 2.4B: © Bettmann/Corbis. 2.5B: NASA, ESA, M. Robberto (Space Telescope Science Institute/ESA) and the Hubble Space Telescope Orion Treasury Project Team. 2.6: Courtesy of NASA and the European Space Agency. 2.7: Courtesy of ESO/S. Guisard. 2.8: Courtesy of Robert Williams, the Hubble Deep Field Team, and NASA. 2.9: Courtesy of NASA/ESA/S, BeckwithSTScI and the HUDF Team. 2.11: Courtesy of J. Bally University of Colorado and H. Throop SWRI, NASA. 2.12A: Courtesy of NASA/JPL/DLR (Deutsche Forschungsanstalt für Luftund Raumfahrt e.V., Berlin, Germany) 2.12B: Courtesy of NASA/JPL/University of Arizona/University of Colorado 2.12C: Courtesy of NASA/JPL/Arizona State University. 2.13: Courtesy of NASA/Roger Arno. 2.14A: Courtesy of NASA/JPL-Caltech/Univer-

sity of Arizona 2.14B: Courtesy of NASA/ JPL-Caltech/University of Arizona/ Texas A&M University. 2.22: Courtesy of the U.S. Geological Survey. 2.31A: © Jim Sugar/Corbis. 2.32: Courtesy of NOAA PMEL Vents Program. 2.33: Courtesy of D. McKay, NASA /JSC; K. Thomas-Keprta, Lockheed-Martin; R. Zare, Stanford; and NASA/APOD. 2.34: Courtesy of University Photography, Cornell University. 2.35: Courtesy of J. Schmidt/National Park Service.

Chapter 3

Opener: © Douglas Peebles Photography/ Alamy. 3.6B: Courtesy of Larry Mayer, University of Maine. 3.9A: Created with GeoMapApp software using U.S. Geological Survey data. Software described in Ryan et al., 2009. Global multi-resolution topography (GMRT) synthesis data set. *Geochem. Geophys. Geosyst.* 10: Q03014, doi:10.1029/2008GC002332. 3.10: After Sverdrup, H. U., Johnson, M. W., and Fleming, R. H., 1942. *The Oceans: Their Physics, Chemistry and General Biology.* New York: Prentice-Hall. 3.11: Based on data collected aboard the RV *Endeavor*, 2010. 3.12: Courtesy of Divins, D. L., NOAA/National Geophysical Data Center. 3.13: Images courtesy of NOAA/National Geophysical Data Center. 3.15A: Courtesy of University of Edinburgh 3.15B: After Arthur Holmes, 1944. *Principles of Physical Geology.* London: Thomas Nelson and Sons, Ltd. 3.16A: Courtesy of NOAA/National Geophysical Data Center. 3.17A: Courtesy of NOAA Photo Library, NOAA Central Library, NOAA Office of NOAA Corps Operations 3.17B: Courtesy of NOAA Photo Library 3.17C: Courtesy of New Zealand American Submarine Ring of Fire 2007 Exploration, NOAA Vents Program, the Institute of Geological & Nuclear Sciences and NOAA-OE. 3.18, map: Courtesy of NOAA 3.18: After Vine, F. J., 1966. *Science* 154: 1405–1415. 3.19A: After Jacquelyne Kious and Robert I. Tilling, 1996. *This Dynamic Earth: The Story of Plate Tectonics.* Washington, DC: U.S. Government Printing Office. 3.20: Data from Muller, R. D., Sdrolias, M., Gaina, C., and Roest, W. R., 2008. Age, spreading rates, and spreading asymmetry of the world's ocean crust. *Geochem-*

istry Geophysics Geosystems 9 (Q04006); courtesy of Elliot Lim, CIRES & NOAA/ NGDC. 3.21: From Bullard, E., Everett, J. E., and Smith, A. G., 1965. A Symposium on Continental Drift. *Philosophical Transactions of the Royal Society of London, A,* 258(1088): 41–51. 3.23: After the U.S. Geological Survey. 3.24A: Reproduced by permission of the American Geophysical Union from Sykes, L. R. 1967. Mechanisms of earthquakes and nature of faulting on the mid-oceanic ridges. *Journal of Geophysical Research* 72: 2131–2153. 1967 American Geophysical Union 3.24B: Courtesy of NOAA/National Geophysical Data Center. 3.25A: After the U.S. Geological Survey. 3.25B: Created with GeoMapApp software using U.S. Geological Survey data. 3.26A: After Kerr, R. A. 1997. Deep-Sinking Slabs Stir the Mantle. *Science* 275: 613–615. 3.28A: Courtesy of NOAA/National Geophysical Data Center 3.28B: Courtesy of Dr. Anthony Philpotts, Philpotts, A. R., and Ague, J. J., 2009. *Principles of Igneous and Metamorphic Petrology,* 2e. Cambridge: Cambridge University Press. 3.30: Created with GeoMapApp software using U.S. Geological Survey data. 3.31: Courtesy of NOAA/National Geophysical Data Center. 3.33: Created with GeoMapApp software using U.S. Geological Survey data. 3.34B: Created with GeoMapApp software using U.S. Geological Survey data. 3.36: Map courtesy of NOAA/National Geophysical Data Center. 3.37B: Created with GeoMapApp software. 3.37, inset: Courtesy of NOAA. 3.38D: Courtesy of NASA/ISS Crew Earth Observations experiment and the Image Science & Analysis Laboratory, Johnson Space Center. 3.39A: Courtesy of Andrew C. Thomas, University of Maine 3.39B: Courtesy of the U.S. National Park Service. 3.42A,B: Courtesy of the SeaWiFS Project, NASA/Goddard Space Flight Center, and ORBIMAGE. 3.43: Image courtesy of the SeaWiFS Project, NASA/Goddard Space Flight Center, and ORBIMAGE. 3.44A: Courtesy of Jeremy Young, University College London. 3.45: © vitek12/ Shutterstock. 3.46: From Townsend, D. W., Keller, M. D., Holligan, P.M., Ackleson, S. G., and Balch, W. M., 1994. Coccolithophore blooms in relation to hydrography in the Gulf of Maine. *Continental Shelf*

Research 14: 979–1000. 3.49A: Courtesy of NOAA Photo Library. 3.50: Courtesy of Aubrey Whymark, www.tektites.co.uk. Box 3C, Figure C, top: Courtesy of NOAA/ National Geophysical Data Center; bottom, courtesy of David T. Sandwell, Scripps Institution of Oceanography. Box 3F, Figure A: Photo by M. Lansing, courtesy of NOAA Photo Library Box 3F, Figure C: Courtesy of IODP-USIO.

Chapter 4

Opener: © Keren Su/China Spa/Alamy. 4.11C: © Sue Robinson/Shutterstock. 4.15: Courtesy of CERES instrument team, NASA. 4.16: Courtesy of NOAA-NASA GOES Project. 4.22: Data courtesy of NOAA World Ocean Data Center. 4.24: Courtesy of the National Oceanographic Data Center.

Chapter 5

Opener: Courtesy of NASA. 5.20F: Courtesy of NOAA. 5.23: Courtesy of NASA. 5.24B: Courtesy of NOAA's National Hurricane Center. 5.25A: Courtesy of NASA 5.25B: Courtesy of Dr. Steven Babin and Ray Sterner, © 2003, the Johns Hopkins University Applied Physics Laboratory, all rights reserved. Box 5D: Photograph courtesy of Derek Houtman. Box 5F, bottom image: Courtesy of NASA. Box 5H, Figure A: Courtesy of the Library of Congress. Box 5H, Figure D: © 2003, the Johns Hopkins University Applied Physics Laboratory, all rights reserved.

Chapter 6

Opener: © Mana Photo/Shutterstock. 6.12: After Knauss, John, 1978. *Introduction to Physical Oceanography*. Englewood Cliffs, NJ: Prentice-Hall. 6.13A: Courtesy of U.S. Navy. 6.17A: Courtesy of the NOAA Photo Library 6.17B: Courtesy of Capt. G. Anderson Chase, Maine Maritime Academy. 6.19C: Courtesy of DigitalGlobe. 6.22A: © AFP/Getty Images 6.22B: Courtesy of the U.S. Geological Survey. 6.23: Data courtesy of NOAA. 6.25: Photographs © Bill Brooks/Alamy. 6.26: Data courtesy of NOAA. 6.34: Data courtesy of NOAA.

Chapter 7

Opener: © Images & Stories/Alamy. Page 235: Photo of Alfred Redfield by Jan Hahn © Woods Hole Oceanographic Institution. 7.1B: Courtesy of NOAA/Steve Fisher. 7.6: Data from NOAA's World Ocean Database. 7.10: After Parsons, T., and Takahashi, M., 1973. *Biological Oceanographic Processes*. Oxford: Pergamon Press. 7.11: Courtesy of the SeaWiFS Project, NASA/ Goddard Space Flight Center and ORBIMAGE, NASA image created by Jesse Allen, Earth Observatory. 7.21A,B: Courtesy of

NASA. 7.23: After Friedrich, 1973. 7.24B: © Corbis. Box 7G, satellite image: Courtesy of the SeaWiFS Project, NASA GSFC, and ORBIMAGE NASA.

Chapter 8

Opener: © Wim van Egmond/Visuals Unlimited/Corbis. 8.1: From Haeckel, Ernst, 1904. *Kunstformen der Natur* (Leipzig and Vienna: Verlag der Bibliographischen Instituts). 8.3: Courtesy of NASA. 8.4: © Christian Uhrig/istock. 8.5A,B: Courtesy of NOAA; C courtesy of Yuuji Tsukii. 8.6A: Courtesy of Sven A. Kranz, Alfred Wegener Institute for Polar and Marine Research. 8.6B: Courtesy of Annette M. Hynes, University of Georgia. 8.6C: Courtesy of Douglas G. Capone, University of Southern California. 8.7: Courtesy of the Image Science and Analysis Laboratory, NASA-Johnson Space Center. 8.8: Courtesy of the Bigelow Laboratory for Ocean Sciences. 8.9: Courtesy of Zackary I. Johnson, Duke University. 8.10: Courtesy of Daniel Vaulot, CNRS, Station Biologique de Roscoff. 8.11: After Fritsch, 1935. 8.12: After Fritsch, 1935. 8.13A: Courtesy of Alan L. Baker, University of New Hampshire. 8.14A: Courtesy of Dr. Dolors Blasco, Institute de Ciencias del Mar, Barcelona, Spain. 8.14B: Courtesy of SeaWiFS Project, NASA/Goddard Space Flight Center and ORBIMAGE. 8.22: After Jacobsen, D. M., and Anderson, D. M., 1986. Thecate heterotrophic dinoflagellates: Feeding behavior and mechanisms. *Journal of Phycology* 22: 249–258. 8.24: Data from Sugandha Shankar, University of Maine. 8.29A: Left, courtesy of Miriam Godfrey, Woods Hole Oceanographic Institution; right, courtesy of Maria Elisabeth Albinsson, CSIRO Marine and Atmospheric Research, Hobart, Tasmania. 8.29B: Left, courtesy of Tomas Sowles, Maine Department of Marine Resources; right, courtesy of John R. Dolan, Laboratoire d'Océanographie de Villefranche, France. 8.31A: Data courtesy of D. M. Anderson, Woods Hole Oceanographic Institution. 8.32: Courtesy of Donald M. Anderson, Woods Hole Oceanographic Institution. 8.33, bottom: Courtesy of Donald M. Anderson, Woods Hole Oceanographic Institution. 8.35: Courtesy of Heather Spalding, University of Hawaii. 8.36B: © David Wrobel/Visuals Unlimited/Corbis. 8.36C: © Wim van Egmond/ Visuals Unlimited/Corbis. 8.37A: Courtesy of NOAA Sanctuaries Collection, NOAA Photo Library, Claire Fackler, CINMS, NOAA. 8.39A,B,D: Courtesy of Islands in the Sea 2002, NOAA/OER, NOAA Ocean Explorer Program. 8.39C: © Masa Ushioda/Alamy. 8.40A: Courtesy of John M. Carroll, Stony Brook University.

Chapter 9

Opener: © Wim van Egmond/Visuals Unlimited/Corbis. 9.4B: Courtesy of Robert S. Steneck, University of Maine. 9.4C: Courtesy of Joel W. Martin, from Martin, J. W., Truisdale, F. M., and Felder, D. L., 1988. The megalopa stage of the gulf stone crab, *Menippe Adina* Williams and Felder, 1986, with a comparison of megalopae in the genus Menippe. *Fishery Bulletin* 86: 289–297. 9.5B: Courtesy of Robert S. Steneck, University of Maine. 9.9: © Wim van Egmond/Visuals Unlimited/Corbis. 9.18A: After Kirk, 1994. 9.18B: Courtesy of NASA. 9.21: Whale photo courtesy of Dr. Lewis S. Incze. 9.22A: © David Liittschwager/National Geographic Society/Corbis. 9.22B: After Hardy, which he based on Morton (1954). 9.25: © David Wrobel/Visuals Unlimited/Corbis. 9.26: © Kevin Raskoff, Monterey Peninsula College. 9.27A: © WaterFrame/Alamy. 9.27B: Courtesy of NOAA Photo Library. 9.28: © Brandon Cole Marine Photography/Alamy. 9.29: Photos courtesy of Dr. Peter Auster, National Undersea Research Program and University of Connecticut. 9.30: © Visual&Written SL/Alamy. 9.31A: © David Wrobel/Visuals Unlimited/Corbis. 9.32B: © D. P. Wilson/FLPA/Photo Researchers, Inc. 9.34A: © David Wrobel. 9.34B: © Gavin Newman/Alamy.

Chapter 10

Opener: © Michele Westmoreland/Science Faction/Corbis. 10.1A: From Brusca and Brusca. 10.1B: Courtesy of Julie Bedford, NOAA. 10.2C: Courtesy of Richard J. Hoffman, University of Nebraska. 10.3A: Courtesy of David Burdick, NOAA Photo Library. 10.3B: Courtesy of George Schmahl, NOAA. 10.4: Courtesy of NOAA Photo Library. 10.5: Courtesy of Julie Bedford, NOAA. 10.6C: © blickwinkel/ Alamy. 10.7: Courtesy of OAR/NOAA National Undersea Research Program and the NOAA Photo Library. 10.8: Courtesy of Stephen Childs. 10.10A: Courtesy of Dr. David Gems, University College, London. 10.10B: After Brusca and Brusca 2003. 10.11A: Courtesy of Rick Brusca and Gary Brusca, also in Brusca, G. J., and Brusca, R. C., 1978. *A Naturalist's Seashore Guide— Common Marine Life of the Northern California Coast and Adjacent Shores*. Eureka, California: Mad River Press. 10.14A: Courtesy of the NOAA Photo Library, Dr. Anthony R. Picciolo, NOAA NODC. 10.14B: From Brusca and Brusca, 2003. 10.15C: Courtesy of Rachael Norris and Marina Freudzon. 10.17A: From Brusca and Brusca, 2003. 10.19A: From Brusca and Brusca, 2003. 10.20B: From the collection of Andrew D. Sinauer. 10.20C: © Volker Steger/Photo Researchers, Inc. 10.22A: Courtesy of

Parent Géry. 10.23: Courtesy of Karen N. Pelletreau and Mary Rumpho, University of Maine. 10.24: © Aurora Photos/Alamy. 10.27: Courtesy of Dann Blackwood, the U.S. Geological Survey, Woods Hole. 10.28: © WaterFrame/Alamy. 10.29: Courtesy of the NOAA Photo Library, NOAA/Monterey Bay Aquarium Research Institute. 10.30: © Prisma Bildagentur AG/Alamy. 10.31B: From Brusca and Brusca 2003. 10.32: Courtesy of Kevin J. Eckelbarger. 10.33: After Meglitsch, P. A., 1972. *Invertebrate Zoology*, 2e. New York: Oxford University Press. 10.34, inset: © Natural Visions/Alamy. 10.35A: Courtesy of Robert R. Steneck, University of Maine 10.35B: Courtesy of NOAA Photo Library 10.35D: © WaterFrame/Alamy. 10.36A: From Brusca and Brusca, 2003. 10.37C: Courtesy of J. Malcolm Shick, University of Maine. 10.38A: © Teguh Tirtaputra/Shutterstock. 10.39: © John Anderson/istock. 10.40A: From Brusca and Brusca, 2003. 10.41: © Gergo Orban/Shutterstock. Box 10A, Figure A: Courtesy of Lee, Harper, 1884. *Sea Monsters Unmasked*. London: William Clowes and Son. Box 10A, Figure B: © Reuters/Corbis. Box 10A, Figure C: Courtesy of Murray, J., and Hjort, J., 1912. *The Depths of the Ocean*. London: MacMillan & Co.

Chapter 11

Opener: © Specialist Stock/Corbis. 11.2C: © Brandon Cole Marine Photography/Alamy. 11.3A: © anne de Haas/istock. 11.3B: © blickwinkel/Alamy. 11.4: Courtesy of Tara Anderson, NOAA Photo Library. 11.5A: Courtesy of Terry Goss, Wikipedia. 11.5B: Courtesy of Brandi Noble, NOAA. 11.6: B © B.W. Folsom/Shutterstock. 11.8: © tororo reaction/Shutterstock. 11.9: Courtesy of Gregory B. Skomal, NOAA Photo Library. 11.11: © Brandon Cole Marine Photography/Alamy. 11.12: © Brandon Cole Marine Photography/Alamy. 11.14: © Daniel Gotshall/Visuals Unlimited/Corbis. 11.21: © Maximilian Weinzierl/Alamy. 11.22A: After Bigelow and Schroeder, 1953. 11.22B: © WaterFrame/Alamy. 11.23: © Niels Poulsen/Alamy. 11.24: After Breder, C. M. Jr., 1959. Studies on social groupings in fishes. *Bulletin of the American Museum of Natural History* 117: 393–482. 11.25A: After Potts, G. W., 1970. The schooling ethology of *Lutianus monostigma* (Pisces) in the shallow reef environment of Aldbra. *Journal of the Zoological Society of London* 161: 223–235. 11.25B: © Rich Carey/Shutterstock. 11.25C: © Felix Hug/Corbis. 11.26: After Castro, P., and Huber, M. E., 2007. *Marine Biology*, 6e. Boston: McGraw Hill. 11.27A: After Barton, Michael, 2007. *Bond's Biology of Fishes*. Belmont, CA: Thomson Brooks. 11.27B: After Moyle, Peter B., and Cech, Joseph

J., 2004. *Fishes: An Introduction to Ichthyology*. Upper Saddle River, NJ: Prentice Hall. 11.28: © Dennis Frates/Alamy. 11.29: © Jeff Rotman/Alamy. 11.30: Courtesy of David Csepp, NOAA Photo Library. 11.31: Steps 1–5 after Laroche, W. A., 1981. Development of larval smooth flounder. *Liopsetta putnami*, with a redescription of development of winter flounder. *Pseudopleuronectes amcricanus* (Family *Pleuroncctidae*). *Fishery Bulletin, U.S.* 78: 897–909; step 6 after Bigelow, H. B., and Schroeder, W. C., 1953. Fishes of the Gulf of Maine. *Fishery Bulletin of the Fish and Wildlife Service*. 53: 1–577. 11.33: After Hjort, J., and Lea, E., 1914. The age of herring. *Nature* 94: 60–61. 11.35A: After Barton, Michael, 2007. *Bond's Biology of Fishes*. Belmont, CA: Thomson Brooks. 11.35B: After Bigelow, H. B., and Welsh, W. W., 1925. Fishes of the Gulf of Maine. *Bulletin of the United States Bureau of Fisheries* XL: 1–567. 11.36B: Courtesy of Loren Grey. Box 11A, photo: © Hoberman Collection/Corbis. Box 11B, sailfish: © Reinhard Dirscherl/Alamy. Box 11B, tuna: © Richard Herrmann/Visuals Unlimited/Corbis. Box 11B, wahoo: © Visuals Unlimited/Corbis. Box 11C, Figure A: After Bigelow and Schroeder, 1953. Box 11C, Figures B,C: Courtesy of Dr. Steven Campana, Bedford Institute of Oceanography, Canada.

Chapter 12

Opener: © William Manning/Corbis. 12.5C: Courtesy of J. Malcolm Shick. 12.13: Courtesy of Linda Healy, Darling Marine Center, University of Maine. 12.16A: Courtesy of the NOAA Photo Library. 12.16B: Courtesy of Sugandha Shankar. 12.17: Courtesy of Robert S. Steneck. 12.19: Courtesy of NASA. 12.21A: Courtesy of Bigelow Laboratory for Ocean Sciences. 12.22A,B: From Brauer, A., 1906. "Die Tiefsee-Fische. I. Systematischer Teil," in C. Chun, *Wissenschaftl. Ergebnisse der deutschen Tiefsee-Expedition 'Valdivia,'* (1898–1899). Germany: Jena. 12.22C: From Regan, C. T., 1914. Fishes. British Antarctic ("Terra Nova") Expedition, 1910. Natural History Report. *Zoology* 1(1): 1–54; D © David Shale/Naturepl.com. 12.23: Courtesy New Zealand American Submarine Ring of Fire 2007 Exploration, NOAA Vents Program, the Institute of Geological and Nuclear Sciences, and NOAA-OE. 12.24: Courtesy of NOAA Okeanos Explorer Program, Galápagos Rift Expedition 2011. Box 12A, Figure A: Courtesy of David Burdick, NOAA Photo Library. Box 12A, Figure B: Courtesy of NOAA/NESDIS.

Chapter 13

Opener: © Visuals Unlimited/Corbis. 13.1: Courtesy of NOAA Photo Library, Julie

Bedford, NOAA. 13.2: Courtesy of NOAA Photo Library, Lieutenant Elizabeth Crapo, NOAA Corps. 13.3: © Susan Flashman/istock. 13.4: Courtesy of the U.S. Fish and Wildlife Service. 13.5: Courtesy Kristy L. Townsend. 13.6: © Mark Conlin/Alamy. 13.8: © Kenneth Jones/Alamy. 13.10: Courtesy of Dr. Robert Ricker, NOAA/NOS/ORR. 13.11: © Steve Bloom Images/Alamy. 13.12A: © Avico Ltd/Alamy. 13.12B: © Megan Whittaker/Alamy. 13.12C: © Green Stock Media/Alamy. 13.13: Courtesy of Andrew D. Sinauer. 13.14B: Courtesy of David McIntyre. 13.16: © Brian & Sophia Fuller/Alamy. 13.17A: Courtesy of NOAA Photo Library, Dr. Steve Lonhart. 13.17B: After Estes, J. A., and Bodkin, J. L., 2002. "Otters," in *Encyclopedia of Marine Mammals*, eds. W. F. Perrin, B. Wursig, and J. G. M. Thewissen (San Diego: Academic Press). pp. 842–858. 13.18: Courtesy of the NOAA Photo Library. 13.19: Courtesy of Nancy Heise. 13.20A: Courtesy of M. Boylan. 13.20B: Courtesy of NOAA. 13.21: Courtesy of Sarah Sonsthagen, U.S. Geological Survey. 13.24: After Berta, A., and Sumich, J. L., 1999. *Marine Mammals: Evolutionary Biology*. San Diego: Academic Press. 13.25: After Berta, A., and Sumich, J. L., 1999. *Marine Mammals: Evolutionary Biology*. San Diego: Academic Press. 13.26: © Hugh Lansdown/Shutterstock. 13.28: After Kooyman, G. L., and Ponganis, P. J., 1997. The Challenges of Diving to Depth: The deepest sea divers have unique ways of budgeting their oxygen supply and responding to pressure. *American Scientist* 85: 530–539. 13.34A: © Mary Evans Picture Library/Alamy. 13.34B: © Jeremy Sutton-Hibbert/Alamy. 13.35: Data from International Whaling Statistics. 13.36: © Karel Havlicek/National Geographic Society/Corbis. 13.37: © Xu Jian/Naturepl.com. 13.38: Courtesy of Beth Josephson/NOAA. 13.39: Courtesy of the NOAA Fisheries, Office of Protected Resources.

Chapter 14

14.3: Data from *The State of World Fisheries and Aquaculture*, 2012. Food and Agriculture Organization, United Nations. 14.4: Data from the Population Reference Bureau, Washington, DC. 14.5: After Akinori, T., Oozeki, Y., Kubota, H., and Lluch-Cota, S. E., 2008. Contrasting spawning temperature optima: Why are anchovy and sardine regime shifts synchronous across the North Pacific? *Progress in Oceanography* 77: 225–232. 14.6: Data from Mayo, R. K., Brodziak, J. K. T., Thompson, M., Burnett, J. M., and Cadrin, S. X., 2002. Biological Characteristics, Population Dynamics, and Current Status of Redfish, *Sebastes fasciatus Storer*, in the Gulf of Maine—Georges

Bank Region. *NOAA Northeast Fisheries Science Center* Reference Document 02–05. 14.7A: Courtesy of Goode, George Brown, 1887. *Fisheries and Fishery Industries of the United States*. Washington, DC: Government Printing Office. 14.7B: © Bettmann/Corbis. 14.9A: Courtesy of NOAA 14.9B: © Natalie Fobes/Science Faction/Corbis. 14.10: Data from NOAA Fisheries Service. 14.11: After NOAA, 1995. 14.12: Data from NOAA, NMFS. 14.13: After Gulland, J. A., 1974. *The Management of Marine Fisheries*. Seattle: University of Washington Press. 14.14A: Data from NOAA Fisheries Service 14.14B: After Cushing, D. H., 1975. *Marine Ecology and Fisheries*. Cambridge, UK: Cambridge University Press. 14.16: After Beverton, Raymond J. H., and Holt, Sidney. J., 1957. On the Dynamics of Exploited Fish Populations. Ministry of Agriculture, Fisheries and Food. *Fisheries Investigations*, Series II, Volume XIX. London. 14.18: Data from *The State of World Fisheries and Aquaculture*, 2012. Food and Agriculture Organization, United Nations. 14.19: After *The State of World Fisheries and Aquaculture*, 2012. Food and Agriculture Organization, United Nations. 14.20A,B: Courtesy of Paul Rawson. 14.21: All photos courtesy of Christopher Bartlett.

Chapter 15

15.1: Courtesy of NOAA. 15.2: After Nienhuis, P. H., 1993. Nutrient cycling and foodwebs in Dutch estuaries. *Hydrobiologia* 265: 15–44. 15.3A: © Sarah Weston/Alamy 15.3, inset: Courtesy of Bigelow Laboratory for Ocean Sciences, National Center for Marine Algae and Microbiota. 15.13B: After Riegman, R., Noordeloos,

A. A. M., and Cadée, G. C., 1992. Phaeocystis blooms and eutrophication of the continental coastal zones of the North Sea. *Marine Biology* 112: 497–484. 15.4A: After Goolsby, D. A., and Battaglin, W. A., 2000. Nitrogen in the Mississippi Basin—Estimating Sources and Predicting Flux to the Gulf of Mexico. *U.S. Geological Survey* Fact Sheet 135–00, December. 15.5: Data from Nancy N. Rabalais, Louisiana Universities Marine Consortium, and R. Eugene Turner, Louisiana State University. 15.6: After Clark, R. B., 1997. *Marine Pollution*. Oxford: Clarendon Press. 15.7A: © Rob Stapleton/AP/Corbis 15.7B: © John Gaps III/AP/Corbis 15.7C: © Accent Alaska.com/Alamy. 15.8: Courtesy of the U.S. Geological Survey, U.S. Coast Guard. 15.9: Data from U.S. Energy Information Administration, Annual Energy Review, 2010. 15.10: Courtesy of the NOAA Photo Library. 15.11: © Jinny Goodman/Alamy. 15.13: Photos courtesy of Dr. Daniel A. Dixon, Climate Change Institute, University of Maine. 15.15: Data from Etheridge, D. M. et al., 1996. Natural and anthropogenic changes in atmospheric CO_2 over the last 1000 years from air in Antarctic ice and firn. *Journal of Geophysical Research* 101: 4115–4128. 15.16A: Data from NOAA. 15.16B: Data from Hansen, J., Ruedy, R., Sato, M., Imhoff, M., Lawrence, W., Easterling, D., Peterson, T., and Karl, T., 2001. A closer look at United States and global surface temperature change. *Journal of Geophysical Research* 106: 23947–23963. 15.17: After Crowley, T. J., 2000. Causes of climate change over the past 1000 years. *Science* 289: 270–277. 15.18: After Pachauri, R. K., and Reisinger, A., eds., 2007. *Climate*

Change 2007: Fourth Assessment Report (AR4). Contribution of Working Groups I, II and III to the Fourth Assessment Report of the Intergovernmental Panel on Climate Change. Geneva, Switzerland: IPCC. 15.20: After Levitus, S., Antonov, J. I., Boyer, T. P., and Stephens, C., 2000. Warming of the World Ocean. *Science* 287: 2225–2229. 15.22: After Doney, S. C., Fabry, V. J., Feely, R. A., and Kleypas, J. A., 2009. Ocean acidification: The other CO_2 problem. *Annual Review of Marine Science* 1: 169–192. 15.23: Courtesy of the U.S. Geological Survey. 15.24: Courtesy of NASA/Goddard Space Flight Center Scientific Visualization Studio. 15.25A: After the National Snow and Ice Data Center. 15.25B: Courtesy of NASA. 15.26A,B: Courtesy of Michael Van Woert, NOAA, NOAA Photo Library; C courtesy of Jeff Schmaltz, MODIS Rapid Response Team, NASA/GSFC.

Appendix A

Figure B: © ITAR-TASS Photo Agency/Alamy. Figure F: Courtesy of NASA.

Appendix C

Figure A: Courtesy of the Wildlife Conservation Society. Figure B: Left and center courtesy of the U.S. Naval Historical Center. Figure C: Left and right courtesy of Dr. Rhian Waller, University of Maine. Figure E: Right, courtesy of NOAA/Institute for Exploration/University of Rhode Island.

Index

Page numbers followed by the letter *f* denote entries that are included in a figure; the letter *b* after a page number indicates that the entry is included in a box.